Time Series for Data Science

CHAPMAN & HALL/CRC
Texts in Statistical Science Series

Joseph K. Blitzstein, *Harvard University, USA*
Julian J. Faraway, *University of Bath, UK*
Martin Tanner, *Northwestern University, USA*
Jim Zidek, *University of British Columbia, Canada*

Recently Published Titles

A First Course in Linear Model Theory, Second Edition
Nalini Ravishanker, Zhiyi Chi, Dipak K. Dey

Foundations of Statistics for Data Scientists
With R and Python
Alan Agresti and Maria Kateri

Fundamentals of Causal Inference
With R
Babette A. Brumback

Sampling
Design and Analysis, Third Edition
Sharon L. Lohr

Theory of Statistical Inference
Anthony Almudevar

Probability, Statistics, and Data
A Fresh Approach Using R
Darrin Speegle and Brain Claire

Bayesian Modeling and Computation in Python
Osvaldo A. Martin, Raviv Kumar and Junpeng Lao

Bayes Rules!
An Introduction to Applied Bayesian Modeling
Alicia Johnson, Miles Ott and Mine Dogucu

Stochastic Processes with R
An Introduction
Olga Korosteleva

Introduction to Design and Analysis of Scientific Studies
Nathan Taback

Time Series for Data Science
Analysis and Forecasting
Wayne A. Woodward, Bivin Philip Sadler and Stephen Robertson

For more information about this series, please visit: https://www.routledge.com/Chapman--HallCRC-Texts-in-Statistical-Science/book-series/CHTEXSTASCI

Time Series for Data Science

Analysis and Forecasting

Wayne A. Woodward, Bivin P. Sadler and
Stephen D. Robertson

CRC Press
Taylor & Francis Group
Boca Raton London New York

CRC Press is an imprint of the
Taylor & Francis Group, an **informa** business

A CHAPMAN & HALL BOOK

First edition published 2022
by CRC Press
6000 Broken Sound Parkway NW, Suite 300, Boca Raton, FL 33487-2742

and by CRC Press
2 Park Square, Milton Park, Abingdon, Oxon, OX14 4RN

CRC Press is an imprint of Taylor & Francis Group, LLC

ISBN: 978-0-367-53794-4 (hbk)
ISBN: 978-0-367-54389-1 (pbk)
ISBN: 978-1-003-08907-0 (ebk)

DOI: 10.1201/9781003089070

Typeset in Times
by Newgen Publishing UK

To Beverly, Ellie, and Melissa

and in memory of

Henry L. (Buddy) Gray

Contents

Preface   xv
Acknowledgments   xix
Authors   xxi

1 Working with Data Collected Over Time   **1**
  1.1 Introduction   1
  1.2 Time Series Datasets   3
       1.2.1 Cyclic Data   3
              1.2.1.1 Sunspot Data   3
              1.2.1.2 DFW Temperature Data   5
              1.2.1.3 Air Passengers Data   6
       1.2.2 Trends   7
              1.2.2.1 Real Datasets That Have Trending Behavior   8
              1.2.2.2 The Problem with Trends   9
  1.3 The Programming Language R   10
       1.3.1 The *tswge* Time Series Package   11
       1.3.2 Base R   12
       1.3.3 Plotting Time Series Data in R   12
       1.3.4 The *ts* Object   13
              1.3.4.1 Creating a *ts* Object   14
              1.3.4.2 More About *ts* Objects   16
       1.3.5 The `plotts.wge` Function in *tswge*   18
              1.3.5.1 Modifying the Appearance of Plots Using the *tswge*
                       `plotts.wge` Function   19
       1.3.6 Loading Time Series Data into R   19
              1.3.6.1 The .csv file   20
              1.3.6.2 The .txt file   21
              1.3.6.3 Other File Formats   21
       1.3.7 Accessing Time Series Data   22
              1.3.7.1 Accessing Data from the Internet   22
              1.3.7.2 Business / Proprietary Data: Ozona Bar and Grill   28
  1.4 Dealing with Messy Data   29
       1.4.1 Preparing Time Series Data for Analysis: Cleaning, Wrangling,
              and Imputation   29
              1.4.1.1 Missing Data   29
              1.4.1.2 Downloading When no .csv Download Option Is Available   33
              1.4.1.3 Data that Require Cleaning and Wrangling   34
              1.4.1.4 Programatic Method of Ingestion and Wrangling Data from
                       Tables on Web Pages   37
  1.5 Concluding Remarks   37
       Appendix 1A   37

2   Exploring Time Series Data   **41**
2.1   Understanding and Visualizing Data   41
2.1.1   Smoothing Time Series Data   41
2.1.1.1   Smoothing Data Using a Centered Moving Average Smoother   42
2.1.1.2   Other Methods Available for Smoothing Data   44
2.1.1.3   Moving Average Smoothing versus Aggregating   44
2.1.1.4   Using Moving Average Smoothing for Estimating Trend in
Data with Fixed Cycle Lengths   47
2.1.2   Decomposing Seasonal Data   49
2.1.2.1   Additive Decompositions   50
2.1.2.2   Multiplicative Decompositions   56
2.1.3   Seasonal Adjustment   58
2.1.3.1   Additive Seasonal Adjustment   58
2.1.3.2   Multiplicative Seasonal Adjustment   59
2.2   Forecasting   60
2.2.1   Predictive Moving Average Smoother   62
2.2.2   Exponential Smoothing   64
2.2.2.1   Forecasting with Exponential Smoothing beyond the
Observed Dataset   65
2.2.3   Holt-Winters Forecasting   66
2.2.3.1   Additive Holt-Winters Equations   66
2.2.3.2   Multiplicative Holt-Winters Equations   67
2.2.4   Assessing the Accuracy of Forecasts   68
2.3   Concluding Remarks   70
Appendix 2A   71

3   Statistical Basics for Time Series Analysis   **75**
3.1   Statistics Basics   75
3.1.1   Univariate Data   75
3.1.2   Multivariate Data   80
3.1.2.1   Measuring Relationships between Two Random Variables
in a Bivariate Random Sample   81
3.1.2.2   Assessing Association from a Bivariate Random Sample   82
3.1.3   Independent vs Dependent Data   85
3.2   Time Series and Realizations   89
3.2.1   Multiple Realizations   90
3.2.1.1   Time Series 1: X_t 92
3.2.1.2   Time Series 2: Y_t (Example 3.3 Continued)   94
3.2.2   The Effect of Realization Length   97
3.3   Stationary Time Series   100
3.3.1   Plotting the Autocorrelations of a Stationary Process   104
3.3.2   Estimating the Parameters of a Stationary Process   105
3.3.2.1   Estimating μ 105
3.3.2.2   Estimating the Variance   109
3.3.2.3   Estimating the Autocovariance and Autocorrelation   109
3.3.2.4   Plotting Sample Autocorrelations   110
3.4   Concluding Remarks   113
Appendix 3A   113
Appendix 3B   114

4   The Frequency Domain   **121**
  4.1   Trigonometric Review and Terminology   122
  4.2   The Spectral Density   125
    4.2.1   Euler's Formula   126
    4.2.2   Definition and Properties of the Spectrum and Spectral Density   127
      4.2.2.1   The Nyquist Frequency   128
      4.2.2.2   Frequency $f = 0$ 129
      4.2.2.3   The Spectral Density and the Autocorrelation Function   131
    4.2.3   Estimating the Spectral Density   132
      4.2.3.1   The Sample Spectral Density   132
      4.2.3.2   Smoothing the Sample Spectral Density   134
      4.2.3.3   Parzen Spectral Density Estimate vs Sample Autocorrelations   135
      4.2.3.4   Why We Plot Spectral Densities in Log Scale   137
  4.3   Smoothing and Filtering   140
    4.3.1   Types of Filters   140
    4.3.2   The Butterworth Filter   143
  4.4   Concluding Remarks   145
  Appendix 4A   145

5   ARMA Models   **151**
  5.1   The Autoregressive Model   151
    5.1.1   The AR(1) Model   152
      5.1.1.1   The AR(1) in Backshift Operator Notation   152
      5.1.1.2   The AR(1) Characteristic Polynomial and Characteristic Equation   153
      5.1.1.3   Properties of a Stationary AR(1) Model   153
      5.1.1.4   Spectral Density of an AR(1)   155
      5.1.1.5   AR(1) Models with Positive Roots of the Characteristic Equation   155
      5.1.1.6   AR(1) Models with Roots Close to +1   159
      5.1.1.7   AR(1) Models with Negative Roots of the Characteristic Equation   160
      5.1.1.8   Nonstationary 1st-order Models   162
      5.1.1.9   Final Comments Regarding AR(1) Models   162
    5.1.2   The AR(2) Model   163
      5.1.2.1   Facts about the AR(2) Model   164
      5.1.2.2   Operator Notation and Characteristic Equation for an AR(2)   164
      5.1.2.3   Stationary AR(2) with Two Real Roots   167
      5.1.2.4   Stationary AR(2) with Complex Conjugate Roots   168
      5.1.2.5   Summary of AR(1) and AR(2) Behavior   173
    5.1.3   The AR(p) Models   173
      5.1.3.1   Facts about the AR(p) Model   174
      5.1.3.2   Operator Notation and Characteristic Equation for an AR(p)   175
      5.1.3.3   Factoring the AR(p) Characteristic Polynomial   176
      5.1.3.4   Factor Tables for AR(p) Models   177
      5.1.3.5   Dominance of Roots Close to the Unit Circle   184
    5.1.4   Linear Filters, the General Linear Process, and AR(p) Models   187
      5.1.4.1   AR(1) in GLP Form   188
      5.1.4.2   AR(p) in GLP Form   190
  5.2   Autoregressive-Moving Average (ARMA) Models   191
    5.2.1   Moving Average Models   192
      5.2.1.1   The MA(1) Model   192
      5.2.1.2   The MA(2) Model   195

		5.2.1.3	The General MA(q) Model	198
		5.2.1.4	Invertibility	198
	5.2.2	ARMA(p,q) Models		201
		5.2.2.1	Stationarity and Invertibility of an ARMA(p,q) Process	201
		5.2.2.2	AR versus ARMA Models	203
5.3	Concluding Remarks			207
Appendix 5A				208
Appendix 5B				211

6 ARMA Fitting and Forecasting — **217**

6.1	Fitting ARMA Models to Data			217
	6.1.1	Estimating the Parameters of an ARMA(p,q) Model		218
		6.1.1.1	Maximum Likelihood Estimation of the ϕ and θ Coefficients of an ARMA Model	218
		6.1.1.2	Estimating μ	220
		6.1.1.3	Estimating σ_a^2	220
		6.1.1.4	Alternative estimates for AR(p) models	223
	6.1.2	ARMA Model Identification		229
		6.1.2.1	Plotting the Data and Checking for White Noise	229
		6.1.2.2	Model Identification Types	230
		6.1.2.3	AIC-type Measures for ARMA Model Fitting	231
		6.1.2.4	The Special Case of AR Model Identification	239
6.2	Forecasting Using an ARMA(p,q) Model			245
	6.2.1	ARMA Forecasting Setting, Notation, and Strategy		245
		6.2.1.1	Strategy and Notation	245
		6.2.1.2	Forecasting X_{t_0+l} for $l \le 0$	246
		6.2.1.3	Forecasting a_{t_0+l} for $l > 0$	246
	6.2.2	Forecasting Using an AR(p) Model		246
		6.2.2.1	Forecasting Using an AR(1) Model	246
	6.2.3	Basic Formula for Forecasting Using an ARMA(p,q) Model		251
	6.2.4	Eventual Forecast Functions		255
	6.2.5	Probability Limits for ARMA Forecasts		255
		6.2.5.1	Facts about Forecast Errors	256
		6.2.5.2	Lack of Symmetry	260
	6.2.6	Assessing Forecast Performance		260
		6.2.6.1	How "Good" Are the Forecasts?	260
		6.2.6.2	Some Strategies for Using RMSE to Measure Forecast Performance	261
6.3	Concluding Remarks			270
Appendix 6A				270

7 ARIMA and Seasonal Models — **275**

7.1	ARIMA(p,d,q) Models			275
	7.1.1	Properties of the ARIMA(p,d,q) Model		276
		7.1.1.1	Some ARIMA(p,d,q) Models	276
		7.1.1.2	Characteristic Equations for Models (a)–(c)	277
		7.1.1.3	Limiting Autocorrelations	277
		7.1.1.4	Lack of Attraction to a Mean	279
		7.1.1.5	Random Trends	280
		7.1.1.6	Differencing an ARIMA(0,1,0) Model	281

		7.1.1.7	ARIMA Models with Stationary and Nonstationary Components	282
		7.1.1.8	The Stationary AR(2) Model: $(1-1.4B+.65B^2)X_t = a_t$	282
	7.1.2	Model Identification and Parameter Estimation of ARIMA(p,d,q) Models		287
		7.1.2.1	Deciding Whether to Include One or More 1–B Factors (That Is, Unit Roots) in the Model	287
		7.1.2.2	General Procedure for Fitting an ARIMA(p,d,q) Model to a Set of Time Series Data	291
	7.1.3	Forecasting with ARIMA Models		298
		7.1.3.1	ARMA Forecast Formula	298
7.2	Seasonal Models			304
	7.2.1	Properties of Seasonal Models		305
		7.2.1.1	Some Seasonal Models	305
	7.2.2	Fitting Seasonal Models to Data		311
		7.2.2.1	Overfitting	312
	7.2.3	Forecasting Using Seasonal Models		321
7.3	ARCH and GARCH Models			327
	7.3.1	ARCH(1) Model		329
	7.3.2	The ARCH(p) and GARCH(p,q) Processes		332
	7.3.3	Assessing the Appropriateness of an ARCH/GARCH Fit to a Set of Data		334
	7.3.4	Fitting ARCH/GARCH Models to Simulated Data		335
	7.3.5	Modeling Daily Rates of Return Data		337
7.4	Concluding Remarks			338
Appendix 7A				339
Appendix 7B				340

8	**Time Series Regression**			**343**
8.1	Line+Noise Models			343
	8.1.1	Testing for Linear Trend		344
		8.1.1.1	Testing for Trend Using Simple Linear Regression	344
		8.1.1.2	A t-test Simulation	348
		8.1.1.3	Cochrane-Orcutt Test for Trend	350
		8.1.1.4	Bootstrap-Based Test for Trend	352
		8.1.1.5	Other Methods for Testing for Trend in Time Series Data	355
	8.1.2	Fitting Line+Noise Models to Data		355
	8.1.3	Forecasting Using Line+Noise Models		358
8.2	Cosine Signal+Noise Models			360
	8.2.1	Fitting a Cosine Signal+Noise Model to Data		361
	8.2.2	Forecasting Using Cosine Signal+Noise Models		364
		8.2.2.1	Using fore.sigplusnoise.wge:	366
	8.2.3	Deciding Whether to Fit a Cosine Signal+Noise Model to a Set of Data		367
		8.2.3.1	A Closer Look at the Cyclic Behavior	369
8.3	Concluding Remarks			376
Appendix 8A				377

9	**Model Assessment**			**381**
9.1	Residual Analysis			381
	9.1.1	Checking Residuals for White Noise		383
		9.1.1.1	Check Residual Sample Autocorrelations against 95% Limit Lines	383
		9.1.1.2	Ljung-Box Test	384
	9.1.2	Checking the Residuals for Normality		389

9.2	Case Study 1: Modeling the Global Temperature Data	390
	9.2.1 A Stationary Model	391
	9.2.1.1 Checking the Residuals	392
	9.2.1.2 Realizations and their Characteristics	393
	9.2.1.3 Forecasting Based on the ARMA(4,1) Model	394
	9.2.2 A Correlation-Based Model with a Unit Root	395
	9.2.2.1 Checking the Residuals	396
	9.2.2.2 Realizations and their Characteristics	397
	9.2.2.3 Forecasting Based on ARIMA(0,1,1) Model	398
	9.2.3 Line+Noise Models for the Global Temperature Data	400
	9.2.3.1 Checking the Residuals, $\hat{a}_t$, for White Noise	401
	9.2.3.2 Realizations and their Characteristics	402
	9.2.3.3 Forecasting Based on the Signal-plus-Noise Model	403
	9.2.3.4 Other Forecasts	405
9.3	Case Study 2: Comparing Models for the Sunspot Data	407
	9.3.1 Selecting the Models for Comparison	408
	9.3.2 Do the Models Whiten the Residuals?	409
	9.3.3 Do Realizations and Their Characteristics Behave Like the Data?	410
	9.3.4 Do Forecasts Reflect What Is Known about the Physical Setting?	412
	9.3.4.1 Final Comments about the Models Fit to the Sunspot Data	412
9.4	Comprehensive Analysis of Time Series Data: A Summary	413
9.5	Concluding Remarks	413
	Appendix 9A	414
10	**Multivariate Time Series**	**417**
10.1	Introduction	417
10.2	Multiple Regression with Correlated Errors	417
	10.2.1 Notation for Multiple Regression with Correlated Errors	418
	10.2.2 Fitting Multiple Regression Models to Time Series Data	419
	10.2.2.1 Including a Trend Term in the Multiple Regression Model	422
	10.2.2.2 Adding Lagged Variables	423
	10.2.2.3 Using Lagged Variables and a Trend Variable	425
	10.2.3 Cross Correlation	426
10.3	Vector Autoregressive (VAR) Models	428
	10.3.1 Forecasting with VAR(p) Models	429
	10.3.1.1 Univariate Forecasts	431
	10.3.1.2 VAR Analysis	432
	10.3.1.3 Comparing RMSEs	436
	10.3.1.4 Final Comments	437
10.4	Relationship between MLR and VAR Models	437
10.5	A Comprehensive and Final Example: Los Angeles Cardiac Mortality	437
	10.5.1 Applying the VAR(p) to the Cardiac Mortality Data	439
	10.5.2 The Seasonal VAR(p) Model	441
	10.5.3 Forecasting the Future	444
	10.5.3.1 Short vs. Long Term Forecasts	445
10.6	Conclusion	446
	Appendix 10A	447
	Appendix 10B	449

11 Deep Neural Network-Based Time Series Models **455**
11.1 Introduction 455
11.2 The Perceptron 455
11.3 The Extended Perceptron for Univariate Time Series Data 457
    11.3.1 A Neural Network Similar to the AR(1) 458
        11.3.1.1 The Architecture 458
        11.3.1.2 Fitting the MLP 458
        11.3.1.3 Forecasting 460
        11.3.1.4 Cross Validation Using the Rolling Window RMSE 463
    11.3.2 A Neural Network Similar to AR(p): Adding More Lags 464
    11.3.3 A *Deeper* Neural Network: Adding a Hidden Layer 467
        11.3.3.1 Differences and Seasonal "Dummies" 470
11.4 The Extended Perceptron for Multivariate Time Series Data 475
    11.4.1 Forecasting Melanoma Using Sunspots 475
        11.4.1.1 Architecture 475
        11.4.1.2 Fitting the Baseline Model 475
        11.4.1.3 Forecasting Future Sunspot Data for Predicting Future Melanoma 476
        11.4.1.4 Forecasting the Last Eight Years of Melanoma 478
        11.4.1.5 Fitting a Competing Model 478
        11.4.1.6 Assessing the Competing Model on the Last Eight Years of
            Melanoma Data 479
        11.4.1.7 Forecasting the Next Eight Years of Melanoma 480
    11.4.2 Forecasting Cardiac Mortality Using Temperature and Particulates 482
        11.4.2.1 General Architecture 482
        11.4.2.2 Train / Test Split 483
        11.4.2.3 Forecasting Covariates: Temperature and Particulates 483
        11.4.2.4 Model Without Seasonal Indicator Variables 484
        11.4.2.5 Model With Seasonal Indicator Variables 486
11.5 An "Ensemble" Model 487
    11.5.1 Final Forecasts for the Next Fifty-Two Weeks 488
    11.5.2 Final Forecasts for the Next Three Years (Longer Term Forecasts) 489
11.6 Concluding Remarks 491
Appendix 11A 491
Appendix 11B 492

References *497*
Index *501*

Preface

We believe this to be a truly unique textbook for teaching introductory time series and feel it is appropriate for master's level and undergraduate students in data science, statistics, mathematics, economics, finance, MBA programs, and any of the sciences. The contents of this book were developed from time series courses taught by the authors in two applied master's programs at Southern Methodist University: an Applied Statistics/Data Analytics program within the Statistical Science Department and an online Master of Science in Data Science.

OUR GOAL

The goal of this text is to provide students with an understandable and "friendly" introduction to time series analysis that will provide them with a fundamental understanding of time series analysis and will equip them with a wide range of tools they can use in the analysis of time series data in their careers. Our experience is that we have been successful in accomplishing these goals. We have learned how to effectively address this audience. Many students have stepped out of this course and begun using the tools in this book to analyze time series data "on the job." We have received encouraging feedback from students, professors, and practitioners, alike. We're excited about the value you will find as well!

TOPICS COVERED

This book covers a variety of time series–related topics. We devote the entirety of Chapter 1 to discussing time series datasets, what they are, characteristic behaviors often seen in time series data, where to find and access such data, as well as methods of wrangling and cleaning time series data. That is, we don't discuss analysis in this chapter, just data. We also introduce the basics of using the statistical software package, R, and in particular the CRAN time series package *tswge*. Chapter 2 covers topics such as data smoothing, data decomposition, seasonal adjustment, exponential smoothing, and Holt-Winters forecasting. Chapter 3 is a unique introduction to fundamental statistical concepts, a discussion of the difference between a simple random sample and a time series realization, why correlation can "be our friend", and an introduction to stationary time series. Chapter 4 provides a brief, understandable introduction to the frequency domain. The frequency domain will be used as we understand and analyze data, thus its placement early in the book. Chapters 5 and 6 cover ARMA modeling, including such topics as the factor table for understanding the underlying components of the model, model identification, parameter estimation, and forecasting. Chapter 7 addresses nonstationary time series analysis using ARIMA, seasonal, and ARCH/GARCH models, while Chapter 8 provides a unique and hands-on look at time series regression along with its features and shortfalls. Case studies are presented in Chapter 9 and include the analysis of global temperature and sunspot data. Chapter 10 facilitates the inclusion of outside information by presenting multivariate topics including multiple regression with correlated errors and the VAR model for modeling and forecasting multivariate time series data. Finally, Chapter 11 presents a clear coverage of increasingly popular neural network/deep learning methods for analyzing time series data.

PREREQUISITES

Prerequisites for the readers of the text are relatively minimal. A calculus background is valuable but not necessary for students using this book. The book is surprisingly advanced without falling back on calculus-based derivations. We believe that the book is accessible to serious students who have not had courses in calculus or a statistics course beyond the introductory level. While we are concerned that this may "turn off" some instructors, we believe that you will be surprised at the mathematical rigor yet applied nature of this book. Give it a try! For instructors or students interested in more mathematical detail, we have provided supplemental videos produced by the authors, in-depth appendices, and references to related resources (textbooks, journal articles, etc.).

PROBLEM SETS

Of course, to truly master these tools, the student must apply the methods and practice, practice, practice. The book contains problem sets using real-world datasets and interesting and relevant questions. In addition, we carefully designed chapter problems to specifically address learning outcomes associated with the chapter. The result is a set of problems that can reasonably be completed in full by a student to assist in efficiently facilitating mastery of the concepts in each chapter. We feel this allows for students, upon completion of their study, to confidently be able to say to themselves and others that they fully grasp the concepts covered in this book.

TSWGE SOFTWARE

For software choice, we have created an R package *tswge*, which is available on CRAN to accompany this book. Extensive discussion of the use of *tswge* functions is given within the chapters and in appendices following each chapter. The *tswge* package currently contains about 70 functions and that number will continue to grow. Consult the book's website http://blog.smu.edu/timeseries/ for updates concerning the software and the book. We have added guidance concerning R usage throughout the entire book, including code for nearly all examples and corresponding figures. The CRAN package *tswge* also contains many datasets, several of them containing real data, including a collection of datasets associated with figures and examples in the book. Because of the clearly described R syntax, students should not be required to have previous knowledge of R. The number of examples and the focus on participation by the reader to enter the provided R code will ground the students with the necessary R coding skills. In fact, this textbook could be a resource for an introductory course in R programming with time series applications.

HANDS-ON EXPERIENCE

Finally, we feel what really distinguishes the book is its strong emphasis on encouraging students to participate as they read through the chapters. The engaged student should type in the code and/or copy and paste code segments available from the website to reproduce examples and corresponding plots and output.

QR CODES

Approximately 80 QR codes, strategically placed throughout the chapters, link to tutorial videos produced by the authors. These videos are key learning tools for the student. The videos typically offer additional detail on a particular topic, carefully explain a particularly important example, or, in a few cases, introduce a related topic or method. It is important to note that the book could be consumed independently of these videos and extra resources; however, the resources available through the QR codes can offer significant extra detail, insight, and perspective to the interested student.

Acknowledgments

Many colleagues contributed time and expertise to this book. We specifically appreciate the advice and encouragement from Darren Homrighausen and Alan Elliott. Many of the contributions have been made from students who have been taught out of a preliminary set of notes that led to the current book. These students did an excellent job of proofreading the text, beta testing the software, and working through the problem sets. The following students of Robertson deserve recognition: John Bookas, Georgios Chatzikyriakidis, Emily Fashenpour, Radford Freel, Alexis Gambino, C.J. Olson, Seth Ozment, Charles Patterson, Kirstin Pruitt, Yifei Wang, Alex White, and Yao Zhang. Sadler has singled out the following students for recognition: Edward Roske, Justin Ehly, Cameron Stewart, Nick Yu, Jorge Olmos, Paul Swenson, Mrinmoy Bhaumik, Morgan Nelson, Sreeni Prabhala, Mel Schwan, Farbod Khomeini, Kebur Fantahun, Michael Mazel, Paritosh Rai, Chance Robinson, David Josephs, Mai Dang, and Sam Coyne.

Authors

Wayne Woodward is Professor Emeritus in the Department of Statistical Science at Southern Methodist University and served as department chair for 16 years. He is a Fellow of the American Statistical Association (ASA), an elected member of the International Statistical Institute, and a recipient of ASA's Don Owen Award. Woodward has been recognized as an SMU Altshuler University Distinguished Teaching Professor, and in 2006 he was named the United Methodist Church Scholar/Teacher of the Year at SMU. Woodward directed or co-directed 21 Ph.D. students and has published over 75 research articles and four previous books, one of which is on the topic of time series analysis.

Bivin Sadler has over 20 years of experience in teaching mathematics and statistics at both the undergraduate and graduate levels. Sadler began teaching in the newly formed Master of Science in Data Science (MSDS) program at Southern Methodist University in 2014. He was promoted to be the Lead Faculty of this program in 2018 and has since created several courses in the curriculum including co-creating the time series course, along with Woodward, in which much of the pedagogy of this book was polished. Bivin serves on SMU's Big Data Advisory Board, regularly presents at local, state, and national venues, and in 2022 he received the "M" Award and the Rotunda Scholar's Teaching and Service Award, both of which are highly regarded awards at SMU.

Stephen Robertson is Senior Lecturer and the Director of Southern Methodist University's Master's in Applied Statistics and Data Analytics (MASDA) program. Prior to his employment at SMU, Robertson spent 12 years working in government and industry. Through the frequent teaching of many undergraduate and graduate courses including master's level time series analysis at SMU, he has extensive experience in the classroom. He is the recipient of a number of teaching awards at SMU, including the prestigious Provost's Teaching Award in February 2020. Robertson is an innovative teacher who seeks creative and accessible teaching methods to explain time series topics in an applied setting.

Working with Data Collected Over Time

1

1.1 INTRODUCTION

Welcome to the fascinating world of Time Series Analysis! Especially for the statistician, data scientist, or quantitative analyst who must analyze a wide variety of data types, time series analysis is an important field of study. The applicability of time series analysis to "real-world" data will be frequently exemplified throughout the text. While the focus of this book is on application, sufficient theoretical detail is provided for the reader to appreciate the underlying techniques.

In recent years, there has been a major increase in the availability of vast amounts of data. This relatively new phenomenon has (fortunately, for you!) created a shortage of individuals who can analyze data, offering attractive employment and career opportunities for those willing to learn how to analyze such data. Furthermore, of the many types of available data, it is important to note that a substantial share is categorized as *time series data*. Consequently, industrial analytics groups have realized the value of utilizing time series analysis techniques for analyzing real data. As this developing pattern continues, it is inevitable that knowledge of time series will become a common expectation of professional data analysts, so learning how to analyze time series data will be beneficial (and lucrative!) for your career.

Why are time series data different?

It is important to understand that time series data cannot be appropriately or accurately analyzed using standard statistical techniques most frequently learned in undergraduate and even graduate statistics courses. In these courses, a key assumption is that the data are randomly sampled, resulting in independent observations. This assumption is violated for time series data, and it is inappropriate to apply classical statistical analysis techniques in such time-dependent settings. Fortunately, important (but different) information is present in time series data, and corresponding adjustments in the methodology take advantage of this information. The goal of this book is to address this very issue; that is, the book is designed so that *readers will be equipped with the tools necessary to analyze time series data* that are encountered in practice.

Various time series applications

To which types of naturally occurring settings does time series analysis apply, and why? Consider the following scenarios and identify a common theme. A cardiologist monitors a patient's cardiogram results to recognize irregular heartbeats over time. A geologist studies a seismograph to identify unusual deviations in recent days which may help predict a future earthquake. A stockbroker closely tracks stock prices over a time period of interest to optimize investment strategies. A major retail chain utilizes historical data to predict consumer demand of a particular product throughout the year to better manage inventory. A professional sports team uses data on annual ticket sales to correlate the delayed time effect of widespread advertising campaigns on ticket sales. A climatologist studies global temperatures over an extended period of time to detect potential shifting patterns in temperatures. What key word do all of these scenarios have

DOI: 10.1201/9781003089070-1

in common? *Time!* In each of these instances, data are measured sequentially in time, and the questions of interest specifically revolve around how the corresponding variables are dependent on time.

Relevance of time series

Why is time series analysis a relevant field of study? As illustrated in the previous examples, many important questions are associated with time series data, and the analysis of such data can help answer these questions. For example, what if the cardiologist discovers, by studying cardiogram data, that lives could be saved by detecting a previously unidentified abnormality in heartbeats over time? Or, what if the retail chain determines with a high degree of certainty how seasonal patterns predict customer demand of a product of interest, suggesting a better inventory strategy to maximize profit? What if the stockbroker is able to identify subtle indicators in the recent daily markets which give evidence that a particular stock purchase is likely to yield high returns? The wide variety of this relatively short list of examples further supports the applicability and relevance of time series data.

Perhaps an even more visible example of relevant time series data occurred (ironically) during the writing of this book in 2020–2021. During that time, the world-wide COVID-19 pandemic raced rampantly across the globe, creating a flood of concerned interest in daily, weekly, and monthly data that would provide crucial information about new cases, cured cases, and deaths caused by COVID-19. This was a unique time frame in which essentially all news stations, radio stations, and other media sources frequently reported variations of these time series data to inform the public about the most up-to-date status of the COVID-19 virus. The daily tracking of virus deaths was a question of interest to all, and because of this widespread interest, the relevance of summarizing and processing data, and in particular, time series data, was suddenly in the world's spotlight.

A final example of time series data that has potentially significant world-wide implications had also gained much momentum around the same time frame as the COVID-19 pandemic. The introduction of Bitcoin in 2009 led, according to Forbes Advisor (2021), to the world's first "cryptocurrency"; that is, Bitcoin is a decentralized digital currency which allows individual investors to buy and sell directly without a bank. Forbes describes Bitcoin as "digital gold" that investors can use to hedge against market volatility and inflation. Interestingly, the price of the first coins was under $150 per coin, but this price had consistently surged and as of April 2021 (the time of the writing of this book) had risen to nearly $50,000. It is only speculative to guess at this time where the price will go, but it is nearly certain it will be a focus of governments and markets for years to come! It is also clear that time series analysis will be used to provide informed forecasts of future behavior.

Where to start?

These examples illustrate that time series data are prevalent and are of valuable interest to a world-wide audience. As a result, several natural questions come to mind.

- How does one gain access to such data?
- What software and coding skills are required to begin analysis of time series data?
- What mathematical and statistical knowledge is necessary to really understand time series data and to have a working knowledge of the techniques used to adequately analyze the data?
- What assumptions must be considered before making conclusions, and what are the limitations and qualifications of these conclusions?

These are the very questions that this book will answer.

Guided by invaluable industrial experience and from teaching many semesters of applied time series students, the authors have devoted much time and consideration to the presentation of topics covered in this book. It will soon become apparent that our strategy is rather different from most textbooks. In addition to presenting a plethora of useful techniques, our goal is to provide enough discussion to help readers understand the techniques without having to take a "deep dive" into the theory. The journey is intended to be a particularly "hands-on experience" – we highly encourage you to participate as you follow along! The reader will frequently be provided with a set of codes to reproduce results from an example; in addition, QR codes throughout the text are provided which are supplementary short teaching videos that allow the interested student to learn more detail about various topics. Such activities and problem sets have been designed and included to enhance your learning experience. We are pleased to be a part of your time series education, so with these thoughts in mind, have fun as you learn about the following time series topics!

1.2 TIME SERIES DATASETS

Time series data exhibit a variety of behaviors which we will discuss in the following sections. We will illustrate these behavior types using "real-world" examples such as the intriguing sunspot data, temperature data, DOW Jones and individual stock prices, sales data (monthly, daily, …), and more. We begin by discussing data that have some sort of cyclic behavior.

1.2.1 Cyclic Data

Many time series datasets have a cyclic pattern, by which we will mean that the data display rises and falls in somewhat of a repetitive fashion. Such data are sometimes referred to as "pseudo-periodic", a term that we will use synonymously with "cyclic".[1] The sunspot data in Figure 1.1 and the Dallas-Ft. Worth (DFW) monthly temperature data in Figure 1.3 are examples of cyclic data. We discuss these two datasets and examine their similarities and differences.

1.2.1.1 Sunspot Data

Figure 1.1 shows annual sunspot data for the years 1700–2020. Sunspots are areas of solar explosions or extreme atmospheric disturbances on the sun. In 1848, the Swiss astronomer Rudolf Wolf introduced a method of counting sunspot activity, and monthly data using his method are available since 1749. See Waldmeier (1961).

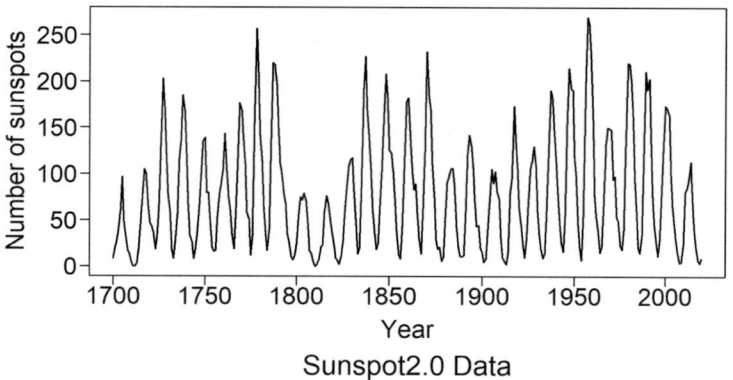

Sunspot2.0 Data

FIGURE 1.1 Annual number of sunspots from 1700 through 2020.

---

1 Data that are *truly periodic* have a behavior that repeats *exactly* over a fixed time frame. A good example of purely periodic data is the sine curve. Thus, pseudo-periodic (or cyclic) data are data that *tend* to repeat behaviors.

Sunspots have created a considerable amount of interest in the scientific community for two main reasons:

(1) Sunspot activity tends to affect us here on earth. For example, high sunspot activity causes interference with radio communication and is associated with higher ultraviolet light intensity and northern light activity.
(2) Sunspot activity has a cyclic behavior that has a cycle length of about 11 years. Examining Figure 1.1 shows that there are 29 cycles in the 321 years for an average cycle length of about 11 years. Actually, cycle lengths tend to randomly vary from 9 to 13 years.

While the cyclic behavior in Figure 1.1 is clear, it is often useful to examine short snippets of the data to better visualize the specific behavior. Figure 1.2 shows the number of sunspots from 1867 through 1950. The vertical lines identify the years at which there was a peak in the sunspot numbers and the horizontal arrows represent the time between the peaks. For the years plotted in Figure 1.2, the cycle lengths were 13, 10, 12, 12, 11, and 9 years, respectively. The cycle lengths seem to vary randomly, and there does not appear to be an "adjustment to a fixed cycle length". In fact, it is the understanding of these authors that scientists do not have a physical explanation for the approximately 11-year cycle. The sunspot data are a classic example of cyclic data with varying cycle lengths. In fact, Yule (1927) developed the autoregressive process as a means of describing the "disturbed" periodic behavior of the sunspot data. Recognizing this behavior will be critical when we forecast data in future chapters.

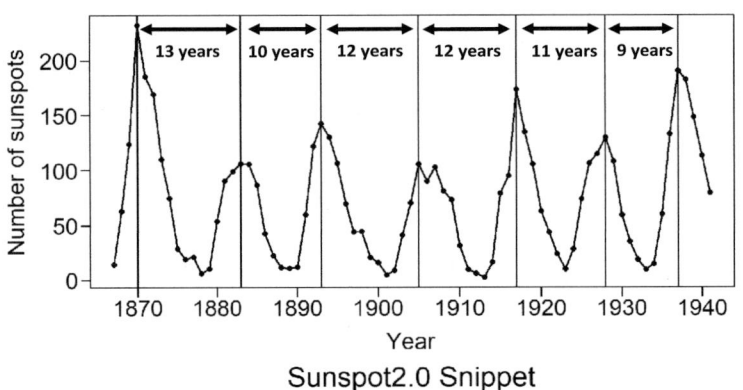

Sunspot2.0 Snippet

FIGURE 1.2   Snippet from Figure 1.1 showing years 1867–1950.

Key Point: Viewing time series realizations over a short time span can improve the interpretation of the data behavior.
— We will refer to this as viewing data over a *snippet* of time

It is important to note that beginning in July, 2015 a new method for enumerating sunspot activity has replaced the Wolfer[2] method, and a revised dataset has been developed. See Clette, Cliver, Lefêvre, Svalgaard, Vaquero, and Leibacher (2016). Using the new method, annual data are available from 1700 to the present, and monthly values exist for the years 1749 to the present. To avoid confusion, the new sunspot counting version is numbered 2.0. The new data are available at http://sidc.oma.be/silso/home which is discussed in Section 1.3.7.[3]

---

2 "Wolfer" is the commonly used name for the famous sunspot data originated by Rudolf Wolf.
3 SILSO data/image, Royal Observatory of Belgium, Brussels.

Note

1. Sunspot data used in this book will always be version 2.0.
   - Although the original version of the sunspot data has been widely published and modeled, future sunspot numbers will be of the 2.0 form.
   - We use version 2.0 to provide the ability to evaluate forecasts of future values which will be available only in 2.0 form.
2. As of the writing of this book, the sunspot data available in Base R are the original Wolfer version, and as such, are not available after July 2015.
3. We caution readers to be alert to the fact that there are now two sets of sunspot data.

1.2.1.2 DFW Temperature Data

Figure 1.3 shows the average monthly temperature for Dallas-Ft. Worth (DFW) (where the authors live) from January 1900 through December 2020. Although it is difficult to see clearly, temperatures follow the expected pattern. That is, they are low in the winter and high (for DFW very high!) in the summer.

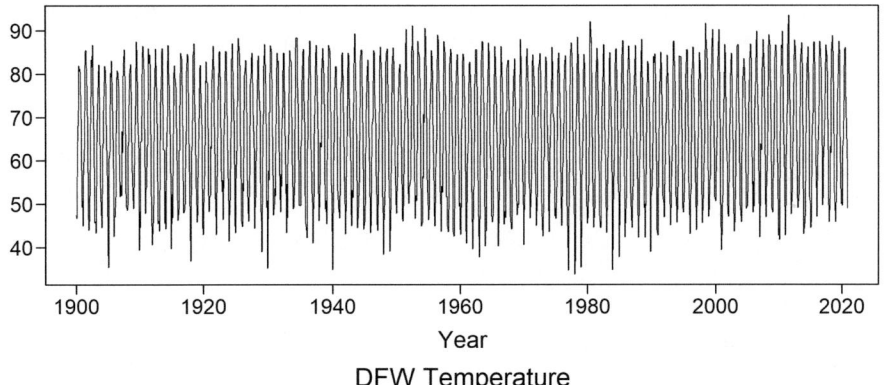

FIGURE 1.3   Average monthly temperature for Dallas-Ft. Worth for January 1900 through December 2020.

By viewing the data in Figure 1.4, we again see the value of focusing on a snippet of time, which in this case is for the years 1979 through 1986. For our purposes, we call a cycle the number of months between the hottest month of each year. We see that, in DFW, the hottest month is either July or August. Note that the temperature data have a smooth progression from summer to winter and again from winter to summer with the average temperatures in the fall and spring being similar. The resulting overall pattern is "sort of" sinusoidal, or pseudo-sinusoidal.

It is useful to note that random variations (noise) will be present in data. The hottest month in 1979 through 1981 was July.[4] The hottest month in 1982 was August, so the third cycle in the plot has length 13 months. August was also the hottest month for 1983 through 1985, so the next three cycles are of length 12 (August to August). Finally, in 1986 the hottest month was July, so that the corresponding cycle length is 11 months. While the cycle lengths were not all equal to 12 months, note that whenever a 13-month cycle occurs, it is always followed by either a 12-month or an 11-month cycle to "stay in or get back in sync". That is, suppose the hottest month in one year is July and then the two succeeding years had temperature cycle lengths of 13. That would indicate that the hottest month in the third year was September, which has never been the hottest month in DFW in the dataset available back to September of 1898. The DFW

---

4 Those living in the DFW area in the summer of 1980 will have that summer "burned" in their memory.

temperature data are an example of cyclic data with a *fixed* cycle length. In this case, the 12-month cycle has a physical cause − regular and predictable motion of the Earth around the Sun. Because the cycles in the temperature are related to the calendar year, the temperature data are referred to as *seasonal* data.

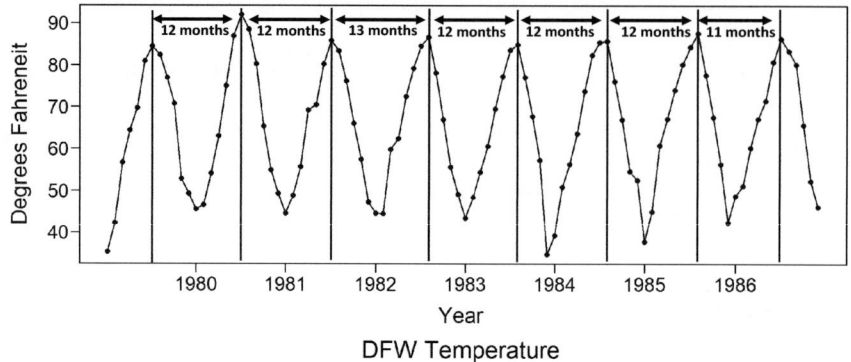

FIGURE 1.4   Average monthly temperature in Dallas, Texas from 1979 to 1986.

Note: The key difference between the sunspot data and the temperature data is that while the sunspot cycle lengths tend to vary randomly between 9 and 13 years, the temperature cycle lengths are fixed to the point that they stay "in sync" with the 12-month year.

> **Key Points:**   The sunspot and temperature data are *cyclic* data. However:
> 1.   The sunspot cycles lengths seem to vary randomly from 9 to 13 years.
> 2.   The temperature data have a *fixed*, physically explainable cycle length.
>   −   Because the cycles in the temperature are related to the calendar year, the temperature data are an example of *seasonal* data.
>   −   The temperature data can also be described as pseudo-sinusoidal.

1.2.1.3 Air Passengers Data

Figure 1.5 is a dataset containing the total number (in thousands) of international airline passengers per month for the 12 years from 1949−1960. This dataset has been extensively analyzed and is a classical dataset in the time series literature. The data go through a 12-month cyclic pattern that is similar from year to year and is based on the calendar year. Thus, the Air Passengers data are another example of *seasonal* data. Additionally, the data tend to be trending upward in time. That is, the number of airline passengers is increasing in time. Trending behavior in time series will be discussed in Section 1.2.2. There is also an expanding within-year variability. This type of behavior, referred to as multiplicative seasonality, will be discussed in Chapter 2.

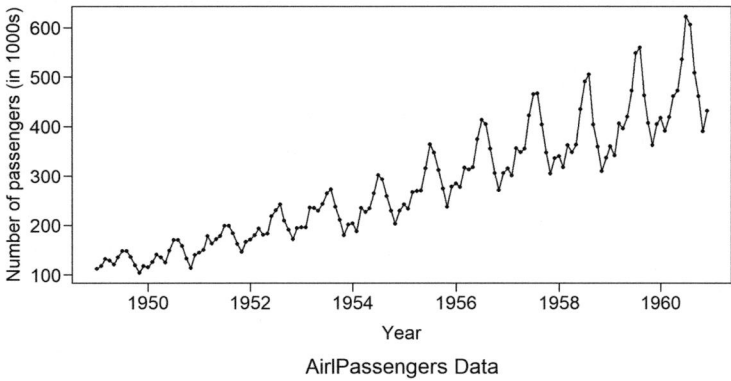

FIGURE 1.5   Number of International Passengers on Airlines from 1949 to 1960.

Figure 1.6 is a snippet of the air passenger data from 1957 to 1960. It can be seen that air travel is light from January through April, is high during the summer months, and begins to drop in September through November with a slight increase in December. This pattern, although not sinusoidal, is repeated from year to year. The cyclic behavior of the air passenger data is repeated on an annual basis, and is an example of *seasonal* data that is not pseudo-sinusoidal.

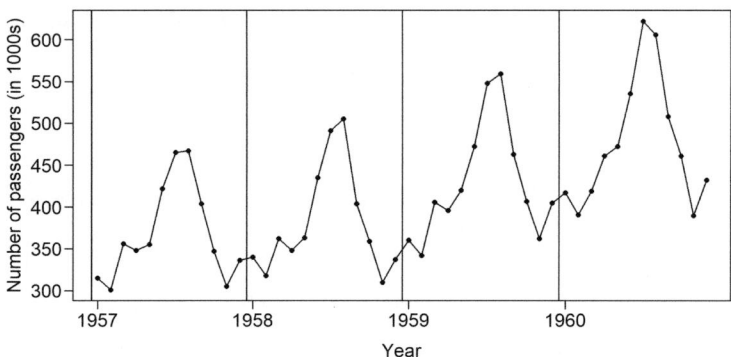

FIGURE 1.6   Number of International Passengers on Airlines from 1957 to 1960.

1.2.2 Trends

A trend is a tendency for data to increase (or decrease) steadily over time. We noted that the Air Passengers data in Figure 1.5 have an increasing trending behavior in addition to the seasonal pattern noted. A *linear* trend would be a "tendency" for the data to increase (or decrease) in a *linear* fashion (see Figure 1.7(a)). Trends might tend to follow a curve such as the exponential trending shape in Figure 1.7(b). Figure 1.7(c) is a time series that has a downward trend, but is more irregular in nature than the trends in Figures 1.7(a) and (b). A typical pattern for datasets is *random trending behavior* such as that in Figure 1.7(d), which has the appearance of aimless *wandering*. That is, there may be a series of short or long trends, sometimes in opposite directions.

Key Point: Data with trending and random wandering behavior are not cyclic in nature. They are sometimes called *aperiodic* (not periodic) because there is no regular rise and fall behavior.

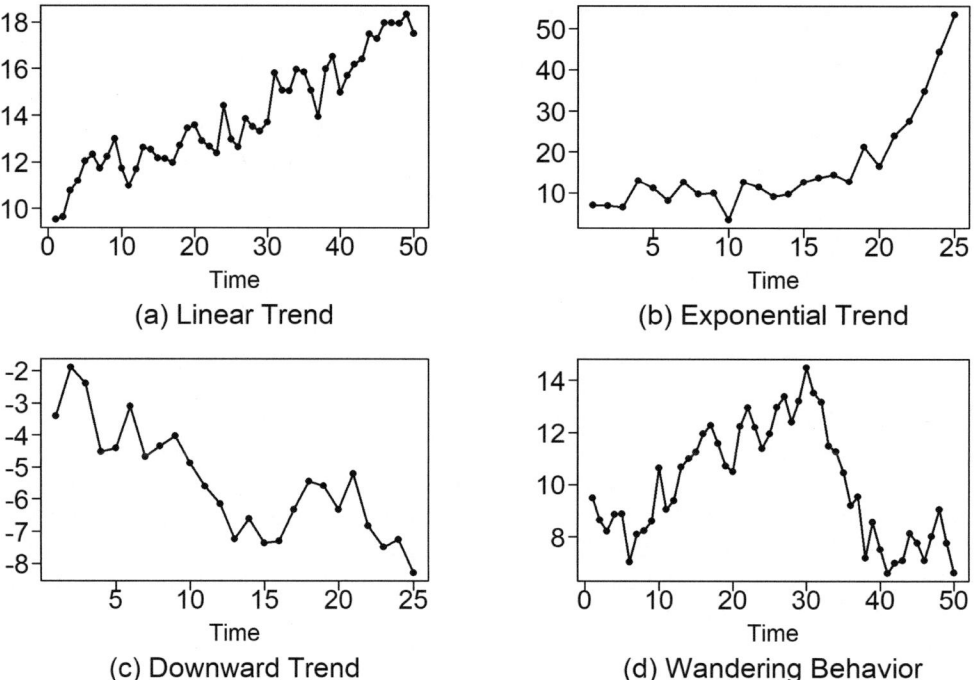

FIGURE 1.7 Plots showing (a) a linear trend, (b) an exponential trend, (c) a downward irregular trend, and (d) wandering pattern.

Trends may be the main feature of a set of time series data such as for the time series in Figure 1.7. However, the air passenger data in Figure 1.5 show that a set of data may have seasonal and trending behavior. The possible variations are endless.

1.2.2.1 Real Datasets That Have Trending Behavior

(1) Monthly Dow Jones Closing Average
The Dow Jones (DOW) has been used as a measure of the health of the US stock market for over 100 years. Figure 1.8(a) shows the monthly closing averages for the years 1985 through 1995. There we see an upward trending behavior with a dip associated with Black Monday.[5] However, the DOW averages recovered and continued to increase far beyond the pre-Black Monday levels.

(2) West Texas Intermediate Crude Oil Price
Figure 1.8(b) is a plot of monthly price of West Texas Intermediate Crude Oil (WTI) from January 1990 through June 2008. The price of this grade of oil (affectionately known as "Texas Light Sweet") is used as a benchmark for fuel prices around the world and can often be seen in professional reports and heard in newscasts. The data shown in Figure 1.8(b) appear to have an even more explosively increasing trend than the DOW data in Figure 1.8(a).

(3) Bitcoin
Figure 1.8(c) shows daily Bitcoin closing prices from May 1, 2021 through April 30, 2021. These data tend to be trending upward and are of interest because the use of cryptocurrency (or virtual money) is in

5 October 19, 1987 is referred to as Black Monday because the DOW dropped almost 22% to become one of the most notorious days in US financial history.

its infancy. The financial experts, at the time of this writing, are split on their predictions concerning the growth (or even the permanence) of this type of currency.

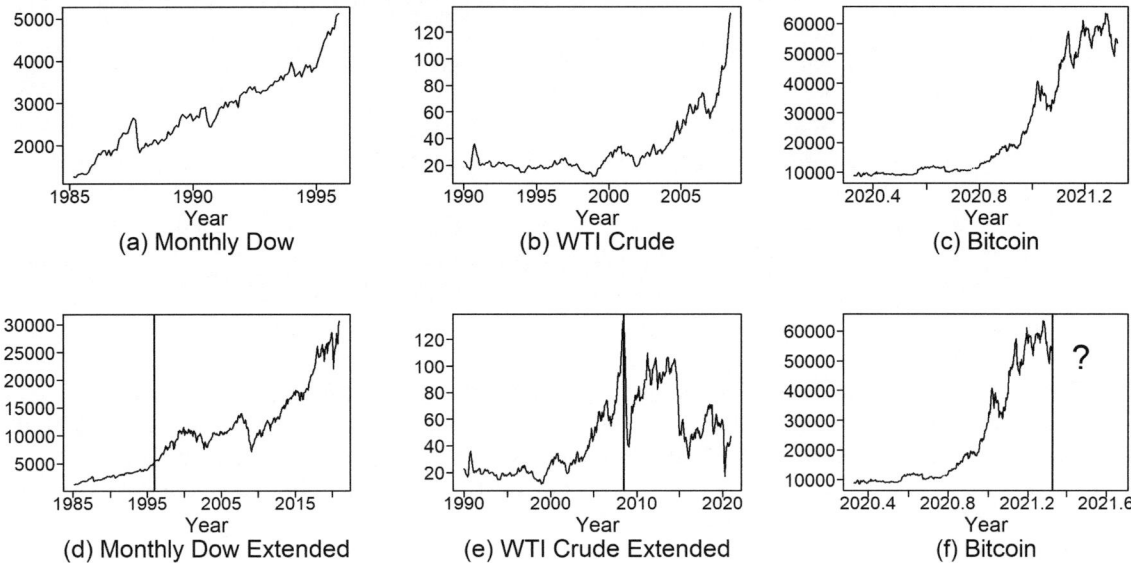

FIGURE 1.8 (a) Monthly DOW closing averages from March 1985 through December 1995, (b) monthly WTI Crude oil prices from January 1990 through June 2008, (c) Daily Bitcoin prices from May 1, 2020 through April 30, 2021, (d) monthly DOW closing averages from March 1985 through December 2020, (e) WTI crude oil prices from January 1990 through December 2020, and (f) Bitcoin prices as in (c) with the next 120 days unknown as of April 30, 2021.

1.2.2.2 The Problem with Trends

Plots of data such as those in Figures 1.8(a)−(c) are informative in the sense that we gain information about the behavior of a time series *within the time frame of the plot*. For example, Black Monday shows up clearly and the plot helps us to understand its impact. However, in most cases the question of interest is, *"Will the current trending behavior continue?"* To illustrate the "flaky" nature of trends, we extend the time frame for the DOW and WTI datasets. Figure 1.8(d) shows the DOW data from 1985 through 2020 with a line drawn at the end of 1995 where Figure 1.8(a) ended. There it is seen that the trending behavior continues after 1995 "on the whole". The increase was not without its temporary "dips". For example, it is easy to see the effects of the Great Recession between 2007−2009 and the COVID effects on the market in 2020. However, the stock market continues to increase, and as of the writing of this book, is above "pre-COVID" levels.

Note that in June 2008, based on the WTI data in Figure 1.8(b), one would probably predict a continued rise in oil prices such as had been seen in the previous 15 years. Banking heavily on such a prediction would not have been wise! Note that, in contrast to the DOW data, the WTI data in Figure 1.8(e) dipped precipitously in late 2008 and then wandered without displaying much evidence of sinusoidal, seasonal, or overall trending behavior. Any perceived trends in such data may vanish at a moment's notice. This is an important characteristic that is prevalent in many time series datasets, for example, stock market data for individual stocks. Again, understanding this behavior will be useful in modeling and forecasting such datasets.

In Figure 1.8(f) we again show the Bitcoin prices for 2021 through April (the time of this writing). The data show a nice upward trend, but time will tell whether the trend continues.

Key Points:

1. Trending behavior in a time series provides information about the time frame *in which the data were collected.*
2. Predictions that trending behavior will continue is analogous to *extrapolating* in regression beyond the range of the predictor variables.
3. We will discuss trending behavior and tests for trend in Chapters 7–8.
4. At this point we note that predictions about future values should involve domain knowledge as well as data analysis.
   - For example, while in the short term, the stock market tends to "bounce around". A longer range look indicates that a continued long-term trend should be expected to continue unless conditions change regarding the strength of the US economy. See Figure 1.9.

Figure 1.9 shows the year-end DOW price for the years 1915 through 2020. A continued upward trend overshadows isolated behaviors like the Great Depression beginning in 1929, Black Monday in 1987, the Great Recession in 2007–2009, and so forth. Data such as that in Figure 1.9 might lead the analyst to predict a long-term increase in the DOW. *Caution:* This obviously does not hold for individual stocks because a corporation or entire industry may cease to be viable.

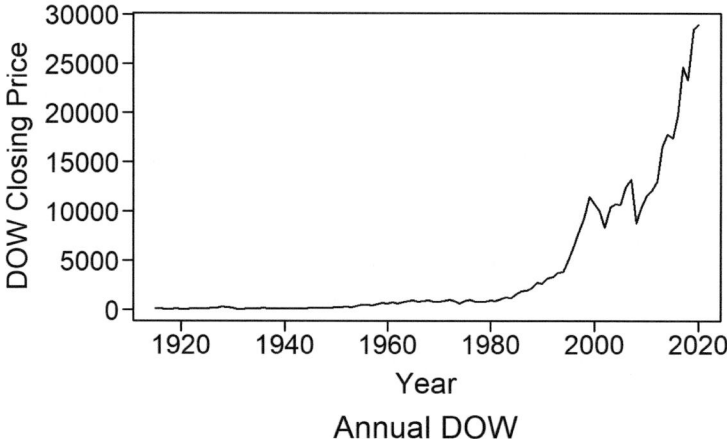

FIGURE 1.9   Annual year-end DOW closing prices for 1915 through 2020.

1.3 THE PROGRAMMING LANGUAGE R

Modern time series analysis requires proficiency in one or more of the readily available software tools such as R, Python, SAS, SPSS, Minitab, and STATA just to name a few. We will focus on R in this text as it is easily available on Mac, PC, and Linux operating systems, is open source with over 12,000 user-supplied packages in the CRAN repository, is widely used by the academic, industrial, and commercial communities, and has a free development environment called RStudio.

There are an increasing number of data science consulting companies. One of the leaders is a company named *Fingent*, which states in its comparison between R and other software (Python, etc.) with respect to time series analysis:

R offers one of the richest ecosystems to perform data analysis. Since there are 12,000 packages in the open-source repository, it is easy to find a library for any required analysis. Business managers will find that its rich library makes R the best choice for statistical analysis, particularly for specialized analytical work.

...

Since R is developed by academicians and scientists, it is designed to answer statistical problems. It is equipped to perform time series analysis. It is the best tool for business forecasting.

—Fingent

A minimal level of R experience is assumed, although we provide video resources to bring those new to R up to this minimal level. The video resources associated with this book are available via QR code as well as by means of a list of URLs on the website. The QR code can be accessed by simply pointing the camera of most smartphones at the QR Code, after which a prompt automatically accesses the video on the web. This is a nice feature because of the ability to access video instruction on a smartphone or PC while working on the computer. We believe that this feature is both quick and intuitive. The first QR code accessible video in this book is a tutorial about loading R and RStudio, which is a great way to get a running start!

QR 1.1 Installing R and RStudio

1.3.1 The *tswge* Time Series Package

The R package *tswge,* which is available both on CRAN and on the book's GitHub site, provides a set of functions and datasets to accompany this text. As mentioned above, time series data may possess a wide variety of characteristics and behaviors. This fact, combined with the various types of questions of interest that may be paired with the data, motivates the methods, models, tests, and techniques presented in this book. These tools have been collected in *tswge*. The *tswge* functions will be introduced in the text, and then details about each function's purpose, input parameters involved in the function statement, and output variables are given in an appendix to the chapter in which the function is introduced.

We believe that both inexperienced and seasoned R users will find the videos and examples in this book easy to follow and hopefully informative. Most of the coding examples are designed so that the readers can (and are very much encouraged to) run the code directly on their own R console and experience the example for themselves!

Note 1: In order to use the R code provided in the book, you will need to load R onto your computer. From this point going forward, we will assume that you have downloaded R, installed *tswge* from CRAN, and loaded *tswge*.

Note 2: If you haven't done this already, NOW is the time to perform these tasks! The following step-by-step instructions will help.

Note 3: The instructions below are applicable to both PCs and Macs.

Step 1: Download R onto your computer. (RStudio is covered in the tutorial and will not be discussed here.)
(a) Go to https://cran.r-project.org/mirrors.html
(b) Select a CRAN mirror site close to you

(c) In the box labeled "Download and Install R", click on the link associated with your operating system.
(d) Click on the link *Download R xx.xx.xx* (The R version will change in time, thus the xx.xx.xx.)
(e) Open the .exe file and answer the questions to install R on your system. (We recommend that you choose "base" and the default options at each step.)

Step 2: Open the R package at which time you should see the command console.

Step 3: Install tswge

(a) Select *Packages* in the top menu on the R screen
(b) Select *Install Package(s)...*
(c) Choose a CRAN mirror closest to you
(d) You will be given a *long* list of packages. Choose *tswge*.

Load tswge
In order to use the *tswge* package you must access it for use. You can use either of the following methods:

(i) From the top menu under Packages, select *Load Package...*
    − you will be given a list of packages available to you. Select *tswge*.
(ii) Enter the command
    library(tswge)
    at the beginning of any code that uses *tswge*.

QR 1.2 Loading tswge

1.3.2 Base R

As you downloaded R onto your computer, you chose "base", which includes what we will refer to in this book as "Base R". Base R is a collection of functions and datasets, many of which relate to time series analysis. The functions and datasets are always available to you in an R session. The **plot** function, which we will discuss in the next section, is a readily accessible Base R function.

1.3.3 Plotting Time Series Data in R

The first thing a time series analyst should do when analyzing a time series is to ***plot the data***. In the following example, we discuss considerations involved in plotting time series data in R.

Example 1.1 Plotting a Time Series Dataset
*A **standard dot plot**:* One dataset in *tswge* is **dfw.mon,** which contains the average monthly temperatures in DFW from 1900 through 2020. This dataset is plotted in Figure 1.3. In this example, we will only use the data for the 10 years from January 2011 through December 2020, which for our purposes is contained in the *tswge* dataset **dfw.2011**. To access the data (once you have loaded the *tswge* package), issue the command

```
data(dfw.2011)
```

By typing the command

```
dfw.2011
```

the dataset's contents are listed as follows:

```
  [1]  42.8 49.5 61.3 70.8 72.8 86.8 91.4 93.4  80.0 68.2  57.9 47.6  50.4  52.5
 [15]  64.3 70.3 77.9 84.3 87.7 86.5 80.0 67.0  59.7 51.2  49.1 52.0  56.4  63.0
 [29]  72.3 82.6 84.5 87.1 82.4 68.2 53.5 43.1  45.3 47.0  55.1 66.3  74.4  82.4
 [43]  83.8 86.2 80.3 71.6 51.5 50.1 44.5 45.7  56.1 65.8  70.9 82.1  87.1  87.3
 [57]  82.7 71.2 58.7 53.7 47.0 55.2 61.2 68.1  72.5 84.0  87.4 85.8  81.5  74.1
 [71]  63.5 49.7 51.2 60.6 65.7 69.3 75.4 82.5  86.6 84.4  80.6 69.6  62.4  49.7
 [85]  45.8 51.1 63.3 61.6 79.0 85.7 88.8 85.2  78.1 66.2  52.5 48.4  45.8  50.2
 [99]  55.0 66.0 73.4 79.9 84.6 87.4 85.5 65.5  53.5 50.0  50.3 49.6  63.4  64.6
[113]  73.8 81.9 85.7 86.0 74.7 65.0 60.4 49.0
```

Using the plot command in Base R, that is,

```
plot(dfw.2011)
```

we obtain the plot in Figure 1.10(a). To put it lightly, this plot is a mess. The plot looks like a random pattern of points. Certainly, the seasonal cyclic behavior of the DFW temperature data is not visible. This is because the R function **plot** produces a dot plot (or scatterplot) by default. The scatterplot is useful for showing the relationship between two variables, like height and weight. However, given the data in **dfw.2011**, the variable on the horizontal axis is the vector index $t = 1, 2, \ldots, n$ where n is the length of the vector, in our case $n = 120$. So, Figure 1.10(a) is a scatterplot between the vector index and the data in the vector, and is clearly inadequate for our purposes.

Connecting the dots: Time series data are collected along a time axis, and it is important to know how the data evolve in time. To see this more clearly, we "connect the dots (points)". Figure 1.10(b) uses the R command

```
plot(dfw.2011,type= 'l')
```

where **type= 'l'** tells the computer to connect the points with lines.

Note: The value for **type** is the letter **l** (for line), not the number 1.

The data in Figure 1.10(b) now show the seasonal pattern previously seen in Figure 1.4 but for a different time period.

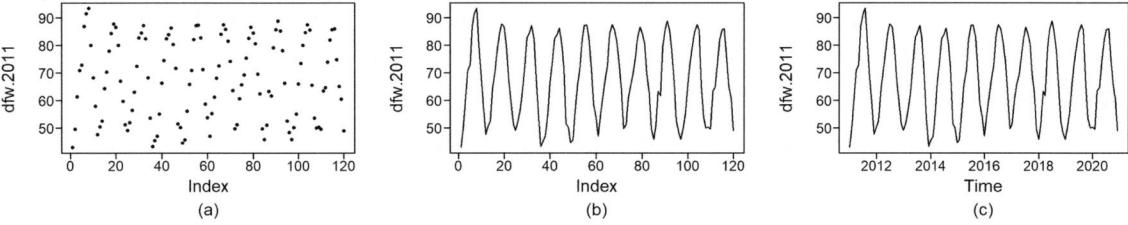

FIGURE 1.10 Monthly DFW temperature data: (a) as a dot plot, (b) the figure in (a) connecting the dots, and (c) the data in (b) with the horizontal axis correctly labeled in time.

The plot still has problems: Specifically, the horizontal axis in Figure 1.10(b) is indexed from time points 1 to 120 instead of years. In Section 1.3.4 we will discuss the *ts* object, which is a special object in R that is designed specifically to store time series data. Note that Figure 1.10(c) is a plot of a *ts* object, and we will discuss it in the next section.

1.3.4 The *ts* Object

We will introduce several sources and examples of time series data in this chapter. We begin our exploration with the **dfw.2011** dataset. The missing information in Figure 1.10(b) results from the fact that the dataset **dfw.2011** does not contain any date information. Examination of the data, which are listed above, shows that the data reside in a vector which is by default indexed from 1 to 120.

1.3.4.1 Creating a ts Object

As mentioned, R makes it easy to work with time series data by facilitating its storage in an aptly named *ts* object.

Example 1.2 Creating a *ts* Object using the dfw.2011 Data

We recall from Example 1.1 that the data in **dfw.2011** are monthly average temperatures at DFW for the years 2011–2020. We will create a *ts* object that incorporates this date information, and call it **dfw.2011.ts**. The command we use to create the *ts* object is the following:

```
dfw.2011.ts = ts(dfw.2011,start=c(2011,1),frequency=12)
```

This command says to place the vector data, **dfw.2011**, into a *ts* object and to associate it with dates that start with January 2011 (**start=c(2011,1)**). It also stipulates that the data are monthly (**frequency=12**). Now by typing the command

```
dfw.2011.ts
```

the contents of this *ts* object are listed as follows:

	Jan	Feb	Mar	Apr	May	Jun	Jul	Aug	Sep	Oct	Nov	Dec
2011	42.8	49.5	61.3	70.8	72.8	86.8	91.4	93.4	80.0	68.2	57.9	47.6
2012	50.4	52.5	64.3	70.3	77.9	84.3	87.7	86.5	80.0	67.0	59.7	51.2
2013	49.1	52.0	56.4	63.0	72.3	82.6	84.5	87.1	82.4	68.2	53.5	43.1
2014	45.3	47.0	55.1	66.3	74.4	82.4	83.8	86.2	80.3	71.6	51.5	50.1
2015	44.5	45.7	56.1	65.8	70.9	82.1	87.1	87.3	82.7	71.2	58.7	53.7
2016	47.0	55.2	61.2	68.1	72.5	84.0	87.4	85.8	81.5	74.1	63.5	49.7
2017	51.2	60.6	65.7	69.3	75.4	82.5	86.6	84.4	80.6	69.6	62.4	49.7
2018	45.8	51.1	63.3	61.6	79.0	85.7	88.8	85.2	78.1	66.2	52.5	48.4
2019	45.8	50.2	55.0	66.0	73.4	79.9	84.6	87.4	85.5	65.5	53.5	50.0
2020	50.3	49.6	63.4	64.6	73.8	81.9	85.7	86.0	74.7	65.0	60.4	49.0

It is clear from the data listing that the time series starts in January 2011 and ends in December 2020.

Now that we have the data in the *ts* object **dfw.2011.ts**, we can issue the command

```
plot(dfw.2011.ts)
```

and the resulting plot is in Figure 1.10(c). This plot "connects the points" and provides the accurate date information.

If we issue the command

```
class(dfw.2011.ts)
```

the output

```
[1] "ts"
```

informs us that **dfw.2011.ts** is a *ts* object.[6] In contrast, by issuing the command

```
class(dfw.2011)
```

---

6 The *ts* objects need not have a ".ts" extension. We have named the file **dfw.2011.ts** as a reminder that it is a *ts* file. However, it could have been named **dfw.2011** or **dfw.last10**.

we obtain the output

```
[1] "numeric"
```

indicating that **dfw.2011** is simply a numeric vector.

Example 1.3 The **lynx** dataset

Another classical dataset is the "lynx" data. This dataset contains the annual number of Canadian lynx that were trapped in the McKenzie River District in Canada between 1821 and 1934. It has been referenced in several seminal time series papers and software vignettes and is so popular that it is available in Base R as the *ts* object **lynx**. It may seem unusual that this dataset would be of such interest, so we will examine it further. We begin our examination by listing and plotting the data using the following commands:

```
data(lynx)
plot(lynx)
```

The **lynx** data are plotted in Figure 1.11.

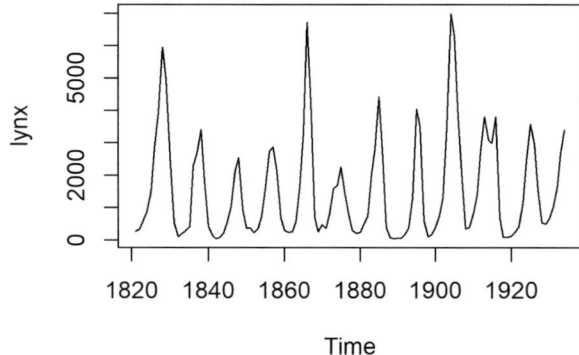

FIGURE 1.11   Annual number of lynx trapped in the McKenzie River district of Canada from 1821 to 1934.

From observing the data, we notice the following:

(a) The data appear to (surprisingly) have a 10-to-11-year cycle (period), and there is some evidence that the variable amplitude of the peaks has a pattern as well. This behavior is an example of cyclic behavior similar to the sunspot data in Figure 1.1(a). The somewhat puzzling 10-11-year cycle is the reason for the interest in this dataset.

(b) Examination of the plot shows that the dataset **lynx** must have some date information associated with it because the horizontal axis is in years, not the simple index 1 to 114. By issuing the command

```
class(lynx)
```

the output

```
[1] "ts"
```

indicates that the dataset **lynx** is already in *ts* form. Furthermore, we can see the information stored in this *ts* object by typing

```
lynx
```

after which we observe the following output:

```
Time Series:
Start = 1821
End = 1934
Frequency = 1
```

```
  [1]   269   321   585   871  1475  2821  3928  5943  4950  2577   523    98   184   279   409
 [16]  2285  2685  3409  1824   409   151    45    68   213   546  1033  2129  2536   957   361
 [31]   377   225   360   731  1638  2725  2871  2119   684   299   236   245   552  1623  3311
 [46]  6721  4254   687   255   473   358   784  1594  1676  2251  1426   756   299   201   229
 [61]   469   736  2042  2811  4431  2511   389    73    39    49    59   188   377  1292  4031
 [76]  3495   587   105   153   387   758  1307  3465  6991  6313  3794  1836   345   382   808
 [91]  1388  2713  3800  3091  2985  3790   674    81    80   108   229   399  1132  2432  3574
[106]  2935  1537   529   485   662  1000  1590  2657  3396
```

Along with the 114 annual counts of the number of lynx trapped, this *ts* object contains the start (1821) and end (1934) years of the time series. Recall that for the monthly data in the *ts* object **dfw.2011.ts**, the **frequency** attribute took on the value 12 indicating that there were 12 equally spaced observations per unit of time (year). For the **lynx** *ts* object, **frequency=1**, indicating that each observation represents a full year.

1.3.4.2 More About ts Objects

Returning our focus to the *ts* object itself, we discuss some features and uses for this powerful tool which is available for storing time series data.

(1) Extracting Vector Data from a ts Object

A *ts* object contains numeric data along with other attributes that assist in time series plotting. Suppose, however, that we have a *ts* object and want to extract the numeric vector containing (only) the data. This can be accomplished, using the *ts* object **dfw.2011.ts** as an example, with the following R command:

dfw.2011.num = as.numeric(dfw.2011.ts)

To verify that **dfw.2011.num** is a numeric vector, we issue the command

class(dfw.2011.num)

and obtain the output

```
[1] "numeric"
```

To actually view the data in the vector, we use the command

dfw.2011.num

and obtain the output

```
  [1] 42.8 49.5 61.3 70.8 72.8 86.8 91.4 93.4 80.0 68.2 57.9 47.6 50.4 52.5 64.3
 [16] 70.3 77.9 84.3 87.7 86.5 80.0 67.0 59.7 51.2 49.1 52.0 56.4 63.0 72.3 82.6
 [31] 84.5 87.1 82.4 68.2 53.5 43.1 45.3 47.0 55.1 66.3 74.4 82.4 83.8 86.2 80.3
 [46] 71.6 51.5 50.1 44.5 45.7 56.1 65.8 70.9 82.1 87.1 87.3 82.7 71.2 58.7 53.7
 [61] 47.0 55.2 61.2 68.1 72.5 84.0 87.4 85.8 81.5 74.1 63.5 49.7 51.2 60.6 65.7
 [76] 69.3 75.4 82.5 86.6 84.4 80.6 69.6 62.4 49.7 45.8 51.1 63.3 61.6 79.0 85.7
 [91] 88.8 85.2 78.1 66.2 52.5 48.4 45.8 50.2 55.0 66.0 73.4 79.9 84.6 87.4 85.5
[106] 65.5 53.5 50.0 50.3 49.6 63.4 64.6 73.8 81.9 85.7 86.0 74.7 65.0 60.4 49.0
```

Looking back at Example 1.1 we see that **dfw.2011.num** is identical to the original vector dataset **dfw.2011**.

(2) More on Creating a ts Object

Consider a time series defined in the vector **x** below:

x = c(10,20,30,40,50,60,70,80,90,100,110,120,130,140,150,160,170)

Assume this is monthly data (**frequency = 12**) which start in June, 2018 (**c(2018,6)**). To construct a *ts* object from this vector form of the time series, we invoke the following R commands (output follows):

```
xTSmonth = ts(x,start = c(2018,6),frequency = 12)
xTSmonth
```

```
      Jan  Feb  Mar  Apr  May  Jun  Jul  Aug  Sep  Oct  Nov  Dec
2018                           10   20   30   40   50   60   70
2019   80   90  100  110  120  130  140  150  160  170
```

Alternatively, if the same data were actually quarterly data (**frequency = 4**) and started in the 3rd quarter of 1986 (**c(1986,3)**), we could construct a *ts* object for the series stored in **x** using the code (output follows):

```
xTSquarter = ts(x,start = c(1986,3),frequency = 4)
xTSquarter
```

```
      Qtr1  Qtr2  Qtr3 Qtr4
1986                10    20
1987    30    40    50    60
1988    70    80    90   100
1989   110   120   130   140
1990   150   160   170
```

While the *ts* object has nice default formats for data recorded in months and quarters, we recognize that many time series will not fit this mold. Suppose that each observation is a day, and the unit of time is a week rather than a year. Furthermore, assume that the first observation was on the 4th day of the first week (**c(1,4)**). In this case the frequency would be 7 and the *ts* object would be constructed accordingly (output follows):

```
xTSweek = ts(x,start = c(1,4),frequency = 7)
xTSweek
Time Series:
Start = c(1, 4)
End = c(3, 6)
Frequency = 7
[1] 10  20 30  40 50  60 70 80  90 100  110  120 130 140  150 160  170
```

While the visual format is not as descriptive as it was for months and quarters, the information available is still useful. The input **start=c(1,4)** indicates that the data start on the 4th day of the first week. Because the series is 17 observations in length, it will take four observations to finish the first week, another seven to fill the second week, and then the remaining six observations will end on the 6th day of the 3rd week, consistent with the output above: **end=(3,6)**. While the daily data in this example are not presented in a convenient visual format, the period of seven in the data is stored in the **frequency** field.

Finally, we will often want to analyze only a subset of a *ts* object. This can be accomplished using the **window** function. With respect to the previous example, suppose we wanted to only work with data from the second week. We could accomplish this with the following code:

```
window(xTSweek,start = c(2,1),end = c(2,7))
```

```
Time Series:
Start = c(2, 1)
End = c(2, 7)
Frequency = 7
[1]  50  60  70 80  90 100  110
```

As another example, assume we want to subset the **AirPassengers** data to yield only the number of airline passengers during 1950:

```
window(AirPassengers,start = c(1950,1),end = c(1950,12))
```

```
      Jan Feb Mar  Apr May Jun Jul Aug Sep Oct  Nov Dec
1950  115 126 141  135 125 149 170 170 158 133  114 140
```

Key Points:

1. The *ts* object is a valuable tool for storing time series and their most important attributes in a single place.
2. The *ts* object can be useful in plotting and subsetting the series, and we will see that it is helpful in modeling as well.
3. There is a *multivariate* version of a *ts* object called an *mts* object for multivariate data structures (see footnote 9). Analysis of multivariate time series will be covered in Chapters 10 and 11.

1.3.5 The `plotts.wge` Function in *tswge*

The **plotts.wge** command in *tswge* extends the Base R plotting function, **plot**. The **plotts.wge** function can plot datasets consisting of numeric vectors as well as *ts* objects. This function makes it easy to produce useful plots that quickly provide the information needed by the analyst without specifying a lot of parameters for the graph. On the other hand, if it is important to obtain a higher quality plot for written reports, presentations, etc., **plotts.wge** allows you to "dress up" your plots as desired with colors, labels, line widths, and more.

Consider again the **lynx** dataset plotted in Figure 1.11. Note the asymmetric behavior of the lynx dataset, specifically that the peaks are much more variable than troughs from cycle to cycle. Most analysts who work with the **lynx** data analyze the logarithm of the **lynx** data instead of the **lynx** data themselves. An analyst may simply want to view this series quickly (Figure 1.12) to get an idea of any periodic behavior. The following code will take the logarithms and plot the log data.

```
data(lynx)
log.lynx=log(lynx)
# Note that log.lynx retains the ts file information contained in file lynx
plotts.wge(log.lynx)
```

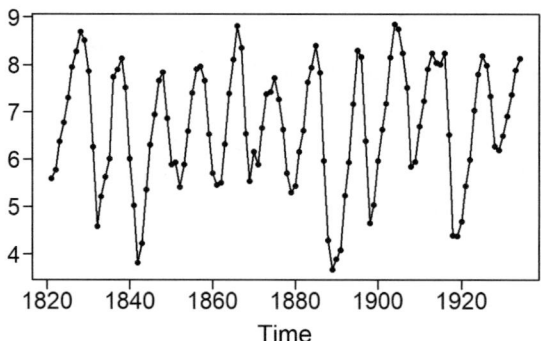

FIGURE 1.12 Basic **plotts.wge** plot of the log-lynx data.

From the plot we notice that the cyclic behavior is more symmetric in terms of variability of peaks versus troughs.

> **Key Point:** The *tswge* function `plotts.wge` produces plots that:
> - include data values as points in the graph if the realization length is less than or equal to 200
> - do not include the points in the graph if the realization length is greater than 200

1.3.5.1 Modifying the Appearance of Plots Using the tswge *plotts.wge* Function

You may want to add customized labels, a title, color, or other plotting options that make the plot more informative and visually appealing. An example of code that would render more detail to the plot is given in Figure 1.13 below. A description of the function can be displayed in R with the command: `?plotts.wge`. The description is also included in Appendix 1A.

```
plotts.wge(log.lynx, style = 1, xlab = "Date", ylab = "Logarithm of Lynx Data",
main = "Natural Logarithm of Number of Lynx Trapped from 1821 and 1934 ", text_
size = 12)
```

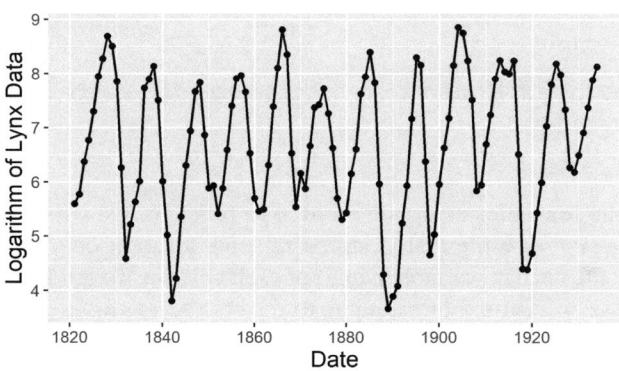

FIGURE 1.13   Example of a plot of the classic log-lynx data with style = 1.

> **Key Point:** *tswge* functions will be described in the appendix of the chapter in which they are introduced.
> - `plotts.wge` is described in Appendix 1A

1.3.6 Loading Time Series Data into R

To this point, we have worked with data that were "native to R"; in practice, of course, we will usually want to load external data into R. The comma separated value (`.csv`) format is the format of many time series datasets we will discuss in Section 1.3.7. Also, external data may be stored in text (`.txt`) files and Excel (`.xlsx`) files. Sections 1.3.6.1 and 1.3.6.2 provide useful information for reading `.csv` and `.txt` files, respectively.

1.3.6.1 The .csv file

In a **.csv** file, each observation is separated by a comma. While *.csv* stands for "comma-separated values", other delimiters such as semicolons, are allowed. On the book's GitHub site, a file named **AirPassengers.csv** is a **.csv** file containing the air passenger data discussed earlier. Download this file to your computer. It is common practice to open this type of file using a spreadsheet software package like Excel, although opening it in a common text editing application will show the separating commas explicitly. Figure 1.14 displays the first 10 values of this dataset in both Excel (left) and raw text format (right). Note that the columns (values) are separated by commas and the rows (observations) are separated by a new line.

	A	B
1	Date	NumPassengers
2	Jan-49	112
3	Feb-49	118
4	Mar-49	132
5	Apr-49	129
6	May-49	121
7	Jun-49	135
8	Jul-49	148
9	Aug-49	148
10	Sep-49	136
11	Oct-49	119

```
Date,NumPassengers
Jan-49,112
Feb-49,118
Mar-49,132
Apr-49,129
May-49,121
Jun-49,135
Jul-49,148
Aug-49,148
Sep-49,136
Oct-49,119
```

(a) Microsoft Excel   (b) Text Editor

FIGURE 1.14 Screen shot of first 10 rows of the AirPassengers data stored in comma separated format in (a) Excel and (b) a standard text editor.

The read.csv Function

R makes it easy to read in **.csv** files using the **read.csv** function. We will use the **read.csv** function to read the **AirPassengers.csv** file that is stored in some location on your computer. Note that the first line of the dataset is the header (variable name) of each column. To tell R to treat this first row as the header we specify **header = TRUE**. For example, to read the **AirPassengers.csv** file you could use the command

```
AirPassengersData = read.csv("Your file location//AirPassengers.csv",header = TRUE)
```

> **Note:** To reiterate, the file path will be specific to your computer.
> For example, in Windows the path might be something like
>   "c:\\Documents and Settings\\My Documents\\My Data Files"

A simple way to use the **read.csv** function is to use it along with the **file.choose** function in order to select the file without having to identify the path in the user's file system. The complete function call is

```
AirPassengersData=read.csv(file.choose(),header=TRUE)
```

This command will bring up a file selection screen in which one can navigate to the file to be loaded. At this point, the user selects **AirPassengers.csv** in the subdirectory in which it resides, and the data will be loaded into R. It is always a good idea to check to make sure the data were read in as expected. The following line of code will print the first 10 lines of the dataset.[7]

---

7 The R function "**head**" prints the first *n* lines of a dataset.

```
head(AirPassengersData,n=10)
```

```
      Date      NumPassengers
1     Jan-49    112
2     Feb-49    118
3     Mar-49    132
4     Apr-49    129
5     May-49    121
6     Jun-49    135
7     Jul-49    148
8     Aug-49    148
9     Sep-49    136
10    Oct-49    119
```

QR 1.3 Import Data
with read.csv

1.3.6.2 The .txt file

As mentioned, sometimes data are saved in standard text (.**txt**) files. Below are screenshots of two .**txt** files: **sample1.txt** and **sample2.txt**.

```
sample1.txt      sample2.txt
34               34  42  55
42               23  36  33
55
23
36
33
```

The Scan Function

To follow this example, create the two files in a text editor (not much typing is involved) and save them in a location of your choice. The **scan** function can be used to read each of these datasets into the same numerical R vector.

```
s1=scan("Your file location//Sample1.txt")
s2=scan("Your file location//Sample2.txt")
```

The numeric vectors **s1** and **s2** are both equal to

```
[1]  34  42  55  23  36  33
```

That is, the scan function reads the data row-wise into a vector and there is no header option. You can also use the **file.choose** function with **scan**.

1.3.6.3 Other File Formats

In addition to the common .**csv** and .**txt** file formats, one may also encounter any number of other formats such as Excel formats (.**xlsx**), **JSON,** or **XML,** just to name a few. Because R is a well-established language, it is likely that an input function exists which accommodates nearly any format. Also, because R is open source, if a new format comes into vogue in the future, it will usually not take long for someone to write a function that is compatible with the new format. While R can read .**xlsx** files, the instructions given in this chapter involve reading .**csv** files. An Excel file is easily converted into a .**csv** file.

1.3.7 Accessing Time Series Data

So far, we have discussed how to read and plot time series data. But who is collecting time series data and where can it be found? We have seen that Base R and individual R packages themselves (including **tswge**) are a great source of time series data. For example, the **AirPassengers** and **lynx** datasets are **ts** objects in Base R which can be made available for use by simply issuing the commands

```
data(AirPassengers)
data(lynx)
```

Oh, if it were always this easy! The datasets available in R are just the tip of the iceberg! In this section, we will discuss the internet as a source of time series data and provide instruction on how to access data from internet websites.

1.3.7.1 Accessing Data from the Internet

There are an incredible number of websites that contain time series data. (Sometimes access requires a fee, but that is not the case for the data in the following examples.) Some of the websites are privately owned and some are government sources of data. In this section we will briefly discuss the following websites and illustrate the procedure for accessing data from them:

(1) FRED: The Federal Reserve Economic Database https://fred.stlouisfed.org
(2) Silso: Sunspot Index and Long-Term Solar Observations   www.sidc.oma.be/silso/
(3) Yahoo! Finance https://finance.yahoo.com/
(4) National Weather Service www.weather.gov
(5) New York City Taxi & Limousine website https://www1.nyc.gov/site/tlc/about/about-tlc.page

Two additional websites containing many time series datasets are https://census.gov and https://epa.gov

(1) FRED: The Federal Reserve Economic Database
At the intersection of government and research lies the Federal Reserve Economic Database, or *FRED*. Created in 1991 by the Research Department at the Federal Reserve Bank of St. Louis, *FRED* is a repository of over *a hundred thousand* economic time series from around the world! The website is https://fred. stlouisfed.org . The homepage is shown in Figure 1.15.

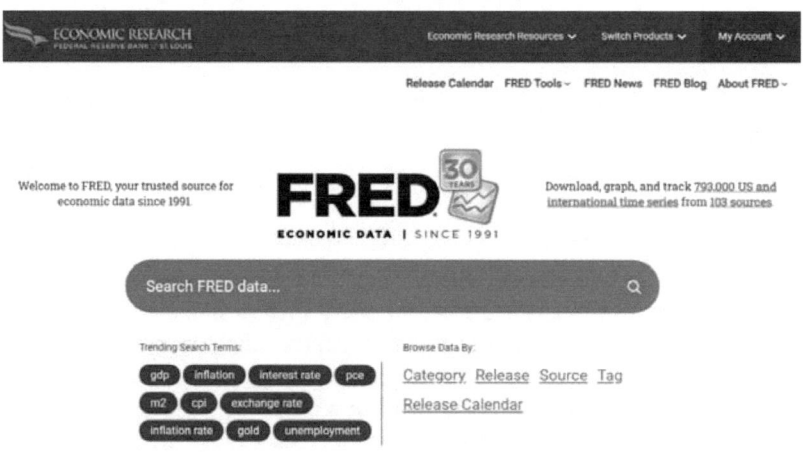

FIGURE 1.15   FRED Homepage.

The various datasets are found by selecting *Category* from the above menu, after which a screen appears which shows various options. These include *Money, Finance, & Banking*, *Production & Business*, and *Prices* among others. We will examine two of the datasets: (a) West Texas crude oil prices and (b) median days houses stay on the market before being sold.

(a) The West Texas Intermediate (WTI) Crude Oil Data
The webpage in Figure 1.16(a) has a plot of the same data as Figure 1.8(e). For comparison, Figure 1.8(e) is reproduced in Figure 1.16(b). This dataset is contained in the *Prices* category under the subcategory *Commodities*. This is accessed on the FRED website by navigating to the WTI data using the following steps:

- *Categories → Prices → Commodities*
- Scroll down and choose: *Global Price of WTI Crude*[8] and select *monthly*
- Click on *Global Price of WTI Crude* after which a plot of the current WTI crude oil data will appear
- To access the data, select *Download* and then *CSV(data)*
- Save the downloaded file, **POILWTIUSDM.csv,** to a subdirectory of choice
- In R, read the **.csv** file using
  WTI=read.csv(file.choose(),header=TRUE)
- A *ts* file can be created using the command
  wtcrude2020=ts(WTI$POILWTIUSDM,start=c(1990,1), end=c(2020,12), frequency=12)

QR 1.4
Download Data
from FRED

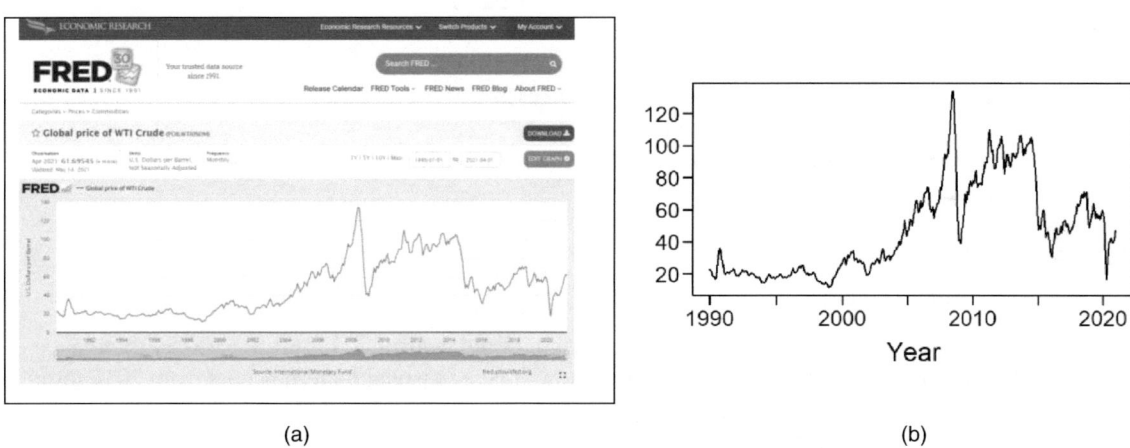

(a)   (b)

FIGURE 1.16   (a) Screenshot of WTI Crude data from January 1990 through April 2021 and (b) R-based plot of the *ts* object wtcrude2020 that extends through December 2020.

(b) Median Days on Market
Another dataset on the FRED website is a monthly tally of the median number of days a house stays on the market before being sold. The webpage is shown in Figure 1.17(a).

---

8 One category under *Commodities* is <u>Crude Oil Prices: West Texas Intermediate (WTI) – Cushing Oklahoma</u>. This is NOT the crude oil data set we are using. You should scroll down to *Global Price of WTI Crude*.

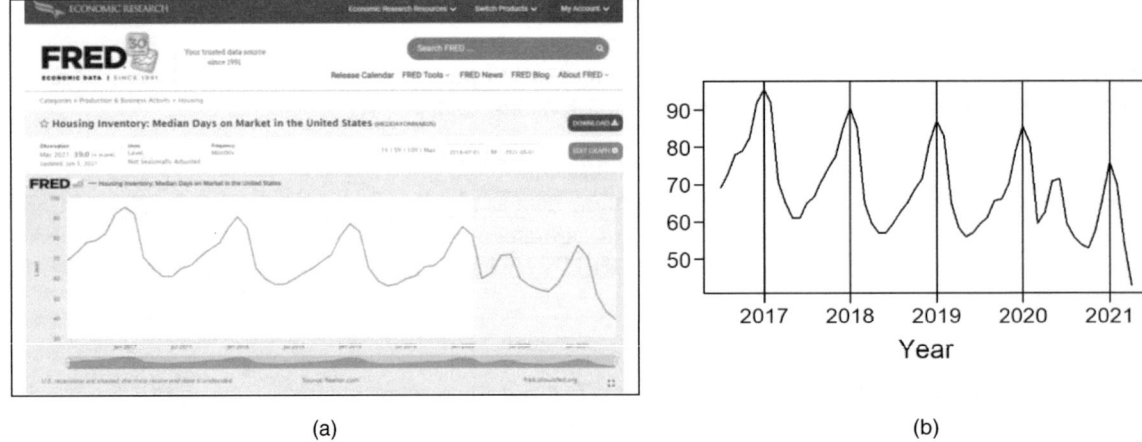

(a)   (b)

FIGURE 1.17 (a) Screenshot of the FRED website for median number of days to sell a house from July 2016 through April 2021 and (b) R plot of the data in (a).

This dataset can be accessed using the following steps:

- *Categories → Production & Business Activity → Housing*
- Scroll down and choose: *Housing Inventory: Median Days on Market in the United States* and select *Monthly, Not Seasonally Adjusted.* Figure 1.17(a) will appear.
- To access the data select *Download* and then *CSV(data)*
- Save the downloaded file **MEDDAYSONMARUS.csv** to a subdirectory of choice
- In R, read the **.csv** file using
  MedDays=read.csv(file.choose(),header=TRUE)
- A *ts* file can be created using the command
  MedDays.ts=ts(MedDays$MEDDAYONMARUS,start=c(2016,7) end=c(2021,4), frequency=12)

To obtain the plot of the data in Figure 1.17(b) use the command

plotts.wge(MedDays.ts)

Figure 1.17(b) is much easier to read than the screenshot of the website in Figure 1.17(a). Although we do not have a lengthy record, examination of this plot shows that a seasonal pattern occurs from 2017 through 2019 during which time houses typically sold faster (fewer days on the market) in May, June, and July while during December, January, and February, the number of days on the market is higher. Notice that the seasonal pattern abruptly changed during 2020 with May and June having higher than usual days on the market. One would suspect that this was due to the COVID outbreak declared a pandemic in March 2020. Notice also that things "turned around" in July through November with fewer than the normal number of days on the market. Also, during March and April of 2021, houses were selling unusually quickly.

(2) Silso: Sunspot Index and Long-Term Solar Observations
This website provides the new Sunspot2.0 numbers that were mentioned in Section 1.2.1. The link is www.sidc.oma.be/silso/ . The steps for retrieving the annual data are given below.

(a) Annual Data (Available for 1700 through the Present)
- *Click: Data → Sunspot Number → Total Sunspot Number*
- Select *Yearly mean total sunspot number [1700 - now]* and choose *CSV*
- The file **SN_y_tot_v2.0.csv** will be downloaded. Save this file to a subdirectory of choice

- *Note:* This `.csv` file uses a semicolon delimiter instead of a comma(default). To read the `.csv` file in R use the command

  `ss=read.csv(file.choose(),';',header=FALSE)`

- The data frame[9] `ss` has four columns, the second of which contains the sunspot data. The second variable has the default name `V2`. A *ts* file can be created using the command

  `sunspot2.0=ts(ss$V2,start=1700,frequency=1)`

A plot of the `sunspot2.0` data was given in Figure 1.1 and is repeated in Figure 1.18(a).

(b) ***Monthly Data (Available for 1749 through the Present)***
- *Click: Data → Sunspot Number → Total Sunspot Number*
- Select *Monthly mean total sunspot* number [1/1749−now] and choose CSV
- The file `SN_m_tot_v2.0.csv` will be downloaded. Save this file to a subdirectory of choice
- *Note:* As in the case of yearly means, this `.csv` file uses a semicolon delimiter instead of a comma. To read the file `SN_m_tot_v2.0.csv` file in R use the command `ss.month=read.csv(file.choose(),';',header=FALSE)`
- The data frame `ss.month` has seven columns. The fourth column contains the sunspot numbers and has the default name `V4`. A *ts* file can be created using the command

  `sunspot2.0.month=ts(ss.month$V4,start=c(1749,1),frequency=12)`

QR 1.5
Download the
Sunspot Data

A plot of the `sunspot2.0.month` is given in Figure 1.18(b).

Notes: In most (but not all) cases the datasets we will analyze in this book extend through 2020. For the sunspot data, in order to replicate the datasets plotted here, remove any data recorded on or after January 1, 2021.

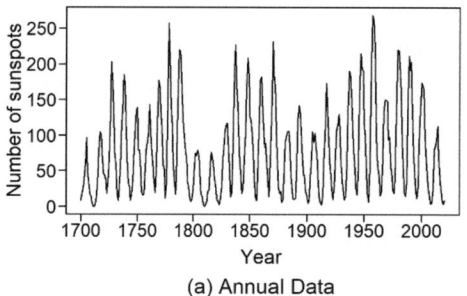 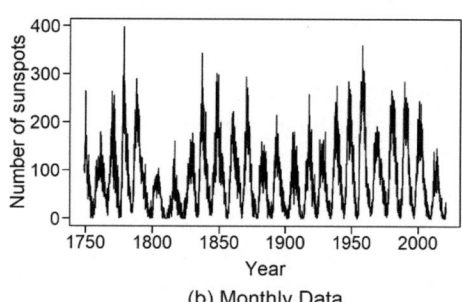

(a) Annual Data   (b) Monthly Data

FIGURE 1.18   (a) Plot of annual sunspot2.0 numbers from 1700 through 2020 and (b) plot of monthly sunspot2.0 numbers from 1749 through 2020.

(3) Yahoo! Finance
Stock market and other financial data are available on numerous sites on the internet. For our purposes we will use *Yahoo! Finance* which can be found at https://finance.yahoo.com/ . We will discuss stock prices for *Tesla*.

---

9 A "data frame" is a data structure in R that is similar to a matrix but can have various object classes rather than only numbers. It is one of, if not the, most common data structures in R. For more information on data frames simply type `?data.frame` into the R console.

(a) Tesla

Figure 1.19 below is a plot of the Tesla stock prices from January 1, 2020 through April 30, 2021. A dramatically increasing trending behavior until January 2021 is observed at which time the stock stayed fairly constant until February 2021 at which time it dropped. At the time of the writing of this section, the stock seems to be making a slight recovery.[10]

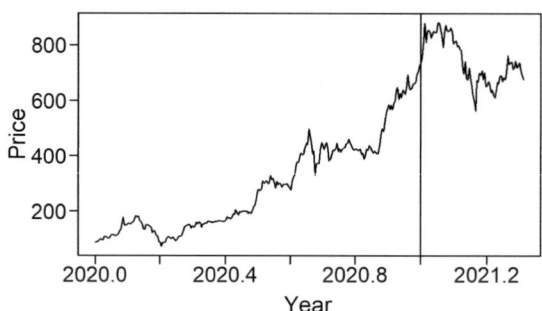

FIGURE 1.19   Tesla stock prices from January 1, 2020 through April 30, 2021.[a]
[a] For example, the date 2020.4 is 40% through year 2020 or late May.

In order to access the Tesla data and prepare it for plotting, use the following steps:

- In the search box at the top of the *Yahoo! Finance* homepage enter *Tesla*; the ticker symbol will appear as *TSLA*.
- Select *TSLA* and on the resulting screen select *Historical Data* and *Daily*
- Fix dates from 01/01/2020 to 04/30/2021
- Select *Apply* and then *Download*
- The file **TSLA.csv** will be downloaded. Save this file to a subdirectory of choice.
- To read the file **TSLA.csv** in R, use the command
  tesla=read.csv(file.choose(),header=TRUE)
- The data frame **tesla** has seven columns. The sixth column contains the adjusted closing price for that particular day with the variable name **Adj.Close**
- A *ts* file can be created using the command
  tesla.ts=ts(tesla$Adj.Close,start=c(2020,1),frequency=254)
  (There were 254 business days in 2020)
- The command **plotts.wge(tesla.ts)** produces Figure 1.19 (without the more descriptive labels and vertical line at January 2021 shown there.)

(4) National Weather Service

This is a website published by the National Weather Service that provides an abundance of information about local forecasts and historical data at www.weather.gov. The monthly average temperature in Dallas Ft. Worth (DFW) from September 1898 through the current month can be obtained from the link www.weather.gov/fwd/dmotemp . In *tswge* there are two files obtained from this link:

- (a) **dfw.mon**: DFW monthly average temperatures from January 1900 through December 2020. *Note:* These data are plotted in Figure 1.3.
- (b) **dfw.yr**: DFW average annual temperatures from 1900 through 2020.

These *ts* files are plotted in Figure 1.20.

---

10 Recall the caution to not assume that observed trends will necessarily continue. The first author went against his own advice and bought Tesla stock in early January 2021.

Note: The above link for the DFW temperatures does not provide the option to download a `.csv` file. In Section 1.4.1.2 we will discuss procedures for obtaining the temperature data from the link.

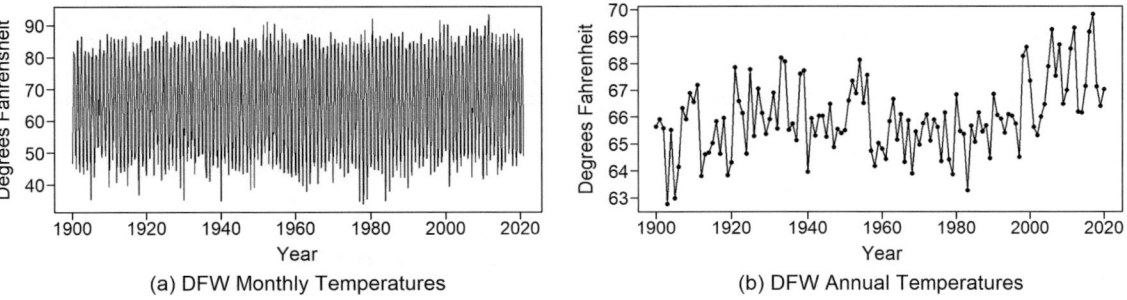

(a) DFW Monthly Temperatures   (b) DFW Annual Temperatures

FIGURE 1.20   (a) DFW monthly temperatures from 1900 through 2020 and (b) DFW annual average temperatures for the same time period.

Notes: Figure 1.20(a) is the plot in Figure 1.3. Figure 1.20(b) has a wandering behavior with an upward trend beginning in the 1980s that suggests a warming trend which is the topic of much discussion. However, in DFW the years 2018–2020 have been cooler.

(5) New York City Taxi & Limousine website

Suppose that an analyst wants to analyze the effect of COVID-19 on taxicab trips in New York City. Beginning in 2009, the city of New York mandated that all taxis report the distance, length, cost, pickup and drop-off location, and other trip information for each taxicab ride in the city. The data can be obtained from the New York City Taxi & Limousine website www1.nyc.gov/site/tlc/about/about-tlc.page shown in Figure 1.21.

FIGURE 1.21   New York Taxi & Limousine Commission homepage.

The file `yellowcab.precleaned.ts` is a *ts* file in *tswge* containing the number of trips per month for Yellow Cabs (a taxicab company), and these data are plotted in Figure 1.22. COVID-19 was declared a pandemic in March 2020, and it is of interest to understand the degree to which this affected demand for taxi service in New York City. Figure 1.22 dramatically shows the devastating impact.

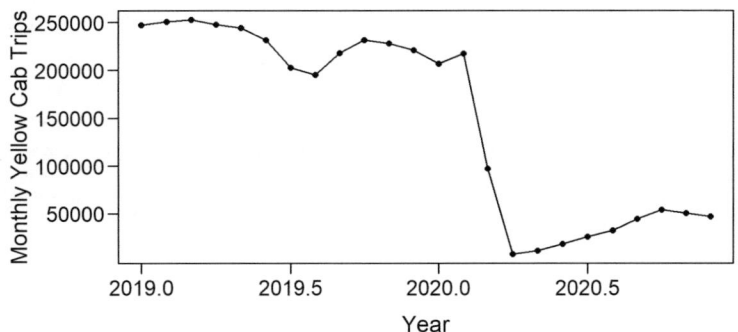

FIGURE 1.22   Number of Yellow Taxicab trips per month from January 2019 through December 2020.

Key Points

1. The file **yellowcab.precleaned** is *not* directly attainable by downloading a .csv or .txt file from the NYC Taxicab website.
2. A considerable amount of data manipulation was required to "pre-clean" the data.
3. Section 1.4.1.3 discusses techniques for pre-cleaning data and specifically for producing the *ts* file **yellowcab.precleaned**.

1.3.7.2 Business / Proprietary Data: Ozona Bar and Grill

Time series data may also be obtained directly from a business or organization. For example, one of the most popular restaurants in Dallas, Texas (and frequented quite often by the authors of this text) is a place near the campus of Southern Methodist University named *Ozona Bar and Grill*. While Ozona has a full menu full of Dallasite favorites, it is famous for its chicken fried steak! We asked the Director of Operations for relevant historical data, and he graciously provided us with a time series of the daily number of chicken fried steaks that were sold in June and July of 2019.[11] The data are provided in the **ozona** data frame in the *tswge* package and are displayed below (Figure 1.23). Stop in next time you are in the neighborhood! Thank you, Director of Operations, Cory Wauson!

```
data(ozona)
ozona.ts=ts(ozona$CFS_Sold)
plot(ozona.ts,type='o',xlab='Day')
```

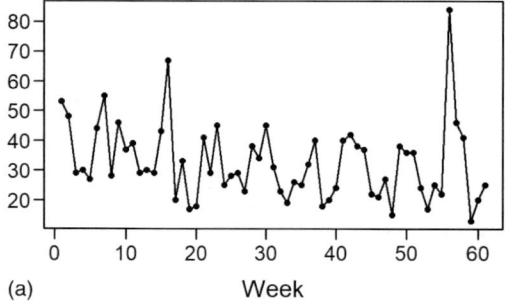
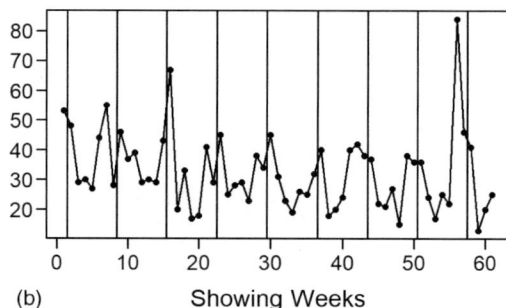

(a)   Week

(b)   Showing Weeks

FIGURE 1.23 (a) Number of chicken fried steaks sold at Ozona Bar and Grill in June and July, 2019 and (b) plot (a) with vertical lines separating weeks

In Figure 1.23(a) a fairly random-looking pattern of sales is observed, with a slight decreasing trend except for a large spike in late July. Figure 1.23(b) separates weeks with vertical lines between Saturday and Sunday for each week. In this plot we see a definite "seasonal" pattern in which weekends, particularly Friday and Sunday, tend to have the most sales.

Key Point: When plotting data from a private source such as Ozona, be sure that permission to use the data has been granted.

---

11 As all good data scientists should do, we sampled the (delicious) chicken fried steak data!

1.4 DEALING WITH MESSY DATA

Despite the wide variety of types and sources of time series data the analyst may encounter, one common denominator is that such data are often not available as data files in R or convenient .**csv** files that are immediately suitable for visualization and analysis. In many cases the data must be manipulated into a suitable form.

After obtaining the data, the next step is to perform any necessary cleaning, wrangling,[12] and/or handling of missing values in order to obtain the data in the form needed for analysis. Nearly all data scientists discover that after they leave the classroom and enter the "real world of data", a large part of analyzing data is *preparing the data for analysis*. This is an important and complex topic that we will only briefly introduce here using a couple of examples.

1.4.1 Preparing Time Series Data for Analysis: Cleaning, Wrangling, and Imputation

In this section, we give two examples in which the data need to be cleaned or otherwise manipulated before analysis can take place. These examples address three types of "problem data".

- Data with missing values.
- Data on websites that do not provide a download option.
- Data that can be obtained from a larger data file but must be assembled and organized before analysis takes place.

1.4.1.1 Missing Data

It is important to note that the data analytic procedures discussed in this book assume that data are obtained at equally spaced time intervals and that there are no missing data values. (For information about analyzing irregularly spaced time series data see Jones, 2016; Wang, Woodward and Gray, 2009). The Bitcoin dataset in ***tswge* (bitcoin)** is a daily ***ts*** object that contains the price of Bitcoin from May 1, 2020 to April 30, 2021 as per Yahoo Finance on April 30, 2021. A quick inspection of the ***ts*** object shows that the data are daily (**frequency**=365) which begin on the 122nd day of 2020 (May 1) and end on the 121st day of 2021 (April 30). At that time the data had three missing values on October 9, 12, and 13 of 2020. Before analysis can proceed, these values need to be "imputed". That is, they need to be approximated and then put into the dataset in the place of the missing values. There are many ways to impute missing data. Suppose x_t is missing in a dataset. Two simple approaches are:

- Set $x_t = x_{t-1}$ where x_{t-1} is the previous data value (which is known). This approach is commonly known as Last Observation Carried Forward (LOCF).
- Linear interpolation: that is, conceptually draw a straight line connecting x_{t-1} and x_{t+1} which are the known values on each side of the missing value. Then assign x_t to be the value on the line at time t. See Figure 1.24(a).

October 12 and 13 represent an instance in which two adjacent values are missing. In this case, linear interpolation is illustrated in Figure 1.24(b) and the procedure is analogous. Connect the two known values x_{t-1} and x_{t+2} with a straight line and find the values on that line at t and $t+1$. This procedure is the basis for the linear interpolation formulas in the code below.

---

12 Wrangling is the processing of transforming and mapping data from one form to another for purposes of preparing it for analysis.

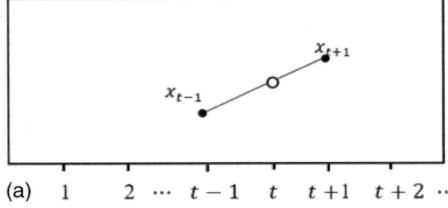
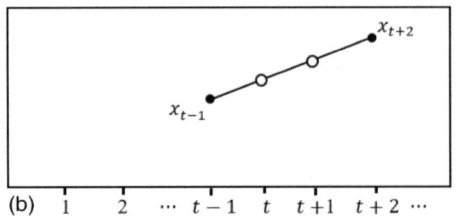

FIGURE 1.24 Illustration of linear interpolation. (a) with one missing value (x_t) and (b) with two adjacent missing values (x_t and x_{t+1}).

Key Points

1. The techniques described in this book for analyzing time series data require that data are equally spaced and that there are no missing values.
2. Methods do exist for analyzing data with unequally spaced and missing data.

(a) Bitcoin Prices

As mentioned above, Bitcoin prices in the **bitcoin** dataset are missing in 2020 for October 9, 12, and 13. We will use linear interpolation to impute the missing values. Figure 1.25(a) shows Bitcoin prices for the year from May 1, 2020 through April 30, 2021. The horizontal axis is **Index**. The variable **Index** begins at 1 for May 1, 2020 and goes through 365 (a full year) on April 30, 2021. Table 1.1 pairs some key dates with the corresponding **Index**.

TABLE 1.1 Date/Index Pairs for Bitcoin data

DATE	INDEX
May 1, 2020	1
October 1, 2020	154
October 9, 2020	162
October 12, 2020	165
October 13, 2020	166
October 22, 2020	174
April 30,2021	365

Dealing with Messy Data

Bitcoin has continued to rise as of April 30, 2021 with some indication of "slowing down". As mentioned, the Bitcoin prices for October 9, 12, 13 in 2020 (**Index** values 162, 165, 166) are missing. In order to better see the missing data time frame, we will focus on the snippet of time from October 1 through October 22, 2020. According to Table 1.1 these dates have **Index** values 154 and 174, respectively. Figure 1.25(b) shows the snippet range within the two vertical lines. If you look very closely at the data between the lines you can see tiny gaps. Figure 1.25(c) shows the data within the focused range and we can see that there are three missing values at Indices 162, 165, and 166. Finally, Figure 1.25(d) shows the snippet data with missing values imputed using linear interpolation.

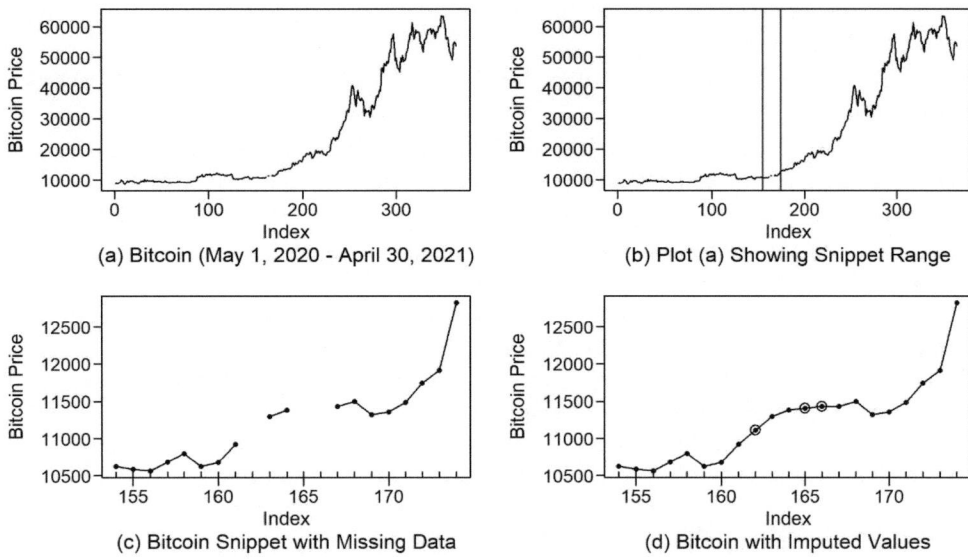

FIGURE 1.25   (a) Bitcoin data from May 1, 2020 through April 30, 2021, (b) Plot in (a) showing the snippet of time from October 1 through October 22, 2020 with the vertical strip, (c) Snippet of Bitcoin data from October 1 through October 22, 2020 showing missing values on October 9, 12, and 13 and (d) Figure (c) with imputed values.

You can use the following steps to impute the missing values using linear interpolation. (Recall that the snippet data are the subset **bitcoin[154:174]**)

Impute missing values using linear interpolation

```
# Linear interpolation with one missing value
bitcoin[162]=bitcoin[161]+(bitcoin[163]-bitcoin[161])/2
# Linear interpolation with two adjacent missing values
bitcoin[165]=bitcoin[164]+(bitcoin[167]-bitcoin[164])/3
bitcoin[166]=bitcoin[164]+2*(bitcoin[167]-bitcoin[164])/3
```

The imputed values can be viewed as:

```
bitcoin[162]
[1]  11106.02

bitcoin[165]
[1]  11399.29

bitcoin[166]
[1]  11414.4
```

Recently, Yahoo Finance has updated the Bitcoin dataset and have published the actual values for October 9, 12, and 13 in 2020. This provides the unique opportunity to check the imputed values versus the actual values! Follow the steps below to download the data from Yahoo Finance and compare the performance of the imputed values:

- In the search box enter *Bitcoin*; the ticker symbol will appear as *BTC-USD*.
- Select *BTC-USD* and on the resulting screen select *Historical Data* and *Daily*.

- Fix dates from 05/01/2020 to 04/30/2021 to collect "12 months" or "a year" of daily data.
- Select *Apply* and then *Download*
- The file **BTC-USD.csv** will be downloaded; Adjusted Close (variable **Adj.Close**) is the variable of interest

  Note: **Adj.Close** values for October 9, 12, and 13 in 2020 are missing. It will be necessary to impute values for these dates before analysis can continue.
- Save this file to a subdirectory of choice.
- To read the file **BTC-USD.csv** in R use the command
  BTC=read.csv(file.choose(),header=TRUE)
- The data frame **BTC** has seven columns. The sixth column contains the adjusted closing price for that day with the variable name **Adj.Close**
- The data in **BTC$Adj.Close** are character values, so enter the command
  Bitcoin=as.numeric(BTC$Adj.Close)
- A warning will be issued that there are missing values, and by examining the file **Bitcoin** it follows that the missing values are at time points $t = 162, 165,$ and 166.

Table 1.2 compares the actual values with the imputed data.

TABLE 1.2   Actual versus Imputed Values for Bitcoin Data

Date	Imputed	Actual
October 9, 2020	$11,106.02	$11,064.46
October 12, 2020	$11,399.29	$11,555.36
October 13, 2020	$11,414.40	$11,425.90

Figure 1.26 shows the Bitcoin data (as dots connected with a solid line) along with the three imputed values as open circles. The plot shows that the imputed values for October 9 and 13 were quite close to the true values while the Bitcoin price for October 12 was higher than the simple interpolation predicted.

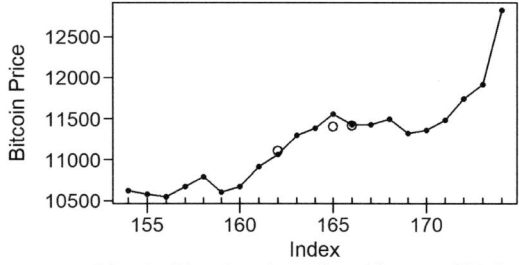

Bitcoin Showing Actual and Imputed Values

QR 1.6 Bitcoin Imputation

FIGURE 1.26   Bitcoin data from October 1 through October 22, 2020 as solid dots connected by solid lines and the imputed values for October 9, 2, and 13 as open circles.

Key Points

1. There are many imputation techniques (we used linear interpolation).
2. While the *mean* is an often-used imputed value for missing data in a random sample, the correlation structure of time series data is such that the mean is rarely a wise imputation choice in a time series setting.
3. It is very important to consider why the value is missing in the first place.
4. For a deeper discussion of these topics please check out the video using the QR code.

1.4.1.2 Downloading When no .csv Download Option Is Available

The previous examples have focused on websites for which the data can be downloaded onto a `.csv` (or other formatted) file. However, the DFW temperature data obtained from the weather.gov link www.weather.gov/fwd/dmotemp   does not have a download option. Figure 1.27 shows the monthly DFW temperatures from September 1898 through the current month (April 2021 as of the writing of this book). In this example, we will utilize the data from January 1900 through December 2020. The following steps allow the user to access these monthly averages for analysis in R even though there is no download option.

- Select the data (selection shown with shading in Figure 1.27) and use Ctrl-C for copying.

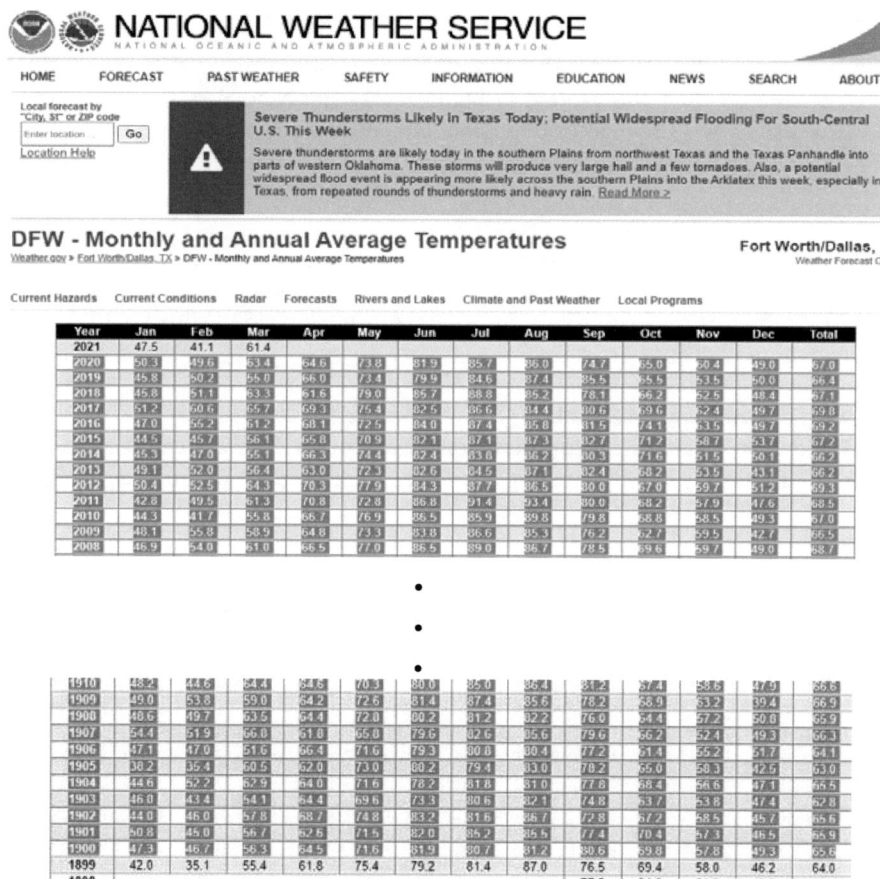

FIGURE 1.27   National Weather Service monthly data for Dallas Ft. Worth.

- Paste (Ctrl-V) the data into a blank Excel file.
- Delete the last column containing annual averages.
- Select all of the columns (A–M).
- Click *Data → Sort*.
- The checkbox "My data has headers" should not be checked.
- For "Sort by" choose *Column A*.
- For "Order" choose *Smallest to Largest*.
  The data should now be sorted from earliest dates to the most recent.
- Delete the first column containing years.
- Select all remaining columns (A–L), copy (Ctrl-C) and paste them (Ctrl-V) into a blank text file, and name it `dallastemp.txt`
- Read the data into R using the scan function

```
dfw.mon=scan("c:\\Your subdirectory\\dallastemp.txt")
```
- Convert **dfw.mon** to a *ts* file
```
dfw.mon=ts(dfw.mon,start=c(1900,1),frequency=12)
```
- Calculate the annual data using the aggregate command
```
dfw.yr=aggregate(dfw.mon,FUN=mean)
```

These *ts* files are plotted in Figure 1.27.

Note: The *ts* files obtained above are identical to the DFW temperatures files by the same name in *tswge*.

1.4.1.3 Data that Require Cleaning and Wrangling

In this section, we discuss the cleaning and wrangling steps that were used to create the "nice" *ts* file **yellowcab.precleaned** that was plotted in Figure 1.22. The New York City Taxicab information is available on the website www1.nyc.gov/site/tlc/about/about-tlc.page which is shown in Figure 1.21. In order to access the data, use the following steps:

- On the homepage select *Data and Research.*
- Select *Data* and then *Aggregated Reports.*
- Select *Monthly Data Reports(CSV)* near the bottom of the screen
  The file **data_reports_monthly.csv** will be downloaded. Save this file to a subdirectory of choice.
- In R, read the csv file using
  NYCabRaw = read.csv(file.choose(), header = TRUE)
- *Note:* The data in data frame **NYCabRaw** are shown below. There it can be seen that the data will not be ready for use by simply creating a *ts* file as we have done in previous examples.[13]

QR 1.7 Wrangling

head(NYCabRaw)

	Month.Year	License.Class	Trips.Per.Day	Farebox.Per.Day	Unique.Drivers	Unique.Vehicles	Vehicles.Per.Day
1	2021-02	FHV - High Volume	414,755	–	47,206	47,267	28,637
2	2021-02	Green	2,305	52,276	904	881	446
3	2021-02	Yellow	48,499	755,115	5,277	4,496	3,182
4	2021-01	FHV - Black Car	10,386	–	3,816	3,806	1,752
5	2021-01	FHV - High Volume	383,932	–	47,622	47,592	28,582
6	2021-01	FHV - Livery	23,890	–	5,132	5,030	3,086

	Avg.Days.Vehicles.on.Road	Avg.Hours.Per.Day.Per.Vehicle	Avg.Days.Drivers.on.Road	Avg.Hours.Per.Day.Per.Driver
1	17.0	7.1	17.1	7.1
2	14.2	4.1	13.9	4.1
3	19.8	7.9	17.6	7.6
4	14.3	4.4	14.3	4.4
5	18.6	6.9	18.7	6.8
6	19.0	5.1	18.9	5.0

	Avg.Minutes.Per.Trip	Percent.of.Trips.Paid.with.Credit.Card	Trips.Per.Day.Shared
1	16	–	–
2	18.8	79%	–
3	12.4	76%	–
4	22	–	–
5	15	–	–
6	15	–	–

"Wrangling" the dataset, in this context, involves filtering and subsetting the data to obtain the appropriate number of columns and rows (months). This will involve:

---

13 Note that if you access the NYC Taxicab dataset using the procedure outlined above, it will be different from the one accessed here which was obtained in early 2021.

    (a) Selecting only the **Month.Year, License.Class** and **Trips.Per.Day** columns

    (b) Filtering the dataset to contain only the Yellow taxicabs

    (c) Reversing the **Trips.Per.Day** column to reorder the data from earliest to latest dates. (The data originate from the website in descending order from latest to earliest dates.)

    (d) Subsetting the dataset to only include the data from January 2019 through December 2020. "Cleaning", on the other hand, will involve converting the trips per day into a numeric format in order to pass them to the plot function. This will involve only one step:

    (e) Deleting the comma from the **Trips.Per.Day** column

The dplyr Package

To wrangle and clean this dataset in R, the ***dplyr*** package will be utilized. This package has become a leading source of functions for cleaning and wrangling data in R. While we will not provide a full tutorial here,[14] a key characteristic of the ***dplyr*** package is that the code is easy to implement and is intuitive to understand; thus, the following examples are intended to be both instructive and easy to follow.

(a) Selecting the Month.Year, License.Class and Trips.Per.Day columns

In order to select only the three columns of interest, use the **select** function in ***dplyr***:

```
NYCabNew = NYCabRaw %>% select(c(Month.Year,License.Class,Trips.Per.Day))
head(NYCabNew)
```

	Month.Year	License.Class	Trips.Per.Day
1	2021-02	FHV - High Volume	414,755
2	2021-02	Green	2,305
3	2021-02	Yellow	48,499
4	2021-01	FHV - Black Car	10,386
5	2021-01	FHV - High Volume	383,932
6	2021-01	FHV - Livery	23,890

(b) Filtering the Data to Retain Only the Yellow Taxicabs

Next, filter the dataset to contain only the trips per day for the Yellow cabs. This can be accomplished using the **filter** function in ***dplyr***:

```
NYCabNew = NYCabNew %>% filter(License.Class == "Yellow")
head(NYCabNew)
```

	Month.Year	License.Class	Trips.Per.Day
1	2021-02	Yellow	48,499
2	2021-01	Yellow	44,052
3	2020-12	Yellow	47,145
4	2020-11	Yellow	50,285
5	2020-10	Yellow	54,221
6	2020-09	Yellow	44,646

We next perform the cleaning step.

(c) Deleting the Comma from the Trips.Per.Day Column

Visually, the **Trips.Per.Day** column already appears to be numeric; however, R considers this column to contain character strings. To verify this fact we issue the Base R **class** command

```
class(NYCabNew$Trips.Per.Day)
```

```
[1] "character"
```

To remove the comma we call the **sub** function in Base R which finds a specified pattern (the first argument), and replaces it with a character value (the second argument) in the given data (the third argument),

---

14 A great resource to learn more about the ***dplyr*** and other useful R packages is Wickham and Grolemund (2017).

and then returns the results as character values. For this reason, the result must be changed (cast) to a numeric value to complete the process. Here is the code to do both:

```
NoCommaTrips = sub(",","",NYCabNew$Trips.Per.Day)
NoCommaTrips = as.numeric(NoCommaTrips)
NoCommaTrips
```

```
  [1]  48499  44052  47145  50285  54221  44646  32491  25816  18325  11237   7928  96993 217216
 [14] 206604 220786 227654 231171 217747 194798 202443 231335 244017 247742 252634 250654 247315
 [27] 263609 271501 284121 267983 253182 253186 290362 297508 310169 304169 303280 282565 306706
 [40] 309471 315084 298163 271676 277042 321877 325857 334865 332075 327451 313229 337071 336737
 [53] 350380 337321 320718 332231 371257 381878 397780 393886 392470 351816 369686 377076 397244
 [66] 374156 359029 372979 410831 424459 435701 430669 444633 411238 419776 440576 459087 445782
 [79] 409305 422771 460393 476569 487275 497661 466522 444537 450634 479527 483921 470147 406246
 [92] 445861 479148 492962 503249 507914 499556 476544 474005 459114 468377 484783 463841 463784
[105] 503139 502097 515848 520764 516585 482811 481410 484122 506532 487470 427740 475475 503152
[118] 501708 490577 518212 507167 434297 445727 463701 457996 517972 404115 472752 494137 499374
[131] 504798 415567 397969 479376
```

(d) Reversing the `Trips.Per.Day` Column to Reorder the Data from Earliest to Latest Dates

Now that we have cleaned the `Trips.Per.Day` column, we need to reverse the order of the data so that the dates run from earliest to latest. We reverse the data `NoCommaTrips` using the Base R `rev` function.

```
NoCommaTrips = rev(NoCommaTrips)
NoCommaTrips
```

```
  [1] 479376 397969 415567 504798 499374 494137 472752 404115 517972 457996 463701 445727 434297
 [14] 507167 518212 490577 501708 503152 475475 427740 487470 506532 484122 481410 482811 516585
 [27] 520764 515848 502097 503139 463784 463841 484783 468377 459114 474005 476544 499556 507914
 [40] 503249 492962 479148 445861 406246 470147 483921 479527 450634 444537 466522 497661 487275
 [53] 476569 460393 422771 409305 445782 459087 440576 419776 411238 444633 430669 435701 424459
 [66] 410831 372979 359029 374156 397244 377076 369686 351816 392470 393886 397780 381878 371257
 [79] 332231 320718 337321 350380 336737 337071 313229 327451 332075 334865 325857 321877 277042
 [92] 271676 298163 315084 309471 306706 282565 303280 304169 310169 297508 290362 253186 253182
[105] 267983 284121 271501 263609 247315 250654 252634 247742 244017 231335 202443 194798 217747
[118] 231171 227654 220786 206604 217216  96993   7928  11237  18325  25816  32491  44646  54221
[131]  50285  47145  44052  48499
```

This listing is in the correct earliest to latest date order.

(e) Subsetting the Dataset to Only Include the Data from January 2019 through December 2020

We complete the wrangling by subsetting the data to contain only the trips for January 2019 to February 2021. To do this we will need to first create a *ts* object for the data so we can then use the **window** function.

```
NYCabNewts=ts(NoCommaTrips,start=c(2010,1),frequency=12)
NYCabNewtsShort=window(NYCabNewts,start=c(2019,1),end = c(2020,12))
NYCabNewtsShort
```

	Jan	Feb	Mar	Apr	May	Jun	Jul	Aug	Sep	Oct	Nov	Dec
2019	247315	250654	252634	247742	244017	231335	202443	194798	217747	231171	227654	220786
2020	206604	217216	96993	7928	11237	18325	25816	32491	44646	54221	50285	47145
2021	44052	48499										

Notice that this is the same as *tswge ts* object `yellowcab.precleaned.ts` .

Note: Now it is clear why the descriptive "pre-cleaned" was included in the *tswge* taxicab *ts* object.[15] We can now plot the data to complete the example.

---

15 As noted previously, if you access the NYC Taxicab dataset using the procedure described above, it will be different from the one accessed here which was obtained in early 2021.

```
plotts.wge(NYCabNewtsShort, xlab = "Data", ylab = "Monthly Yellow Taxi Trips")
```

This code produces the plot previously shown in Figure 1.22.

1.4.1.4 Programatic Method of Ingestion and Wrangling Data from Tables on Web Pages

Earlier, it was shown that time series may be obtained from tables on web pages by copying and pasting the data into a spreadsheet, saving it as a `.csv` file and then loading it into R. While this is a very useful method, it has two significant drawbacks: a) It may not always work or may take significant wrangling to get it into the right form. b) An actual person will need to be available to perform the steps. If the data are updated often this may create a significant strain on human and other resources. The video below demonstrates a programatic method that may work when the copy and paste method presents challenges and, even when the other method does work, this method may be preferred as it automates the process thus freeing up valuable resources.

QR 1.8 Web Scraping

1.5 CONCLUDING REMARKS

In this opening chapter, the importance of the field of time series data has been motivated. Time series data are widely available, and the analysis of such data is in demand. The first step in analyzing time series data is having the skillset to access and manipulate time series data in a way that is compatible for statistical software, in our case, R. Common, basic R code has been introduced that will serve as a great starting place in the beginning steps of the analysis of a time series dataset. In this text, the authors' R package *tswge* will be the primary programming tool for analyses. Now that you have the tools to produce a time series dataset that is in a form that can be analyzed, you are ready to learn the next steps in the process! In the chapters ahead, a comprehensive array of methods will be introduced that will be beneficial in providing *useful* forecasts and inferences to a wide variety of time series settings.

APPENDIX 1A

TSWGE FUNCTION

```
plotts.wge = function (x,style = 0,ylab = "",xlab = "Time", main = "", col =
"black", text_size= 12, lwd = .75, cex = .5, cex.lab = .75, cex.axis = .75, xlim
= NULL, ylim = NULL)
```

is a function designed to produce time series plots. It has two uses: (1) using default options it plots a dataset as a time series, connecting the dots, etc. with no parameter specifications needed (2) it creates custom plots using the parameters in Base R function **plot** along with the **style** parameter described below.

style specifies the appearance of the output. **style=0** produces a plot similar to Figure 1.12, and Figure 1.13 was obtained using **style=1** which uses *ggplot2*. (See the related discussion in Section 1.3.5)

Notes: Color, axis labeling and numbering, and other plot-related options can be modified using the plot parameters in the call statement. For more information on plot parameters, use the command **?plot**.

TSWGE DATASETS INTRODUCED IN THIS CHAPTER

Some of the following datasets were downloaded from websites. In order to avoid the necessity to perform the download for future use, we have included the following datasets in *tswge*.

   bitcoin – Bitcoin data from May 1, 2020 through April 30, 2021 (includes imputed values from Section 1.4.1.1)

   dfw.mon – DFW monthly temperature data from January 1900 through December 2020

   dfw.yr – DFW annual temperature data from January 1900 through December 2020

   dfw.2011 – DFW monthly temperatures from January 2011 through December 2020

   dow1985 – Monthly DOW closing averages from March 1985 through December 2020

   MedDays – Median days a house stayed on the market between July 2016 and April 2020

   ozona – Daily number of chicken-fried steaks sold at Ozona Bar and Grill during June and July, 2019

   sunspot2.0 – Sunspot2.0 annual data from 1700 through 2020

   sunspot2.0.month – Sunspot2.0 monthly data from January 1749 through December 2020

   tesla – Tesla stock prices from January 1, 2020 through April 30, 2021

   wtcrude2020 – Monthly WTI crude oil prices from January 1990 through December 2020

   yellowcab.precleaned – Number of trips per month for Yellow Cabs in New York City from January 2019 through December 2020

PROBLEMS

1.1 Google search "sunspot data" and find the classical (Wolfer) *annual* sunspot data from 1749 through 2014. Plot a subset of the sunspot2.0 data from 1749 through 1914. On the same graph plot the classical sunspot data (using R function **points** (or **lines**)). Label the axes and compare the similarities and differences between the two time series.

1.2 Repeat Problem 1.1 using *monthly* sunspot data from 1749 through 2014.

1.3 The US Bureau of Transportation records monthly data on the number of air passengers in the US. Use these data to create a time series plot of the number of air passengers in the US from 2002 to the present.

1.4 From the FRED website, access the data set containing monthly Total Vehicle Sales **(TOTALNSA)** in the United States from January 1st 1976 to July 1st 2021. The data are available at https://fred. stlouisfed.org/series/TOTALNSA

   (a) Download the **.csv file TotalNSA.csv** to a location of choice.

   (b) Using the techniques discussed in Section 1.3.7 read the .csv file using the commands
   NSA.read=read.csv(file.choose(),header=TRUE)
   x=NSA.read$TOTALNSA

   (c) Use the command **head(x)** to list the first six data values in **x**

(d) Plot the data in x using the command `plotts.wge(x)`

(e) Convert the vector **x** to a *ts* file named `NSA`

(f) Plot the dataset `NSA` using `plotts.wge`

(g) Describe the behavior of the data (cyclic, seasonal, wandering,…)

1.5 From the FRED website, access the monthly data set containing the number of new houses sold between 1965-2020. The data are available at https://fred.stlouisfed.org/series/HSN1F

(a) Download the `.csv` file `HSN1F.csv` to a location of choice.

(b) Using the techniques in Problem 1.4 create a vector x contatining discussed in Section 1.3.7 read the .csv file using the commands

```
HSN1F.read=read.csv(file.choose(),header=TRUE)
x=HSN1F.read$HSN1F
```

(c) Plot the data in x using the command `plotts.wge(x)`

(d) Convert the vector **x** to a *ts* file named `HSN1F`

(e) Plot the dataset `HSN1F` using `plotts.wge`

(f) Describe the behavior of the data (cyclic, seasonal, wandering,…)

1.6 Kaggle is a website for data scientists that contains various examples of real-world time series data. Kaggle also includes code and contains data analysis projects designed for competition among data scientists. This problem involves data containing the monthly sales for the North American Industry Classification System (NAICS) code 44X72: Retail Trade and Food Services. The data are U.S. Total sales in billions of dollars from Jan 1992 to the present date. This dataset is from a Kaggle Competition and is found at https://www.kaggle.com/landlord/usa-monthly-retail-trade?select=2017+NAICS+Definition.csv

(a) Download the data. The downloaded data are in a .zip file named archive.zip. Extract the data

```
SeriesReport-Not Seasonally Adjusted Sales - Monthly (Millions of
Dollars).csv
```

(b) Using the techniques in Problem 1.4 create a vector x containing total US sales in millions of dollars using the commands

```
NAICS.read=read.csv(file.choose(),header=TRUE)
x=NAICS.read$Value
```

(c) Plot the data in **x** using the command `plotts.wge(x)`

(d) Convert the vector **x** to a *ts* file named `NAICS` containing only data from January 2000 through December 2019.

(e) Convert `NAICS` to billions using the command `NAICS=NAICS/1000`
Plot the dataset `NAICS` using plotts.wge

(f) Describe the behavior of the data (cyclic, seasonal, wandering,…)

1.7 *Medium* is a website that provides a publishing platform. Among the data on the website are 10 good time series datasets. These can be accessed using the command https://medium.com/analytics-vidhya/10-time-series-datasets-for-practice-d14fec9f21bc

(a) The resulting page shows 10 datasets. Click on the file `Cali Emissions.csv` which contains CO2 emissions data for North America in million metric tons. The data are shown for the years 1980 through 2017.

(b) There is no automatic download provision, so highlight the emissions data, and copy it into the text file CO2.txt in your subdirectory of choice.

(c) Open Excel with a blank spreadsheet.

(d) Under the *Data* heading select *Get data from Text/CSV* and choose the file `CO2.txt` that you created. Save the resulting file as `CO2.csv` inn your subdirectory.

(e) Issue the commands

```
CO2.read=read.csv(file.choose(),header=TRUE)
x=CO2.read$Column2
```

(f) Plot the data in **x** using the command `plotts.wge(x)`

(g) Convert the vector **x** to a *ts* file named `CO2` containing annual data from 1980 through 2017.

(h) Plot the dataset **CO2** using `plotts.wge`

(f) Describe the behavior of the data (cyclic, seasonal, wandering,…)

1.8 Use the World Bank API (WDI package) to gather data that will allow you to create a time series plot of the total worldwide railroad passengers. Compare any trends you see in this plot to that of the air passenger data displayed in Problem 1.3.

1.9 Use the World Bank API (WDI package) to gather data that will allow you to construct a time series plot that compares the GDP of the US, Mexico, and Canada over the last 30 years.

1.10 Using the New York Taxi and Limousine Commission website, construct and plot a time series of the number of Yellow Taxis per month for 2010 to 2020 and label the axes appropriately.

Exploring Time Series Data

2

In Chapter 1 a variety of time series datasets were presented, and basic interpretations were made of the available information. We showed how to extract time series data from various sources and how to import data into R. In the current chapter, the next step in the process will be considered; that is, it is now time to begin the early stages of analyzing time series data.

Analogous to the utilization of visual representations in traditional data analysis, plotting data will play a central role in the time series setting. As mentioned in Chapter 1, plots of time series can reveal information about the data that would not otherwise be obvious. While plotting the data is the important first step in any time series analysis, it is possible that such plots may not be sufficient to provide an enhanced understanding of valuable underlying information. Time series data often have patterns that are noisy. Another issue is that the data may be so dense that they appear as a compact cloud of points and may hide or mask any possible patterns that exist. In this chapter, strategies will be discussed that assist in understanding data and forecasting future values.

2.1 UNDERSTANDING AND VISUALIZING DATA

A first step in any data analysis is understanding the data. For example, it is informative to calculate summary statistics, such as the mean, median, standard deviation, range, quartiles, etc. While these statistics can be calculated for time series data, there are many methods and statistics that are designed specifically for understanding the way time series data evolve in time. In Chapter 1 we discussed data that were cyclic, trending, wandering, and seasonal or a combination of these types. We also noted that viewing data over small snippets of time is helpful to better visualize the finer details. The types of behavior discussed above are often accompanied by unexpected spikes and dips and general noisiness.

It is often the case that such noisy behavior tends to hide the underlying fundamental patterns in the time series. In this section we will discuss methods designed specifically for enhancing the understanding of the data. These methods include smoothing methods for "smoothing out" the anomalies that obscure the "big-picture". Also, given a time series that contains, say seasonal and trending behavior, it may be that the repetitive seasonal behavior is covering up some important trending movement in the data that can be seen better after removing the seasonal effects. Seasonal adjustment is a widely used method in time series analysis. For example, many of the datasets in the FRED website discussed in Chapter 1 are seasonally adjusted, or the website provides an option to download either seasonally adjusted or non-seasonally adjusted data.

2.1.1 Smoothing Time Series Data

Several methods exist for "smoothing out" the noisy (possibly unimportant) behavior from a time series so that an important underlying "signal" can be better understood. We begin by discussing the most basic of smoothing methods: the *centered moving average smoother*.

DOI: 10.1201/9781003089070-2

2.1.1.1 Smoothing Data Using a Centered Moving Average Smoother

The centered moving average smoother is a method for replacing data values in a time series with an average of data values surrounding (and including) that data point. For example, a centered moving average smoother of order three replaces a data value x_t at time t with $s_t = (x_{t-1} + x_t + x_{t+1})/3$. That is, the average value is assigned to the middle (2nd) time point. Therefore, a centered moving average smoother of order three cannot be assigned to the first or last time point of a time series. It follows that the higher the order, the more values will be missing at the beginning and end of the smoothed dataset. For a 3rd-order centered moving average, the averaging formula moves along the time series dataset, considering three consecutive data values together until arriving at the last three time points.

Definition 2.1: Centered Moving Average Smoother

Let $x_t, t = 1,\ldots,n$ be a set of time series data. The centered moving average smoother is defined as follows:

Case 1: *m* is an odd number: Let $k = (m-1)/2$. For $k < t < n - k$, the smoothed data value, s_t, at time t is given by

$$s_t = \frac{1}{m} \sum_{i=t-k}^{t+k} x_i \tag{2.1}$$

Case 2: *m* is an even number: Let $k = m/2$. For $k < t < n - k$, the smoothed data value, s_t, at time t is given by

$$s_t = \frac{x_{t-k}}{2m} + \frac{1}{m} \sum_{i=t-k+1}^{t+k-1} x_i + \frac{x_{t+k}}{2m} \tag{2.2}$$

Examples:

(a) For a 5th-order centered moving average at times $2 < t < n - 2$, we have
$$s_t = (x_{t-2} + x_{t-1} + x_t + x_{t+1} + x_{t+2})/5. \tag{2.3}$$

(b) A 4th-order moving average smoother at times $2 < t < n - 2$, is given by
$$s_t = \frac{x_{t-2}}{8} + \frac{x_{t-1} + x_t + x_{t+1}}{4} + \frac{x_{t+2}}{8}$$

The centered moving average smoother has two basic uses:

(1) Smoothing designed to remove (potentially meaningless) fluctuations from the data.
(2) Removing cyclic behavior from seasonal or other cyclic data with fixed cycle lengths.

Example 2.1 shows the use of centered moving average smoothing for the purpose of detecting or better understanding underlying, fundamental signals in the data.

Example 2.1 Smoothing the Tesla and DFW Temperature Data

The Tesla stock prices from January 1, 2020 through April 30, 2021 are shown in Figure 1.19. We discussed the fact that there has been a steady increase until early 2021 at which time the price leveled off and decreased. However, there is considerable day-to-day fluctuation, especially in 2021. Figure 1.20(b) shows the DFW average annual temperatures from 1900 through 2020. There we see considerable year-to-year fluctuation but somewhat of an increase beginning in about 2000.

Figures 2.1(a) and (d) are plots of the Tesla and DFW temperature data, respectively. Figures 2.1(b), (c), (e), and (f) show smoothed versions of Figures 2.1(a) and (d). By use of the centered moving average smoother, the day-to-day fluctuations are smoothed out; note that the 8th order produces more smoothing than the 3rd order. The fundamental behavior including the leveling off and decline in early 2021 are more clearly seen by minimizing the noisy day-to-day changes. The effect of smoothing is more obvious in the DFW temperature data. Year-to-year fluctuation is quite dramatic in the original dataset in Figure 2.1(d). A 3rd-order smoothing provides some clarity, but the 8th-order smoothing shown in Figure 2.1(f) clearly shows an almost level behavior from 1900 through about 1985. Since that time there has been a rise that may or may not have leveled off in the last few years. Particularly noticeable in Figure 2.1(f) is that the smoothing has removed the extremes. Specifically, the extremely high temperatures in 2012, 2016, and 2017 are moderated by the lower temperatures in surrounding years.

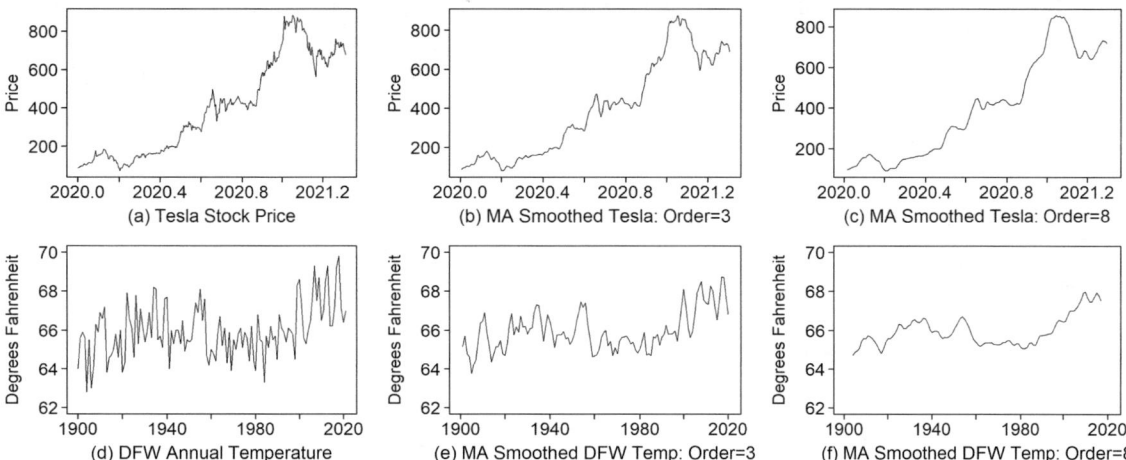

FIGURE 2.1   (a) Tesla stock prices, (b) and (c) data in (a) after applying 3rd- and 8th-order moving average smoothers, respectively. (d) DFW annual temperature data, (e) and (f) data in (d) after applying 3rd- and 8th-order moving average smoothers, respectively.

The *tswge* function `ma.smooth.wge` performs a centered moving average smoothing procedure on a set of data. Suppose **x** is a time series dataset; then the code

```
x.s=ma.smooth.wge(x,order=5,plot=TRUE)
```

plots the data in **x** and overlays it with a plot of the smoothed data (which resides in **x.s$smooth**). The command

```
x.s=ma.smooth.wge(x,order=5,plot=FALSE)
```

simply performs the smoother and retains the smoothed data in **x.s$smooth**. Figures 2.1(a)−(c) were obtained using the code below which performs centered moving average smoothers of orders 3 and 8 on the dataset **tesla**. [1]

```
par(mfrow=c(1,3))
data(tesla)
tesla.3=ma.smooth.wge(tesla,order=3,plot=FALSE)
tesla.8=ma.smooth.wge(tesla,order=8,plot=FALSE)
plot(tesla)
```

1 Code such as that given above often produces the plots without careful axis labeling and numbering. For example, **tesla** is a *ts* object (and the horizontal axis of a plot reflects the given dates). However, **tesla.3$smooth** is not a *ts* object. Creating a plot like Figure 2.1(b) would first require converting **tesla.3$smooth** to a *ts* file. Such steps will typically not be given in short clips of code throughout the book.

```
plot(tesla.3$smooth)
plot(tesla.8$smooth)
# the plots based on the tswge ts object dfw.yr are obtained analogously
```

Example 2.2 Smoothing a simulated cosine + trend + noise dataset

Figure 2.2(a) shows a cosine curve (c_t) with period 10, Figure 2.2(b) is a trend line $\ell_t = -2 + .05t$, and Figure 2.2(c) is the sum of the data in Figures 2.2(a) and (b) forming the full underlying signal, $c_t + \ell_t$, which has the appearance of "ascending cosine curves". Figure 2.2(d) is a random noise sequence, z_t, and Figure 2.2(e) is the addition of the noise in Figure 2.2(d) to the underlying signal in Figure 2.2(c) to form the "simulated observed signal", x_t, in Figure 2.2(e) given by $x_t = c_t + \ell_t + z_t$. The "observed data" in Figure 2.2(e) are quite noisy and have the appearance of "noisy" data trending upward, with some hint of cyclic behavior. Figure 2.2(f) shows the results of applying a 5th-order centered moving average smoother to the data in Figure 2.2(e). The output from the 5th-order smoother, shown in Figure 2.2(f) superimposed on the data, does a good job of removing the noise and revealing the signal (cosine + trend line). The plots in Figures 2.2(d)−(f) are obtained using the code[2]

```
set.seed(6946)
t=1:60
cosine=cos(2*pi*t/10)
line=-2+.05*t
z=rnorm(n=60,sd=1)
x=cosine+line+z
plot(x)
ma.smooth.wge(x,order=5)
```

2.1.1.2 Other Methods Available for Smoothing Data

A topic closely related to smoothing is the more general topic of *filtering*, which will be discussed in more detail in Chapter 4. In that chapter, we will introduce the Butterworth filter which is a popular filtering technique among engineers and scientists. Other smoothing methods include LOESS (locally estimated scatterplot smoothing), (Cleveland and Devlin, 1988). In Section 4.3 we introduce the *first difference filter*. While the methodologies behind the filtering and smoothing techniques are sometimes different, they share the common goal of either identifying or removing an underlying signal. In the current chapter we will use also filtering to remove seasonal components to isolate a trend. As will be discussed in the following chapters, smoothing (and more generally filtering) is also useful for transforming data to a form suitable for forecasting.

2.1.1.3 Moving Average Smoothing versus Aggregating

In Chapter 1 we noted that the new annual **sunspot2.0** data are available from 1700 to the present and monthly from 1749 through the most recent full year.[3] The annual and monthly datasets are available in *tswge* as **sunspot2.0** and **sunspot2.0.month**, respectively. Figure 2.3(a) is a plot of **sunspot2.0** containing data for the 272 years between 1749 and 2020. Figure 2.3(b) is a plot of **sunspot2.0.month** containing 3264 monthly observations between 1749 and 2020. Specifically,

---

2 The Base R function **rnorm** generates n random normals. The default values are **mean=0** and **sd=1**.
   Also, note that **plot=TRUE** is the default option in **ma.smooth.wge.**
3 As of the writing of this book, the latest full year is 2020.

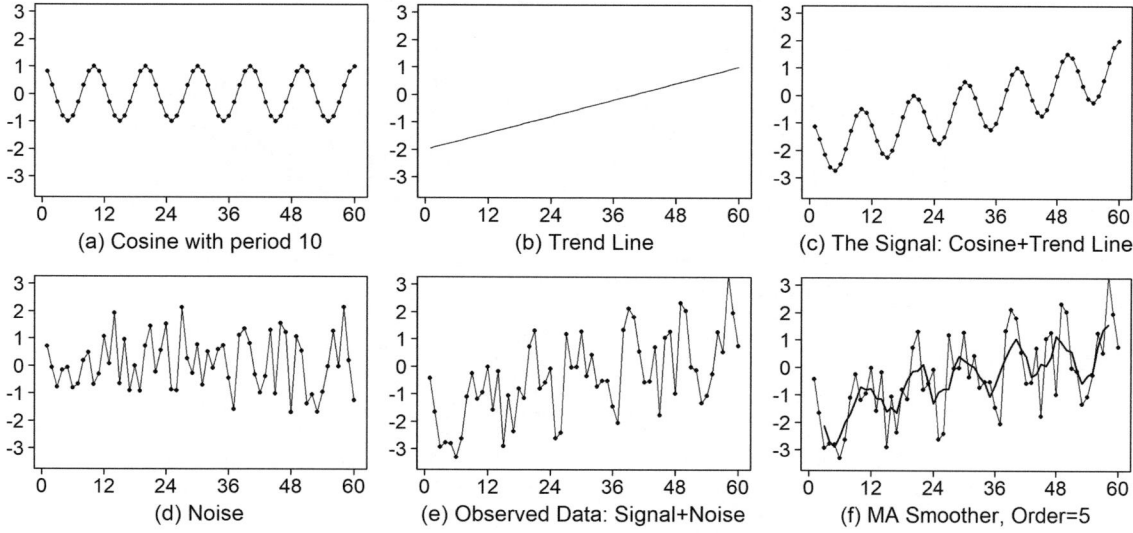

FIGURE 2.2 (a) Cosine curve with period 10, (b) the trend line, (c) cosine+line (the underlying signal), (d) the noise component, (e) signal+noise (the observed data), and (f) observed data in (e) with fifth order smoothing.

sunspot2.0.month is a *ts* object with **frequency=12**. The data in Figure 2.3(a) are for the years 1749 −2020 for which monthly data are available, and they are contained in the file

sunspot2.0.1749=window(sunspot2.0,start=1749,end=2020)

The annual data in **sunspot2.0.1749** can be calculated from the monthly data using a procedure called *aggregating*. We have used the Base R function **aggregate** to average the 12 monthly observations in each year, which results in one observation per year. Specifically, the command

ss.yr=aggregate(sunspot2.0.month,FUN=mean)

instructs R to use the sunspot monthly data, group it into subsets of 12 months (because **frequency= 12** in the *ts* file) and summarize the information in each subset using the mean function (**FUN=mean**). Figure 2.3(c) is a plot of the aggregated data, and it consists of 272 data values almost identical to the data in **sunspot2.0** for the years 1749−2020.[4] Other summary statistics can be used. For more about aggregation methods see R help for the **aggregate** function. The annual data in **ss.yr** obtained using the **aggregate** command above are plotted in Figure 2.3(c). Note that **ss.yr** is a *ts* file, the first few lines of which are

```
Time Series:
Start = 1749
End = 2020
Frequency = 1
   [1]   134.875000  139.000000   79.441667   79.666667
   [5]    51.125000   20.358333   15.933333   16.983333
   [9]    54.041667   79.341667   89.941667  104.775000
```

---

4 **ss.yr** and **sunspot2.0.1749** differ slightly because *ss.yr* is simply the mean of the 12 monthly values while **sunspot2.0.1749** takes into account the number of days per month.

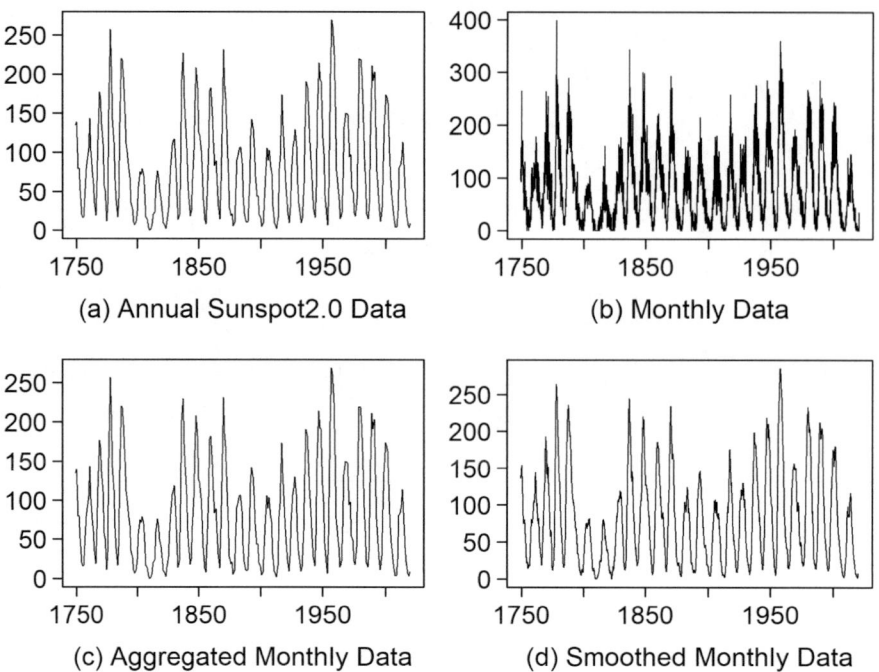

FIGURE 2.3   (a) Annual sunspot data for the years 1749–2020, (b) monthly data for the same time period, (c) annual data obtained by aggregating the monthly data, (d) smoothed monthly data using a 12th-order moving average smoother.

To create the plot in Figure 2.3(d) we applied a centered moving average smoother of order 12 to the monthly data in **sunspot2.0.month**. The plots in Figure 2.3 can be obtained using the following code.

```
data(sunspot2.0)
data(sunspot2.0.month)
sunspot2.0.12=ma.smooth.wge(sunspot2.0.month,order=12,plot=FALSE)
sunspot2.0.sm12=ts(sunspot2.0.12$smooth,start=c(1749,1),
frequency=12)
par(mfrow=c(2,2))
plotts.wge(sunspot2.0.1749)
plotts.wge(ss.yr)
plotts.wge(sunspot2.0.month)
plotts.wge(sunspot2.0.sm12)
```

Note: The aggregated data in Figure 2.3(c) and the smoothed data in Figure 2.3(d) appear to be very similar. However,

(i)   Figure 2.3(c) is a plot of 272 points of annual data for years 1749 through 2020 obtained by aggregating the monthly data for each year into an annual value by averaging the monthly data.

(ii) Figure 2.3(d) is the result of a moving average smoother. It contains $3264 - 12 = 3252$ *monthly* data values.[5]

---

5   Note that six data values are "lost" at each end of the series because of the smoother. The rolling window calculates an "annualized" value for each consecutive 12 months.

2.1.1.4 Using Moving Average Smoothing for Estimating Trend in Data with Fixed Cycle Lengths

We discussed the centered moving average smoother in Section 2.1.1.1 and demonstrated that it can be used to "uncover" an underlying signal that was "covered up" by noise. It can also be used to remove cyclic content with fixed cycle length.

(1) Cosine+Line+Noise Data

Figure 2.2 illustrated the use of moving average smoothing to uncover a "trending cosine" curve of period 10 that is embedded in noise. Figure 2.4(a) shows the noisy data previously shown in Figure 2.2(e). Figure 2.4(b) shows the result of a 5th-order moving average smoother shown earlier in Figure 2.2(f). As mentioned previously, the 5th-order smoother does a good job of removing the noise from the data and revealing the underlying cosine+line signal. Figure 2.4(c) shows the effect of increasing the smoothing order to the *period length*, 10. The resulting smoothed data estimates the trend (in this case a line) and removes the cyclic component altogether. The following is an important point.

> **Key Point:** If a centered moving average smoother of order k is applied to fixed period cyclic data with fixed cycle length *where k is the cycle length*, then the smoother removes the cyclic component and "reveals" (or estimates) the trend.

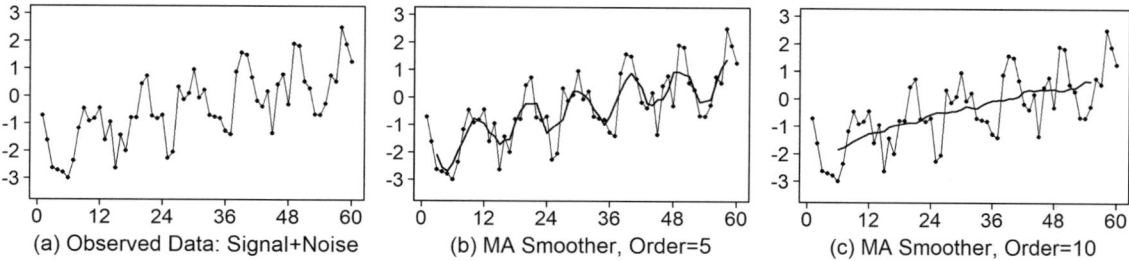

(a) Observed Data: Signal+Noise   (b) MA Smoother, Order=5   (c) MA Smoother, Order=10

FIGURE 2.4   (a) Cosine+line+noise data, (b) and (c) observed data in (a) with 5th and 10th order smoothing, respectively.

(2) Air Passengers Data

The monthly **AirPassengers** data in Figure 2.5(a) were discussed in Chapter 1 where it was noted that there is a repetitive (seasonal but non-sinusoidal) behavior each year. The moving average smoothed data of order 12 in Figure 2.5(b) show no evidence of the within-year seasonal behavior, but instead it is a curve that tracks the trend in airline travel. For example, there were fairly "flat" spots around 1953 and 1958 where there was not much increase in the number of passengers over the previous year. These were both temporary in that air travel continued to increase after each of these "flat" spots. The point is that there are annual patterns that are more easily seen in Figure 2.5(b) than in the **AirPassengers** data (although close examination of Figure 2.5(a) shows the annual patterns in question). Thus, the 12th-order moving average smoother behaves like the 10th-order moving average smoother on the period 10 sinusoidal data in Figure 2.4(c) in that it removes the cyclic behavior. It is interesting to note the effect of a 15th-order smoother on the **AirPassengers** data shown in Figure 2.5(c). The 15th order introduces some wiggle in the smoothed data reinforcing the fact that the cyclic component is most effectively removed when the order of the smoother is equal to the period length of the cyclic data.

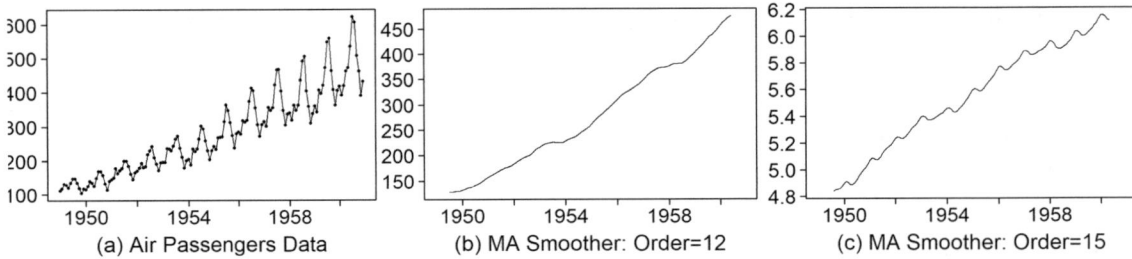

FIGURE 2.5 (a) Air Passengers data, (b) and (c) observed data in (a) with 12th and 15th -order smoothing, respectively.

> **Key Point:** While a moving average smoother of order k applied to cyclic data with a fixed cycle length k will remove the cyclic behavior and retain the trend, smoothers of orders greater than k or less than k will retain some cyclic behavior.

QR 2.1 Smoothing Orders

(3) Texas Unemployment Data

Figure 2.6(a) displays the monthly unemployment rate in Texas for the years 2000–2019, obtained from the Texas Workforce Commission website https://twc.texas.gov. The data are in *tswge ts* file **tx.unemp. unadj**. The data show a seasonal behavior (which is not very clear in Figure 2.6(a)) but shows up better in the short time snippet in Figure 2.6(b). The seasonal pattern within a year appears to be that unemployment rates are lower in January through May than they are in the summer months, after which time they decline.

Figure 2.6(c) is a plot of the original unemployment data after applying a 12th-order moving average smoother. As previously discussed, the smoother removed the seasonal behavior and retained the "trend" which in this case was anything but linear. The trending behavior is important, and there is much to be learned from it. Unemployment rose to a peak of about 7.5% in 2004, dropped to about 4%, then increased rapidly to about 8% during the Great Recession, and stayed at that level from June 2009 through September 2011. The unemployment rate then consistently declined through 2019 when unemployment was at about 3.3%.[6] Figure 2.6(c) was obtained using the commands

```
data(tx.unemp.unadj)
tx.unemp.sm=ma.smooth.wge(ts.unemp.unadj,order=12, plot=FALSE)
tx.unemp.sm12=ts(tx.unemp.sm$smooth,start=c(2000,1),frequency=12)
plotts.wge(ts.unemp.sm12)
```

---

6 Although data are available through 2020, the COVID pandemic caused very different and unpredictable unemployment behavior.

Key Point: A moving average smoother of order k applied to cyclic data with a fixed cycle length, k, will remove the cyclic behavior and retain the trend even when the trend follows a wandering path as in the Texas Unemployment data.

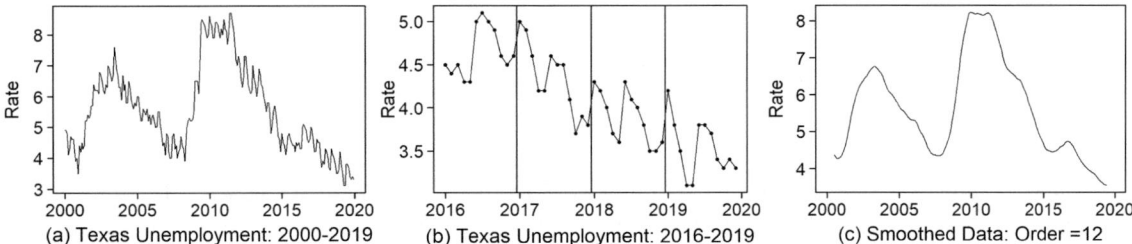

FIGURE 2.6 (a) Monthly Texas unemployment rates from 2000 through 2019 and (b) unemployment rates from 2016 through 2019, (c) original data after applying a 12th-order moving average smoother.

2.1.2 Decomposing Seasonal Data

We defined seasonal data in Chapter 1 as cyclic data that have fixed periods and having a pattern related to the calendar. The **AirPassengers** dataset in Figure 1.5 and Figure 2.5(a) is a classical seasonal dataset. The dataset not only has a within-year seasonal behavior but also has a (near linear) increasing trend. It is common practice to consider seasonal data, x_t, to have: (a) a within-year seasonal component, s_t, (b) a longer-term trend component, tr_t, and (c) a random noise component, z_t. For example, we have seen these behaviors in the Air Passengers and Texas Unemployment data. Time series analysts focus on two types of seasonal models:

Additive Seasonal Data
The data, x_t, at time t can be thought of as a sum given in (2.4)

$$x_t = s_t + tr_t + z_t.$$
(2.4)

Multiplicative Seasonal Data
The data, x_t, at time t can be expressed as the product given in (2.5)

$$x_t = s_t \times tr_t \times z_t.$$
(2.5)

Example 2.3 Air Passengers Data
To illustrate the difference between the data types that are best fit with an additive and with a multiplicative model, we use the **AirPassengers** dataset. As previously noted, the **AirPassengers** data, plotted in Figure 2.7(a), have a seasonal and trend component but also the within-year variability increases in time.

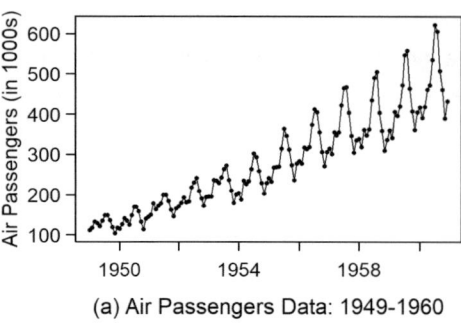
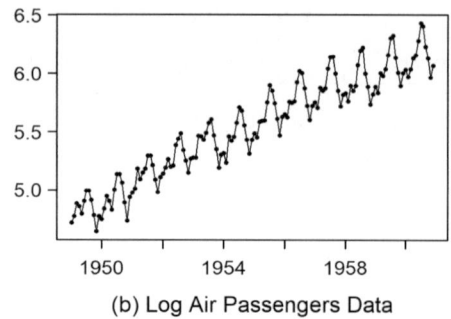

(a) Air Passengers Data: 1949-1960   (b) Log Air Passengers Data

FIGURE 2.7   (a) Air Passengers data and (b) logarithm of Air Passengers data.

Datasets with this type of behavior are often modeled using multiplicative models. To eliminate the variability increase, analysts often take the logarithms of the data and use the "log data" for analysis. The R commands below create the **logAirPassengers** data which are plotted in Figure 2.7(b).

```
data(AirPassengers)
logAirPassengers=log(AirPassengers)
```

Recall that the **AirPassengers** data are in the form of a *ts* object, so **logAirPassengers** is also a *ts* object. The **logAirPassengers** data in Figure 2.7(b) show no increase in the within-year variability and are a classic example of data that are modeled using the additive model in (2.4). We will discuss the differences in modeling strategies for these two datasets below.

The additive and multiplicative decompositions follow similar implementation steps:

1. Estimate the trend component.
2. Remove the trend component which results in a dataset that is primarily made up of the seasonal fluctuations in the data.
3. Calculate an "average" within-year seasonal component.
4. Find the remaining noise.

We begin by discussing the additive model which is the more intuitive of the two.

Key Points

1. Data with increasing within-year variability, i.e. multiplicative data, are common in economics and other fields.
2. Analysts model multiplicative data in two ways:
   – Using the multiplicative model in (2.5)
   – Taking the logarithm of the data and modeling the log data using an additive model.

2.1.2.1 Additive Decompositions

In this section, we will obtain additive decompositions of the **log Air Passengers** and the Texas Unemployment data.

(1) Log Air Passengers Data
When data are analyzed using the additive model in (2.4), the assumption is made that the data are the sum of seasonal, trend, and random noise components. The analysis steps involved in decomposing

the **logAirPassengers** data are discussed. In practice, the components in (2.4) are estimated and an estimated model can be described as

$$x_t = \hat{s}_t + tr_t + \hat{z}_t. \tag{2.6}$$

Additive Decomposition: Step-by-step

The decomposition of the logAirPassengers data is accomplished using the following steps.

(a) Estimate the Trend Component
Figure 2.8 is a plot of the **logAirPassengers** dataset overlaid with the result of applying a 12th-order centered moving average smoother to the data. These plots were obtained using the commands:

```
data(AirPassengers)
logAirPassengers=log(AirPassengers)
logair.12=ma.smooth.wge(logAirPassengers,order=12)
logair.sm12=ts(logair.12$smooth,start=c(1949,1),frequency=12)
```

The **logair.12** object contains the data plotted in the overlaid trend in Figure 2.8. These data are listed below:

Year	Jan	Feb	Mar	Apr	May	Jun	Jul	Aug	Sep	Oct	Nov	Dec
1949	NA	NA	NA	NA	NA	NA	4.8373	4.8411	4.8466	4.8512	4.8545	4.8600
1950	4.8698	4.8814	4.8934	4.9043	4.9128	4.9237	4.9405	4.9574	4.9744	4.9919	5.0131	5.0338
1951	5.0478	5.0609	5.0738	5.0884	5.1069	5.1243	5.1383	5.1528	5.1637	5.1715	5.1784	5.1894
1952	5.2039	5.2181	5.2316	5.2437	5.2574	5.2707	5.2829	5.2921	5.3041	5.3233	5.3436	5.3574
1953	5.3677	5.3783	5.3884	5.3978	5.4038	5.4072	5.4104	5.4103	5.4084	5.4068	5.4062	5.4106
1954	5.4196	5.4283	5.4351	5.4422	5.4507	5.4611	5.4737	5.4897	5.5040	5.5164	5.5294	5.5427
1955	5.5579	5.5727	5.5875	5.6027	5.6167	5.6312	5.6459	5.6598	5.6742	5.6876	5.7008	5.7147
1956	5.7272	5.7389	5.7507	5.7607	5.7708	5.7804	5.7887	5.7965	5.8048	5.8141	5.8231	5.8327
1957	5.8427	5.8535	5.8649	5.8755	5.8857	5.8945	5.9016	5.9070	5.9100	5.9107	5.9116	5.9138
1958	5.9174	5.9229	5.9261	5.9276	5.9297	5.9305	5.9330	5.9384	5.9462	5.9564	5.9678	5.9773
1959	5.9853	5.9941	6.0040	6.0149	6.0266	6.0407	6.0545	6.0662	6.0731	6.0807	6.0919	6.1020
1960	6.1125	6.1212	6.1284	6.1374	6.1457	6.1515	NA	NA	NA	NA	NA	NA

In terms of the estimated model in (2.6), $tr_t =$ **logair.sm12** is the near linear curve in Figure 2.8.

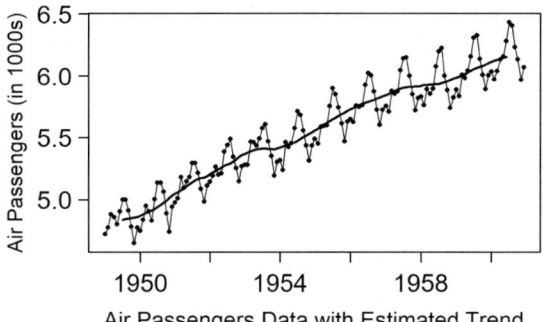

Air Passengers Data with Estimated Trend

FIGURE 2.8 Log Air Passenger data overlaid with 12th-order centered moving average smoother.

(b) Remove the Trend Component from the Data
The next step is to subtract the estimated trend component from the data (**logAirPassengers**). This can be accomplished using the code:

```
seas.logair=logAirPassengers-logair.sm12
round(seas.logair,4) # rounding to 4 decimal places for listing below
```

The data in **seas.logair** are plotted in Figure 2.9(a) and the *ts* file is shown below:

Year	Jan	Feb	Mar	Apr	May	Jun	Jul	Aug	Sep	Oct	Nov	Dec
1949	NA	NA	NA	NA	NA	NA	0.1599	0.1561	0.0661	-0.0721	-0.2101	-0.0893
1950	-0.1249	-0.0451	0.0553	0.0010	-0.0844	0.0802	0.1953	0.1784	0.0882	-0.1016	-0.2769	-0.0922
1951	-0.0710	-0.0503	0.1080	0.0054	0.0406	0.0575	0.1550	0.1406	0.0512	-0.0839	-0.1948	-0.0774
1952	-0.0622	-0.0251	0.0311	-0.0452	-0.0479	0.1138	0.1552	0.1968	0.0383	-0.0711	-0.1961	-0.0896
1953	-0.0896	-0.1002	0.0754	0.0618	0.0299	0.0858	0.1656	0.1955	0.0597	-0.0549	-0.2133	-0.1073
1954	-0.1015	-0.1919	0.0245	-0.0173	0.0047	0.1148	0.2368	0.1905	0.0529	-0.0826	-0.2162	-0.1090
1955	-0.0689	-0.1217	-0.0002	-0.0080	-0.0182	0.1214	0.2512	0.1895	0.0688	-0.0745	-0.2327	-0.0871
1956	-0.0782	-0.1148	0.0082	-0.0145	-0.0088	0.1438	0.2347	0.2074	0.0673	-0.0905	-0.2210	-0.1091
1957	-0.0901	-0.1464	0.0101	-0.0233	-0.0135	0.1505	0.2405	0.2393	0.0914	-0.0614	-0.1913	-0.0967
1958	-0.0884	-0.1608	-0.0345	-0.0754	-0.0353	0.1449	0.2635	0.2862	0.0552	-0.0730	-0.2312	-0.1572
1959	-0.0992	-0.1593	0.0024	-0.0335	0.0137	0.1163	0.2518	0.2600	0.0646	-0.0719	-0.2003	-0.0981
1960	-0.0794	-0.1524	-0.0905	-0.0040	0.0112	0.1307	NA	NA	NA	NA	NA	NA

The year-to-year seasonal behavior is much easier to visualize in Figure 2.9(a) without the "interference" of the trend. Specifically, the pattern in each year is similar (high travel in summer, low in November, still low but slightly up in December, continued low in January and February, up in March, and so forth). However, there are year-to-year variations: air travel in November was unusually low in 1950 and then abnormally high for July and August of 1958.

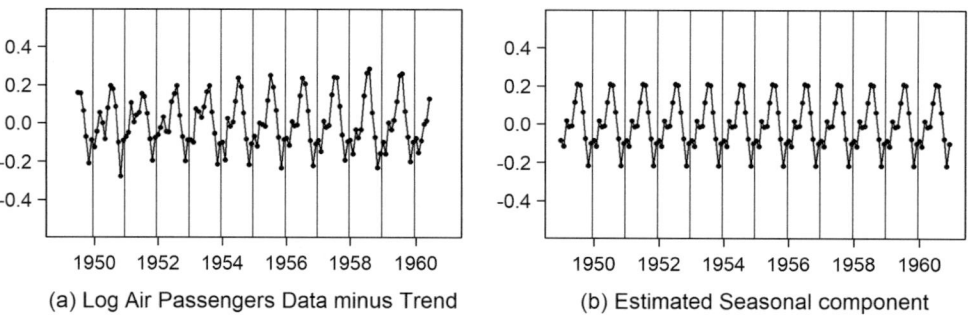

(a) Log Air Passengers Data minus Trend   (b) Estimated Seasonal component

FIGURE 2.9   (a) Log Air Passengers data minus the trend component in Figure 2.8 and (b) estimated seasonal component.

(c) Calculate an "Average" Within-year Seasonal Component

Note that the seasonal component in the model (2.4) is an overall pattern that is the same from year to year. That is, $\{s_t, t = 1,...,12\} = \{s_{t+12}, t = 1,...,12\} = \{s_{t+2(12)}, t = 1,...,12\} = \cdots$. The noise component, z_t, adjusts for the year-to-year variations from the overall seasonal pattern. The seasonal pattern, $s_t, t = 1,...,12$, is estimated by averaging across years, and the estimated seasonal component, $\hat{s}_t$, (which is the same for each year) is plotted in Figure 2.9(b). The following code can be used to calculate the monthly means:

```
# convert the ts file to a vector
seas.logair.numeric=as.numeric(seas.logair) # convert ts file to a vector
# convert this vector to a matrix with ncol=number of years
seas.logair.matrix=matrix(seas.logair.numeric,ncol=12) #12 years and 12 months
```

	[,1]	[,2]	[,3]	[,4]	[,5]	[,6]	[,7]	[,8]	[,9]	[,10]	[,11]	[,12]
[1,]	NA	-0.1249	-0.0710	-0.0622	-0.0896	-0.1015	-0.0689	-0.0782	-0.0901	-0.0884	-0.0992	-0.0794
[2,]	NA	-0.0451	-0.0503	-0.0251	-0.1002	-0.1919	-0.1217	-0.1148	-0.1464	-0.1608	-0.1593	-0.1524
[3,]	NA	0.0553	0.1080	0.0311	0.0754	0.0245	-0.0002	0.0082	0.0101	-0.0345	0.0024	-0.0905
[4,]	NA	0.0010	0.0054	-0.0452	0.0618	-0.0173	-0.0080	-0.0145	-0.0233	-0.0754	-0.0335	-0.0040

[5,]	NA	-0.0844	0.0406	-0.0479	0.0299	0.0047	-0.0182	-0.0088	-0.0135	-0.0353	0.0137	0.0112
[6,]	NA	0.0802	0.0575	0.1138	0.0858	0.1148	0.1214	0.1438	0.1505	0.1449	0.1163	0.1307
[7,]	0.1599	0.1953	0.1550	0.1552	0.1656	0.2368	0.2512	0.2347	0.2405	0.2635	0.2518	NA
[8,]	0.1561	0.1784	0.1406	0.1968	0.1955	0.1905	0.1895	0.2074	0.2393	0.2862	0.2600	NA
[9,]	0.0661	0.0882	0.0512	0.0383	0.0597	0.0529	0.0688	0.0673	0.0914	0.0552	0.0646	NA
[10,]	-0.0721	-0.1016	-0.0839	-0.0711	-0.0549	-0.0826	-0.0745	-0.0905	-0.0614	-0.0730	-0.0719	NA
[11,]	-0.2101	-0.2769	-0.1948	-0.1961	-0.2133	-0.2162	-0.2327	-0.2210	-0.1913	-0.2312	-0.2003	NA
[12,]	-0.0893	-0.0922	-0.0774	-0.0896	-0.1073	-0.1090	-0.0871	-0.1091	-0.0967	-0.1572	-0.0981	NA

Comparing **seas.logair.matrix** with **seas.logair** it can be seen that the rows and columns are reversed. The matrix with columns representing months as in the *ts* file is obtained by transposing the matrix using the Base R function **t**.

```
seas.logair.matrix.t=t(seas.logair.matrix)
```

[1,]	NA	NA	NA	NA	NA	NA	0.1599	0.1561	0.0661	-0.0721	-0.2101	-0.0893
[2,]	-0.1249	-0.0451	0.0553	0.0010	-0.0844	0.0802	0.1953	0.1784	0.0882	-0.1016	-0.2769	-0.0922
[3,]	-0.0710	-0.0503	0.1080	0.0054	0.0406	0.0575	0.1550	0.1406	0.0512	-0.0839	-0.1948	-0.0774
[4,]	-0.0622	-0.0251	0.0311	-0.0452	-0.0479	0.1138	0.1552	0.1968	0.0383	-0.0711	-0.1961	-0.0896
[5,]	-0.0896	-0.1002	0.0754	0.0618	0.0299	0.0858	0.1656	0.1955	0.0597	-0.0549	-0.2133	-0.1073
[6,]	-0.1015	-0.1919	0.0245	-0.0173	0.0047	0.1148	0.2368	0.1905	0.0529	-0.0826	-0.2162	-0.1090
[7,]	-0.0689	-0.1217	-0.0002	-0.0080	-0.0182	0.1214	0.2512	0.1895	0.0688	-0.0745	-0.2327	-0.0871
[8,]	-0.0782	-0.1148	0.0082	-0.0145	-0.0088	0.1438	0.2347	0.2074	0.0673	-0.0905	-0.2210	-0.1091
[9,]	-0.0901	-0.1464	0.0101	-0.0233	-0.0135	0.1505	0.2405	0.2393	0.0914	-0.0614	-0.1913	-0.0967
[10,]	-0.0884	-0.1608	-0.0345	-0.0754	-0.0353	0.1449	0.2635	0.2862	0.0552	-0.0730	-0.2312	-0.1572
[11,]	-0.0992	-0.1593	0.0024	-0.0335	0.0137	0.1163	0.2518	0.2600	0.0646	-0.0719	-0.2003	-0.0981
[12,]	-0.0794	-0.1524	-0.0905	-0.0040	0.0112	0.1307	NA	NA	NA	NA	NA	NA

The following code creates the vector **months** containing the column means excluding the missing data.

```
months=colMeans(seas.logair.matrix.t, na.rm=TRUE)
round(months,4)
```

```
[1] -0.0867 -0.1153 0.0172 -0.0139 -0.0098 0.1145  0.2100 0.2036 0.0640 -0.0761 -0.2167 -0.1012
```

```
seas.means=rep(months,12) # replicates the 12 monthly means for each year (12)
seas.means=ts(seas.means,start=c(1949,1),frequency=12)
```

The *ts* file **seas.means** is plotted in Figure 2.9(b).

(d) Find the Remaining Noise Component
We have computed the following:

Trend estimate: **logair.sm12**
Seasonal component estimate: **seas.means**

The estimated noise in (2.6), $\hat{z}_t$, is calculated using the code:

```
logair.noise=logAirPassengers-logair.sm12-seas.means
```

Plot the Decomposition
Figure 2.10(d) is a plot of the **logair.noise** along with the other parts of the decomposition procedure:

(a) log **AirPassengers** data
(b) estimated trend component
(c) estimated seasonal component
(d) noise

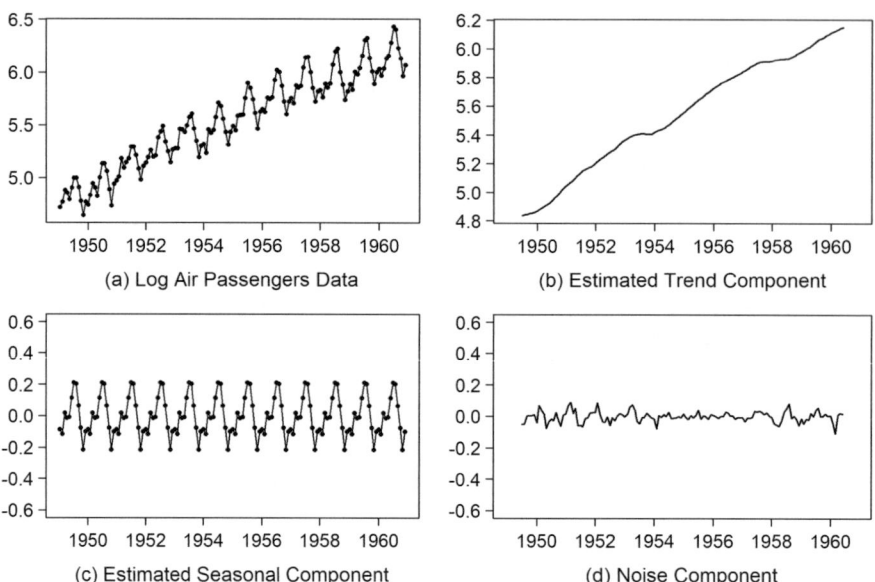

FIGURE 2.10   Additive Decomposition of Log Air Passengers Data.

(2) Texas Unemployment Data

We will repeat the steps used in the additive decomposition of the **logAirPassengers** dataset, but without as much detail.

(a) Estimate the Trend Component

Figure 2.11(a) shows the Texas Unemployment data (not seasonally adjusted) along with the trend estimate using a centered moving average smoother of order 12. This plot can be obtained using the commands:

```
data(tx.unemp.unadj)
tx.unemp.sm=ma.smooth.wge(tx.unemp.unadj,order=12,plot=FALSE)
tx.unemp.sm12=ts(tx.unemp.sm$smooth,start=c(2000,1), frequency=12)
plotts.wge(tx.unemp.sm12)
```

(b) Remove the Trend Component from the Data

Remove the estimated trend component from the data (**tx.unemp.unadj**) using the command

```
seas.tx.unemp=tx.unemp.unadj-tx.unemp.sm12
```

Figure 2.11(b) is a plot of **seas.tx.unemp** where it can be seen that there is considerable year-to-year variability in the seasonal behavior, but the general behavior is for the unemployment in summer months to be high. This was observed in the time snippet in Figure 2.6(b).

(c) Calculate an "Average" Within-year Seasonal Component

The estimated seasonal pattern, $\hat{s}_t, t = 1,\dots,12$, is found by averaging across years, and the seasonal component consists of repeating this pattern for all years in the dataset. Figure 2.11(c) shows the estimated seasonal component. The "average" seasonal pattern is for high unemployment in the summer months with the best (lowest) unemployment in April and May.

```
seas.tx.unemp.numeric=as.numeric(seas.tx.unemp) #convert ts file to a vector
# convert this vector to a matrix with ncol=number of years
seas.tx.unemp.matrix=matrix(seas.tx.unemp.numeric,ncol=20) # 20 years
seas.tx.unemp.matrix.t=t(seas.tx.unemp.matrix) # transpose matrix
months=colMeans(seas.tx.unemp.matrix.t,na.rm=TRUE) # colMeans are monthly means
```

```
seas.means=rep(months,20)   # replicates the 12 monthly means for each year (20)
seas.means=ts(seas.means,start=c(2000,1),frequency=12)
```

The data plotted in Figure 2.11(c) are in the vector **seas.means**.

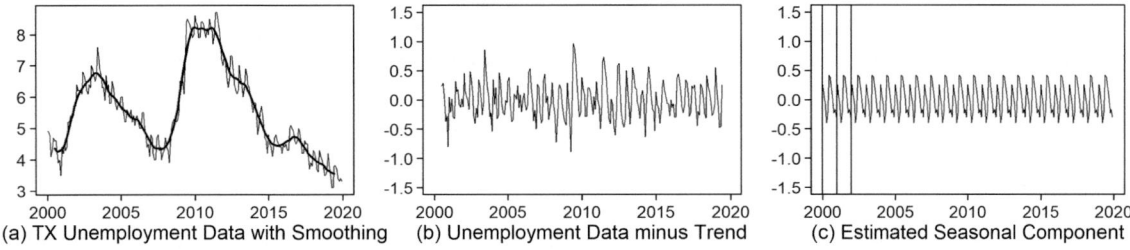

(a) TX Unemployment Data with Smoothing   (b) Unemployment Data minus Trend   (c) Estimated Seasonal Component

FIGURE 2.11   (a) Texas unemployment rates for 2000 through 2019 and showing moving average smoothed data of order 12, (b) Texas unemployment data minus trend, and (c) estimated seasonal component.

(d) Find the Remaining Noise Component
We have computed the following:

> Trend estimate: **tx.unemp.sm12**
> Seasonal component estimate: **seas.means**

The estimated noise in (2.6), $\hat{z}_t$, is calculated using the code:

```
tx.unemp.noise=tx.unemp.unadj-tx.unemp.sm12-seas.means
```

Plot the Decomposition
Figure 2.12(d), similar to that of Figure 2.10(d), is a plot of **tx.unemp.noise** along with the other parts of the decomposition procedure:

(a) Texas unemployment data
(b) estimated trend component
(c) estimated seasonal component
(d) noise

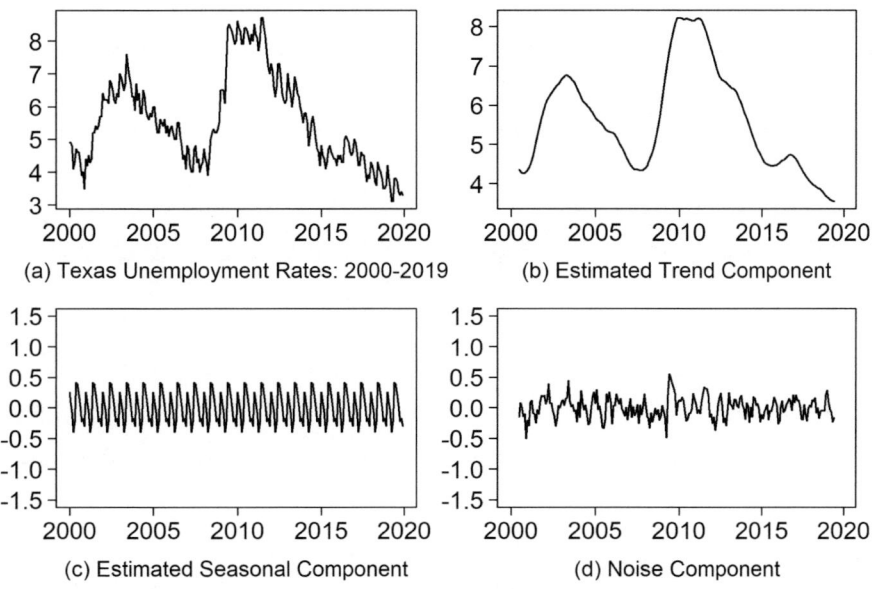

(a) Texas Unemployment Rates: 2000-2019   (b) Estimated Trend Component

(c) Estimated Seasonal Component   (d) Noise Component

FIGURE 2.12   Additive Decomposition of Texas Unemployment Data.

2.1.2.2 *Multiplicative Decompositions*

We will obtain a multiplicative decomposition of the **AirPassengers** dataset shown in Figures 2.5(a) and 2.7(a). Recall that these data have a seasonal behavior that is increasing in time along with within-year variability that is also increasing in time. While such data can be modeled by taking the logarithm and then modeling the log data using an additive model, in this section we use a multiplicative model to analyze the "raw" **AirPassengers** data. When data are analyzed using multiplicative decomposition, the assumption is made that the data are the product of seasonal, trend, and noise components. The estimated model is

Multiplicative Decomposition: Step-by-step

$$x_t = \hat{s}_t \times tr_t \times \hat{z}_t.$$ (2.7)

(1) Estimate the Trend Component

As with the additive model, the first step is to use a centered moving average smoother, again in this case of order 12. We previously computed and plotted the moving average smoother in Figure 2.5(b). The steps involved are as follows:

```
data(AirPassengers)
AirPass.sm12=ma.smooth.wge(AirPassengers,order=12)
AirPass.sm12=ts(AirPass.sm12$smooth,start=c(1949,1),frequency=12)
```

Recall that, in terms of the estimated model in (2.7), $tr_t =$ **AirPass.sm12**. This near linear curve is shown as part of the full decomposition in Figure 2.14(b).

(2) Remove the Trend Component from the Data

The next step is to remove the estimated trend component from the dataset, **AirPassengers**. This can be accomplished by *division* (instead of subtraction) using the code:

```
seas.AirPass=AirPassengers/AirPass.sm12
```

The year-to-year seasonal behavior is much easier to visualize in Figure 2.13(a) after removing the "interference" of the trend and increasing within-year variability. Note also that the within-year variability is not increasing as much as in Figure 2.14(a). The increase in variability in the final model (2.7) is caused by the increasing trend being multiplied by the seasonal data in Figure 2.13(a). The seasonal patterns in Figure 2.13(a) are similar to those for the additive data shown in Figure 2.9(a).

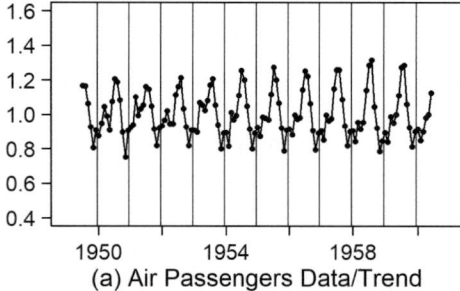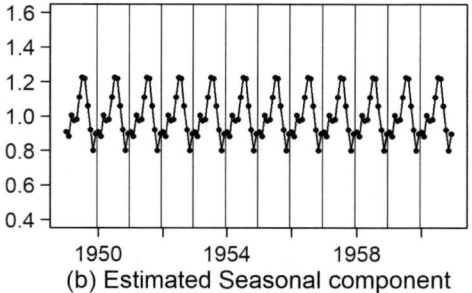

(a) Air Passengers Data/Trend   (b) Estimated Seasonal component

FIGURE 2.13   (a) Air Passenger data with trend component removed (by division) and (b) estimated seasonal component.

(3) Calculate an "Average" Within-year Seasonal Component.

As in the additive model, the seasonal component in the model (2.5) is an overall pattern that is assumed to be the same from year to year, and the noise component, z_t, adjusts for the year-to-year variations from the overall seasonal pattern. The estimated seasonal component, $\hat{s}_t$, (which is the same for each year) is plotted in Figure 2.13(b). Note the similarity between Figure 2.13(b) and Figure 2.9(b) which was the seasonal component for additive decomposition of the **logAirPassengers** data.

The following code can be used to calculate the monthly means:

```
seas.AirPass.numeric=as.numeric(seas.AirPass) #convert ts file to a vector
# convert this vector to a matrix with ncol=number of years
seas.AirPass.matrix=matrix(seas.AirPass.numeric,ncol=12) # 12 years
seas.AirPass.matrix.t=t(seas.AirPass.matrix) # transpose matrix
months=colMeans(seas.AirPass.matrix.t,na.rm=TRUE) # colMeans are monthly means
seas.means=rep(months,12)  # replicates the 12 monthly means for each year (12)
seas.means=ts(seas.means,start=c(1949,1),frequency=12)
```

The *ts* file **seas.means** is plotted in Figure 2.13(b).

(4) Find the Remaining Noise Component

We have computed the following:

> Trend estimate: **AirPass.sm12**
> Seasonal component estimate: **seas.means**

The estimated noise in (2.6), $\hat{z}_t$, is calculated using the code:

```
Air.Pass.noise=AirPassengers/(AirPass.sm12*seas.means)
```

Figure 2.14 is a plot of the component parts of the multiplicative decomposition of the **AirPassengers** data.[7]

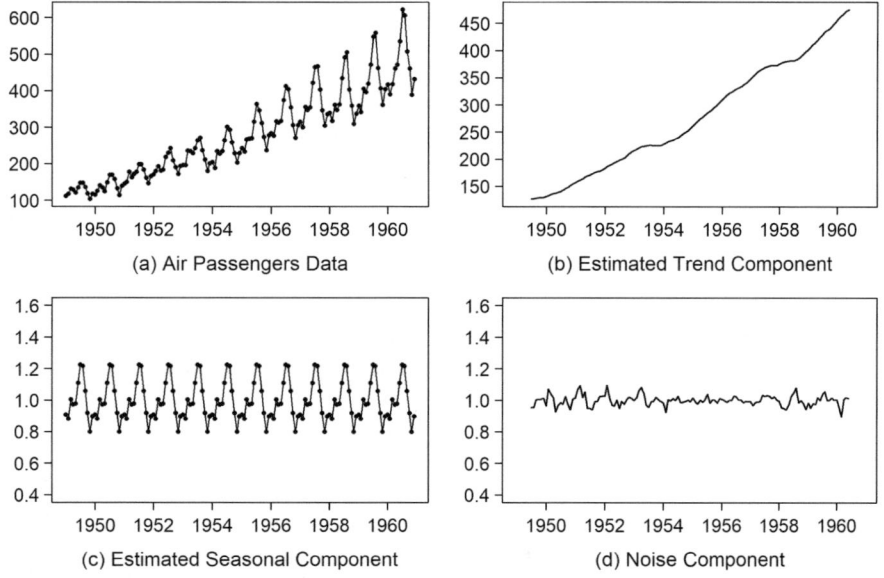

(a) Air Passengers Data

(b) Estimated Trend Component

(c) Estimated Seasonal Component

(d) Noise Component

FIGURE 2.14 Multiplicative Decomposition of Air Passengers Data.

---

7 Base R function **decompose** can be used for time series decomposition. (See Appendix 2A)

2.1.3 Seasonal Adjustment

As you looked through the FRED website you probably noticed that many (even most) datasets are *seasonally adjusted*. Other government websites such as the Census (https://census.gov) and the Bureau of Labor Statistics (www.bls.gov/web/laus/laumstrk.htm) provide seasonally adjusted data. Seasonal adjustment techniques can be quite complex and have been widely studied and developed by data scientists at the Census Bureau and elsewhere. In this section we briefly discuss the basic ideas and goals of seasonal adjustment, go through a step-by-step rudimentary seasonal adjustment procedure, apply seasonal adjustment to the Texas Unemployment and Air Passengers data, and then discuss software that is available for state-of-the-art seasonal adjustment procedures.

The question is: "Why are so many economic-type datasets seasonally adjusted?" The Texas Unemployment data in Figure 2.12(a) contain a lot of information to be understood. There is year-to-year trending along with within-year seasonal patterns. These within-year seasonal patterns are fairly predictable, but they do have (possibly important) variations from year to year and month to month. One method for seeing the "big picture" is to perform a centered moving average smoother to obtain a trend estimate as shown in Figure 2.12(b). For example, the smoothed data in Figure 2.12(b) show the year-to-year changes such as the Great Recession which was followed by a consistent decline in unemployment until the COVID pandemic in 2020.[8] For this reason, the trend estimate obtained by moving average smoothing can be thought of as a type of "seasonal adjustment" because it removes the within-year seasonal changes and shows the overall behavior. However, notice in Figure 2.11(b) that within-year seasonal patterns differed quite a bit from year to year. If an important economic event occurred in May of some year, that information is smeared out in the trend estimate. For example, the patterns within the year 2009 were "more extreme" than those in other years. What is needed is a seasonal adjustment that removes the normal seasonal patterns, shows the overall trending behavior, but also retains month-to-month information that can highlight unusual activity.

2.1.3.1 Additive Seasonal Adjustment

Seasonal adjustments are related to the decompositions previously discussed. If an additive decomposition is appropriate, then an additive seasonal adjustment is used. A similar pairing applies in the multiplicative case.

Because an additive decomposition was appropriate for the Texas Unemployment data, an additive-based seasonal adjustment will be used for this dataset. The most straightforward additive seasonal adjustment method that has some ability to accomplish the goals discussed in the preceding paragraph is to obtain the seasonally adjusted data, sa_t, using the formula

$$sa_t = x_t - \hat{s}_t, \tag{2.8}$$

which subtracts the seasonal component (shown in Figure 2.12(c)) from the data.

Figure 2.15(a) shows the seasonally adjusted Texas Unemployment data using this simple method. The command to create the seasonally adjusted data is

```
tx.unemp.adj.1 = tx.unemp.unadj - seas.means
```

Figure 2.15(a) shows the Texas Unemployment data superimposed with the trend estimate obtained using a centered moving average smoother of order 12. Figure 2.15(b) shows the seasonally adjusted data computed using (2.8). That is, it is a plot of `tx.unemp.adj.1`. This plot is similar to the estimated trend but with more detail concerning monthly changes that might be important. The website https://twc.texas. gov contains seasonally adjusted data, which is available in *tswge* data file `tx.unemp.adj`. The seasonal

---

8 As stated in a previous footnote, although data are available through 2020, the COVID pandemic caused very different and unpredictable unemployment behavior.

adjusted data on the https://twc.texas.gov website is much smoother than the data computed using (2.8). Finally, the mathematically sophisticated seasonal adjustment techniques developed at the Census Bureau are contained in the CRAN package *seasonal*.[9] The function **seas** in the *seasonal* package performs the Census-Bureau seasonal adjustment. Although **seas** has numerous options, for purposes of this discussion we simply used **seas** with all default options. The following commands were used:

```
library(seasonal) # assuming that "seasonal" has been installed
census=seas(tx.unemp.unadj)
# the seasonally adjusted data are in the ts file census$data[,3]
```

Figure 2.15(d) is a plot of **census$data[,3]**. Examination of Figure 2.15 shows that the seasonal adjustment using **seas** with default options provides the same information as the one using (2.8).[10]

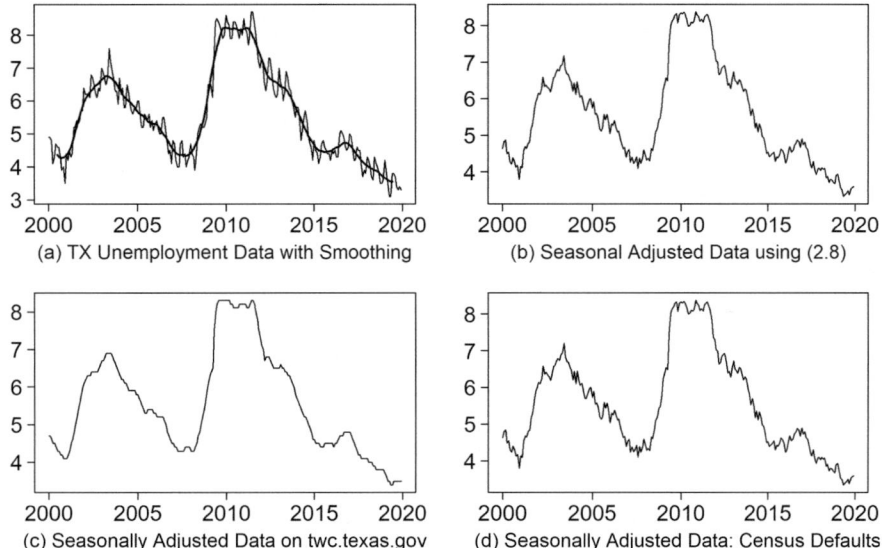

FIGURE 2.15   (a) Texas unadjusted unemployment data superimposed with the output from a 12th-order moving average smoother, (b) seasonally adjusted data using (2.8), (c) seasonally adjusted data from the twc.texas.gov website, and (d) seasonally adjusted data using the default options of the function seas.

2.1.3.2 Multiplicative Seasonal Adjustment

Because multiplicative decomposition was appropriate for the **AirPassengers** data, we will use multiplicative seasonal adjustment for this dataset. Analogous to the method used for the additive seasonal adjustment, the seasonally adjusted data uses the formula

$$\widehat{sa}_t = x_t / \hat{s}_t, \tag{2.9}$$

to *divide* the data by the seasonal component (shown in Figure 2.14(c)). The command to create the seasonally adjusted data is

---

9 The seasonal adjustment developed at the Census Bureau is referred to as X13-ARIMA-SEATS. Information about the Census Bureau seasonal adjustment can be found at www.census.gov/data/software/x13as.html

10 The *seasonal* package has a variety of seasonal adjustment methods including methods using seasonal ARIMA models. These models will be discussed in Chapter 7.

```
AirPassengers.adj = AirPassengers/seas.means
```

Figure 2.16(a) shows the **AirPassengers** data superimposed with the trend estimate obtained using a centered moving average smoother of order 12. Figure 2.16(b) is a plot of the seasonally adjusted data computed using (2.9). That is, it is a plot of **AirPassengers.adj**. Again, the seasonally adjusted data are similar to the estimated trend but with more detail concerning monthly changes. The following code uses the **seas** function with default options.

```
library(seasonal) # assuming that seasonal has been installed
census=seas(AirPassengers)
# the seasonally adjusted data are in the ts file census$data[,3]
```

Figure 2.16(c) is a plot of **census$data[,3]**. Examination of Figure 2.16(c) shows that the seasonal adjustment using **seas** is smoother and is less affected by the larger within-year variability increase in later years. Overall, the results are similar to those seen in Figure 2.16(b) with the seasonal effects removed.

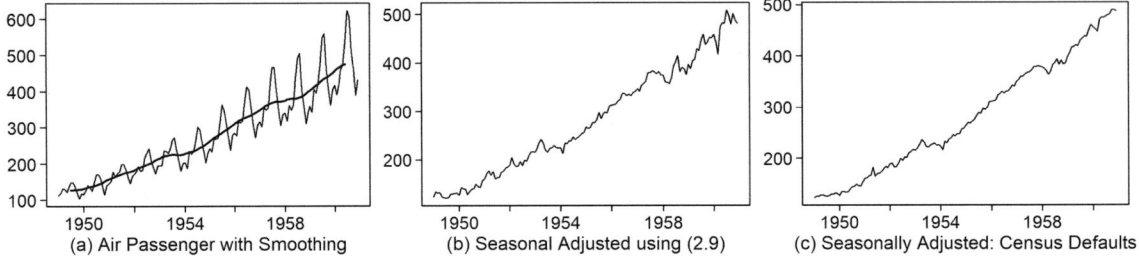

FIGURE 2.16   (a) Air Passengers data superimposed with the output from a 12ᵗʰ-order moving average smoother, (b) seasonally adjusted data using (2.9), (c) seasonally adjusted data using the default options of the function seas in the CRAN package *seasonal.*

In this section, we have shown how smoothing, decomposition, and seasonal adjustment are useful techniques to help the analyst understand and visualize data. These techniques have been illustrated using various examples. For both additive and multiplicative time series data, the separation of the original time series data into trend,[11] seasonality, and residual components enhances understanding of the individual components and the dataset as a whole. In many cases, unless a time series dataset is decomposed into its component parts, important information is hidden and unrecognized.

2.2 FORECASTING

Obviously, a primary application of time series analysis is that of forecasting. It was previously stated that one of the purposes of smoothing is to forecast future values. Indeed, if evidence exists that previous patterns in a dataset will continue into the future, various techniques can be used to make forecasts. The techniques discussed in this chapter are "non-model" based approaches to forecasting, but later chapters will be devoted to "model-based" forecasting.

---

11 It should be emphasized that the "trend" recognized by the decomposition method is not necessarily a deterministic trend that can be assumed to continue indefinitely. Instead, it should be considered to be "the current trend". In later chapters we introduce a different methodology that also dissects (decomposes) time series data into various components and relevant details will be provided related to theory and useful application.

There are many situations for which businesses, organizations, and individuals are interested in the ability to make forecasts into the future. For example, a business owner may wish to forecast the future demand of a certain product so that the appropriate amount of inventory is ensured to be in stock. A city making decisions about infrastructure might need to forecast its population in ten years. A difficult problem but one of interest to many in the financial sector (and to most of us, actually) is forecasting the ebb and flow of the stock market and of individual stocks so that sound investing decisions can be made. Each of these examples illustrates the applicability and need for forecasting techniques. Without a better option, such forecasts are often made somewhat subjectively, based on past memory of similar events, rumor, or perhaps by intuitive but presumptuous calculations which provide educated guesses and estimates.

Fortunately, time series analysis techniques provide a mathematical, theory-based alternative if the underlying mathematical assumptions are appropriate. This results in algorithms that calculate forecasts along with corresponding prediction limits at a given level of confidence, analogous to calculating the sample mean plus or minus a margin of error in the non-time series, random sample setting. The typical scenario is that a forecasting algorithm is developed which is then used to predict an outcome of interest. The forecast comprises parameter estimates which can be found using software in an effort to achieve optimal predictive ability. Notice that this is similar to linear regression methodology used for predicting a dependent variable. The difference is that linear regression modeling is based on independent errors and is thus inappropriate for time-dependent data. While some of the mathematical theory supporting time series methodology is quite complex, software is available that performs the heavy lifting. Also, it is convenient that a proposed time series forecasting model can often be "checked" (validated). In Chapters 6–11 these issues will be discussed in detail.

> **Key Point:** We will use the terms *prediction* and *forecast* synonymously.

We will first introduce moving average forecasting, an intuitively simple method which can give reasonable 1-step ahead forecasts. A 1-step ahead forecast at time t is a forecast of x_{t+1} given data up to and including x_t. Exponential smoothing is a somewhat more sophisticated method of forecasting that also uses smoothing techniques. Finally, the Holt-Winters forecasting technique, an extension of exponential smoothing designed to account for trend and/or seasonality, is introduced.

> **Key Points**
>
> 1. In later chapters, we will provide a comprehensive analysis of various "model-based" forecasting methodologies.
>   - Model-based forecasting methods produce forecasts based on a particular mathematical model (for example, the autoregressive model) that has been fit to the data to describe the underlying physical process generating the data over time and involves certain distributional assumptions about residuals.[12]
> 2. The forecasting methods in this chapter are smoothing methods which are simply functions of the data that make few assumptions about the underlying physical process.

---

12 We will begin the discussion of model-based methods in Chapter 5.

2.2.1 Predictive Moving Average Smoother

In Section 2.1.1.1 we explained how the centered moving average smoother is used for visualizing a smoothed version of the data for purposes of "recovering" underlying signals or removing noise or seasonal effects. We can also use moving averages for prediction. Instead of the centered moving average discussed in Section 2.1.1.1, we will use the *predictive (rolling or trailing)* moving average for forecasting future values. If the data do not exhibit seasonality or trend, then to predict the value of the time series at time $t+1$, it makes sense to "predict" x_{t+1} to be the average of the previous k data values for some k. That is, letting $\tilde{x}_{t+1}$ denote the 1-step ahead prediction of x_{t+1} given data to time t, then a reasonable and very simple predictor would be $\tilde{x}_{t+1} = \left(x_t + x_{t-1} + \cdots + x_{t-k+1}\right)/k$. That is, the predictor of x_{t+1} is the average of the last k data values. Suppose, for example, that we want to use a 3-point moving average predictor for a dataset of length $t = 20$. Note that the 1-step ahead prediction of x_4 using this predictor would be $\tilde{x}_4 = \left(x_3 + x_2 + x_1\right)/3$. Because the dataset has 20 data values, there is a known value for x_4, so we can compare the predictor $\tilde{x}_4$ with the actual value, x_4, to evaluate the accuracy of the predictor. Using this procedure, we can predict the values x_4 up to x_n for which the prediction is $\tilde{x}_n = \left(x_{n-1} + x_{n-2} + x_{n-3}\right)/3$. Again, all of these 1-step ahead predictors can be compared with actual values to ascertain the quality of each prediction. Notice also, that we have observed the data values necessary to calculate $\tilde{x}_{n+1} = \left(x_n + x_{n-1} + x_{n-2}\right)/3$. In this case $\tilde{x}_{n+1}$ is a predictor of a future value which we presumably do not know. We can make an assessment about the quality of this prediction based on the predictions of $x_4, \ldots, x_n$, all of which can be "checked". These forecasts can be obtained using *tswge* function **ma.pred.wge**. The procedure is illustrated in the following example.

Key Point: A 1-step ahead forecast at time t is the forecast of x_{t+1} given data up to and including x_t.

Example 2.4 Moving Average Prediction of Monthly WTI Crude Prices 2018-2020
Figure 2.17(a) shows the monthly price of West Texas Intermediate crude oil for the years 2018−2020. During that time period the price varied dramatically from a low of $20 per /barrel to a high of about $70. The function **ma.pred.wge** below calculates 1-step ahead predictions using a fifth order moving average smoother. In the third command below, the **wtc36** *ts* file is changed to a numeric vector so that the data will simply be indexed by 1−36 (months) for convenience.

```
data(wtcrude2020)
wtc36=window(wtcrude2020,start=c(2018,1))
wtc36=as.numeric(wtc36)
ma5=ma.pred.wge(wtc36,order=5,n.ahead=4)
```

The **wtc36** data and the 1-step ahead predictions, **ma5$pred**, are shown below (rounded to two decimal places)

```
round(wtc36,2)
```

```
 [1]   63.56  62.15  62.86  66.32  69.89  67.52  70.99 67.99  70.19  70.75 56.57 48.64
[13]   51.36  54.99  58.15  63.88  60.73  54.68  57.51 54.84  56.86  53.98 57.11 59.86
[25]   57.71  50.60  29.88  16.81  28.79  38.30  40.75 42.36  39.61  39.53 41.52 47.09
```

```
round(ma5$pred,2)
```

```
 [1]     NA     NA     NA     NA     NA 64.96  65.75  67.52  68.54  69.32 69.49 67.30
[13]   62.83  59.50  56.46  53.94  55.40  57.82  58.49  58.99  58.33  56.92 55.57 56.06
[25]   56.53  57.10  55.85  51.03  42.97  36.76  32.88  30.91  33.40  37.96 40.11 40.75
[37]   42.02  41.95  42.42  43.00
```

The **wtc36** data are plotted again in Figure 2.17(b) along with the moving average predictors listed above.

Notes

(a) Because the moving average predictor is fifth order, the data begin at $t = 6$.

(b) The prediction at $t = 6$, obtained using the first five data values is \$64.96 while the actual value at $t = 6$ was \$67.52. Additionally, at time $t = 12$, the predicted price was \$67.30 while the actual price was \$48.64, a drop in prices that was not anticipated by the moving average. In general, the moving average predictor lags behind dramatic changes in the behavior of the process.

(c) Predictions for $t = 6$ through $t = 36$ are 1-step ahead predictions.

(d) The **ma.pred.wge** command above specifies forecasts up to four steps *beyond* the end of the dataset (n.ahead=4). The prediction $\tilde{x}_{37}$ is the average of the last five data values, x_{32} through x_{36}, which are observed. Data are only available up through $t = 36$, so the forecast for time $t = 38$ is the average

$$\tilde{x}_{38} = \frac{x_{33} + x_{34} + x_{35} + x_{36} + \tilde{x}_{37}}{5}$$

The forecast $\tilde{x}_{38}$, which is based on data up to and including time $t = 36$, is called a 2-step ahead forecast. The predictions for $t = 38$ through $t = 40$ proceed in a similar manner, gradually increase, but will eventually level off.

Key Points

1. In general, a forecast of $x_{t+\ell}$ given data up to and including time t is called an ℓ-step ahead forecast.

2. Figure 2.17(b) shows 1-step ahead forecasts for times $t = 6$ through $n + 1 = 41$.

3. Forecasts for $t = 42, 43,$ and 44 are also shown. These forecasts are two, three, and four-step ahead forecasts, respectively.

4. Note that the parameter **n.ahead** in the **ma.pred.wge** function specifies how many steps beyond the observed data we want to forecast.

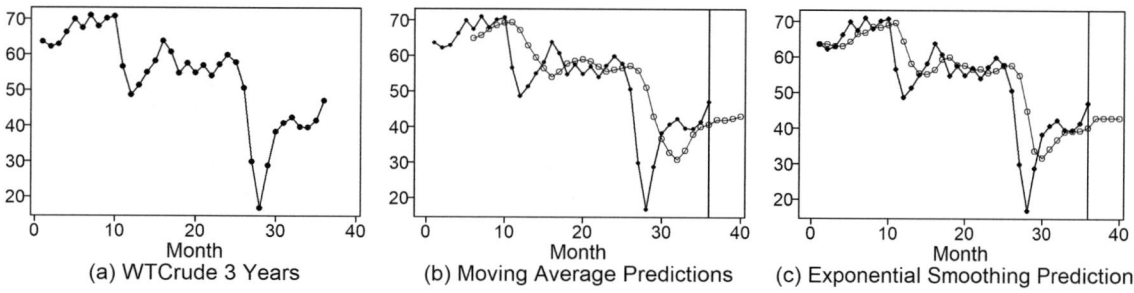

(a) WTCrude 3 Years   (b) Moving Average Predictions   (c) Exponential Smoothing Prediction

FIGURE 2.17 (a) Monthly price of West Texas Intermediate crude oil for the years 2018–2020, (b) crude oil prices along with 1-step ahead predictors using a 5th-order smoother, and (c) 1-step ahead predictions using exponential smoothing with $\alpha = .4$.

Key Points

1. The moving average predictions are based on "smoothing".
2. Thinking of the data value x_t as a sum $x_t = \mu_t + e_t$ (that is, a mean value plus a noise term), then at time t, instead of predicting the next *data value*, x_{t+1}, the prediction can be viewed as an estimator of the mean μ_{t+1}.

2.2.2 Exponential Smoothing

While moving average prediction is easy to conceptualize and calculate, it is unrealistic to assume that all preceding data values will have an equal amount of influence on future values. It is more intuitive to believe that in many cases more recent data should have more influence on future values than data in the more distant past because it is more reflective of the current state of reality. A smoothing method that takes this into consideration is called *exponential smoothing*. Exponential smoothing was first introduced by R.G. Brown (1956). To motivate the methodology, we again consider that each data value x_t in a time series is comprised of a mean value at each time point t and an independent error term with zero mean and constant variance. That is, $x_t = \mu_t + e_t$. For $0 \le \alpha \le 1$, the exponential smoothing recursion for $t = 1, 2, \ldots, n$ is[13]

$$u_{t+1} = \alpha x_t + (1-\alpha)u_t \tag{2.10}$$

The expression $\alpha x_t + (1-\alpha)u_t$ in (2.10) is a linear combination of the current value x_t along with u_t, which is the estimate of μ_t based on data up to but not including time t. "Standing at time t" the weighted average given in (2.10) is the predictor of μ_{t+1}. Note that the estimate u_t is heavily dependent on the smoothing parameter α which ranges from 0 to 1. The closer α is to 1, the more weight is given to most recent data while the closer α is to 0, the less weight is given to most recent data. Of course, the optimal value of α will depend on the particular application of the dataset and can be tuned accordingly.

Because the above formula is recursive, each new estimate u_t is dependent on the previous estimate u_{t-1}, which is dependent on the previous estimate u_{t-2}, and so on. This formula does not immediately appear to be "exponential", so how does the method get its name?

Letting $t = 1, 2, 3, 4$, we see that the recursive formula yields the following:

$$u_1 = x_1$$

$$u_2 = \alpha x_1 + (1-\alpha)u_1$$

$$= \alpha x_1 + (1-\alpha)x_1$$

$$= x_1$$

$$u_3 = \alpha x_2 + (1-\alpha)u_2$$

$$= \alpha x_2 + (1-\alpha)x_1$$

$$u_4 = \alpha x_3 + (1-\alpha)u_3$$

$$= \alpha x_3 + (1-\alpha)\left[\alpha x_2 + (1-\alpha)x_1\right]$$

QR 2.2 Exponential Smoothing Derivation

---

13 It is typical practice to set $u_1 = x_1$ to initiate the recursion.

$$= \alpha x_3 + \alpha(1-\alpha)x_2 + (1-\alpha)^2 x_1$$

$$u_5 = \alpha x_4 + (1-\alpha)u_4$$

$$= \alpha x_4 + (1-\alpha)\left[\alpha x_3 + \alpha(1-\alpha)x_2 + (1-\alpha)^2 x_1\right]$$

$$= \alpha x_4 + \alpha(1-\alpha)x_3 + \alpha(1-\alpha)^2 x_2 + (1-\alpha)^3 x_1$$

and in general,

$$u_k = \sum_{j=1}^{k-2} \alpha(1-\alpha)^{j-1} x_{k-j} + (1-\alpha)^{k-2} x_1. \tag{2.11}$$

The exponential nature of the exponential smoother is apparent in (2.11).

It is clear from (2.11) that since α is between zero and one, the recursive formula gives less weight to earlier observations and more weight to the most recent observations, and that the magnitude of the weighting is exponential.

The *tswge* function **expsmooth.wge(x,alpha,n.ahead)** performs exponential smoothing where **x** is the dataset, **alpha** is the value for α, and **n.ahead** is the number of steps beyond the end of the dataset you want to forecast (default=0). If you do not specify a value for **alpha**, the function will "pick" one based on a minimization of the one--step-ahead forecast errors. Because the goal is to forecast the mean level μ_t, and because the μ_ts are likely to be smoother than the data, minimizing one-step-ahead forecast errors may not be the best solution. Values of **alpha** between .2 and .4 are often used.

Example 2.4 (Revisited)

Figure 2.17(c) shows the **wtc36** data along with the u_t values. The figure shows one-step-ahead forecasts using exponential smoothing with $\alpha = .4$. The resulting predictors are similar to those in Figure 2.17(b) obtained using a 5th-order moving average predictor. Using the file **wtc36** that was created in Example 2.4, Figure 2.17(c) can be obtained using the command

```
wtc36.exp=expsmooth.wge(wtc36,alpha=.4,n.ahead=4)
```

2.2.2.1 Forecasting with Exponential Smoothing beyond the Observed Dataset

A comment about exponential smoothing forecasts is in order. Recalling that for t within the time frame of the data, u_{t+1} is an estimate of the mean μ_{t+1} given data up to and including time t. When $t = n$, the end of the dataset, then u_{n+1} can be calculated using data up to and including $t = n$. That is, the prediction equation gives an estimate of μ_{n+1}. Because there is no assumed trend or seasonal effect in the data, exponential smoothing prediction beyond the end of the data is based on the assumption that, based on what we know at time $t = n$, the "best" estimate of $\mu_{n+k} = u_{n+1}$ for all $k \geq 1$. Figure 2.17(c) is a plot of forecasts for times $n+1,\ldots,n+4$. Note that the prediction for $t = n+1$ is slightly different from the prediction for time $t = n$. However, the predictions for times $t = n+2, n+3$, and $n+4$ are all the same as the prediction for $t = n+1$.

While these forecasts are unexciting and not appealing intuitively, the "conservative" nature of the forecasts can be the "best thing to do" with randomly wandering data like the West Texas crude oil prices. That is, predicting an upward trend to continue may not be a good idea if the trend is not persistent. Having said that, these predictions have a tendency to track poorly with upward or downward trends even when visual inspection, and possibly other information, suggest that the trend is continuing.[14] If a time series

---

14 The exponential smoothing forecasts are the optimal forecasts for data from "ARIMA(0,1,1)" models, which will be defined in Chapter 7. In future chapters we will make model identification decisions *before* using a particular model for prediction.

dataset exhibits trend or seasonality, simple exponential smoothing should not be used as a forecasting model. The Holt-Winters forecasting approach, to be discussed in Section 2.2.3, is designed to account for trend and seasonality.

Example 2.5 DFW Annual Temperature Data

Dataset **dfw.yr** contains the annual temperature data for the Dallas-Ft. Worth area for the years 1900 through 2020. As you can see, there is considerable year-to-year variability. Figures 2.18(a) and 2.18(b) show the 7th-order moving average predictor and the exponential smoothing predictor using $\alpha = .2$. These plots provide a better view of the fact that the predictors are predicting mean values. These smoothing-based predictors show that there seems to be a recent increase in temperatures.

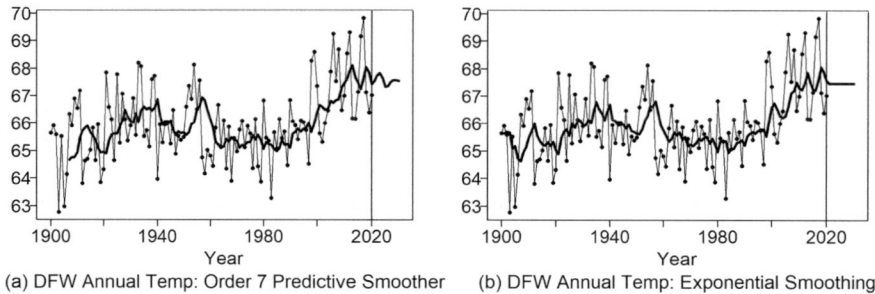

(a) DFW Annual Temp: Order 7 Predictive Smoother   (b) DFW Annual Temp: Exponential Smoothing

FIGURE 2.18   (a) DFW Annual Temperature data along with 7th-order moving average predictor and (b) DFW data with exponential smoothing predictor with $a = 2$.

The plots were obtained using the code

```
data(dfw.yr)
expsmooth.wge(dfw.yr,alpha=.2,n.ahead=10)
ma.pred.wge(dfw.yr,n.ahead=10)
```

QR 2.3 Exponential Smoothing Example

Notice the forecasts beyond the end of the series. The moving average forecasts show some pattern but will eventually become horizontal while the exponential smoothing forecasts are horizontal from the beginning since all forecasts are equal to the initial forecast for $n+1$.

2.2.3 Holt-Winters Forecasting

The Holt-Winters approach is a technique developed by economists Holt (1957) and Winters (1960). This method was designed to generalize exponential smoothing to account for trend and seasonality. As mentioned above, this method is much more adept at tracking time series with trend and/or seasonality than is simple exponential smoothing. As with decomposition and seasonal adjustment, there are additive and multiplicative versions of Holt-Winters forecasting.

2.2.3.1 Additive Holt-Winters Equations

In this section we consider prediction when an additive model (see (2.4)) is appropriate. The formulas for Holt-Winters forecasting are generalizations of exponential smoothing equations, and the Holt-Winters equations are:

$$u_t = \alpha\left(x_t - s_{t-m}\right) + \left(1 - \alpha\right)\left(u_{t-1} + v_{t-1}\right)$$

$$v_t = \beta\left(u_t - u_{t-1}\right) + \left(1 - \beta\right)v_{t-1} \tag{2.12}$$

$$s_t = \gamma\left(x_t - u_{t-1}\right) + \left(1 - \gamma\right)s_{t-m}$$

where $0 \le \alpha, \beta, \gamma \le 1$, and where m is the period length (for example, frequency in a *ts* file). For monthly data, $m = 12$, and for quarterly data, $m = 4$. The u_ts are related to simple exponential smoothing and provide a baseline. The v_ts and s_ts relate to trend and seasonal effects, respectively.

For times $t = m + 1, \ldots, n$, the 1-step ahead forecasts, $\hat{x}_t$, for the mean at time t, are given by

$$\hat{x}_t = u_{t-1} + v_{t-1} + s_{t-m}.$$

Forecasts for $x_{n+\ell}, \ell = 1, \ldots, K$ (that is, up to K steps beyond the end of the observed data), are given recursively by

$$\hat{x}_{n+\ell|n} = u_n + mv_n + s_{n+\ell-m\ell'}.$$

where $\ell' = [(\ell - 1)/m] + 1$ with $[(\ell - 1)/m]$ denoting the greatest integer less than or equal to $(\ell - 1)/m$. Here, α, β, and γ are smoothing parameters, and can be obtained using R commands. Notice that each of the three Holt-Winters equations follow the general exponential smoothing form. The only reasonable method for implementing the Holt-Winters computations is by using a computer. R has several packages that contain functions for performing the calculations. We will use the function **HoltWinters** which is in Base R.[15]

Example 2.6 Texas Unemployment Rates

In this example we will apply Holt-Winters forecasting to the non-seasonally adjusted Texas Unemployment data shown in Figure 2.12(a). We earlier obtained an "additive" decomposition and performed additive seasonal adjustment on this dataset. In this example additive Holt-Winters forecasts will be obtained. Figure 2.19(a) shows the data overlayed by the 1-step ahead Holt-Winters forecasts. The two curves essentially lie on top of each other and are indistinguishable. Figure 2.19(b) shows forecasts for five years beyond the end of the dataset. There it can be seen that the seasonal and trending behavior are forecast to continue. For more information see Holt (1957) and Winters (1960).

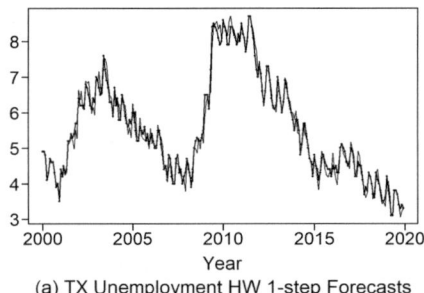
(a) TX Unemployment HW 1-step Forecasts

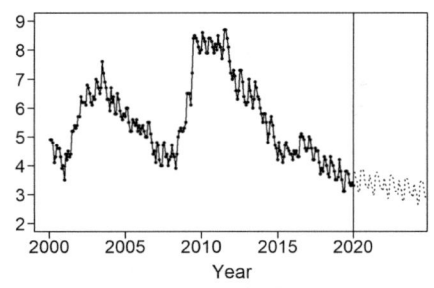
(b) TX Unemployment HW Forecast 5 Years

FIGURE 2.19   (a) Texas Unemployment data with 1-step ahead Holt-Winters forecasts and (b) Texas Unemployment data for 2000–2019 and Holt-Winters forecasts for the next five years.

2.2.3.2 Multiplicative Holt-Winters Equations

As the term implies, the multiplicative Holt-Winters equations are applicable to data for which the multiplicative model in (2.5) is appropriate. In this case, the Holt-Winters equations are:

$$u_t = \alpha(x_t / s_{t-m}) + (1 - \alpha)(u_{t-1} + v_{t-1})$$

---

15 The values of α, β, and γ are chosen by the **HoltWinters** function and are provided as output.

$$v_t = \beta\left(u_t - u_{t-1}\right) + \left(1-\beta\right)v_{t-1} \tag{2.13}$$

$$s_t = \gamma\left(x_t / u_{t-1}\right) + \left(1-\gamma\right)s_{t-m}$$

where $0 \le \alpha, \beta, \gamma \le 1$, and where again, m is the frequency.

Predictions for x_t values are given by

$$\hat{x}_t = (u_{t-1} + v_{t-1})s_{t-m.}$$

Forecasts for $x_{n+\ell}, \ell = 1,\ldots,K$ (that is, up to K steps beyond the end of the observed data) are given recursively by

$$\hat{x}_{n+\ell\mid n} = \left(u_n + \ell v_n\right)s_{n+\ell-m\ell'.} \tag{2.14}$$

where $\ell' = [(\ell-1)/m] + 1$.

Example 2.7 Air Passengers data

Consider once again the **AirPassengers** data, first plotted in this chapter in Figure 2.7(a). Because later time points reveal increasing magnitudes of the cyclic peaks and troughs, the time series is multiplicative rather than additive. Figure 2.20(a) shows the one-step-ahead Holt-Winters forecasts for years 1950–1960 (superimposed on actual data) given by the formula using the estimated smoothing parameters and coefficients above. The one-step-ahead forecasts (solid line) are quite accurate and are barely distinguishable from the data (points on the line). Figure 2.20(b) shows the **AirPassengers** data along with Holt-Winters forecasts for the next three years (dotted line). These forecasts appear to accurately extend the preceding pattern. The following code produces the plots in Figure 2.20.

```
# Figure 2.20(a)
ap.hw=HoltWinters(AirPassengers,seasonal="mult")
plot(ap.hw)
# Figure 2.20(b)
ap.pred=predict(ap.hw,n.ahead=36)
plot(ap.hw,ap.pred,lty=1:2)
```

QR 2.4 Holt
Winters Example

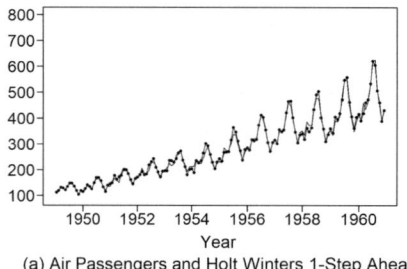

(a) Air Passengers and Holt Winters 1-Step Ahead

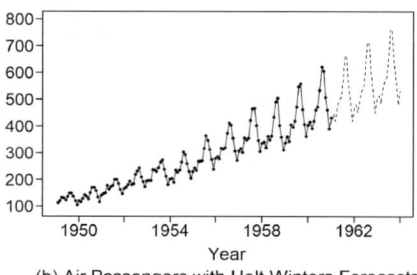

(b) Air Passengers with Holt Winters Forecasts

FIGURE 2.20 (a) Air Passengers data with 1-step ahead Holt-Winters predictions and (b) Air Passengers data with Holt-Winters forecasts for years 1961–1963.

2.2.4 Assessing the Accuracy of Forecasts

The forecasts in Figures 2.19(b) and 2.20(b) "look" like they are good predictors. However, the "acid" test is to compare forecasts with actual values. Consider the **AirPassengers** data, which are for the 144 months from January 1949 through December 1960. (And we do not know the actual values for the following years.) Instead of forecasting 36 months "beyond the known" data, a common method for assessing accuracy in this situation is, for example, to base the forecasts on the first nine years (108 months) and "forecast" the next 36 months. In this example, the first nine years would be considered to be the dataset from which forecasts were produced. The forecasts can then be compared with the known values that were "held out". Common terminology is to refer to the first nine years in this example as the *training data* and the last three years as the *test data*. The code

```
AP9=window(AirPassengers,start=c(1949,1),end=c(1957,12))
AP9.hw=HoltWinters(AP9,seasonal="mult")
AP9.pred=predict(AP9.hw,n.ahead=36)
plot(AirPassengers,type='l')
points(AP9.pred,type='l',lty=2)
```

produces Figure 2.21(a), which gives a visual comparison between the forecasts (dashed lines) and actual values for the last three years.

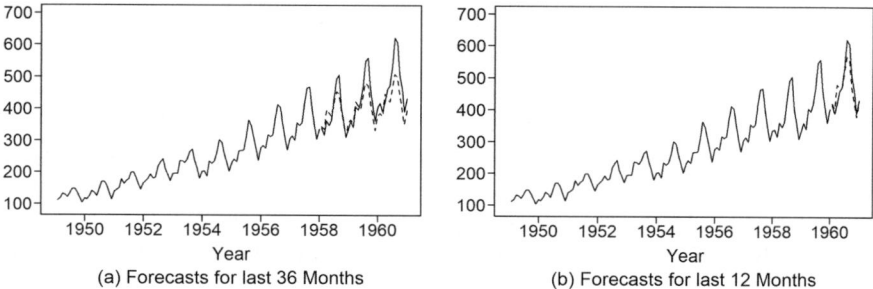

(a) Forecasts for last 36 Months   (b) Forecasts for last 12 Months

FIGURE 2.21 (a) Forecasts and actual values for the last 36 months of the Air Passenger data and (b) same as (a) for the last 12 months.

The forecasts and actual values for months 109–144 are given below:

```
Actual=window(AirPassengers,start=c(1958,1))
Actual
```

Year	Jan	Feb	Mar	Apr	May	Jun	Jul	Aug	Sep	Oct	Nov	Dec
1958	340	318	362	348	363	435	491	505	404	359	310	337
1959	360	342	406	396	420	472	548	559	463	407	362	405
1960	417	391	419	461	472	535	622	606	508	461	390	432

```
Pred=AP9.pred
round(Pred,0)
```

Year	Jan	Feb	Mar	Apr	May	Jun	Jul	Aug	Sep	Oct	Nov	Dec
1958	344	336	398	386	376	419	454	447	398	353	314	357
1959	365	357	422	409	399	444	480	473	421	374	332	377
1960	386	377	446	432	421	469	507	500	444	395	351	398

Creating the R vector **Error** we obtain the following:

```
Error=Pred-Actual
round(Error,0)
```

Year	Jan	Feb	Mar	Apr	May	Jun	Jul	Aug	Sep	Oct	Nov	Dec
1958	4	18	36	38	13	-16	-37	-58	-6	-6	4	20
1959	5	15	16	13	-21	-28	-68	-86	-42	-33	-30	-28
1960	-31	-14	27	-29	-51	-66	-115	-106	-64	-66	-39	-34

Some differences are positive and some are negative. A common method of obtaining an "overall" quantification of the quality of the predictions is to find the mean square error (MSE), which eliminates the issue with signs. Letting x_t and f_t represent the actual and predicted values at time t, respectively, then the MSE for the forecasts above is

$$MSE = \frac{1}{36} \sum_{t=1}^{36} \left(f_t - x_t \right)^2 . \tag{2.15}$$

The MSE in this case can be found using the R command

`MSE=mean(Error^2)`

which is 2010.37. This value is in squared units and is not intuitively meaningful. For this reason, it is useful to take the square root of the MSE, which is called the *root mean square error* (RMSE). The RMSE is in the units of the problem. For the forecasts above, the RMSE is given by

`RMSE=sqrt(MSE)`

which is 44.84. Examining the differences, this is sort of an average amount by which forecasts differ from true values. Figure 2.21(b) shows the forecasts for the last year (12 months). Visually the forecasts look quite good, and in this case the RMSE is 31.98.

In this book we will use the RMSE to assess forecast performance and to compare forecasts from different models or methods. Another measure of forecast quality that has intuitive appeal is the mean absolute deviation, which is the average of the absolute values of the errors. Specifically,

$$MAD = \frac{1}{36} \sum_{t=1}^{36} \left| f_t - x_t \right| . \tag{2.16}$$

In Problem 2.5 you are asked to find the MAD for the above forecasts.

2.3 CONCLUDING REMARKS

In this chapter some preliminary time series data analysis techniques were introduced and illustrated. It was shown that applying various smoothing techniques can make time series data easier to visualize, interpret, and analyze. A thorough description of the decomposition of a time series dataset was presented, and the resulting applications related to decomposition were considered. It is common for a data scientist to encounter seasonally adjusted data; this term and related methods were defined and illustrated using examples. Finally, one of the most important applications of time series data analysis is that of forecasting. While in the chapter we introduced non model-based forecasting concepts using smoothing, we will consider a more theoretical approach in Chapters 5–10 when we study model-based forecasting. In Chapter 11 we will return to non model-based methods in the discussion of neural network analysis of time series data.

APPENDIX 2A

TSWGE FUNCTIONS

(a) `exp.smooth.wge(x,alpha=NULL,n.ahead=0,plot=TRUE):`   Computes   exponential smoothing forecasts

x is a vector containing the time series
alpha is the α parameter.
n.ahead is the number of steps beyond the end of the dataset for which you want to forecast (default=0)
plot=TRUE (default) produces a plot of the data overlayed with the 1-step ahead forecasts.
if **n.ahead** > 0, the forecasts beyond the dataset are plotted with a dashed line.

Output
> **alpha**
> **forecasts:** Calculates the exponential smoothing 1-step ahead forecasts for $t = 1,\ldots,n$ and forecasts beyond
> n, if **n.ahead>0**.

Note: You can specify **alpha**, or if you leave **alpha** out of the call statement, the function will pick α by minimizing 1-step ahead forecast errors. This is not usually the goal of exponential smoothing; selecting values of **alpha** somewhere in the range from .2 to .4 is common practice.

(b) `ma.pred.wge(x,order=3,k.ahead=1,plot=TRUE):`   computes   a   predictive   (rolling) moving average on the data in **x**

x is a vector containing the time series
order is the order of the predictive moving average (default=3)
plot=TRUE (default) produces a plot of the data along with the predictions.

Example

```
data(AirPassengers)
ma.pred.wge(AirPassengers,order=5,k.ahead=20)
```

provides 1-step ahead predictions for the data up to x_{n+1} where n is the length of the **AirPassengers** data. Also, using the **AirPassengers** data, the function extends the predictions to x_{n+20}. The data and the predictions will be plotted.

(c) `ma.smooth.wge(x,order=3,plot=TRUE):`   computes   a   centered   moving   average smoother on the data in **x**
x is a vector containing the time series
order is the order of the centered moving average smoother (default=3)
plot=TRUE (default) produces a plot of the data along with the smoothed version

Example

```
data(AirPassengers)
ma.smooth.wge(AirPassengers,order=5)
```

TSWGE DATASETS INTRODUCED IN CHAPTER 2

tx.unemp.unadj– Monthly Texas Unemployment Rates from 2000 through 2019 (not seasonally adjusted)

tx.unemp.adj – Monthly Texas Seasonally Adjusted Unemployment Rates from 2000 through 2019

us.retail – quarterly US retail sales (in $millions) from the fourth quarter of 1999 through the second quarter of 2021

NAICS - Monthly US total sales in retail trade and food services in billions of dollars from Jan 2000 through December 2019

NSA: Monthly Total Vehicle Sales in the United States from January 1976 through July 2021.

PROBLEMS

1. Use a 5th-order moving average smoother to smooth the following. In each case plot the smoothed data as an overlay on the original data:
    (a) **bitcoin**
    (b) **wtcrude2020**
2. Using the *tswge ts* object **dfw.mon**, do the following:
    (a) Use the Base R window function to create a *ts* object consisting of the monthly temperatures from January 2000 through December 2021. Name the file **dfw.mon.2000**.
    (b) Apply a 5th-order moving average smoother to the data in **dfw.mon.2000**. Plot the monthly data and overlay the smoothed data. Describe the appearance of the smoothed data.
    (c) Apply an additive decomposition to **dfw.mon.2000** using the techniques in Section 2.1.2.1 and plot the four plots as in Figure 2.10.
    (d) Base R has a function called **decompose** that performs an additive or multiplicative decompostion. Issue the following commands:

    ```
    dfw.decomp=decompose(dfw.mon.2000,type='additive')
    plot(dfw.decomp)
    Compare the plots with those you produced in (c).
    ```

3. Using the **decompose** function, produce a plot corresponding to the multiplicative decomposition shown in Figure 2.14 for the **AirPassengers** data. Use the command

    ```
    decompose(AirPassengers,type='multiplicative')
    ```

4. Consider the following three *tswge* datasets.
    (1) **us.retail**: quarterly US retail sales (in $millions) from the fourth quarter of 1999 through the second quarter of 2021.
    (2) **NSA**: Monthly Total Vehicle Sales (**NSA**) in the United States from January 1976 through July 2021.
    (3) **NAICS**: Monthly US total sales in retail trade and food services in billions of dollars from Jan 2000 through December 2019.

For each of these datasets:

(a) Plot the data.

(b) Discuss behavior of the time series.

(c) Decompose the data (additive or multiplicative – you decide).

(d) These data are not seasonally adjusted. Seasonally adjust the data using either (2.8) or (2.9) and plot your results.

(e) Seasonally adjust the data using the default options in the Census Bureau seasonal command.

(f) Plot the raw data, and the seasonal adjustments obtained in (d) and (e). Compare the results. Does the seasonal adjustment highlight behavior not noticed in the non-seasonally adjusted data?

- **NAICS.adj** contains seasonally adjusted data from the Kaggle website. For the NAICS dataset, include a plot of **NAICS.adj** in (e).

5. Use MAD to measure the accuracy of the **AirPassengers** forecasts obtained in Section 2.2.4.

6. For each of the following datasets, use the strategy in Section 2.2.3 to find Holt-Winters forecasts for the last k data values in the dataset, where you choose k. In each case

(i) Fit an *additive model* to the data and use it to forecast the last k values.

(ii) Fit a *multiplicative model* to the data and use it to forecast the last k values.

(iii) For each dataset calculate the RMSEs for your forecasts and use this as a measure to assess which forecasting technique (additive or multiplicative) performs best.

    (a) **AirPassengers**

    (b) log **AirPassengers**

    (c) **lynx**

    (d) log **lynx**

    (e) **tx.unempl.unadj**

    (f) **us.retail**

Statistical Basics for Time Series Analysis

3

In Chapter 2 we discussed the topic of observing, smoothing, decomposing, seasonal adjusting, and forecasting time series data. We refer to the methodology used in that chapter as non-model based or data analytic methods. Most time series analysis techniques are based on statistical models. In this chapter we discuss statistical fundamentals that are basic to analysis of time series, and this material will be used in future chapters as the basis for our time series modeling. In Section 3.1 we review some basic concepts from classical (not time series) statistical analysis.

3.1 STATISTICS BASICS

The following will be a quick review of basics for some readers and a brief introduction for others. We begin by discussing univariate statistical analysis. That is, we discuss the concepts involved in the analysis of a single variable.

3.1.1 Univariate Data

Fundamental to statistical analysis is the *random variable,* which is defined in Definition 3.1.

Definition 3.1: *Random Variable*: A *random variable* is a variable whose values are the (numerical) outcomes of a random phenomenon associated with some probability distribution. There are two basic types of random variables:

(a) discrete random variables: possible outcomes can be counted
(b) continuous random variables: there is a *continuum* of possible outcomes

Key Point: The terms *random phenomenon* and *random process* will be used interchangeably from this point forward

DOI: 10.1201/9781003089070-3

Examples of Random Processes, Random Variables, and Probability Distributions

(a) *Random Process:* Tossing a coin twice.
   Random Variable: The *number* of heads. (discrete)
   Probability Distribution: Prob(0)=.25, Prob(1)=.5, and Prob(2)=.25[1]

(b) *Random Process:* Randomly selecting male students at a large university
   Random variable: Student *height* in inches. (continuous: actual height is a *number* along a *continuum*)
   Probability Distribution: The true probability distribution would be the heights of all male students at the university. Assuming that the data on all male students are unavailable, we could randomly select several male students and plot a histogram of heights. It would probably tend to be bell-shaped with most heights between 64 inches and 76 inches, a few taller than 76 inches and a few shorter than 64 inches.

(c) *Random Process:* Customers entering a store during a given day.
   Random Variable: The *number* of customers who enter the store (discrete)
   Probability Distribution: We would not know the true probability distribution, but it could be visually estimated using a histogram of daily number of customers over a period of several days.

Key Point: Histograms can be used to "visually estimate" a probability distribution.

The underlying probability distribution may be theoretical (such as in the case of the coin tossing example), or it may be the collection of all "things" about which we want to know information (such as the student height example). We often want to know characteristics about the underlying population or probability distribution. This information is often summarized using parameters that are estimated from observed data.

Definition 3.2: *Parameters* are numerical characteristics or descriptives of a population or probability distribution.

Definition 3.3: The *expected value* is a theoretical (population) average.[2] Parameters of interest related to the random variable X include the following, which we state in terms of expected values:

(i) Mean, μ_X: Measures the "center" of the distribution and is denoted $\mu_X = E[X]$.

(ii) Variance, σ_X^2: A measure of how "spread out" the values of the underlying distribution are. The variance measures the average squared amount by which observations tend to differ from the mean. Variance is denoted $\sigma_X^2 = E\left[(X - \mu_X)^2\right]$.

(iii) Standard Deviation, σ_X: We often use σ_X as a measure of spread because it can be interpreted as the typical amount by which values in the distribution differ from the mean (in the units of the problem). The standard deviation is simply the square root of the variance.

Random variables are typically denoted by capital letters, such as $X, Y,$ or Z. A particular occurrence of the random variable produces a number. This leads to the following definition.

---

1 The four possible, equally likely outcomes of two tosses are HH, HT, TH, TT.
2 The calculation of expected values for continuous random variables involves the use of calculus.

Definition 3.4: *Observed Values* are specific outcomes of the underlying random phenomenon (process).

Observed values of random variables are numbers, and they are denoted by lower case letters, for example, *x*. Suppose our random variable, *X*, is the number of heads after two tosses of a coin. If after tossing the coin twice the outcome "HT" is observed, then the particular outcome is $x = 1$. The classical study of random variables makes extensive use of collections of these observed values called a *sample*. A particularly useful (and often assumed) type of sample is known as a *random sample* and is defined next.

Definition 3.5: A *random sample*, say of size *n*, is the result of *n* *independent* instances of the underlying random phenomenon (process).

Random samples are often denoted, $x_j, j = 1,\ldots,n$. They provide the data scientist with *n* independent (unrelated) observations of the random variable. Consequently, a random sample provides information about the underlying probability distribution of the phenomenon of interest.

Definition 3.6: A *statistic* is an estimate of a population parameter based on a sample.

Given the values, $x_i, i = 1,\ldots,n$, in a random sample, the following are useful statistics:

(i) Sample Mean: $\bar{x} = \dfrac{1}{n}\sum_{i=1}^{n} x_i$. The symbol $\bar{x}$ is read as "x-bar", and it is simply the average of the observed data values.

(ii) Sample Variance: $s_X^2 = \dfrac{1}{n-1}\sum_{i=1}^{n}(x_i - \bar{x})^2$.[3] The sample variance is approximately the average of the squared differences from the sample mean. This interpretation leaves most people "cold" regarding what is being measured, primarily because it is in squared units. For this reason we often use the standard deviation, the square root of the variance, for a more interpretable measure.

(iii) Sample Standard Deviation: $s_X = \sqrt{s_X^2}$. The standard deviation is in the units of the observed data (not in squared units) and can be thought of as the typical amount by which data values differ from the sample mean.

QR 3.1 Unbiasedness

Key Points

1. Random variables are variables whose (numerical) values are determined by a random process.
2. Probability distributions are characterized by parameters such as the mean (μ_X), variance (σ_X^2), and standard deviation (σ_X).
3. We use random samples to obtain sample-based estimates of parameters
    - a single observation of a random variable gives limited information about the true mean and no information about the true variance or standard deviation.

---

3 It is common to use the denominator $n-1$, which produces an unbiased estimator of σ_X^2. Some practitioners use the denominator n, which produces the maximum likelihood estimator. These concepts are discussed in standard statistics textbooks, but will not be reviewed here.

Theorem 3.1: Central Limit Theorem Let $X_i, i = 1, \ldots, n$ denote a random sample from a population with mean μ and variance σ^2. Then for "large n" the random variable

$$Z = \frac{\overline{X} - \mu_X}{\sigma_X / \sqrt{n}} \tag{3.1}$$

is approximately distributed as a Normal random variable with mean zero and variance one (denoted N(0,1)) even if the random variables, X_i are not normally distributed.[4]

Definition 3.7: The standard error of $\overline{X}$ (which is the standard deviation of the population of $\overline{X}s$) is given by

$$SE(\overline{X}) = \sigma_X / \sqrt{n}. \tag{3.2}$$

An important interpretation of Definition 3.7 is that $\sigma_X / \sqrt{n}$ is the typical amount by which $\overline{X}$ tends to differ from the true mean. Consequently, as n increases, $\overline{X}$ tends to get closer and closer to the mean μ_X. We say that $\overline{X}$ "converges to μ_X" as n increases.

"Large" often simply means $n > 30$. Because we usually do not know the population variance, we often calculate the statistic

$$\tilde{Z} = \frac{\overline{X} - \mu_X}{s_X / \sqrt{n}}, \tag{3.3}$$

QR 3.2
R Simulation of CLT

which has an approximate normal distribution for large n (see Moore, McCabe, and Craig, 2017).[5]

Key Points

1. As n increases, the standard error of $\overline{X}$ that is, $\left(\sigma_X / \sqrt{n}\right)$ decreases and $\overline{X}$ converges to μ_X. That is, $\overline{X}$ taken from a large sample should be a good estimate of μ and is probably a better estimate than $\overline{X}$ calculated from a sample with smaller sample size.

2. According to the Central Limit Theorem, for "large n" (typically greater than 30) the random variable

$$Z = \frac{\overline{X} - \mu_X}{\sigma_X / \sqrt{n}}$$

---

4 We are assuming that readers are familiar with the Normal (Gaussian) distribution that is characterized by the classic bell-shaped curve. If not, any introductory statistics text will provide a discussion of this distribution.

5 If the $X_i s$ have a normal distribution, then the statistic in (3.3) has a Student's t distribution. For large n the t distribution approaches the normal distribution, so we denote the statistic with a "$\tilde{Z}$" (suggesting that it is approximately N(0,1)). Data scientists use this "normal approximation" even when the $X_i s$ are not normally distributed.

has an approximate N(0,1) distribution even if the random variables, X_i, are not normally distributed.

3. The random variable

$$\tilde{Z} = \frac{\bar{X} - \mu_X}{s_X / \sqrt{n}}$$

is approximately N(0,1) if n is large.

Example 3.1 Consider the data in Table 3.1, which represent a sample of 30 students at a large university who were randomly assigned to an Introductory Computer Science course. Table 3.1 shows the scores on a ten-point quiz given early in the semester.

The true mean and other parameters would be based on the population of students at the university who took the Introductory Computer Science course. The histogram of the scores is shown in Figure 3.1 where it is seen that no students scored below five or greater than ten (a perfect score). The histogram is approximately bell-shaped, and thus we can conclude that the distribution is approximately normal. Also, from the histogram it is reasonable to guess that the average score is about 7.5. To calculate the sample average, we find $\bar{x} = \frac{1}{n}\sum_{i=1}^{n} x_i = \frac{1}{30}(222.6) = 7.42$. Letting **x** denote the R vector containing the quiz scores, we can find the sample average using the R command **mean(x)** which yields 7.42.

TABLE 3.1   Quiz Scores

STUDENT	QUIZ	STUDENT	QUIZ	STUDENT	QUIZ
1	7.4	11	7.6	21	8.0
2	8.4	12	8.8	22	9.0
3	8.8	13	6.1	23	8.9
4	6.4	14	7.2	24	7.5
5	10.0	15	6.6	25	5.5
6	5.5	16	7.0	26	8.5
7	7.3	17	5.3	27	7.4
8	5.9	18	7.9	28	6.3
9	7.1	19	8.1	29	7.7
10	7.9	20	7.6	30	6.9
				Sum	222.6

The mean and standard deviation could be calculated by hand, but for datasets of this magnitude we would again use R, and the command **sd(x)** gives the result 1.147531. The interpretation of the standard deviation is that the values in the list tend to differ from the average ($\bar{x} = 7.42$) by about 1.15.

For a bell-shaped (normal) distribution, about 95% of the observations fall within two standard deviations of the mean and about 68% fall within one standard deviation of the mean. For more details, see any introductory text in statistics. Because the histogram for this example is approximately bell-shaped, we will assume that the distribution is approximately normal and thus that about 95% of the observations should be between $7.4 - 2(1.15) = 5.1$ and $7.4 + 2(1.15) = 9.7$. For the dataset in Table 3.1, 29 out of 30 or 96.7% fall within this range.

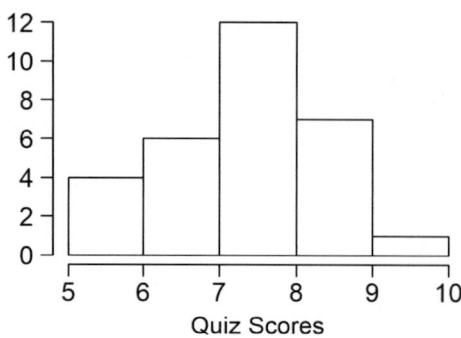

QR 3.3 Example 3.1

FIGURE 3.1   Quiz scores for a sample of 30 students selected randomly from the population of all students enrolled in Introductory Statistics.

3.1.2 Multivariate Data

When two or more random variables are observed on members of a random sample, the resulting data are called *multivariate data*. The special case of two variables is referred to as *bivariate* data.

Table 3.2 is a generic illustration of a typical tabular presentation of a bivariate random sample. We see that Subject 1 has the value x_1 for the random variable X and the value y_1 for the random variable Y. The two observations for Subject 1 are considered as an ordered pair, (x_1, y_1), and the bivariate random sample is represented by $(x_i, y_i), i = 1 .., n$.

TABLE 3.2   Random Sample of Ordered Pairs

SUBJECT	X	Y
1	x_1	y_1
2	x_2	y_2
⋮	⋮	⋮
n	x_n	y_n

Example 3.1 (revisited): Bivariate Data Consider the data in Table 3.1 which represent a sample of 30 students at a large university who were randomly assigned to an Introductory Computer Science course. In Table 3.3 we show not only the quiz scores from Table 3.1 but also the final exam score for each student.

TABLE 3.3   Quiz and Final Exam Scores

STUDENT	QUIZ	FINAL	STUDENT	QUIZ	FINAL	STUDENT	QUIZ	FINAL
1	7.4	79.8	11	7.6	80.7	21	8.0	84.2
2	8.4	82.0	12	8.8	94.5	22	9.0	87.8
3	8.8	76.1	13	6.1	50.1	23	8.9	94.1
4	6.4	62.7	14	7.2	68.3	24	7.5	78.2
5	10.0	98.2	15	6.6	64.4	25	5.5	62.4
6	5.5	43.0	16	7.0	67.2	26	8.5	85.1
7	7.3	76.5	17	5.3	53.9	27	7.4	77.8
8	5.9	61.4	18	7.9	78.8	28	6.3	67.6
9	7.1	78.5	19	8.1	85.7	29	7.7	70.2
10	7.9	88.7	20	7.6	81.7	30	6.9	73.6
						Sum	222.6	2253.2

The means and standard deviations are given below (again obtained using R functions **mean** and **sd**). The quiz scores are in R vector **x** and the final exam scores are in **y**. We have the following results:

```
mean(x)   [1]  7.42
sd(x)     [1]  1.15
mean(y)   [1]  75.11
sd(y)     [1]  13.15
```

The histograms for these two variables are shown in Figure 3.2. There it can be seen that the histograms for both variables are approximately bell shaped with the final exam scores being slightly skewed to the left. By examination of the histogram in Figure 3.2(b) we see that the mean of the final exam scores appears to be about about 75 which is consistent with the results above. The majority of the scores are between 60 and 90 with a few above 90 and a few below 60.

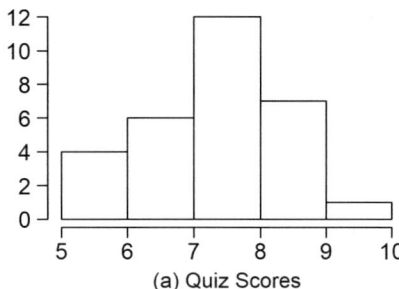

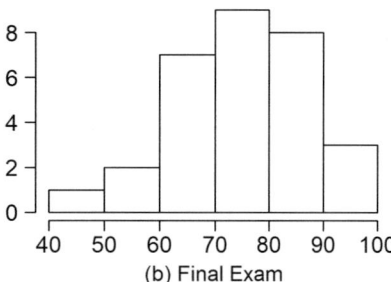

(a) Quiz Scores   (b) Final Exam

FIGURE 3.2 (a) Quiz and (b) Final Exam scores for a random sample of 30 students enrolled in Introductory Computer Science.

Key Point: While the histograms and statistics derived from the sample discussed to this point are interesting and informative, they do not provide information about the relationship between the two random variables.

3.1.2.1 Measuring Relationships between Two Random Variables in a Bivariate Random Sample

The Key Point above emphasizes the fact that histograms, means, and standard deviations do not provide information about how two variables are related to each other. As an example, an instructor would probably like to know whether students who did well on the quiz also tended to do well on the final exam, and vice versa. She might also want to know if some students who did poorly on the quiz dramatically raised their grade on the final exam. The histograms and sample statistics shown above do not answer these questions.

What is needed are measures of the relationship between the two variables. The most common population parameters used to measure such relationships are the covariance, γ_{XY}, and the correlation, ρ_{XY}.

Definition 3.8: The covariance, γ_{XY}, and correlation, ρ_{XY}, are parameters that measure the linear[6] relationship between two variables. These are defined as follows:

(a) $\gamma_{XY} = E\left[\left(X - \mu_X\right)\left(Y - \mu_Y\right)\right]$

---

6 We will discuss the manner in which the correlation measures linear association as we discuss the sample correlation coefficient.

(b) $\rho_{XY} = \dfrac{\gamma_{XY}}{\sigma_X \sigma_Y}$

Technically, the covariance is the expected value (or theoretical average) of the cross-product $(X - \mu_X)(Y - \mu_Y)$. It is a measure of how two variables "move together". To facilitate the interpretation, we usually use the correlation, which is a "standardized" version of the covariance that has the property $-1 \le \rho_{XY} \le 1$ for any two random variables X and Y.

Notes

(a) $\gamma_{XX} = E\left[(X - \mu_X)(X - \mu_X)\right] = E\left[(X - \mu_X)^2\right] = \sigma_X^2$

(b) $\rho_{XX} = \dfrac{\sigma_X^2}{\sigma_X \sigma_X} = 1$

3.1.2.2 Assessing Association from a Bivariate Random Sample

Definition 3.8 and the notes that follow discuss (unknown) parameters. In a practical setting we estimate the association using the sample correlation coefficient.

Sample correlation coefficient: The *sample statistic used to estimate* ρ_{XY} based on a random sample of bivariate data, $(x_i, y_i), i = 1, \ldots, n$, is the sample correlation coefficient (sometimes called the Pearson product-moment correlation coefficient). Given a bivariate dataset, the sample correlation coefficient, r_{XY}, is defined by

$$r_{XY} = \frac{\sum_{i=1}^{n}(X_i - \bar{X})(Y_i - \bar{Y})}{\sqrt{\sum_{i=1}^{n}(X_i - \bar{X})^2}\sqrt{\sum_{i=1}^{n}(Y_i - \bar{Y})^2}}. \tag{3.4}$$

Values of r_{XY} share the property with the population correlation coefficient, ρ_{XY}, that $-1 \le r_{XY} \le 1$.

Scatterplots: Another tool for evaluating the association between two variables is the scatterplot. A scatterplot is a graph in which the ordered pairs, $(x_1, y_1), (x_2, y_2), \ldots, (x_n, y_n)$ are plotted as n points on an X-Y plane. Figure 3.3 shows scatterplots for several pairs of variables. Figure 3.3(a) shows a strong positive linear relationship between X and Y. Specifically, as X increases, Y tends to increase and vice versa. Referring to the definition of r_{XY}, note that r_{XY} is a function of the products of the differences $x_i - \bar{x}$ and $y_i - \bar{y}$. In Figure 3.3(a) we see that nearly all of the points are in the upper right or lower left quadrants. Points in the upper right quadrant are such that $x_i > \bar{x}$ and $y_i > \bar{y}$, so $(x_i - \bar{x})$ and $(y_i - \bar{y})$ are both positive, and consequently, so is their product. Likewise, for points in the lower left quadrant, it follows that $x_i < \bar{x}$ and $y_i < \bar{y}$, so $(x_i - \bar{x})$ and $(y_i - \bar{y})$ are both negative, and again their product, $(x_i - \bar{x})(y_i - \bar{y})$, is positive. Therefore, since most of the products will be positive (and the ones that are negative will have relatively small magnitude), the overall sum $\sum_{i=1}^{n}(x_i - \bar{x})(y_i - \bar{y})$ and thus r_{XY} will tend to be positive since the denominator is

always positive. Based on the data in Figure 3.3(a), if you are told that an x-y pair has $x = 2$, then you would be fairly confident that the y value is about $y = 1.5$. In general, you could predict another value for y if you know the corresponding value for x. However, note that we only know that this relationship exists when x is between -3 and 3. This leads to the following definition.

QR 3.4 Visualizing Autocovariance

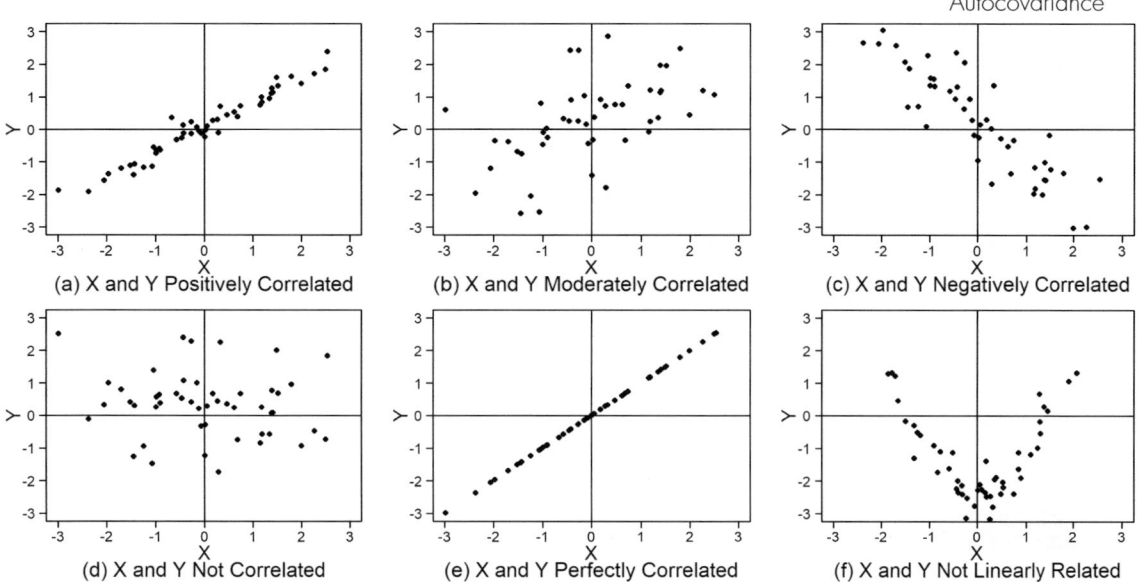

FIGURE 3.3   (a)–(f) are scatterplots with a variety of correlation structures.

Definition 3.9: Prediction of a y value based on an x value that is outside the range of observed x values is an *extrapolation*.

For example, to predict y if $x = 5$, which is outside the range of observed x values $(-3, 3)$, is an example of extrapolation. Every introductory statistics book known to humankind discourages extrapolation because the fact that a linear relationship exists between X and Y in the range of the variables in the sample (in this case between -3 and 3) does not guarantee that the same relationship extends beyond the range of the original data.

Similar to Figure 3.3(a), the data in Figure 3.3(b) suggest a positive, although now moderate, correlation between X and Y. However, unlike Figure 3.3(a), for a given value of x, there is quite a range of plausible y values. Suppose you have sampled a new observation, $x = 2$, of the random variable X. Based on the plot, you would not be as confident that the corresponding y value is close to 1.5 although you would probably be willing to predict that it would be greater than zero. That is, "knowing x" helps in predicting y (but not as much as for the data in Figure 3.3(a)).

In contrast to Figures 3.3(a) and (b), Figure 3.3(c) shows an example of a fairly strong negative correlation, where large values of X are associated with small values of Y, and vice versa. We see that nearly all points are now in the upper left or lower right quadrants. Points in the upper left quadrant are such that $x_i < \bar{x}$ and $y_i > \bar{y}$, so $(x_i - \bar{x})$ is negative, $(y_i - \bar{y})$ is positive, and thus the product $(x_i - \bar{x})(y_i - \bar{y})$ is negative. Likewise, for points in the lower right quadrant, it follows that that $x_i > \bar{x}$ and $y_i < \bar{y}$, so $(x_i - \bar{x})$ is positive, $(y_i - \bar{y})$ is negative, and again the product is negative. Therefore, because most of the products will be negative (and the ones that are positive will have relatively small magnitude), we expect the overall sum $\sum_{i=1}^{n}(x_i - \bar{x})(y_i - \bar{y})$, and thus r_{XY} to be negative this time. This negative relationship suggests that if you have a new randomly selected ordered pair with $x = 2$, then you can be relatively confident that the associated y value will be negative and likely less than -1.

Figure 3.3(d) shows an example in which there seems to be no detectable correlation between X and Y. "Knowing x" does not help in predicting y. Think about what this implies in terms of $(x_i - \bar{x})$, $(y_i - \bar{y})$, $(x_i - \bar{x})(y_i - \bar{y})$, $\sum_{i=1}^{n}(x_i - \bar{x})(y_i - \bar{y})$, and thus r_{XY}. See Problem 3.9.

Figure 3.3(e) shows an example of a "perfect" positive linear relationship. The points all lie on a line whose equation you could derive using elementary algebra. Given the equation, if you "know x," then you can calculate y (again, within the range of the observed x values).

Finally, Figure 3.3(f) shows another situation in which there is a relationship between X and Y. Specifically, y values tend to be high when x values are either low or high. On the other hand, for x values toward the middle of the range of x (that is, near zero), the y values tend to be low. Consequently, there is a relationship, but it is not linear. Think again about what this implies for $(x_i - \bar{x})$, $(y_i - \bar{y})$, $(x_i - \bar{x})(y_i - \bar{y})$, $\sum_{i=1}^{n}(x_i - \bar{x})(y_i - \bar{y})$, and thus r_{XY}. See Problem 3.10.

Calculating Correlation using R: Suppose the data for x are contained in the R vector **x** and that y values are in **y**. Then the correlation coefficient is calculated using the R command `cor(x,y)`. The sample correlation was calculated for the data in each plot in Figure 3.3, and the following values were obtained: (a) $r_{XY} = .976$, (b) $r_{XY} = .518$, (c) $r_{XY} = -.883$, (d) $r_{XY} = -.158$, (e) $r_{XY} = 1$, (f) $r_{XY} = -.211$.

Key Points

1. The data in Figure 3.3(d) were generated from a distribution for which the true correlation, $\rho_{XY} = 0$. The sample correlation coefficient r_{XY} is estimating this value and is doing a good job as it is small ($r_{XY} = -.158$).
2. The sample correlation coefficient for the data in Figure 3.3(f) is near zero. This emphasizes the fact that r_{XY} is measuring *linear* association. Again, for these data there is a relationship, but it is not linear and is not "detected" by the sample correlation coefficient.

Example 3.1 (revisited) Figure 3.4 shows a scatterplot of the quiz-final exam data in Example 3.1 along with vertical and horizontal lines specifying $\bar{x} = 7.42$ and $\bar{y} = 75.11$, respectively. Given the example scatterplots in Figure 3.3 it appears that the scatterplot is a "positively correlated" version of Figure 3.3(c), and, in fact, $r_{XY} \approx .9$. If the quiz scores are in R vector **x** and the final exam scores are in **y**, then the correlation coefficient can be obtained by issuing the command

```
cor(x,y)
[1] 0.9006451
```

As would be expected, there is convincing evidence of a strong positive linear relationship between quiz scores and final exam scores. Knowing the early quiz score helps predict performance on the final exam. From the scatterplot we see that there does not seem to be much evidence of students greatly improving on the final exam scores nor did many students underperform on the final.

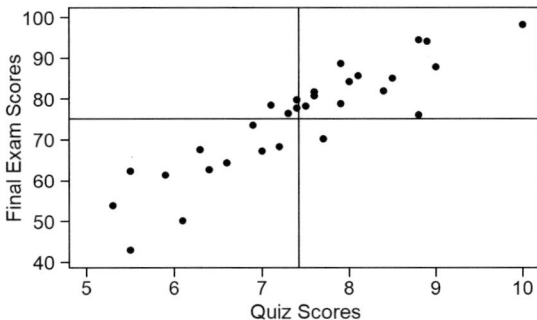

FIGURE 3.4   Scatterplot of quiz scores and final exam scores for the data in Table 3.3.

3.1.3 Independent vs Dependent Data

The main difference between a random sample and time series data is that a random sample is composed of independent observations while time series data have a dependence structure. To illustrate, we provide the following example.

Example 3.2

(a) Independent Sum Data: Consider the experiment of rolling two dice (a white one and a black one) and finding the sum of the rolls. In this case, the random variable is:

$X =$ the sum of the dots showing after rolling two dice.

Suppose the experiment of rolling two dice and finding the sum was performed and the result of the rolls was ⚃ and ⚄, so that the sum of the rolls was $x = 9$. If, after obtaining this single observation, you had to "guess" what the mean of the distribution is, your best guess would be 9. While a single observation of a random variable provides some information about the mean or center of the distribution, it makes sense that the sample mean of n unrelated observations is a better estimate of the mean. To illustrate further, the random process of rolling two dice and finding the sum of the rolls was repeated 30 times. The results are given in Table 3.4

To start with, we consider the first 10 outcomes (the first column). For the first data column, $\bar{x} = 8.2$. It makes sense that 8.2 is a reasonable estimate of the (to this point unknown) mean, and is likely to be more accurate than the estimate based simply on the first observation. Continuing the procedure, we expanded the sample until we obtained 30 repetitions; that is, we obtain a sample of size $n = 30$. Letting **x** denote the R vector containing the 30 "sum of the rolls" dataset, we calculated $\bar{x}$ using the base R command **mean(x)**. In this case $\bar{x} = 7.17$. The actual mean is $\mu = 7$ (see Appendix 3A), so the additional sample size improved the estimate. This will tend to be the case in general (although it is not a certainty).[7]

---

7 For example, if our original observation was $x = 7$, then there is no sample (no matter how large the sample size) for which the sample average is a better estimate of the true mean ($\mu_X = 7$).

TABLE 3.4   Sample Results for Dice-Rolling Process

REPETITION	SUM OF ROLLS	REPETITION	SUM OF ROLLS	REPETITION	SUM OF ROLLS
1	9	11	4	21	4
2	7	12	3	22	8
3	8	13	9	23	9
4	10	14	6	24	7
5	5	15	8	25	6
6	7	16	6	26	3
7	9	17	6	27	8
8	10	18	7	28	7
9	10	19	11	29	8
10	7	20	8	30	5

Again, letting **x** denote the R vector containing the data in Table 3.4, the sample variance, s_X^2, can be obtained using R command **var(x)**. For these data we obtained $s_X^2 = 4.35$. In our case, the sample standard deviation is $s_X = 2.09$ which has the interpretation that individual sums of the rolls of two dice tend to differ from the sample mean ($\bar{x} = 7.17$) by about 2. The sample standard deviation, s_X, can be obtained using R command **sd(x)**, or by using **stdev(x)=sqrt(var(x))**. In Appendix 3A we show that for the dice rolling distribution, $\mu_X = 7$, $\sigma_X^2 = 5.833$, and $\sigma_X = 2.415$. Note that, although a single observation gives us limited information about the true mean, μ_X, a single observation gives *no informa-tion* about the variability. This is because with a single observation, x, we have $\bar{x} = x$ and $n = 1$ but the

$$\text{sample standard deviation } s_X = \sqrt{\frac{1}{n-1}\sum_{i=1}^{n}(x_i - \bar{x})^2} = \frac{0}{0}, \text{ which is undefined.}[8]$$

Before discussing the statistical basics of time series analysis, we show Figure 3.5(a), which is a plot of the 30 outcomes in Table 3.4, indexed by time (that is, ordered by repetition number). The experiments were run sequentially in time, so that the first run of the experiment produced a sum of 9, the second time a sum of 7, and so forth. That is, $x_1 = 9$, $x_2 = 7$, $x_3 = 8$, etc., and we plot $x_i, i = 1, \ldots, 30$. In this plot we see the randomness associated with a random sample. If we connect the points, as we have done in Figure 3.5(b), we get a better look at how the sums "progress through time". We note that the random-ness expresses itself in that the tth sum appears to have no predictive ability concerning the value of the $(t+1)^{st}$ sum.

As an illustration of this randomness, Figure 3.5(c) shows the mean line at 7. There are 14 instances in which the sum is greater than 7. In these cases, the next sum had the following characteristics:

(i)   in 5 cases, the next sum is also above 7
(ii)   in 4 cases, the next sum is equal to 7
(iii)   in 5 cases, the next sum is less than 7.

Consequently, if the sum at "time" t is greater than 7, there is no information in the dataset regarding whether the sum at $t+1$ is above, equal to, or less than 7. This is the essence of a random sample in a "time series" framework. Data such as that shown in Figure 3.5 are called *white noise*, for which a precise def-inition will be given in Definition 3.14.

---

8   It also makes intuitive sense that if only one observation is available, then you know nothing about how much different observations tend to differ from each other.

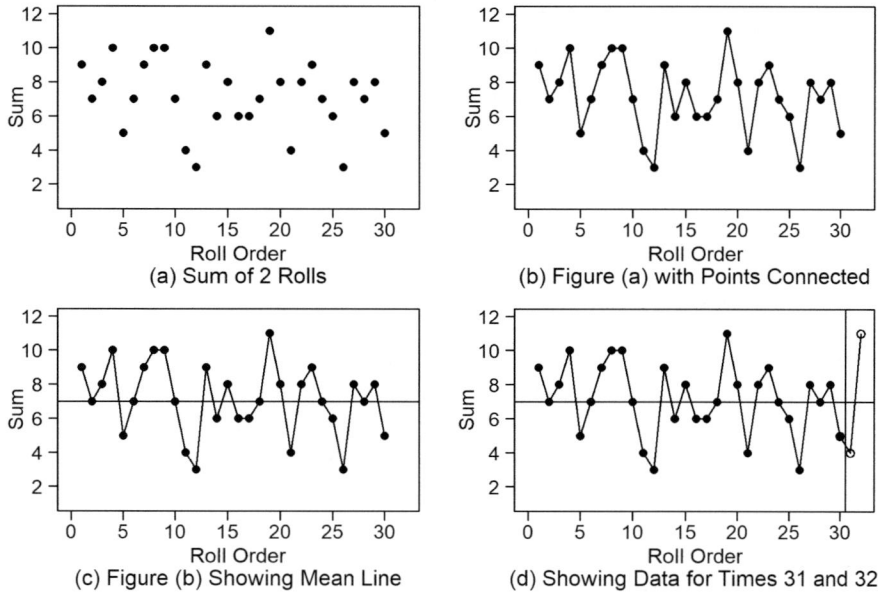

FIGURE 3.5 (a) Plot of the sum of two dice data in Table 3.4, (b) data in (a) with points connected, (c) plot (b) showing mean line at 7, and (d) plot showing data values for 31ˢᵗ and 32ⁿᵈ instances of the random process.

Figure 3.5(d) shows the original sum data for $t = 1,\ldots,30$ along with outcomes at $t = 31$ and $t = 32$. Note that the last observed outcome (at $t = 30$) is 5. However, this fact provides no information about the next two observations, which were 4 and 11, respectively. In fact, if you knew that the mean of the process is $\mu = 7$ (or if you observed that the average of the first 30 observations is $\bar{x} = 7.17$), then your best guess for the outcomes at times $t = 31$ and $t = 32$ would be 7 in both cases, regardless of the value at time $t = 30$.

We are now in position to make the following Key Points.

Key Points:

1. Random samples arranged in sequence produce a *white noise* time series (which will be formally defined in Definition 3.14)
2. The independent nature of a random sample sequenced by time leads to the fact that the value of a random variable at time t gives no information concerning the value at time $t + k$ for $k \geq 1$.

This leads to the following Key Points, which we have separated from the Key Points above for emphasis.

Key Points

1. In classical statistics, we typically work with random samples in which the observations are independent, and we avoid cases in which a sample is not truly random.
2. In time series analysis, we *focus* on data that *are dependent (that is, related)*.
   - this is not a "bad thing" – in fact, it is quite the opposite!
   - we will use the dependence structure to our advantage for forecasting and other analyses.

(b) Dependent Sum Data: To emphasize Key Point Number 2 regarding the advantage of dependence, we simulated a "coin tossing experiment" using a random process producing the random variable, Y, which does not protect the independent nature of the data that occurs from simply rolling two dice and finding the sum.[9] We denote this random process the "dependent sum" process, and from Footnote 9, we see that $\mu_Y = 7$. We will refer to the random process that produced the data in Table 3.4 as the "original (independent) sum" process. The "dependent sum" data are shown in Figure 3.6(a) and again in Figure 3.6(b) with the points connected by straight lines to better show the pattern in time along with a horizontal line showing the mean line $\mu_Y = 7$. There is something very different between the data in Figure 3.5 and Figure 3.6. We see in Figure 3.6 that "what happens at time $t+1$ is related to what happened at time t". For example, between time $t = 10$ and time $t = 22$, all dependent sum outcomes are greater than the mean $\mu_Y = 7$. Similarly, the four outcomes from $t = 24$ through $t = 27$ are all below the mean. The outcome at time $t = 30$ is 9. If you were asked to predict the 31st and 32nd outcomes of this process, because of the correlation structure of the data, you would probably guess the values to be somewhere in the neighborhood of 9 (and likely above the mean). We repeated the dependent dice tossing experiment two more times. Figure 3.6(c) shows that the outcomes at times $t = 31$ and $t = 32$ are both 10, which are consistent with our conjecture. That is, the dependence structure of the data provided information about the expected future outcomes.

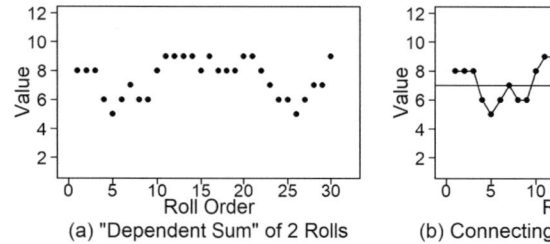
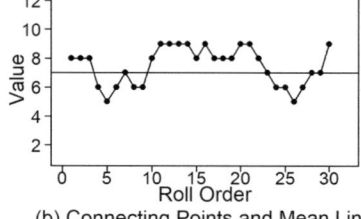
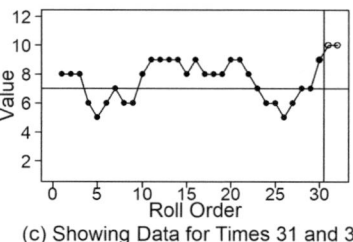

(a) "Dependent Sum" of 2 Rolls (b) Connecting Points and Mean Line (c) Showing Data for Times 31 and 32

FIGURE 3.6   (a) Plot of the sum of "dependent sum" data, (b) data in (a) with points connected and showing mean line at 7, and (c) plot showing data values for 31st and 32nd instances of the random process.

(c) The "Oscillating" Process: As another example, consider the data in Figure 3.7(a). We denote the data with the notation, w_t. Figure 3.7(a) shows the data, w_t, $t = 1,\ldots, 30$ while Figure 3.7(b) connects the points and plots the mean line, $\mu_w = 7$.[10] The sample mean of the data in Figure 3.7(a) is $\bar{w} = 6.87$. Figure 3.7(b) shows the data connected by straight lines along with the mean line, which better illustrates the general oscillating behavior of the data, typically back and forth across the mean line. We note that the value of the random variable at $t = 30$ is 2. Consequently, the correlation structure of the data suggests that the 31st observation will probably be above 7 and the 32nd below 7. In Figure 3.7(c) we see that indeed these predictions are accurate. Again, the dependence structure aided us in predicting future outcomes.

---

9 To introduce dependence, the "dependent sum" process (denoted here by the random variable Y_t, was obtained as follows: at time $t = 1$, Y_1 is the sum of the rolls of the two dice as before. So $E(Y_1) = 7$. (See Appendix 3A). For $t = 2$, we roll two dice again and obtain a "new sum" random variable, Z_2. Again $E(Z_2) = 7$. Then, we define $Y_2 = .6Y_1 + .4Z_2$, and using the properties of expectation to be discussed in Footnote 15, it follows that $E(Y_2) = .6(7) + .4(7) = 7$. Continuing this process we see that at time t for $t = 2,\ldots,20$, $Y_t = .4Y_{t-1} + .4Z_t$, and using the logic above, we see that $E(Y_{t-1}) = 7$ and $E(Z_t) = 7$. It follows that for $t = 2,\ldots,20$, we have $E(Y_t) = .6E(Y_{t-1}) + .4E(Z_t) = .6(7) + .4(7) = 7$. Consequently, for $t = 1,\ldots,20$, $E(Y_t) = \mu_Y = 7$. Note that we round the value of y_t to the nearest integer to be consistent with the original sum-of-the-rolls variable, x_t.

10 The formula for W_t is $W_t = 7 - .97(W_{t-1} - 7) + a_t$ where the a_ts are random normal variables with mean 0 and variance 1. The population mean of the random variable W_t is 7. This will be discussed in Chapter 5. Please be patient! Again, we round the value of W_t to the nearest integer to be consistent with the original sum-of-the-rolls variable, x_t.

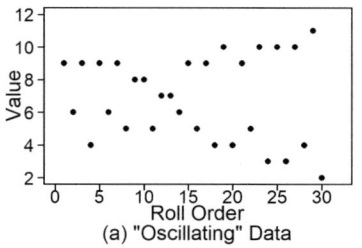
(a) "Oscillating" Data

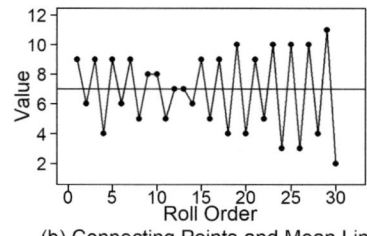
(b) Connecting Points and Mean Line

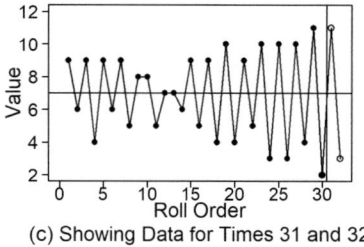
(c) Showing Data for Times 31 and 32

FIGURE 3.7 (a) Plot of the oscillating data, (b) data in (a) with points connected and a horizontal line at 7, and (c) plot showing data values for $t = 31$ and $t = 32$.

Key Points

1. For the "dependent sum" and "oscillating" data, an understanding of the dependence between consecutive runs of the "experiment" is helpful in predicting outcomes at $t = 31$ and $t = 32$.
2. Because of the independence (that is, lack of dependence structure) in the "original sum" data in Figure 3.5, knowing that the sum at time $t = 30$ is 5 provides no assistance in predicting the sums at $t = 31$ and $t = 32$.

3.2 TIME SERIES AND REALIZATIONS

Using the preceding review as a lead-in, we are now in a position to discuss time series as a collection of random variables. Informally, we can think of a time series as a collection of observations made sequentially in time. The datasets in Figures 1.1, 1.3, and so forth, are examples of time series data using this informal definition. However, to be more precise, we need to define time series in terms of random variables. We have the following definition.

Definition 3.10: A *time series* is a collection of random variables, $\{X_t\}$, indexed on time.[11]

A collection of random variables is generically referred to as a *stochastic process*. When the index set is time, we refer to the stochastic process as a *time series*. We will focus on time series for which the index, t, takes on integer values. Definition 3.10 says that each time, t, is associated with "its own" random variable, X_t. Recall that it is common to denote a random variable using a capital letter.

Some Time Series Parameters: Because each time, t, has "its own" random variable, X_t, these random variables have their own distribution and thus possess their own parameters. Among these are the following:

(a) Random variable X_t has a mean, μ_{X_t} (just as the random variable X has mean μ_X).
(b) Random variable X_t has a variance, $\sigma^2_{X_t}$ (just as the random variable X has variance σ^2_X).
(c) There is a covariance, $\gamma_{X_{t_1} X_{t_2}}$, and correlation, $\rho_{X_{t_1} X_{t_2}}$, between random variables X_{t_1} and X_{t_2}. The covariance and correlation between two random variables in the same time series are called the *autocovariance* and *autocorrelation*, respectively. Again, these correspond to the covariance and correlation between random variables X and Y.

---

11 We usually omit the curly brace and simply refer to the time series X_t when there is no confusion.

Notes

(a) If $t_1 \neq t_2$, it may follow (but not necessarily) that $\mu_{X_{t_1}} \neq \mu_{X_{t_2}}$ or $\sigma^2_{X_{t_1}} \neq \sigma^2_{X_{t_2}}$, or both.

(b) If $t_1 = t_2$, then $\rho_{X_{t_1} X_{t_2}} = \rho_{X_{t_1} X_{t_1}} = 1$.

In Section 3.1, we defined a random variable to be a variable whose values are the numerical outcomes of a random phenomenon, and we defined specific outcomes to be observed values. This leads to the following definition.

Definition 3.11: A *realization* of the time series $\{X_t\}$, is a set of specific outcomes of the random phenomenon (process).

We typically denote a realization of n outcomes using the notation, $x_t, t = 1, \ldots, n$. Earlier we referred to the data sets plotted in Figures 1.1 among others as a time series. These are actually realizations of the underlying time series. For example, for the original sum of two dice example, if you decided to repeat the experiment (that is, roll the dice 30 more times and find the sums), this would produce another realization from the underlying time series, but it would be almost a certainty that the new realization would differ from the one plotted in Figure 3.5(b). Figure 3.8(a) shows the realization in Figure 3.5(b) along with two other realizations from the same random process. Although different from each other, the realizations exhibit similar behavior. This leads to the following definition.

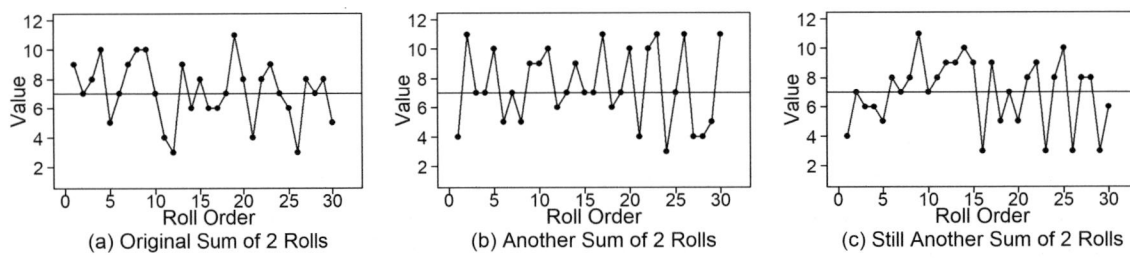

FIGURE 3.8   Three realizations from the random process of rolling two dice 30 times and finding the sums.

Definition 3.12: The collection of all possible realizations from a time series is called an *ensemble*.[12]

3.2.1 Multiple Realizations

Consider the case of a time series for which, as mentioned, each t "has its own" random variable, X_t. For example, X_3 has mean μ_{X_3} and standard deviation, σ_{X_3}. If we have a realization of length, say $n = 100$ (or $n = 1,000$), from the time series, we only have *one* observation, x_3, on the random variable X_3. Lacking other knowledge about the time series, our best estimate of μ_{X_3} is the single value of x_3, and there is no information about σ_{X_3}. Multiple realizations would be needed to obtain reasonable estimates of the mean and standard deviation of X_3.[13]

Suppose we have a time series, $\{X_t\}$, from which we have m (independent) realizations. Call these realizations $x_t^{(1)}, x_t^{(2)}, \ldots, x_t^{(m)}, t = 1, \ldots, n$. Then, $x_1^{(i)}, i = 1, \ldots, m$ are the observed values of a random sample from the random variable X_1 (the first observation of each realization). We can estimate the mean, $\mu_{X_{t_1}}$,

---

12 The ensemble is often a theoretical construct achievable only in the imagination.

13 Stationarity of a time series provides the ability to obtain information from a single realization. This will be discussed in Section 3.3.

variance, $\sigma_{X_{t_1}}^2$, and standard deviation, $\sigma_{X_{t_1}}$, along with the autocovariance, $\gamma_{X_{t_1}X_{t_2}}$, and autocorrelation, $\rho_{X_{t_1}X_{t_2}}$, using the following calculations:

(a) Sample Mean: $\bar{x}(t_1) = \dfrac{1}{m}\sum_{i=1}^{m} x_{t_1}^{(i)}$

(b) Sample Variance: $s_{X_{t_1}}^2 = \dfrac{1}{m-1}\sum_{i=1}^{m}\left(x_{t_1}^{(i)} - \bar{x}(t_1)\right)^2$ and standard deviation: $s_{X_{t_1}} = \sqrt{s_{X_{t_1}}^2}$.

(c) Sample Autocovariance and Autocorrelation between Random Variables X_{t_1} and X_{t_2}:

$$\hat{\gamma}_{X_{t_1}X_{t_2}} = \frac{1}{m-1}\sum_{i=1}^{m}(x_{t_1}^{(i)} - \bar{x}(t_1))\left(x_{t_2}^{(i)} - \bar{x}(t_2)\right)$$

$$r_{X_{t_1}X_{t_2}} = \frac{\sum_{i=1}^{m}(x_{t_1}^{(i)} - \bar{x}(t_1))\left(x_{t_2}^{(i)} - \bar{x}(t_2)\right)}{(m-1)s_{X_{t_1}}s_{X_{t_2}}}$$

(3.5)

QR 3.5 Multiple Realizations

Note that in this case, we are assuming that we have multiple observations $x_{t_1}^{(1)}, x_{t_1}^{(2)}, \ldots, x_{t_1}^{(m)}$ and $x_{t_2}^{(1)}, x_{t_2}^{(2)}, \ldots x_{t_2}^{(m)}$ for each of the two random variables, X_{t_1} and X_{t_2}.

Example 3.3 Multiple Realizations

Figure 3.9(a) shows a realization, x_t, $t = 1,\ldots,20$ from a time series. For this realization, $x_1 = 17$, $x_2 = 19$, $\ldots$, and $x_{20} = 60$. The realization has a somewhat wandering behavior, but the general tendency is to increase in time. Given the notation above, there are 20 random variables, X_t, $t = 1,\ldots,20$. The random variable at time t, that is, X_t, has a mean, μ_{X_t}. For example, X_2 is a random variable which has a mean, μ_{X_2}, and we have observed a single observation, 19, from its distribution. Similarly, the random variable X_{18} has a mean, $\mu_{X_{18}}$, and we have observed the single observation, $x_{18} = 47$.

Figure 3.9(b) shows a realization, y_t, $t = 1,\ldots,20$ from another time series, Y_t. This realization has characteristics that are similar to the one in Figure 3.9(a). The appearance again suggests that the time series involves a tendency for data values to increase in time. As in Figure 3.9(a), for a given value of t, the realization in Figure 3.9(b) gives us only one observation. For example, for the random variable X_5 we have the single observed value $x_5 = 23$ while X_{15} has the single observed value $x_{15} = 39$.

As noted in Section 3.1, only a limited amount of information can be obtained from one observation from a random variable. In this example, we consider the information obtained by observing multiple realizations from each of these time series.

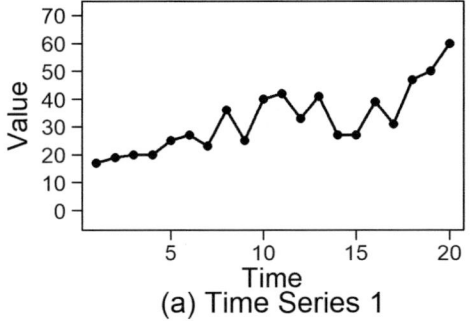
(a) Time Series 1

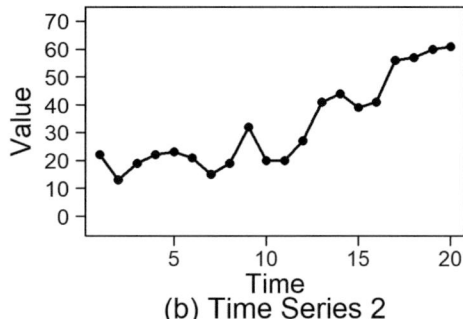
(b) Time Series 2

FIGURE 3.9 Realizations from two time series.

3.2.1.1 Time Series 1: X_t

The increase in time for the realization in Figure 3.9(a) suggests that the true means, μ_t, increase in time. To further examine this speculation about the increase, we randomly generated nine more realizations from the same random process (phenomenon) that generated the realization in Figure 3.9(a).[14] In Figure 3.10(a) we show the realization in Figure 3.9(a), while Figure 3.10(b) shows this realization (in bold) along with nine new realizations from the same time series. In Figure 3.10(b) we see that indeed there is a tendency for realizations to increase in time (as was the case in the original realization). That is, the multiple realizations reinforce the conjecture that the true means increase over time. We estimate the population mean, μ_{X_t} and standard deviation, σ_{X_t} at time t, by finding the sample average $\left(\bar{x}(t)\right)$ and sample standard deviation (s_{X_t}) of the 10 observations at time t. The sample data for times $t = 5$, 10, and 15, highlighted in Figure 3.10(b) (using vertical strips), are shown in Table 3.5. This table and Figure 3.10(b) suggest that both the means and standard deviations of the random variables increase with time. To get a better understanding of the pattern of the means, we found the average $\bar{x}_t$ for each t, where $t = 1,\ldots,20$. These sample averages are plotted in Figure 3.10(c) as open squares, and we see that they increase in a fairly linear manner. Also plotted are the true means, which follow a line with positive slope.

TABLE 3.5   Data from Multiple Realizations for $t = 5, 10$, and 15 Shown in Figure 3.10(b)

REALIZATION	X_5	X_{10}	X_{15}
1	25	40	27
2	21	27	48
3	25	35	49
4	21	34	33
5	21	31	51
6	23	27	39
7	20	26	48
8	25	34	42
9	23	37	46
10	22	47	31
$\bar{x}_t$	22.6	33.8	41.4
s_{X_t}	1.9	6.5	8.5

---

14 We generated the realizations from a time series model. A typical analysis would proceed knowing only the data, and we would attempt to understand the underlying "true model". We will "disclose" the true model after we have discussed the analysis based on the data alone.

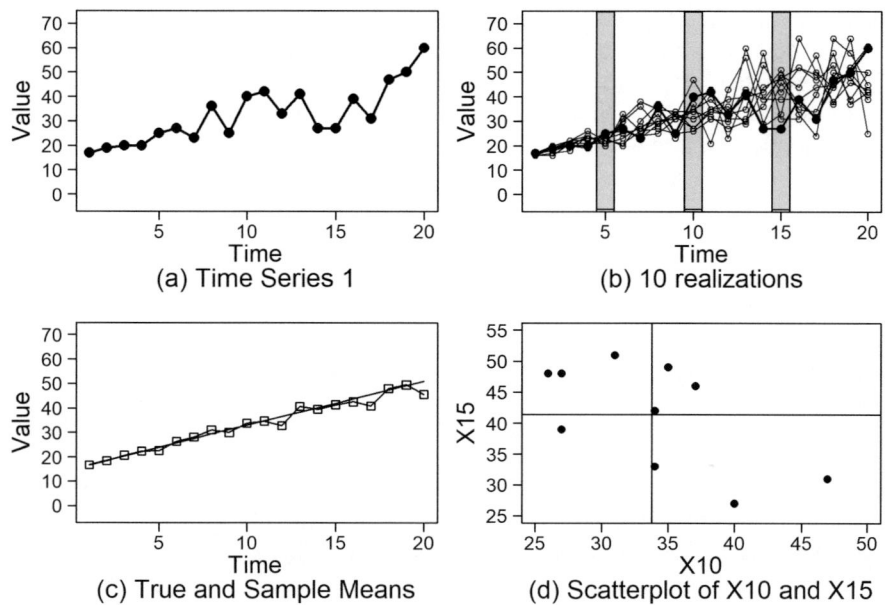

FIGURE 3.10   (a) A realization from Time Series 1, (b) multiple realizations from Time Series 1, highlighting $t = 5,10,15$, (c) plot of sample means (as open squares) and the true mean line, (d) scatterplot between X_{10} and X_{15} based on data in columns two and three in Table 3.5.

Estimating autocorrelations from multiple realizations. For any integers t_1 and t_2 between 1 and 20, we can also estimate the autocorrelation between random variables X_{t_1} and X_{t_2} because these are simply two random variables in the same sense that quiz and final exam scores are two random variables. For example, to estimate the autocorrelation between X_{10} and X_{15}, we use the associated columns in Table 3.5 and perform the following calculation:

$$r_{X_{10}X_{15}} = \frac{(40-33.8)(27-41.4)+(27-33.8)(48-41.4)+\cdots+(47-33.8)(31-41.4)}{9(6.5)(8.5)}$$

$$= -0.62$$

These results can also be obtained using the following R code:

```
x10=c(40,27,35,34,31,27,26,34,37,47)
x15=c(27,48,49,33,51,39,48,42,46,31)
cor.x10x15=cor(x10,x15)
```

Figure 3.10(d) shows a scatterplot of the ten ordered pairs $(40,27),(27,48),\ldots,(37,46),(47,31)$ along with a vertical line at the sample mean of X_{10} (33.8) and a horizontal line at the sample mean of X_{15} (41.4). Although these two random variables are uncorrelated by construction, there seems to be a negative correlation for this sample due to the "cloud of points" in Figure 3.10(d) slopes downward to the right. X_{10} values above its mean (33.8) seem to be paired with X_{15} values below its mean (41.4), and vice versa.

Inside Information:
The statistical model used to generate Time Series 1 has the form

$$X_t = 15+1.8t+.6ta_t, \tag{3.6}$$

where for each t, a_t is a randomly chosen observation from a normal (Gaussian) distribution with mean zero and standard deviation one. Consequently, μ_{X_t} follows the straight line, $\mu_{X_t} = 15 + 1.8t$.[15] Also, at time t, $\sigma_{X_t} = .6t$.[16] Thus, we have $\sigma_{X_5} = 3, \sigma_{X_{10}} = 6$, and $\sigma_{X_{15}} = 9$, indicating that the standard deviations also increase in time. These population standard deviations are well estimated by the sample standard deviations shown in Table 3.5.

Keep this in mind:
In the analysis of real data, *we will not in all likelihood know the "inside information"*. We have the following Key Points based on the analysis of Time Series 1.

QR 3.6
Explanation of
Footnotes 16

Key Points

1. The original realization appears to be upward trending, leading us to suspect that the true means, μ_{X_t}, are increasing in time.
2. If the original realization was our only realization, then while we would suspect that the time series has a general tendency to increase, our evidence would be weak.
3. The multiple realizations allow us to estimate the mean, μ_{X_t}, and standard deviation, σ_{X_t}, for each t using the sample mean, $\bar{x}_t$, and sample standard deviation, s_{X_t}, respectively.
4. The general tendency for each realization to trend upward in time and the fact that the sample means at each t tend to increase both suggest that the population means, μ_{X_t}, trend upward in time.
5. The fact that visually, the variability in the realizations seems to increase as t increases along with the fact that the sample standard deviations increase with time both suggest that the population standard deviations, σ_{X_t}, are increasing with time.
6. In this case, we can obtain several realizations from the time series only because we know its "formula" (Equation (3.6)). In other situations (for example, the shampoo sales data to be discussed in Example 3.5), multiple realizations can be obtained without a formula (different stores, different time periods, and so forth).
7. By using the formula (inside information), we have shown that the true means actually trend linearly upward ($\mu_{X_t} = 15 + 1.8t$) and that the standard deviations $\sigma_{X_t} = .6t$ also increase with time.

3.2.1.2 Time Series 2: Y_t (Example 3.3 Continued)

As was the case with Figure 3.9(a), the increase in time for the realization in Figure 3.9(b) suggests that the true means, μ_{Y_t}, increase in time. To examine this supposition as we did for Time Series 1, we generated nine more realizations from the same random process that generated the realization in Figure 3.9(b). In Figure 3.11(a) we show the realization in Figure 3.9(b), and Figure 3.11(b) is a plot of the original realization (in bold) along with the nine new randomly selected realizations from the same time series. In Figure 3.11(b) we see that additional realizations from the underlying and "unknown" model go "all over the place". There is no uniform tendency for the data to trend up or follow any other pattern.

---

15 Two facts about expectation are used here: (1) if c is a constant, then $E[c] = c$ and (2) expectation is additive, that is, $E[aX + bY] = aE[X] + bE[Y]$. Thus, in our case, $X_t = 15 + 1.8t + .6ta_t$, so $\mu_t = E[X_t] = E[15 + 1.8t + .6ta_t] = E[15] + E[1.8t] + E[.6ta_t] = 15 + 1.8t + .6tE[a_t] = 15 + 1.8t$ because for each t it follows that $15 + 1.8t$ is deterministic constant and $E[a_t] = 0$ (a_t has zero mean).
16 The variance at time t is $Var(t) = E[(X_t - \mu_{X_t})^2] = E[(15 + 1.8t + .6ta_t - (15 + 1.8t))^2] = E[(.6ta_t)^2] = E[(.36)t^2 a_t^2] = .36t^2 E[a_t^2] = .36t^2$ because $E[a_t] = 0$ and $1 = Var[a_t] = E[(a_t - 0)^2] = E[a_t^2]$. Also, $SD(X_t) = \sqrt{Var(X_t)} = .6t$.

Note: Although Figure 3.11(b) simply looks like an uninterpretable "mess", there is much to learn from a closer examination of these realizations.

Figure 3.11(b) highlights the multiple values at times $t = 1$ and 2 and at times $t = 9$ and 10. Table 3.6 shows the actual values along with the sample means ($\bar{x}_1 = 28.4$, $\bar{x}_2 = 28.9$, $\bar{x}_9 = 28.1$, $\bar{x}_{10} = 30.0$), sample standard deviations ($s_{X_1} = 18.3$, $s_{X_2} = 17.8$, $s_{X_9} = 14.9$, $s_{X_{10}} = 12.5$). The sample autocorrelations, $r_{X_1 X_2} = .90$ and $r_{X_9 X_{10}} = .79$ are also shown in the table.

While not following a recognizable pattern, the realizations from Time Series 2 show positive correlation between adjacent values. That is, x_t tends to be fairly close to x_{t-1} (or x_{t+1} This is seen in Table 3.6 where the sample autocorrelations, $r_{X_1 X_2} = .90$ and $r_{X_9 X_{10}} = .79$ are fairly high. The scatterplots associated with both of these random variable pairs are shown in Figures 3.11(d) and (e). There it can be seen that (i) there is a strong positive autocorrelation in each case and (ii) the scatterplots look similar. We will return to Figure 3.11(f) in Section 3.3, but it is not too early to begin thinking about it.

TABLE 3.6   Data from Multiple Realizations for $t = 1$ and 2 and Again for $t = 9$ and 10 Shown in Figure 3.11

REALIZATION	X_1	X_2	X_9	X_{10}	
1	23	15	32	21	
2	13	24	31	25	
3	46	43	43	36	
4	35	25	8	26	
5	8	4	30	38	
6	35	33	9	10	
7	63	63	41	50	
8	3	7	6	16	
9	36	36	38	36	
10	22	39	43	42	
$\bar{X}_1, \bar{X}_2$	28.4	28.9	28.1	30.0	$\bar{X}_9, \bar{X}_{10}$
s_{X_1}, s_{X_2}	18.3	17.8	14.9	12.5	$s_{X_9}, s_{X_{10}}$
$r_{X_1 X_2}$		.90		.79	$r_{X_9 X_{10}}$

Table 3.7 shows the sample means and standard deviations for the ten observations from each random variable X_t, $t = 1, \cdots, 20$.[17] Examination of Table 3.7 shows that the sample means are quite similar across time, and that they tend to be between 28.1 and 34.3. The sample means are plotted in Figure 3.11(c) along with the true mean line at 30. Likewise, the sample standard deviations range from 7.4 to 18.3 with 7.4 being unusually low.

---

17 The actual observations are shown in Table 3.6 for $X_1, X_2, X_9,$ and X_{10}, whereas Table 3.7 gives only the sample statistics for all 20 of the time periods.

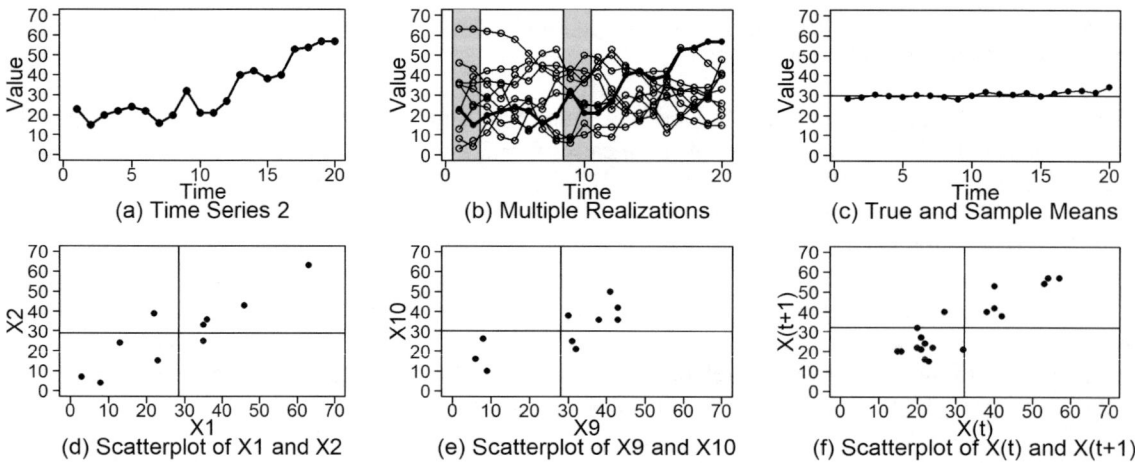

FIGURE 3.11   (a) A realization from Time Series 2, (b) multiple realizations from Time Series 2, (c) sample and true means, (d) and (e) scatterplots between X_1 and X_2 and between X_9 and X_{10}, respectively, and (f) scatterplot of X_t and X_{t+1}, $t = 1,…,19$ for Time Series 2 realization in (a).

In addition to the sample statistics for each random variable, the estimated autocorrelations between X_t and X_{t+1} are given in the table for $t = 1,…,19$. Note that the sample autocorrelations between X_1 and X_2 and between X_9 and X_{10} were already given in Table 3.6 (along with the data from which they were calculated). We noted that the estimated autocorrelations (.90 and .79), and scatterplots (Figures 3.11(d) and (e)) were quite similar for the random variable pairs X_1, X_2 and X_9, X_{10} respectively. This behavior is not unique to these two random variable pairs, and in fact the estimated autocorrelations between random variables X_t and X_{t+1} (that is, random variables that differ by one on the time axis) tended to range from about .75 to .95 (with one unusually small estimate of .56). Figure 3.11(f) is a scatterplot of the 19 pairs of data values plotted in Figure 3.11(a) that are separated by one time unit. That is, $(x_1, x_2), (x_2, x_3),…, (x_{19}, x_{20})$. These ordered pairs (which are members of the same realization) will be listed in Table 3.8, and are shown in Figure 3.11(f) as a scatterplot. This scatterplot is consistent with a correlation coefficient of about .85.

Key Point: In Section 3.3 we will discuss *stationary* processes, which are members of a broad class of time series that share the behavior exhibited by Time Series 2.

TABLE 3.7   Means and Standard Deviations for Table 3.6 Data $x_t^{(i)}, i = 1,…, 10$ for Time t

t	1	2	3	4	5	6	7	8	9	10
$\bar{x}_t$	28.4	28.9	30.5	29.6	29.3	30.2	29.9	29.2	28.1	30.0
s_{X_t}	18.3	17.8	14.6	14.9	14.9	13.8	14.5	16.6	14.9	12.5
$r_{X_t X_{t+1}}$	.90	.95	.90	.89	.88	.85	.90	.91	.79	***

	11	12	13	14	15	16	17	18	19	20
$\bar{x}_t$	31.7	30.9	30.5	31.3	29.8	31.1	32.3	32.4	31.6	34.3
s_{X_t}	14.5	14.3	11.3	10.1	7.4	10.7	12.6	12.3	12.5	13.3
$r_{X_t X_{t+1}}$	*	.84	.80	.86	.82	.56	.75	.94	.91	.83

* $r_{X_{10} X_{11}} = .93$

Inside Information:
The realizations in Figure 3.11(b) are generated using the following:

$$X_t = 3 + .9X_{t-1} + a_t,$$ (3.7)

where for each t, a_t is a randomly selected normal random variable with zero mean and variance 40. This is our first example of an "autoregressive model". Model (3.7) is known as an autoregressive process of order 1 (denoted AR(1)) and will be discussed in detail in Chapter 5. For the process in (3.7), it can be shown that the true means and standard deviations are constant across time ($\mu_{X_t} = 30$ and $\sigma_{X_t} = 14.5$ for each t, where $t = 1,\ldots,20$). Also, the autocorrelation $\rho_{X_t X_{t+1}} = .9$ for each t, for $t = 1,\ldots,19$.

Key Points

1. Both realizations in Figure 3.9 show a trending-type behavior.
2. The availability of multiple realizations revealed the fact that the trending in Figure 3.9(a) is "real" while the trending in Figure 3.9(b) is not repeatable in other realizations from the same model.
3. Points 1 and 2 above show that it is difficult to establish whether an observed trend should be predicted to continue based on a single realization, especially if n is small.
   – We will return to this topic in Chapter 8.

3.2.2 The Effect of Realization Length

The apparent trending in Time Series 1 and 2 in Example 3.3 was more fully understood because of the availability of multiple realizations. The two realizations in Figure 3.9 seemed to be trending upward and were almost indistinguishable in terms of an interpretation. Recall however that the use of multiple realizations revealed the following:

(a) *Time Series 1:* Realizations tended to consistently increase in time and to have increasing variability.
(b) *Time Series 2:* Realizations wandered around aimlessly. There was a tendency for neighboring data values to be similar and the average of the ten multiple realizations at each time period was approximately 30.

Although we have spent some time discussing the information that can be obtained from multiple realizations, the fact is that we will usually have only one realization. (This will be a focus of future discussion.) Time series analysis usually involves longer realizations, and it is not uncommon for realizations of length $n = 20$ (as was the case in Example 3.3) to contain insufficient information for a thorough analysis. In Figure 3.12, we show Time Series 1 and 2 extended to realization lengths $n = 100$. In Figure 3.12, the first 20 data values in each plot are the values of the realizations in Figure 3.9, and the following 80 data values are obtained by "following the formula" that produced the first 20. From single realizations of length $n = 100$, we see the following:

(a) *Time Series 1:* The realization continues to increase in time and the increase in variability is very obvious.
(b) *Time Series 2:* The realization continues to increase until about $n = 30$, then declines for about the next 10 time points, but in general, just wanders around. The tendency for neighboring data values to be similar continues and the realizations wander around, never getting very far from 30.

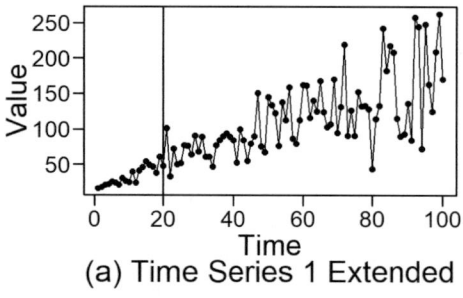 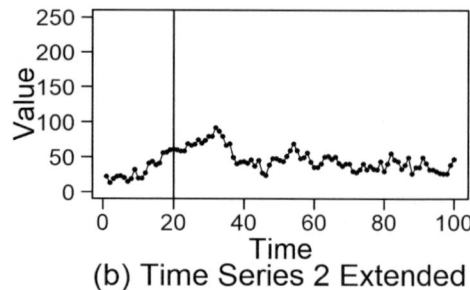

(a) Time Series 1 Extended   (b) Time Series 2 Extended

FIGURE 3.12   Figures 3.9 (a) and (b) extended to realizations of length $n = 100$.

Key Points

1. The results in this example are somewhat "sobering" because, in most time series applications these authors have encountered, *only one realization is available.*
2. Drawing conclusions from a single realization must be done with caution.
3. Individual time series of length as small as $n = 20$ rarely provide enough information for an informed analysis.

Example 3.4 December Sales Data with Multiple Realizations

Based on the analysis of the two datasets in Figure 3.9, it is clear that the behavior of the time series data can be misleading if not properly understood. For example, suppose the realizations in Figure 3.9 represent daily sales (in hundreds of dollars) of a certain product in two different stores for the days December 1–20, 2009. Let's assume that Figures 3.9(a) and (b) refer to sales data for Store 1 and Store 2, respectively. In each case the store owner is led to believe that sales of the product will increase during the first 20 days of December. We let the nine additional realizations represent the sales data for December 1–20, for the years 2010–2018. Although not identified as such in the plots, we now assume that each of the realizations in Figure 3.10(b) and Figure 3.11(b) corresponds to one of the years between 2009 and 2018. For Store 1, the multiple realizations shown in Figure 3.10(b) give the store owner some confidence that the first 20 days of December 2019 will behave in a similar manner. However, given the data in Figure 3.11(b) for Store 2, the owner should see that the increase in sales in 2009 was more of a "random occurrence", and an increase in sales during the first 20 days of December should not be expected to occur on an annual basis.

Forecasting Using Multiple Realizations

In many applications, data (such as the hypothetical sales data in Figure 3.9) are collected for the purpose of making "forecasts". We consider two situations:

(a) In November 2019, the store owner might want to forecast sales for the first 20 days of December using the information from the previous years.
   (i) *Store 1*: The multiple realizations should provide the owner of Store 1 with some confidence that sales will increase in a similar manner *given that conditions have not changed.*
       – For example, in November 2020 (the COVID pandemic year) it would be advisable for the store owner to proceed cautiously and not assume similar behavior.
   (ii) *Store 2:* For the owner of Store 2, about the only information suggested by the multiple realizations is that sales seem to average about $3,000 per day with no discernible pattern.
       – Again, this prediction assumes that conditions do not change (which would likely not have been true in pandemic years).

(b) Data are often collected for purposes of predicting into the immediate future. That is, in the sales data example, store owners might want to predict sales in 2018 for December 21–24 supposing that historical data are only available for December 1–20.

   (i) *Store 1*: The sales for December 1–20 seem to follow a linearly increasing pattern. Fitting a line to the data (which seems justified for days 1–20 of each year) and extending it to the next four days is a type of *extrapolation*. As we have discussed, extrapolation is discouraged (or should at least be done very cautiously). In this case, extrapolation is a bad idea unless there is additional information suggesting that conditions will not change. However, one might speculate that sales would increase more rapidly in these days due to last-minute Christmas shopping.

   (ii) *Store 2*: It seems that the best guess for the future is to predict the average of observed data (that is, $3,000 per day). However, last minute Christmas shopping might alter this expectation.

Key Points
1. Forecasting strategies will be a main focus of this book.
2. The forecasting techniques we will discuss are based on forecasting the immediate future given a *single realization*.
3. Forecasts based on time series models will be accompanied by prediction limits that acknowledge the inherent uncertainty.

Example 3.5 Shampoo Sales Data

Figure 3.13 is a plot of shampoo sales at a hypothetical store with sales at days 10 and 33 in bold. In time series terminology, the data in Figure 3.13 represent a realization, $x_t, t = 1,\ldots,50$. We note that $x_{10} = 9$ and $x_{33} = 20$. For this particular store, sales seemed to be higher on day 33 than on day 10.

FIGURE 3.13   Shampoo sales recorded at a particular store.

We may question whether sales at day 33 are typically higher on average than at day 10, that is, whether $\mu_{X_{33}}$ is greater than $\mu_{X_{10}}$. In order to better understand the situation, we obtained corresponding sales data for four randomly selected, similar stores. Figure 3.14 shows shampoo sales data for these four stores. The sales data for times $t = 10$ and $t = 33$ are highlighted. In this case, these realizations represent a sample from the ensemble of corresponding sales data for all the stores of interest. Averaging the data over the five realizations at times $t = 10$ and $t = 33$, we obtain $\bar{x}_{10} = (9 + 10 + 7 + 16 + 9)/5 = 10.2$ and $\bar{x}_{33} = (21 + 23 + 21 + 17 + 20)/5 = 20.4$. The evidence suggests that $\mu_{X_{33}} > \mu_{X_{10}}$. We cannot be sure of this fact because we only have a sample of five observations and not the entire ensemble itself. Given that we only have a sample of five stores at each value of t, we do not know either $\mu_{X_{10}}$ or $\mu_{X_{33}}$ for certain. However, we do have evidence (the data).

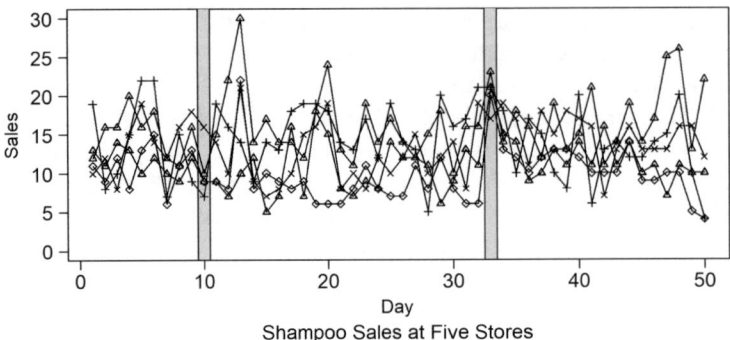

FIGURE 3.14   Shampoo sales recorded over the same time period at the store in Figure 3.13 and four similar stores.

3.3 STATIONARY TIME SERIES

In Section 3.2 we discussed the fundamentals of a statistical approach to time series analysis. We noted that in most time series scenarios these authors have encountered, there is only one realization available for analysis. For example, the data for monthly DOW closing averages, the price of West Texas Intermediate crude oil, Texas unemployment data, Dallas-Ft. Worth temperature data, and sunspots are only available as single realizations. We cannot somehow turn back the clock and obtain other realizations. In Example 3.3 we discovered the fact that the availability of only single, admittedly short ($n = 20$) realizations caused problems with interpretation. So, what do we do? We discussed the fact that longer realizations provide increased information. Also, *stationary time series* are a type of time series for which we can make meaningful analysis given only a single realization. This statement is sufficiently important to state separately as a Key Point.

> **Key Point:** Meaningful analysis of a *stationary time series* can often be made given only a single realization.

Time Series 2 in Example 3.3 is an example of a realization from a stationary time series process. In Figure 3.11(b) and Table 3.7, we can see that the ten realizations from Time Series Model 2, although seemingly dissimilar, actually share the following properties:

(a) The sample means for each t were all in the neighborhood of 30 and showed no tendency to increase or decrease with time.
(b) The sample standard deviations were similar for $t = 1,\ldots,20$.
(c) The autocorrelations $r_{X_t X_{t+1}}, t = 1,\ldots,19$ were similar and showed no tendency to change with time.

Another way to express the behavior of Time Series 2 is that it seems to be in a sort of *equilibrium*. That is, the basic behavior of the time series does not change with time. Recall that we have said that a time series is a sequence of random variables X_t with population parameters, mean, μ_{X_t} , and standard deviation, σ_{X_t} . Also, the population correlation between random variables X_{t_1} and X_{t_2} is denoted $\rho_{X_{t_1} X_{t_2}}$, and is called the population autocorrelation. With this notation in mind, a stationary process is defined in Definition 3.13.

Definition 3.13 (Stationarity):[18] The time series $\{X(t)\}$; is said to be *covariance stationary* if

1. $\mu_{X_t} = \mu$ (constant mean for all t)
2. $\sigma^2_{X_t} = \sigma^2 < \infty$ (i.e., a finite, constant variance for all t)
3. $\gamma_{X_{t_1} X_{t_2}}$ and $\rho_{X_{t_1} X_{t_2}}$ depend only on $t_2 - t_1$.

QR 3.7 Conditions of Stationarity

Condition 1: $\mu_{X_t} = \mu$ **(constant mean for all t)**
Figure 3.10(b) strongly suggests that for Time Series 1 in Example 3.3, the means for each t are not constant. In fact, we showed (using our "inside information") that μ_{X_t} is linearly increasing in time, which is supported by the plot of sample means in Figure 3.10(c). Consequently, Time Series 1 is *not* a stationary process. However, the realizations from Times Series 2 shown in Figure 3.11(b) wander around 30 but show no uniform tendency to trend up or down. This behavior is consistent with Condition 1 of stationarity. In Example 3.3 we stated that the realizations in Figure 3.11(b) were from an AR(1) model with $\mu = 30$. We will study AR(1) processes in Chapter 5 where we will discuss the fact that these processes are stationary.

The Air Passengers data (Figure 2.7(a)) and annual DOW data (Figure 1.9) increase in time which makes the constant mean questionable. The global temperature data show an increase, and the recent debate about global warming is in essence a debate about whether there is an increasing "population mean".

Condition 2: $\sigma^2_{X_t} = \sigma^2 < \infty$ **(constant and finite variance)**
Processes that satisfy this condition seem to have a variability that does not change with time. Time Series 1 is an example of a time series with increasing variability. The changing variability is not apparent in Figure 3.10(a). However, by obtaining multiple realizations shown in Figure 3.10(b), the changing variability becomes clear. Additionally, examining a longer realization, shown in Figure 3.12(a), the changing variability became obvious. While the variability change is not very apparent in the original realization, by examining the ten realizations in Figure 3.10(b) we see that the variability is increasing with time.

As mentioned in Chapter 2, the Air Passengers data (Figure 2.7(a)) show more variability in later years, and one method for adjusting for this increase in variability is to analyze the "log airline data" (Figure 2.7(b)) which has a more stable variability across time. Also, the bat echolocation data in Figure 3.15(a)[19] and the seismic data in Figure 3.15(b)[20] show a distinct decrease in variability across time toward the end of the realizations.[21]

---

18 Time series satisfying the conditions in Definition 3.13 are sometimes referred to as covariance stationary, weakly stationary, or second-order stationary. There is another more restrictive type of stationarity called strict stationarity. In this book, when we refer to a stationary process, we will be referring to one that satisfies the conditions of Definition 3.13.
19 Figure 3.15(a) shows big brown bat echolocation data furnished by Al Feng of the Beckman Center at the University of Illinois. The data consist of 381 points sampled at 7-microsecond intervals.
20 Figure 3.15(b) is a seismic Lg wave from an earthquake known as Massachusetts Mountain earthquake (August 5, 1971) which was recorded at the Mina, Nevada station.
21 We stated that a random sample of size $n = 1$ gives no information about variability. Because a realization represents "a single time series observation" it may be confusing that we are using the "single observation" to discuss variability change. However, in the seismic data, for example, by looking "down the realization" we can obtain the following information: (a) The variability changes with time. (b) In this particular case, the variability change down the timeline is gradual, so that some information about the variability at time t_k may be gleaned from the time-to-time variability of observations close to t_k.

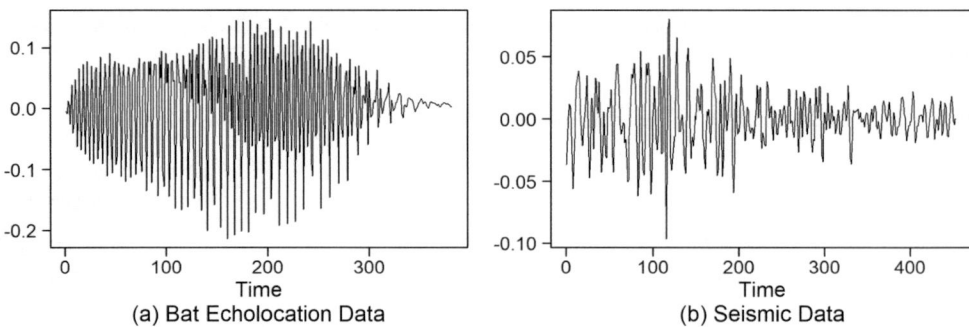

(a) Bat Echolocation Data   (b) Seismic Data

FIGURE 3.15   (a) Bat Echolocation data and (b) Seismic Lg wave from an earthquake.

Condition 3: $\gamma_{X_{t_1}X_{t_2}}$ **and** $\rho_{X_{t_1}X_{t_2}}$ **depend only on** $t_2 - t_1$.
This condition can be restated as follows:

> "The autocorrelation between X_{t_1} and X_{t_2} depends on how far apart t_1 and t_2 are, not where they are in time."

If Condition 3 is satisfied, there is no confusion in replacing the notation, $\gamma_{X_{t_1}X_{t_2}}$ with $\gamma_{t_2-t_1}$, and similarly denoting $\rho_{X_{t_1}X_{t_2}}$ with $\rho_{t_2-t_1}$. Letting $t_2 - t_1 = k$, we refer to γ_k and ρ_k as the *autocovariance and autocorrelation at lag k*, respectively. Also, it follows that $\gamma_k = \gamma_{-k}$ and $\rho_k = \rho_{-k}$. See Key Point 1 below.

Key Points

1. $\gamma_{-k} = E\left[\left(X_t - \mu\right)\left(X_{t-k} - \mu\right)\right] = E\left[\left(X_{t-k} - \mu\right)\left(X_t - \mu\right)\right] = \gamma_k$
2. $\rho_{-k} = \rho_k$
3. $\gamma_0 = E\left[\left(X_t - \mu\right)\left(X_t - \mu\right)\right] = E\left[\left(X_t - \mu\right)^2\right] = \sigma^2$
4. In the notations γ_k and ρ_k, the integer k is referred to as the ***lag***.

A realization for which Condition 3 seems reasonable is Time Series 2, multiple realizations of which are shown in Figure 3.11(b). Table 3.7 shows the correlation estimates between X_t and X_{t+1} for $t = 1, 2, \ldots, 19$. These estimated correlations were obtained using the ten realizations that were available. The autocorrelations tended to be in the range .7 to .9. That is, the autocorrelations between adjacent variables did not seem to change across the timeline.

As mentioned, the typical situation is that only one realization is available (and for our example, this is Time Series 2 in Figure 3.11(a)). Table 3.8 lists the available pairs of observations in Time Series 2 that are one time unit apart. These are plotted in Figure 3.11(f) and in the discussion following Figure 3.11 we noted that the associated correlation coefficient is in the neighborhood of .85. Section 3.3.2.3 will provide a formula for estimating the "lag 1" autocorrelation based on the scatterplot data in Figure 3.11(f).

TABLE 3.8   19 Pairs of Observations in Time Series 2 that Differ by One Time Unit

QR 3.8 Positive and Negative Autocorrelations

t	X_t	X_{t+1}	t	X_t	X_{t+1}
1	22	13	11	20	27
2	13	19	12	27	41
3	19	22	13	41	44
4	22	23	14	44	39
5	23	21	15	39	41
6	21	15	16	41	56
7	15	19	17	56	57
8	19	32	18	57	60
9	32	20	19	60	61
10	20	20			

While it may be difficult to "tell by looking" that a correlation structure stays constant across time, there are obvious examples in which this does not happen. An example of a time series realization that clearly violates this condition is the linear chirp data in Figure 3.16(a). Examination of the data shows a strong positive autocorrelation between X_t and, say X_{t+51} early in the data set. This positive autocorrelation appears to become smaller in time to the point that toward the end of the realization the autocorrelation between X_t and X_{t+1} is near zero or even negative. The Nyctalus noctula hunting bat echolocation signal sampled at 44×10^{-5} seconds shown in Figure 3.16(b) has a varying correlation structure that is not quite so obvious. Time series associated with DFW temperatures, lynx trappings, and Texas Unemployment have realizations for which it is not apparent that there is a violation of Condition 3.

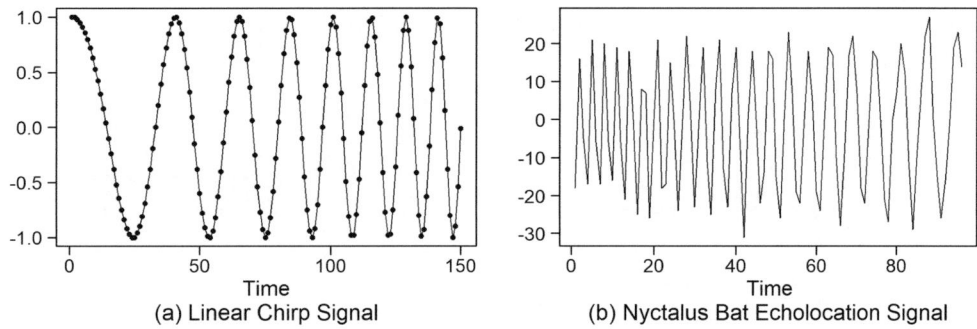

FIGURE 3.16   (a) Linear chirp signal and (b) echolocation signal from a Nyctalus noctula bat.

Definition 3.14: White Noise Process
A process X_t is said to be white noise if the following conditions hold.

1.   Each X_t has zero mean and finite variance, σ_X^2
2. X_{t_1} and X_{t_2} are uncorrelated if $t_1 \neq t_2$.

Note: From Definition 3.14 it follows that $\gamma_k = \rho_k = 0$ if $k \neq 0$ and $\gamma_{X_{t_1} X_{t_2}} = \sigma_X^2$ when $t_1 = t_2$.

In words, whenever lag $k = 0$, the autocovariance is equal to the process variance and the autocorrelation is equal to one.

In a white noise process, each observation is uncorrelated with all other observations. It is analogous to a random sample indexed in time.[22] An important fact is that white noise processes are stationary. (See Problem 3.8.)

Key Points

1. Stationary time series models, such as autoregressive or autoregressive moving average models, are associated with model-based parameters such as the:
   - mean (μ)
   - variance $\sigma^2 (\gamma_0)$
   - autocorrelations and autocovariances $(\rho_k$ and $\gamma_k)$
2. It is common practice to plot the autocorrelations, $\rho_k, k = 0, 1, \ldots, K$ for some integer K.
   - There is no need to plot autocorrelations for negative lags since $\rho_{-k} = \rho_k$, for all lags.
3. In Section 3.3.1 we will discuss the estimation of these parameters for a given realization.

3.3.1 Plotting the Autocorrelations of a Stationary Process

It is important to understand the extent to which a plot of the autocorrelations describes the behavior of a time series realization. Figure 3.17 shows realizations from four stationary, autoregressive time series models, while Figure 3.18 shows the corresponding model-based autocorrelations for lags 0–30.

Realization 1 in Figure 3.17(a) has a random wandering behavior. Note that it is typical for x_t and x_{t+1} to be relatively close to each other with a few exceptions, e.g. around $t = 50$. That is, the value of the random variable X_{t+1} is usually not very far from the value of X_t, and, as a consequence, there is a noticeably strong positive autocorrelation between the random variables X_t and say X_{t+1} (ρ_1 is at least .9). Note also that the autocorrelation between X_t and X_{t+k} decreases as the lag k increases. This decrease leads to the fact that by lag 30 there is very slight correlation between X_t and X_{t+30}. By examining Figure 3.18(a) it appears that $\rho_{30} \approx .10$.

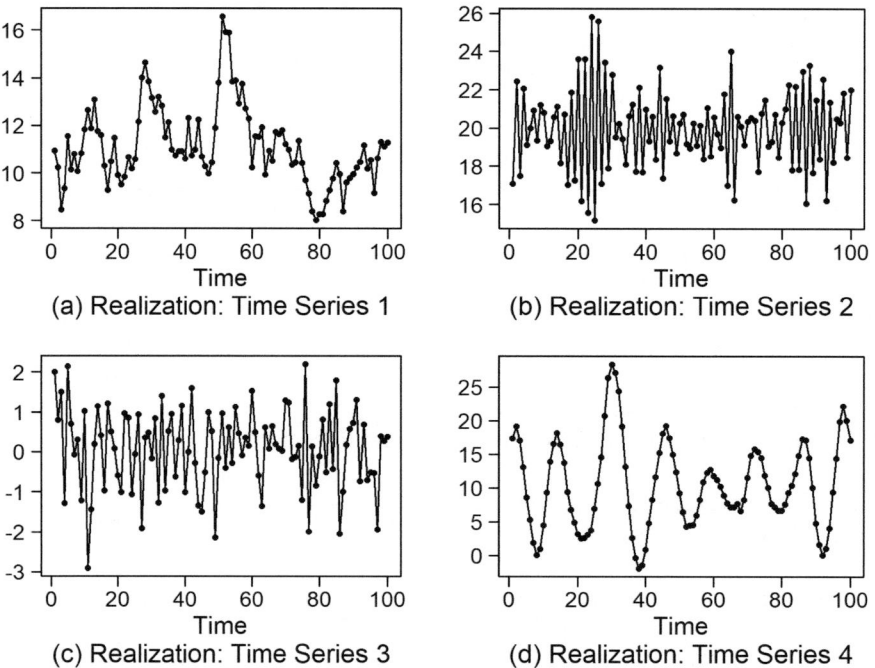

(a) Realization: Time Series 1

(b) Realization: Time Series 2

(c) Realization: Time Series 3

(d) Realization: Time Series 4

FIGURE 3.17 Four realizations from stationary processes. (a) Realization 1 (b) Realization 2 (c) Realization 3 (d) Realization 4.

---

22 From a mathematical perspective, we know that "independence" and "uncorrelated" are equivalent concepts if the data are normally distributed. In general, independence implies uncorrelated, but not vice versa.

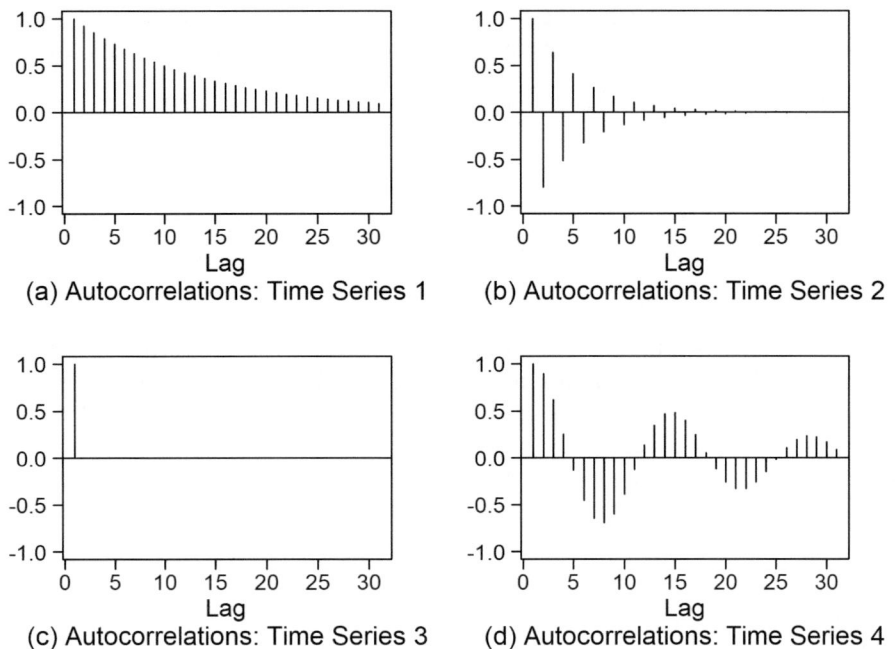

FIGURE 3.18   True autocorrelations from models associated with realizations in Figure 3.17 (a) Realization 1 (b) Realization 2 (c) Realization 3 (d) Realization 4.

Realization 2 in Figure 3.17(b) shows a highly oscillatory behavior. In fact, if x_t is above average, then x_{t+1} tends to be below average, x_{t+2} tends to be above average, and so forth. The autocorrelations in Figure 3.18(b) describe this behavior where we see that $\rho_1 \approx -0.8$ while $\rho_2 \approx 0.6$, and $\rho_3 \approx -0.5$, and so on. Note also that the up-and-down pattern is sufficiently imperfect that the autocorrelations damp to near zero by about lag 15.

Realization 3 in Figure 3.17(c) shows an absence of pattern. That is, there appears to be no relationship between X_t and X_{t+1} and as a matter of fact, there is seemingly no correlation between X_t and X_{t+k} for any $k \neq 0$. In fact, this is a realization from a white noise model, and for this model, $\rho_k = 0$ whenever $k \neq 0$, as can be seen in Figure 3.18(c). (Notice that in all autocorrelation plots, $\rho_0 = 1$.)

Realization 4 in Figure 3.17(d) is characterized by cyclic behavior with an average cycle-length of about 14 time points. The corresponding autocorrelations in Figure 3.18(d) show a damped sinusoidal behavior. Note that, not surprisingly, because of the cycle length of about 14, there is a positive correlation between X_t and X_{t+14}. Also, there is a substantial negative correlation at lags 7 and 8, because within the sinusoidal cycle of length 14, if x_t is above the average, then x_{t+7} would be expected to be below average and vice versa. Note also that this autocorrelation structure holds up fairly "solidly" as far as $k = 28$.

3.3.2 Estimating the Parameters of a Stationary Process

In this section we discuss the estimation of the parameters of a stationary process given a realization of length n.

3.3.2.1 Estimating μ

Note that if the means do not depend on time (that is, Condition 1 is satisfied), then the given realization provides n observations from random variables that share the same population mean (μ). Thus, because each observation is estimating the same mean, it makes sense that we can estimate μ with the sample mean of $x_1, \ldots, x_n$, that is, by "averaging down the realization time line". The formula looks familiar: $\bar{x} = \frac{1}{n} \sum_{t=1}^{n} x_t$. This looks like a sample average from a random sample.

> **Key Point:** For a stationary time series, we can use the data across time to estimate the mean because the mean is assumed to be the same for each time, t.

The difference, however, is that the observations are *not independent*, and instead are related. The sample mean from a random sample is based on n independent observations of the phenomenon. However, if observations in a time series are highly dependent, then a realization of length 30 does not provide 30 independent pieces of information.

Recall that when data are from a random sample of size n with variance σ_X^2, then we have the classical result that $Var(\bar{X}) = \sigma_X^2/n$ (and $SE(\bar{X}) = \sigma_X/\sqrt{n}$). However, if X_t is a stationary time series, then the variance of $\bar{X}$ based on a realization of length n is given by

$$Var(\bar{X}) = \frac{\sigma_X^2}{n} + 2\frac{\sigma_X^2}{n} \sum_{k=1}^{n-1} \left(1 - \frac{|k|}{n}\right)\rho_k \tag{3.8}$$

Equation 3.8 is written in the form $Var(\bar{X}) = \frac{\sigma^2}{n} + Q$ where clearly the standard error of $\bar{X}$ increases over that of a random sample of size n (that is, $Q > 0$) when the autocorrelations are positive. The following example illustrates the effect of dependence on the variability of $\bar{X}$.

Example 3.6 Consider the data plotted in Figure 3.19 where, in each case, the true mean is 30 and the population variance is $\sigma_X^2 = 50$. Figure 3.19(a) is a plot of a random sample (with data values connected). Figure 3.19(b) is a "sample" (actually a realization) from the AR(1) model that generates data values using the formula

$$X_t = .3 + .99 X_{t-1} + a_t, \tag{3.9}$$

where the $a_t s$ are white noise with mean zero and $\sigma_a^2 = 1$ and $\sigma_X^2 = 50$. In Figure 3.19(a), it can be seen that the independent observations in Figure 3.19(a) provide unrelated observations "hovering" around the mean $\mu = 30$. On the other hand, note that in Figure 3.19(b), the first observation is about 40 and subsequent observations are correlated with (sort of "tied to") this observation. In fact, the correlation structure is such that all of the 50 data values are above 30.

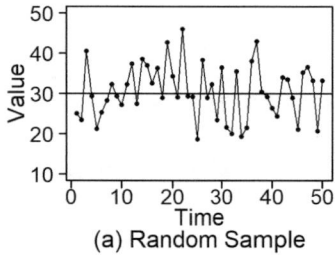

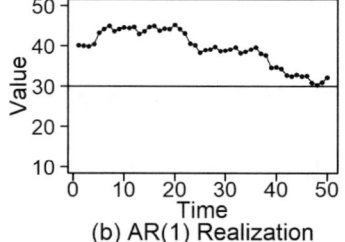

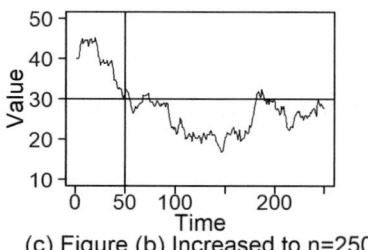

(a) Random Sample   (b) AR(1) Realization   (c) Figure (b) Increased to n=250

FIGURE 3.19 (a) Random sample (white noise) of length $n = 50$ (b) AR(1) realization of length $n = 50$ from the model $X_t = 9 + .99X_t + a_t$ with $\sigma_a^2 = 1$. In both cases, the theoretical mean variance = 30 and variance = 50 (c) the realization in (b) extended to $n = 250$ observations.

Analogy: The following trivial analogy may be helpful. (Take it or leave it.) Consider the situation in which two sets of 50 testers are attempting to estimate the weight of a metal object. They are allowed to pick up and carefully examine the object but are not allowed to place it on a scale.

Figure 3.19(a) Data: The situation in Figure 3.19(a) is analogous to each of the 50 testers holding the object and independently and carefully using their common sense and expertise to come up with their "best guesses".

Figure 3.19(b) Data: The data in Figure 3.19(b) is analogous to the situation in which we assume that Tester 1 uses the same carefully thought-out techniques employed by *each* of the 50 testers represented in Figure 3.19(a). However, Tester 2 holds the object, finds out the first tester's estimate, and instead of coming up with an independent well thought-out estimate, estimates the weight to be fairly close to the estimate given by Tester 1. Similarly, after holding the object but before giving an estimate, Tester 3 learns the second tester's estimate and gives a similar guess. The 49 testers following Tester 1 use the same strategy on down the line.

Question: Which estimate, (a) or (b), would you put the most faith in?[23] With regard to estimate (b), the only tester who gave an independent and seriously thought-out estimate was the first one. The others "just went along".

The sample mean of the data in Figure 3.19(a) is 30.6 (quite close to 30), while the mean for the data in Figure 3.19(b) is 34.7, leaving the impression that the true mean of this process is really above $\mu = 30$. Figure 3.19(c) shows the realization in Figure 3.19(b) extended to 250 data values. There we continue to see a solid connection between neighboring data values, but the realization does eventually "wander" back toward $\mu = 30$. Because the testers did hold the object, the estimates never strayed extremely far from the true value.

Calculating $Var(\bar{X})$ *(and SE* $(\bar{X}))$: We have obtained $\bar{X}$ from three realizations:

(a)   From the random sample of size $n = 50$
(b)   From the highly dependent realization of length $n = 50$
(c)   From the highly dependent realization of length $n = 250$

We calculate $Var(\bar{X})$ (and $(SE(\bar{X}))$ in each case.

(a) $Var(\bar{X}) = \dfrac{\sigma^2}{n} = \dfrac{50}{50} = 1$ $\left(SE(\bar{X}) = 1\right)$

(b) $Var(\bar{X}) = \dfrac{50}{50} + 2\sum_{k=1}^{49}\left(1 - \dfrac{k}{50}\right).99^k = 42.6$ $(SE(\bar{X}) = 6.5)$

(c) $Var(\bar{X}) = \dfrac{50}{250} + 2\dfrac{50}{250}\sum_{k=1}^{249}\left(1 - \dfrac{k}{250}\right).99^k = 25.2$ $\left(SE(\bar{X}) = 5.0\right)$

Two other bits of information are of interest:

(a)   In order for the "dependent group" to achieve an $SE(\bar{X}) = 1$, it would require a realization length of $n = 9750$.[24]
(b)   Suppose there was only one observation in the "random sample" group. In this case, $\bar{X} = X_1$, in which case $Var(\bar{X}) = 50/1 = 50$ with $SE(\bar{X}) = 7.1$. Surprisingly, this is not much greater than the corresponding quantities for a realization of length $n = 50$ in the "dependent group".

---

23 Testers in group (a) would probably have the better overall estimate of the mean.
24 The correct reaction at this point is "WOW".

In the previous example, we generated realizations from a model that introduced a strong autocorrelation structure, $\rho_k = .99^{|k|}$, that dies out very slowly. For comparison, we consider the following autocorrelation structures (that are associated with AR(1) models). Each of these models is designed to have mean $\mu = 30$ and variance $\sigma^2 = 50$.[25]

(a) $X_t = .3 + .99X_t + a_t$, $\sigma_a^2 = 1$

(b) $X_t = 1.5 + .95X_t + a_t$, $\sigma_a^2 = 4.9$

(c) $X_t = 3 + .9X_t + a_t$, $\sigma_a^2 = 9.5$

(d) $X_t = 6 + .8X_t + a_t$, $\sigma_a^2 = 18$

The autocorrelation plots for these four models are given in Figure 3.20.

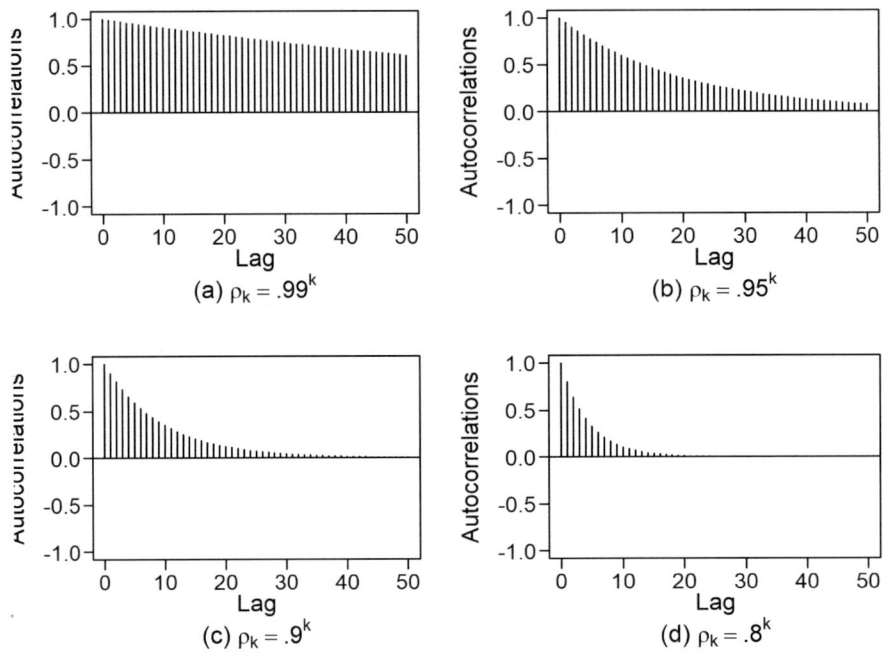

FIGURE 3.20 Model-based autocorrelations for AR(1) models $X_t = \beta + \phi_1 X_{t-1} + a_t$ with (a) $\phi_1 = .99$, (b) $\phi_1 = .95$, (c) $\phi_1 = .9$, and (d) $\phi_1 = .8$.

The graphs in Figure 3.20 show various levels of exponential damping associated with AR(1) models given by $X_t = \beta + \phi_1 X_{t-1} + a_t$, where ϕ_1 is positive and less than 1. As previously mentioned, Figure 3.20(a) shows that the autocorrelations for .99 die out very slowly. The correlation between adjacent values is $\rho_1 = .99$, while the autocorrelation between variables 25 time units apart is $\rho_{25} = .99^{25} = .78$ (still a substantial correlation). At lag 49, we have $\rho_{49} = .99^{49} = .61$. Consequently, in the previous example with a realization of length $n = 50$, the correlation between the observation made by Tester 1 and by Tester 50 would be .61. That is, all observations in the realization of length 50 were markedly related. The autocorrelation structure in Figure 3.20(b) shows that the first and last observations are only very slightly correlated. Finally, in Figures 3.20(c) and (d), the correlation dies out by lags 30 and 15, respectively. That is, if the testers had autocorrelations similar to those in Figure 3.20(d), then Tester 20's estimate would

---

25 Autoregressive processes and their properties will be discussed in Chapter 5.

essentially be independent of that of Tester 1. Table 3.9 provides analogous results to those shown earlier for the $\phi_1 = .99$ case.

TABLE 3.9 $SE(\overline{X})$ for Realizations with $\rho_k = \phi_1^k$ and Realization Lengths Necessary to Obtain $SE(\overline{X}) = 1$, which is Associated with a White Noise Realization of Length $n = 50$

MODEL	$n = 50$	$n = 250$	$n = 500$	$n = 1000$	n required for $SE(\overline{X}) = 1$
(a) $\phi_1 = .99$	6.53	5.02	4.00	2.99	9750
(b) $\phi_1 = .95$	5.00	2.68	1.94	1.38	1925
(c) $\phi_1 = .90$	3.93	1.91	1.37	0.97	940
(d) $\phi_1 = .80$	2.86	1.33	0.94	0.67	445

Even with autocorrelations such as those in Figure 3.20(d) that "die out" by lag 20, a realization of length $n = 445$ is "similar" (in terms of standard error of $\overline{X}$) to a white noise realization (random sample) of length $n = 50$. The take-away from this example is summarized in the following Key Points.

Key Points

1. When analyzing a random sample, we are often able to consider $n = 30$ as a "large sample" for purposes of normal approximation, and so forth.
2. The presence of autocorrelations, especially those such as shown most dramatically in Figures 3.20(a) and (b), reduces the amount of information about the mean that is available in a realization.
3. Time series data with autocorrelation structures such as those illustrated in Figure 3.20 may require n to be 100, 1000, 5000, or more to provide the same amount of information about the population mean that is available in a random sample of size $n = 50$.

3.3.2.2 Estimating the Variance

The estimate of the variance of a stationary process is $\hat{\gamma}_0$ where $\hat{\gamma}_k$ is defined in Section 3.3.2.3.

3.3.2.3 Estimating the Autocovariance and Autocorrelation

As mentioned, because of stationarity, $\gamma_k = E\left[(X_t - \mu_X)(X_{t+k} - \mu_X)\right]$ does not depend on t. Consequently, it seems reasonable to estimate γ_k from a single realization by "moving down the time axis and finding the average of all crossproducts of observations in the realization separated by k time units". There are $n - k$ such pairs, and these are shown in Table 3.10.

TABLE 3.10   Lag k Data Pairs

LAG k PAIRS	
X_t	X_{t+k}
X_1	X_{1+k}
X_2	X_{2+k}
$\vdots$	$\vdots$
X_{n-k-1}	X_{n-1}
X_{n-k}	X_n

QR 3.9 Sample Autocovariance and Autocorrelation

The estimator or γ_k is denoted by $\hat{\gamma}_k$, and is defined by

$$\hat{\gamma}_k = \frac{1}{n}\sum_{t=1}^{n-k}\left(X_t - \bar{X}\right)\left(X_{t+k} - \bar{X}\right), \quad 0 \le k \le n \tag{3.10}$$

$$= 0, \quad k \ge n$$

$$= \hat{\gamma}_{-k}, \quad k < 0$$

Notes:

(1) Viewing Table 3.10 makes it easy to see that if $k = n-1$, there is only one pair of data values that differ by $n-1$ time units, namely X_1 and X_n. In this case, the "sum" in estimator $\hat{\gamma}_{n-1}$ has only a single term: $\hat{\gamma}_{n-1} = \frac{1}{n}(X_1 - \bar{X})(X_n - \bar{X})$. Consequently, the estimator of γ_{n-1} would be expected to be of poorer quality than, for example, the estimator of γ_1 which is estimated using an "average" of $n-1$ cross-products of data pairs separated by one time unit.

(2) From Figure 3.20, we see that model-based autocorrelations damp toward zero (at differing rates). Because the autocorrelations are simply standardized versions of the autocovariances, it follows that both $\rho_k \to 0$ and $\gamma_k \to 0$ as n gets large.[26]

(3) Using (3.10) it follows that

$$\hat{\gamma}_0 = \frac{1}{n}\sum_{t=1}^{n}\left(X_t - \bar{X}\right)^2 \quad (= \hat{\sigma}^2). \tag{3.11}$$

(4) The estimator of the autocorrelation, ρ_k is given by

$$\hat{\rho}_k = \hat{\gamma}_k / \hat{\gamma}_0. \tag{3.12}$$

This estimator is called the *sample autocorrelation*. From examination of (3.10), it is clear that $\hat{\gamma}_k$ (and $\hat{\rho}_k$) values will tend to be "small" when k is large relative to n. For example, in Note (1) above we showed that $\hat{\gamma}_{n-1}$ is simply one term divided by n. If n is large, then $\hat{\gamma}_{n-1}$ (and $\hat{\rho}_{n-1}$) are likely to be small, which mimics the behavior of the true values.

3.3.2.4 *Plotting Sample Autocorrelations*

Figures 3.17 and 3.18 showed plots of four realizations from autoregressive time series models and their associated model-based (theoretical) autocorrelations. In the discussion regarding Figures 3.17 and 3.18, we described how patterns in the autocorrelations provide information about the basic nature of realizations from time series models. Of course, given a set of data, we will not know the underlying model, so we will need to estimate the autocorrelations. Figure 3.21 shows plots of the sample autocorrelations, $\hat{\rho}_k, k = 0,...,40$ calculated from the four realizations in Figure 3.17. These sample autocorrelations are estimates of the true autocorrelations, shown in Figure 3.18, and they approximate the behavior of the true autocorrelations. In particular, Figures 3.18(d) and 3.21(d) are very similar and show a damped sinusoidal behavior with a period of about 14. Figure 3.21(b) has a damped oscillating behavior similar to Figure 3.18(b) but the damping is more extreme in the true autocorrelations. In Figure 3.21(a) we note that there is correlation among the $\hat{\rho}_k$ values. For example, notice that when a

---

26 Remember that in time series analysis, "large" n may be several thousand time units.

value of $\hat{\rho}_k$ tends to underestimate ρ_k, values of $\hat{\rho}_j$ for j "near" k also tend to underestimate the true value. This can produce a cyclic behavior in sample autocorrelation plots when no such behavior is present in the plot of ρ_k or the data. Also, it should be noted that sample autocorrelations for a white noise realization will of course not be exactly equal to zero. In Sections 6.1.2 and 9.1.1 we will discuss tests for white noise that will help decide whether sample autocorrelations such as those shown in Figure 3.21(c) are larger in magnitude than would be expected for a realization from a white noise process.

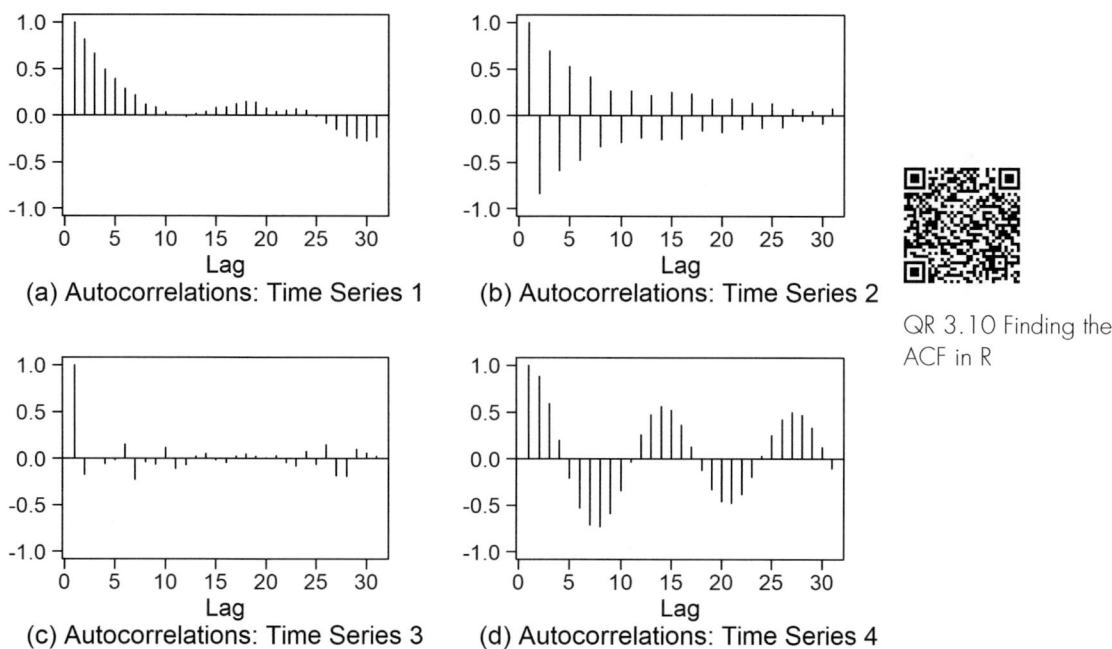

(a) Autocorrelations: Time Series 1

(b) Autocorrelations: Time Series 2

QR 3.10 Finding the ACF in R

(c) Autocorrelations: Time Series 3

(d) Autocorrelations: Time Series 4

FIGURE 3.21 Sample autocorrelations, $\hat{\rho}_k$, calculated from realizations in Figure 3.17 (a), (b), (c), and (d), respectively.

Key Points

1. Given a realization of length n, the estimators of γ_k and ρ_k are 0 for $k \geq n$.
2. $-1 \leq \hat{\rho}_k \leq 1$
3. The $\hat{\rho}_k s$ are referred to as the *sample autocorrelations*.
4. Model-based or "true" autocorrelations such as those in Figure 3.18 and 3.20 are only available when we know the "inside information" concerning the model used to generate the realization.
5. In practice, given a set of time series data such as sales data or crude oil price data, we will not know the "true" autocorrelations, and will only be able to calculate the sample autocorrelations such as those in Figure 3.21.
6. An analysis of a set of time series data almost always begins by examining plots of the data and the sample autocorrelations.

Example 3.7 Why we use $\hat{\rho}_k$ to estimate ρ_k

You may have noticed that the estimator $\hat{\gamma}_k$ in (3.10) always has a divisor of n regardless of the number of terms in the sum. This seems odd; admittedly, a more natural estimate of γ_k would have been to replace μ_X by $\bar{X}$ and find the *average* of the $n - k$ (instead of n) cross products

$$\tilde{\gamma}_k = \frac{1}{n-k}\sum_{t=1}^{n-k}\left(X_t - \bar{X}\right)\left(X_{t+k} - \bar{X}\right), \ 0 \le k < n$$
$$= 0, \quad k \ge n$$
$$= \tilde{\gamma}_{-k}, \ k < 0. \tag{3.13}$$

Then, the corresponding autocorrelation estimator would be $\tilde{\rho}_k = \tilde{\gamma}_k / \tilde{\gamma}_0$. Figures 3.22(a), (b), and (c) show a realization of length $n = 75$, the true autocorrelations associated with the model that generated the data in Figure 3.22(a), and the sample autocorrelations (defined by (3.10)) up to lag $k = 74$, respectively. We see that the true autocorrelations damp fairly rapidly and that the sample autocorrelations damp some-what more rapidly and then "hover" around zero. In Figure 3.22(d), we plot the alternatively proposed autocorrelation estimates, $\tilde{\rho}_k$, based on (3.13). In the plot, we see that $\hat{\rho}_k$ and $\tilde{\rho}_k$ are similar for small k, but as k gets closer to n, the sample autocorrelations, $\hat{\rho}_k$, begin to wander around zero while the estimates, $\tilde{\rho}_k$, become very erratic. Even more disturbing is the fact that nearly all $\tilde{\rho}_k$ for lag $k = 57$ and above are less than -1. This is particularly concerning because autocorrelations (which are correlations) must fall between -1 and 1. For these reasons, the alternate formula $\tilde{\rho}_k$, which is based on an autocovariance $\tilde{\gamma}$ that divides by $n - k$, *is not used to estimate* $\tilde{\rho}_k$.

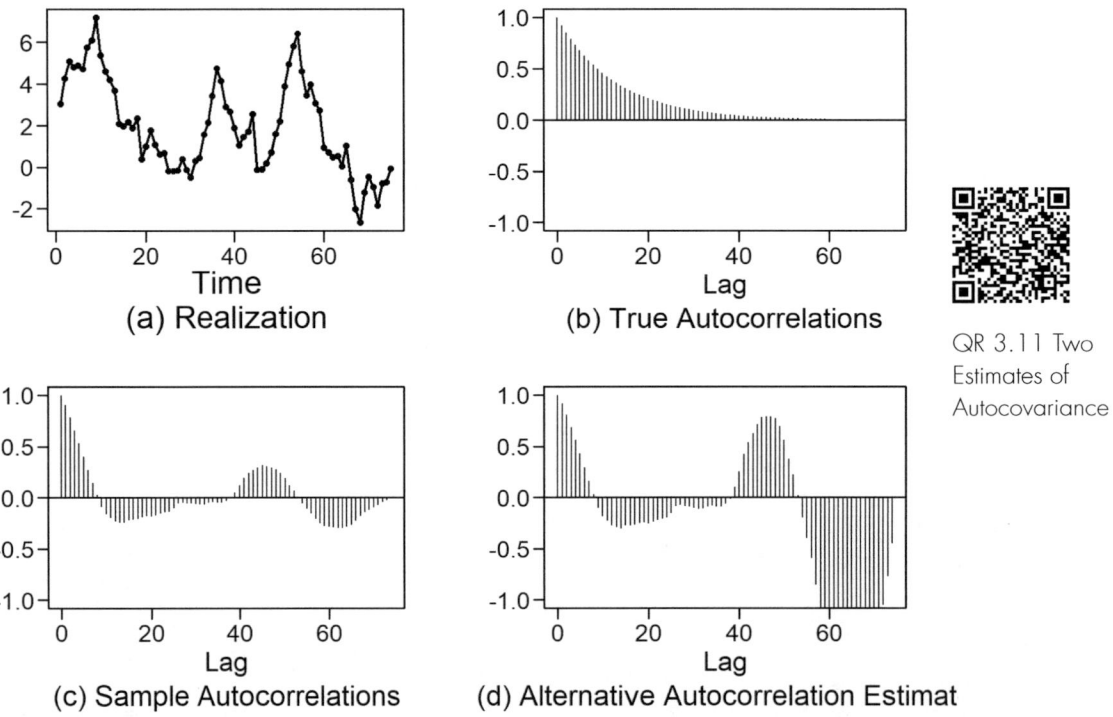

(a) Realization

(b) True Autocorrelations

QR 3.11 Two Estimates of Autocovariance

(c) Sample Autocorrelations

(d) Alternative Autocorrelation Estimat

FIGURE 3.22   (a) Realizations from a time series model, (b) and (c) theoretical (ρ_k) and sample $(\hat{\rho}_k)$ autocorrelations, respectively, and (d) sample autocorrelations using proposed estimates, $\hat{\rho}_k$.

3.4 CONCLUDING REMARKS

In this chapter, several pivotal concepts which are influential in the analysis of time series data were introduced. The importance of understanding the dependence structure in a time series realization was emphasized. Comparisons and contrasts were made between dependent and independent data structures, and interesting relationships were explored. Also, of relevance to future chapters are the autocovariance and autocorrelation, which are the time series versions of the familiar covariance and correlation, respectively. These terms were defined and considered both in mathematical form and by graph. A significant amount of time series methodology has been developed for stationary time series, which will become apparent very soon! The term "stationary" was defined and illustrated in this chapter, and stationary datasets were compared to those for which data are nonstationary. Several examples of both simulated and actual data were presented to illustrate this wide array of important topics.

APPENDIX 3A

In this appendix we obtain the mean, variance, and standard deviation associated with the random process of rolling two dice and finding the sum of the rolls. Table 3A.1 shows the 36 equally likely possible outcomes when we roll a white die and a black die and find the sum.

TABLE 3A.1   Probability Distribution for the Sum of the Rolls of Two Dice

	•	••	•••	•• ••	•• • ••	•• •• ••
•	2	3	4	5	6	7
••	3	4	5	6	7	8
•••	4	5	6	7	8	9
•• ••	5	6	7	8	9	10
•• • ••	6	7	8	9	10	11
•• •• ••	7	8	9	10	11	12

Consequently, we see that the probability of a sum of 12 is 1/36 (which can only happen if the white die and black die both land on 6) while the probability of 6 is 5/36, and so forth. Table 3A.2 shows the possible values and associated probabilities.

TABLE 3A.2   Possible Outcomes and Probabilities for the Random Process of Rolling Two Dice and Finding the Sum

SUM	PROBABILITY
2	1/36
3	2/36
4	3/36
5	4/36
6	5/36
7	6/36

SUM	PROBABILITY
8	5/36
9	4/36
10	3/36
11	2/36
12	1/36

This is an example of a discrete probability distribution in which there are 11 possible outcomes. Suppose we have a discrete probability distribution with possible values $y_1, y_2, \ldots, y_k$ and the probability of y_i is p_i, $i = 1, \ldots, k$. Then from introductory statistics, we know that the mean is given by $\mu = \sum_{i=1}^{k} p_i y_i$ and the variance is given by $\sigma^2 = \sum_{i=1}^{k} p_i (y_i - \mu)^2$. So, for the "sum" distribution we have

$$\mu = \frac{1}{36}(2) + \frac{2}{36}(3) + \frac{3}{36}(4) + \frac{4}{36}(5) + \frac{5}{36}(6) + \frac{6}{36}(7) + \frac{5}{36}(8) + \frac{4}{36}(9) + \frac{3}{36}(10) + \frac{2}{36}(11) + \frac{1}{36}(12)$$
$$= 7$$

This seems reasonable because 7 is the most likely sum and the distribution is symmetric about 7. The variance is

$$\sigma^2 = \frac{1}{36}(2-7)^2 + \frac{2}{36}(3-7)^2 + \frac{3}{36}(4-7)^2 + \frac{4}{36}(5-7)^2 + \frac{5}{36}(6-7)^2 + \frac{6}{36}(7-7)^2 + \frac{5}{36}(8-7)^2$$

$$+ \frac{4}{36}(9-7)^2 + \frac{3}{36}(10-7)^2 + \frac{2}{36}(11-7)^2 + \frac{1}{36}(12-7)^2$$

$$= 5.833,$$

and the standard deviation is $\sigma = 2.415$.

APPENDIX 3B

TSWGE FUNCTION

`plotts.sample.wge(x,lag.max,trunc,arlimits,periodogram):` plots a realization, sample autocorrelations, and Parzen spectral density (both plotted in dB). Note: See Chapter 4 for discussion of the Parzen spectral density. (The periodogram is not covered in this book but is discussed in QR4.4 video.) The periodogram is plotted as an option. See Woodward et al. (2017).

x is a vector containing the time series realization
lag.max is the maximum lag at which to plot the sample autocorrelations (default is 25)
trunc (integer ≥0) specifies the truncation point for the Parzen window

trunc=0 (default) (or if no value for M specified) indicates default truncation point $M = 2\sqrt{n}$

trunc>0 is a user-supplied truncation point, i.e. calculations will be based on the user-supplied value of **M**

arlimits (default=**FALSE**) is a logical variable specifying whether 95% limit lines will be included on sample autocorrelation plots

periodogram (default=**FALSE**) specifies whether the periodogram will be plotted

Example:

```
data(AirPassengers);airlog=log(AirPassengers);
sp=plotts.sample.wge(x=airlog,lag.max=50)
```

plots the realization(**x=airlog**), sample autocorrelations (lags 0 to **lag.max=50**), and the Parzen spectral estimator (in dB) using truncation point $M = 2\sqrt{n}$. The vector **$autplt** contains the sample autocorrelation values while **$freq** and **$pgram** contain the frequencies and Parzen-based spectral estimates in dB, respectively.

BASE R COMMANDS

(a) **mean(x):** calculates the mean of a time series realization where **x** is a vector containing the time series realization.

   Example: The basic R command

```
data(AirPassengers);mean.air=mean(AirPassengers)
```

   calculates the sample mean of the air passengers data in Base R dataset **AirPassengers** and places it in vector **mean.air**. In this case **mean.air** is 280.2986.

(b) **var(x):** calculates the variance of a time series realization where **x** is a vector containing the time series realization.

   Example: The basic R command

```
data(AirPassengers);var.air=var(AirPassengers)
```

   calculates the sample variance (using denominator $n-1$) of the **AirPassengers** data and places it in variable **var.air**. In this case **var.air** is 14391.92.

(c) **sd(x):** calculates the standard deviation of a time series realization where **x** is a vector containing the time series realization.

   Example: The basic R command

```
data(AirPassengers);sd.air=sd(AirPassengers)
```

   calculates the sample standard deviation (using denominator $n-1$) of the **AirPassengers** data in Base R dataset **AirPassengers** and places it in variable **sd.air**. In this case **sd. air.** is 119.9663, which is the square root of **var.air**.

(d) **acf(x, lag.max, type, plot)** calculates (and optionally plots) the sample autocorrelations (or sample autocovariances and sample variance) of a time series realization where

x is a vector containing the time series realization

lag.max is the maximum lag at which to calculate the sample autocorrelations (or autocovariances)

type='correlation' (default) outputs the sample autocorrelations ($\hat{\rho}_k$ in our notation)

    ='covariance' specifies the autocovariances

    ='partial' designates the partial autocorrelations (discussed in Section 6.1)

plot is a logical variable. **plot='TRUE'** (default) produces a plot of the sample autocorrelations (or autocovariances)

Example:

```
data(lynx); aut=acf(x=lynx, lag.max=25)
```

calculates and plots the first 25 sample autocorrelations for the lynx data in *ts* object **lynx** shown in Figure 1.11. The sample autocorrelations are placed in the vector **aut$acf** (which has 26 elements counting **aut$acf[1]=1**, i.e. the sample autocorrelation at lag 0). If **type= 'covariance'** is selected, then the vector **aut$acf** contains the sample autocovariances, and **aut$acf[1]** contains the sample variance (using the divisor n).

TSWGE DATASETS RELATED TO THIS CHAPTER

---

bat – Bat echolocation signal shown in Figure 3.15(a)
mm.eq – Seismic signal shown in Figure 3.15(b)
linearchirp – Linear chirp signal shown in Figure 3.16(a)
noctula – Echolocation signal for the Nyctalus noctula hunting bat in Figure 3.16(b)

PROBLEMS

3.1 The following data are annual sales of a hypothetical company in millions of dollars:

Period	Sales
1	76
2	70
3	66
4	60
5	70
6	72
7	76
8	80

Compute by hand (i.e., calculator) the estimates $\hat{\gamma}_0$, $\hat{\gamma}_1$, $\hat{\rho}_0$, and $\hat{\rho}_1$.

3.2 Figure 1.16(b) is a plot of the *tswge* dataset **wtcrude2020** which contains the monthly West Texas intermediate crude oil prices from January 1990 through December 2020. Figure 1.3 is a plot of the monthly average temperatures in the Dallas-Ft. Worth area from January 1900 through December 2020. These data are stored in *tswge* file **dfw.mon**. For each of these datasets, plot the realization and the sample autocorrelations. Explain how these plots describe (or fail to describe) the behavior in the data.

3.3 Dice-Rolling

    (A) The following R code can be used to repeat the random process of rolling two dice, finding the sum, and plotting the resulting data. This generates *another* set of 30 "sum-of-two-rolls" data and plots the new data in the format of Figure 3.5(c).

```
set.seed(8327)
x=rep(0,30)
for(i in 1:30) {
roll=sample(1:6,2)
x[i]=roll[1]+roll[2]}
plotts.wge(x)
abline(h=7)
```

    Note: The command **roll=sample(1:6,2)** instructs R to randomly choose two numbers (**roll[1]** and **roll[2]**) between 1 and 6, inclusive. The following line adds the two rolls.

    (a) Using the above commands, plot the resulting 30 new "sum of rolls".

    (b) Do these realizations have the same random appearance as Figure 3.5? Explain.

    (c) In the plot in (a) (using seed 8327), there are 11 instances in which the sum of the rolls is greater than the true mean of 7. In these situations, the next sum was

        (i)   above 7 _____ times

        (ii) below 7 ____ times

        (iii) equal to 7 ____ times

    (B) Change the seed from 8327 to another positive integer and repeat the process. Answer (a) and (b) for the data using the seed you selected. How many instances were there in which the sum of the rolls is greater than 7? In each of these situations answer (i)–(iii) above.

    (C) What do your results in (A) and (B) tell you about the ability to predict the sum at time $t+1$ if you know the sum at time t?

    (D) Use **plotts.sample.wge** to plot the datasets in (A) and (B) along with their sample autocorrelations. What do the autocorrelations say about the relationship among outcomes in the "sum of the rolls" data? Note: The Parzen spectral density estimate will be discussed in Chapter 4.

3.4 Following are displayed two sets of figures, each containing four plots. The first set shows four realizations of length n = 100 each generated from a time series model. The sample mean is also plotted for each realization. The second set contains four autocorrelation functions based on the models used to generate the four realizations (in random order). Match each realization with its corresponding autocorrelation plot. Explain your answers.

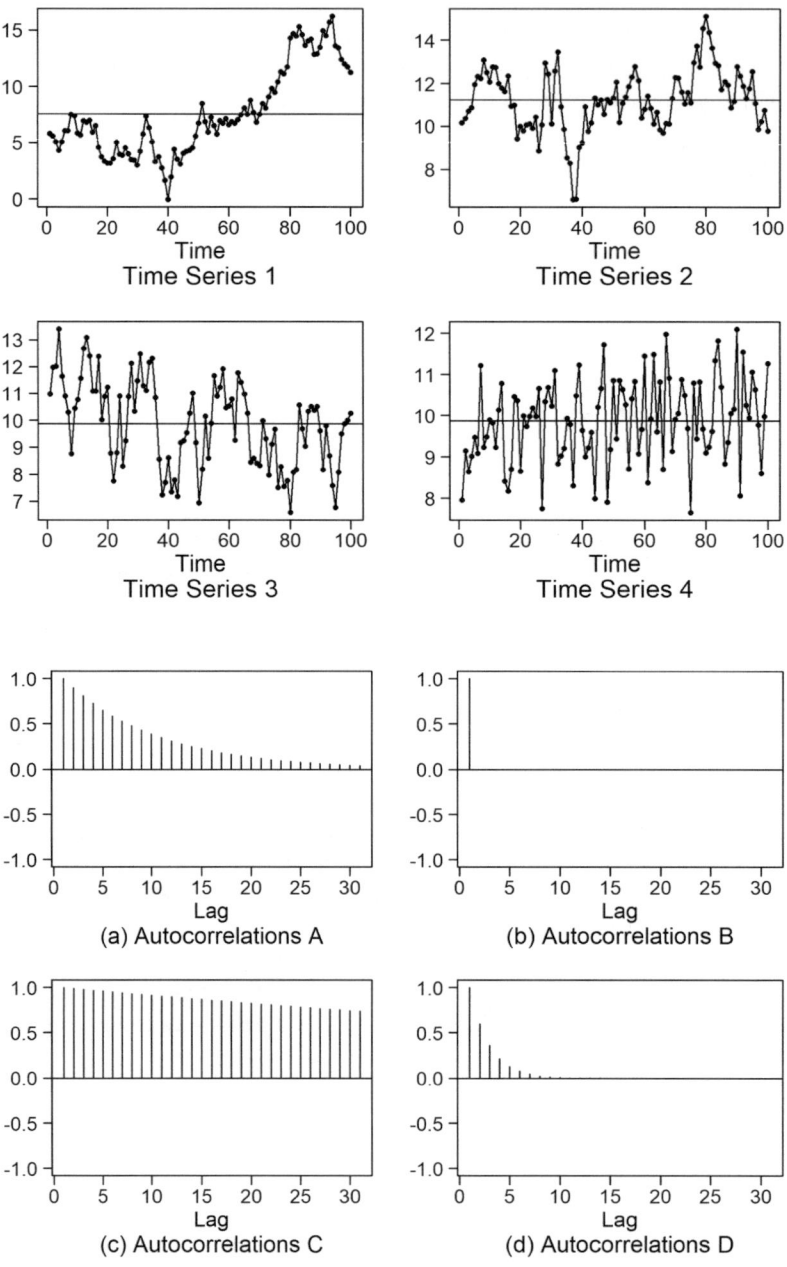

3.5 Following are displayed two sets of figures, each containing four plots. The first set shows four realizations of length n =50 each generated from a time series model. The second set contains four autocorrelation functions based on the models used to generate the four realizations (in random order). Match each realization with its corresponding autocorrelation plot. Explain your answers.

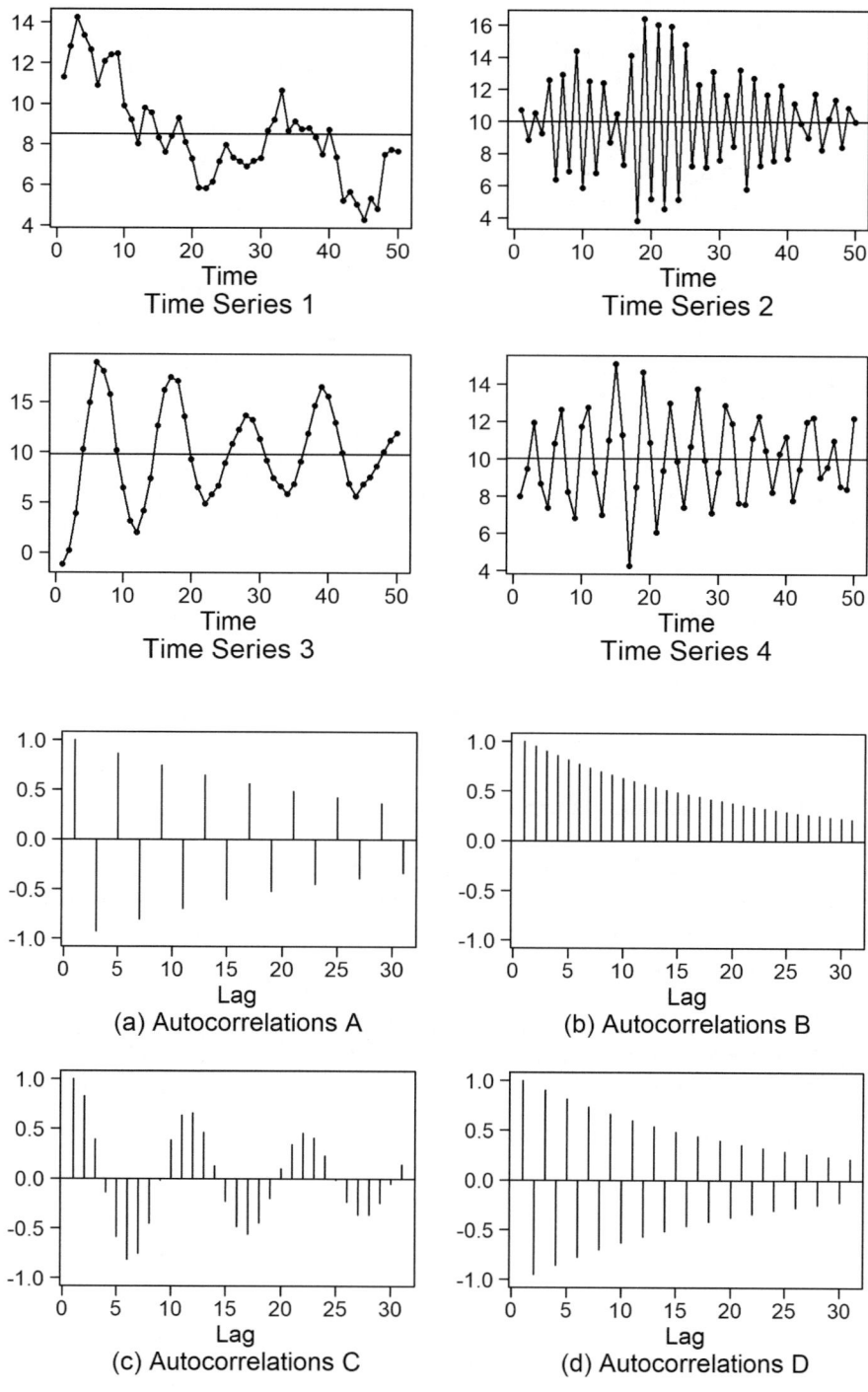

3.6 The following plots are realizations of length 100. One of these realizations is white noise and the other three are not. Which is the white noise realization?

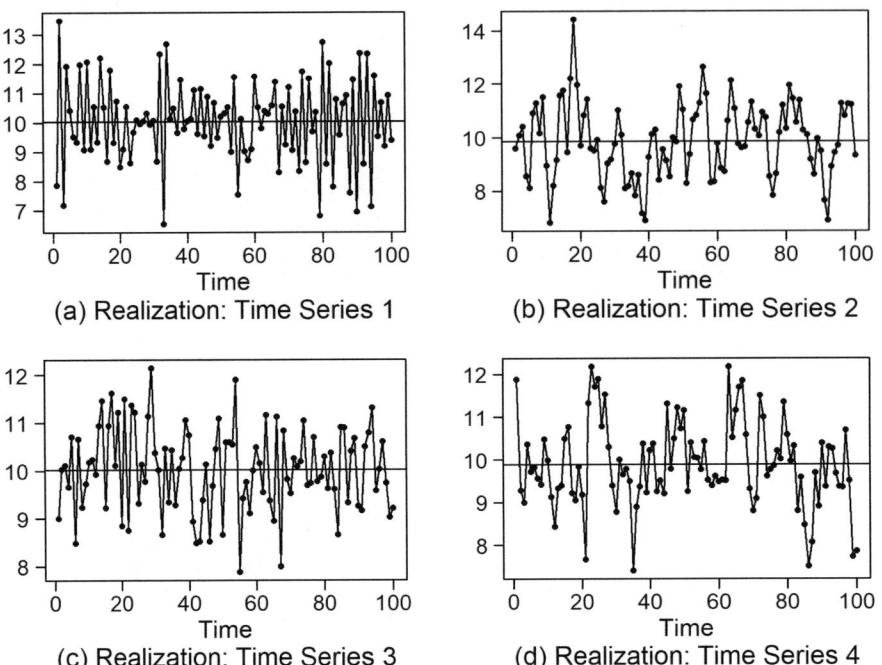

(a) Realization: Time Series 1

(b) Realization: Time Series 2

(c) Realization: Time Series 3

(d) Realization: Time Series 4

3.7 Find a time series dataset (on the internet, etc.) not discussed in the book or in class. For this time series:
   (a) Plot the time series realization and sample autocorrelations.
   (b) Describe the behavior (wandering, periodic, etc.) in the realization and how that behavior manifests itself in the sample autocorrelations.

   Hint: Use the *tswge* function `plotts.sample.wge`.

3.8 Show that the three conditions of stationarity are satisfied by a white noise process.

3.9 Recall that Figure 3.3(d) is a plot in which there is no discernible linear relationship between X and Y, and "knowing x" does not help in predicting y. Discuss what this means in terms of $(x_i - \overline{x})$, $(y_i - \overline{y})$, $(x_i - \overline{x})(y_i - \overline{y})$, $\sum_{i=1}^{n}(x_i - \overline{x})(y_i - \overline{y})$, and thus r_{XY}.

3.10 Figure 3.3(f) is a plot in which there is a relationship between X and Y, but it is not a linear relationship. However, in this case, "knowing x" does help in predicting y, although the correlation coefficient is very close to $r_{XY} = 0$. Explain the reason for the near zero correlation by discussing the terms $(x_i - \overline{x})$, $(y_i - \overline{y})$, $(x_i - \overline{x})(y_i - \overline{y})$, and $\sum_{i=1}^{n}(x_i - \overline{x})(y_i - \overline{y})$.

The Frequency Domain

4

In Chapter 1 we defined cyclic behavior to indicate that a time series realization rises and falls in somewhat of a repetitive fashion. We refer to such data as cyclic or pseudo-periodic. An example of cyclic data is the sunspot data (Figure 1.1) which have an intriguing cyclic behavior of about 11 years. We noted that sunspot cycle lengths varied randomly from cycle to cycle. The DFW monthly temperature data (Figure 1.3) are cyclic data with a fixed 12-month cycle length that can be explained by the earth's rotation around the sun. Such fixed-period cyclic data for which the time axis is tied to the calendar year are referred to as seasonal data. The Air Passengers data (Figure 1.5) are another example of seasonal data. In Chapter 1 we also discussed data for which there was no obvious cyclic behavior. The monthly Dow Jones index (Figure 1.8(d)), West Texas Intermediate crude oil prices (Figure 1.8(e)), and DFW annual temperature data (Figure 1.20(b)) are examples of data with trending or wandering behavior but for which there is no detectable cyclic behavior. The Air Passengers data have both seasonal and trending patterns.

The statistical analysis of time series was discussed in Chapter 3 and the concept of stationarity was introduced. We showed that for stationary time series, it "makes sense" to estimate the mean, variance, and autocorrelations using one realization. Plots of time series realizations and sample autocorrelations provide us with information about the time series behavior. Time series realizations are indexed by "time" while sample autocorrelations are indexed by "lag", which is the difference between time points. Both of these indices have *time* as their basis, and it is natural to think of time series in the *time domain*.

In this chapter, we examine time series from the perspective of their cyclic (or frequency) content. This type of study is said to be based on the *frequency domain*. Frequency domain analysis is based on the fact, shown originally by Joseph Fourier in 1807, that a broad class of functions, $f(x)$, can be written as a linear combination of sines and cosines. See the video with the QR code below for an incredible demonstration of the power of these representations.

QR 4.1 Fourier Representation

The spectral density, discussed in Section 4.2, is the main tool we will use for analyzing the frequency content in data. This type of frequency domain analysis is commonplace in science and engineering and is the topic of this chapter. It is our experience that this will likely be the first introduction to the frequency domain analysis for data science students taking their first time series course. We use the frequency domain as a key part of our analyses of time series data throughout the remainder of this book.

DOI: 10.1201/9781003089070-4

4.1 TRIGONOMETRIC REVIEW AND TERMINOLOGY

We are assuming some familiarity with trigonometric functions. Consequently, we first go over some introductory material, much of which will be a review. Figures 4.1(a) and (b) show the familiar sine and cosine waves. We will use the following terminology:

Periodic Function: A function, $g(t)$, is said to be periodic with period (or cycle length) $p > 0$ if p is the smallest value such that $g(t) = g(t + kp)$ for all integers k. Clearly, functions $\sin(t)$ and $\cos(t)$ are periodic with period (cycle length) 2π. Figures 4.1(a) and (b) show two complete cycles (periods) with the two cycles separated by the vertical lines at 2π. Although Fourier analysis is based on sines and cosines, other periodic functions exist, such as the function plotted in Figure 4.1(c), which is a periodic function of period 10.[1]

Key Point: The terms "cycle length" and "period" will be used interchangeably.

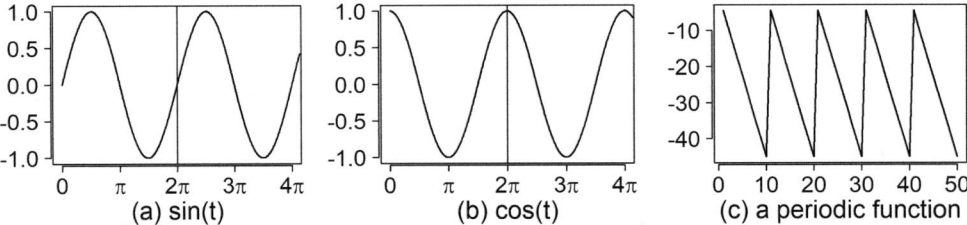

FIGURE 4.1　(a) sin(t) and (b) cos(t), and (c) another periodic function.

Phase Shift: The function $\sin(t+\ell)$ is said to be a sine wave with phase shift ℓ. Such a function still has a cycle length of 2π, but it is "shifted" along the horizontal axis. Figure 4.2(a) overlays the functions $\sin(t)$ and $\sin(t+1)$. At time $t = 0$, $\sin(t+1)$ takes on the value of $\sin(t)$ at $t = 1$. Figure 4.2(b) overlays $\sin(t)$ and $\sin(t + \pi)$. Visually, the two curves are mirror images of each other across a horizontal line at zero.

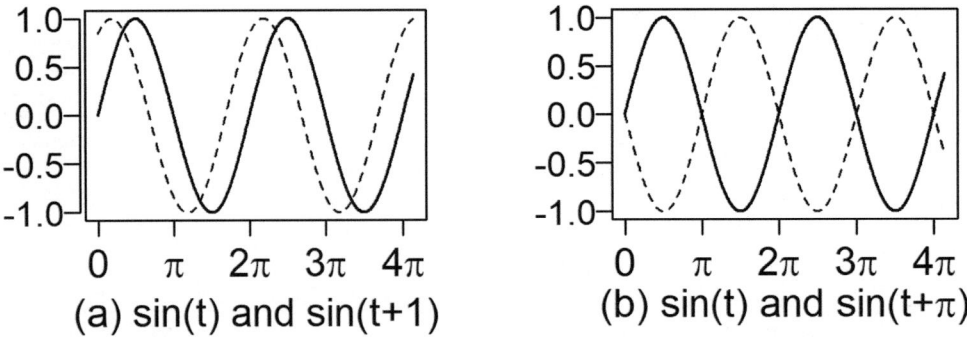

FIGURE 4.2　(a) Overlay showing sin(t) from Figure 4.1(a) (solid) and the phase-shifted sin (t+1) (dashed) lines; (b) Overlay showing sin(t) from Figure 4.1(a) (solid) and the phase-shifted sin (t+π) (dashed) lines.

1　Actual time series data will almost never be "truly periodic". See datasets in the first three chapters. We give this definition as the fundamental behavior which cyclic data approximates.

Frequency (denoted by f): Related to period, another descriptive measure of the behavior of a periodic function is the *frequency*, which can be described in two ways. Frequency is

(a) 1/period (cycle length)
(b) the number of cycles the function goes through in a unit of time

Figure 4.1(a) and Figure 4.3(a) show that $\sin(t)$ has a period of 2π and hence a frequency of $1/(2\pi)$. Figure 4.3(b) indicates that $\sin(2\pi t)$ has period and frequency of one. In Figures 4.3(c) and 4.3(d), we see that $\sin 2\pi(2)t$ has period $1/2$ and frequency two and that $\sin(2\pi(.5)t)$ has period two and frequency $1/2$. Table 4.1 summarizes the information obtained from Figure 4.3.

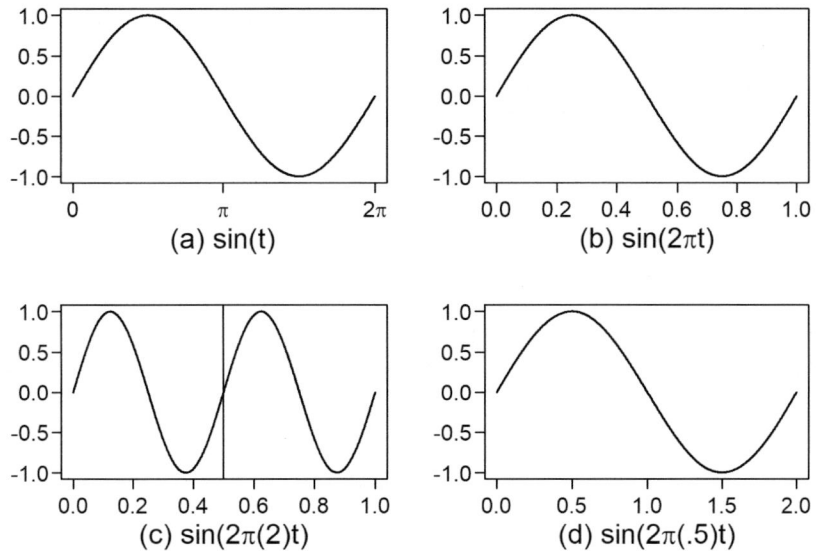

FIGURE 4.3   (a) $\sin(t)$, (b) $\sin(2\pi t)$, (c) $\sin 2\pi(2)t$ and (d) $\sin(2\pi(.5)t)$.

TABLE 4.1   Periods and Frequencies for Selected Sinusoidal Functions

	PERIOD	FREQUENCY (f)
(a) $\sin(t)$	2π	$1/(2\pi)$
(b) $\sin(2\pi t)$	1	1
(c) $\sin(2\pi(2)t)$	.5	2
(d) $\sin(2\pi(.5)t)$	2	.5
In General		
$\sin(2\pi ft)$	$1/f$	f

Key Points

1. The function $\sin(2\pi ft)$ has frequency f. That is, it goes through f cycles per unit.
2. If the data are collected annually, monthly, or hourly then the corresponding frequencies measure the number of cycles per year, month, or hour, respectively.
3. In scientific and engineering literature, it is common to measure frequency in Hertz (Hz), which is the number of cycles per second.

Cyclic/Pseudo-periodic Data: As mentioned in Chapter 1 and in the introduction to this chapter, many time series data that are likely to be encountered will be "sort of" periodic which we refer to as cyclic or pseudo-periodic data. The sunspot and log-lynx datasets are shown in Figure 4.4(a) and (b), respectively. These datasets have previously been discussed and we have noted that they both have fairly repeatable cycles (but do not satisfy the exacting conditions of a periodic function). The average cycle length for the sunspot data is 11 years so that the sunspot data have a frequency of about $1/11 = .09$. The log-lynx data have a similar cycle length and frequency. Figure 4.4(c) is a plot of a simulated realization that goes through about seven cycles in the realization of length $n = 100$. Consequently, the cycle length is about $100/7 = 14.3$ and the associated frequency is about $f = 7/100 = .07$. The following ***tswge*** code plots the data below:

```
data(sunspot2.0)
data(lynx)
llynx=log10(lynx)
x3=gen.arma.wge(n=100,phi=c(1.75,-.95),sn=203)
par(mfrow=c(1,3),mar=c(4,3,.2,.5))
plotts.wge(sunspot2.0)
plotts.wge(llynx)
plotts.wge(x3)
```

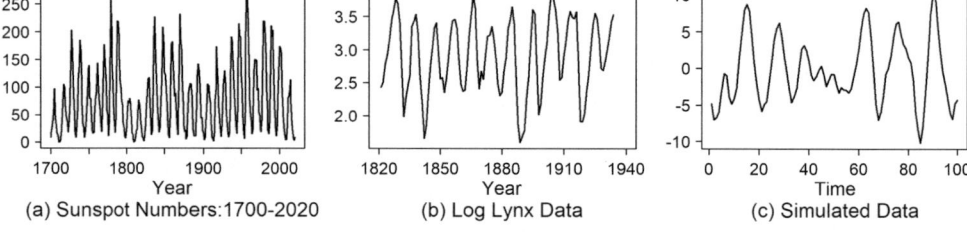

(a) Sunspot Numbers:1700-2020   (b) Log Lynx Data   (c) Simulated Data

FIGURE 4.4   (a) Sunspot data (1800–present), (b) Log Lynx data, and (c) Simulated data.

Non-cyclic (aperiodic) Data: Consider the three datasets plotted in Figure 4.5. As we will see, these figures share the property that there is no discernible cyclic (or even pseudo-cyclic) behavior. Monthly West Texas Intermediate crude oil prices from January 1990 through December 2020 are shown in Figure 4.5(a). As mentioned in Sections 1.2.2.1 and 1.2.2.2, the price of oil has moved around erratically during this time period with no discernible pattern. Figure 4.5(b) shows the Dow Jones monthly closing averages from March 1985 through December 2020. While the DOW seemed to generally trend upward during that period, there were "dips" and pattern changes "along the way". See Sections 1.2.2.1 and 1.2.2.2. Specifically, there is no discernible pseudo-cyclic behavior. Finally, Figure 4.5(c) shows a simulated realization which has a behavior that in places looks pseudo-cyclic (eg. from $t = 75$ through $t = 150$). However, the behavior before or after this time window does not carry on any such pattern.[2]

> **Key Point:** We will often refer to the non-cyclic behavior seen in Figure 4.5 as *aimless wandering*.

---

2 We use the term "aperiodic" loosely. Non-cyclic behavior is correctly described as aperiodic. Technically, $g(t)$ is an aperiodic function if there does not exist a period $p > 0$ such that $g(t) = g(t + kp)$ for all integers k. Therefore, based on this precise definition, the pseudo-periodic functions in Figure 4.4 are aperiodic. However, we will use the term aperiodic to indicate functions such as those in Figure 4.5 that show no cyclic tendencies. We will discuss this further in Section 4.2.

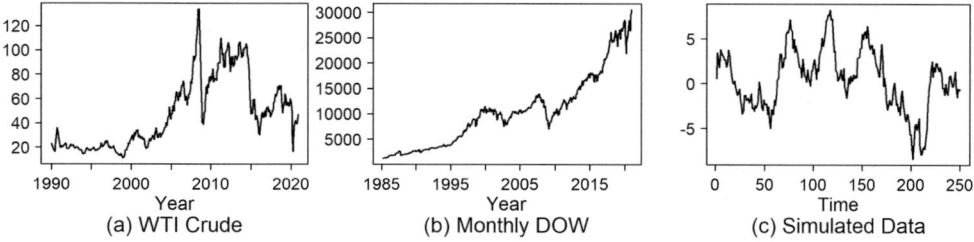

FIGURE 4.5 (a) Monthly West Texas Intermediate Crude price from January 1990 through December 2020, (b) Dow Jones monthly closing averages from March 1985 through December 2020, and (c) simulated data.

4.2 THE SPECTRAL DENSITY

Figure 4.6(a) has the general appearance of time series realizations we will encounter in this text. This signal is the sum of sine waves with different cycle lengths and phase shifts. Close examination of Figure 4.6(a) shows a period of about length 25 (notice that the data seem to go through about four cycles in the realization of length $n = 100$). Shorter term periodic behavior is also visible in the dataset. The question remains, "What frequencies (or cycle lengths) are present in the data?"

Inside Information: Figure 4.6(a) is the sum

$$f(x) = 4\sin\left(2\pi(.04)t\right) + 1.25\sin\left(2\pi(.125)t + 1\right) + .5\sin\left(2\pi(.25)t + 2.5\right).\qquad(4.1)$$

That is, $f(x)$ is the sum of the three components. Note that multiplier weights on the three components are 4, 1.25, and .5, respectively. The components are described in Table 4.2.

TABLE 4.2 Characteristics of Components that Sum to Give Figure 4.6(a)

COMPONENT	FREQUENCY	PERIOD	PHASE SHIFT	MULTIPLIER
1	.040	25	0.0	4.00
2	.125	8	1.0	1.25
3	.250	4	2.5	.50

What is needed is a tool to identify the frequency content of a stationary time series. The good news is that such a tool exists. The *spectrum* (or in standardized form, the *spectral density*) is the main tool in our tool kit for identifying underlying frequency behavior in stationary time series data.

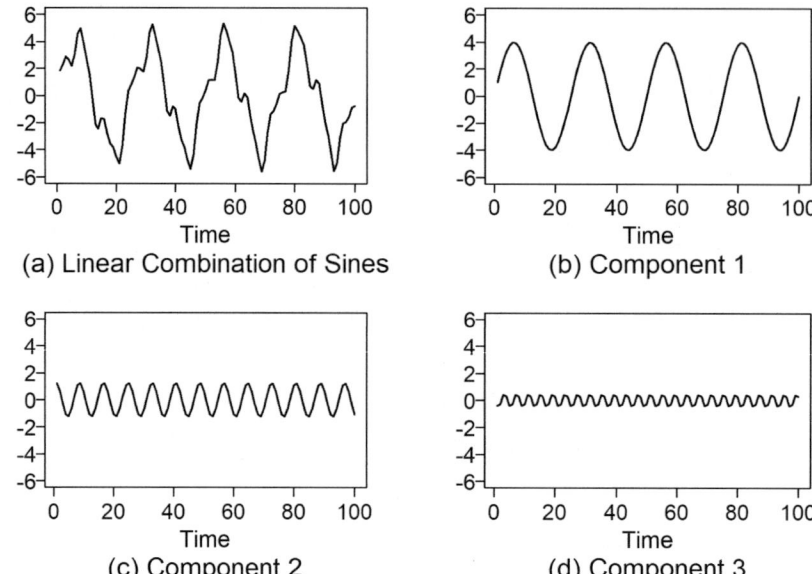

FIGURE 4.6   (a) Sum of the sine waves in (b), (c), and (d).

The graphs in Figure 4.6 can be obtained using the following code:

```
t=1:100
y=rep(0,100)
y1=rep(0,100)
y2=rep(0,100)
y3=rep(0,100)
for(i in 1:100){y1[i]= 4*sin(2*pi*.04*i)}
for(i in 1:100){y2[i]=1.25*sin(2*pi*.125*i+1)}
for(i in 1:100){y3[i]=.5*sin(2*pi*.25*i+2.5)}
y=y1+y2+y3
par(mfrow=c(2,2),mar=c(4,2.5,1.5,.5))
plot(y,type='l',xlab='(a) Linear Combination of Sines',ylim=c(-6,6))
plot(y1,type='l',xlab='(b) Component 1',ylim=c(-6,6))
plot(y2,type='l',xlab='(c) Component 2',ylim=c(-6,6))
plot(y3,type='l',xlab='(d) Component 3',ylim=c(-6,6))
```

4.2.1 Euler's Formula

The spectrum and spectral density of a time series are defined as functions of frequency. Some trigonometric and complex variables formulas of importance are stated here for completeness:

(a) $\sin(-\theta) = -\sin(\theta)$

(b) $\cos(\theta) = \cos(-\theta)$

(c) $e^{i\theta} = \cos(\theta) + i\sin(\theta)$ (Euler's Formula)

(d) $e^{-i\theta} = \cos(\theta) - i\sin(\theta)$

4.2.2 Definition and Properties of the Spectrum and Spectral Density

We are now in position to define the spectrum and spectral density.

Definition 4.1: Let X_t be a stationary time series with autocovariance γ_k and autocorrelation ρ_k. Then for $|f| \le .5$:

(1) The *spectrum* of X_t is defined by

$$P_X(f) = \sum_{k=-\infty}^{\infty} e^{-2\pi i f k} \gamma_k. \tag{4.2}$$

(2) The *spectral density* of X_t is defined by

$$S_X(f) = \sum_{k=-\infty}^{\infty} e^{-2\pi i f k} \rho_k. \tag{4.3}$$

Using Euler's formula, we obtain the "more pleasing" formulas:

$$P_X(f) = \sigma_X^2 + 2\sum_{k=1}^{\infty} \gamma_k \cos(2\pi f k), \tag{4.4}$$

and

$$S_X(f) = 1 + 2\sum_{k=1}^{\infty} \rho_k \cos(2\pi f k). \tag{4.5}$$

These formulas emphasize that the spectrum and spectral density are real-valued functions which is not apparent from (4.2) and (4.3). The condition that $|f| \le .5$ will be addressed in Section 4.2.2.1 when we discuss the Nyquist frequency.

Key Point: In this text, we will focus on the spectral density.

Important Properties of Spectral Densities:

1. $S_X(f) \ge 0$
2. $S_X(f) = S_X(-f)$
3. $S_X(f) = 1 + 2\sum_{k=1}^{\infty} \rho_k \cos(2\pi f k)$, where $|f| \le .5$ (Equation 4.5 above)
4. $\sum_{-.5}^{.5} S_X(f) e^{2\pi i f k} df = \rho_k$

Properties 1 and 2 show that $S_X(f)$ is a non-negative, even function.

Key Point: We plot spectral plots only for $f \geq 0$ because $S_X(f)$ is an even function and the plot for negative values would simply be a mirror image.

Property 3 above says that if the autocorrelation function, ρ_k, $k = 0, 1, 2, \ldots$ is known, then the spectral density can be calculated. On the other hand, Property 4 says that if we know the spectral density, $S_X(f)$, then we can calculate the autocorrelation function.[3]

Key Points

1. Properties 3 and 4 show that ρ_k and $S_X(f)$ contain equivalent mathematical information
    − if you know one, you can calculate the other.
2. ρ_k and $S_X(f)$ are called *Fourier transform pairs*.

4.2.2.1 The Nyquist Frequency

Definition 4.1 has an interesting restriction on the range of f. Specifically, the definitions for the spectrum and spectral density are subject to the constraint that $|f| \leq .5$. For time series data sampled at unit increments, the shortest period that can be observed is a period of 2. (Think about it.) That is, the highest observable frequency is $f = 0.5$, which is called the *Nyquist* frequency. To better understand the situation, consider the two signals $X_t = \sin(2\pi(.2)t)$ and $Y_t = \sin(2\pi(1.2)t)$, which have frequencies 0.2 and 1.2, respectively. These two signals are defined along a continuum, but we consider the case in which they are sampled at the integers. Figure 4.7 shows an overlay of these two signals, and it is interesting to note that the two signals are equal at integer values of t (shown as dots on the graph). For example, letting $t = 2$,

$$\sin\left(2\pi(.2)2\right) = \sin\left(.8\pi\right)$$
$$\sin\left(2\pi(1.2)2\right) = \sin\left(2\pi(1+.2)2\right) = \sin(4\pi + .8\pi) = \sin\left(.8\pi\right)$$

because $\sin\left(\theta + k\pi\right) = \sin\left(\theta\right)$ if k is an even integer. Consequently, the two curves are indistinguishable when sampled at the integers. Consider again the curve $\sin\left(2\pi(1.2)t\right)$ (bold in Figure 4.7). If we sample at the integers, we have no way of knowing that this curve passed through more than a full cycle between, for example, $t = 1$ and $t = 2$. This phenomenon is called *aliasing*, because the frequency $f = 1.2$ "falsely appears" to be the same as $f = 0.2$. For data collected at the integers, it would be impossible to detect a period less than two sample units. Consequently, we must restrict frequencies to $|f| \leq .5$. This leads to the following definition.

Definition 4.2: Nyquist Frequency
For data $x_t, t = 1, 2, \ldots, n$, then the Nyquist frequency is $1/2$, and the shortest observable period is 2.[4]

---

3 In this book, you will not be expected to calculate the integral-based formula in Property 4 or compute the infinite sum in Property 3.

4 If data are sampled at the increments of Δ, then the Nyquist frequency is $1/(2\Delta)$, and the shortest observable period is 2Δ.

Key Point: In order to recover frequencies higher than $1/(2\Delta)$, it is necessary to sample more rapidly.

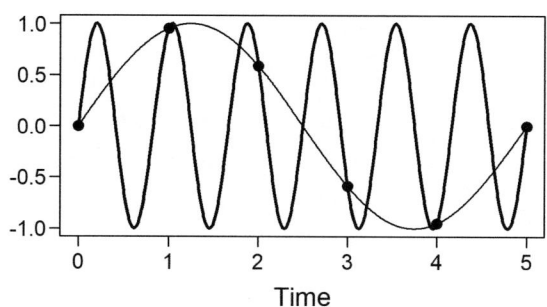

FIGURE 4.7   Plots of $\sin(2\pi(.2)t)$ (thin line) and $\sin(2\pi(1.2)t)$ (bold line).

Figure 4.7 can be obtained using the following code:

```
t=1:1000
tm=t/200
y1=sin(2*pi*.2*tm)
z1=sin(2*pi*1.2*tm)
tp=c(0,1,2,3,4,5)
yp=sin(2*pi*.2*tp)
par(mfrow=c(1,1),mar=c(2.5,2.5,1,1.5))
plot(tm,y1,type='l',xlab='Time',xlim=c(0,5))
points(tm,z1,type='l',lwd=2)
points(tp,yp,pch=19)
```

4.2.2.2 Frequency f=0

Notice that the restriction $|f| \leq .5$ includes the frequency $f = 0$. Before proceeding, we need to discuss the meaning of $f = 0$. Consider the following list of frequencies and associated cycle lengths.

FREQUENCY	CYCLE LENGTH
.1	1/.1=10
.01	100
.001	1,000
.0001	10,000

We see that the closer the frequency gets to $f = 0$, the longer the cycle length. Consequently, $f = 0$ is associated with an "infinite" cycle length. That means there is no periodic content in the data "at all". For example, the realization in Figure 4.8(a) has no periodic content (exhibits "$f = 0$" behavior) and can be characterized as having "aimless wandering" behavior.

If a realization of length $n = 200$ has only aimless wandering, all we can really say is that, if there is a period, it is longer than the realization length, $n = 200$. Figure 4.8(b) shows the realization in Figure 4.8(a) extended to 800 time points. The realization of length $n = 200$ is shown in the shaded region. For this longer realization we see that there is a period of length 250 (marked off by vertical lines on the plot). However, because in Figure 4.8(a) we only observed the first 200 points, we were unable to detect this period. Note that the frequency associated with a period of 250 is $f = 1 / 250 = .004$ which is quite close to $f = 0$.

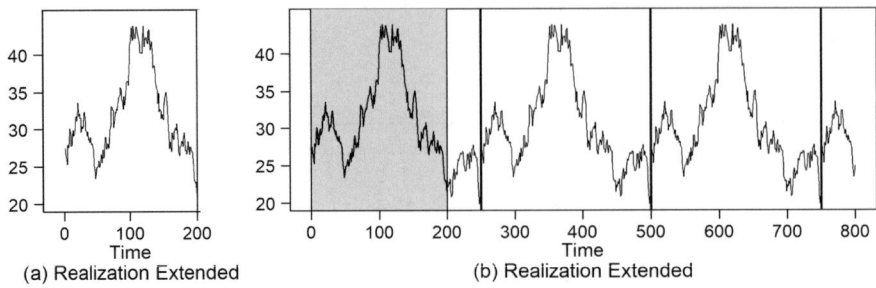

(a) Realization Extended   **(b) Realization Extended**

FIGURE 4.8   (a) Realization showing aimless wandering behavior and (b) Realization extended to 800 points showing periodic behavior.

Key Points

1. Data associated with frequency $f = 0$ never have a repeating behavior and are characterized by "aimless wandering" behavior.
2. Given a realization of length n that shows only "aimless wandering", all we can confidently say is that if there is a repeating period in the data, the period length is longer than n.

Example 4.1 Spectral Density for Data with Two Dominant Frequencies

We have claimed that the spectral density is designed to identify the underlying frequency content of a time series realization. In Figure 4.9(a) we show a realization from an *autoregressive* model (these models will be discussed in Chapter 5). Examination of the realization shows that there is a dominant cyclic behavior and that the data go through about 10 cycles in the realization of length $n = 200$. That is, the period length is about 20 so that the associated frequency is about $f = 1 / 20 = .05$. Further examination of the realization shows that there is a "jagged wiggle" in the data. This behavior is associated with a very short period that corresponds to high frequency (nearly oscillating up and down) behavior. Recall that $f = 1 /(\text{period})$, so that the shorter the period length, the greater (higher) the frequency. Figure 4.9(c) shows the "true" spectral density. For the autoregressive model from which the realization is generated, the spectral density in (4.5) has a simple closed-form solution. This solution, which will be given in (5.30), was used to calculate the spectral density shown in Figure 4.9(c).[5]

Recall that the spectral density is designed to identify frequency content in a time series realization. In Figure 4.9(c) we see that the spectral density has a peak at about $f = .05$ and another peak at about $f = .45$. The peak at $f = .05$ is associated with the previously mentioned period lengths of about 20. The peak at $f = .45$ is associated with cycle lengths of approximately $1 / .45 = 2.2$. These short cycles are associated with the "high frequency" wiggle in the data. The true autocorrelations in Figure 4.9(b) show a dominant slowly damping sinusoidal behavior with period of about 20. There is very little indication of the high frequency behavior, but close observation shows that ρ_1 appears to be slightly smaller than ρ_2, which is not consistent with the smooth sinusoidal behavior associated with the longer cycle.

---

5 Spectral densities are typically plotted in log scale. We will discuss the reason for this in Section 4.2.3.4. Spectral densities could be calculated using natural logs or log base 10. However, in this book, we will follow the practice used widely in the scientific community of plotting spectral densities in decibels (dB), which are the standard units for measuring acoustical energy and are calculated as $10 \log_{10} S_X (f)$.

Key Points

1. The spectral density is designed to identify frequency content in stationary time series data.
2. Frequency content in a realization is represented by "peaks" in the spectral density.

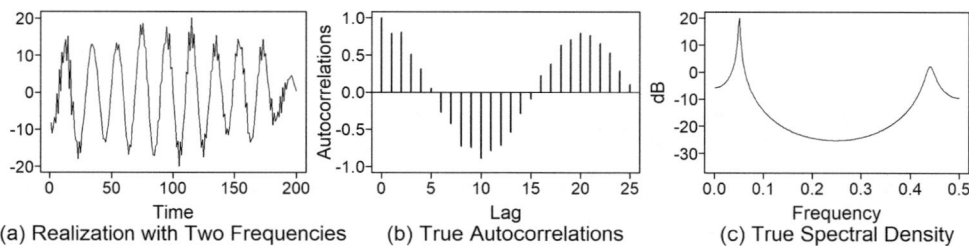

FIGURE 4.9 (a) Realization with two dominant frequencies, (b) true autocorrelations, and (c) the true spectral density for the model from which realization (a) was generated.

The graphs in Figure 4.9 are obtained using the single *tswge* command

```
x=plotts.true.wge(n=200,phi=c(0.1300,1.4414,-0.0326,-0.8865),sn=9310)
```

Example 4.2 Spectral Density of White Noise

An example in which the formula for the spectral density in (4.5) can be easily calculated is the case of white noise. Recall that for white noise, $\rho_0 = 1$ and $\rho_k = 0$ for $k \neq 0$. In this case, $S_X(f) = 1 + 2\sum_{k=1}^{\infty} \rho_k \cos(2\pi f k)$ simplifies to $S_X(f) = 1$ for $|f| \geq .5$. The fact that the spectral density is flat indicates that all frequencies are equally present in white noise. Specifically, there are no peaks in the spectral density. Figure 4.10(a) shows the plot for $f \in [-.5, .5]$ while Figure 4.10(b) shows the plot for only $f \in [0, .5]$ as will be our practice in this book.

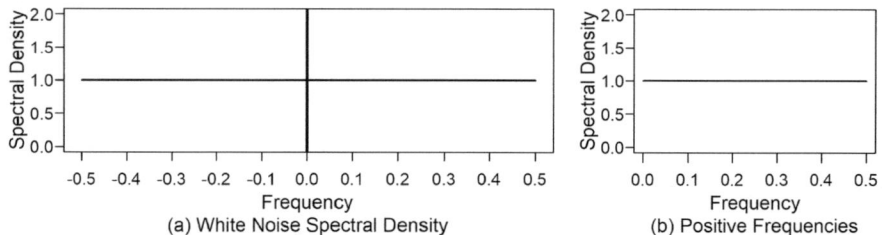

FIGURE 4.10 (a) White noise spectral density and (b) white noise spectral density restricted to positive frequencies.

QR 4.2 Spectral Density for White Noise

4.2.2.3 *The Spectral Density and the Autocorrelation Function*

Recall that the spectral density, $S_X(f)$, and the autocorrelation function, $\rho_k, k = 0, 1, 2, \dots$ contain equivalent mathematical information. That is, if you know one of the functions, you can mathematically compute the other. Despite their "equivalence", we illustrated in Example 4.1 the fact that the spectral density is a better tool for identifying underlying cyclic behavior.

> **Key Point:** The spectral density is far superior to the autocorrelations as a tool for identifying underlying cyclic behavior (even though they contain essentially equivalent mathematical information).

4.2.3 Estimating the Spectral Density

In practice, given a realization of length n from a time series, we do not have sufficient information to calculate the "true" spectral density in (4.5) which involves an infinite sum of autocorrelations. For one thing, given an actual set of time series data, we do not know the "true" autocorrelations. That is, although (4.5) involves an infinite sum of autocorrelations, ρ_k, we only "know" the estimates, $\hat{\rho}_k$, defined in (3.12), and from (3.10) it follows that for a realization of length n, only $\hat{\rho}_k, k = 1, \ldots, n-1$ can be calculated.

> **Key Point:** In practice, given a realization of time series data, we do not know the true autocorrelations necessary to compute $S_X(f)$ in (4.5). However,
>
> - we can estimate ρ_k using the sample autocorrelations, $\hat{\rho}_k$, defined in (3.12).
> - we can only calculate $\hat{\rho}_k, k = 1, \ldots, n-1$.

QR 4.3 Sample
Spectral Density

4.2.3.1 The Sample Spectral Density

Based on the above Key Point, the "natural estimate" of the spectral density is obtained by substituting the sample autocorrelations into the formula for $S_X(f)$. Using this strategy, we obtain

$$\hat{S}_X(f) = 1 + 2\sum_{k=1}^{n-1}\hat{\rho}_k\cos(2\pi fk), \ |f| \le .5. \tag{4.6}$$

The estimate in (4.6) is called the *sample spectral density*. Figure 4.11(a) shows the realization in Figure 4.9(a) with a vertical line at $t = 100$. Figure 4.11(b) shows the sample spectral density calculated using (4.6) based on the first 100 time points in Figure 4.11(a) (that is, those that occur before the vertical line at $n = 100$). The two peaks are visible, but the behavior of the sample spectral density is very erratic (variable). While the true spectral density is a smooth curve, the sample spectral density is not at all smooth. We next calculate the sample spectral density based on the full realization of length $n = 200$ in Figure 4.11(a). The resulting sample spectral density for the longer realization length is plotted in Figure 4.11(c). There is no apparent improvement over the plot in Figure 4.11(b). Most estimators, like the sample mean and sample variance, tend toward the true values as the sample size increases. The behavior of the sample spectral density is unusual and disappointing.

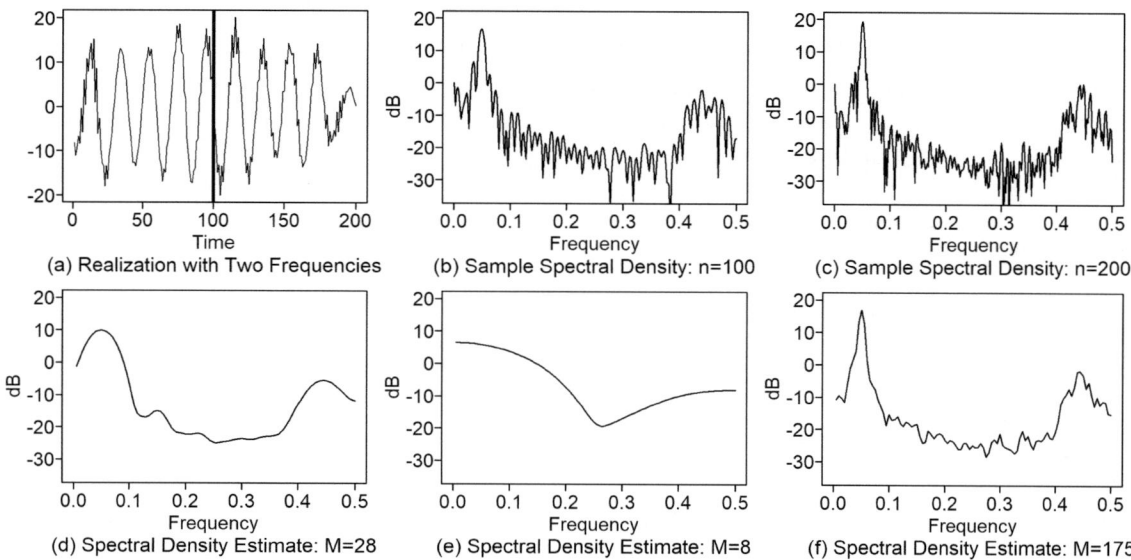

FIGURE 4.11 (a) Realization in Figure 4.9(a), (b) sample spectral density based on the first 100 points (to the left of the vertical line) in (a), (c) sample spectral density for the entire realization in (a); (d), (e), and (f): Parzen spectral density estimates for $M = 28$ (default value), 8 and 175, respectively, for the 200 point realization in (a).

The plots in Figure 4.11 can be obtained using the following R code.

```
x=gen.arma.wge(n=200,phi=c(0.1300,1.4414,-0.0326,-0.8865),sn=9310)
# this generates an AR(4) realization (details in Chapter 5)
par(mfrow=c(2,3))
plotts.wge(x)
abline(v=100)
sample.spec.wge(x[1:100])
sample.spec.wge(x)
parzen.wge(x)
parzen.wge(x,trunc=8)
parzen.wge(x,trunc=175)
```

For a realization, **x**, of length n, the sample spectral density can be obtained using the *tswge* command `sample.spec.wge(x)`.

Key Points

1. The sample spectral density does not "improve" as the realization length increases.
   − that is, it does not converge to the true spectral density as n increases.
2. Another spectral estimate related to the sample spectral density is the periodogram. See Woodward, et al. (2017) and QR4.4 for discussions of the periodogram.

QR 4.4
Periodogram

4.2.3.2 Smoothing the Sample Spectral Density

To understand the problem associated with the non-convergence of the sample spectral density, we recall that the quality of $\hat{\rho}_k$ decreases as k increases. Note that from (3.12), given a realization of length n, the sample autocorrelation at lag k is given by

$$\hat{\rho}_k = \frac{\dfrac{1}{n}\displaystyle\sum_{t=1}^{n-k}\left(x_t - \bar{x}\right)\left(x_{t+k} - \bar{x}\right)}{\hat{\sigma}_X^2}$$

As discussed in Section 3.3.2.3, using this general formula for $\hat{\rho}_k$, it follows that

$$\hat{\rho}_1 = \frac{\dfrac{1}{n}\displaystyle\sum_{t=1}^{n-1}\left(x_t - \bar{x}\right)\left(x_{t+1} - \bar{x}\right)}{\hat{\sigma}_X^2},$$

because there are $n-1$ data pairs in the realization separated by one time period. The $n-1$ pairs are x_1 and x_2, x_2 and $x_3,\dots$, x_{n-1} and x_n. Thus, $\hat{\rho}_1$ is "a sort of average" of $n-1$ cross products $\left(x_t - \bar{x}\right)\left(x_{t-1} - \bar{x}\right)$. On the other hand,

$$\hat{\rho}_{n-1} = \frac{\dfrac{1}{n}\left(x_n - \bar{x}\right)\left(x_1 - \bar{x}\right)}{\hat{\sigma}_X^2},$$

because x_1 and x_n are the only pair of data values separated by $n-1$ time points. Consequently, we do not expect $\hat{\rho}_{n-1}$ to be a very good estimator of ρ_{n-1}. This is analogous to estimating the population mean, μ, of a population based on a sample of size $n = 1$. And notice that the problem with estimating ρ_{n-1} based on one cross-product occurs whether $n = 100$ or $n = 500{,}000$.

This issue with the inability to obtain reliable estimates of ρ_{n-1} (and any ρ_k where k is close to n) gives us a clue regarding how to improve the sample spectral density. Through the years, several "smoothing" methods have been developed that have two characteristics:

(1) The sum in (4.6) is truncated at some value $M < n-1$.
(2) A "smoothing" or "window" function, λ_k, is used to minimize the impact of $\hat{\rho}_k$ as k increases by the fact that, $\lambda_k \to 0$ as $k \to \infty$.

The general formula for a "smoothed" spectral density estimator is

$$\hat{S}_X\left(f\right) = \lambda_0 + 2\sum_{k=1}^{M}\lambda_k\hat{\rho}_k\cos\left(2\pi fk\right), \left|f\right| \le .5. \tag{4.7}$$

Implementation involves a choice for M and a window function. A common choice for M is $2\sqrt{n}$. Although there are many window functions, we have chosen to implement the Parzen window in *tswge*. The Parzen window is defined as

$$\lambda_k = 1 - 6\left(\frac{k}{M}\right)^2 + 6\left(\frac{k}{M}\right)^3, \quad 0 \le k \le M/2$$

$$= 2\left[1 - \left(\frac{k}{M}\right)\right]^3, \quad M/2 < k \le M \tag{4.8}$$

$$= 0, \; k > M,^6$$

and is implemented in *tswge* functions **parzen.wge** and **plotts.sample.wge**. Using **parzen. wge(x)**, the Parzen spectral density estimate is obtained using the default value $M = 2\sqrt{n}$. You can also specify a truncation value, say $M = 8$, by using the command **parzen.wge(x,trunc=8)**.

Example 4.1 (revisited) Figure 4.11(d) shows the Parzen spectral density estimate for the dataset considered in Example 4.1 (shown originally in Figure 4.9(a)) using the default value $M = 2\sqrt{200} = 28$. In this plot the two peaks at $f = .05$ and $f = .45$ are visible, although not as sharp as in the true spectral density in Figure 4.9(c). Recall that Figure 4.11(c) is the sample spectral density for the realization of length $n = 200$ shown in Figure 4.9(a). The Parzen window has "smoothed" the sample spectral density dramatically. Figures 4.11(e) and (f) show Parzen spectral density estimates for $M = 8$ and $M = 175$, respectively. The spectral estimate for $M = 8$ uses only the first eight sample autocorrelations, and consequently loses much of the detail in the spectral density. In this case, there is a peak in the spectral density at $f = 0$, with a "plateau" from about $f = .4$ to $f = .5$. Clearly, $M = 8$ is not an optimal choice for M. The Parzen spectral density estimate for $M = 175$ shown in Figure 4.11(f) has sharp peaks but with much of the "jagged" appearance of the sample spectral density. The use of $M = 2\sqrt{n}$ is simply a guide. For this particular dataset, $M = 50$ seems to produce a smooth spectral estimate with sharper peaks than for $M = 28$. Try it!

> **Key Point:** The windowed spectral estimators (properly used) can do an excellent job of smoothing the sample spectral density and providing information about frequency content in the data.

4.2.3.3 Parzen Spectral Density Estimate vs Sample Autocorrelations

Figure 4.12(a) is a plot of the "two-frequency" realization considered in Example 4.1, while Figures 4.12(b) and (c) are plots of the associated sample autocorrelations and Parzen spectral estimator (with default truncation point $M = 28$), respectively. We noted that the model-based spectral density is a better tool than the autocorrelations for identifying underlying frequency content. Figure 4.12 shows that similarly, the Parzen spectral density estimate does a better job of detecting frequency content than do the sample autocorrelations. The two frequencies show up clearly in Figure 4.12(c), while there is absolutely no indication of the high frequency behavior in the sample autocorrelations! The plots in Figure 4.12 can be obtained using the following R code:

```
x=gen.arma.wge(n=200,phi=c(0.1300,1.4414,-0.0326,-0.8865),sn=9310)
# this generates an AR(4) realization (details in Chapter 5)
plotts.sample.wge(x)
```

---

6 The Parzen window has the property that the weights, λ_k, damp to zero as k gets large. Parzen's window uses a cubic damping function. Other windows include the Bartlett window (triangular damping),Tukey's window (a tapered cosine), and a host of others, mostly named after the researcher. We have chosen to use the Parzen window, although there are a variety of alternatives that might be preferable in certain circumstances. Emmanuel Parzen was a giant in the history of time series analysis and a friend of the first author. For those interested in investigating spectral windows, see Jenkins and Watts (1968) as a good starting point.

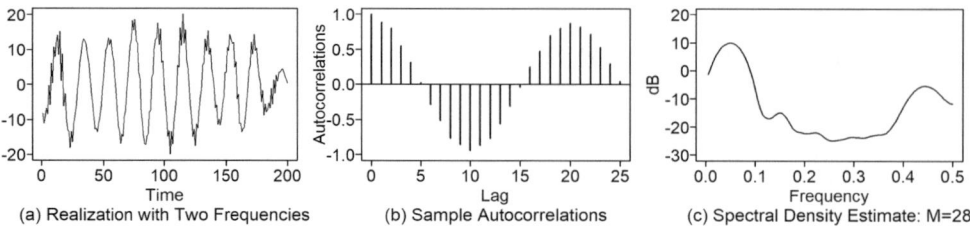

FIGURE 4.12 (a) Realization in Figure 4.9(a), (b) sample autocorrelations, and (c) Parzen spectral density estimate with $M = 28$ calculated from the data in (a).

Example 4.3 Parzen spectral density estimate for the "sum-of-sines data" in Figure 4.6(a).
We return to the "sum-of-sines" dataset in Figure 4.6(a), which is shown again in Figure 4.13(a). As mentioned earlier, this is actually the deterministic curve specified in (4.1) which is a sum of three sinusoidal components with frequencies .04, .125, and .25, respectively. Using the following commands, we obtain the plot in Figure 4.13.

```
y1=rep(0,100)
y2=rep(0,100)
y3=rep(0,100)
y=rep(0,100)
for(t in 1:100){y1[t]= 4*sin(2*pi*.04*t)}
for(t in 1:100){y2[t]=1.25*sin(2*pi*.125*t+1)}
for(t in 1:100){y3[t]=.5*sin(2*pi*.25*t+2.5)}
y=y1+y2+y3
par(mfrow=c(1,3))
plotts.wge(y)
parzen.wge(y,trunc=20)
parzen.wge(y,trunc=30)
```

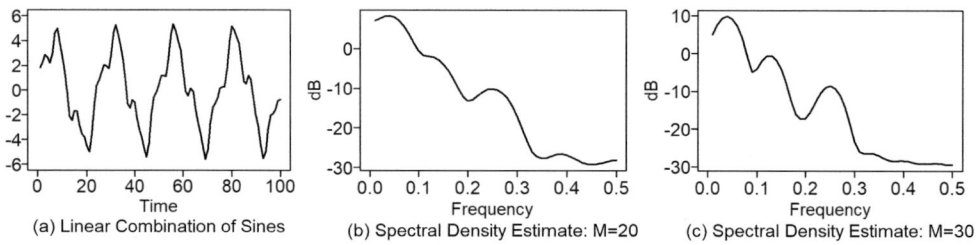

FIGURE 4.13 (a) Sum-of-sines curve previously shown in Figure 4.6(a); (b) and (c) Parzen spectral density estimates for the data in (a) using $M = 20$ (default) and $M = 30$, respectively.

Note that the Parzen spectral density estimator does not "know" the inside information about the frequencies in (4.1). Based on the data, for which it was difficult to visually discern the various frequencies in the data, the Parzen spectral density estimate in Figure 4.13(b) has a weak peak at about $f = .04$, more like an inflection point at about $f = .125$, and a peak at about $f = .25$. For $n = 100$, the "common" choice for M is $2\sqrt{100} = 20$, and this is the truncation point, M, which the command **parzen.wge(y)** uses by default. Trying another value for M, (say M=30) we issued the command:

```
parzen.wge(y,trunc=30)
```

and obtained Figure 4.13(c) which shows the three frequency components more clearly than in Figure 4.13(b). Notice also that the multipliers in (4.1) for frequencies .04, .125, and .25 decreased from 4 to 1.25 to .5, respectively. Correspondingly, the peaks in the spectral density in Figure 4.13(c) decrease in height (and again, the Parzen spectral estimator was not "provided" with that information).

Key Point: Higher peaks tend to indicate more promine nt behavior at the associated frequency.

Parzen spectral density estimate vs sample autocorrelations: Before leaving this example, we show the sample autocorrelations and the Parzen spectral density estimate in Figures 4.14(b) and (c) for the sum-of-sines data, which are plotted again in Figure 4.14(a). We note that the sample autocorrelations in Figure 4.14(b) have the form of a damped sinusoid with a period of about 25 which corresponds to a period of $f = 1/25 = .04$. There is *absolutely no indication in the sample autocorrelations* of the other two frequencies ($f = .125$ and $f = .25$). However, all three frequency components are identified by the spectral estimate in Figure 4.14(c). This re-emphasizes the fact that the spectral density is the proper tool for identifying frequency content in data. Letting **y** denote the vector obtained using the above code for the sum-of-sines data, the plots in Figure 4.14 can be obtained using the ***tswge*** command:

```
plotts.sample.wge(y,trunc=30)
```

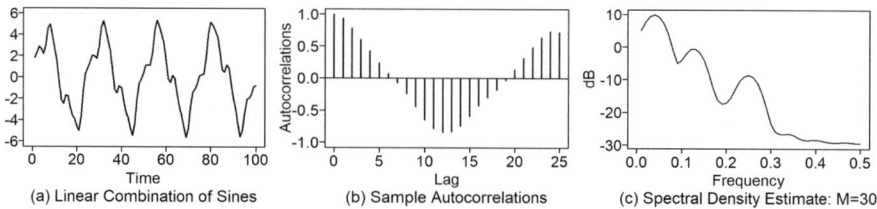

FIGURE 4.14   (a) Sum-of-sines data and corresponding (b) sample autocorrelations and (c) Parzen spectral estimate ($M = 30$).

4.2.3.4 Why We Plot Spectral Densities in Log Scale

Recall that the previous spectral densities have been plotted in log scale (in our case, dB). To understand the reasoning for this, consider Figure 4.15. Figure 4.15(a) is the Parzen spectral density estimate using $M = 30$ for the sum-of-sines data (shown previously in Figure 4.14(a)) and plotted in dB. Figure 4.15(b) shows the Parzen spectral density estimate with $M = 30$ plotted in non-log scale. In this plot, we see that the main peak (at $f = .04$) completely dominates the plot, and the other two peaks are barely visible.

Key Point: Plotting spectral densities in log scale allows secondary peaks to be more visible.

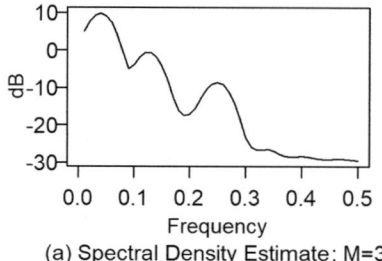

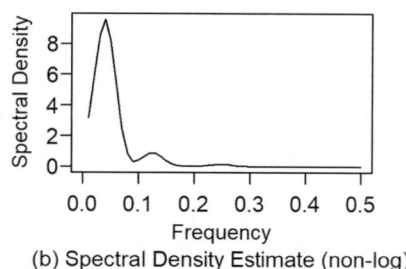

FIGURE 4.15   (a) Parzen spectral density estimate using $M = 30$ for the sum-of-sines data in Figure 4.14(a) plotted in dB, and (b) Parzen spectral density estimate in (a) but plotted in non-log scale.

Example 4.4   Spectral density estimates for real datasets.
In this example, we will find spectral estimates for some datasets of interest.

(1) Datasets Shown in Figure 4.4
Figures 4.4(a) and (b) show plots of the sunspot data and the log-lynx data. These were given as examples of cyclic/pseudo-periodic data with cycle lengths of about 11 and 10, respectively. That is, the data have some cyclic-type behavior, but the data are not technically periodic. In Figure 4.16, we show these datasets along with their sample autocorrelations and Parzen spectral density estimates, using the default truncation point $= 2\sqrt{n}$. As expected, both plots of the Parzen spectral density estimate in Figure 4.16(c) and (f) have peaks at about $f = .10$. The spectral density for the sunspot data also shows a peak at $f = 0$. Examination of the sunspot data shows that there is some longer cyclic or wandering behavior related to the peaks. The log-lynx data have a peak only at $f = .10$, consistent with the observation that there seems to be no other pattern in the data. For both datasets, the sample autocorrelations show a damped sinusoidal behavior with period of about 10. The plots in Figure 4.16(a)–(c) can be obtained using the commands:

```
data(sunspot2.0)
plotts.sample.wge(sunspot2.0)
```

while those in Figures 4.16(d)–(f) are obtained using the code

```
data(llynx)
plotts.sample.wge(llynx)
```

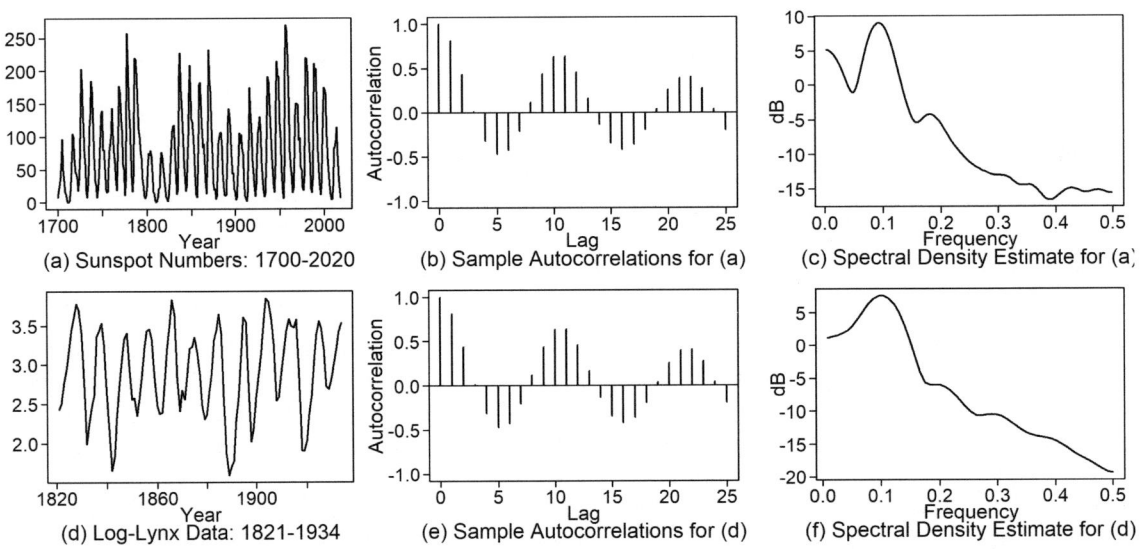

FIGURE 4.16   (a) Sunspot data (1700–present), (b) sample autocorrelations, and (c) Parzen spectral density for (a), (d) log-lynx data, (e) sample autocorrelations and (f) Parzen spectral density estimate for (c).

Key Point: It is worth repeating the fact that frequency content in a realization is represented by "peaks" in the spectral density.

(2) Datasets Shown in Figure 4.5
Figure 4.5 shows datasets for the monthly West Texas intermediate crude oil price from January 1990 through December 2020, and the monthly DOW closing price from March 1985 through December 2020.

These plots were shown to illustrate aperiodic data which have a wandering behavior. The crude oil prices have an unpredictable wandering behavior with upward trending behavior ending in a dramatic peak of about $120/barrel in the summer of 2008 followed by an equally dramatic drop to $34 dollars/barrel toward the end of that year. The prices increased and then remained at about $100/barrel in the years 2012–2014 but as of the writing of this book, have remained low since that time with a catastrophic drop to about $15/barrel in the spring of 2020. The general picture is one of aperiodic and volatile behavior. The DOW data are also aperiodic with a general tendency to increase but with "dips along the way". Figure 4.17 shows these datasets along with the Parzen spectral density estimates using the default value for truncation point, M, in each case. The spectral density for each of the datasets has a distinct peak at $f = 0$, which would be expected. Additionally, the sample autocorrelations have a slowly damping exponential appearance for both datasets. Specifically, there is a strong positive correlation between data values close to each other in time. This is true in the crude oil data despite a few dramatic changes in oil price. A horizontal line is drawn at the mean of the data values in Figures 4.17(a) and (d), where it can be seen that there tend to be long sections of time at which the data values stay above or below the mean.

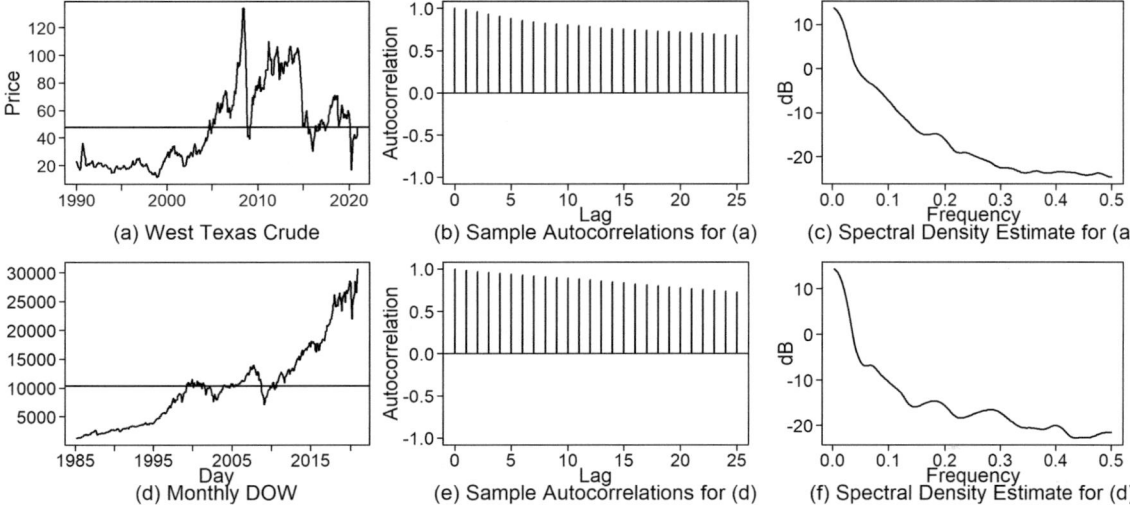

FIGURE 4.17 (a) Monthly West Texas Intermediate Crude price from January 1990 through December 2020 (b) sample autocorrelations, and (c) Parzen spectral density estimate for (a). (d) Dow Jones monthly closing price from March 1985 through December 2020, (e) sample autocorrelations, and (f) Parzen spectral density estimate for (c).

Key Point: A spectral density whose only peak is at $f = 0$, as in Figures 4.17(c) and (f), indicates that the realization shows no cyclic behavior associated with a period of less than n.

(3) A White Noise Dataset

In Figure 4.10(b) we plotted the spectral density for white noise and showed it to be the constant $S_X(f) = 1, 0 \le f \le .5$. We note that if we had plotted this spectral density in dB, then the constant horizontal line would have been at $10 log_{10}(1) = 0$. Figure 4.18(a) shows a white noise realization of length $n = 100$. This realization has the typical random behavior associated with white noise. It is important to note that frequency behavior is difficult to recognize in Figure 4.18(a), but this is not because there is no cyclic at all (as in Figures 4.17(a) and (c)). On the contrary, all frequencies are present in white noise with equal intensities.

Figure 4.18(b) is a plot of the Parzen spectral density estimate in dB (using the default value $M = 20$). At first glance, the spectral density plot is disappointing in that it does not look flat at all. However, notice that the plot is centered around zero and that there seem to be about as many dips as peaks. However, note the vertical scale on Figure 4.18(b). In this plot, dB ranges from about −3.5 to 2. Contrast this with the spectral plots in Figure 4.17(c) and (f), where dB ranges from −20 to 10, which is similar to the dB range for the spectral density plots in Figure 4.16. That is, the spectral density in Figure 4.18(b) does not vary much from zero. In Figure 4.18(c), we show this spectral density plot with dB ranging from −10 to 10. Given this rescaling of the spectral density, it is seen that the Parzen spectral density estimate is fairly flat and is centered around the horizontal line at dB= 0.

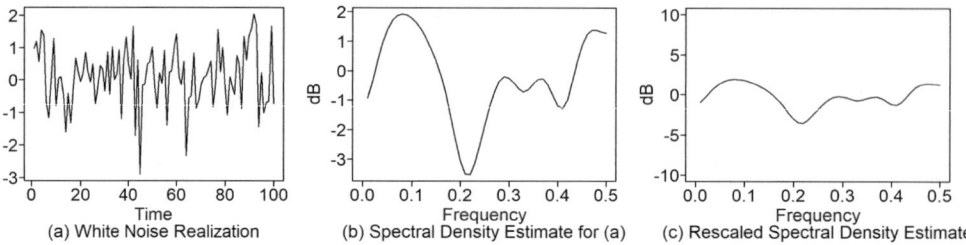

(a) White Noise Realization   (b) Spectral Density Estimate for (a)   (c) Rescaled Spectral Density Estimate

FIGURE 4.18 (a) A white noise realization, (b) the Parzen spectral density estimate for the data in (a), and (c) Figure 4.18(b) with expanded vertical scale.

Key Point: All frequencies are present in white noise with equal intensities.

4.3 SMOOTHING AND FILTERING

Filters are used in many fields of science and engineering to "filter out" certain types of frequencies from a dataset. The centered moving average smoother discussed in Chapter 2 is a type of filter. In Chapter 2, we discussed smoothing time series data to remove "noisy" parts of a realization and retain (or pass through) the true "signal". The main smoother discussed in that section is the *centered moving average smoother*. Equations 2.1 and 2.2 give the general formulas for centered moving average smoothers of odd and even orders, respectively.

4.3.1 Types of Filters

Figure 4.19(a) shows the monthly West Texas intermediate crude oil price shown in Figure 4.17(a), among other places. Figure 4.19(b) shows the result of a 9th-order moving average smoother applied to the crude oil data. There we see that the data have been smoothed (that is, the high frequency "wiggle" has been removed) while the major movements of oil prices described in the discussion of Figure 4.17(a) are "passed through". Because the moving average smoother passed through the low-frequency part of the data and removed the high-frequency part of the data, we say that the moving average smoother is an example of a *low-pass filter*.

In Figure 4.19(c) we show the differenced crude oil data. That is, we calculate the differences $x_t - x_{t-1}$, where x_t is the crude oil price at month t. Note that $x_1 - x_0$ cannot be computed, so we are only able to calculate the differences $x_t - x_{t-1}$ for $t = 2,\ldots,n$. Taking a slight liberty with the time index notation, we define the difference $x_t - x_{t-1}$ to be d_{t-1}, so that we have the differenced data $d_t, t = 1,2,\ldots,n-1$. The differenced data are interesting and appear to have done the exact opposite of smoothing. That is, differencing the data retained (or passed through) the "high-frequency wiggle" and tended to remove the low-frequency wandering behavior. Because of this fact, we say that differencing the data is a form of *high-pass filtering*.

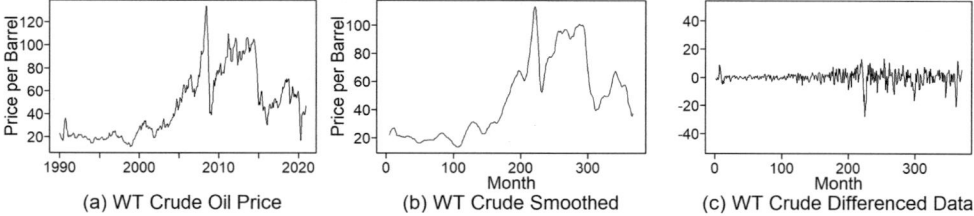

FIGURE 4.19   (a) Monthly West Texas Intermediate Crude oil price from January 1990 through December 2020, (b) Smoothed version of (a) using a centered moving average smoother of order 9, and (c) Differenced crude oil price data.

```
# Note: This code is for WT Crude from Jan 1990 through December 2020
data(wtcrude2020)
n=length(wtcrude2020)
parz.wt=parzen.wge(wtcrude2020,plot=FALSE)
s9=ma.smooth.wge(wtcrude2020,order=9,plot=FALSE)
parz.s9=parzen.wge(s9$smooth[5:114],plot=FALSE)
d1.wt=rep(NA,n)
for(t in 2:n) d1.wt[t-1]=wtcrude2020[t]-wtcrude2020[t-1]
parz.d1=parzen.wge(d1.wt[1:n-1],plot=FALSE)
par(mfrow=c(1,3),mar=c(4.5,3,2,.5))
plot(wtcrude2020,type='l',xlab='(a) WT Crude Oil Price')
plot(s9$smooth,type='l',xlab='(b) WT Crude Smoothed')
plot(d1.wt,type='l',xlab='(c) WR Crude Differenced data',ylim=c(-50,50))
```

Key Points

1. *Low-pass filters* pass through lower frequencies and remove higher frequencies
   - low-pass filtered data have a smoother appearance than the original data
   - a centered moving average smoother is an example of a low-pass filter.
2. *High-pass filters* pass through higher frequencies and remove lower frequencies
   - high-pass filtered data have a noisy/high-frequency appearance
   - a difference is an example of a high-pass filter
   - high-pass filters are often used to remove low-frequency noise, eliminate humming sounds in audio signals, and so forth
3. *Band-pass filters* pass through behavior in a certain band (or range) of frequencies.
4. *Band-stop filters* remove behavior in a particular band of frequencies and retain the frequencies outside this band.

The two filters we have discussed to this point are the centered moving average smoother which is a *low-pass filter*, and the first difference which is a high pass filter.

Example 4.5 Smoothing and differencing the two-frequency data in Example 4.1

In this example, we consider filtering the data in Figure 4.12, which consist of distinct low-frequency behavior at $f = .05$ and high-frequency behavior at $f = .45$. Figures 4.20(a) and (b) are plots of the data and Parzen spectral density estimate, which have been shown previously but are included here for purposes of comparison. In Example 4.1 we noted the behavior associated with the ten cycles, each of approximate length 20. This is seen as a peak in the Parzen spectral density estimate at about $f = .05$. Similarly, the high-frequency behavior is evidenced by the "noisy wiggle" in the data and a peak in the Parzen spectral density estimate at about $f = .45$.

Figures 4.20(c) and (d) show the smoothed data after applying a 6th-order centered moving average smoother. The effect of the smoothing (low-pass filter) is to remove the "wiggle" from the realization and essentially eliminate the peak in the Parzen spectral density estimate at $f = .45$.

Figure 4.20(e) shows the differenced data $(x_t - x_{t-1})$. In this figure, the high-frequency behavior is accentuated but there is still some indication of the ten cycles that were present in the original data. The Parzen spectral density estimate in Figure 4.20(f) shows that the peak at $f = .45$ is higher than the peak at $f = .05$ (indicating that the peak at $f = .45$ is more dominant), but there is still a distinct peak at $f = .05$.

Plots in Figure 4.20 can be obtained using the following commands:

```
x=gen.arma.wge(n=200,phi=c(0.1300,1.4414,-0.0326,-0.8865),sn=9310)
n=length(x)
par(mfrow=c(3,2),mar=c(3,2.5,1,1.5))
plot(x,type='l',ylim=c(-20,20))
parzen.wge(x)
s6=ma.smooth.wge(x,order=6,plot=FALSE)
plot(s6$smooth,type='l',ylim=c(-20,20))
parzen.wge(s6$smooth[4:197])
for(t in 2:n) d1[t-1]=x[t]-x[t-1]
plot(d1,type='l',ylim=c(-20,20))
parzen.wge(d1[1:n-1])
```

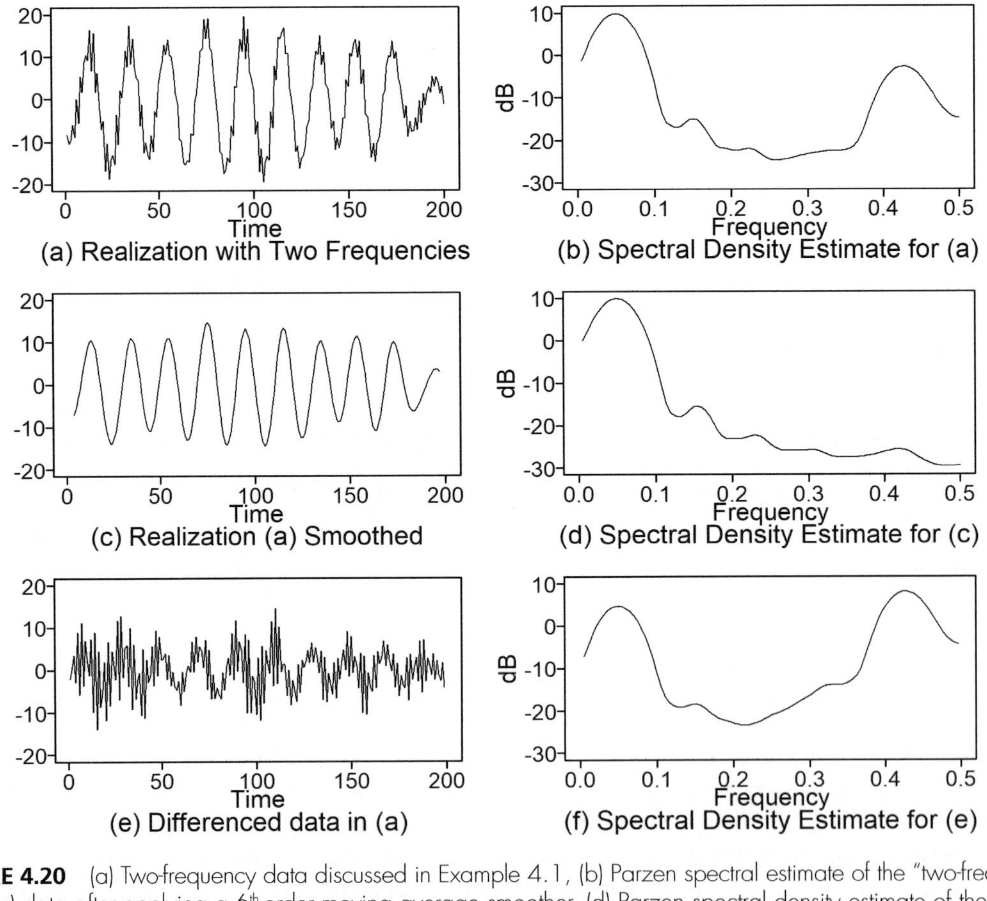

FIGURE 4.20 (a) Two-frequency data discussed in Example 4.1, (b) Parzen spectral estimate of the "two-frequency" data, (c) data after applying a 6th-order moving average smoother, (d) Parzen spectral density estimate of the data in (c), (e) data in (a) differenced, and (f) Parzen spectral density estimate of the data in (e).

> **Key Points:** Regarding the data in Figure 4.20(a):
> 1. The 6th-order moving average smoother did a good job of removing the high-frequency behavior and retaining a "smoother" version of the low-frequency behavior.
> 2. The difference acted as a high-pass filter, but it did not sufficiently remove the low-frequency behavior that was dominant in the realization.

4.3.2 The Butterworth Filter

In Example 4.5, we noted that while differencing is a method of high-pass filtering, it is not a very good one. That is, while it did accentuate the high-frequency behavior in the data in Example 4.5, it did not really remove the low-frequency signal. We say that the $f = .05$ behavior "leaked" through the difference filter. What is needed is a method of filtering that does a better job of retaining a certain range of frequency behavior while at the same time removing parts of the signal related to unwanted frequencies. One of the more common filters used by scientists and engineers for this purpose is the Butterworth filter (see Butterworth, 1930; Porat, 1997).[7]

The low-pass Butterworth filter is designed to remove frequencies above a specified "cutoff" frequency, and a high-pass Butterworth filter works in an analogous manner. While the Butterworth filter can implement a variety of filtering methods, the *tswge* function **butterworth.wge** is designed to produce a low-pass or high-pass Butterworth filter on an R vector. The **butterworth.wge** function has the general form

butterworth.wge(x,order,type,cutoff,plot)

where type is either **'l'** for low-pass or **'h'** for high-pass. The parameter **cutoff** specifies the cutoff frequency, and **order** specifies the order of the Butterworth filter. That is, if we apply a Butterworth filter to remove frequencies above $f = .2$, we could apply a low-pass Butterworth filter with **cutoff=.25**. We have found that a good choice is **order=4**, which is the default. The function also provides the option to plot the original and filtered data, and **plot=TRUE** is the default.

Example 4.5 (revisited) Using the Butterworth filter on the two-frequency data
In this example, we apply low-pass and high-pass Butterworth filters in an attempt to recover a smooth pseudo-sinusoidal signal via a low-pass Butterworth filter along with a signal representing the high-frequency component of the data without the low frequency sinusoidal component. The plots in Figure 4.21 were obtained using the following code:

```
x=gen.arma.wge(n=200,phi=c(0.1300,1.4414,-0.0326,-0.8865),sn=9310)
par(mfrow=c(3,2),mar=c(3,2.5,1,1.5))
plot(x,type='l',ylim=c(-20,20))
parzen.wge(x)
low=butterworth.wge(x,type='l',cutoff=.25,plot=FALSE)
plot(low$x.filt,type='l',ylim=c(-20,20))
parzen.wge(low$x.filt)
high=butterworth.wge(x,type='h',cutoff=.25,plot=FALSE)
plot(high$x.filt,type='l',ylim=c(-20,20))
parzen.wge(high$x.filt)
```

QR 4.5
Butterworth
Filter –
Example 4.5

---

7 Letting x_t and y_t denote the original and filtered data, respectively, the Butterworth filter is a recursive filter defined by $y_t = \sum_{k=0}^{N} \alpha_k x_{t-k} - \sum_{k=0}^{N} \beta_k y_{t-k}$, where the α_ks and β_ks are selected by the function to provide the appropriate filter and N is the order of the filter.

Figure 4.21 shows that the Butterworth filter did precisely what it was intended to do. That is, the low-pass filter smoothed the data and eliminated the peak in the spectral density at $f = .45$ while retaining the peak at $f = .05$. Similarly, the high-pass filter eliminated the low-frequency pseudo-sinusoidal behavior and retained the high-frequency wiggle. The peak in the spectral density at $f = .05$ is removed, while the one at $f = .45$ is retained. In particular, the high-pass filter performed much better than the difference filter.

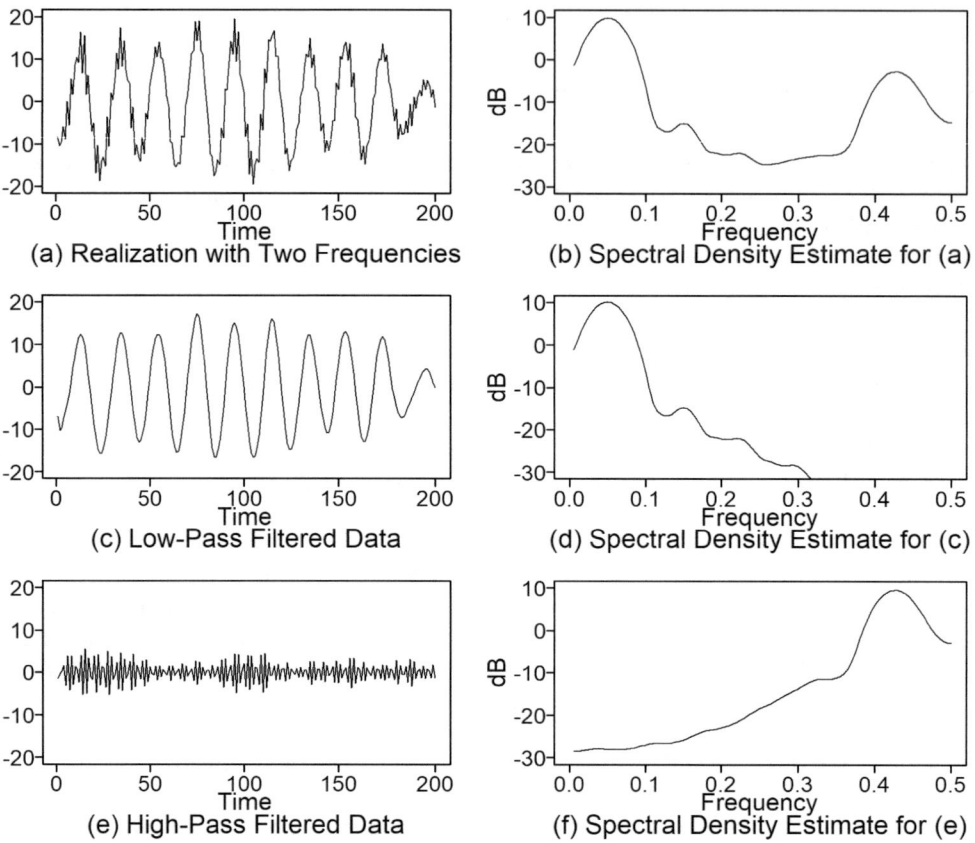

FIGURE 4.21 (a) Two-frequency data discussed in Example 4.1, (b) Parzen spectral estimate of the "two-frequency" data, (c) data in (a) after applying low-pass Butterworth filter, (d) Parzen spectral density estimate of the data in (c), (e) data in (a) after applying high-pass Butterworth filter, and (f) Parzen spectral density estimate of the data in (e).

Key Points

1. The Butterworth filter can be "fine-tuned" to retain and remove certain frequencies.
2. Through use of the cutoff frequency, the Butterworth filter can be designed to remove (or retain) specified frequency ranges.
   - the centered moving average smoother and the difference do not have this type of "frequency control".
3. The Butterworth filter has "end-effects" issues. The *tswge* function **butterworth.wge** does not control for these, so data values toward the ends of the filtered data should not be viewed as reliable.

- The larger the filter order, the more values are subject to end-effects problems.
- This is like the centered moving average filter, which cannot be applied to ends of the data.
- The number of undefined values increases as the order increases.
4. The Butterworth filter (either low-pass or high-pass) can be best implemented by understanding the frequency behavior in the original data.

4.4 CONCLUDING REMARKS

In this chapter we have provided tools for understanding the frequency content of data. We formalized the concepts of frequency and period, and we discussed cyclic, pseudo-periodic, and aperiodic data. Our treatment is atypical in introductory applied time series texts in that we discuss the frequency domain early and then use the associated tools throughout the remainder of the book to better understand models, forecasts, and so forth.

We discussed the information available in the spectral density, and we presented the Parzen spectral density estimator as a tool for estimating the spectral density given a set of time series data. As we discuss time series models in the following chapters, we will often use information provided by the spectral density to understand and formulate models.

Smoothing is commonly used in business, economic, and scientific/engineering applications to isolate key features of a dataset. For example, centered moving average smoothing was discussed in Chapter 2 as a technique for removing noisy components of data to better observe the dominant behavior. In this chapter we discussed filtering as a generalization of smoothing. We introduced the Butterworth filter, which also provides the ability to smooth data (that is, it is a low-pass filter), but in this case, smoothing is achieved by controlling which "high-frequency" behavior is to be removed. Similarly, the Butterworth high-pass filter was introduced as an effective tool that could be used for emphasizing subtle high-frequency behavior that is obscured by the dominating low-frequency features of the data.

APPENDIX 4A

TSWGE FUNCTIONS

(a) `parzen.wge(x, dbcalc, trunc, plot)` calculates and alternatively plots the spectral density estimate based on the Parzen window.

x is a vector containing the time series realization $x_t, t = 1,\ldots,n$
`dbcalc` is a logical variable. `dbcalc = "TRUE"` (default) calculates the smoothed spectral estimator on the log scale (in dB)
`trunc` (integer ≥ 0) specifies the truncation point.
`trunc = 0` (default) (or no value for `trunc` is specified) indicates that the default truncation point is $M = 2\sqrt{n}$
`trunc > 0` is a user-specified truncation point, i.e. calculations will be based on $M = trunc$

> **plot** is a logical variable. **plot = "TRUE"** (default) produces a plot of the smoothed spectral density estimate (in log or non-log scale depending on **dbcalc**).

Example:

```
data(sunspot2.0); ssparz=parzen.wge(ss2.0)
```

calculates and plots the smoothed spectral estimator (in dB) for the sunspot data (**ss2.0**) using the default truncation point, $M = 2\sqrt{319} = 36$. The vectors **ssparz$freq** and **ssparz$pgram** contain the frequencies and associated Parzen-based spectral estimates.

(b) **butterworth.wge(x,order,type,cutoff)** is a *tswge* R function that requires functions in the CRAN package *signal* which should be installed on your machine before calling **butterworth.wge**.

x is a vector containing the time series realization $x_t, t = 1,\ldots,n$

order is the order of the Butterworth filter you want to use (default = 4). **type** should be either **"low"**, **"high"**, **"pass"**, or **"stop"** for a low-pass, high-pass, band-pass, or band-stop (notch) filter, respectively.

cutoff
- if **type = "high"** or **"low"**, then **cutoff** is the scalar cutoff frequency
- if **type = "stop"** or **"band"**, then **cutoff** is a 2-component vector containing the lower and upper cutoff frequencies

Example: The dataset **wages** is contained in *tswge*. The *tswge* R commands to perform the low-pass filter described in Example 2.4 are

```
data(wages)
wages05 = butterworth.wge(wages,order = 4,type = "low", cutoff = 0.05)
```

The filtered data set is **wages05$x.filt**.

(c) **gen.sigplusnoise.wge(n,b0,b1,coef,freq,psi,phi,vara,plot,sn)** is a *tswge* R function that generates a realization from the signal-plus-noise model

$$x_t = b0 + b1*t + coef[1]*\cos(2\pi freq[1]t + psi[1]) + coef[2]*\cos(2\pi freq[2]t + psi[2]) + z_t$$

n= length of realization to be generated (*t* in the above formula specifies the integers from 1 to *n*)
b0 is the y-intercept
b1 is the slope of a linear regression component (default = 0)
coef is a 2-component vector specifying the coefficients (if only one cosine term is desired, define **coef[2]** = 0) (default = **c(0,0)**)
freq is a 2-component vector specifying the frequency components (0 to .5) for the cosine terms (default = **c(0,0)**)
psi is a 2-component vector specifying the phase shift (0 to 2π) (default =**c(0,0)**)
phi is a vector containing the AR model for the noise process (see Chapter 5). (default = 0)
vara is the variance of the noise (default = 1).

Notes

(1) The AR process, z_t, serves as the "noise" in the above signal-plus-noise model. We will see in Chapter 5 that an AR process is "driven" by a white noise process, a_t. In *tswge*, we assume the white noise process to be normally distributed with zero mean. The variance of the white noise process is specified using **vara**.

(2) If you want the "noise" in the signal-plus-noise model to be white noise, then use the default value for **phi** (=0).

plot is a logical variable. **plot = TRUE** (default) produces a plot of the generated realization.

sn determines the seed used in the simulation. (default = 0) (see note below)

Note: **sn = 0** (default) produces results based on a randomly selected seed. If you want to reproduce the same realization on subsequent runs of **gen.sigplusnoise.wge**, then set **sn** to the same positive integer value on each run.

Example: The command

```
x = gen.sigplusnoise.wge(n = 100,coef = c(1.5,3.5), freq = c(.05,.2),psi =
c(1.1,2.8), phi = .7,vara = 2)
```

calculates

$$x_t = 1.5\cos\left(2\pi(.05)t + 1.1\right) + 3.5\cos\left(2\pi(.2)t + 2.8\right) + z_t,$$

where z_t is a realization from the AR(1) process $z_t = .7z_{t-1} + a_t$, where a_t is a zero-mean white noise process with variance $\sigma_a^2 = 2$. The realization is based on a randomly selected seed (based on **sn**) and the resulting signal-plus-noise realization is placed in the vector **x**.

PROBLEMS

4.1 For each of the following datasets, use **plotts.sample.wge** to plot the data, sample autocorrelations, and Parzen spectral density estimate (using the default truncation point $M = 2\sqrt{n}$). For each dataset, explain how the sample autocorrelations and Parzen spectral density estimate describe (or fail to describe) the behavior in the data:
   (a) **dfw.mon**
   (b) **wtcrude2020**
   (c) **sunspot2.0**
   (d) **llynx** (log of the lynx data)
   (e) **doppler**
   (f) Simulated data **f=gen.arma.wge(n=200,phi=c(2.55,-2.42,.855),mu=10, sn=7285)**
   (g) Simulated data **g=gen.arma.wge(n=200,phi=c(.65,.62,-.855),mu=50, sn=5698)**
   (h) Simulated data **h=gen.sigplusnoise.wge(n=250, coef=c(2,1),freq= c(.05,.28), psi=c(1.2,2),phi=.9,sn=278)**
4.2 Following are displayed three sets of figures, each containing four plots. The first set shows four realizations of length $n = 200$, each generated from a time series model. The second and third sets contain four autocorrelation functions and four spectral densities, respectively. Match each realization with the corresponding autocorrelations and spectral density. Explain your answers.

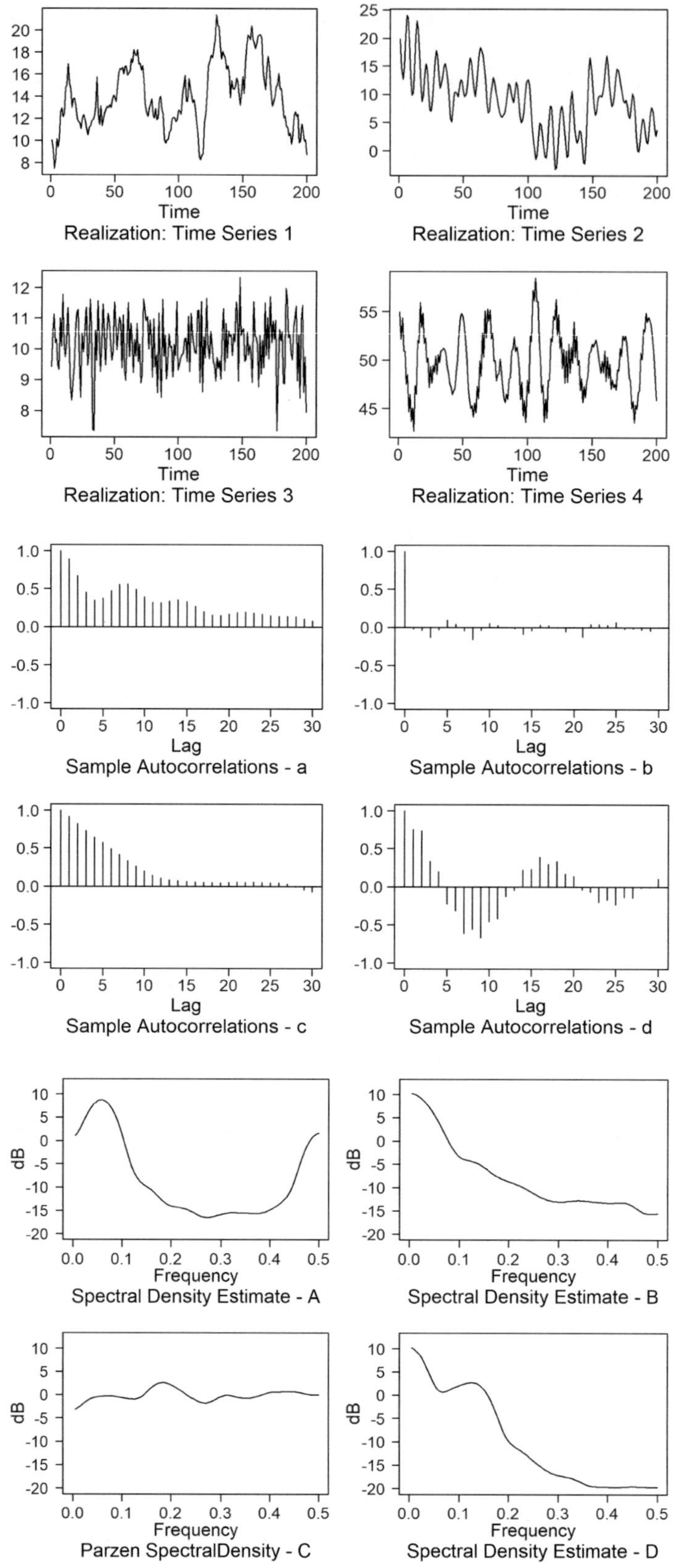

Realization: Time Series 1

Realization: Time Series 2

Realization: Time Series 3

Realization: Time Series 4

Sample Autocorrelations - a

Sample Autocorrelations - b

Sample Autocorrelations - c

Sample Autocorrelations - d

Spectral Density Estimate - A

Spectral Density Estimate - B

Parzen SpectralDensity - C

Spectral Density Estimate - D

4.3 Using the *tswge* command below, generate a time series realization.

```
y=gen.sigplusnoise.wge(n=100,coef=c(2,0),freq=c(.1,0),psi=
c(0,0),phi=0,sn=17)
```

(a) Using the description for `gen.sigplusnoise.wge` in the appendix to this chapter, write down the equation satisfied by **y**.
(b) Plot the data in vector **y**.
(c) Filter the data in **y** with a high-pass Butterworth filter with cutoff=.05. Plot the filtered data, and describe and explain the results.
(d) Filter the data in **y** with a low-pass Butterworth filter with cutoff=.05. Plot the filtered data, and describe and explain the results.
(e) Filter the data in **y** with a high-pass Butterworth filter with cutoff=.15. Plot the filtered data, and describe and explain the results.
(f) Filter the data in **y** with a low-pass Butterworth filter with cutoff=.15. Plot the filtered data, and describe and explain the results.
(g) Apply a centered moving average smoother of order 5 to the data in **y**. Plot the smoothed data, and describe and explain the results.

4.4 Using the *tswge* command below, generate a time series realization.

```
x=gen.arma.wge(n=50,phi=c(1.83000,-1.97500,1.82085,-0.97510),sn=2787)
```

In the following, you will be asked to discuss the frequency behavior in the data and Parzen spectral density estimate. In each case, we recommend that you use `plotts.sample.wge` on the dataset (or filtered dataset) under discussion.

(a) Discuss the frequency behavior visible in the realization (**x**) and the Parzen spectral density estimate.
(b) Use the Butterworth filter to perform a low-pass filter using cutoff=.12. Discuss the effect of the filter by discussing the frequency behavior visible in the filtered data and associated Parzen spectral density estimate.
(c) Use the Butterworth filter to perform a high-pass filter using cutoff=.12. Discuss the effect of the filter by discussing the frequency behavior visible in the filtered data and associated Parzen spectral density estimate.
(d) Use the Butterworth filter to perform a low-pass filter using cutoff=.23. Discuss the effect of the filter by discussing the frequency behavior visible in the filtered data and associated Parzen spectral density estimate. How does the performance of this filter compare with the low-pass filter using cutoff=.12, and why?
(e) Use the Butterworth filter to perform a high-pass filter using cutoff=.07. Discuss the effect of the filter by discussing the frequency behavior visible in the filtered data and associated Parzen spectral density estimate. How does the performance of this filter compare with the high-pass filter using cutoff=.12, and why?
(f) Use the Butterworth filter to perform a low-pass filter using cutoff=.3. Discuss the effect of the filter by discussing the frequency behavior visible in the filtered data and associated Parzen spectral density estimate. How would you explain the effect of the filter?
(g) Use the Butterworth filter to perform a high-pass filter using cutoff=.3. Discuss the effect of the filter by discussing the frequency behavior visible in the filtered data and associated Parzen spectral density estimate. How would you explain the effect of the filter? (Notice the end-effects in the filtered data. How might you adjust for the end-effects?)

ARMA Models

5

All models are wrong, but some are useful.
George Box (The Father of Modern Time Series)

A great place to start when using models is to understand that we are most likely ***not*** going to be able to identify or define the exact process that is generating the data we are analyzing. In fact, we will probably not be able to define the process mathematically because it is just too complex. However, the process can often be approximated in a very *useful* way using statistical models! If we can find a "reasonable" model, we will find that we can calculate probabilities, forecasts, and other items of interest. In addition, models often offer insight into outside forces that may be driving changes in the response, which may be our principal interest. A big take-away here is to not stress about finding the "right" model out of tens, hundreds, thousands or more potential models… it is likely true that they are all wrong! Our aim is to find a *useful* model, and that is largely the focus of the remainder of this text. Box and Jenkins (1970) (yes, the Box who was quoted above) is *the* landmark book that introduced a practical and integrated approach to model building that has been in use for the last half century. While this book (that is, the one you are reading) provides updated, and in many cases, more accessible approaches to model-building, the "Box-Jenkins" procedure is at the heart of what we do.

Stationary time series play a fundamental role in time series data analysis. Even when seasonal, trending, or other (nonstationary) features are present in the data, most time series models include a stationary component. Section 5.1 introduces a very important tool in the time series practitioner's tool kit for analyzing stationary time series data, the autoregressive, AR (p), model. Section 5.2 discusses the autoregressive-moving average (ARMA (p,q)) model which is an important generalization of the AR model.

5.1 THE AUTOREGRESSIVE MODEL

The $AR(p)$ model is similar to a multiple regression model in which the "independent variables" are lagged versions of X_t.

Definition 5.1: The process X_t is said to satisfy an $AR(p)$ model if

$$X_t = \beta + \phi_1 X_{t-1} + \phi_2 X_{t-2} + \cdots + \phi_p X_{t-p} + a_t \tag{5.1}$$

where ϕ_k, $k = 1, \ldots, p$ are real constants, $\beta = \left(1 - \phi_1 - \phi_2 - \cdots - \phi_p\right)\mu$, $\phi_p \neq 0$, and a_t is a white noise process with zero mean and finite variance σ_a^2.

The formula in (5.1) says that the value of the process at time t is a linear combination of the p previous values plus a random noise component a_t. We begin our discussion of AR models by discussing their properties, including stationarity conditions and the behavior of autocorrelations and spectral densities for specific models. Before discussing the general AR(p) model, we begin our discussion with the simplest AR model, the AR(1) model.

5.1.1 The AR(1) Model

The time series, X_t, is said to satisfy an AR(1) model if

$$X_t = \beta + \phi_1 X_{t-1} + a_t,\tag{5.2}$$

where ϕ_1 is a real, nonzero constant, and a_t is a white noise process with finite variance σ_a^2. The constant $\beta = (1 - \phi_1)\mu$, is called the *moving average constant*. In essence, the AR(1) model specifies that the value of the process at time t depends on the value of the process at time $t-1$, plus a random noise component, a_t, and a constant, β. This seems like a reasonable way to describe how a time series might progress in time. In fact, (5.2) has the appearance of a simple linear regression, but it differs from the standard regression model because the independent variable, X_{t-1}, is a "lagged" version of the dependent variable X_t. Theorem 5.1 states a key result concerning AR(1) processes.

Theorem 5.1: An AR(1) process is stationary if and only if $|\phi_1| < 1$.
Proof: See Woodward et al. (2017)

As a result of Theorem 5.1, the models in the left list below are stationary and those on the right are nonstationary.

Stationary Models	Nonstationary Models
$X_t = .6X_{t-1} + a_t$	$X_t = 1.2X_{t-1} + a_t$
$X_t = 10 - .8X_{t-1} + a_t$	$X_t = 10 - 3.1X_{t-1} + a_t$
$X_t = 50 + .99X_{t-1} + a_t,$	$X_t = X_{t-1} + a_t.$

Thus, by simple examination of an AR(1) model, it is very easy to determine whether it describes a stationary process. We will see that for higher order AR(p) models, that is, where $p > 1$, assessing stationarity by simple examination of the coefficients can be difficult or impossible. See Example 5.5.

5.1.1.1 The AR(1) in Backshift Operator Notation

The AR(1) model is sometimes expressed using the *backshift operator* defined by $BX_t = X_{t-1}$. Note that $Bc = c$ for a constant c. The AR(1) model in (5.2), that is, the model $X_t = (1 - \phi_1)\mu + \phi_1 X_{t-1} + a_t$, can be written as

$$X_t - \mu - \phi_1(X_{t-1} - \mu) = a_t,$$

or, in backshift operator notation

$$(1 - \phi_1 B)(X_t - \mu) = a_t,\tag{5.3}$$

QR 5.1
Backshift Notation

or,

$$\phi(B)(X_t - \mu) = a_t,$$

where $\phi(B)$ is the operator

$$\phi(B) = 1 - \phi_1 B.$$

Key Points

1. In this book, we often work with the simplified zero-mean forms of the models in (5.2) and (5.3).
2. If an AR(1) process, X_t, has mean μ, then we can create a zero-mean process, $\tilde{X}_t$, by setting $\tilde{X}_t = X_t - \mu$.
3. Given a set of data, we typically analyze it as a zero-mean process by first estimating μ by $\bar{X}$, subtracting $\bar{X}$ from each observation, and modeling the remaining data using the "zero-mean" model.

5.1.1.2 The AR(1) Characteristic Polynomial and Characteristic Equation

Associated with the operator $\phi(B) = 1 - \phi_1 B$ in (5.3) is the *characteristic polynomial*, $\phi(z) = 1 - \phi_1 z$, which is not an operator but is an algebraic expression where z is a real number. The equation $\phi(z) = 1 - \phi_1 z = 0$ is called the *characteristic equation*. Notice that the root of the characteristic equation obtained by solving $1 - \phi_1 z = 0$ is $z = 1/\phi_1$. Because the root of the characteristic equation of an AR(1) model is the reciprocal of the coefficient, ϕ_1, the result in Theorem 5.1 can be restated as follows.

Theorem 5.1 (restatement): An AR(1) process is stationary if and only if the root, r, of the characteristic equation, satisfies $|r| = \left| \dfrac{1}{\phi_1} \right| = |\phi_1^{-1}| > 1.$

Key Point: While Theorem 5.1 applies to AR(1) processes, Theorem 5.3 will give stationary conditions for general AR(p) models. These conditions will also be based on the roots of the characteristic equation.

5.1.1.3 Properties of a Stationary AR(1) Model

In this section, we discuss the properties of a stationary AR(1) process of the form $X_t = (1 - \phi_1)\mu + \phi_1 X_{t-1} + a_t$ with $|\phi_1| < 1$. See Appendix 5B for a derivation of the properties listed below. Let X_t be a stationary AR(1) process satisfying (5.3).

(a) The mean of X_t is given by $E[X_t] = \mu$ (a constant)

Note that if $\mu = 0$, then (5.2) simplifies to

$$X_t = \phi_1 X_{t-1} + a_t, \tag{5.4}$$

which, in operator notation, is

$$(1 - \phi_1 B) X_t = a_t. \tag{5.5}$$

The AR(1) model expressed in (5.4) and (5.5) is called the *zero-mean form* of the AR(1) process which, as stated earlier, we will often use for simplicity.

(b) The variance of X_t is given by

$$\text{var}[X_t] = \frac{\sigma_a^2}{1 - \phi_1^2}. \tag{5.6}$$

By (5.6), $\text{var}[X_t]$ is a constant, which we will denote σ_X^2, and is finite because σ_a^2 is assumed to be finite and $|\phi_1| < 1$.

(c) The autocovariance between X_t and X_{t+k}, is given by

$$\text{cov}[X_t, X_{t+k}] = \phi_1^k \sigma_X^2. \tag{5.7}$$

From (5.7) it follows that the autocovariance between X_t and X_{t+k} depends only on k, and thus can be denoted by γ_k. Recall that if X_t is a stationary process then $\gamma_0 = \text{cov}[X_t, X_t] = \sigma_X^2$, so (5.7) can be written as

$$\text{cov}[X_t, X_{t+k}] = \phi_1^k \gamma_0.$$

Also, because $\rho_k = \gamma_k / \gamma_0$, it follows that

$$\rho_k = \phi_1^k, \quad k > 0. \tag{5.8}$$

Note that because $\gamma_k = \gamma_{-k}$ and consequently, $\rho_k = \rho_{-k}$, then for $k = 0, \pm 1, \pm 2, \ldots$, (5.7) and (5.8), respectively, become

$$\gamma_k = \phi_1^{|k|} \gamma_0$$

and

$$\rho_k = \phi_1^{|k|}, \tag{5.9}$$

for $= 0, \pm 1, \pm 2, \ldots$.

QR 5.2
Properties

5.1.1.4 Spectral Density of an AR(1)

The true (model-based) autocorrelations for an AR(1) model with parameter ϕ_1 are given in (5.9). Recall from Section 4.2 that if the true autocorrelations are known, then the true spectral density can be calculated. In the case of the AR(1) model, the spectral density is given by

$$S_X(f) = \frac{\sigma_a^2}{\gamma_0} \frac{1}{\left|1 - \phi_1 e^{-2\pi i f}\right|^2}$$

$$= \frac{1 - \phi_1^2}{\left|1 - \phi_1 e^{-2\pi i f}\right|^2}. \tag{5.10}$$

See Woodward et al. (2017). Using Euler's formula, $e^{-2\pi i f} = \cos 2\pi f - i \sin 2\pi f$, the denominator of (5.10) becomes

$$\left|1 - \phi_1 e^{-2\pi i f}\right|^2 = \left|1 - \phi_1\left(\cos 2\pi f - i \sin 2\pi f\right)\right|^2$$

$$= \left|\left(1 - \phi_1 \cos 2\pi f\right) - i\phi_1 \sin 2\pi f\right|^2$$

$$= \left(1 - \phi_1 \cos 2\pi f\right)^2 + (\phi_1 \sin 2\pi f)^2,$$

because $|a + bi|^2 = a^2 + b^2$. Thus, (5.10) can be written as

$$S_X(f) = \frac{1 - \phi_1^2}{\left(1 - \phi_1 \cos 2\pi f\right)^2 + (\phi_1 \sin 2\pi f)^2}. \tag{5.11}$$

In (5.11) we have simplified the formula for $S_X(f)$ in (5.10) to emphasize the fact that $S_X(f)$ is a real number that is easy to calculate. However, in practice we will use *tswge* functions **plotts.true.wge** and **true.arma.spec.wge** to calculate model-based AR spectral densities.

5.1.1.5 AR(1) Models with Positive Roots of the Characteristic Equation

Consider the AR(1) models

$$X_t - .9X_{t-1} = a_t \tag{5.12}$$

and

$$X_t - .7X_{t-1} = a_t. \tag{5.13}$$

where a_t is zero-mean white noise with variance $\sigma_a^2 = 1$. Using the backshift operator, these models can be written as $(1 - .9B)X_t = a_t$ and $(1 - .7B)X_t = a_t$, with characteristic equations $1 - .9z = 0$ and $1 - .7z = 0$, respectively. Both models are stationary because the roots, 1.11 and 1.43, are greater than one in absolute value. (This could also be determined by simple examination of the models using Theorem 5.1). Figure 5.1(a) shows the model-based autocorrelations for the AR(1) model with $\phi_1 = 0.9$. From (5.9), the model-based autocorrelations for this AR(1) model should be of the form of a damped exponential curve defined by $\rho_k = .9^k$. The autocorrelations in Figure 5.1(a) exhibit the expected behavior. The model-based spectral density plotted in Figure 5.1(b) is simply a plot of the function in (5.10) with $\phi_1 = .9$. The dominant feature of the spectral density is a peak at $f = 0$.

Figures 5.2(a) and (b) show that the AR(1) model with $\phi_1 = .7$ has exponentially damping autocorrelations that damp more quickly than for the $\phi_1 = .9$ case. Also, the peak in the spectral density is again at $f = 0$, but the peak is not as "sharp".

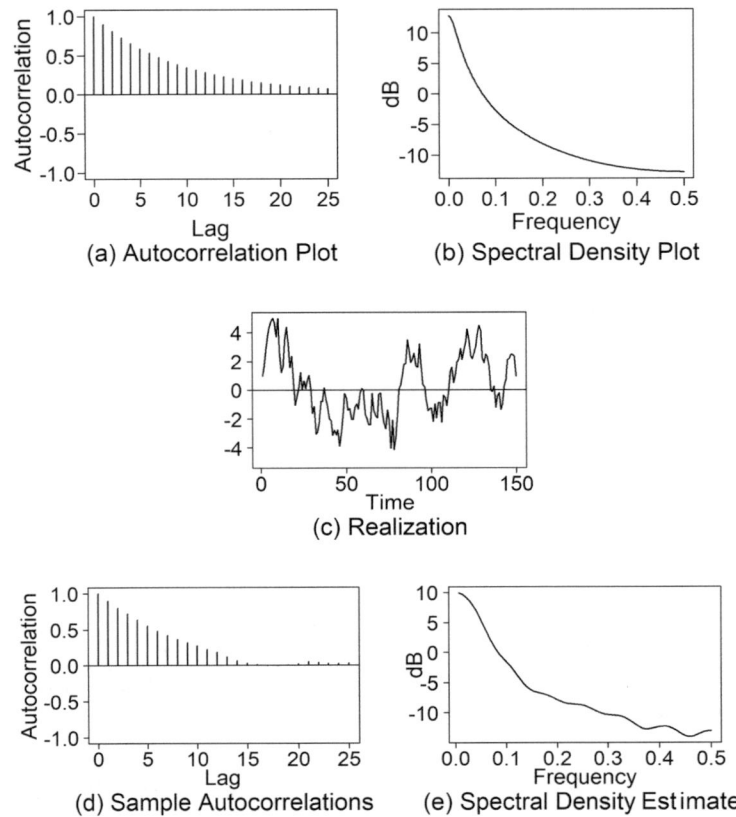

FIGURE 5.1 (a) Model-based autocorrrelations and (b) spectral density for $(1 - .9B)X_t = a_t$, (c) realization of length $n = 150$ from this AR(1) model, (d) sample autocorrelations and (e) Parzen spectral density estimate based on the realization in (c).

QR 5.3
Characteristic
Equation of
AR(1)

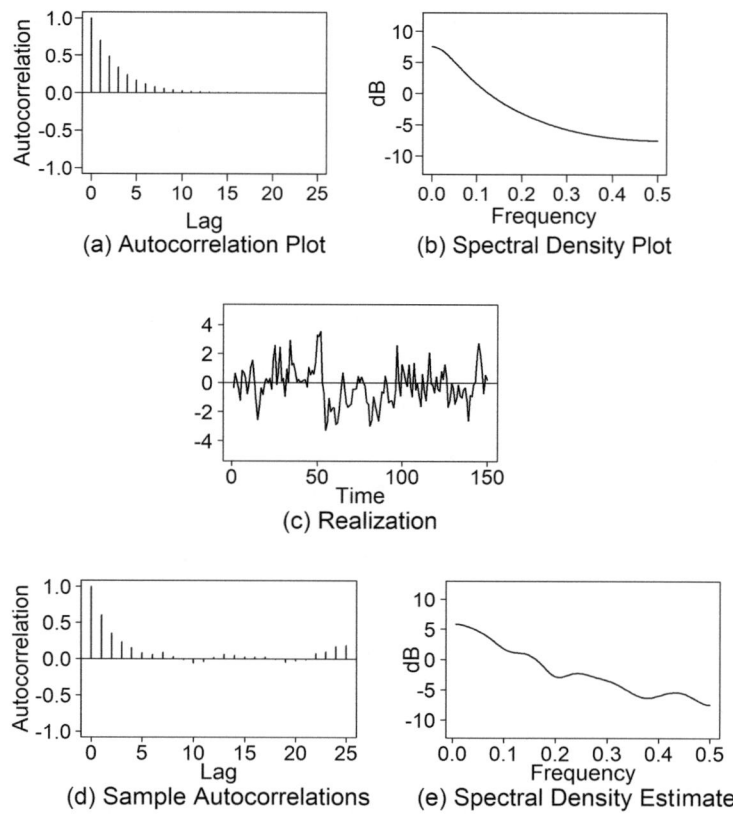

FIGURE 5.2 (a) Model-based autocorrelations and (b) spectral density for $(1 - .7B)X_t = a_t$, (c) realization of length $n = 150$ from this AR(1) model, (d) sample autocorrelations and (e) Parzen spectral density estimate based on the realization in (c).

Figures 5.1(a)-(b) and 5.2(a)-(b) were obtained using the commands

```
plotts.true.wge(phi=.9,plot.data=FALSE)
plotts.true.wge(phi=.7,plot.data=FALSE)
```

Key Points

1. It is important to note that Figures 5.1(a)−(b) and Figures 5.2(a)−(b) show the "theoretical" autocorrelations and spectral densities based on the mathematical models in (5.12) and (5.13), respectively.
2. In practice, given a set of data, we estimate the autocorrelations using the sample autocorrelations defined in (3.12), and we estimate the spectral density using a spectral density estimate such as the Parzen spectral density estimate (see Section 4.2.3).

Figures 5.1(c)−(e) and Figures 5.2(c)−(e) show realizations generated from these AR(1) models along with sample-based estimates of the autocorrelations and spectral density. The sample-based estimates in Figures 5.1(d)−(e) and Figures 5.2(d)−(e) have the same general appearance as the model-based quantities in Figures 5.1(a)−(b) and Figures 5.2(a)−(b), respectively. However, they are not exact replicas.

The realizations, sample autocorrelations, and Parzen spectral densities in Figures 5.1(c)–(e) were obtained using the commands

```
x=gen.arma.wge(n=150,phi=.9,sn=20)
plotts.sample.wge(x)
```

while those in Figures 5.2(c)–(e) were produced by

```
x=gen.arma.wge(n=150,phi=.7,sn=130)
plotts.sample.wge(x)
```

Key Points

1. Although the realizations in Figure 5.1(c) and Figure 5.2(c) were generated from the AR(1) models in (5.12) and (5.13), respectively, the sample autocorrelations and spectral density estimates were calculated without reference to a given model.
   - The sample autocorrelations and spectral density estimates are similar to, but not exactly the same as, the model-based versions.
2. A particular model may turn out to be "useful" if the sample autocorrelations and spectral density estimates calculated from the data have the general appearance of the corresponding model-based quantities.

One consequence of the autocorrelation structure for the $\phi_1 = .9$ case is that there will be a strong positive correlation, $\rho_1 = .9$, between adjacent observations in realizations from this model. That is, realizations should exhibit a tendency for the value of the realization at time t to be fairly close to the value at time $t+1$. This positive correlation associated with $\phi_1 = 0.9$ remains fairly strong (holds up) for observations at time t and time $t+k$ for values of k larger than $k = 1$. For example, the correlation between observations separated by seven time points is $\rho_7 = .9^7 = .48$. This implies that, for example, if the process takes on a value substantially below the mean at time t, then the values of the process at times in the neighborhood of t are also likely to be below the mean. As noted in Section 3.3.1, this type of correlation structure induces a random wandering behavior in the realizations. By construction, the mean of the process is zero. Notice in Figure 5.1(c), for example, that several x_t values for t near $t = 125$ tend to remain above the mean. The random, aperiodic wandering behavior is reflected by a peak in the spectral density at $f = 0$. For $\phi_1 = .9$, this peak is fairly "sharp" in both the model-based values and the sample-based Parzen estimates.

Now notice in Figure 5.2(a) that the correlation structure for $\phi_1 = .7$ dies out rather quickly. For example, while the correlation between values of the process at times t and $t+1$ is $.7^1 = .7$, the correlation between times t and $t+7$ is $.7^7 = .08$. That is, there is almost no correlation between values separated by seven units of time, and the "lingering" behavior above and below the mean that was seen for $\phi_1 = .9$ would not be expected to occur. The realization in Figure 5.2(c) reflects this fact. Additionally, the process variance is also noticeably smaller for $\phi_1 = .7$ than for $\phi_1 = .9$. This can be verified mathematically from (5.6). That is, because $\sigma_a^2 = 1$ in the generated realizations, the variance for $\phi_1 = .7$ is $1/(1-.7^2) = 1.96$ and for $\phi_1 = .9$ is $1/(1-.9^2) = 5.26$. Figures 5.2(b) and (e), i.e. the model-based and the sample-based spectral densities in the $\phi_1 = .7$ case, both have mild peaks at or about $f = 0$.

Key Points

1. The realizations in Figures 5.1(c) and 5.2(c) represent an "artificial case" in which the realization is actually generated from the underlying AR(1) model.
2. Recalling the George Box quote in the opening paragraph to this chapter, "real" data will in all likelihood not be a realization from any AR (or ARMA, ARIMA, …) model.

Try This: Submit the following commands *several times* to get an idea of the variability in realizations, sample autocorrelations, and spectral density estimates that you might expect to see from realizations generated from a model.[1]

```
x=gen.arma.wge(n=150,phi=.9)
plotts.sample.wge(x)
```

Because no seed was set by specifying a value for **sn**, you will get a new randomly selected realization each time you submit the **gen.arma.wge** command above.

Repeat the procedure by replacing **phi=.9** with **phi=.7**.

QR 5.4 Investigating
AR(1) Realizations

5.1.1.6 AR(1) Models with Roots Close to +1

Figure 5.3 shows realizations from two AR(1) models with roots close to one in absolute value. Specifically we show realizations from AR(1) models $(1 - \phi_1 B)(X_t - 10) = a_t$, with $\phi_1 = .95$ and $\phi_1 = .99$. Noting that the theoretical mean of the model from which these realizations were generated is $\mu = 10$ (which is plotted as a horizontal line in the plots), it can be seen that there is considerable aimless, aperiodic wandering above and below the mean, especially for $\phi_1 = .99$. We describe this by saying that for $\phi_1 = 0.95$ and 0.99, the autocorrelations, $\rho_k = \phi_1^k$, are strong (or *persistent*). That is, the autocorrelations remain substantial even when k is moderately large. As an example, for $\phi_1 = .99$, amazingly $\rho_{50} = .99^{50} = .61$. That is, there is a correlation of .61 between observations 50 time periods apart! The sustained wandering behavior increases as values ϕ_1 increase toward $\phi_1 = 1$. At $\phi_1 = 1$ the model is not a stationary AR(1) model because the absolute value of the root is equal to one instead of greater than one. We will discuss this model later in Section 5.1.1.8 and in Chapter 7. The data for Figure 5.3 were generated using the commands

```
gen.arma.wge(n=150,phi=.95,sn=305)
gen.arma.wge(n=150,phi=.99,sn=404)
```

---

1 See Appendix 5A for a discussion of *tswge* function **gen.arma.wge**

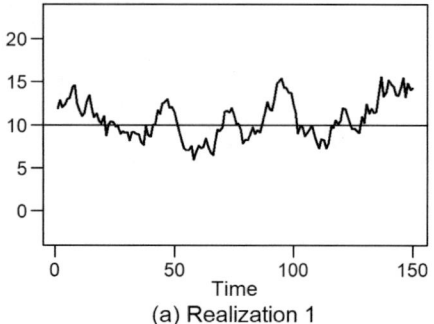
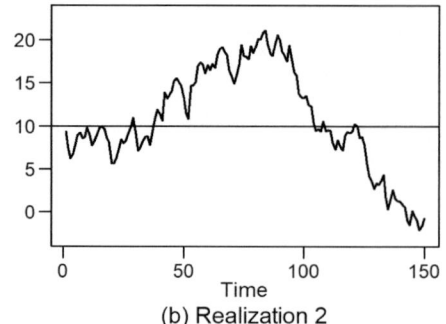

(a) Realization 1   (b) Realization 2

FIGURE 5.3   Realizations from two AR(1) models with roots close to one in absolute value.

Example 5.1 Dow Jones Stock Market Data

Figure 5.4(a) shows the monthly Dow Jones closing averages for March 1985 through December 2020 (dataset **dow1985.ts** in *tswge*). (Source: Federal Reserve Bank). Of course, in this case we do not know the "actual" model, the model-based autocorrelations, or spectral density. Consequently, we must analyze the data based on the realization along with the corresponding sample-based estimates of autocorrelations and spectral densities. It can be seen that the time series has the typical wandering appearance of an AR(1) process with positive ϕ_1. The sample autocorrelations and Parzen spectral density estimate are shown in Figures 5.4(b) and (c), where it can be seen that the autocorrelations are slowly damping, and the spectral density has a peak at zero. By comparing Figure 5.4(a) with Figure 5.3 it can be seen that ϕ_1 appears to be close to one, that is, the model is nearly nonstationary (or maybe nonstationary with $\phi_1 = 1$). We will return to the analysis of data such as the Dow Jones data when we discuss the topics of nonstationarity, model identification, and parameter estimation.

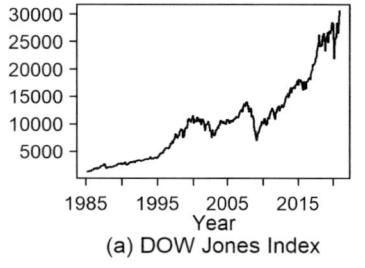
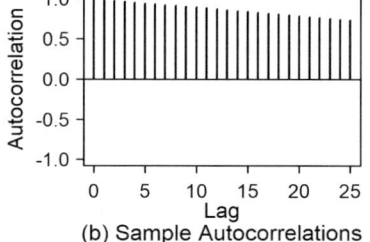
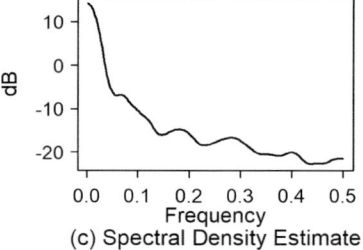

(a) DOW Jones Index   (b) Sample Autocorrelations   (c) Spectral Density Estimate

FIGURE 5.4   (a) Dow Jones monthly closing averages from March 1985 through December 2020, (b) sample autocorrelations, and (c) Parzen spectral density estimate.

5.1.1.7 AR(1) Models with Negative Roots of the Characteristic Equation

To describe the behavior of an AR(1) model associated with a negative root of the characteristic equation, we consider the AR(1) model

$$X_t + .7X_{t-1} = a_t. \tag{5.14}$$

Note that $\phi_1 = -.7$ and the characteristic equation is $1 + .7z = 0$ which has root $-1/.7 = -1.43$. Thus the root is negative and since $|-1.43| > 1$, the process is stationary. Using (5.8) the model-based autocorrelations for positive k are given by $\rho_k = (-.7)^k = (-1)^k (.7)^k$. So,

$$\rho_1 = -.7$$

$$\rho_2 = (-1)^2(.7)^2 = .49$$

$$\rho_3 = (-1)^3(.7)^3 = -.343$$

$$\vdots$$

That is, the autocorrelations follow a damped, oscillating behavior, and if a value of the process at time t is above the mean, then the value at time $t+1$ will likely be below the mean, and vice versa. Similarly, because $\rho_2 = .49$, the value at time $t+2$ will tend to be on the same side of the mean as the value at time t. This oscillating behavior describes a period of length two (that is, the frequency is .5, the Nyquist frequency). Consequently, we expect the spectral density to have a peak at $f = .5$. The model-based autocorrelations in Figure 5.5(a) illustrate the expected damped oscillating behavior. The dominant feature of the model-based spectral density plotted in Figure 5.5(b) (a plot of the function in (5.10) with $\phi_1 = -.7$) is a peak at $f = .5$.

Figure 5.5(c) shows a realization of length $n = 100$ generated from $(1 + .7B)X_t = a_t$, along with sample-based estimates of the autocorrelations and spectral density. The expected oscillatory behavior in the realization is quite apparent in Figure 5.5(c). The sample autocorrelations oscillate, and the Parzen spectral density estimate has some "wiggle" but its main feature is a peak at $f = .5$. As in the case of the AR(1) with positive roots, whenever $|\phi_1|$ gets closer to one, these behaviors become more pronounced (sample autocorrelations damp more slowly but are still oscillatory, the peak in the spectral density is higher, etc.) See Problem 5.1(b). The following *tswge* commands create the plots in Figure 5.5.

```
plotts.true.wge(n=100,phi=-.7,plot.data=FALSE)
x=gen.arma.wge(n=150,phi=-.7,sn=878)
plotts.sample.wge(x)
```

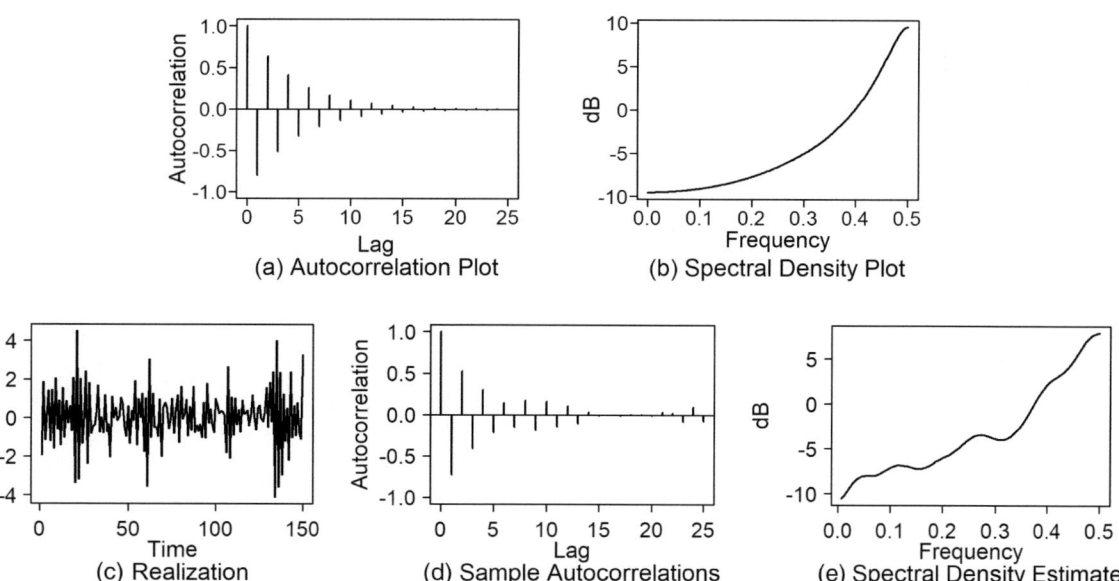

FIGURE 5.5 (a) Model-based autocorrrelations and (b) spectral density for $(1 + .7B)X_t = a_t$, (c) realization of length $n = 100$ from this model, (d) sample autocorrelations and (e) Parzen spectral density estimate based on the realization in (c).

5.1.1.8 Nonstationary 1st-order Models

In Figure 5.6, realizations are shown from the 1st-order model in (5.3) with $\phi_1 = 1$ and $\phi_1 = 1.1$ where $\sigma_a^2 = 1$. In neither case is it true that $|\phi_1| < 1$, so these are realizations from *nonstationary* models. Examination of the realizations shows that for $\phi_1 = 1$, the "wandering or quasi-linear behavior" is similar to, but more pronounced than, that in the realizations shown in Figure 5.3 for both $\phi_1 = 0.95$ and 0.99. In Chapter 7 we will find that the model,

$$(1-B)(X_t - \mu) = a_t,$$

is useful for modeling real datasets. The realization in Figure 5.6(a) is generated from the model (5.3) with $\phi_1 = 1$ and $\mu = 0$. Note that there does not seem to be an attraction to the mean value $\mu = 0$, which does not even show in the plot.

The realization associated with $\phi_1 = 1.1$ demonstrates what Box et al. (2008) refer to as "explosive" behavior. The random wandering seen in Figure 5.6(a) associated with $\phi_1 = 1$ is typical of certain real data series such as stock prices as mentioned in Section 5.1.1.6. However, the explosive realization associated with $\phi_1 = 1.1$ is not similar to that of any datasets that we will attempt to analyze, being better modeled as a deterministic process such as $X_t = -e^{\left(at + bt^2 + ct^3\right)}$. The plot in Figure 5.6(a) was obtained using the *tswge* command

```
x=gen.arima.wge(n=150, d=1,sn=734)
```

Because *tswge* will not generate realizations associated with roots inside the unit circle, the plot in Figure 5.6(b) was obtained by straightforward computation of the generating equation. The following code produces the plot in Figure 5.6(b).

```
a=gen.arma.wge(n=500,sn=10,plot=FALSE) #generates 500 N(0,1) white noise
xx.spin=rep(0,500)
xx=rep(0,150)
for(i in 1:499) {xx.spin[i+1]=1.1*xx.spin[i]+a[i]}
xx[1:150]=xx.spin[350:499] # generates 500 and keeps the last 150
plot(xx[1:150])
```

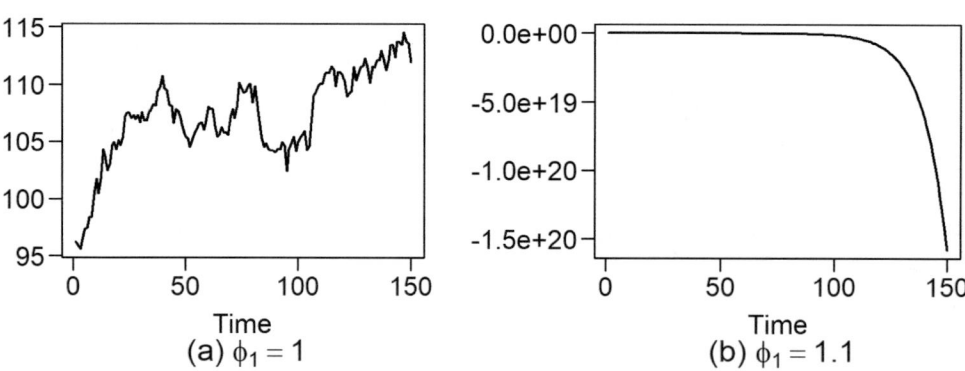

FIGURE 5.6   Realizations from 1st-order nonstationary models, (a) $(1-B)X_t = a_t$ and (b) $(1-1.1B)X_t = a_t$.

5.1.1.9 Final Comments Regarding AR(1) Models

While the AR(1) is a useful model, it has its limitations. The peak in the spectral density is either at $f = 0$ or $f = .5$, implying that realizations will either display aimless (aperiodic) wandering or will oscillate back and forth. Many datasets, for example, data like the DFW monthly average temperatures and business-related data such as the **AirPassengers** data shown previously and in Figure 5.7, clearly require more complex models to explain their behavior.

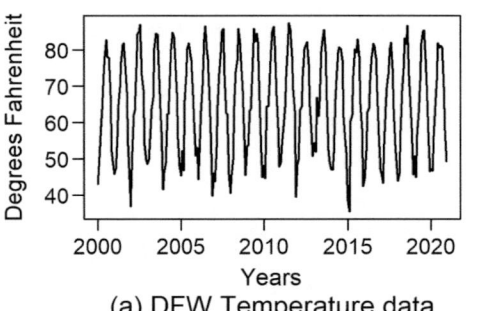

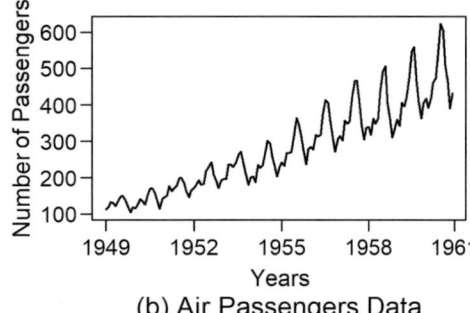

(a) DFW Temperature data

(b) Air Passengers Data

FIGURE 5.7   (a) DFW monthly temperature data (January 1990–December 2020) and (b) Monthly number of international airline passengers (January 1949–December 1960).

Key Points

1. As $\phi_1 \uparrow 1$, realizations from an AR(1) model begin to appear similar to realizations from a nonstationary model.

2. As $\phi_1 \to 0$, the AR(1) model approaches white noise.

5.1.2 The AR(2) Model

Understanding the features of the AR(1) and AR(2) models is critical for gaining insight into the more general AR(p) model. We have already discussed the AR(1) model, so, in this section, we consider the AR(2) model

$$X_t = \beta + \phi_1 X_{t-1} + \phi_2 X_{t-2} + a_t, \tag{5.15}$$

where ϕ_1 and ϕ_2 are real constants, $\beta = (1 - \phi_1 - \phi_2)\mu$, $\phi_2 \neq 0$, and a_t is a white noise process with zero mean and finite variance σ_a^2.

Analogous to the AR(1) model, the AR(2) model in (5.15) has the appearance of a multiple regression model with two independent variables. Again, the "independent variables" are values of the dependent variable at the two previous time periods. In essence, the model specifies that the value of the process at time t is a linear combination of values at the two previous time periods plus a random noise component that enters the model at time t.

As in the case of the AR(1) model, it will often be useful to use the zero-mean form, and express the model as

$$X_t - \phi_1 X_{t-1} - \phi_2 X_{t-2} = a_t. \tag{5.16}$$

5.1.2.1 Facts about the AR(2) Model

(a) $E[X_t] = \mu$, for the "non-zero mean" form of the AR(2) model in (5.15).

(b) The process variance is given by $\sigma_X^2 = \gamma_0 = \dfrac{\sigma_a^2}{1 - \phi_1\rho_1 - \phi_2\rho_2}$, which is a finite constant.

(c) The autocorrelations satisfy the equations

$$\rho_k = \phi_1\rho_{k-1} + \phi_2\rho_{k-2}, \quad k > 0. \tag{5.17}$$

See Appendix 5B.

Setting $k = 1$ and $k = 2$ in (5.17), we obtain the 2nd-order *Yule-Walker equations*

$$\rho_1 = \phi_1\rho_0 + \phi_2\rho_1$$
$$\rho_2 = \phi_1\rho_1 + \phi_2\rho_0 \tag{5.18}$$

where, of course, $\rho_0 = 1$. Knowing ϕ_1 and ϕ_2 allows us to solve this 2×2 system of equations for ρ_1 and ρ_2. Autocorrelations ρ_k, $k > 2$ can be calculated using the recursion $\rho_k = \phi_1\rho_{k-1} + \phi_2\rho_{k-2}$. These autocorrelations are also computed by *tswge* functions `true.arma.aut.wge` and `plotts.true.wge`.

The model-based spectral density of an AR(2) model is given by

$$S_X(f) = \frac{\sigma_a^2}{\gamma_0} \frac{1}{\left|1 - \phi_1 e^{-2\pi i f} - \phi_2 e^{-4\pi i f}\right|^2}.$$

5.1.2.2 Operator Notation and Characteristic Equation for an AR(2)

Recall that when discussing the AR(1) model, we defined the backshift operator $BX_t = X_{t-1}$. More generally, the backshift operator is defined as

$$B^k X_t = X_{t-k}$$

Using the more general form of the backshift operator, we can write the AR(2) model in (5.16) as

$$X_t - \phi_1 B X_t - \phi_2 B^2 X_t = a_t$$

or

$$(-\phi_1 B - \phi_2 B^2) X_t = a_t. \tag{5.19}$$

We often use the "shorthand" notation $\phi(B)X_t = a_t$ where $\phi(B)$ is the operator $\phi(B) = 1 - \phi_1 B - \phi_2 B^2$. Converting the operator $\phi(B)$ to the algebraic quantity $\phi(z)$ as in the AR(1) case, we obtain the AR(2) characteristic polynomial $\phi(z) = 1 - \phi_1 z - \phi_2 z^2$. The corresponding AR(2) characteristic equation is given by

$$\phi(z) = 1 - \phi_1 z - \phi_2 z^2 = 0. \tag{5.20}$$

It should be noted that the characteristic equation has two roots, r_1 and r_2. These roots are either both real or are complex conjugate pairs, i.e. $r_1 = a + bi$ and $r_2 = r_1^* = a - bi$, where z^* denotes the complex conjugate of the complex number z.

> **Key Point:** The behavior of the autocorrelations and spectral density of an AR(2) model depend on whether the roots of the characteristic equation are real or complex. See Woodward et al. (2017).

Theorem 5.2 gives the stationarity conditions for an AR(2) model and is an extension of the result given in Theorem 5.1(restatement) for AR(1) models.

Theorem 5.2 An AR(2) is stationary if and only if all of the roots of the characteristic equation are greater than one in absolute value.

Complex numbers that satisfy the condition $|z| = |a + bi| = \sqrt{a^2 + b^2} = 1$ are said to fall on the *unit circle* on the complex plane. A common way to express the condition in Theorem 5.2 is that *the roots of the characteristic equation fall outside the unit circle.*

> **Key Point:** Theorem 5.2 can be restated as:
>
> An AR(2) model is stationary if and only if all of the roots of the characteristic equation fall *outside the unit circle.*

Consider the following AR(2) models. According to the results of Theorem 5.2, which of these models are stationary?

(a) $X_t - .2X_{t-1} - .63X_{t-2} = a_t$

(b) $X_t - .7X_{t-1} + 1.2X_{t-2} = a_t$

(c) $X_t - 1.6X_{t-1} + .8X_{t-2} = a_t$

Unlike in the AR(1) case, by only examining the size of the coefficients it is not straightforward to determine whether the models are stationary. In the following example we will check each of these models.

Example 5.2 Checking AR(2) Models for Stationarity

Model (a) $X_t - .2X_{t-1} - .63X_{t-2} = a_t$

This model can be written in operator notation as $(1 - .2B - .63B^2)X_t = a_t$ and the resulting characteristic equation is $1 - .2z - .63z^2 = 0$. In order to find the roots, we factor this 2nd-order polynomial as $(1 - .9z)(1 + .7z) = 0$. It follows that the roots are $r_1 = 1/.9 = 1.11$ and $r_2 = 1/(-.7) = -1.43$, and because both of these roots are greater than one in absolute value, Model(a) is a stationary AR(2) model. The *tswge* function **unit.circle.wge** displays points in the complex plane along with the unit circle. Figure 5.8 is obtained using the *tswge* command

```
unit.circle.wge(real=c(1.11,-1.43),imaginary=c(0,0))
```

The figure shows that both roots are real (fall along the horizontal real axis) and that both roots fall outside the unit circle (which is shown as the circle in the plot).

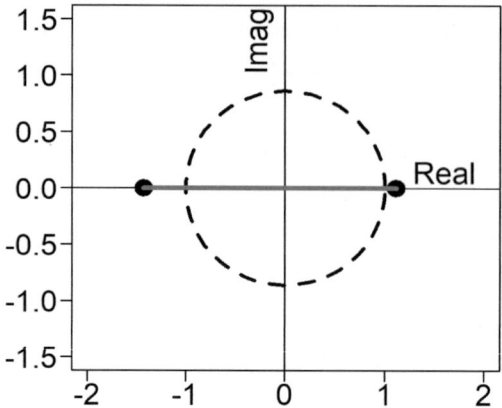

FIGURE 5.8   Plots of $r_1 = 1.11$. and $r_2 = -1.43$ in the complex plane.

Model (b) $X_t - .7X_{t-1} + 1.2X_{t-2} = a_t$

The operator form for this model is $\left(1 - .7B - 1.2B^2\right)X_t = a_t$ and the resulting characteristic equation is $1 - .7z - 1.2z^2 = 0$. By factoring, we get $\left(1 - 1.5z\right)\left(1 + .8z\right) = 0$, for which the roots are thus $r_1 = 1/1.5 = .67$ and $r_2 = -1/.8 = -1.25$. Since $r_1 = .67 < 1$, this is *not* a stationary AR(2) model. Figure 5.9, obtained using the command

```
unit.circle.wge(real=c(-1.25,.67),imaginary=c(0,0))
```

is a plot of the roots which illustrates that $r_1 = .67$ falls inside the unit circle and thus the model exhibits "explosive" behavior such as that seen in Figure 5.6(b), even though $r_2 = -1.25$ lies outside the unit circle.

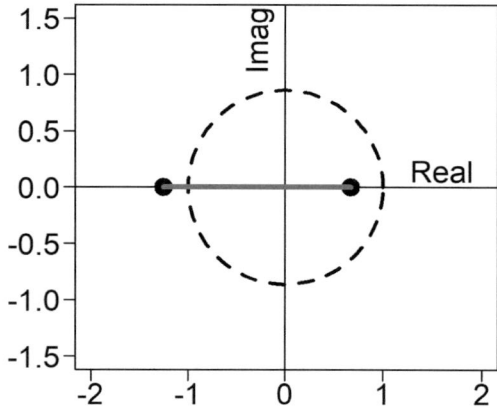

QR 5.5
Characteristic
Equation of AR(2)

FIGURE 5.9   Plots of $r_1 = .67$ and $r_2 = -1.25$ in the complex plane.

Model (c) $X_t - 1.6X_{t-1} + .8X_{t-2} = a_t$

The model in operator form is $\left(1 - 1.6B + .8B^2\right)X_t = a_t$ and the characteristic equation is $1 - 1.6z + .8z^2 = 0$. This 2nd-order polynomial cannot be factored into two 1st-order factors with real coefficients, so we need to use the quadratic formula, from which we know that the roots of $az^2 + bz + c = 0$ are $r_1, r_2 = r_1^* = \dfrac{-b \pm \sqrt{b^2 - 4ac}}{2a}$. In our case, $a = .8$, $b = -1.6$, and $c = 1$. Going through the mathematics, we find that

$$r_1, r_1^* = \frac{1.6 \pm \sqrt{(-1.6)^2 - 4(.8)(1)}}{2(.8)} = 1 \pm .5i, \text{ and in both cases, } |r| = |r^*| = \sqrt{1^2 + .5^2} = 1.12. \text{ Thus, by Theorem}$$

5.2, Model (c) is a stationary AR(2) model. Figure 5.10, obtained using the command

```
unit.circle.wge(real=c(1,1),imaginary=c(.5,-.5))
```

shows that both roots are outside the unit circle.

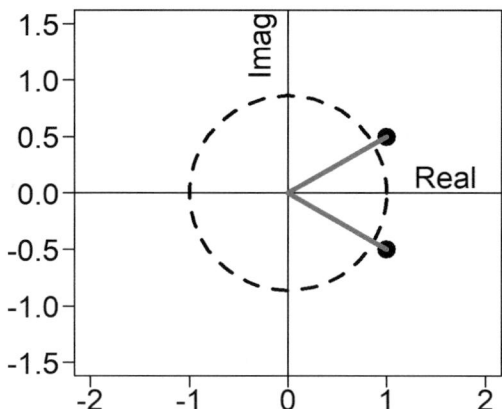

QR 5.6
Characteristic
Equation with
Complex Roots

FIGURE 5.10   Plots of complex conjugate roots $1 \pm .5i$ in the complex plane.

5.1.2.3 Stationary AR(2) with Two Real Roots

Model (a) is an example of a stationary AR(2) model for which the characteristic equation has two real roots, 1.11 and −1.33. In this case the roots have opposite signs, but they could also both be positive or both be negative. The factored characteristic polynomial for Model (a) is $(1 - .9z)(1 + .7z)$, and each of these 1st-order factors are typical of AR(1) models. For example, the AR(1) model $(1 - .9B)Y_t = a_t$, for some Y_t, has the characteristic equation $1 - .9z = 0$. We examined such a model in Section 5.1.1.5 where we saw that the model $(1 - .9B)Y_t = a_t$ will have realizations with wandering behavior, exponentially damping autocorrelations, and a spectral density with a peak at $f = 0$. In Section 5.1.1.7, we discussed the AR(1) model with characteristic equation $1 + .7z = 0$. There we saw oscillating behavior in the realization, oscillating autocorrelations, and a peak in the spectral density at $f = .5$.

Interestingly, the AR(2) model (Model (a))

$$X_t - .2X_{t-1} - .63X_{t-2} = (1 - .9B)(1 + .7B)X_t = a_t \tag{5.21}$$

will show a "combination" of these two 1st-order behaviors. The model-based autocorrelations and spectral density are shown in Figure 5.11(a) and (b), respectively. As a result of the discussion above, it follows that the spectral density in Figure 5.11(b) should show peaks at both $f = 0$ and $f = .5$. (Compare with Figures 5.1(b) and 5.5(b)). Consistent with the discussion in Section 5.1.1.5, because the 1st-order factor, $1 - .9B$, is associated with a root closer to the unit circle than $1 + .7B$, the peak of the spectral density at $f = 0$ is higher than it is at $f = .5$. The autocorrelations in Figure 5.11(a) show a damped exponential with a slight indication of oscillating behavior associated with the $(1 + .7B)$ factor.

Figure 5.11(c) shows a realization of length $n = 150$ generated from $(1-.9B)(1+.7B)X_t = a_t$ along with sample-based estimates of the autocorrelations and spectral density. The realization displays some "wandering behavior" with superimposed "oscillatory behavior". (Compare with Figure 5.1(c) and Figure 5.5(c).) The sample autocorrelations at early lags are similar to the model-based autocorrelations, and the dominant features of the Parzen spectral density estimate in Figure 5.11(e) are peaks at $f = 0$ and at $f = .5$. The plots in Figure 5.11 are obtained using the ***tswge*** commands

```
plotts.true.wge(phi=c(.2,.63),plot.data=FALSE)
x=gen.arma.wge(n=150,phi=c(.2,.63),sn=13)
plotts.sample.wge(x)
```

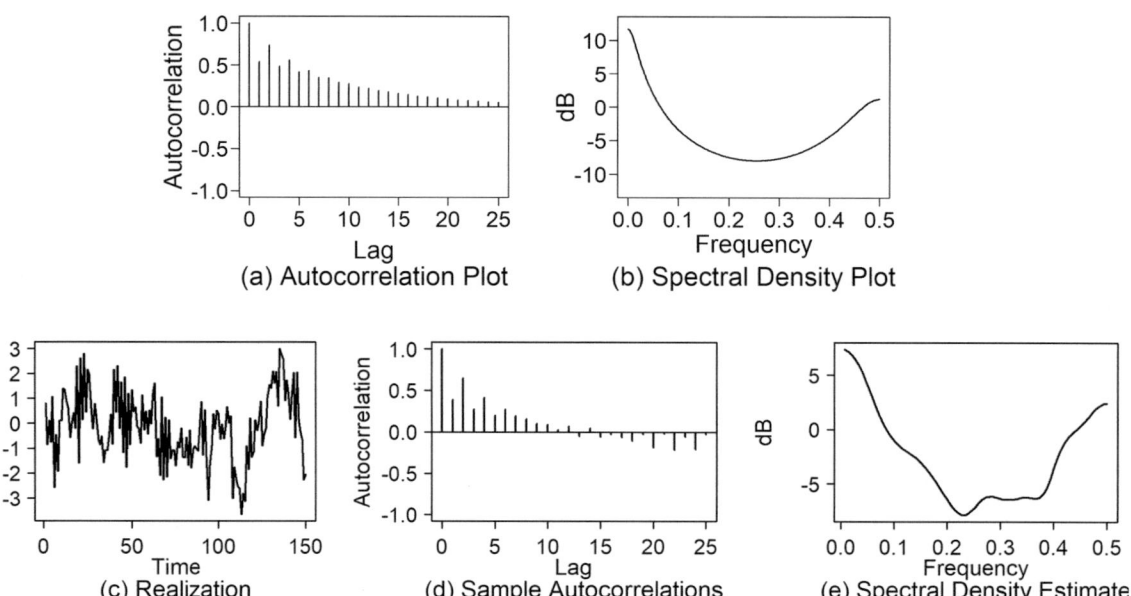

FIGURE 5.11   (a) Model-based autocorrelations and (b) spectral density for the model $(1-.9B)(1+.7B)X_t = a_t$, (c) realization of length $n = 150$ from this model, (d) sample autocorrelations and (e) Parzen spectral density estimate based on the realization in (c).

In the AR(2) model (5.21), the positive real root $(1/.9)$ is closer to one in absolute value than is the negative real root $(1/(-.7))$. The consequence of this, which we will re-visit in the case of the AR(p) model in Section 5.1.3.5, is that the behavior associated with the positive real root is "stronger". The model $(1-1.6B+.63B^2)X_t = (1-.9B)(1-.7B)X_t = a_t$ has two positive real roots while the model $(1+1.6B+.63B^2)X_t = (1+.9B)(1+.7B)X_t = a_t$ has two negative real roots. The behavior of these two models is examined in Problem 5.2.

5.1.2.4 Stationary AR(2) with Complex Conjugate Roots

Returning to Model (c) in Example 5.2, we note that Model (c) is stationary because the associated characteristic equation for the model, $1-1.6z+.8z^2 = 0$, has the complex conjugate roots $1 \pm .5i$ which are both greater than one in absolute value.

The following are facts concerning AR(2) models associated with complex roots.

(a) A stationary AR(2) model $(1 - \phi_1 B - \phi_2 B^2)X_t = a_t$ with complex conjugate roots has a model-based autocorrelation function, ρ_k, which has the appearance of a damped sinusoidal curve with system frequency[2]

$$f_0 = \frac{1}{2\pi} \cos^{-1} \left(\frac{\phi_1}{2\sqrt{-\phi_2}} \right).^3 \tag{5.22}$$

(b) Realizations will tend to be pseudo-sinusoidal and associated with frequency behavior at approximately f_0.

(c) The model-based spectral density

$$S_X(f) = \frac{\sigma_a^2}{\sigma_X^2 \left| 1 - \phi_1 e^{-2\pi i f} - \phi_2 e^{-4\pi i f} \right|^2} \tag{5.23}$$

will have a peak at about f_0. The spectral density in (5.23) can be calculated using *tswge* functions `true.arma.spec.wge` and `plotts.true.wge`.

AR(2) Model (c):
For Model (c), $\phi_1 = 1.6$ and $\phi_2 = -.8$ (again – watch the signs!), so f_0 in (5.22) is given by

$$f_0 = \frac{1}{2\pi} \cos^{-1} \left(\frac{\phi_1}{2\sqrt{-\phi_2}} \right)$$

$$= \frac{1}{2\pi} \cos^{-1} \left(\frac{1.6}{2\sqrt{.8}} \right)$$

$$= .0738. \tag{5.24}$$

Figures 5.12(a)−(b) display the model-based autocorrelations and spectral density for Model (c). The model-based autocorrelations in Figure 5.12(a) have a damped sinusoidal appearance with period length of about 14, which is consistent with the expected system frequency $f_0 = 1/14 \approx .07$. Note that the true spectral density in Figure 5.12(b) has a peak slightly below $f = .1$.

Figure 5.12(c) shows a realization of length $n = 150$ generated from Model (c). Figures 5.12(d) and (e) show the sample-based estimates of the autocorrelations and spectral density, respectively. We first note that the sample autocorrelations are similar to the model-based autocorrelations in that both show a damped sinusoidal behavior with a period of about 14 lags, which is consistent with a frequency of about .07. The Parzen spectral density estimate based on the realization in Figure 5.12(c) shows a distinct peak at about $f = .07$, which is not as sharp as the peak in Figure 5.12(b). The realization has the appearance of a pseudo-sinusoidal curve that goes through about 11−12 periods in the realization of length $n = 150$.

---

2 When we use the term "system frequency", we are referring to f_0 as given in (5.22). Specifically, as we discuss data in terms of their frequency behavior, we will use the notation f, unless reference is specifically made to the frequency calculated in (5.22).

3 Note that (5.22) states that $\cos(2\pi f_0) = \phi_1 / (2\sqrt{-\phi_2})$. Also note that $2\pi f_0$ is in radians.

That is, the period length is about $150 / 11.5 = 13$, which corresponds to a frequency of $f = 1 / 13 = .077$. Figures 5.12(a)–(e) were obtained using the commands

```
plotts.true.wge(phi=c(1.6,-.8),plot.data=FALSE)
x=gen.arma.wge(n=150,phi=c(1.6,-.8),sn=19)
plotts.sample.wge(x)
```

Another AR(2) model associated with complex roots:

Figures 5.13(a) and (b) show model-based autocorrelations and spectral density, respectively, for the model $\left(1 + .5B + .8B^2\right)X_t = a_t$. For this model, the roots of the characteristic equation $1 + .5z + .8z^2 = 0$ are $-.3 \pm 1.1i$, which are complex and outside the unit circle. Consequently, this is also a stationary model. Using (5.22) we obtain $f_0 = \dfrac{1}{2\pi}\cos^{-1}\left(\dfrac{-.5}{2\sqrt{.8}}\right) = .3$. The model-based autocorrelations in Figure 5.13(a) have a damped sinusoidal appearance with period length about 3, which is consistent with the expected system frequency $f_0 \approx .33$. Because of the short period length, the sinusoidal nature of the autocorrelations is not apparent and has an oscillating appearance. The spectral density in Figure 5.13(b) has a peak at about $f = .3$.

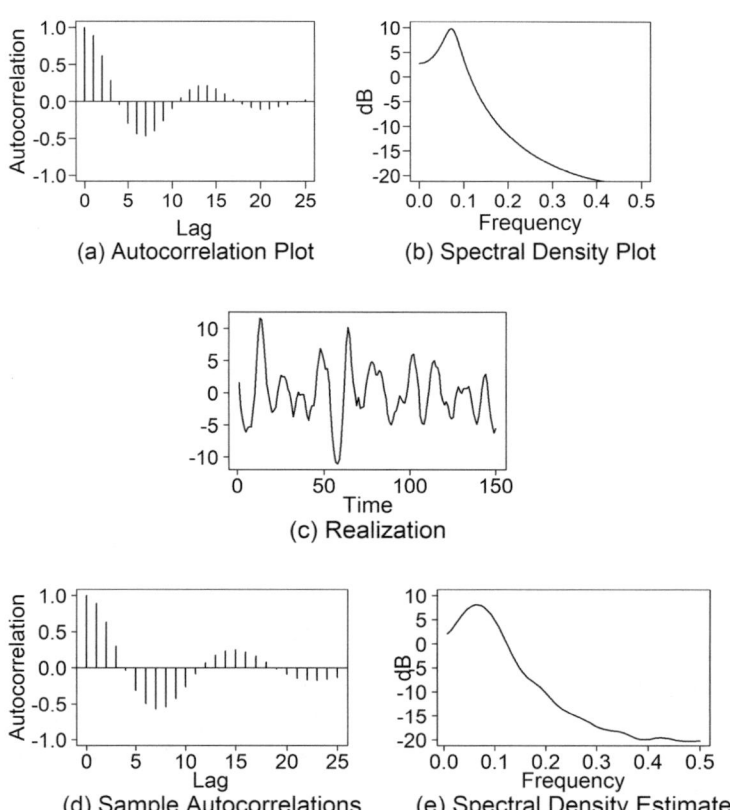

(a) Autocorrelation Plot   (b) Spectral Density Plot

(c) Realization

(d) Sample Autocorrelations   (e) Spectral Density Estimate

FIGURE 5.12 (a) Model-based autocorrrelations and (b) spectral density for $\left(1 - 1.6B + .8B^2\right)X_t = a_t$, (c) realization of length $n = 150$ from $\left(1 - 1.6B + .8B^2\right)X_t = a_t$, (d) sample autocorrelations and (e) Parzen spectral density estimate based on the realization in (c).

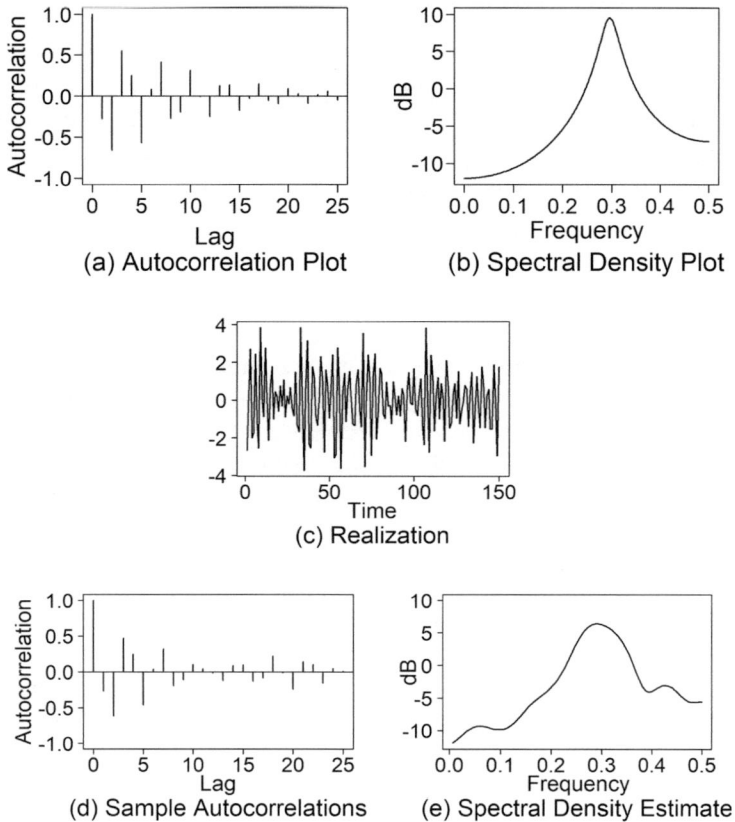

FIGURE 5.13   (a) Model-based autocorrrelations and (b) spectral density for $\left(1+.5B+.8B^2\right)X_t = a_t$, (c) realization of length $n = 150$ from $\left(1+.5B+.8B^2\right)X_t = a_t$, (d) sample autocorrelations and (e) Parzen spectral density estimate based on the realization in (c).

Figures 5.13(c)–(e) show a realization of length $n = 150$ generated from $\left(1+.5B+.8B^2\right)X_t = a_t$ along with sample-based estimates of the autocorrelations and spectral density, respectively. The realization has high-frequency behavior, and there are about 49 periods (do you agree?). That is, the period length is about 150/49=3.1 and $f \approx 1/3.1 \approx .3$, which is consistent with the calculated f_0. The sample autocorrelations in Figure 5.13(d) have a damped oscillating appearance (again, the sinusoidal nature is not obvious) with period length of about 3, and the Parzen spectral density estimate has a peak at about $f = .3$, which is consistent with the model-based spectral density in Figure 5.13(b). Figures 5.13(a)–(e) were obtained using the *tswge* commands

```
plotts.true.wge(phi=c(-.5,-.8),plot.data=FALSE)
x=gen.arma.wge(n=150,phi=c(-.5,-.8),sn=19)
plotts.sample.wge(x)
```

When discussing AR(1) and AR(2) models, we have typically plotted the model-based autocorrelations and spectral densities. We have then plotted realizations from the model of interest, sample autocorrelations, and Parzen spectral density estimates based on the generated realizations. (See, for example, Figure 5.12 and Figure 5.13.) We went through this detail to show what the "theoretical values" are and how these were estimated from data. We have seen that the sample autocorrelations and Parzen spectral density estimates have done a good job of estimating these "theoretical/model-based" quantities.

> **Key Points**
>
> 1. When analyzing real data, we will not "know" model-based or theoretical autocorrelations and spectral densities.
> 2. From this point on when discussing models, we will typically not plot model-based autocorrelations and spectral densities unless they are needed for clarity.

Example 5.3 Canadian Lynx Data

Figure 5.14(a) (dataset **lynx** in *tswge*) shows the annual number of Canadian Lynx trapped in the Mackenzie River district of the Northwest Canada for the period 1821–1934 and Figure 5.14(b) shows the log (base 10) of the annual numbers trapped. These plots were previously given in Figures 1.11 and 1.12, respectively. As mentioned previously, this dataset that interested researchers because of the cyclic behavior with cycle lengths of about 10 years. See, for example, Tong (1977), Bhansali (1979), Woodward and Gray (1983), and Woodward et al. (2017). The log of the lynx data is usually used for analysis because the resulting peaks and troughs of the cycles behave in similar manners, and thus the data are more AR-like. Also shown in Figure 5.14(c) and (d) are the sample autocorrelations and the Parzen spectral density estimate of the log-lynx data. The sample autocorrelations have a damped sinusoidal appearance with period of about 10, and the spectral density estimate in Figure 5.14(d) has a peak at about $f = .10$. The log-lynx data have the appearance of AR(2) data with a pair of complex conjugate roots associated with system frequency of about $f_0 = .10$. If an AR(2) model is fit to the data, the model is $X_t - 1.38X_{t-1} + .75X_{t-2} = a_t$. Using (5.22), the system frequency associated with this AR(2) model is $f_0 = .10$.

The plots in Figure 5.14 can be obtained using the commands

```
data(tswge)
plotts.wge(lynx)
llynx=log10(lynx)
plotts.sample.wge(llynx)
```

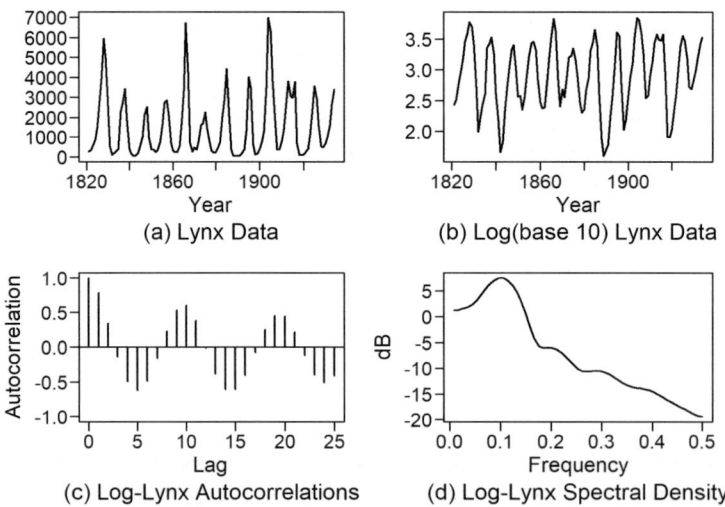

FIGURE 5.14 (a) Number of lynx trapped, (b) natural log of the data in (a), (c) sample autocorrelations of the data in (b), and (d) Parzen spectral density estimate of the data in (b).

5.1.2.5 Summary of AR(1) and AR(2) Behavior

Before proceeding to further examination of general AR(p) models, we summarize the information we have learned about AR(1) and AR(2) models:

(a) *Facts about the AR(1) model $X_t - \phi_1 X_{t-1} = a_t$.*

    (i) An AR(1) process is stationary if and only if $|\phi_1| < 1$, or equivalently, if the root $r_1 = \phi_1^{-1}$ of the characteristic equation is greater than one in absolute value.

    (ii) The model-based autocorrelation function of a stationary AR(1) process is given by $\rho_k = \phi_1^{|k|}$. This autocorrelation function is a damped exponential if $\phi_1 > 0$ and an oscillating damped exponential if $\phi_1 < 0$.

    (iii) Realizations from an AR(1) model with $\phi_1 > 0$ tend to be aperiodic with a general "wandering" behavior. When $\phi_1 < 0$, the realizations will tend to oscillate back and forth across the mean.

    (iv) The spectral density $S_X(f)$ has a peak at $f = 0$ if $\phi_1 > 0$ and at $f = 0.5$ if $\phi_1 < 0$.

(b) *Facts about an AR(2) model $X_t - \phi_1 X_{t-1} - \phi_2 X_{t-2} = a_t$.*

An AR(2) process is stationary if and only if the roots of the characteristic equation $1 - \phi_1 z - \phi_2 z^2 = 0$ lie outside the unit circle.

The features of the AR(2) process depend on the nature of the roots of the characteristic equation.

Case 1: The roots of the characteristic equation are real.

Each real root corresponds to "1st-order behavior" in the realizations, autocorrelations, and spectral density.

Case 2: The roots of the characteristic equation are complex.

    (i) The model-based autocorrelation function is a damped sinusoidal with system frequency

$$f_0 = \frac{1}{2\pi} \cos^{-1}\left(\frac{\phi_1}{2\sqrt{-\phi_2}} \right) \text{ given in (5.22).}$$

    (ii) Realizations from an AR(2) model with complex conjugate roots will tend to be pseudo-sinusoidal with frequency f_0 given in (5.22), that is, with period $1/f_0$.

    (iii) The spectral density will have a peak near f_0 given in (5.22).

Key Point: As we will see in Section 5.1.3, knowing the behavior of AR(1) and AR(2) models is the key to understanding the behavior of an AR(p) model.

5.1.3 The AR(p) Models

We are now ready to discuss the general AR(p) model

$$X_t = \beta + \phi_1 X_{t-1} + \phi_2 X_{t-2} + \cdots + \phi_p X_{t-p} + a_t \tag{5.25}$$

defined in Definition 5.1. Note again that the AR(p) "looks" like a multiple regression equation, where in this case, the "independent variables" are the p previous values of the "dependent variable" X_t.

Another way to write (5.25), after rearranging terms, is

$$X_t - \mu - \phi_1\left(X_{t-1} - \mu\right) - \phi_2\left(X_{t-2} - \mu\right) - \cdots - \phi_p\left(X_{t-p} - \mu\right) = a_t. \tag{5.26}$$

As in the case of AR(1) and AR(2) models, we will frequently express the AR(p) in the zero-mean form

$$X_t - \phi_1 X_{t-1} - \phi_2 X_{t-2} - \cdots - \phi_p X_{t-p} = a_t. \tag{5.27}$$

Equations (5.25)–(5.27) give the impression that an AR(p) model will be much more complicated to deal with than an AR(1) or AR(2) model. While this is somewhat true, as mentioned in the above Key Point, an understanding of the characteristics of AR(1) and AR(2) models leads directly to understanding the behavior of an AR(p) model. We will discuss this shortly.

5.1.3.1 Facts about the AR(p) Model

(a) $E\left[X_t\right] = \mu$, for the "non-zero mean" form of the AR(p) model in (5.25) and (5.26).

(b) The process variance is $\sigma_X^2 = \gamma_0 = \dfrac{\sigma_a^2}{1 - \phi_1\rho_1 - \phi_2\rho_2 - \cdots - \phi_p\rho_p}$, which is constant and finite when X_t is stationary.

(c) The autocorrelations of an AR(p) process satisfy

$$\rho_k = \phi_1\rho_{k-1} + \phi_2\rho_{k-2} + \cdots + \phi_p, \quad k > 0. \tag{5.28}$$

Equation (5.28) is a generalization of (5.17) for the AR(2) case, and it leads to the $p \times p$ *Yule-Walker Equations*

$$\rho_1 = \phi_1 + \phi_2\rho_1 + \cdots + \phi_p\rho_{p-1}$$

$$\rho_2 = \phi_1\rho_1 + \phi_2 + \cdots + \phi_p\rho_{p-2}$$

$$\vdots$$

$$\rho_p = \phi_1\rho_{p-1} + \phi_2\rho_{p-2} + \cdots + \phi_p. \tag{5.29}$$

Analogous to the AR(2) case, knowing ϕ_1, $\phi_2, \cdots, \phi_p$ allows us to solve this $p \times p$ system of equations for ρ_k, $k = 1, \ldots, p$. Model-based autocorrelations, ρ_k, $k > p$, can be computed using the recursion $\phi_1\rho_{k-1} + \phi_2\rho_{k-2} + \cdots + \phi_p\rho_{k-p}$. Not surprisingly, we use computer functions to perform these calculations. The *tswge* functions `true.arma.aut.wge` and `plotts.true.wge` use the Durbin Levinson algorithm to calculate model-based autocorrelations, ρ_k, $k = 1, 2, \ldots$. See Durbin (1960), Levinson (1947), and Woodward et al. (2017).

(d) The spectral density of an AR$\left(p\right)$ model is given by

$$S_X\left(f\right) = \frac{\sigma_a^2}{\gamma_0} \frac{1}{\left|1 - \phi_1 e^{-2\pi i f} - \phi_2 e^{-4\pi i f} - \cdots - \phi_p e^{-2p\pi i f}\right|^2}. \tag{5.30}$$

Key Points

1. As in the AR(1) and AR(2) cases, the behavior of the realizations, autocorrelations, and spectral density depend on the roots of the characteristic equation defined below for the AR(p) model.
2. If the coefficients $\rho_1, \rho_2, \ldots, \rho_p$ of an AR(p) model are known, then the Yule-Walker equations can be used to solve for the model-based autocorrelations, $\phi_1, \phi_2, \ldots, \phi_p$.
3. Key Point 2 indicates that, using (5.30), the spectral density of an AR(p) process can be calculated if $\rho_1, \rho_2, \ldots, \rho_p$ are known.

5.1.3.2 Operator Notation and Characteristic Equation for an AR(p)

The AR(p) model in (5.26) can be written in operator notation as

$$\left(1 - \phi_1 B - \phi_2 B^2 - \cdots - \phi_p B^p\right)\left(X_t - \mu\right) = a_t \tag{5.31}$$

or by using a "shorthand" notation, $\phi(B)(X_t - \mu) = a_t$, where $\phi(B)$ is the pth order operator $\phi(B) = 1 - \phi_1 B - \phi_2 B^2 - \cdots - \phi_p B^p$. Converting the operator $\phi(B)$ to the algebraic quantity $\phi(z)$ results in the general AR(p) characteristic polynomial $\phi(z) = 1 - \phi_1 z - \phi_2 z^2 - \cdots - \phi_p z^p$. The corresponding AR($p$) characteristic equation is

$$\phi(z) = 1 - \phi_1 z - \phi_2 z^2 - \cdots - \phi_p z^p = 0. \tag{5.32}$$

The characteristic equation has p roots $r_1, r_2, \ldots, r_p$ which are real and/or complex, where the complex roots appear as conjugate pairs and some roots may be repeated.

Theorem 5.3 is the fundamental result regarding the stationarity of AR(p) processes. This result is the extension of Theorems 5.1 and 5.2 to AR(p) processes.

Theorem 5.3 An AR(p) process is stationary if and only if all of the roots of the characteristic equation are greater than one in absolute value.

Example 5.4 An AR(4) Model
Consider the AR(4) model

$$X_t - .13X_{t-1} - 1.4414X_{t-2} + .0326X_{t-3} + .8865X_{t-4} = a_t, \tag{5.33}$$

where $\sigma_a^2 = 1$. The operator notation for this model is

$$\left(1 - .13B - 1.4414B^2 + .0326B^3 + .8865B^4\right)X_t = a_t, \tag{5.34}$$

and the corresponding characteristic equation is

$$1 - .13z - 1.4414z^2 + .0326z^3 + .8865z^4 = 0. \tag{5.35}$$

Figure 5.15(a) is a realization of length $n = 200$ from (5.33), and Figures 5.15(b) and (c) are the associated sample autocorrelations and Parzen spectral density estimate, respectively. These plots were obtained using the commands

```
x=gen.arma.wge(n=200,phi=c(0.1300,1.4414,-.0326,-.8865),sn=9310,plot=FALSE)
plotts.sample.wge(x)
```

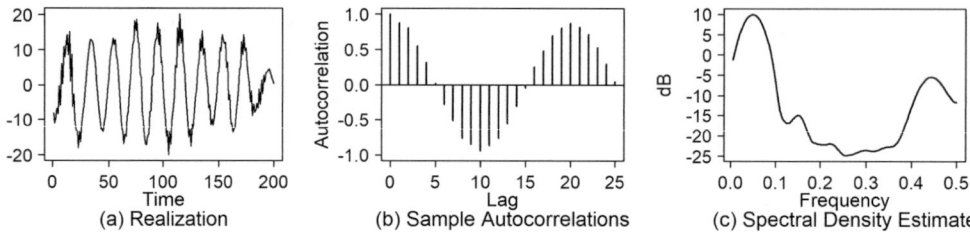

FIGURE 5.15   (a) Realization from the AR(4) model in (5.33), (b) sample autocorrelations, and (c) Parzen spectral density estimate calculated from the data in (a).

Note that Figure 5.15(a) was previously shown in Figure 4.20(a). The **gen.arma.wge** code specifies that $\phi_1 = .13$, $\phi_2 = 1.4414$, $\phi_2 = -.0326$, and $\phi_4 = -.8865$. These correspond to the coefficients in (5.33) (again −be careful about signs). The seed is set to 9310, and we select default values, $\mu = 0$ and $\sigma_a^2 = 1$. As noted in Chapter 4, this realization has a low-frequency behavior of about $f = .05$ and a high-frequency content at approximately $f = .45$.

Section 5.1.2.5 summarizes information about how the behaviors of realizations, autocorrelations, and spectral densities are related to the coefficients of AR(1) and AR(2) models. This leads us to the following question.

> **Question:** What is it about the coefficients in (5.33) that created the behavior of the realization, sample autocorrelations, and spectral density in Figure 5.15?

We begin to answer this question by factoring the AR(p) characteristic polynomial.

5.1.3.3 Factoring the AR(p) Characteristic Polynomial

The roots of a quadratic equation can be found by using the quadratic formula. However, things get more "sticky" for polynomial orders greater than two. The cubic equation

$$1 - 2.1z + 1.6z^2 - .4z^3 = 0$$

can be factored into the form $(1 - .5z)(1 - 1.6z + .8z^2) = 0$. (Do you remember how to do this?) Based on this factorization, the roots are found by setting $(1 - .5z) = 0$, in which case $r_1 = 1/.5 = 2$, and by setting $(1 - 1.6z + .8z^2) = 0$, in which case the roots are $r_2 = 1 + .5i$ and $r_3 = 1 - .5i$. That is, this AR(3) will have 1st-order behavior associated with $1 - .5B$ (that is, a frequency of zero), 2nd-order cyclic behavior with system frequency $f_0 = .07$, and the process is stationary because all roots are outside the unit circle.

The 5th-order polynomial equation

$$1 - 2z + 1.94z^2 - 1.32z^3 + .72z^4 - .16z^5 = 0,$$

can be factored as $(1-.4z)(1-1.6z+.8z^2)(1+.5z^2)=0$. (Good luck with this one!) The resulting roots are $r_1 = 1/.4 = 2.5$, $r_2 = 1+.5i$, $r_3 = 1-.5i$, $r_4 = \sqrt{2}i$, and $r_5 = -\sqrt{2}i$. Realizations will have behavior that is a combination of 1st-order factor $1-.5B$ and cyclic behavior associated with the system frequencies of the 2nd-order factors $1-1.6B+.8B^2$ ($f_0 = .07$) and $1+.4B^2$ ($f_0 = .25$). The process is stationary because all roots are outside the unit circle. (It is a useful exercise to check this.)

Higher order polynomial equations, such as the 5th-order equation above, must be solved using numerical methods. See for example, Press, Teukolsky, Vetterling, and Flannery (2007).

Note that in each case above, we

- (a) factored the polynomial (into first- and/or second-order factors), and
- (b) found the roots based on setting each of these factors equal to zero.

Theorem 5.4 below provides a generalization of these observations.

Theorem 5.4: The pth order polynomial $1-\phi_1 z - \phi_2 z^2 - \cdots - \phi_p z^p$ can always be factored as a product of

- (a) 1st-order (linear) factors which are associated with real roots
- (b) 2nd-order (quadratic) factors for which the roots are complex conjugate pairs[4]

Key Points

1. $\phi(z) = 1 - \phi_1 z - \phi_2 z^2 - \cdots - \phi_p z^p = 0$ can *always be* factored as a product of
   (a) 1st-order (linear) factors – associated with the real roots
   (b) 2nd-order polynomial (quadratic) factors – associated with complex conjugate pairs
2. Roots of $\phi(z) = 0$ can be found by factoring the polynomial into the 1st- and 2nd-order factors and then setting each factor equal to zero.
3. A 2nd-order factor associated with a complex conjugate pair is referred to as an *irreducible 2nd-order factor.*
4. While 1st- and 2nd-order equations can be solved easily, polynomial equations of 3rd-order and above are difficult or impossible to solve algebraically (that is, by using something similar to the quadratic formula). For this reason, such higher order polynomials are solved numerically using a computer.

5.1.3.4 Factor Tables for AR(p) Models

Theorem 5.4 states that any pth-order polynomial can be expressed as a product of 1st-order and/or irreducible 2nd-order factors. Understanding the 1st- and 2nd-order factors is the "key" to understanding the AR(p) model.

---

4 A 2nd-order factor associated with complex conjugate root pairs can be factored further to a product of two linear factors, but these two factors have complex (not real) coefficients.

> **Key Points**
>
> 1. 1st-order and 2nd-order factors like $(1 - \alpha_1 B)$ and $(1 - \alpha_1 B - \alpha_2 B^2)$ serve as building blocks (or the DNA) of an AR(p) model.
> 2. Features of an AR(p) model are simply a combination of 1st- and 2nd-order features.
>   - That is, there is no new "third-order feature."

Example 5.4 (Revisited)

Consider again the AR(4) model in (5.33):

$$\left(1 - .13B - 1.4414B^2 + .0326B^3 + .8865B^4\right)X_t = a_t.$$

The associated characteristic equation is

$$1 - .13z - 1.4414z^2 + .0326z^3 + .8865z^4 = 0.$$

The factored form (obtained numerically) is

$$\left(1 - 1.89B + .985B^2\right)\left(1 + 1.76B + .9B^2\right) = 0.$$

The *tswge* function **factor.wge** uses numerical methods to produce these factors and outputs them in a "Factor Table". The factor table is a very useful tool for quickly summarizing the "DNA" of an AR(p) model regarding the 1st- and 2nd-order factors. The command

```
factor.wge(phi=c(.13,1.4414,-.0326,-.8865))
```

produces the output:

```
                    AR Factor Table
Factor              Roots          Abs Recip  System Freq
1-1.8900B+0.9850B^2 0.9594+-0.3079i  0.9925     0.0494
1+1.7600B+0.9000B^2 -0.9778+-0.3938i 0.9487     0.4391
```

Table 5.1 shows the format that will be used to display factor tables in this book.

QR 5.7
Factored Models

TABLE 5.1 Factor Table for $X_t - .13X_{t-1} - 1.4414X_{t-2} + .0326X_{t-3} + .8864X_{t-4} = a_t$

FACTOR	ROOTS	$|r|^{-1}$	f_0
$1 - 1.89B + .985B^2$	$.96 \pm .31i$	.99	.05
$1 + 1.76B + .9B^2$	$-.98 \pm .39i$	.95	.44

The first thing we notice is that the factor table shows the (1st- and 2nd-order) factors of the model. In this case, there are two 2nd-order factors, $1 - 1.89B + .985B^2$ and $1 + 1.76B + .9B^2$, as noted above. In the following, we summarize the wealth of information contained in the factor table.

Factor Table Format

For each 1ˢᵗ-order or irreducible 2ⁿᵈ-order factor of the model, the factor table is displayed in four columns:

Column 1: The first and/or irreducible 2ⁿᵈ-order factors.

Column 2: Roots of the characteristic equations associated with the factors.

Column 3: The absolute reciprocal of the roots. This measure provides two key pieces of information:

    (a) *Stationarity*: If the roots are all outside the unit circle, the process is stationary. Because we table the absolute values of the *reciprocal* of the roots, the check for stationarity is whether all of these values are *less* than one.

    (b) *A measure of how close the roots are to the unit circle*. (The closer the absolute value of the reciprocal is to one, the closer the root is to the unit circle.) We will see that roots closest to the unit circle dominate the behavior of a stationary AR process.

Column 4: System frequencies f_0. For stationary models:

    (a) $f_0 = 0$ for 1ˢᵗ-order factors with positive real roots
    (b) $f_0 = .5$ for 1ˢᵗ-order factors with negative real roots
    (c) f_0 is given by (5.22) for irreducible 2ⁿᵈ-order factors

---

Example 5.5 Consider the following AR(p) models.

    (A) $X_t - 1.95X_{t-1} + 1.85X_{t-2} - .855X_{t-3} = a_t$
    (B) $X_t - 2.6X_{t-1} + 3.34X_{t-2} - 2.46X_{t-3} + .9024X_{t-4} = a_t$
    (C) $X_t - 2.85X_{t-1} + 3.24X_{t-2} - 2.03X_{t-3} + .6X_{t-4} = a_t$

Figure 5.16 shows realizations from these three models given in random order. The code that generated the realizations will not be given here because that would defeat the purpose of the following question. Sorry!

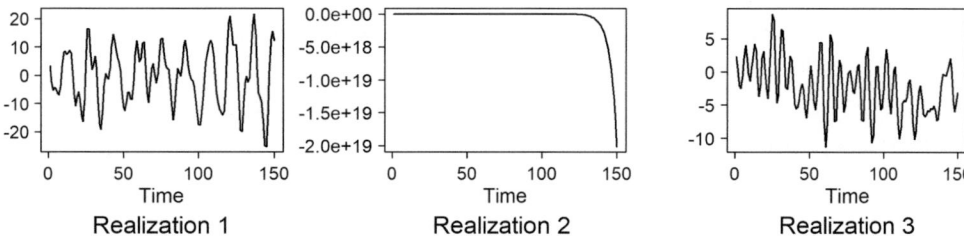

FIGURE 5.16 Realizations from AR(p) Models (A), (B), and (C) in random order.

Question: By examining the coefficients of the models, can you identify which models are stationary and which realizations correspond to which model?

You Can't? Don't Worry. Neither Can We! The coefficients themselves are not very informative.

Our Solution: Use `factor.wge` to solve the associated characteristic equations and provide key information.

Model (A): $X_t - 1.95X_{t-1} + 1.85X_{t-2} - .855X_{t-3} = a_t$

The command

```
factor.wge(phi=c(1.95,-1.85,.855))
```

produces the following factor table.

TABLE 5.2   Factor Table for Model (A): $X_t - 1.95X_{t-1} + 1.85X_{t-2} - .855X_{t-3} = a_t$

FACTOR	ROOTS	$|r|^{-1}$	f_0
$1 - .95B$	1.05	.95	.00
$1 - B + .9B^2$	$.55 \pm .90i$	.95	.16

The factorization shows that the properties of this AR(3) model will be a combination of 1st-order behavior associated with $1 - .95B$ (that is, aperiodic wandering behavior with a "system frequency" of $f_0 = 0$) and 2nd-order   behavior associated with $1 - B + .9B^2$, that is, pseudo-sinusoidal realizations associated with system frequency $f_0 = \dfrac{1}{2\pi} cos^{-1}\left(\dfrac{1}{2\sqrt{.9}}\right) = .16$. The root associated with the 1st-order factor is $1/.95 = 1.05$, and the roots associated with the 2nd-order factor are $.55 \pm .90i$. The factor table shows that the absolute value of the reciprocal of all three roots is .95, which indicates stationarity because the reciprocals are all less than one in absolute value.

Based on the factor table in Table 5.2 and the above discussion, this model should produce realizations that show random wandering (associated with $1 - .95B$) and pseudo-sinusoidal behavior with frequency of about .16 (or period $1/.16 = 6$). Realization 3 in Figure 5.16 has pseudo-sinusoidal behavior along a wandering path. By counting, we see that there are about 23 cycles in the realization of length $n = 150$, implying that the period length is about $150/23 = 6.5$, which is consistent with a system frequency of $f_0 = .16$.

Conclusion: Realization 3 is from Model (A).

Figure 5.17 shows Realization 3, along with its sample autocorrelations and Parzen spectral density estimate. In addition to the realization behavior noted above, the sample autocorrelations have a sinusoidal behavior (2nd-order) that is not symmetric about zero, but instead seems to follow a damped exponential path (1st-order). Also the spectral density estimate has a peak at zero (associated with a 1st-order factor with positive real root) and a peak slightly below $f = .2$. These are consistent with the information in the factor table. Figure 5.17 can be obtained using the commands

```
x=gen.arma.wge(n=150,phi=c(1.95,-1.85,.855),sn=129,plot=FALSE)
plotts.sample.wge(x)
```

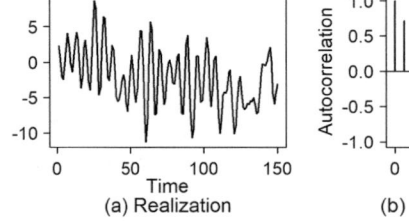 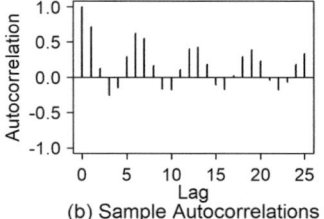 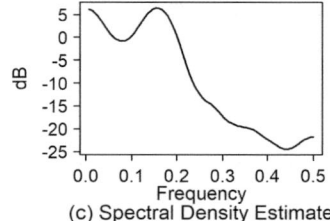

FIGURE 5.17   (a) Realization 3 in Figure 5.16, (b) sample autocorrelations, and (c) Pzen spectral density estimate calculated from this realization.

Model (B): $X_t - 2.6X_{t-1} + 3.34X_{t-2} - 2.46X_{t-3} + .9024X_{t-4} = a_t$

The command

```
factor.wge(phi=c(2.6,-3.34,2.46,-.9024))
```

produces the factor table in Table 5.3.

TABLE 5.3   Factor Table for Model (B): $X_t - 2.6X_{t-1} + 3.34X_{t-2} - 2.46X_{t-3} + .9024X_{t-4} = a_t$

FACTOR	ROOTS	$|r|^{-1}$	f_0
$1 - 1.8B + .96B^2$	$.94 \pm .40i$	.98	.065
$1 - .8B + .94B^2$	$.43 \pm .94i$	.97	.182

In this case, the 4th-order polynomial factors into two irreducible 2nd-order polynomials. That is, the characteristic equation has two pairs of complex conjugate roots. The factor table shows that the factored form is $(1 - 1.8B + .96B^2)(1 - .8B + .94B^2)X_t = a_t$. The model is s*tationary* because all values in column 3 are less than one. The factor table indicates that the data will have cyclic behavior associated with frequencies $f_0 = .065$ and $f_0 = .182$. Realization 1 in Figure 5.16 goes through about 10 cycles in the 150 time points, so the period is about 15 and the associated frequency is about $f = 1/15 = .067$. Also, there appears to be a higher frequency behavior, but it is difficult to assess the period length. Realization 1 in Figure 5.16 is the only realization showing characteristics consistent with Model (B).

***Conclusion:* Realization 1 is from Model (B).**

Examining the components: In order to further understand AR(4) Model (B), Figure 5.18(a) shows a realization from an AR(2) model associated with the factor $1 - 1.8B + .96B^2$, which as can be seen from the factor table, is associated with a frequency $f = .065$ (or period length of about 15). The realization is cyclic and goes through about 10 cycles across 150 data points; that is, the period is about 15. The sample autocorrelations in Figure 5.18(b) are a smooth, slowly damping sinusoid again with period of about 15, while the Parzen spectral density estimate in Figure 5.18(c) has a peak slightly below $f = .10$.

Figure 5.18(d) shows a realization from the AR(2) model with the factor $1 - .8B + .94B^2$, which is associated with a frequency $f = .182$ (or period length of about 5.5). The realization has higher frequency behavior, going through about 28 cycles in 150 data values, indicating a period length of about $150/28 = 5.36$ and a frequency of $f = .19$. The sample autocorrelations have a damping sinusoidal behavior with period of about 5. The Parzen spectral density estimate has a peak at about $f = .20$.

Figure 5.18(g) displays Realization 1 from Figure 5.16 which we concluded was from AR(4) Model (B). The Parzen spectral density estimate has two peaks: (1) one slightly below $f = .1$ and (2) one at about $f = .2$. The sample autocorrelations in Figure 5.18(h) are very similar to those in Figure 5.18(b) for the AR(2) model associated with the lower frequency but show no indication of the higher frequency.

The plots in Figure 5.18 can be obtained using the following commands

```
xa=gen.arma.wge(n=150,phi=c(1.8,-.96,0,.001),mu=0,sn=3233)
plotts.sample.wge(xa) # plots (a-c)
xd=gen.arma.wge(n=150,phi=c(.8,-.94,0,.001),mu=0,sn=3233)
plotts.sample.wge(xd) # plots (d-f)
xg=gen.arma.wge(n=150,phi=c(2.6,-3.34,2.46,-.9024),mu=0,sn=3233)
plotts.sample.wge(xg) # plots (g-i)
```

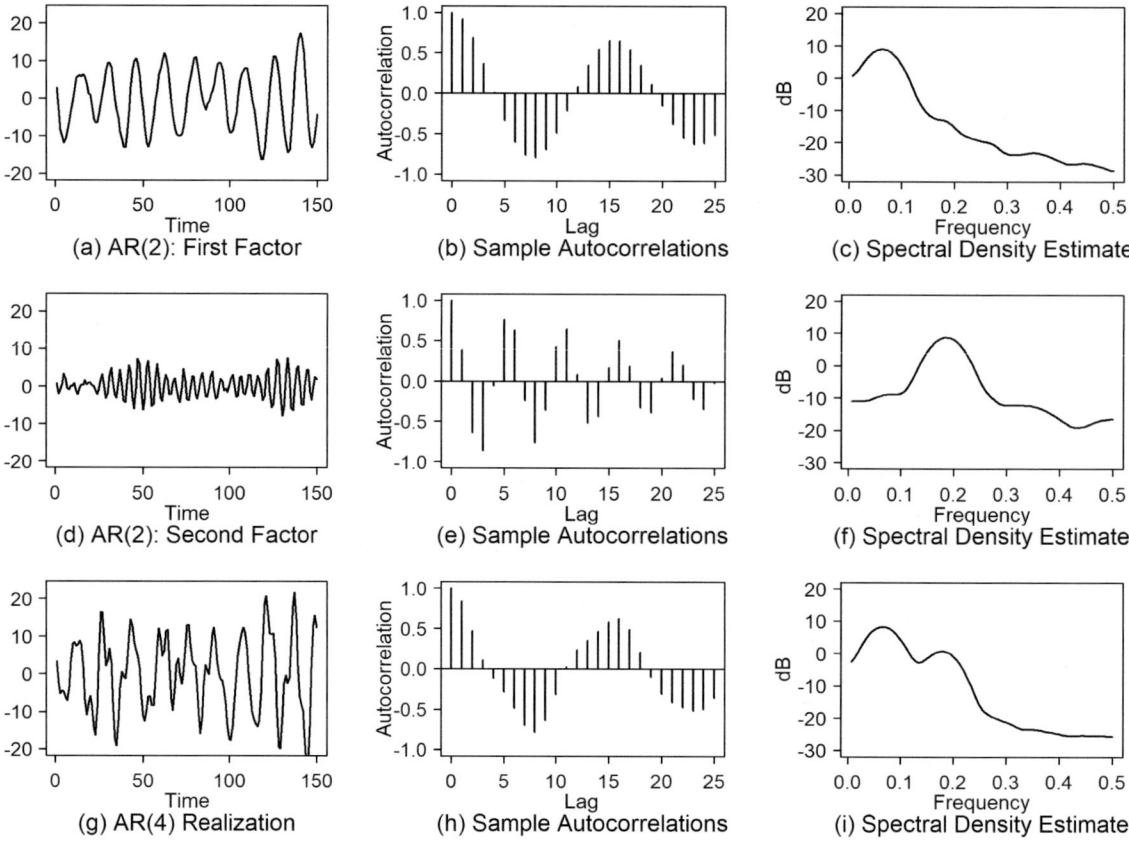

FIGURE 5.18 (a) Realization of length $n = 150$ from AR(2) model $\left(1 - 1.8B + .96B^2\right)X_t = a_t$, (b) sample autocorrelations and (c) Parzen spectral density estimate based on the realization in (a). (d) realization of length $n = 150$ from AR(2) model $\left(1 - .8B + .94B^2\right)X_t = a_t$, (e) sample autocorrelations and (f) Parzen spectral density estimate based on the realization in (d). (g) realization of length $n = 150$ from AR(4) model $\left(1 - 1.8B + .96B^2\right)\left(1 - .8B + .94B^2\right)X_t = a_t$, (h) sample autocorrelations and (i) Parzen spectral density estimate based on the realization in (g).

Model (C): $X_t - 2.85X_{t-1} + 3.24X_{t-2} - 2.03X_{t-3} + .6X_{t-4} = a_t$
The command

```
factor.wge(phi=c(2.85,-3.24,2.03,-.6))
```

produces the factor table in Table 5.4.

TABLE 5.4   Factor Table for Model (C): $X_t - 2.85X_{t-1} + 3.24X_{t-2} - 2.03X_{t-3} + .6X_{t-4} = a_t$

AR-FACTOR	ROOTS	$|r|^{-1}$	f_0
$1 - 1.25B$	.80	1.25	0
$1 - .80B$	1.25	.80	0
$1 - .80B + .60B^2$	$.67 \pm 1.11i$	.77	.16

As a first check for stationarity, we scan the third column and see a value of 1.25, which is not less than 1. Consequently, the model is nonstationary. (Equivalently, the first-order factor $1 - 1.25B$ is associated with the root 1/1.25=.8, which is inside the unit circle.) We see that Realization 2 has "explosive behavior".[5]

Conclusion: **Realization 2 is from Model (C)** (No further analysis is appropriate.)

---

mult.wge: A useful *tswge* command

The *tswge* function **factor.wge** has been shown to be valuable for revealing the underlying factors and associated behavior of an AR model. The vector **phi** is input to the **factor.wge** function, which factors the model into first- and/or irreducible 2nd-order factors, and prints a factor table.

Consider the case in which you have an AR(5) model in factored form, for example,

$$\left(1 - 1.8B + .95B^2\right)\left(1 + .6B\right)\left(1 + .8B^2\right)X_t = a_t,$$

and you want to generate a realization from it. The function **gen.arma.wge** requires the autoregressive model to be specified by its model coefficients in the vector **phi**. Before generating a realization from this model, the AR coefficients must be available *in unfactored form*.

The *tswge* package provides a function, **mult.wge**, for multiplying the factors and outputting the model coefficients. Considering the AR(5) model above, the **mult.wge** command that produces the model coefficients is

```
cf=mult.wge(fac1=c(1.8,-.95),fac2=-.6, fac3=c(0,-.8))
```

The output from this command includes the vector

```
cf$model.coef [1] 1.200 -0.670 0.390 0.104 -0.456
```

which contains the parameters of the model. The model can now be written as

$$X_t - 1.2X_{t-1} + .67X_{t-2} - .39X_{t-3} - .104X_{t-4} + .456X_{t-5} = a_t.$$

Either of the two commands

```
x=gen.arma.wge(n=200,phi=cf$model.coef,sn=10)
```

or

```
x1=gen.arma.wge(n=200,phi=c(1.2,-.67,.39,.104,-.456),sn=10)
```

can be used to generate the same realization from the desired model.

Notes:

(1) See Appendix 5A for more information on the **mult.wge** command.
(2) Several *tswge* commands require model parameters as input.

---

5 **gen.arma.wge** will not generate realizations from a nonstationary process. The realization in Figure 5.16(c) was generated directly from the equation $x_t = 2.85\,x_{t-1} - 3.24\,x_{t-2} + 2.03\,x_{t-3} - .6x_{t-4} + a_t$, where a_t is Normal(0,1) white noise.

5.1.3.5 Dominance of Roots Close to the Unit Circle

Factors associated with roots closer to the unit circle tend to be more dominant in the combined behavior of an AR(p) model. In order to further illustrate this concept, we consider the following models given in factored form:

Model (A): Previously studied Model (A) (complex and real roots equidistant from unit circle)

$$X_t - 1.95X_{t-1} + 1.85X_{t-2} - .855X_{t-3} = a_t, \text{ or in factored form:}$$
$$(1 - .95B)(1 - B + .9B^2)X_t = a_t.$$

Model (A_real): Real root closer to the unit circle than complex root

$$X_t - 1.71X_{t-1} + 1.22X_{t-2} - .475X_{t-3} = a_t, \text{ or in factored form:}$$
$$(1 - .95B)(1 - .76B + .5B^2)X_t = a_t.$$

Model (A_complex): Complex root closer to the unit circle than real root

$$X_t - 1.70X_{t-1} + 1.60X_{t-2} - .63X_{t-3} = a_t, \text{ or in factored form:}$$
$$(1 - .70B)(1 - B + .9B^2)X_t = a_t.$$

In order to understand these models, we compare their factor tables given in Table 5.5.

TABLE 5.5   Factor Tables for Models (A), (A_real), and (A_complex)

	FACTOR	ROOTS	$|r|^{-1}$	f_0
Model (A)	$1 - .95B$	1.05	.95	0
	$1 - B + .9B^2$	$.55 \pm .90i$	.95	.16
Model (A_real)	$1 - .95B$	1.05	.95	0
	$1 - .76B + .5B^2$	$.76 \pm 1.20i$	.70	.16
Model (A_complex)	$1 - .70B$	1.43	.70	0
	$1 - B + .9B^2$	$.55 \pm .90i$	.95	.16

While examining Table 5.5 notice the following:

(a) All three models are AR(3) models with a 1st- and 2nd-order factor
(b) All three models have system frequencies 0 and .16

(c) Model (A): $|r|^{-1} = .95$ for both real and complex roots
    – consequently, both real and complex roots are equally close to the unit circle

(d) Model (A_real): $|r|^{-1} = .95$ for the real root, and $|r|^{-1} = .70$ for the complex roots
    – the real root is closer to the unit circle than are the complex roots

(e) Model (A_complex): $|r|^{-1} = .70$ for the real root, and $|r|^{-1} = .95$ for the complex roots
    – the complex roots are closer to the unit circle than is the real root

To illustrate the fact that roots closer to the unit circle dominate, we generate realizations from each of the three models. These realizations, sample autocorrelations, and Parzen spectral density estimates are shown in Figure 5.19. The plots in Figure 5.19 can be obtained using the following commands:

```
A=gen.arma.wge(n=150,phi=c(1.95,-1.85,.855), sn=3847)
plotts.sample.wge(A) # plots (a-c)
A_real=gen.arma.wge(n=150,phi=c(1.71,-1.22,.475), sn=327)
plotts.sample.wge(A_real) # plots(d-f)
A_complex=gen.arma.wge(n=150,phi=c(1.70,-1.60,.63), sn=2813)
plotts.sample.wge(A_complex) # plots(g-i)
```

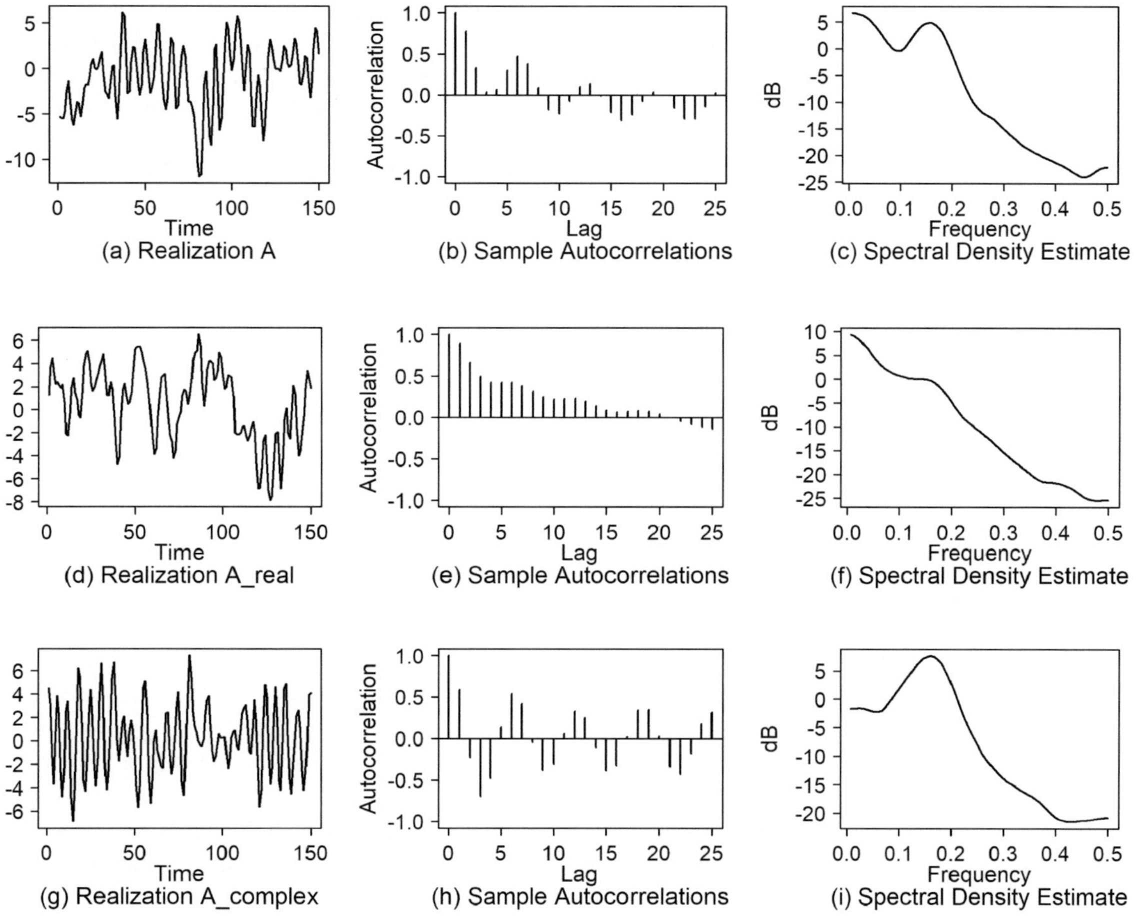

FIGURE 5.19 (a) realization, (b) sample autocorrelations, and (c) Parzen spectral density estimate from Model (A); (d) realization, (e) sample autocorrelations, and (Parzen spectral density estimate from Model (A_real); (g) realization, (h) sample autocorrelations, and (i) Parzen spectral density estimate from Model (A_complex).

We begin with Figures 5.19(a)−(c), which are based on a realization generated from Model (A): $(1-.95B)(1-B+.9B^2)X_t = a_t$ using *tswge* command **gen.arma.wge**. The data in Figure 5.19(a) are based on a different seed than the one used to generate the data from Model (A) shown, among other places, in Figure 5.17(a). Take a few minutes to compare Figures 5.17 and Figure 5.19(a)−(c). In both cases, the realization shows a pseudo-sinusoidal behavior associated with period of about 6, along a wandering path. The sample autocorrelations show damped sinusoidal behavior along a damped exponential path. Finally, the Parzen spectral density estimates in Figure 5.17(c) and Figure 5.19(c) both show peaks at zero and at some frequency slightly below $f = .2$. An important take-away from this example is that the real and complex roots are equidistant from the unit circle, and consequently both behaviors are equally apparent in the figures.

Figure 5.19(d) shows a realization from Model(A_real). It was noted that the real root in Model(A_real) is substantially closer to the unit circle than are the complex roots. Consequently, in Figure 5.19(d) we see wandering behavior with only slight periodic behavior. The sample autocorrelations in Figure 5.19(e) show only a damped exponential behavior and the Parzen spectral density estimate in Figure 5.19(f) has a peak at $f = 0$ with a slight inflection point (but not really a peak) at $f = .16$. This illustrates the fact that the real root (which is the root closest to the unit circle) dominates the behavior.

Figure 5.19(g) shows a realization from Model (A_complex) in which the complex roots are substantially closer to the unit circle than is the real root. Figure 5.19(g) shows pseudo-sinusoidal behavior, but instead of this behavior being along a wandering path as in Figure 5.19(a), the pseudo-sinusoidal behavior is basically centered along the horizontal axis. The sample autocorrelations in Figure 5.19(h) show a damped sinusoidal behavior that is fairly symmetric around the zero axis. That is, there is not much evidence of the damped exponential behavior. Finally, the spectral density estimate in Figure 5.19(i) has a peak at about $f = .16$ and only a weak upward tendency, but not really a peak, around $f = 0$.

QR 5.8
Dominance

Key Points

1. Factors associated with roots closer to the unit circle tend to be more dominant in the combined behavior of an AR(p) model.
2. Much of the behavior of an AR process can be traced to its dominant factors, that is, those with roots closest to the unit circle.
3. The factor table is an invaluable tool for inspecting the factors of an AR model, for understanding its composition, and ultimately for interpreting the physical process generating the data.

As we discuss the analysis of actual time series data in later chapters, the validity of these Key Points will become apparent. In the following example, we illustrate the use of the factor table to learn about the underlying features of an AR(9) model fit to the sunspot data.

Example 5.6 Sunspot Data
The sunspot data (dataset **sunspot2.0** in *tswge*) is based on the new counting procedure for the years 1700 through 2020, Clette et al. (2016). Figures 5.20(a)−(c) show the sunspot data from 1700−2020, the sample autocorrelations, and the Parzen spectral density estimate, respectively. As mentioned previously, the cyclic behavior with periods of about 10−11 years has been the topic of extensive investigation. A reasonable model for these data is an AR(9) (take our word for it for now). We will discuss model identification of the sunspot data in Example 6.8.

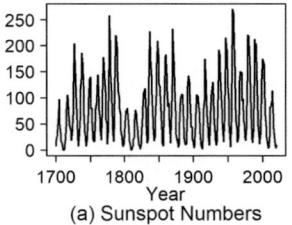 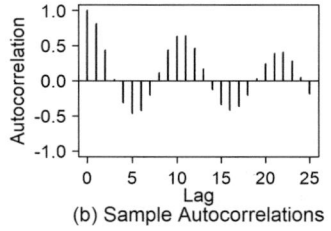 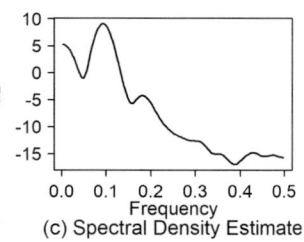

(a) Sunspot Numbers   (b) Sample Autocorrelations   (c) Spectral Density Estimate

FIGURE 5.20   (a) sunspot data, (b) sample autocorrelations, and (c) Parzen spectral density estimate.

The AR(9) model fit to the (zero-mean form) of the sunspot data is[6]

$$\left(1-1.17B+.41B^2+.13B^3-.10B^4+.07B^5-.01B^6-.02B^7+.05B^8-.22B^9\right)\left(X_t-78.52\right)=a_t \quad (5.36)$$

While no information can be gained from the actual coefficients of the AR(9) model, we use the factor table in Table 5.6 to understand the underlying features of the model.

TABLE 5.6   Factor Table for the AR(9) Model Fit to `sunspot2.0` Data

FACTOR	ROOTS	$|r|^{-1}$	f_0
$1-1.61B+.95B^2$	$0.85\pm0.58i$	0.97	0.094
$1-0.94B$	1.06	0.94	0.000
$1-0.60B+0.72B^2$	$0.42\pm1.10i$	0.85	0.192
$1+1.44B+0.59B^2$	$-1.21\pm0.46i$	0.77	0.442
$1+0.56B+0.59B^2$	$-.48\pm1.22i$	0.76	0.309

The factor table shows, not surprisingly, that the dominant component is a 2nd-order factor associated with $f_0=.094$ (e.g. periods of about $1/.094=10.6$ years). The second component (which also has a root fairly close to the unit circle) is associated with $f_0=0$, which is based on the long term or aperiodic behavior of the amplitude heights. The other factors are associated with more subtle features of the data. Note that none of these interpretations are possible by examination of the model coefficients alone.

5.1.4 Linear Filters, the General Linear Process, and AR(*p*) Models

Linear filters play a major role in time series data analysis. The linear filter produces an output process which is a linear combination of lagged values of an input process. Given a set of input data, x_t, the linear filter is defined as

$$y_t=\sum_{j=-\infty}^{\infty}h_j x_{t-j}, \qquad (5.37)$$

where x_t is the observed (input) data, y_t is the output, and the h_js are real-valued constants. A centered moving average is an example of a linear filter. Specifically, a 5th-order centered moving average smoother

---

6 We will discuss techniques for estimating the parameters and identifying the order, *p*, in Section 6.1.2. The estimates shown here are maximum likelihood estimates. See Section 6.1.1.

is defined as $s_t = \left(x_{t-2} + x_{t-1} + x_t + x_{t+1} + x_{t+2}\right)/5$, which can be expressed in the form of (5.37) where $h_j = 1/5$ for $j = -2, -1, 0, 1,$ and 2, and $h_j = 0$, elsewhere.

Key Points

1. If the linear filter "starts" at $j = 0$, it is called a *realizable* filter and has the property that X_t at time t depends only on present and past values of a_t.
2. A centered moving average is *not* a realizable filter because its value at time t depends on past, present, and future values of t.

A stationary AR(p) process is a special case of a particular type of realizable linear filter called a *general linear process* which is defined in Definition 5.2.

Definition 5.2: (*General Linear Process*) The process X_t given by

$$X_t - \mu = \sum_{j=0}^{\infty} \psi_j a_{t-j}, \tag{5.38}$$

is called a *general linear process* (GLP) if a_t is a white noise process with zero mean and finite variance and if $\sum_{j=0}^{\infty} |\psi_j| < \infty$.

Theorem 5.5

(1) A GLP satisfying Definition 5.2 is a stationary process.
(2) The variance of a GLP as defined in Definition 5.2 is given by $\sigma_X^2 = \sigma_a^2 \sum_{j=0}^{\infty} \psi_j^2$.

Key Points

1. A GLP as defined by Definition 5.2 is a *stationary process*.
2. For a general discussion of the GLP, see Woodward et al. (2017) and Brockwell and Davis (1991).

As mentioned above, stationary AR(p) models are special cases of the general linear process (GLP). In this section we illustrate this concept using an AR(1) model and then give a general result for AR(p) processes.

5.1.4.1 AR(1) in GLP Form

For the AR(1) process $(1 - \phi_1 B)(X_t - \mu) = a_t$, the operator $(1 - \phi_1 B)$ is associated with an inverse operator, $(1 - \phi_1 B)^{-1}$, for which $(1 - \phi_1 B)^{-1}(1 - \phi_1 B)X_t = X_t$. Consequently,

$$(1 - \phi_1 B)^{-1}(1 - \phi_1 B)(X_t - \mu) = (1 - \phi_1 B)^{-1} a_t,$$

or

$$X_t - \mu = (1 - \phi_1 B)^{-1} a_t.$$

A natural question is, "How is $(1 - \phi_1 B)^{-1}$ defined?" Brockwell and Davis (1991) show that inverse operators are defined analogously to their algebraic counterparts. The algebraic counterpart in this case is $(1 - \phi_1 z)^{-1}$, where z is a real number. Note that $(1 - \phi_1 z)^{-1} = \dfrac{1}{1 - \phi_1 z}$, which can be solved by long division,

$$
\begin{array}{r}
1 + \phi_1 z + \phi_1^2 z^2 + \phi_1^3 z^3 + \cdots \\
\hline
1 - \phi_1 z)\ 1 \\
\underline{1 - \phi_1 z} \\
+\ \phi_1 z \\
\underline{+\ \phi_1 z - \phi_1^2 z^2} \\
+\ \phi_1^2 z^2 \\
\underline{+\phi_1^2 z^2 + \phi_1^3 z^3} \\
+\ \phi_1^3 z^3
\end{array}
$$

That is,

$$(1 - \phi_1 z)^{-1} = \frac{1}{1 - \phi_1 z}$$

$$= 1 + \phi_1 z + \phi_1^2 z^2 + \phi_1^3 z^3 + \cdots$$

$$= \sum_{k=0}^{\infty} \phi_1^k z^k.$$

Converting back to operator notation (that is, substituting the operator, B, for the numerical quantity, z) it follows that the operator $(1 - \phi_1 B)^{-1} = \sum_{k=0}^{\infty} \phi_1^k B^k$. Consequently, the AR(1) model, $(1 - \phi_1 B) X_t = a_t$, can be written as

$$X_t - \mu = (1 - \phi_1 B)^{-1} a_t,$$

$$= \left(\sum_{k=0}^{\infty} \phi_1^k B^k \right) a_t$$

$$= \sum_{k=0}^{\infty} \phi_1^k a_{t-k}$$

$$= \sum_{k=0}^{\infty} \psi_k a_{t-k} \qquad (5.39)$$

where $\psi_k = \phi_1^k$. The last equation in (5.39) is a GLP, by Definition 5.2, because $\sum_{k=0}^{\infty} |\psi_k| = \sum_{k=0}^{\infty} |\phi_1|^k < \infty$ for $|\phi_1| < 1$ (see Theorem 2.3, Woodward et al. (2017)).[7]

5.1.4.2 AR(p) in GLP Form

Equation (5.39) shows that a stationary AR(1) model, $\phi(B)(X_t - \mu) = a_t$, where $\phi(B) = 1 - \phi_1 B$, can be expressed as $X_t - \mu = \phi^{-1}(B)a_t = \sum_{k=0}^{\infty} \psi_k a_{t-k}$, where the ψ_ks are defined as the coefficients of the inverse operator $\phi^{-1}(B) = 1/\phi(B)$. In the AR(1) case, the ψ_ks are simply given by $\psi_k = \phi_1^k$ and $1/\phi(B) = 1/(1 - \phi_1 B) = \sum_{k=0}^{\infty} \phi_1^k B^k$. The generalization of this result applies to stationary AR(p) models. It can be shown (see Brockwell and Davis, 1991 and Woodward, et al., 2017) that if $\phi(B) = (1 - \phi_1 B - \cdots - \phi_p B^p)X_t = a_t$ is a stationary AR(p) process, that is, if all roots of $\phi(z) = 0$ are outside the unit circle, then X_t can be written in GLP form. In other words, for stationary AR(p) processes, X_t, it follows that $X_t = \phi^{-1}(B)a_t = \sum_{k=0}^{\infty} \psi_k a_{t-k}$ where $\sum_{k=0}^{\infty} |\psi_k| < \infty$. Again, the ψ-weights can be found by long division as shown in Example 5.7.

Example 5.7 GLP form for $(1 - 1.6B + .8B^2)X_t = a_t$
In order to find the ψ-weights associated with this AR(2) model, we note that

$$X_t = (1 - 1.6B + .8B^2)^{-1} a_t$$

$$= \frac{1}{1 - 1.6B + .8B^2} a_t.$$

As in the AR(1) case, the operator $1/(1 - 1.6B + .8B^2)$ is defined analogously to its algebraic counterpart, $1/(1 - 1.6z + .8z^2)$. This ratio is an infinite-order polynomial, the coefficients of which are the ψ-weights. Specifically, the ratio $\dfrac{1}{1 - 1.6z + .8z^2}$ can be obtained by long division as follows:

$$
\begin{array}{r}
1 + 1.6z + 1.76z^2 + \cdots \\
\hline
1 - 1.6z + .8z^2) \, 1 \\
\underline{1 - 1.6z + .8z^2} \\
+ 1.6z - .8z^2 \\
\underline{+ 1.6z - 2.56z^2 + 1.28z^3} \\
+ 1.76z^2 - 1.28z^3
\end{array}
$$

QR 5.9
Polynomial
Division

---

7 Alternatively, we could have used the fact that $\dfrac{1}{1 - \phi_1 z}$ is the sum of a geometric series $\sum_{k=0}^{\infty} (\phi_1 z)^k$.

From this long division procedure, we see that $\psi_0 = 1$, $\psi_1 = 1.6$, $\psi_2 = 1.76$, ... A much easier way to find, say the first five ψ-weights beginning with ψ_1, is to issue the *tswge* command

```
psi.weights.wge(phi=c(1.6,-.8),lag.max=5)
```

which gives $\psi_1 = 1.6$, $\psi_2 = 1.76$, $\psi_3 = 1.536$, $\psi_4 = 1.0496$, and $\psi_5 = .4056$. Recall that it is always true that $\psi_0 = 1$, so this is not included in the output list of ψ-weights from `psi.weights.wge`.

5.2 AUTOREGRESSIVE-MOVING AVERAGE (ARMA) MODELS

"Our goal will be to derive models possessing maximum simplicity and the minimum number of parameters consonant with representational adequacy."

–Box and Jenkins (1970)

In Section 5.1 we discussed the autoregressive (AR) models, which are useful models for analyzing stationary time series data. Box and Jenkins (1970) popularized the use of a broader class of models, the autoregressive moving average (ARMA(p,q)) model, of which the autoregressive model is a special case. Specifically, the AR(p) model is the case of an ARMA(p,q) model where $q = 0$. However, we are getting ahead of ourselves because we haven't yet defined the ARMA(p,q) model. While the AR models can be used successfully to provide useful models, the best fitting AR model may have a large number of parameters; that is, p may be large. The quote at the beginning of this section implies that Box and Jenkins preferred models with the fewest parameters necessary to represent the data. They referred to this as a preference for *parsimonious* models. Given a set of stationary data, it is often the case that the "best" ARMA model will have fewer parameters than the "best" AR model. The large number of parameters is not always bad, and some time series giants (among them, Emmanuel Parzen) preferred the exclusive use of AR models for modeling stationary data.

Recall that the AR(p) model,

$$X_t = \beta + \phi_1 X_{t-1} + \cdots + \phi_p X_{t-p} + a_t,$$

expresses the value of the process at time t as a linear combination of the process at times $t-1$, $t-2,\ldots, t-p$, plus a random noise component, a_t, that enters the model at time t. However, the ARMA(p,q) model specifies that the value of the process at time t is a linear combination of the process at times $t-1, t-2,\ldots, t-p$, plus a random noise component at time t, and a linear combination of random noise components that entered the model at times $t-1$, $t-2$, ..., $t-q$. Specifically, we have the following definition.

Definition 5.3: Autoregressive-Moving Average Model
A time series X_t is said to satisfy an ARMA(p,q) model if

$$X_t = \beta + \phi_1 X_{t-1} + \cdots + \phi_p X_{t-p} + a_t - \theta_1 a_{t-1} - \cdots - \theta_q a_{t-q}, \tag{5.40}$$

where ϕ_k, $k = 1,\dots, p$ and θ_k, $k = 1,\dots,q$ are real constants, $\beta = \left(1 - \phi_1 - \phi_2 - \dots - \phi_p\right)\mu$, $\phi_p \neq 0$, $\theta_q \neq 0$, and a_t is a white noise process with zero mean and finite variance σ_a^2.[8]

Notes:

(1) The ARMA(p,q) model can be written in operator notation as $\phi(B)(X_t - \mu) = \theta(B)a_t$ where $\phi(B) = 1 - \phi_1 B - \dots - \phi_p B^p$ and $\theta(B) = 1 - \theta_1 B - \dots - \theta_q B^q$ and where $\phi(B)$ and $\theta(B)$ have no common factors. See Example 5.11.

(2) An AR(p) model is a special case of an ARMA(p,q) model with $q = 0$. Another special case of an ARMA(p,q) model is the *moving average* (MA) model which is the case with $p = 0$. We discuss MA(q) models next.

Caution: Be careful with the signs! We have adopted the notation used by Box, Jenkins, and Reinsel (2008), while other authors may change the signs, *especially for the MA coefficients*. Pay close attention to how the authors have defined the ARMA model as you read different books, journal articles, and websites.

5.2.1 Moving Average Models

Moving average (MA) models are used to model stationary data but are not as useful as AR models. MA models are most useful in combination with the AR model to create the ARMA models.

Definition 5.4: Moving Average Model
A time series X_t is said to satisfy an MA(q) model if

$$X_t - \mu = a_t - \theta_1 a_{t-1} - \dots - \theta_q a_{t-q}, \tag{5.41}$$

where the θ_j's are real constants, $\theta_q \neq 0$, and a_ts are white noise with mean zero and variance σ_a^2.

Recall that the autoregressive model says that the current value of X_t is a linear combination of past values of the process, plus noise. This is a natural way to describe how time series data might evolve in time. Note, however, that the MA(q) model specifies that X_t at time t is a linear combination of present and past noise components. This is not intuitive to many. Nevertheless, the MA(q) model can be used together with the AR(p) model to create a more parsimonious (fewer parameters) ARMA model. We begin with a discussion of the MA(1) (or ARMA(0,1)) model.

5.2.1.1 The MA(1) Model

The MA(1) model is the special case of (5.41) with $q = 1$, that is,

$$X_t - \mu = a_t - \theta_1 a_{t-1}. \tag{5.42}$$

---

8 An additional requirement for X_t to be ARMA(p,q) is given in Note (1).

The MA(1) can be expressed in operator notation as $X_t - \mu = \theta(B)a_t$ where $\theta(B) = 1 - \theta_1 B$. The corresponding MA(1) characteristic function and characteristic equation are given by $\theta(z) = 1 - \theta_1 z$ and $\theta(z) = 1 - \theta_1 z = 0$, respectively. The following are facts about the MA(1) model, X_t.

(a) $E[X_t] = \mu$

(b) The process variance is given by $\sigma_X^2 \left(= \gamma_0\right) = \sigma_a^2 \left(1 + \theta_1^2\right)$

(c) The autocorrelations are given by

QR 5.10
Properties of MA(1)

$$\rho_0 = 1$$

$$\rho_1 = \frac{-\theta_1}{\left(1 + \theta_1^2\right)}$$

(5.43)

$$\rho_k = 0, \; k > 1$$

It can be shown that $|\rho_1| \leq .5$ for an MA(1) process.

The model-based spectral density of an MA(1) process is given by

$$S_X(f) = \frac{\sigma_a^2}{\sigma_X^2} \left|1 - \theta_1 e^{-2\pi i f}\right|^2 .$$

See Woodward et al. (2017).

Example 5.8 Two MA(1) Models
We consider two MA(1) models below.

(a) $X_t = a_t - .99a_{t-1}$. For this model we have

$\theta_1 = .99$ (be careful about signs!)

$$\rho_1 = \frac{-\theta_1}{1 + \theta_1^2} = \frac{-.99}{1 + .99^2} = -.49997$$

$$\rho_k = 0, \; k > 1$$

TABLE 5.7   MA-Factor Table for $X_t = a_t - .99a_{t-1}$

FACTOR	ROOTS	$|r|^{-1}$	f_0
$1 - .99B$	1.01	.99	0

Figure 5.21(a) and (b) show the model-based autocorrelations and spectral density for the MA(1) model $X_t = a_t - .99a_{t-1}$. The autocorrelation plot shows a single nonzero autocorrelation at lag one

($\rho_1 \approx -.5$) with all others being zero (except, of course, at lag zero).[9] It is important to note that system frequencies in an MA factor table (see Table 5.7) are represented by "dips" in the spectral density rather than peaks. Specifically, the spectral density in Figure 5.21(b) has a dip at $f = 0$. The plots below can be plotted using the **tswge** command

```
plotts.true.wge(phi=0,theta=.99,lag.max=25,vara=1,plot.data=FALSE)
```

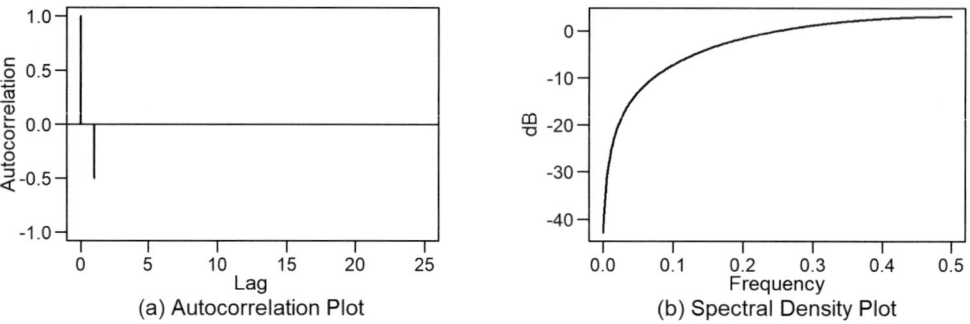

(a) Autocorrelation Plot   **(b) Spectral Density Plot**

FIGURE 5.21   Model-based (a) autocorrelations and (b) spectral density for the MA(1) model $X_t = a_t - .99a_{t-1}$.

> **Key Point:** System frequencies in an MA factor table are represented by "dips" in the spectral density rather than peaks.

Figure 5.22(a) shows a realization from the MA(1) model $X_t = a_t - .99a_{t-1}$, while Figures 5.22(b) and (c) display sample autocorrelations and Parzen spectral density estimates for this realization. Notice that $\hat{\rho}_1 \approx -.5$, which is consistent with the model-based autocorrelation. All other model-based autocorrelations are zero, and the sample autocorrelations for lags greater than one are all small in magnitude. The realization shows an oscillatory behavior consistent with a negative sample autocorrelation at lag one, and near-zero sample autocorrelations for lags greater than one. The main feature of the Parzen spectral density estimate in Figure 5.22(c) is the "dip" at $f = 0$. Figure 5.22 can be obtained using the command

```
x=gen.arma.wge(n=150,theta=.99,sn=53)
plotts.sample.wge(x)
```

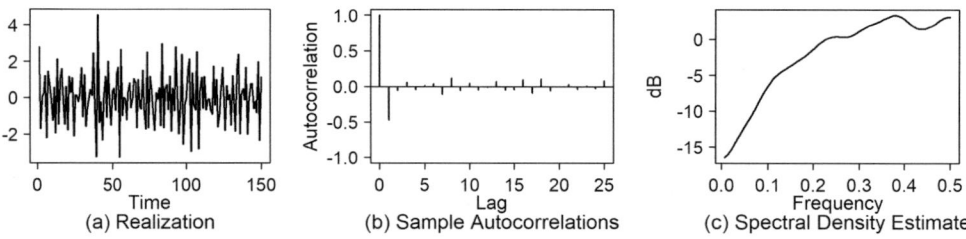

(a) Realization   **(b) Sample Autocorrelations**   **(c) Spectral Density Estimate**

FIGURE 5.22   (a) Realization of $n = 150$ from $X_t = a_t - .99a_{t-1}$, (b) sample autocorrelations and (c) Parzen spectral density estimate based on the realization in (a).

---

9 If you are having trouble visualizing a situation in which $\rho_1 \neq 0$ while $\rho_k = 0$ for $k > 1$, you are not alone!

(b) $X_t = a_t + .8a_{t-1}$. For this model, we have

$\theta_1 = -.8$ (again - be careful about signs!)

$$\rho_1 = \frac{-\theta_1}{1+\theta_1^2} = \frac{.8}{1+.8^2} = .488$$

$\rho_k = 0, \; k > 1$

TABLE 5.8   MA-Factor Table for $X_t = a_t + .8a_{t-1}$

FACTOR	ROOTS	$|r|^{-1}$	f_0
$1+.8B$	-1.25	.80	.50

Figure 5.23(a) shows a realization from the MA(1) model $X_t = a_t + .8a_{t-1}$. Figures 5.23(b) and (c) show sample autocorrelations and Parzen spectral density estimates for this realization. Notice in Figure 5.23(b) that $\hat{\rho}_1$ is slightly less than .5, which is consistent with the model-based correlation of $\rho_1 = 0.488$. All other sample autocorrelations are near zero, and the main feature of the Parzen spectral density estimate in Figure 5.23(c) is the "dip" at $f = .5$ which is the associated system frequency in the factor table (see Table 5.8). Figure 5.23 can be produced using the commands

```
x=gen.arma.wge(n=150,theta=-.8,sn=363)
plotts.sample.wge(x)
```

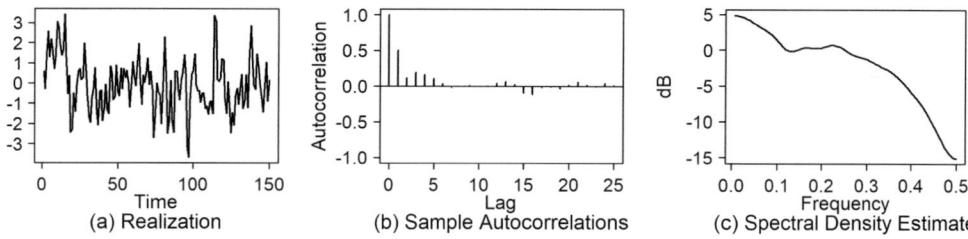

FIGURE 5.23   (a) Realization of $n = 150$ from $X_t = a_t + .8a_{t-1}$, (b) sample autocorrelations and (c) Parzen spectral density estimate based on the realization in (a).

5.2.1.2 The MA(2) Model

The MA(2) model is the special case of (5.41) with $q = 2$, that is,

$$X_t - \mu = a_t - \theta_1 a_{t-1} - \theta_2 a_{t-2}. \tag{5.44}$$

The MA(2) model can be expressed in operator notation as $X_t - \mu = \theta(B)a_t$, where the operator $\theta(B)$ is defined by $\theta(B) = 1 - \theta_1 B - \theta_2 B^2$. The corresponding MA characteristic equation is given by $\theta(z) = 1 - \theta_1 z - \theta_2 z^2 = 0$. The following are facts about the MA(2) model, X_t.

(a) $E[X_t] = \mu$

(b) The process variance is given by $\sigma_X^2 = \sigma_a^2(1 + \theta_1^2 + \theta_2^2)$

(c) The autocorrelations are given by

$$\rho_0 = 1$$

$$\rho_1 = \frac{-\theta_1 + \theta_1\theta_2}{\left(1 + \theta_1^2 + \theta_2^2\right)}$$

$$\rho_2 = \frac{-\theta_2}{\left(1 + \theta_1^2 + \theta_2^2\right)}$$

$$\rho_k = 0, \ k > 2$$

QR 5.11
Properties
of MA(2)

(d) The model-based spectral density of an MA(2) process is given by

$$S_X(f) = \frac{\sigma_a^2}{\sigma_X^2}\left|1 - \theta_1 e^{-2\pi i f} - \theta_2 e^{-4\pi i f}\right|^2.$$

See Woodward et al. (2017).

Example 5.9 An MA(2) Model
We consider the MA(2) model below.

$$X_t = a_t - .4a_{t-1} + .9a_{t-2}.$$

For this model, we have

$$\theta_1 = .4, \ \theta_2 = -.9$$

$$\rho_1 = \frac{-\theta_1 + \theta_1\theta_2}{1 + \theta_1^2 + \theta_2^2} = \frac{-.4 + (-.9)(.4)}{1 + .4^2 + (-.9)^2} = -.3858$$

$$\rho_2 = \frac{-\theta_2}{\left(1 + \theta_1^2 + \theta_2^2\right)} = \frac{.9}{1 + .4^2 + (-.9)^2} = .4569$$

$$\rho_k = 0, \ k > 2$$

TABLE 5.9   MA-Factor Table for $X_t = a_t - .4a_{t-1} + .9a_{t-2}$

FACTOR	ROOTS	$|r|^{-1}$	f_0
$1 - .4B + .9B^2$	$.22 \pm 1.03i$	.95	.22

Figures 5.24(a) and (b) show the model-based autocorrelations and spectral density, respectively, for the MA(2) model $X_t = a_t - .4a_{t-1} + .9a_{t-2}$. The autocorrelation plot shows an autocorrelation a little greater than $-.5$ at lag one and an autocorrelation a little less than .5 at lag two, and $\rho_k = 0, k > 2$. Note that the

factor table has a system frequency of $f = .22$ and there is a *dip* in the spectral density at about this frequency. Output from the command

```
ma2=plotts.true.wge(phi=0,theta=c(.4,-.9),plot.data=FALSE)
```

includes

```
ma2$aut1 [1] 1.0000000 -0.3857868 0.4568528 0.0000000 0.0000000 0.0000000
```

The zeroes continue indefinitely, so it follows that $\rho_0 = 1, \rho_1 = -.386, \rho_2 = .457$, and $\rho_k = 0, k > 2$.

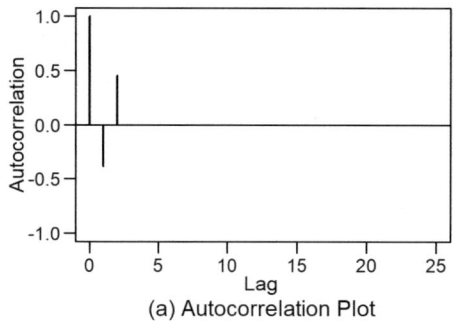 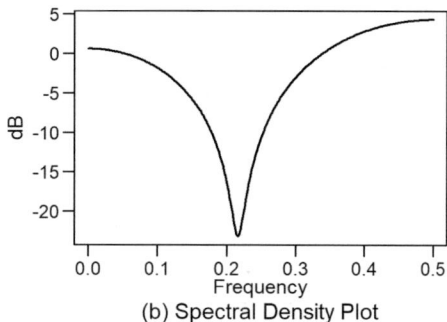

(a) Autocorrelation Plot   (b) Spectral Density Plot

FIGURE 5.24 Model-based (a) autocorrelations and (b) spectral density for the MA(2) model $X_t = a_t - .4a_{t-1} + .9a_{t-1}$.

Figure 5.25(a) shows a realization from the MA(2) model $X_t = a_t - .4a_{t-1} + .9a_{t-2}$. Figures 5.25(b) and (c) show sample autocorrelations and Parzen spectral density estimates for this realization. Notice that $\hat{\rho}_1 \approx -.4$ and $\hat{\rho}_2$ is slightly less than .5 which is consistent with the model-based autocorrelation. All other model-based autocorrelations are zero, and the sample autocorrelations for lags greater than two are all small in magnitude. The realization shows oscillatory behavior and some wandering, but is difficult to describe. That is, what does "the lack of frequency behavior at $f = .22$" look like? *Hmm*. The main feature of the Parzen spectral density estimate in Figure 5.25 is the *dip* at about $f = .22$. Figure 5.25 can be obtained using the commands

```
x=gen.arma.wge(n=150,theta=c(.4,-.9),sn=65)
plotts.sample.wge(x)
```

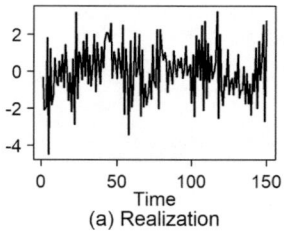 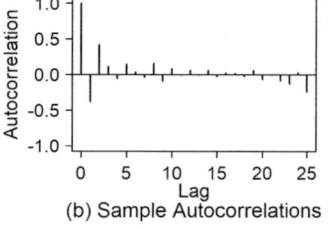 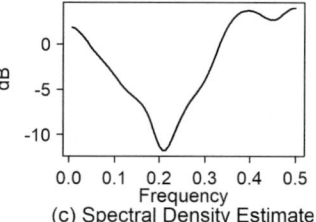

(a) Realization   (b) Sample Autocorrelations   (c) Spectral Density Estimate

FIGURE 5.25 (a) Realization of $n = 150$ from $X_t = a_t - .4a_{t-1} + .9a_{t-2}$, (b) sample autocorrelations and (c) Parzen spectral density estimate based on the realization in (a).

5.2.1.3 The General MA(q) Model

It should be noted that an MA(q) is a finite general linear process (GLP) and is *always stationary*. Recall that a GLP is given by $X_t - \mu = \sum_{j=0}^{\infty} \psi_j a_{t-j}$ so that for an MA(*q*), $\psi_0 = 1, \psi_1 = -\theta_1, \ldots, \psi_q = -\theta_q, \psi_j = 0, j > q$. For example, the MA(2) model $X_t = a_t - .4a_{t-1} + .9a_{t-2}$ is already in GLP form with $\psi_0 = 1$, $\psi_1 = -.4, \psi_2 = .9$, and $\psi_j = 0, j > 2$.

(1) Mean, Variance, and Autocorrelations of an MA(q)

  (a) $E[X_t] = \mu$
  (b) The process variance is given by $\sigma_X^2 (= \gamma_0) = \sigma_a^2 \left(1 + \theta_1^2 + \cdots + \theta_q^2\right)$
  (c) The autocorrelations have the property that $\rho_q \neq 0$ and $\rho_k = 0, k > q$. See Woodward et al. (2017) for a general expression.

(2) Operator Notation and the MA Characteristic Equation
The MA(q) model in (5.41) can be written as

$$X_t - \mu = \left(1 - \theta_1 B - \cdots - \theta_q B^q\right)a_t, \tag{5.45}$$

or by using the shorthand notation, $X_t - \mu = \theta(B)a_t$ where $\theta(B) = 1 - \theta_1 B - \cdots - \theta_q B^q$.

Converting $\theta(B)$ to the algebraic quantity $\theta(z)$, we obtain the MA characteristic polynomial given by $\theta(z) = 1 - \theta_1 z - \cdots - \theta_q z^q$ and the MA characteristic equation

$$\theta(z) = 1 - \theta_1 z - \cdots - \theta_q z^q = 0. \tag{5.46}$$

The MA-factor table can be obtained using *tswge* function, `factor.wge`. For example, for the MA(2) example discussed above, the MA-factor table shown in Table 5.9 can be obtained using `factor.wge(theta=c(.4,-.9))`. Recall that the system frequencies in an MA-factor table are frequencies at which there are *dips* in the spectral densities. That is, these frequencies tend to **not be** in the realization.

Key Points

1. The authors have encountered few real data-sets that are well modeled using an MA(q) model.
2. In making an analogy in which the ARMA(p,q) model is a meal, then the AR component is the entrée while the MA part is the salt and pepper.
3. The general ARMA(p,q) model, which has both AR and MA components, is quite useful and is the topic of the remainder of this chapter.

5.2.1.4 Invertibility

We have seen that when AR(p) processes are stationary, they can be written in GLP form. That is, the model $\phi(B)(X_t - \mu) = a_t$ can be written as $X_t - \mu = \phi^{-1}(B)a_t = \sum_{k=0}^{\infty} \psi_k a_{t-k}$ where $\sum_{k=0}^{\infty} |\psi_k| < \infty$. Note that essentially this says that a stationary AR process can be written as an *infinite order MA process*. A question then arises concerning whether there are conditions under which an MA process can be written as an infinite order AR process. In operator notation, the question is whether $X_t - \mu = \theta(B)a_t$ can be written as

$$\theta^{-1}(B)(X_t - \mu) = a_t. \tag{5.47}$$

Proceeding as in the AR case, we expand $\theta^{-1}(z) = 1/\theta(z)$ in powers of z usually written as $\theta^{-1}(z) = \sum_{j=0}^{\infty} \pi_j z^j$, and formally replace z by the operator B. It follows that

$$\theta^{-1}(B)(X_t - \mu) = \left(\sum_{k=0}^{\infty} \pi_j B^j \right)(X_t - \mu)$$

$$= \sum_{j=0}^{\infty} \pi_j \left(X_{t-j} - \mu \right), \tag{5.48}$$

and from (5.48) we obtain

$$\sum_{j=0}^{\infty} \pi_j \left(X_{t-j} - \mu \right) = a_t, \tag{5.49}$$

which is an *infinite order AR process*. The coefficients are, not surprisingly, called the π-weights. To find the π-weights for the MA(2) model $X_t = a_t - .4a_{t-1} + .9a_{t-2}$, the *tswge* command is

```
pi.weights.wge(theta=c(.4,-.9),lag.max=6)
```

A process that can be written as an infinite order AR is called an *invertible* process. Definition 5.5 gives a more precise definition of invertibility.

Definition 5.4: If an MA(q) process, X_t, can be expressed as in (5.49) where $\sum_{j=0}^{\infty} |\pi_j| < \infty$, then X_t is said to be *invertible*.

The following theorem provides conditions for an MA(q) process to be invertible.

Theorem 5.6 An MA process, X_t, is invertible if and only if the roots of the characteristic equation, $\theta(z) = 1 - \theta_1 z - \cdots - \theta_q z^q = 0$ all lie outside the unit circle.

Key Points

1. To check an MA(q) process for invertibility, use the factor table.
2. For example, Table 5.9 shows that the model $X_t = a_t - .4a_{t-1} + .9a_{t-2}$ is invertible because the two roots (complex conjugate roots, $.22 \pm 1.030$) are outside the unit circle since each root has $|r|^{-1} = .95$.

Reasons for Imposing Invertibility
When using an MA(q) model to analyze time series, we will restrict our focus to invertible models. There are two basic reasons for doing this.

(a) Imposing Invertibility Removes Model Multiplicity
Consider the MA(1) model $X_t = (1 - \theta_1 B) a_t$. From (5.43), we know that $\rho_1 = -\theta_1 / (1 + \theta_1^2)$ and $\rho_k = 0$, when $|k| > 1$. Now consider the MA(1) model

$$X_t = \left(1 - \frac{1}{\theta_1} B\right) a_t. \tag{5.50}$$

For the model in (5.50), ρ_1 is given by

$$\rho_1 = \frac{\dfrac{-1}{\theta_1}}{\left(1 + \dfrac{1}{\theta_1^2}\right)}$$

$$= \frac{-\theta_1}{(1 + \theta_1^2)}.$$

Also, we know that $\rho_k = 0$ for $k > 1$ for model (5.50). In other words, the two MA(1) models $X_t = (1 - \theta_1 B) a_t$ and $X_t = \left(1 - \frac{1}{\theta_1} B\right) a_t$ have the *exact same autocorrelations*.

For example, the models $X_t = (1 - .5B) a_t$ and $X_t = (1 - 2B) a_t$ have the same autocorrelations. However, $X_t = (1 - .5B) a_t$ is invertible, but $X_t = (1 - 2B) a_t$ is not. In general, if $\rho_k, k = 1, 2, \ldots$ are the autocorrelations of an MA(q) process, there are more than one MA(q) model with these same autocorrelations, but only one of them is invertible. By requiring an MA(q) model to be invertible, a *unique* model is associated with a given set of MA(q) autocorrelations.

(b) Invertibility Assures that Present Events Are Associated with the Past in a Sensible Manner
Consider the MA(1) model $X_t = a_t - \theta_1 a_{t-1}$, where for convenience we will assume $\mu = 0$. Rewriting this equation as $a_t = X_t + \theta_1 a_{t-1}$, it follows that $a_{t-1} = X_{t-1} + \theta_1 a_{t-2}$, so that

$$a_t = X_t + \theta_1 \left(X_{t-1} + \theta_1 a_{t-2}\right)$$

$$= X_t + \theta_1 X_{t-1} + \theta_1^2 a_{t-2}$$

$$= X_t + \theta_1 X_{t-1} + \theta_1^2 X_{t-2} + \theta_1^3 a_{t-3}.$$

QR 5.12
Invertibility

Continuing, we see that

$$a_t = X_t + \theta_1 X_{t-1} + \theta_1^2 X_{t-2} + \cdots + \theta_1^k X_{t-k} + \theta_1^{k+1} a_{t-k-1},$$

and rearranging terms,

$$X_t = -\theta_1 X_{t-1} - \theta_1^2 X_{t-2} - \cdots - \theta_1^k X_{t-k} + a_t - \theta_1^{k+1} a_{t-k-1}. \tag{5.51}$$

If $|\theta_1| > 1$, then the associated MA(1) process is not invertible, and (5.51) shows that X_t is *increasingly dependent on the distant past*. In most cases, this is not a physically acceptable model.

For example, for the noninvertible model $X_t = (1 - 2B)a_t$ and letting $k = 20$ in (5.51), we have

$$X_t = -2X_{t-1} - 4X_{t-2} - \cdots - 1048576 X_{t-20} + a_t - 2097152 a_{t-21},$$

and the increasing dependence on terms in the distant past is dramatic.

Key Points

1. All MA(q) models are stationary (because $\sum_{j=0}^{\infty} |\psi_j| = 1 + \sum_{j=0}^{q} |\theta_j| < \infty$).
2. All AR(p) models are invertible (because $\sum_{j=0}^{\infty} |\pi_j| = 1 + \sum_{j=0}^{p} |\phi_j| < \infty$).

5.2.2 ARMA(*p*,*q*) Models

We now study the general ARMA(p,q) process in (5.40) where $p \geq 0$ and $q \geq 0$. For the nonzero mean formula in (5.40), it follows that $E[X_t] = \mu$. However, expressions for the autocorrelations and variance of ARMA(p,q) processes are more complicated (see Woodward et al. (2017)). The *tswge* functions **plotts. true.wge** and **true.arma.aut.wge** calculate these quantities.

We will often express the ARMA(p,q) model in (5.40) in operator notation as

$$(1 - \phi_1 B - \cdots - \phi_p B^p)(X_t - \mu) = (1 - \theta_1 B - \cdots - \theta_q B^q)a_t,$$

or more compactly, $\phi(B)(X_t - \mu) = \theta(B)a_t$.

5.2.2.1 Stationarity and Invertibility of an ARMA(p,q) Process

We will only be interested in processes of the defined in Definition 5.3 that are *stationary* and *invertible*. In this section, we discuss conditions under which an ARMA(p,q) process is stationary and invertible.

(1) Stationarity Conditions for an ARMA (p,q) Process
Theorem 5.7 specifies the conditions for an ARMA(p,q) process to be stationary.

Theorem 5.7 The ARMA(p,q) process, $\phi(B)X_t=\theta(B)a_t$ with $p>0$, is stationary if and only if all the roots of $\phi(z)=0$ fall outside the unit circle.

Theorem 5.7 says that the stationarity of an ARMA(p,q) process depends only on the roots of the characteristic equation associated with the autoregressive side of the equation. In this case, we can write the model in the infinite order moving average (GLP) form $X_t = \sum_{k=0}^{v} \psi_k a_{t-k}$, where $\sum_{j=0}^{\infty} |\psi_j| < \infty$. The ψ-weights can be found by operating on the left-hand and right-hand sides of $\phi(B)X_t=\theta(B)a_t$ by $\phi^{-1}(B)$, to obtain $\phi^{-1}(B)\phi(B)X_t=\phi^{-1}(B)\theta(B)a_t$. It follows that $X_t = \phi^{-1}(B)\theta(B)a_t$; that is, $X_t = \sum_{k=0}^{\infty} \psi_k a_{t-k}$, where the ψ_ks are obtained from the polynomial division $\theta(z)/\phi(z)$, as illustrated in Example 5.7 for the case in which $\theta(z)=1$.

As in the AR(p) case, we recommend finding the ψ-weights using *tswge* function `psi.weights.wge`. For example, consider stationary ARMA(2,1) model $(1-1.6B+.8B^2)X_t = (1+.9B)a_t$. Then the command `psi.weights.wge(phi=c(1.6,-.8),theta=-.9,lag.max=6)` calculates the first six ψ-weights from this model. The output from the above command is

```
[1] 2.50000 3.20000 3.12000 2.43200 1.39520 0.28672
```

QR 5.13 ARMA as GLP

Key Points

1. The `psi.weights.wge` function outputs $\psi_1,\ldots,\psi_{lag.max}$. The user should recall that in all cases, $\psi_0 = 1$.
2. The ψ-weights are useful for obtaining prediction limits for forecasts. This will be discussed in Section 6.3.1.

---

(2) Invertibility Conditions for an ARMA(p,q) Process

Theorem 5.8 gives the conditions for an ARMA(p,q) process to be invertible.

Theorem 5.8 The ARMA(p,q) model, $\phi(B)X_t=\theta(B)a_t$ with $q>0$, is invertible if and only if all the roots of $\theta(z)=0$ fall outside the unit circle.

Theorem 5.8 states that the condition of invertibility of an ARMA(p,q) model depends only on the roots of the MA characteristic equation. If the ARMA(p,q) process is invertible, then again using properties of inverse operators, we can write $\theta^{-1}(B)\phi(B)X_t = a_t$, which is an infinite autoregressive model, usually written as $\sum_{j=0}^{\infty} \pi_j X_{t-j} = a_t$, where $\sum_{j=0}^{\infty} |\pi_j| < \infty$. To find π-weights for the ARMA(2,1) model $(1-1.6B+.8B^2)X_t = (1+.9B)a_t$, the command is

```
pi.weights.wge(phi=c(1.6,-.8),theta=-.9,lag.max=6)
```

QR 5.14 ARMA as AR

Key Points

1. In this book, whenever X_t is said to satisfy an ARMA(p,q) process, the tacit assumption is made that the process is *both stationary* and *invertible*. That is, all the roots of $\phi(z) = 0$ and $\theta(z) = 0$ lie outside the unit circle.

2. A stationary ARMA(p,q) model can be written as an infinite order MA model where $\sum_{j=0}^{\infty} |\pi_j| < \infty$.

3. An invertible ARMA(p,q) model can be written as an infinite order AR model where $\sum_{j=0}^{\infty} |\psi_j| < \infty$.

5.2.2.2 AR versus ARMA Models

In Chapter 6, we will discuss fitting models to data. In some cases, we may find more than one model that appears to be an appropriate fit. For example, it may be that either an AR(10) or an ARMA(3,1) would be appropriate models for a given dataset. As mentioned, some time series analysts prefer to restrict their attention to AR models even if it means the use of a large order of p, such as $p = 10$ or above. If you, however, prefer to use the model with the fewest number of parameters needed to appropriately model the data, then you might choose the ARMA(3,1) model. The fact that a (stationary and invertible) ARMA(3,1) model might be similar to an AR(10) model follows from the fact that the stationary and invertible ARMA(p,q) model can be expressed as an infinite order autoregressive using the the the π-weight formula given above. Because $\sum_{j=0}^{\infty} |\pi_j| < \infty$, it follows that $|\pi_j| \to 0$ as $j \to \infty$. That is, the coefficients of the "infinite order AR" become negligible beyond a point.

Example 5.10 ARMA(3,1) Model $\left(1 - 2.57B + 2.50B^2 - .92B^3\right)X_t = \left(1 - .92B\right)a_t$

We consider the ARMA(3,1) model $\left(1 - 2.57B + 2.50B^2 - .92B^3\right)X_t = \left(1 - .92B\right)a_t$ and illustrate the contributions of the 3rd-order AR and 1st-order MA factors to the final ARMA(3,1) model. The factor table for this model is shown below.

TABLE 5.10   Factor Table for Model $\left(1 - 2.57B + 2.50B^2 - .92B^3\right)X_t = \left(1 - .92B\right)a_t$

AR-FACTOR	ROOTS	$|r|^{-1}$	f_0
$1 - 1.6B + .95B^2$	$.844 \pm .586i$	.973	.097
$1 - .97B$	1.0295	.970	.000

MA-FACTOR	ROOTS	$|r|^{-1}$	f_0
$1 - .92B$	1.111	.900	.000

From the factor table, we see that the irreducible 2nd-order AR-factor is associated with frequency $f_0 = .097$ (or a period length of about 10). The second factor is a 1st-order factor associated with a positive real root. Roots associated with both factors are close to the unit circle. We expect cyclic behavior in the realizations with periods of about 10 units and a wandering behavior associated with the positive real root. The realization from this AR(3) model shown in Figure 5.26(a) has the expected behaviors. The sample autocorrelations in Figure 5.26(b) have a damped sinusoidal behavior along a damped exponential path. Note that most of the autocorrelations are positive. The spectral density shown in Figure 5.26(c) has a peak at $f = 0$ and at about $f = .1$, as expected.

The MA-factor is a 1st-order factor with a positive real root. Recall that the "system frequency" of $f_0 = 0$ in an MA model is associated with a dip in the spectral density at $f_0 = 0$. Figure 5.26(d) shows a realization from $X_t = (1 - .92B)a_t$. Figure 5.26(e) shows that the lag-one autocorrelation is about $-.5$ and all other autocorrelations are quite small. The true lag-one autocorrelation is $\rho_1 = -.92 / (1 + .92^2) = -.498$, and the other true autocorrelations for this MA(1) process are zero. Also, the Parzen spectral density estimate in Figure 5.26(f) has a dip at $f = 0$ with a rise in power (but not a peak) at the higher frequencies, which is consistent with spectral densities shown in Figures 5.21(b) and 5.22(c) for MA(1) processes with positive real roots. This suggests that realizations from the MA(1) model will lack low frequency behavior and tend to be higher frequency. We see this behavior in the MA(1) realization in Figure 5.26(d).

DON'T LOOK at Figures 5.26(g), (h), and (i) yet. They show a realization, the sample autocorrelations, and Parzen spectral density estimate based on a realization from the full ARMA(3,1) model $(1 - 2.57B + 2.50B^2 - .9215B^3)X_t = (1 - .92B)a_t$. First, let's discuss what we would expect for the behavior of the ARMA(3,1) model. This model should incorporate the features of both the AR and MA parts. That is, we expect the realization to be cyclic with a period of about 10. We also expect that the peak at $f = 0$ for the AR(3) part will be partially cancelled out by the dip at $f = 0$ in the MA(1) component. See Examples 5.17 and 5.18.

NOW LOOK: The realization from the ARMA(3,1) model in Figure 5.26(g) has a cyclic behavior associated with a frequency of about $f = .1$ (period of about 10), but the wandering behavior is missing and the cyclic behavior centers around a horizontal center line. The sample autocorrelations in Figure 5.26(h) have a damped sinusoidal behavior similar to that of an AR(2) model. Finally, the Parzen spectral density estimate in Figure 5.26(i) has a peak at about $f = .10$ but only a small hint of a peak at $f = 0$.

The plots in Figure 5.26 can be obtained using the commands

```
fig5.26a=gen.arma.wge(n=150,phi=c(2.57,-2.50,.92),sn=65)
plotts.sample.wge(fig5.26a) # this plots Figure 5.26(a-c)
fig5.26d=gen.arma.wge(n=150,phi=0,theta=.92,sn=65)
plotts.sample.wge (fig5.26d) # this plots Figure 5.26(d-f)
fig5.26g=gen.arma.wge(n=150,phi=c(2.57,-2.50,.92),,theta=.92,sn=65)
plotts.sample.wge(fig5.26g) # this plots Figure 5.26(g-i)
```

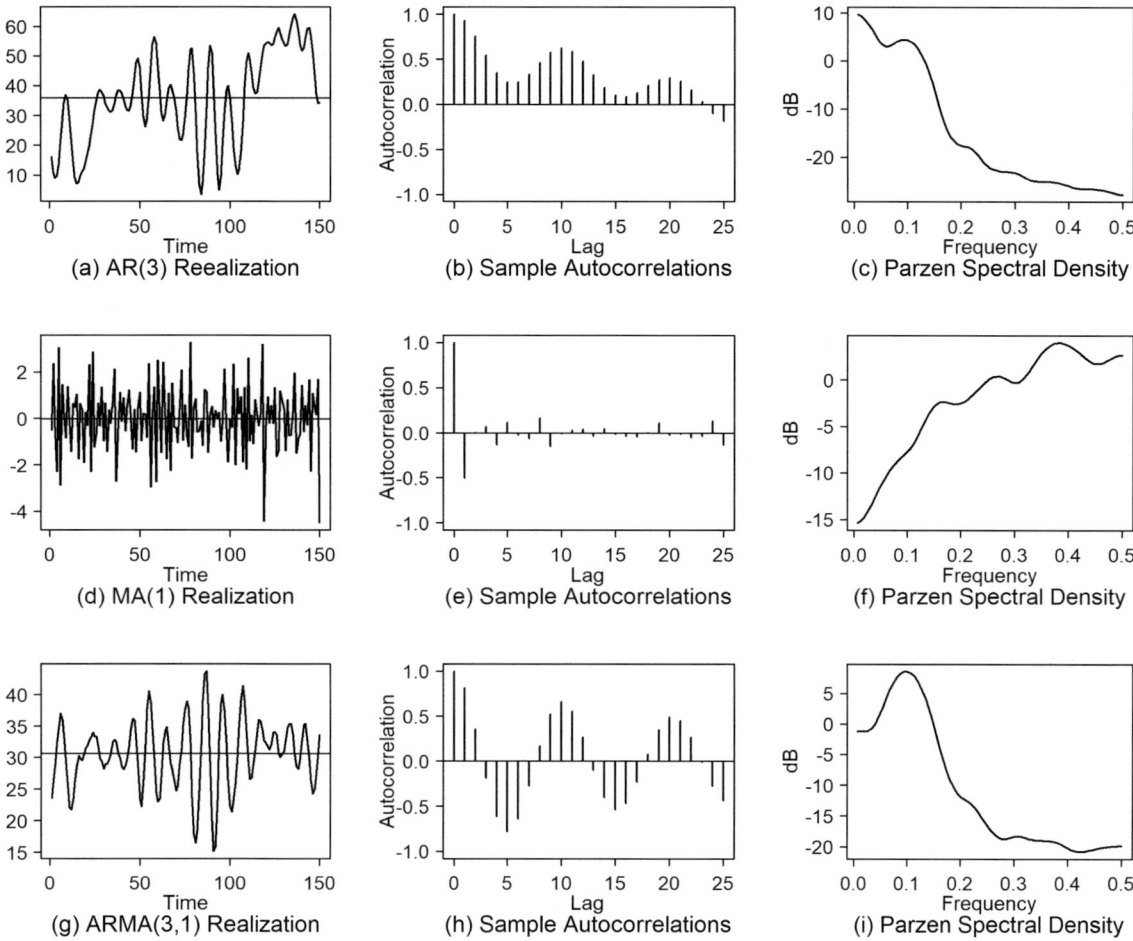

FIGURE 5.26 (a) Realization, (b) sample autocorrelations, and (c) Parzen spectral density estimate for ARMA(3,0): $\left(1-2.57B+2.50B^2-.92B^3\right)X_t = a_t$; (d) realization, (e) sample autocorrelations, and (f) Parzen spectral density estimate for MA(1): $X_t = (1-.92B)a_t$; (g) realization, (h) sample autocorrelations, and (i) Parzen spectral density estimate for ARMA(3,1): $\left(1-2.57B+2.50B^2-.92B^3\right)X_t = (1-.92B)a_t$.

Example 5.11 $\left(1-1.3B+.4B^2\right)X_t = \left(1-.8B\right)a_t$

We next examine the model $\left(1-1.3B+.4B^2\right)X_t = \left(1-.8B\right)a_t$. Using the *tswge* command **true.arma. aut.wge(phi=c(1.3,-.4),theta=.8)**, we obtain the autocorrelations for this model that are viewed in Table 5.11 below:

TABLE 5.11 Autocorrelations for the model: $\left(1-1.3B+.4B^2\right)X_t = (1-.8B)a_t$

LAG(K)	ρ_k
0	1
1	.5
2	.25
3	.125
⋮	⋮

Note that $\rho_1 = .5 = .5^1$, $\rho_2 = .25 = .5^2$, $\rho_3 = .125 = .5^3, \ldots$, and clearly $\rho_k = .5^k$, which we recognize as the model-based autocorrelations of the AR(1) process $(1 - \phi_1 B)X_t = a_t$ with $\phi_1 = .5$. Consequently, the autocorrelations of $(1 - 1.3B + .4B^2)X_t = (1 - .8B)a_t$ are the same as those for the AR(1) model $(1 - .5B)X_t = a_t$. So, what's going on? In a situation like this, we *recommend checking the factor table*. (By now, surely you have noticed that this is a common recommendation!) The *tswge* command `factor.wge(phi=c(1.3,-.4),theta=.8)` gives Table 5.12.

TABLE 5.12   Factor Table for ARMA(2,1) Model $(1 - 1.3B + .4B^2)X_t = (1 - .8B)a_t$

AR-FACTOR	ROOTS	$\lvert r \rvert^{-1}$	f_0
$1 - .8B$	1.25	.8	.00
$1 - .5B$	2.00	.5	.00
MA-FACTOR	ROOTS	$\lvert r \rvert^{-1}$	f_0
$1 - .8B$	1.25	.8	.00

From the table, we see that the factor $(1 - .8B)$ appears on the autoregressive and moving average sides of the equation, and thus violates the condition that $\phi(B)$ and $\theta(B)$ share no common factors in Definition 5.3. Note that if we factor the model as $(1 - .8B)(1 - .5B)X_t = (1 - .8B)a_t$ and pre-multiply both the left-hand and right-hand sides by $(1 - .8B)^{-1}$, we obtain

$$(1 - .8B)^{-1}(1 - .8B)(1 - .5B)X_t = (1 - .8B)^{-1}(1 - .8B)a_t$$

or $(1 - .5B)X_t = a_t$. Consequently, since $(1 - .8B)^{-1}$ and $(1 - .8B)$ are canceling operators, the model specified in the example is *not an* ARMA(2,1) but is actually an AR(1).

> **Key Point:** It is worth repeating that we recommend checking the factor table (for both the AR and MA parts) of an ARMA model in order to fully understand it.
>   − for example, the factor table can identify potential canceling or nearly canceling factors.

Example 5.12 Nearly Canceling Operators

Consider the AR(1) model $(1 - .95B)X_t = a_t$. A realization, sample autocorrelations, and Parzen spectral density estimate are shown in Figures 5.27(a)−(c). There we see wandering behavior in the realization, exponentially damping sample autocorrelations with $\hat\rho_k$ relatively strong (and positive) for small k, and a peak in the Parzen spectral density estimate at $f = 0$.

We next consider the ARMA(1,1) model $(1 - .95B)X_t = (1 - .8B)a_t$. This is clearly an ARMA(1,1) model because the factors do not cancel. However, both factors are associated with positive real roots, and Figures 5.27(d)−(f) show a realization, sample autocorrelations, and Parzen spectral density estimate for the ARMA(1,1) model given by $1 - 95B + X_t = a_t$; we see that the moving average factor has "dampened" the effect of the AR part. In fact, the realization, sample autocorrelations, and Parzen spectral estimator appear similar to white noise, which would have been the case if the model were $(1 - .95B)X_t = (1 - .95B)a_t$. So, in this case, $(1-.8B)$ on the MA side "nearly" canceled the effect of the autoregressive part. Figure 5.27 was plotted using the commands

```
fig5.27a=gen.arma.wge(n=150,phi=.95,sn=20)
plotts.sample.wge(fig5.27a) # this plots Figure 5.27(a-c)
fig5.27d=gen.arma.wge(n=150,phi=.95,theta=.8,sn=20)
plotts.sample.wge (fig5.27d) # this plots Figure 5.27 (d-f)
```

Key Point: The take-away from this example is that "near cancellation" may "hide" some of the autoregressive characteristics.

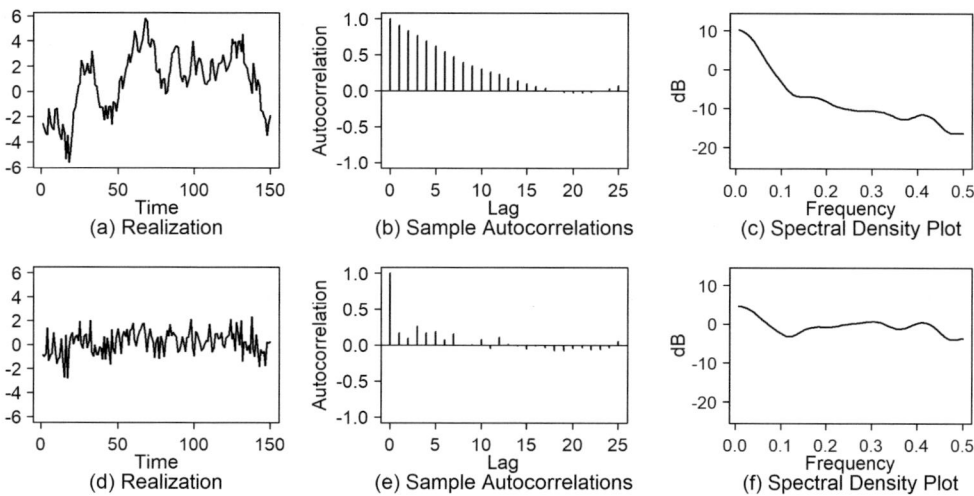

FIGURE 5.27   (a) Realization, (b) sample autocorrelations, and (c) Parzen spectral density estimate for AR(1) model $(1-.95B)X_t = a_t$; (d) realization, (e) sample autocorrelations, and (f) Parzen spectral density estimate for ARMA(1,1) model $(1-.95B)X_t = (1-.8B)a_t$.

5.3 CONCLUDING REMARKS

What a crucial chapter! As was stated in the introduction, the concepts, definitions, and methodologies described in this chapter are at the heart of the analysis of time series data.

The autoregressive AR(1) and AR(2) models were introduced; realizations, autocorrelation function plots, and spectral density plots from both types were compared. These models were generalized to the AR(p) model, for which it was illustrated that by factoring, an AR(p) is (fortunately!) comprised of the simpler AR(1) and AR(2) components.

The AR(p) process was then extended to the autoregressive moving average ARMA(p,q) process, which considers the addition of a moving average component, MA(q). While not commonly used by itself to model real data, the MA(q) component is a valuable addition which interestingly often enhances an AR(p) model.

A novel idea that assists the analyst in extracting and organizing these components of an AR(p) model is the factor table, which will be a frequent consideration in various types of analyses in future chapters.

APPENDIX 5A

TSWGE FUNCTIONS

(a) `factor.wge(phi,theta)` is a function that takes AR, MA, or ARMA polynomials, factors them into their 1st-order and irreducible 2nd-order polynomial factors, and prints a factor table.

`phi` is a vector specifying the coefficients of the AR part of the model
`theta` is a vector specifying the coefficients of the MA part of the model

Example: Consider the model

$$X_t + .3X_{t-1} - .44X_{t-2} - .29X_{t-3} + .378X_{t-4} + .648X_{t-5} = a_t + .5a_{t-1} + .7a_{t-2}.$$

Then define `phi5=c(-.3,.44,.29,-.378,-.648)` and `theta2=c(-.5,-.7)`, and issue the command

`factor.wge(phi=phi5,theta=theta2)`

You will obtain the following factor tables:

AR Factor Table

Factor	Roots	Abs Recip	System Freq
1-1.6000B+0.9000B^2	0.8889+-0.5666i	0.9487	0.0903
1+0.9000B	-1.1111	0.9000	0.5000
1+1.0000B+0.8000B^2	-0.6250+-0.9270i	0.8944	0.3444

MA Factor Table

Factor	Roots	Abs Recip	System Freq
1+0.5000B+0.7000B^2	-0.3571+-1.1406i	0.8367	0.2983

(b) `gen.arma.wge(n,phi=0,theta=0,mu=0,vara=1,plot=TRUE,sn=0)` is a function that generates a realization of length n from a given stationary AR, MA, or ARMA model.

`n` is the realization length
`phi` is a vector of AR parameters (default is `phi=0`)
`theta` is a vector of MA parameters (using signs as defined in this text) (default is `theta=0`)
`mu` is the process mean (default=0)
`vara` is the white noise variance (default=1)
`plot=TRUE` (default) produces a plot of the generated realization
`sn` determines the seed number used in the simulation. (See Note (5) below)

Notes

(1) This function finds p and q as the length of **phi** and **theta**, respectively. For example, if either **phi=0** or **theta=0** (also the default), the values for p and q are set to zero, respectively.

(2) **gen.arma.wge** calls the Base R function **arima.sim**, which uses the same signs as in this text for the AR parameters but opposite signs as in this text for MA parameters. The appropriate adjustments are made within **gen.arma.wge**, so that input parameters **phi** and **theta** for **gen.arma.wge** should contain parameters using the same signs as in this text. However, if you use **arima.sim** directly (which has options not employed in **gen.arma.wge**), then you must remember that the signs needed for the MA parameters have opposite signs as those in **gen.arma.wge**.

(3) It will be useful to call the utility function **mult.wge** if you only know the factors of the model and not the coefficients.

(4) By default, the white noise is zero-mean, normal noise with variance **vara=1**.

(5) **sn=0** (default) produces results based on a randomly selected seed. If you want to reproduce the same realization on subsequent runs of **gen.arma.wge**, then set **sn** to the same positive integer value on each run.

Example: The command

```
x=gen.arma.wge(n=100,phi=c(1.6,-.9),theta=.9,mu=30,vara=1)
```

generates and plots the realization from the model $X_t - 1.6X_{t-1} + .9X_{t-2} = a_t - .9a_{t-1}$ and stores it in the vector **x**. The mean of the process is $\mu = 30$ and the white noise is normally distributed with mean zero and variance one, and the seed is randomly selected.

(c) **mult.wge(fac1,fac2,fac3,fac4,fac5,fac6)**. The above functions dealing with ARMA processes require the coefficients phi and theta as input. If you only know your model in factored form, say for example,

$$\left(1-1.6B+.9B^2\right)\left(1+.9B\right)\left(1+B+.8B^2\right)X_t = a_t,$$

then without multiplying the associated polynomials you do not know the components of the vector **phi** used in the function calls. The *tswge* function **mult.wge** can be used to multiply up to six factors to obtain **phi** and/or **theta**.

fac1 through **fac6** are vectors specifying the factors of the AR or MA part of the model. *Note:* In the above example with three AR factors, the call statement is

```
mult.wge(fac1=c(1.6,-.9),fac2=-.9,fac3=c(-1,-.8))
```

The function outputs the coefficients of the polynomial resulting from the multiplication of the 1st- and 2nd-order factors. The output is in the form of a vector, and also included is the associated characteristic polynomial. For the example above, the output is

```
$char.poly
1 + 0.3*x - 0.44*x^2 - 0.29*x^3 + 0.378*x^4 + 0.648*x^5
$model.coef
[1] -0.300 0.440 0.290 -0.378 -0.648
```

Note This function uses the CRAN package **PolynomF**

(d) **plotts.true.wge(n,phi,theta,lag.max,vara,sn,plot.data)** For a given ARMA(p,q) model, this function generates a realization (optional), calculates the true autocorrelations and spectral density (in dB), and plots the graphs

n is the realization length ($x(t)$, t=1, …, n)
phi is a vector of AR parameters (using signs as in this text, default=0)

theta is a vector of MA parameters (using signs as in this text) (default=0)

lag.max is the maximum lag at which the autocorrelations and autocovariances will be calculated (default=25)

vara is the white noise variance (default=1)

plot.data specifies whether to plot a realization from the model **(default=TRUE)**

sn and **plot.data**. If **plot.data=TRUE**, then the function generates and plots a realization from the specified model. The parameter **sn** plays the same role as in **gen.arma.wge**

Notes

(1) $\max(p, q+1) \leq 25$

(2) This function uses a call to the Base R function **arima.sim**, which uses the same signs as in this text for the AR parameters but opposite signs for MA parameters. The appropriate adjustments are made here so that **phi** and **theta** should contain parameters using the signs as in this text. See discussion of **arima.sim** in the description of **gen.arma.wge**

Example: The command

```
test=plotts.true.wge(n=100,phi=c(1.6,-.90),theta=.9,lag.max=25,vara=1)
```

generates a realization, and computes the true autocorrelations (and autocovariances) and spectral density, and plots the results.

test$data contains the realization if **plot.data=TRUE**

test$autl contains the true autocorrelations (lags 0, 1, ..., **lag.max**). Note:

test$autl[1]=1, that is, the autocorrelation at lag 0.

test$acv contains the true autocovariances (lags 0, 1, ..., **lag.max**). Note:

test$acv[1] is the process variance, i.e. **gvar=g[1]**

test$sd is the process standard deviation

test$spec contains the 251 spectral density values associated with **f** = 0, 1/500, 2/500, ..., 250/500 used in the plot.

(e) **true.arma.aut.wge(phi,theta,lag.max,vara,plot)** calculates the true autocorrelations and autocovariances associated with a given AR, MA, or ARMA model.

phi is a vector of AR parameters (using signs as in this text)

theta is a vector of MA parameters (using signs as in this text)

lag.max is the maximum lag at which the autocorrelations and autocovariances will be calculated (default=25)

vara is the white noise variance (default=1)

plot is a logical variable: **TRUE**=plot autocorrelations, **FALSE**=no plot

Note: $\max(p, q+1) \leq 25$

Example: The command

```
traut=true.arma.aut.wge(phi=c(1.6,-.9),theta=.9,lag.max=25,vara=1)
```

calculates and plots the true autocorrelations, $\rho_k, k = 0,1,\ldots 25$ associated with the ARMA model $X_t - 1.6X_{t-1} + .9X_{t-2} = a_t - .9a_{t-1}$. The autocorrelations are stored in the vector **traut$acf** (which has 26 elements). Similarly the autocovariances $\gamma_k, k = 0,1,\ldots,25$ are stored in the vector **traut$acv**. The process variance is given by **traut$acv[1]**.

(f) **true.spec.wge(phi,theta,plot)** calculates the spectral density associated with a given AR, MA, or ARMA model.

phi is a vector of AR parameters (using signs as in this text)
theta is a vector of MA parameters (using signs as in this text)
plot is a logical variable: **TRUE**=plot spectral density, **FALSE**=no plot

Note: $\max(p, q+1) \leq 25$

Example: The command

trspec=true.arma.spec.wge(phi=c(1.6,-.9),theta=.9)

calculates and plots the true spectral density associated with the ARMA model $X_t - 1.6X_{t-1} + .9X_{t-2} = a_t - .9a_{t-1}$. The vector **trspec** contains the 251 spectral density values associated with $f = 0$, $1/500$, $2/500$, ..., $250/500$ used in the plot.

APPENDIX 5B

STATIONARITY CONDITIONS OF AN AR(1)

In the following, we derive the following properties of the stationary AR(1) model

$$X_t = (1 - \phi_1)\mu + \phi_1 X_{t-1} + a_t,$$

which, by Theorem 6.1, satisfies $|\phi_1| < 1$.

(a) $E[X_t] = \mu$ (a constant)

Taking expectations of the terms in (5.2), we have

$$
\begin{aligned}
E[X_t] &= E[\beta + \phi_1 X_{t-1} + a_t] \\
&= E[(1 - \phi_1)\mu] + \phi_1 E[X_{t-1}] + E[a_t] \\
&= (1 - \phi_1)\mu + \phi_1 E[X_{t-1}] + E[a_t] \\
&= (1 - \phi_1)\mu + \phi_1 E[X_{t-1}],
\end{aligned}
$$

since $E[a_t] = 0$. Now, since X_t is stationary, $E[X_t] = E[X_{t-1}]$, so we will rewrite the equation above as $E[X_t] = (1 - \phi_1)\mu + \phi_1 E[X_t]$. So the following computations follow:

$$E[X_t] - \phi_1 E[X_t] = (1 - \phi_1)\mu,$$
$$E[X_t](1 - \phi_1) = (1 - \phi_1)\mu,$$

$$E[X_t] = (1 - \phi_1) / (1 - \phi_1) \mu$$

$$= \mu \text{, since } (1 - \phi_1) \neq 0 \text{, by the assumption of stationarity.}$$

(b) $Var[X_t] = \sigma_a^2 / (1 - \phi_1^2)$. See (c) below.
(c) The autocorrelation between X_t and X_{t+k} depends only on k.

Without loss of generality, we use the zero-mean form of the AR(1) model. In order to find ρ_k for an AR(1) process, we first find γ_k by evaluating the expectation $E[X_t X_{t+k}] = E[X_{t-k} X_t]$. Recall that we are using the zero-mean form, that is, $\mu = E[X_t] = 0$, for all t. Multiplying both sides of the equality in (5.4) by X_{t-k}, where $k > 0$, yields

$$X_{t-k}[X_t = \phi_1 X_{t-1} + a_t].$$

Taking the expectation of both sides, we obtain

$$E[X_{t-k} X_t] = \phi_1 E[X_{t-k} X_{t-1}] + E[X_{t-k} a_t].$$

The fact that $k > 0$ implies that X_{t-k} occurs prior to time t, and is thus uncorrelated with the noise term a_t that enters the model at time t. That is, for $k > 0$, $E[X_{t-k} a_t] = E[X_{t-k}]E[a_t] = 0$, because $E[X_{t-k}] = E[a_t] = 0$. Because $E[X_{t-k} X_t] = \gamma_{t-(t-k)} = \gamma_k$ and $E[X_{t-k} X_{t-1}] = \gamma_{t-1-(t-k)} = \gamma_{k-1}$, it follows that $\gamma_k = \phi_1 \gamma_{k-1}$ for $k > 0$, and

$$\gamma_1 = \phi_1 \gamma_0$$
$$\gamma_2 = \phi_1 \gamma_1 = \phi_1^2 \gamma_0,$$

so that, in general,

$$\gamma_k = \phi_1^k \gamma_0. \tag{6B.1}$$

It follows that $\rho_k = \phi_1^k \gamma_0 / \gamma_0 = \phi_1^k$ for $k > 0$.

The variance $\sigma_X^2 = \gamma_0$ can be obtained by pre-multiplying both sides of (5.4) by X_t and taking expectations as before, to obtain

$$E[X_t^2] = \phi_1 E[X_t X_{t-1}] + E[X_t a_t].$$

Thus,

$$\gamma_0 = \phi_1 \gamma_1 + E[X_t a_t].$$

The term $E[X_t a_t]$ is found using an analogous procedure, this time by post-multiplying both sides of the zero-mean AR(1) model by a_t and taking expectations, which gives

$$E[X_t a_t] = \phi_1 E[X_{t-1} a_t] + E[a_t^2]$$

Previously, it was shown that $E[X_{t-1} a_t] = 0$, and therefore, $E[X_t a_t] = E[a_t^2] = \sigma_a^2$. Substituting into (5.4) yields

$$\gamma_0 = \phi_1 \gamma_1 + \sigma_a^2$$

so that using (6.6), we have

$$\gamma_0 = \frac{\sigma_a^2}{1 - \phi_1^2},$$

which is a constant and is finite, since $|\phi_1| < 1$.

PROBLEM SET

5.1 (a) Generate realizations of length $n = 80$ from the following AR(1) models, where a_t is zero-mean white noise with $\sigma_a^2 = 1$.

$$(1 - .5B) X_t = a_t$$
$$(1 - .8B) X_t = a_t$$
$$(1 - .95B) X_t = a_t$$

  (i)   For each model, print the factor table.
  (ii) For each model, plot the realization, sample autocorrelations, and Parzen spectral density estimate. Describe the behavior of the model shown in these plots.
  (iii) Explain how the behaviors of the plots in (i) change as the root of the characteristic equation approaches the unit circle.

  (b) Repeat (a) for the models

$$(1 + .5B) X_t = a_t$$
$$(1 + .8B) X_t = a_t$$
$$(1 + .95B) X_t = a_t$$

5.2 Generate realizations of length $n = 150$ from the following AR(2) models, where a_t is zero-mean white noise with $\sigma_a^2 = 1$.

  (a) $\left(1 - 1.6B + .63B^2\right) X_t = a_t$
  (b) $\left(1 + 1.6B + .63B^2\right) X_t = a_t$
  (i)   For each model, print the factor table.
  (ii) What are the roots of the characteristic equations?
  (iii) Write the model in factored form.
  (iv) For each model, plot the realization, sample autocorrelations, and Parzen spectral density estimate. Describe the behavior of the model shown in these plots and the relationship between the factor table information and the plots.

5.3 Determine whether the following models are stationary, and explain your answers.

  (a) $X_t - 1.55X_{t-1} + X_{t-2} - .25X_{t-3} = a_t$
  (b) $X_t - 2X_{t-1} + 1.76X_{t-2} - 1.6X_{t-3} + .77X_{t-4} = a_t$
  (c) $X_t - 1.9X_{t-1} + 2.3X_{t-2} - 2X_{t-3} + 1.2X_{t-4} - .4X_{t-5} = a_t$
  (c) $X_t - 2.5X_{t-1} + 3.14X_{t-2} - 2.97X_{t-3} + 2.016X_{t-4} - .648X_{t-5} = a_t$

5.4 Consider the following realizations:

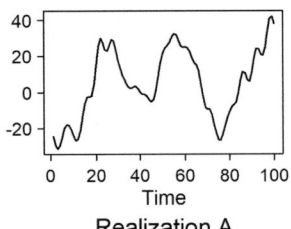

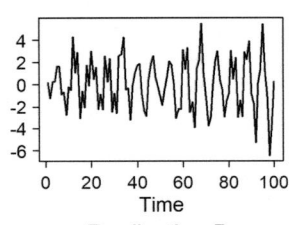

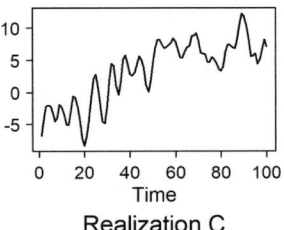

Realization A   Realization B   Realization C

This problem is similar to Example 5.5. Each of these realizations was generated from one of the three models below, given in random order.

(a) $X_t - 2.9X_{t-1} + 3.72X_{t-2} - 2.63X_{t-3} + .828X_{t-4} = a_t$

(b) $X_t - 1.99X_{t-1} + 1.69X_{t-2} - .693X_{t-3} = a_t$

(c) $X_t - .28X_{t-1} - .134X_{t-2} + .8924X_{t-3} = a_t$

Work this problem in the stated order.

(i)   Find factor tables for Models (a)−(c), and use this information to match the realizations to the models from which they were generated. Explain your answers.

(ii)   Generate the realizations for each of the three models above, using the code below:
       For model (a) use: `a=gen.arma.wge(n=100,phi=c(2.9,-3.72,2.63, -.828),sn=21)`
       For model (b): Use the `gen.arma.wge` function, inputting the appropriate coefficients with `sn=229`
       For model (c): Use the `gen.arma.wge` function, inputting the appropriate coefficients with `sn=26`

       For each realization, use **plotts.sample.wge** to display the realization, sample autocorrelations, and spectral density. How does the information in these plots compare with the information in the factor tables? Are there instances in which information is better presented in the factor tables versus the sample plots, or vice-versa?

(iii) For each model, generate another realization (that is, use a different value for **sn**) and again create the **plotts.sample.wge** plots. Do these plots reinforce your answers in (i)? Why or why not?

5.5 Consider the models $X_t = (1 - .8B)a_t$ and $X_t = (1 - 1.25B)a_t$, and compare the autocorrelations. Also note that $X_t = (1 - .5B)a_t$ is invertible. What are the implications of this?

5.6 Determine whether the following models are stationary and/or invertible and explain your answers.

(a) $X_t - 1.6X_{t-1} + X_{t-2} = (1 - .9B)a_t$

(b) $X_t - 2.6X_{t-1} + 1.5X_{t-2} - .38X_{t-3} - .72X_{t-4} = a_t - 1.9a_{t-1} + 1.7a_{t-2} - .72a_{t-3}$

(c) $X_t = a_t - 2.9a_{t-1} + 2.7a_{t-2} - 1.52a_{t-3}$

(d) $X_t - 1.9X_{t-1} + 2.5X_{t-2} - 2.24X_{t-3} + 1.36X_{t-4} - .576X_{t-5} = a_t - .9a_{t-1}$

5.7 The figure below is the true spectral density from one of the three models below. Use the corresponding factor tables to identify the appropriate model and explain your answers.

(a) $X_t - 1.9X_{t-1} + 2.5X_{t-2} - 2.24X_{t-3} + 1.36X_{t-4} - .576X_{t-5} = a_t - .9a_{t-1}$
(b) $X_t - .3X_{t-1} - .9X_{t-2} - .1X_{t-3} + .8X_{t-4} = a_t + .9a_{t-1} + .8a_{t-2} + .72a_{t-3}$
(c) $X_t - .6X_{t-1} + .2X_{t-2} - .54X_{t-3} + .81X_{t-4} = a_t - .99a_{t-1}$

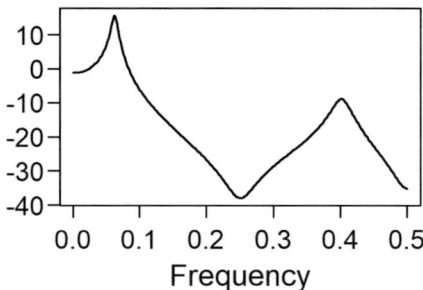

5.8 Generate realizations from the following two models.

(a) $(1 - 1.6B + .9B^2)(X_t - 20) = a_t$
(b) $(1 - 1.6B + .9B^2)(X_t - 20) = (1 - .9B)a_t$.

For each realization, display the results of **plotts.sample.wge**. Discuss the effect of the factor $(1 - .9B)a_t$.

ARMA Fitting and Forecasting

6

In Chapter 5 we discussed the properties of $\text{ARMA}(p,q)$ models, which are a broad class of models frequently used by researchers to model stationary data. We will generically use the term *ARMA model* to refer to a model which might be an $\text{AR}(p)$, $\text{MA}(q)$, or $\text{ARMA}(p,q)$ model where $p > 0$ and $q > 0$. As we begin the discussion of fitting ARMA models to data, it is important to keep in mind the following points.

Key Points

1. Models are used to facilitate the understanding of underlying characteristics of times series data and assist in obtaining forecasts. The George Box quote at the beginning of Chapter 5 reminds us that we are looking for "models that are useful". ARMA models have been found to be useful for modeling stationary data.

2. Many datasets of interest do not satisfy the properties of stationarity. For example, whenever we fit a trend line or include a seasonal component in the model, we have implicitly assumed that the data are not stationary. However, as previously discussed, a final (possibly nonstationary) model will often include a stationary component that will be modeled using an ARMA model.

In this chapter we explain how to fit an $\text{ARMA}(p,q)$ model to a set of data and perform analyses such as forecasting future values. Note, for example, in Example 5.6, we *assumed* a model for the sunspot data for purposes of illustration. But,

*   *Where did the decision to use an AR(9) model come from?*
*   *How were the parameters estimated?*
*   *How are these models used to forecast future values?*

In Section 6.1 we discuss the issue of *fitting* an $\text{ARMA}(p,q)$ model to a time series realization, and in Section 6.2 we show how to use the fitted models to forecast future values.

6.1 FITTING ARMA MODELS TO DATA

In this section we discuss the issue of fitting an $\text{ARMA}(p,q)$ process to a given set of (apparently stationary) data. Fitting an $\text{ARMA}(p,q)$ model to a set of data involves the two steps given in the following Key Points.

DOI: 10.1201/9781003089070-6

Key Points: Fitting an ARMA(p,q) model to data involves two steps:

1. Identifying the orders p and q.
2. Estimating the parameters associated with the chosen ARMA(p,q) model.

Although we must identify p and q *before* finding the estimates of the final model, we address the two steps in the above Key Points in *reverse order*. That is, in Section 6.1.1 we consider the topic of estimating the parameters of a model for which we "know p and q". In Section 6.1.2 we discuss the identification of p and q.

Note:
The main reason for the discussion in "reverse order" is the fact that the strategies for identifying p and q involve the estimation of parameters.

6.1.1 Estimating the Parameters of an ARMA(p,q) Model

Suppose an ARMA(p,q) model is fit to (an apparently stationary) time series realization for purposes of forecasting future values. Recall that the ARMA model has the form

$$\left(1-\phi_1 B-\cdots-\phi_p B^p\right)\left(X_t-\mu\right)=\left(1-\theta_1 B-\cdots-\theta_q B^q\right)a_t, \tag{6.1}$$

where a_t is a zero mean white noise process with finite variance σ_a^2. In this setting, to "fit" an ARMA(p,q) model the following need to be estimated:

(1) p and q
(2) $\phi_1,\ldots,\phi_p$ and $\theta_1,\ldots,\theta_q$
(3) μ and σ_a^2

When we think of estimating the parameters in model (6.1) our minds immediately think of the estimation of the ϕs and θs. However, μ and σ_a^2 are also unknown parameters that must be estimated as part of a completely specified final model. As noted above, the "estimation" of p and q, the *model identification* step, is actually the first decision made in practice. We will discuss model identification in Section 6.1.2.

6.1.1.1 Maximum Likelihood Estimation of the ϕ and θ Coefficients of an ARMA Model

The basic ("gold standard") method used for estimating the parameters $\phi_1, \phi_2,\ldots,\phi_p$ and $\theta_1,\theta_2,\ldots,\theta_q$ of an ARMA(p,q) process is *maximum likelihood*, which is widely used throughout statistical analysis. Maximum likelihood (ML) estimation involves maximizing the likelihood function

$$L = f\left(x_1, x_2 \ldots, x_n\right),$$

which is the joint distribution of the time series realization. ML estimates are obtained using iterative procedures, and involve distributional assumptions concerning the white noise error, a_t. In this book we assume normal white noise with zero mean. Some problematic issues concerning ML estimation include the facts that they can be computationally expensive and have the potential to converge to model estimates for which the roots are inside the unit circle.

Example 6.1 Estimating the parameters of the ARMA(3,1) model in Example 5.10

In Example 5.10 we considered the ARMA(3,1) model

$$\left(1 - 2.57B + 2.50B^2 - .92B^3\right)X_t = \left(1 - .92B\right)a_t, \tag{6.2}$$

where $\sigma_a^2 = 1$. The *tswge* code below generates the realization from this model previously examined in Example 5.10 with the exception that now we use a mean of 30.

```
arma31=gen.arma.wge(n=150,phi=c(2.57,-2.50,.92),theta=.92,mu=30,sn=65)
```

Figures 6.1(a)–(c) are plots of the realization in R vector **arma31**, sample autocorrelations, and Parzen spectral density, shown previously in Figures 5.26(g)–(i), respectively.

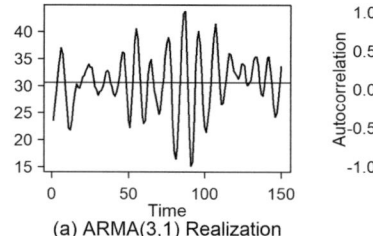
(a) ARMA(3,1) Realization

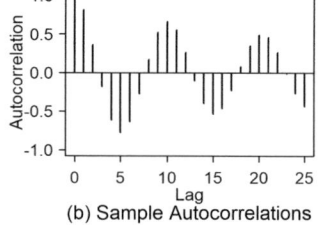
(b) Sample Autocorrelations

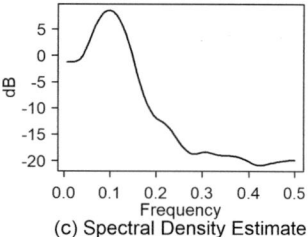
(c) Spectral Density Estimate

FIGURE 6.1 (a) Realization from the ARMA(3,1) model in (6.2), (b) and (c) sample autocorrelations and Parzen spectral density, respectively, of the data in (a).

In order to find the maximum likelihood estimates of the parameters of the ARMA(3,1) model fit to this realization, we use the *tswge* code[1]

```
est.arma.wge(arma31,p=3,q=1)
```

Among the output are the following:

```
$phi    [1]  2.4587424  -2.3304790  0.8286207
$theta  [1]  0.7749636
$avar   [1]  1.035746
$xbar   [1]  30.61015
```

Thus, the final model fit by the ML estimates (to three decimal places) is

$$\left(1 - 2.459B + 2.330B^2 - .829B^3\right)\left(X_t - 30.610\right) = \left(1 - .775B\right)a_t, \tag{6.3}$$

with $\hat{\sigma}_a^2 = 1.036$. While the estimates of the AR and MA parameters are fairly close to the true value, they are "a little off".

Using the factor table to assess estimation performance: If the true model is known, it is natural to compare the coefficient estimates with the true values. *However, we recommend that the quality of the estimates should be evaluated by comparing the associated factor tables*. Table 6.1 is the factor table associated with true model (6.2) while Table 6.2 shows the factor table for the estimated model in (6.3). Comparing Tables 6.1 and 6.2 we see that both models contain a 2nd-order AR factor associated with system frequency of about

---

1 We note again that in this example we are presupposing that we *know* $p = 3$ and $q = 1$.

$f = .1$ with roots similarly close to the unit circle $\left(|r|^{-1} = .973 \text{ vs. } 970\right)$. Both models also contain a 1st-order AR factor and a 1st-order MA factor with positive real roots in which the AR factor has a root slightly closer to the unit circle than the MA factor. Consequently, the MA factor dampens the effect of the 1st-order AR factor. Also, $\bar{x} = 30.610$ and $\hat{\sigma}_a^2 = 1.036$ are close to the true values of 30 and one, respectively. We address the estimation of the mean and white noise variance in the next section. Based on these observations, the basic properties of the estimated model are quite similar to those of the true model.

TABLE 6.1   Factor Table for true model in (6.2)

AR-FACTOR	ROOTS	$|r|^{-1}$	f_0
$1 - 1.599B + .947B^2$	$.844 \pm .586i$	.973	.097
$1 - .971B$	1.030	.971	.000

MA-FACTOR	ROOTS	$|r|^{-1}$	f_0
$1 - .920B$	1.087	.920	.000

TABLE 6.2   Factor Table for fitted model in (6.3)

AR-FACTOR	ROOTS	$|r|^{-1}$	f_0
$1 - 1.578B + .940B^2$	$.838 \pm .600i$	.970	.099
$1 - .881B$	1.135	.881	.000

MA-FACTOR	ROOTS	$|r|^{-1}$	f_0
$1 - .775B$	1.290	.775	.000

6.1.1.2 Estimating μ

As noted, after estimating the ϕ_j and θ_j coefficients, we are not finished. We still need to estimate μ and σ_a^2. We estimate μ by the sample mean $\bar{x}$ in all of the estimation routines used in *tswge*. In the output from the **est.arma.wge** function, the value for the sample mean is given in **$xbar**.

6.1.1.3 Estimating σ_a^2

Given a fitted model with estimates $\hat{\phi}_i, i = 1, \ldots, p$, $\hat{\theta}_i, i = 1, \ldots, q$, and $\bar{X}$, the final goal is to obtain an estimate, denoted by $\hat{\sigma}_a^2$, of the variance of the unknown white noise sequence, $a_t, t = 1, \ldots, n$. The white noise variance estimate is obtained by first estimating the a_ts which, as in regression analysis, are referred to as residuals. The residuals are denoted by $\hat{a}_t, t = 1, \ldots, n$, and $\hat{\sigma}_a^2$ is obtained as an estimate of the variance of these residuals.

Writing the fitted model in the form

$$X_t - \hat{\phi}_1 X_{t-1} - \cdots - \hat{\phi}_p X_{t-p} = \hat{a}_t - \hat{\theta}_1 \hat{a}_{t-1} - \cdots - \hat{\theta}_q \hat{a}_{t-q} + \bar{x}\left(1 - \hat{\phi}_1 - \cdots - \hat{\phi}_p\right), \tag{6.4}$$

and solving for the residuals, $\hat{a}_t$, we obtain

$$\hat{a}_t = X_t - \hat{\phi}_1 X_{t-1} - \cdots - \hat{\phi}_p X_{t-p} + \hat{\theta}_1 \hat{a}_{t-1} + \cdots + \hat{\theta}_q \hat{a}_{t-q} - \bar{x}\left(1 - \hat{\phi}_1 - \cdots - \hat{\phi}_p\right). \tag{6.5}$$

Equation 6.5 appears to provide a straightforward formula for calculating the residuals and ultimately the white noise variance estimate. However, this is not the case, and we will discuss two techniques for calculating residuals and estimating the white noise variance.

Calculating Conditional Residuals
Note, for example, that

$$\hat{a}_1 = X_1 - \hat{\phi}_1 X_0 - \cdots - \hat{\phi}_p X_{1-p} + \hat{\theta}_1 \hat{a}_0 + \cdots + \hat{\theta}_q \hat{a}_{1-q} - \overline{x}\left(1 - \hat{\phi}_1 - \cdots - \hat{\phi}_p\right),$$

which cannot be calculated because $X_0, \ldots, X_{1-p}, \hat{a}_0, \ldots, \hat{a}_{1-q}$, are unknown. It follows that the first residual that can be calculated is

$$\hat{a}_{p+1} = X_p - \hat{\phi}_1 X_{p-1} - \cdots - \hat{\phi}_p X_1 + \hat{\theta}_1 \hat{a}_p + \cdots + \hat{\theta}_q \hat{a}_{p+1-q} - \overline{x}\left(1 - \hat{\phi}_1 - \cdots - \hat{\phi}_p\right).$$

Notice that even though $X_1, \ldots, X_p$ are available, the formula still involves unknown $\hat{a}_k$ values. In order to continue the calculation, we set $\hat{a}_1 = \cdots = \hat{a}_p = 0$ (their unconditional mean). Computation continues for $\hat{a}_{p+2}, \ldots, \hat{a}_n$.

For a fitted $\mathrm{ARMA}(p,q)$ the estimate of the white noise variance based on conditional residuals is

$$\hat{\sigma}_a^2 = \frac{1}{n-p} \sum_{t=p+1}^{n} \hat{a}_t^2. \tag{6.6}$$

Example 6.1 (Revisited)
Writing the fitted ARMA(3,1) model in (6.3) as in (6.4), we obtain

$$X_t - 2.459 X_{t-1} + 2.330 X_{t-2} - .829 X_{t-3} = a_t - .775 a_{t-1} + 30.610\left(1 - 2.459 + 2.330 - .829\right)$$
$$= \hat{a}_t - .775 \hat{a}_{t-1} + 30.610(.042)$$
$$= \hat{a}_t - .775 \hat{a}_{t-1} + 1.286. \tag{6.7}$$

In this example, Equation 6.5 for calculating residuals becomes

$$\hat{a}_t = X_t - 2.459 X_{t-1} + 2.330 X_{t-2} - .829 X_{t-3} + .775 \hat{a}_{t-1} - 1.286. \tag{6.8}$$

Note, for example

$$\hat{a}_1 = X_1 - 2.459 X_0 + 2.330 X_{-1} - .829 X_{-2} + .775 \hat{a}_0 - 1.286,$$

which cannot be calculated because $X_0, X_{-1} X_{-2}$, and $\hat{a}_0$ are unknown.

In order to avoid data values X_t with zero or negative subscripts, the calculation begins with $\hat{a}_4$. The first five values of the realization in Figure 6.1(a) are $X_1 = 23.522$, $X_2 = 26.698$, $X_3 = 28.877$, $X_4 = 32.561$, and $X_5 = 35.089$. The calculation for $\hat{a}_4$ is

$$\hat{a}_4 = X_4 - 2.459 X_3 + 2.330 X_2 - .829 X_1 + .775 \hat{a}_3 - 1.286.$$

As noted, even though $X_1, \ldots, X_4$ are available, the formula still involves the unknown $\hat{a}_3$, so we continue calculation by setting $\hat{a}_3 = 0$. Computation proceeds as follows:

$$\hat{a}_4 = 32.561 - 2.459(28.877) + 2.330(26.698) - .829(23.522) + .775(0) - 1.286 = 2.973,$$

$$\hat{a}_5 = X_5 - 2.459X_4 + 2.330X_3 - .829X_2 + .775\hat{a}_4 - 1.286,$$

$$= 35.089 - 2.459(32.561) + 2.330(28.877) - .829(26.698) + .775(2.973) - 1.286$$

$$= 1.190$$

etc.

Note that the conditional estimate of the white noise variance is

$$\hat{\sigma}_a^2 = \frac{1}{n-3}\sum_{t=4}^{150}\hat{a}_t^2$$

$$= 1.095. \tag{6.9}$$

Key Points: Comments about conditional residuals:
1. Only $n - p$ residuals can be calculated.
2. The residuals $\hat{a}_p, \ldots, \hat{a}_{p+1-q}$ may be poor estimates of the true values, $a_p, \ldots, a_{p+1-q}$. The residuals $\hat{a}_{p+1}, \ldots, \hat{a}_n$ are called *conditional residual estimates* because of their dependence on the starting values, $\hat{a}_p = \cdots = \hat{a}_{p+1-q} = 0$.

Because of the issues with the conditional residuals, we recommend an alternative procedure for calculating residuals that involves *backcasting*.

Calculating Residuals using Backcasting

Recall that a fitted ARMA(p,q) model, such as that in (6.4), is based on an assumption that the current value of the random variable X_t is a linear combination of past values plus a linear combination of present and past noise components. In Section 6.2 we will discuss methods for forecasting future values based on a model of the form (6.4) fit to a dataset. An unusual "backward" model using the same parameter estimates is given in (6.10)

$$X_t - \hat{\phi}_1 X_{t+1} - \cdots - \hat{\phi}_p X_{t+p} = \hat{a}_t - \hat{\theta}_1 \hat{a}_{t+1} - \cdots - \hat{\theta}_q \hat{a}_{t+q} + \bar{x}\left(1 - \hat{\phi}_1 - \cdots - \hat{\phi}_p\right), \tag{6.10}$$

This model has the same parameter estimates as (6.4), but the backward model specifies that the random variable, X_t, is a linear combination of *future* values plus a linear combination of *present and future* noise components. That is, this model literally "goes backward". Given a realization $x_1, x_2, \ldots, x_n$, the natural "thing to do" is to find forecasts $\hat{x}_{n+1}, \hat{x}_{n+2}, \ldots$ and so forth as was done in Chapter 2 with exponential smoothing and Holt-Winters forecasting, and which we will also do with model-based forecasting procedures throughout the remainder of this book. However, given the same realization, the backward model in (6.10) can be used to "forecast backward", that is backcast $\hat{x}_0, \hat{x}_{-1}, \ldots$ and so forth.

Having fit an ARMA(p,q) model to the data, a backcast procedure for obtaining backcast residual estimates and subsequently an estimate of the white noise variance is as follows:

1. Given the data, backcast to obtain $\hat{x}_0, \hat{x}_{-1}, \ldots, \hat{x}_{-K}$ (where $K = 49$ in *tswge*).
2. Use the forward model and (6.5) to begin estimating residuals, $\hat{a}_{-49+p+1}, \hat{a}_{-49+p+2}, \ldots, \hat{a}_0, \hat{a}_1, \hat{a}_2, \ldots, \hat{a}_n$. Note that the starting values are set to zero as before, that is, $\hat{a}_{-49} = \hat{a}_{-49+1} = \hat{a}_{-49+p} = 0$, but the effect of the starting values has diminished by the time $t = 1$.

3. Retain $\hat{a}_1$, $\hat{a}_2$,...,$\hat{a}_n$ (this is a full set of n residuals that are not very dependent on the starting values).

4. Compute $\hat{\sigma}_a^2 = \dfrac{1}{n}\sum_{t=1}^{n}\hat{a}_t^2$ based on the new set of n residuals.

Example 6.1 (Backcast Residuals)
The backcast procedure is computationally intensive. All estimation procedures implemented in *tswge* use a backcasting procedure for obtaining residuals and the estimated white noise variance. In the case of the ARMA$(3,1)$ data in Example 6.1, the backcast-based estimate of the white noise variance is $\hat{\sigma}_a^2 = 1.0356$. Recall that **$avar 1.035746** was one of the output values shown earlier from the command

est.arma.wge(arma31,p=3,q=1)

> **Key Fact:** The *tswge* routines calculate the residuals using *backcasting*.

6.1.1.4 *Alternative estimates for AR(p) models*

While ML estimates are the "gold standard" for all ARMA models, the ML estimates involve (sometimes lengthy) iterative procedures. Two non-iterative estimation techniques, *Yule-Walker(YW)* and *Burg*, have historically been used to estimate the parameters of an AR(p) model, at least in part because of the computing time sometimes required for the ML estimates to converge. Although increased computing power and more efficient ML routines have somewhat alleviated many of the computation issues, we believe it is useful to familiarize readers with these alternatives for AR models. The *tswge* function **est.ar.wge** optionally calculates the ML, YW, or Burg estimates for AR(p) models.

Yule-Walker Estimation for an AR Model
As the name suggests, the Yule-Walker estimates are based on the Yule-Walker equations in (5.29). By replacing the true autocorrelations in these equations with sample autocorrelations and solving the resulting $p \times p$ system

$$\hat{\rho}_1 = \phi_1 + \phi_2\hat{\rho}_1 + \cdots + \phi_p\hat{\rho}_{p-1}$$

$$\hat{\rho}_2 = \phi_1\hat{\rho}_1 + \phi_2 + \cdots + \phi_p\hat{\rho}_{p-2}$$

$$\vdots$$

$$\hat{\rho}_p = \phi_1\hat{\rho}_{p-1} + \phi_2\hat{\rho}_{p-2} \cdots + \phi_p$$

(6.11)

QR 6.1
Yule-Walker
Estimation

for the unknown $\phi_1, \phi_2,...,\phi_p$, the solutions are called the *Yule-Walker estimates*. An appealing feature of the YW estimates is that they *always* result in a *stationary* model.

(2) Burg Estimation for an AR Model
Burg estimates make use of the interesting fact that there is no preference in time direction in AR models (recall $\rho_k = \rho_{-k}$) and consequently that the forces "driving the process forward" also drive it backward. For example, consider the realization in Figure 6.2(a) previously shown in Figure 5.19(d), referred to as Realization A_real, and from the model

$$X_t - 1.71X_{t-1} + 1.22X_{t-2} - .475X_{t-3} = a_t, \qquad (6.12)$$

where $\sigma_a^2 = 1$. Figure 6.2(b) is a plot of the realization in Figure 6.2(a) in "reversed time".

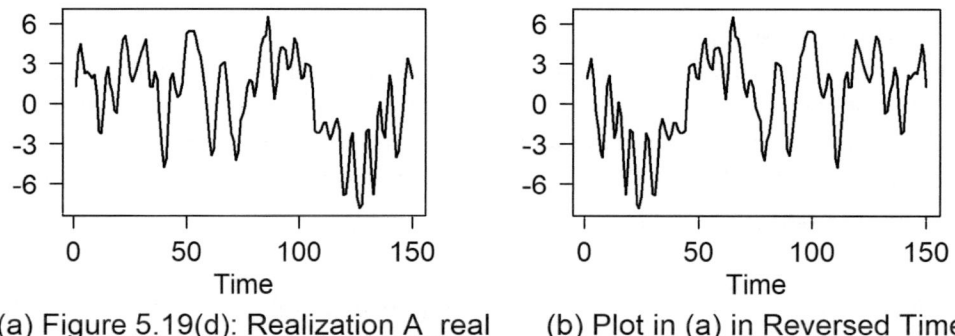

(a) Figure 5.19(d): Realization A_real   (b) Plot in (a) in Reversed Time

FIGURE 6.2   (a) Realization 1 in Figure 5.19(d) and (b) realization in (a) in reversed time.

The following commands generate the plots in Figure 6.2.

```
par(mfrow=c(1,2))
a=gen.arma.wge(n=150,phi=c(1.71,-1.22,.475),sn=327,plot=FALSE)
b=rev(a)
plotts.wge(a)
plotts.wge(b)
```

In Figures 6.2(a) and (b) we see that "flipping" the time (horizontal) axis does not change the general appearance or apparent correlation structure of the realizations. The data in Figure 6.2(b) can be considered to be a realization from the "backward model" (notice the subscripts)

$$X_t - 1.71X_{t+1} + 1.22X_{t+2} - .475X_{t+3} = \delta_t, \qquad (6.13)$$

where δ_t is white noise with zero mean and the same white noise variance, σ_a^2.[2] Because each of these datasets are realizations from models with the same ϕ_k values, Burg (1975) showed how to use this "additional" information to obtain improved least-squares type estimates of the ϕ_ks. For more information, see Woodward et al. (2017). For a discussion of least-squares and Burg estimation, check out QR6.2 and QR6.3, respectively.

QR 6.2 Least   QR 6.3 Burg
Squares Estimation   Estimation

The following Key Points discuss the *tswge* functions that can be used for estimating the parameters of an AR(p) model.

---

2 Readers will recognize (6.13) as another example of a "backward" model analogous to the one shown in (6.10) in the context of backcasting.

> **Key Point:** The *tswge* command **est.ar.wge** fits an AR(p) model to a set of time series data.
> 1. The order p is specified by the user.
> 2. **est.ar.wge** produces the following estimates:
>   (a) ML (default) **(type=mle)** Also, **mle** is the default option.
>   (b) Yule-Walker **(type=yw)**
>   (c) Burg **(type=burg)**

Example 6.2 Estimating the coefficients of the AR(3) Model A_real discussed in Section 5.1.3.5

Consider again the realization in Figure 6.2(a) (and Figure 5.19(d)) from the model

$$\left(1 - 1.71B + 1.22B^2 - .475B^3\right)X_t = a_t, \tag{6.14}$$

with $\sigma_a^2 = 1$, which was referred to as Model A_real. The factored form of the model is

$$\left(1 - .95B\right)\left(1 - .76B + .5B^2\right)X_t = a_t.$$

This model was used to show that the positive real root associated with the factor $\left(1 - .95B\right)$ dominates the behavior of the process because the real root is substantially closer to the unit circle than the complex conjugate roots associated with the factor $\left(1 - .76B + .5B^2\right)$ and system frequency $f_0 = .16$. The following code re-generates the data in Figure 5.19(d) and obtains the ML estimates for an AR(3) fit to the data.

```
A_real=gen.arma.wge(n=150,phi=c(1.71,-1.22,.475),sn=327)
A_real.ml=est.ar.wge(A_real,p=3,type='mle')
# The command A_real.ml=est.ar.wge(A_real,p=3) produces the same results
# since 'mle' is the default type.
```

Among the output we obtain

```
A_real.ml$phi 1.7614 −1.4010 .5605
A_real.ml$avar [1] 0.9093326
```

Consequently, the AR(3) model based on the ML estimates of the ϕs is given by

$$\left(1 - 1.761B + 1.401B^2 - .561B^3\right)X_t = a_t, \tag{6.15}$$

with $\hat{\sigma}_a^2 = .909$, which is quite close to the "true" model in (6.14). Tables 6.3 and 6.4 show factor tables for the true AR(3) model and for the fitted model using ML estimates, respectively. We see that the factors of the fitted model are very similar to those of the true model. That is, there is a 1st-order factor associated with a positive real root and a 2nd-order factor associated with system frequency .16. In both cases the positive real root is closer to the unit circle than are the complex conjugate roots associated with the 2nd-order factor.

TABLE 6.3   Factor Table for true model in (6.2)

AR-FACTOR	ROOTS	$|r|^{-1}$	f_0
$1 - .95B$	1.05	.95	.00
$1 - .76B + .50B^2$	$.76 \pm 1.20i$	.71	.16

TABLE 6.4 Factor Table for ML fit to Model A_real

AR-FACTOR	ROOTS	$\|r\|^{-1}$	f_0
$1 - .90B$	1.12	$.90$	$.00$
$1 - .87B + .63B^2$	$.69 \pm 1.06i$	$.79$	$.16$

Yule-Walker and Burg Estimates of AR(3) fits to Realization A_real.

The Yule-Walker and Burg estimates can be obtained using the commands:

```
A_real.yw=est.ar.wge(A_real,p=3,method='yw')
A_real.burg=est.ar.wge(A_real,p=3,method='burg')
```

The MLE, YW, and Burg estimates of $\phi_1, \phi_2,$ and ϕ_3 are shown in Table 6.5 to three decimal places. It can be seen that all three estimation techniques give estimates close to the true values.

TABLE 6.5 ML, YW, and Burg estimates for AR(3) models fit to the realization in Figure 5.19(d)

	ϕ_1	ϕ_2	ϕ_3	σ_a^2
True	1.710	−1.220	.475	1.000
MLE	1.761	−1.401	.561	0.909
YW	1.757	−1.389	.554	0.910
Burg	1.767	−1.407	.564	0.909

At this point, you may be asking the following question.

> **Question:** Since we can calculate the "gold standard" ML estimates for AR models, why do we bother discussing the other two methods?

One of the answers to the above question is that the YW and Burg estimates are used in practice. However, that's not a compelling explanation. It turns out that while the AR(3) models fit to the **A_real** dataset in Example 6.2 are quite similar for each of the estimation types, the methods have advantages and disadvantages as we will see in the following.

Example 6.3 Problems with Yule-Walker Estimates

Consider the AR(4) model

$$\left(1 - 2.76B + 3.76B^2 - 2.6B^3 + .89B^4\right)X_t = a_t, \tag{6.16}$$

where $\sigma_a^2 = 1$. The following command generates a realization from the model in (6.16) and plots the realizations, sample autocorrelations, and Parzen spectral density estimate. Figure 6.3 shows the resulting plots. The following code will produce the plots in Figure 6.3.

```
ar4=gen.arma.wge(n=100,phi=c(2.76,-3.76,2.6,-.89),sn=468)
plotts.sample.wge(ar4)
```

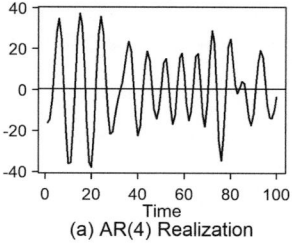
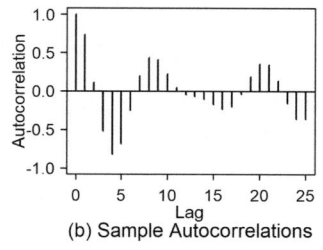
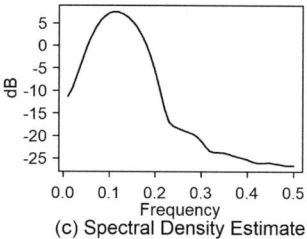

(a) AR(4) Realization   (b) Sample Autocorrelations   (c) Spectral Density Estimate

FIGURE 6.3 (a) Realization from model (6.16), (b) sample autocorrelations, and (c) Parzen spectral density estimate of the data in (a).

Although the data are based on an AR(4) model, the plots have the behavior of an AR(2) with system frequency around $f = .10$. Assuming an AR(4) model, we used the following commands to estimate the parameters.

```
ar4.ml=est.ar.wge(ar4,p=4)
ar4.yw=est.ar.wge(ar4,p=4,type= 'yw')
ar4.bg=est.ar.wge(ar4,p=4,type= 'burg')
```

Table 6.6(a) shows the true and estimated model parameters for the AR(4) model in (6.16). We see from the table that the ML and Burg estimates appear to be similar and quite good. However, the Yule-Walker estimates are surprisingly poor. To better understand the situation, and in particular, to answer the question, "What happened to the Yule-Walker estimates?" we examine (you guessed it) the *factor tables*. Table 6.6(b) shows factor tables for the true model and the three estimated models.

TABLE 6.6   Fitting an AR(4) Model to the Realization from (6.16)

(a) True model parameters and ML, YW, and Burg parameter estimates for AR(4) model fits to the realization in Figure 6.3

	ϕ_1	ϕ_2	ϕ_3	σ_a^2
True	2.76	−3.76	2.60	1.00
MLE	2.77	−3.78	2.63	0.85
YW	1.90	−1.76	.66	5.07
Burg	2.78	−3.81	2.65	0.85

(b) Factor tables for model (6.16) along with ML, YW, and Burg AR(4) fits.

FACTOR	ROOTS	$|r|^{-1}$	f_0
(a) True parameter values			
$1-1.56B+.95B^2$	$.82 \pm .62i$	0.977	0.103
$1-1.20B+.93B^2$	$.65 \pm .81i$	0.966	0.143
(b) ML Estimates			
$1-1.58B+.96B^2$	$.83 \pm .60i$	0.978	0.100
$1-1.19B+.94B^2$	$.63 \pm .81i$	0.971	0.145
(c) Yule-Walker Estimates			
$1-1.38B+.91B^2$	$.76 \pm .72i$	0.954	0.121
$1-0.52B+.14B^2$	$1.90 \pm .194i$	0.368	0.126
(d) Burg Estimates			
$1-1.58B+.96B^2$	$.83 \pm .60i$	0.977	0.100
$1-1.21B+.95B^2$	$.63 \pm .81i$	0.976	0.144

In the factor table for the original model in (6.16) there are two sets of complex conjugate roots, one associated with $f_0 = .103$ and the other with $f_0 = .143$. Points to note about this model are the following:

(a) The two system frequencies are quite close together ($f_0 = .103$ and $f_0 = .143$).
(b) Both pairs of complex conjugate roots are near the unit circle.
(c) The root associated with $f_0 = .103$ is slightly closer to the unit circle than is the root associated with $f_0 = .143$.

The factor tables associated with the ML and Burg estimates do an excellent job of showing the two frequency components present in the factor table for the true model. However, the YW factor table is "another story". We see that the first factor, $1 - 1.38B + .91B^2$, is "in the ballpark" since the factor is associated with a system frequency of $f_0 = .121$ and the roots are fairly close to the unit circle although not as close as those in the original model, as well as in the ML and Burg estimated models. The second factor is associated with essentially the same system frequency $f_0 = .126$, and the roots are quite far from the unit circle. In essence the YW estimates could not "separate" the two system frequencies in the model. Note also that the plots in Figure 6.3 do not separate the two system frequencies because of their close proximity to each other. The *only places* in which we saw the two separate system frequencies were in the factor tables of the true model and of the ML and Burg estimated models. If it is important to detect that there are two system frequencies "close to each other" then certainly the ML and Burg fits are superior to the YW (and to information in the plots in Figure 6.3).

In this example the ML and Burg estimates performed better than the YW estimates. The following summarizes the properties of the three methods for estimating the parameters of an AR(p) model.

PROPERTIES OF THE THREE ESTIMATION TECHNIQUES

(1) *ML Estimates:*
   (a) Are generally quite good.
   (b) Can be computationally intensive. While this is not usually a serious problem with today's computer speeds, simulations or applications that require "real time" parameter estimates can be slowed down considerably when using ML estimates.
   (c) ML estimates can produce models with roots inside the unit circle (beware!)
(2) *Yule-Walker estimates*:
   (a) Are very intuitive and were historically an early technique used for estimating the parameters of an AR process.
   (b) Can be calculated rapidly.
   (c) Always produce a stationary model.
   (d) Are sometimes still used in practice (which is generally a bad idea). It is important to be aware of their shortcomings.
(3) *Burg Estimates:*
   (a) Are usually very similar to ML estimates.
   (b) Always produce a stationary model.
   (c) Can be computed rapidly.
   (d) Because of (a), (b), and (c) they are a good choice when
      – ML estimates are associated with roots inside the unit circle
      – estimating parameters in real time or in extensive simulations
      – analyzing very large datasets.

Key Points

1. We would not have understood the underlying issues with the YW estimates of the AR(4) data in Example 6.3 if we had not *factored* the models.
2. Yule-Walker and Burg estimates *are not applicable to ARMA(p,q) models where $q > 0$*
3. Commands `est.arma.wge(x,p=4,q=0)` and `est.ar.wge(x,p=4)` give the same (MLE) results.
4. Yule-Walker estimates can be very poor when roots are near the unit circle, or when there are repeated roots. We recommend *against* use of Yule-Walker estimates.

6.1.2 ARMA Model Identification

In Section 6.1.1 we discussed the issue of estimating the parameters of an ARMA(p,q) model fit to a set of stationary data where p and q are known. We now address the more general issue of determining *which* ARMA model best fits the data. That is, we need to "identify" the order p before we finalize the estimation of the parameters in the final model. The final models in (6.3) and (6.15) are based on ARMA(3,1) and AR(3) fits, respectively, because we "know" the realizations were generated from ARMA(3,1) and AR(3) models.

Key Point: It is unrealistic to assume that we *know* the orders of an ARMA model that should be fit to a real time series dataset.

6.1.2.1 Plotting the Data and Checking for White Noise

Whenever we fit a model to a set of data we should first plot the data and sample autocorrelations. One thing to check in these plots is whether the data are simply white noise or whether they show obvious nonstationary structure. A quick test for white noise is based on the following facts regarding sample autocorrelations of white noise data:[3]

$$E\left[\hat{\rho}_k\right] \approx 0 \text{ for } k \neq 0$$

- the standard error of $\hat{\rho}_k$ is approximately $1/\sqrt{n}, k \neq 0$ so for moderately large n the $\hat{\rho}_k s$ should be small
- $\hat{\rho}_{k_1}$ and $\hat{\rho}_{k_2}$ are essentially uncorrelated when $k_1 \neq k_2$
- the $\hat{\rho}_k s$ are approximately normal

Based on the above facts, we reject $H_0 : \rho_k = 0$ vs. $H_a : \rho_k \neq 0$ at the 5% level of significance if $|\hat{\rho}_k| > 1.96\left(1/\sqrt{n}\right)$. It is common to accompany plots of sample autocorrelations that might be from white noise with (95%) limit lines at $\pm 2/\sqrt{n}$.[4] Figure 6.4(a) shows a realization of length $n = 150$ from a white noise process generated by the *tswge* command

---

3 See Bartlett (1946).

4 Although the 97.5[th] percentile of the normal distribution is 1.96, it is common practice to round 1.96 to 2 and state the limits as $\pm 2/\sqrt{n}$.

```
x=gen.arma.wge(n=150,phi=0,theta=0,sn=147)
```

Figure 6.4(b) shows the sample autocorrelations. Figure 6.4(a) is consistent with white noise in that there are no cyclic patterns, no wandering, and the data simply seem to be random with no correlation with other values. The sample autocorrelations in Figure 6.4(b) appear to be small (except for $\hat{\rho}_0 = 1$). Figure 6.4(c) shows the same sample autocorrelations as in Figure 6.4(b) but includes the 95% limit lines $\pm 2 / \sqrt{150} = \pm.16$. The sample autocorrelations stay within the limit lines so there is no reason to reject white noise.

> **Key Point:** The 5% applies separately for each k, so it will not be unusual for about 5% of the sample autocorrelations to be outside the 95% limit lines even if the data are white noise.

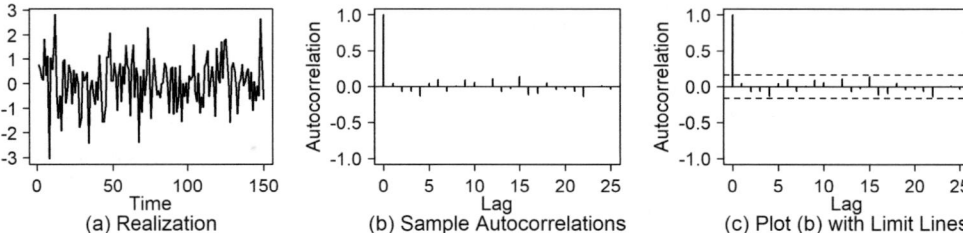

FIGURE 6.4   (a) White noise realization, (b) sample autocorrelations for the data in (a), and (c) sample autocorrelations showing the 95% limit lines.

We next take a look at Figures 6.5(b) and (c) and see that the sample autocorrelations for Figure 6.5(a) have the appearance of sample autocorrelations of white noise data. However, to decide on the basis of the sample autocorrelations alone that the corresponding data (in Figure 6.5(a)) are white noise would clearly be a *major mistake*. The time series in Figure 6.5(a) appears to be higher frequency at the beginning of the realization than it is toward the end. Techniques for analyzing data with time-varying frequencies are discussed in Chapter 13 of Woodward, et al. (2017).

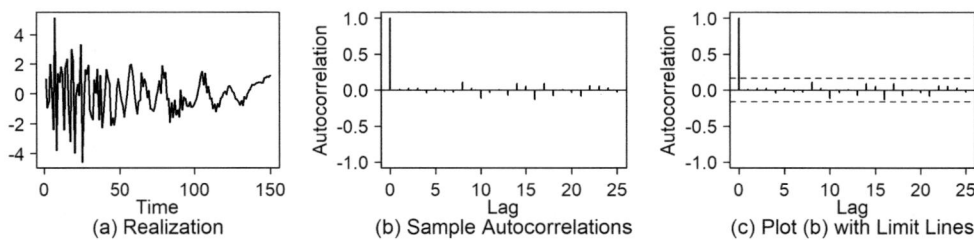

FIGURE 6.5   (a) A realization which is clearly not white noise, (b) sample autocorrelations for the data in (a), and (c) sample autocorrelations showing the 5% limit lines.

> **Key Point:** Always plot the data!

6.1.2.2 Model Identification Types

There are two major strategies commonly used for identifying the order of an ARMA(p,q):

(1) *AIC-type Measures:* Akaike's Information Criterion (AIC) and its variations such as BIC and AICC are popular in a number of applications for deciding among competing models for a set

of data. In the ARMA setting, AIC-type measures are used to "identify" p and q for purposes of fitting an ARMA model to a time series realization. These AIC-type techniques involve fitting ARMA models for a range of values for p and q, and selecting the optimal p and q based on the criterion being used.

(2) ***Pattern Recognition Methods:*** In their ground-breaking 1970 book, G.E.P. Box and G.M. Jenkins used plots of partial autocorrelations to identify the order of an AR process. We will define the partial autocorrelation and discuss its role in identifying the order of an AR model in Section 6.1.2.4.

Key Points

1. In this book we focus on AIC-type model identification which is introduced in Section 6.1.2.3.
2. AIC-type measures are easily automated.
3. Partial autocorrelation methods, to be discussed in Section 6.1.2.4, involve a decision regarding the order of an AR model that is based on observing graphical patterns of partial autocorrelations.
4. AIC-type procedures extend easily to ARMA model identification, while pattern recognition methods are difficult to use for ARMA modeling.

6.1.2.3 AIC-type Measures for ARMA Model Fitting

AIC is a general criterion for model identification that you may have seen before in other settings. For example, it is often used for selecting the independent variables to be used in a multiple regression model. It is also applicable for identifying the order p and q of an ARMA(p,q) model. AIC is actually one of a number of information-based criteria for model selection. Others include BIC and AICC. In the following, the term "AIC" will sometimes be used generically to represent the information-based techniques.

Multiple Regression Review
Consider a hypothetical model selection problem in multiple regression in which dependent variable y is to be predicted using either the single variable x_1 or the two independent variables, x_1 and x_2. In multiple regression, the goal is to reduce the unexplained residual variability, that is, to reduce the mean square error (MSE). Suppose $\hat{y} = 6 + 3x_1$ is the best prediction equation using only x_1, and it has an unexplained variability MSE_1. If independent variable x_2 is added to the model, then there are a multitude of choices for the coefficients b_0, b_1, and b_2 in the prediction equation $\hat{y} = b_0 + b_1 x_1 + b_2 x_2$. Suppose the best prediction equation has unexplained variability MSE_2. Note that one of the choices of b_0, b_1, and b_2 results in the model $\hat{y} = 6 + 3x_1 + 0x_2$ (which has unexplained variability MSE_1). That is, the unexplained variability, MSE_2, of the best prediction equation using the two independent variables, must be at most MSE_1. The key point is that adding a variable to an existing equation will nearly always reduce (and will never increase) the unexplained residual variability. Consequently, the fact that the unexplained residual variability is reduced by adding a new variable is not sufficient evidence to conclude that the new variable is important to the prediction. The goal in multiple regression is to reduce the unexplained residual variability but hopefully to only use the independent variables that are *actually useful*. Clearly, the most useful model will be one with a relatively small MSE. However, because the inclusion of additional variables will never increase (and will nearly always decrease) the MSE, using the MSE for model selection will always yield the model with the most variables (whether they are truly related to the response or not). Techniques are available, for example, stepwise, backward elimination, AIC, etc., for choosing which independent variables are useful. For a discussion of multiple regression, see Moore, McCabe, and Craig (2021). In time series analysis it is common to use AIC, which protects against always selecting a higher order model by adding a "penalty" (see (6.17)) for each new variable added.

ARMA model identification is analogous to model selection in multiple regression. Increasing p and/or q in the ARMA setting indicates the addition of variables (and thus parameters) to the model. When fitting an ARMA(p,q) model to a set of data, a goal is to reduce the unexplained variability, as measured by $\hat{\sigma}_a^2$. Suppose, for example, we fit an ARMA(1,1) model, say $X_t - .8X_{t-1} = a_t + .5a_{t-1}$, to a realization, and associated with this fit is an estimate, $\hat{\sigma}_a^2$, of the white noise variance. Then if we fit an ARMA(2,1) model to the data, one choice for the ARMA(2,1) model is $X_t - .8X_{t-1} - 0X_{t-2} = a_t + .5a_{t-1}$. Consequently, when an ARMA(2,1) model is fit to the realization, the estimated white noise variance must be at most $\hat{\sigma}_a^2$ from the ARMA(1,1) fit.

Key Point: What is needed is a strategy for increasing the order p of the AR model *only if the increase in order is sufficiently helpful.*

As mentioned above, AIC imposes a penalty for adding terms to the model. Theoretically, AIC involves the maximized likelihood. We will use an approximation to AIC that is easy to implement and makes intuitive sense. We define AIC in (6.17) and the strategy is to select the orders p and q such that

$$AIC = \ln(\hat{\sigma}_a^2) + 2(p+q+1)/n \tag{6.17}$$

is minimized. AIC is implemented by selecting the maximum orders we want to allow for p and q, call them P and Q, respectively. Then we fit all models with $0 \le p \le P$ and $0 \le q \le Q$, and choose the order combinations for which AIC is minimized. For simplicity, let $P = 2$ and $Q = 1$ in which case we would fit the models ARMA(0,0), ARMA(0,1), ARMA(1,0), ARMA(1,1), ARMA(2,0), and ARMA(2,1), and select the p and q for which *AIC* is minimized.[5]

Key Point: The AIC procedure for ARMA model identification is basically to reduce the natural log of the estimated white noise variance subject to a penalty for increasing the number of parameters.

BIC and AICC are variations of AIC that are based on *modifications of the penalty*. Specifically, we define BIC and AICC as

$$BIC = \ln(\hat{\sigma}_a^2) + (p+q+1)\frac{\ln(n)}{n} \tag{6.18}$$

and

$$AICC = \ln(\hat{\sigma}_a^2) + (n+p+q+1)/(n-p-q-3). \tag{6.19}$$

AIC tends to select larger orders for longer realization lengths and thus tends to select orders p and q that are too large, for long realizations. BIC adjusts for this tendency by imposing a stronger penalty. AICC tends to serve as a compromise between AIC and BIC.

5 Note that the ARMA(0,0) model is white noise, ARMA(0,1) is an MA(1), ARMA(1,0) is an AR(1), and ARMA(2,0) is an AR(2).

Key Points

1. AIC and its variations are designed to select a *stationary* model. We will discuss nonstationary models in Chapters 7 and 8.
2. When AIC and its alternatives search for $0 \le p \le P$ and $0 \le q \le Q$, one of the options is $p = q = 0$, that is, white noise. Thus, AIC-type methods provide another check for white noise.
3. AIC, BIC, and AICC may produce different "competing" models. The wise analyst does not simply select the order that, for example, AIC selects, but instead selects from among the reasonable models.
4. We will introduce additional strategies and methods for choosing from among these "competing models", not the least of which is *domain knowledge*.

AIC-type Methods for ARMA Model Identification in tswge

The basic *tswge* functions for AIC-type ARMA model identification are `aic.wge` and `aic5.wge`. Given a realization in vector **x**, then to fit an ARMA(p,q) model to this realization where p is allowed to range between 0 and 10, and q ranges from 0 to 4, we use the command

```
aic.wge(x,p=0:10,q=0:4,type= 'aic')
```

To restrict our choice to AR models, we could use

```
aic.wge(x,p=0:10,q=0:0,type= 'aic')
```

Note that `type= 'aic'` is the default and could have been omitted. Other values for `type` are `type= 'bic'` and `type='aicc'`.

QR 6.4 ARMA
Model
Identification

Key Point: *tswge* function, `aic5.wge`, is very useful for identifying alternative ARMA model choices. While the command `aic.wge` selects a single model (based on the criterion selected), `aic5.wge` shows the top five models as selected by the chosen selection criterion.

Example 6.4 Using AIC-type Methods to Select the Order of the Data in `arma31` Which was Generated from the ARMA(3,1) Model in (6.2)

Recall that Figure 6.1(a) shows a realization of length $n = 150$ from the ARMA(3,1) model in (6.2)

$$\left(1 - 2.57B + 2.50B^2 - .92B^3\right)(X_t - 30) = (1 - .92B)a_t,$$

with $\sigma_a^2 = 1$.

The following code generates the realization in Figure 6.1(a).

```
arma31=gen.arma.wge(n=150,phi=c(2.57,-2.50,.92),theta=.92,mu=30,sn=65)
```

Clearly from the data and sample autocorrelations shown in Figures 6.1(a) and (b) respectively, it can be seen that the data in Figure 6.1(a) are not from a white noise model. Assuming we do not know the order of the model that generated the data, we use AIC to select the orders p and q from among the options, $p = 0,1,\ldots,12$ and $q = 0,\ldots,4$ using the *tswge* command

```
a=aic.wge(arma31,p=0:12, q=0:4)
```

Among the output are

```
a$p 3
a$q 1
```

which indicates that AIC has correctly chosen the ARMA(3,1) model.[6] Also included in the output are the parameter estimates

```
a$phi 2.4587424 -2.3304790 0.8286207
a$theta 0.7749636
a$xbar 30.61015
a$vara 1.035746
```

which were previously given in Example 6.1 as output from **est.arma.wge**. By default the criterion used in the above **aic.wge** command was AIC. To identify the orders using BIC and AICC we use the commands

```
b=aic.wge(arma31,p=0:12,q=0:4,type= 'bic')
cc=aic.wge(arma31,p=0:12,q=0:4,type= 'aicc')
```

Among the output from these commands are

```
b$p 2
b$q 0
cc$p 3
cc$q 1
```

The output shows that AICC also selects an ARMA(3,1) model while BIC chooses an AR(2).

> **Key Point:** While **aic.wge** selects the "best" model associated with the chosen criterion, we recommend using the function **aic5.wge** which outputs the top five choices for a given criterion.

For example, the following commands were executed, and the output is shown in Table 6.7.

```
aic5.wge(arma31,p=0:12,q=0:4)
aic5.wge(arma31,p=0:12,q=0:4,type= 'aicc')
aic5.wge(arma31,p=0:12,q=0:4,type= 'bic')
```

---

6 Notice the delay when issuing the AIC command with $p = 0:12$ and $q = 0:4$. In this case, **est.arma.wge** had to find ML estimates for 65 models. In earlier days this might have taken several hours (or overnight) to compute. Recall that Yule-Walker and Burg estimates are only designed to estimate the parameters of an AR(p) model, so because the range of values includes cases in which $q \neq 0$, we cannot use YW or Burg estimates to "speed things up". As mentioned previously, analysts sometimes choose to simplify things by finding the "best" AR model. Choosing the best AR model is the topic of Section 6.1.2.4.

TABLE 6.7   Output Showing the Top 5 AIC, AICC, and BIC Choices for the `arma31` Dataset

p	q	AIC	p	q	AICC	p	q	BIC
3	1	0.1017890	3	1	1.119038	2	0	0.1789293
4	0	0.1056942	4	0	1.122944	3	0	0.1899721
5	1	0.1059744	3	0	1.125800	2	1	0.1948918
2	2	0.1095763	5	1	1.126116	3	1	0.2021435
3	0	0.1096885	2	2	1.126826	4	0	0.2060487

This table shows, as was already noted, that AIC and AICC selected an ARMA(3,1) as the top choice, and BIC selected an AR(2). The table also shows that an ARMA(3,1) was the fourth choice for BIC. Also, an AR(2) model was not one of the top five choices for either AIC or AICC. At this point an obvious question of interest is how the ARMA(3,1) and AR(2) fits compare. The models under consideration are

(a) ARMA(3,1) (This fitted model is already shown in (6.3) and will be repeated here.)

$$\left(1 - 2.459B + 2.330B^2 - .829B^3\right)\left(X_t - 30.610\right) = \left(1 - .775B\right)a_t,$$

with $\hat{\sigma}_a^2 = 1.036$.

(b) AR(2)

$$\left(1 - 1.591B + .938B^2\right)\left(X_t - 30.610\right) = a_t,$$

where $\hat{\sigma}_a^2 = 1.082$.

Comparing the coefficients does not provide information about the characteristics of the two models. Consequently, we examine the factor tables. Table 6.8 shows the factor table for the ARMA(3,1) and AR(2) fitted models.

TABLE 6.8   Factor Tables for the ARMA(3,1) and AR(2) Fits to the `arma31` Data

(a) Factor Table for ARMA(3,1) Model

AR-FACTOR	ROOTS	$\lvert r \rvert^{-1}$	f_0
$1 - 1.578B + .940B^2$	$.838 \pm 600i$	.970	.099
$1 - .881B$	1.135	.881	.000

MA-FACTOR	ROOTS	$\lvert r \rvert^{-1}$	f_0
$1 - .775B$	1.290	.775	.000

(b) Factor Table for AR(2) Model

AR-FACTOR	ROOTS	$\lvert r \rvert^{-1}$	f_0
$1 - 1.591B + .938B^2$	$.848 \pm 589i$	.968	.097

The factor tables show that the AR(2) model has essentially the same 2nd-order factor as the ARMA(3,1) model. Notice that the 1st-order AR and MA factors almost cancel each other, and consequently we conclude that the AR(2) fit to the data is a logical choice for BIC.

Example 6.5 Using AIC-type Methods to Fit ARMA Models to the Log-Lynx Data

In Example 5.3 we examined the log(base 10) of the classical Canadian lynx dataset that exhibits an interesting 10-year cycle. Tong (1977) fit an AR(11) to these data. As we examine a model fit using AIC, we will include an AR(11) as one of the possible model orders. We issue the *tswge* commands:

```
data(lynx)
llynx=log10(lynx)
aic5.wge(llynx,p=0:12,q=0:4)
aic5.wge(llynx,p=0:12,q=0:4,type='aicc')
aic5.wge(llynx,p=0:12,q=0:4,type='bic')
```

The results are shown in Table 6.9. AIC and AICC both select an AR(12) as the top pick with AR(11) as third and second choice for AIC and AICC, respectively. However, BIC, which often selects a model with fewer parameters, chooses an ARMA(2,3), ARMA(3,3), and AR(2) as the top three choices. We will examine the factor tables for the AR(12), ARMA(2,3), and because of its simplicity, the AR(2) model.

TABLE 6.9   Output Showing the Top 5 AIC, AICC, and BIC Choices for the Log-Lynx Data

p	q	AIC	p	q	AICC	p	q	BIC
12	0	−3.128705	12	0	−2.073947	2	3	−2.904979
11	1	−3.121965	11	0	−2.072181	3	3	−2.875738
11	0	−3.121655	11	1	−2.067207	2	0	−2.855197
11	2	−3.115068	11	2	−2.054560	4	3	−2.850560
12	1	−3.112605	12	1	−2.052097	2	1	−2.837161

The three models under consideration are:

(a) AR(12)

$$\left(1-1.116B+.514B^2-.288B^3+.313B^4-.161B^5+.165B^6-.076B^7+.070B^8-.170B^9\right.$$
$$\left.-.138B^{10}+.191B^{11}+.135B^{12}\right)\left(X_t-2.904\right)=\hat{a}_t \tag{6.20}$$

with $\hat{\sigma}_a^2=.035$.

(b) ARMA(2,3)

$$\left(1-1.555B+.953B^2\right)\left(X_t-2.903\right)=\left(1-.453B-.149B^2+.563B^3\right)a_t, \tag{6.21}$$

with $\hat{\sigma}_a^2=.043$.

(c) AR(2)

$$\left(1-1.377B+.740B^2\right)\left(X_t-2.903\right)=a_t, \tag{6.22}$$

with $\hat{\sigma}_a^2=.051$.

If only comparing the coefficients of the three models, it is difficult to see how they are similar to each other and how they are different. In order to understand the models better, we examine the factor tables. The factor tables for the three models are given in Table 6.10.

TABLE 6.10   Factor Tables for AR(12), ARMA(2,3), and AR(2) Models Fit to Log -Lynx Data

(a) Factor Table for AR(12) Model

AR-FACTOR	ROOTS	$|r|^{-1}$	f_0
$1-1.578B+0.977B^2$	$0.81\pm0.61i$	0.988	0.103
$1-0.592B+0.877B^2$	0.34 ± 1.01	0.937	0.199
$1-1.825B+0.877B^2$	$1.04\pm0.24i$	0.936	0.036
$1+0.499B+0.780B^2$	$-0.32\pm1.09i$	0.883	0.296
$1+1.245B+0.660B^2$	$-0.94\pm0.79i$	0.813	0.389
$1+1.135B+0.348B^2$	$-1.63\pm0.46i$	0.590	0.456

(b) Factor Table for ARMA(2,3) Model

AR-FACTOR	ROOTS	$|r|^{-1}$	f_0
$1-1.555B+.953B^2$	$.82\pm.62i$	.976	.103

MA FACTOR	ROOTS	$|r|^{-1}$	f_0
$1-1.203B+.752B^2$	$.80\pm.83i$	.867	.128
$1+.749B$	-1.34	.749	.500

(c) Factor Table for AR(2) Model

AR-FACTOR	ROOTS	$|r|^{-1}$	f_0
$1-1.378B+0.740B^2$	$0.93\pm0.69i$	0.860	0.102

From the factor tables we see that the dominant factor in all three models is a 2nd-order factor associated with a frequency of about $f = 0.10$. However, the roots associated with that factor are much closer to the unit circle in the ARMA(2,3) and AR(12) models than they are in the AR(2) model.

Example 6.6 The Effect of Increasing Realization Length

Recall the ARMA(3,1) model in (6.2) from which the realization **arma31** of length $n = 150$ was generated. To illustrate the effect of increasing realization length on the performance of AIC, AICC, and BIC, we extend the realization in **arma31** to $n = 500$ using the following command:

```
arma31.500=gen.arma.wge(n=500,phi=c(2.57,-2.50,.92),theta=.92,mu=30,sn=65)
```

We then use AIC, AICC, and BIC to select the model orders for the longer series:

```
aic5.wge(arma31.500,p=0:12, q=0:4)
aic5.wge(arma31.500,p=0:12, q=0:4,type= 'bic')
aic5.wge(arma31.500,p=0:12, q=0:4,type= 'aicc')
```

The output (not shown here) reveals that AIC and AICC select an ARMA(7,1) model while BIC picks an AR(4). These models are shown in (6.23) and (6.24), respectively. The ARMA(7,1) model is

$$\left(1-2.521B+2.315B^2-.598B^3-.182B^4-.157B^5+.273B^6-.111B^7\right)\left(X_t-30.217\right)$$
$$=\left(1-.870B\right)a_t, \tag{6.23}$$

where $\hat{\sigma}_a^2 = 0.987$. (It is not surprising that the higher order fit has a lower white noise variance estimate.)

The AR(4) model is given by

$$\left(1-1.671B+.898B^2+.196B^3-.164B^4\right)\left(X_t-30.217\right)=a_t,\tag{6.24}$$

where $\hat{\sigma}_a^2=1.013$.

Recall that the true mean for the ARMA(3,1) model (6.2) is $\mu=30$. The sample mean for the dataset **arma31** (of length $n=150$) is $\bar{x}=30.610$ while the sample mean for **arma31.500** (of length $n=500$) is $\bar{x}=30.217$. That is, as the realization length increased, the sample mean moved closer to the true mean. This behavior fits our understanding of the effect of increasing sample size. However, it is disconcerting that AIC and AICC correctly selected an ARMA(3,1) model for the 150-point dataset but picked an ARMA(7,1) model for the longer realization, **arma31.500**. This illustrates the earlier comment that increasing the realization length is often accompanied by larger model orders selected by AIC. Note that for **arma31.500**, BIC selected an AR(4) model which has the same number of parameters as the original ARMA(3,1) model. We note that the **aic5.wge** output for the **arma31.500** dataset did not include an ARMA(3,1) as one of the top choices by any of the three selection criteria.

We next examine the new models selected by AIC and BIC for the dataset **arma31.500**. Comparing the coefficients in (6.23) and (6.24) to those of the true model (6.2) or the fitted ARMA(3,1) model in (6.3) provides no information concerning the relationship. So, we examine factor tables. Table 6.11(a) is the factor table for the fitted ARMA(3,1) model based on **arma31** previously shown in Table 6.2. As noted earlier the ARMA(3,1) model is dominated by a 2nd-order factor, $1-1.578B+.940B^2$, with roots fairly close to the unit circle and associated system frequency about $f_0=.1$. The model also includes 1st-order AR and MA factors with positive real roots for which the root of the AR factor is closer to the unit circle and the MA factor tends to dampen its effect, leaving an overall effect of a weak 1st-order behavior.

The factor table shown in Table 6.11(b) for the ARMA(7,1) model fit to **arma31.500** includes a 2nd-order factor, $1-1.594B+.937B^2$, very similar to the one in the ARMA(3,1) fit and the true model. The ARMA(7,1) model fit to **arma31.500** contains 1st-order AR and MA factors similar to those in the ARMA(3,1) model fit to **arma31**. Additionally, the ARMA(7,1) model contains two 2nd-order factors whose roots are not close to the unit circle. Knowing that the true model is an ARMA(3,1), these two 2nd-order factors in the ARMA(7,1) model seem to be superfluous.

TABLE 6.11 Factor Table for ARMA(3,1) fit to arma3,1 and the ARMA(7,1) and AR(4) Fits to arma31.500

(a) Factor Table for ARMA(3,1) Model Fit to ARMA31

AR-FACTOR	ROOTS	$|r|^{-1}$	f_0
$1-1.578B+.940B^2$	$.84\pm.60i$	.970	.099
$1-.881B$	1.14	.881	.000

MA-FACTOR	ROOTS	$|r|^{-1}$	f_0
$1-.775B$	1.29	.775	.000

(b) Factor Table for ARMA(7,1) Model Fit to arma31.500

AR-FACTOR	ROOTS	$|r|^{-1}$	f_0
$1-1.594B+.937B^2$	$.85\pm.59i$	.968	.096
$1-.949B$	1.05	.949	.000
$1-.876B+.370B^2$	$1.19\pm1.14i$	.608	.122
$1-.898B+.337B^2$	$-1.33\pm1.09i$	.581	.391

MA-FACTOR	ROOTS	$|r|^{-1}$	f_0
$1-.870B$	1.15	.870	.000

GUIDELINES FOR SELECTING MAXIMUM ORDERS P AND Q IN AN AIC-TYPE SEARCH

You may be wondering how to choose the maximum orders P and Q when using `aic.arma.wge` and `aic5.arma.wge`. Admittedly there are no hard-and-fast rules but the following are considerations:

(1) Use larger orders P and Q than you think you will need.
(2) If the selected order, p is equal to P or q is equal to Q, then enlarge the search.
(3) Let known characteristics of the data help you decide.
  – For example, if you have monthly data, then we recommend choosing P to be larger than 12 to take into consideration any 12th-order monthly behavior. This suggestion will be clearer after discussing seasonal models in Chapter 7.

An Additional Suggestion:

Compare models selected by AIC, BIC, and AICC. If there are major discrepancies in the identified model orders, compare models using factor tables to determine whether the larger order model contains useful information that was missing from the simpler model.

Key Points

1. AIC and AICC have a tendency to select models with more parameters as the realization length increases.
2. Because we know that the true model in Example 6.6 was an ARMA(3,1), then we understand that the fitted ARMA(7,1) model had superfluous factors.
  – This "knowledge" will not be available to us in practical situations.
3. Fitted models selected by AIC, AICC, or BIC will in all likelihood have characteristics similar to those of the true model.
4. In a practical situation with real data and no "true model" it is likely that more than one model can successfully be used for analysis of the data.

6.1.2.4 The Special Case of AR Model Identification

In Section 6.1.1.4 we noted that alternative estimates, Burg and Yule-Walker, were available for estimating the parameters of an AR model. These estimation procedures can also be used in conjunction with AIC-type model identification when you are willing to restrict your search to AR models only. Additionally, as mentioned earlier, the partial autocorrelation function provides a pattern recognition method for AR model identification popularized by Box and Jenkins.

(1) AIC-type AR Model Identification

As previously discussed, the AR model is a widely used model for time series analysis. As noted in Chapter 5, the ARMA model has the advantage of often producing a model with fewer parameters than would be present in a purely AR model fit. As also mentioned, some analysts prefer to deal strictly with AR models.[7]

For AR model identification using AIC-type methods, Equations 6.17−6.19 are modified by setting $q = 0$. The *tswge* functions `aic.ar.wge` and `aic5.ar.wge` perform AR model identification, and in the

---

7 The functions `aic.wge` and `aic5.wge` can be used to restrict the search to AR models by setting Q equal to zero. However, when using `aic.wge` and `aic5.wge`, the parameter estimation is restricted to maximum likelihood.

AR case the identification can be based on ML, Burg, or YW estimates. The functions **aic.ar.wge** and **aic5.ar.wge** use the white noise variance estimate, $\hat{\sigma}_a^2$, in (6.17)–(6.19) associated with the estimation method selected. In the function calls

```
aic.ar.wge(x,p=0:P,type=aic mode, method=estimation method)
aic5.ar.wge(x,p=0:P,type=aic mode, method=estimation method)
```

the parameter options are **type** = **'aic'**, **'bic'**, or **'aicc'** and **method** = **'mle'**, **'burg'**, or **'yw'**. Note that **type** = **'aic'** and **method**=**'mle'** are defaults.

Example 6.7 AR Model Identification for the A_real realization from the AR(3) Model in (6.14)

Let **A_real** be the realization of length $n = 150$ previously generated from AR(3) model (6.14) using the command

```
A_real=gen.arma.wge(n=150,phi=c(1.71,-1.22,.475),sn=327)
```

The data, sample autocorrelations, and Parzen spectral density estimate are shown in Figure 6.6. From the plots in Figure 6.6, it is clear that the data are not white noise.

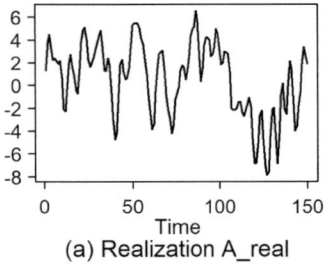
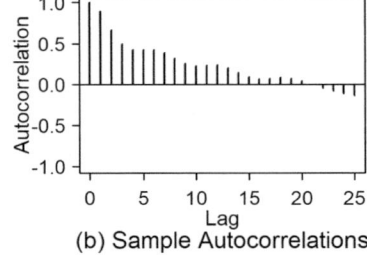
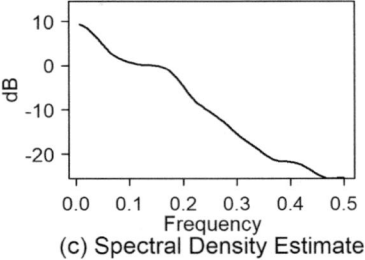

(a) Realization A_real   (b) Sample Autocorrelations   (c) Spectral Density Estimate

FIGURE 6.6 (a) Realization A_real, (b) and (c) sample autocorrelations and Parzen spectral estimate, respectively, of data in (a).

Assuming we do not know the order of the model that generated the data, we select the order p from among the options, $p = 0,1,\ldots,12$ using the *tswge* command **aic.ar.wge(x,p=0:12)**. Among the output is

```
$p 3
```

That is, **aic.ar.wge** has "correctly" chosen an AR(3) model. By default the criterion type is AIC and the estimation method is MLE. The output (not shown here) also includes the ML estimates of the parameters of the AR(3) model (already shown in Table 6.5). Additional output, shown below, includes the value of the minimized AIC, the estimates $\hat{\phi}_j, j = 1, 2,$ and 3, the sample average, and the white noise variance.

```
$type "aic"
$value -0.041711
$p 3
$phi 1.7613987 -1.4010223 0.5605187
$xbar 0.5514184
$vara 0.9093326
```

The command

```
aic5.ar.wge(A_real,p=0:12,type= 'bic',method='burg')
```

yields the following output:

```
Five Smallest Values of bic
Method= burg
```

```
p   bic
3   0.03836811
4   0.06384600
5   0.09125547
6   0.12209338
7   0.15453973
```

Interestingly the top orders were $p = 3, 4, 5, 6$, and 7, respectively. In Problem 6.3 you are asked to identify the order of the AR model using each of the nine type/method combinations.

Example 6.8 Fitting an AR Model to the Sunspot Data

In Example 5.6 we fit an AR(9) model to the sunspot data in **sunspot2.0**. These data are annual averages from 1700 through 2020 using the sunspot2.0 method of counting. Using the commands

```
data(sunspot2.0)
ss=aic.ar.wge(sunspot2.0,p=0:12)
ss
```

we see that indeed AIC selects an AR(9) model. Output from the **aic.ar.wge** command is given below:

```
ss$type "aic"
ss$method "mle"
ss$value 6.361889
ss$p 9
ss$phi 1.165113596 -0.407848045 -0.133850375 0.104019738 -0.068796255
0.007175265 0.023075536 -0.050308201 0.222699470
ss$xbar 78.51651
ss$vara 544.3455
```

Examination of the estimates in **$phi** shows that the model in (5.36) is the fitted ML model rounded to two decimal places. Using

```
aic5.ar.wge(sunspot2.0,p=0:12)
```

we obtain the following output:

```
Five Smallest Values of aic
Method= mle

P   aic
9   6.361889
10  6.367760
11  6.373801
12  6.380013
8   6.407677
```

It can be seen that $p = 10$ is a close second choice (there is very little difference in the AIC values). See Problem 6.2.

Try This:

Let p range from 0 to 20 and continue to use the type/method combination AIC/MLE on the dataset **sunspot2.0**. Before issuing the following command be ready to see how long it takes for the command to run on your computer.

```
aic5.ar.wge(sunspot2.0,p=0:20)
```

It takes my computer over 10 seconds (yours is probably faster). The top five orders are 9, 18, 10, 19, and 11. Unless the model with $p = 18$ is useful in capturing details in the model missed by the AR(9) model, most analysts would prefer the model with 9 coefficients to the one with 18 because it is a much simpler model.

Now modify the above command to compute Burg estimates. That is, issue the following command and again check to see how long it takes to run.

```
aic5.ar.wge(sunspot2.0,p=0:20,method='burg')
```

On my computer the results were almost instantaneous. This is an example of the fact that ML estimates can be quite slow to converge. We have seen this dramatically in Examples 6.4 and 6.6. (Did you run the code?) Unfortunately, in the ARMA(p,q) case where q is allowed to take on a value greater than zero, we are "stuck with" ML estimates and convergence may take a long time.

If speed is an issue, then restricting the search to AR models using Burg estimation may be preferable.[8] The top five orders using Burg estimates are 9, 18, 10, 19, and 20–very nearly the same results.

Now, let's see what orders BIC selects. Issue the following command:

```
aic5.ar.wge(sunspot2.0,p=0:20,type='bic')
```

This command computes the "slow" ML estimates, but notice that the top five orders are 9, 10, 8, 11, 12. That is, the BIC tends to select orders that are smaller than AIC (because there is a greater penalty for increasing the order p). Using BIC as above but this time with Burg estimates yields the same order selections.

Conclusions:
(1) The AR(9) model is a winner with AIC and BIC using MLE and Burg estimates. AR(9) also "wins" using AICC. Yule-Walker estimates also select AR(9) as the top model.
(2) Box and Jenkins (1970) select an AR(2) model, but we will show in Section 9.3 that the AR(9) is the better model.

(2) Pattern Recognition Methods for AR Model Identification Using the Partial Autocorrelation Function

Box, Jenkins, and Reinsel (2008) use the partial autocorrelations as their key tool for identifying the order p of an autoregressive model. There are two definitions of the partial autocorrelation function. We give them both in Definition 6.1.

Definition 6.1: Partial Autocorrelations

Let X_t be a stationary process with autocorrelations $\rho_j, j = 0,1,\dots$.

(a) The partial autocorrelation at lag k, denoted ϕ_{kk}, is the correlation between X_t and X_{t+k} conditional on "knowing" the intervening variables X_{t+1}, X_{t+2}, and X_{t+k-1}.[9]

(b) Consider the following Yule-Walker equations where ϕ_{kj} denotes the jth coefficient associated with the kth order Yule-Walker equations.

---

8 In earlier days of the microcomputer, computing speeds were much slower. Consequently, when using the MLE along with AIC (or its variants) this computation would have taken much longer. In such a case, the first author used to tell his students that they might want to go get a cup of coffee while the AIC routine was running (or in the case of an ARMA setting as in Example 6.6 – check back tomorrow).

9 Recall that the coefficient β_k in the multiple regression equation, $Y = \beta_0 + \beta_1 X_1 + \dots + \beta_k X_k + e_t$ can be interpreted as the partial correlation between Y and X_k after controlling for $X_1,\dots, X_{k-1}$. Analogously, the coefficient ϕ_k in the autoregressive equation $X_t = \beta + \phi_1 X_{t-1} + \phi_2 X_{t-2} + \dots + \phi_k X_{t-k} + a_t$ can be interpreted as the partial correlation between X_t and X_{t+k} after controlling for $X_{t+1},\dots, X_{t+k-1}$.

$k = 1$

$\rho_1 = \phi_{11}$

$k = 2$

$\rho_1 = \phi_{21} + \phi_{22}\rho_1$

$\rho_2 = \phi_{21}\rho_1 + \phi_{22}$

<u>General k</u>

$\rho_1 = \phi_{k1} + \phi_{k2}\rho_1 + \cdots + \phi_{kk}\rho_{k-1}$

$\rho_2 = \phi_{k1}\rho_1 + \phi_{k2} + \cdots + \phi_{kk}\rho_{k-2}$

$\vdots$

$\rho_k = \phi_{k1}\rho_{k-1} + \phi_{k2}\rho_{k-2} + \cdots + \phi_{kk}$

The partial autocorrelation function is defined to be ϕ_{kk}, $k = 1, 2, \ldots$.

Definition 6.1(a) shows that ϕ_{kk} is actually an autocorrelation for each k, so that $|\phi_{kk}| < 1$ for each k. However, Definition 6.1(b) is the definition that is useful in AR model identification. Definition 6.1(b) is straightforward, but may be confusing at first glance. Recall the Yule-Walker equations given in (5.29) and repeated in (6.25):

$\rho_1 = \phi_1 + \phi_2\rho_1 + \cdots + \phi_p\rho_{p-1}$

$\rho_2 = \phi_1\rho_1 + \phi_2 + \cdots + \phi_p\rho_{p-2}$

$\vdots$

$$\rho_p = \phi_1\rho_{p-1} + \phi_2\rho_{p-2} + \cdots + \phi_p. \tag{6.25}$$

The equations in (6.25) are for the case in which the model *is* an AR(p). The ϕ_js, $j = 1, \ldots, p$ are the *actual* coefficients if the model is an AR(p). In Definition 6.1(b) we assume that we know the autocorrelations but not the order p.

- Start by *assuming* $p = 1$, in which case (6.25) says $\rho_1 = \phi_1$. (This is only true if $p = 1$.) To keep things straight, we use the notation $\rho_1 = \phi_{11}$ where the first subscript denotes the AR order assumed and the second designates that ϕ_{11} is the first (and only coefficient) in the AR(1) case. (To be clear, $\phi_{11} = \phi_1$ if the model really is an AR(1).)

- *Assuming* the model *is* an AR(2) then

  $\rho_1 = \phi_1 \quad + \phi_2\rho_1$

  $\rho_2 = \phi_1\rho_1 + \phi_2,$

  Again, this is true only if the model is an AR(2). If we only *assume* the model *is* an AR(2), then we write these equations as

  $\rho_1 = \phi_{21} \quad + \phi_{22}\rho_1$

  $\rho_2 = \phi_{21}\rho_1 + \phi_{22},$

  That is, the coefficient ϕ_{21} is the first coefficient in the AR(2) case while ϕ_{22} is the corresponding second coefficient under the assumption of an AR(2). (If the model is an AR(2) then $\phi_{22} = \phi_2$.)

- Continue this process up to some value P that we believe is at least as large as the true order. and in each case keep the last coefficient, that is $\phi_{11}, \phi_{22}, \ldots$.

> **Key Point:** ϕ_{kk} is the last (kth) coefficient assuming X_t is an AR(k) process.

Consider an AR(p) model with white noise variance σ_a^2 and without loss of generality, zero mean. Then, the AR(p) model, $X_t - \phi_1 X_{t-1} - \cdots - \phi_p X_{t-p} = a_t$, is fully described by $\phi_1, \ldots, \phi_p$. If we decided to express this model as an AR$(p+1)$, then the model would be

$$X_t - \phi_1 X_{t-1} - \cdots - \phi_p X_{t-p} - 0 X_{t-(p+1)} = a_t.$$

That is, the lagged variable $X_{t-(p+1)}$ is not involved in the model, and said another way, $\phi_{p+1} = 0$. Using the same logic, $\phi_{kk} = 0$ if $k > p$.

PACF Model Identification Strategy

Consecutively fit AR(1), AR(2), ..., AR(P) models to the data (where P is a maximum order similar to that used in the AIC-type methods). For the kth fitted model, retain and plot $\hat{\phi}_{kk}$, $k = 1,2,\ldots,P$ where $\hat{\phi}_{kk}$ is the last (kth) coefficient of the fitted AR(k) model. If the model is an AR(p), then it should follow that $\hat{\phi}_{p+1,p+1}$, $\hat{\phi}_{p+2,p+2}, \ldots$ will be near zero compared to $\hat{\phi}_{pp}$. The *tswge* function **pacf.wge** computes and plots the partial autocorrelations.

Example 6.7 (continued)

We return to the AR(3) dataset **A_real** in Example 6.7. A realization from this model, sample autocorrelations, and Parzen spectral density are shown in Figure 6.6. The following code generates the realization **A_real** and calls the *tswge* **pacf.wge** function.

```
A_real=gen.arma.wge(n=150,phi=c(1.71,-1.22,.475),sn=327)
pacf.wge(A_real)
```

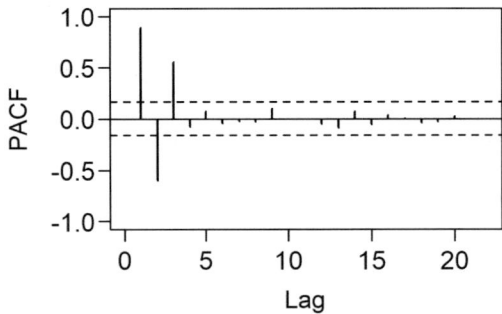

FIGURE 6.7   Base R pacf output for AR(3) data in Figure 6.6(a).

Figure 6.7 shows that $\hat{\phi}_{33}$ is a little larger than .5 and for $k \geq 4$, the sample partial autocorrelations are near zero. The sample partial autocorrelations, $\hat{\phi}_{kk}$, for $k > p$ are asymptotically normal and approximately independent with zero mean and variance $1/n$.[10] The dashed lines in the plot are at $\pm 2(1/\sqrt{n})$, and these limits are often used as a guide for testing $\phi_{kk} = 0$. The plot in Figure 6.7 strongly suggests an AR(3). See problem 6.13 for practice using the partial autocorrelations

---

10 See Quenouille (1949).

Key Points

1. The 95% limit lines for $\hat{\phi}_{kk}$ for $k > p$ where p is the actual order of the AR model are similar to those for sample autocorrelations of white noise data. That is, the 95% applies separately for each $k(> p)$ so it will not be surprising for about 5% of the sample partial autocorrelations (for $k > p$) to fall outside the limit lines.

2. The estimated partial autocorrelations, $\hat{\phi}_{kk}$, calculated using Cramer's rule are the estimates of the kth autoregressive coefficient using the Yule-Walker method.
   - Base R function **pacf** uses Yule-Walker based partial autocorrelations and will be subject to the problems associated with Yule-Walker estimates such as those mentioned in Example 6.3.
   - *tswge* function **pacf.wge** provides the ability to base the calculation of the partial autocorrelations on ML and Burg estimates as well as the traditional YW estimates.

6.2 FORECASTING USING AN ARMA(p,q) MODEL

One of the main purposes for fitting a time series model is to forecast (or predict) future behavior of the time series given a finite realization of its past. We may want to forecast sales for the next quarter based on sales for the past few years, or we may want to address the more challenging problem of forecasting sales for the next four quarters. Based on historical data we could forecast monthly home sales in a particular city for the next year. Other examples include prices of oil, gross national product, Dow Jones Index, and of course, a topic of current interest and concern is to forecast the global temperature for the next few decades. In chapters that follow we will discuss forecasting time series data based on models that include trends, seasonal components, multiple explanatory variables, and so forth.

In Section 6.2.1 we focus on the problem of forecasting the future values of data that satisfy the conditions of stationarity. ARMA-based forecasting was popularized by G.E.P. Box and G.M. Jenkins in their classic 1970 book. These forecasts are based on the underlying assumption that the future is guided by its correlation to the past.

6.2.1 ARMA Forecasting Setting, Notation, and Strategy

In the following it is assumed that a realization has been observed from the time series X_t , for times $t = 1, 2, \ldots, t_0$. The goal is to forecast a future value (or values), say X_{t_0+1} or $X_{t_0+1}, \ldots, X_{t_0+4}$.

Note that the typical case is $t_0 = n$, that is, $X_{t_0} = X_n$ is the last value in the observed realization. In some cases we may, for example, want to "forecast" the last four values in the realization (which are known values) to assess the performance of our forecasts. In this case, $t_0 = n - 4$.

6.2.1.1 Strategy and Notation

$\hat{X}_{t_0}(\ell)$ is the forecast of $X_{t_0+\ell}$ given data up to time t_0

- t_0 is the *forecast origin*
- ℓ is the *lead time* or *horizon*, that is, the number of units (*steps ahead*) that we want to forecast

For example, suppose we observe a realization from the time series X_t for times $t = 1, 2, \ldots, t_0 = 10$, and we want to forecast $X_{12} \left(= X_{t_0+2} \right)$.

Then $\hat{X}_{10}(2)$ is the forecast of X_{12}. Also,

$\hat{X}_{10}(3)$ is the forecast of X_{13}
$\hat{X}_{10}(7)$ is the forecast of X_{17}
$\hat{X}_{10}(\ell)$ is the forecast of $X_{10+\ell}$.

6.2.1.2 Forecasting $X_{t_0+\ell}$ for $\ell \leq 0$

$$\hat{X}_{t_0}(\ell) = X_{t_0+\ell} \text{ if } \ell \leq 0. \tag{6.26}$$

Note that equation (6.26) says that $\hat{X}_{10}(0) = x_{10}$, which has already been observed. Also,

$\hat{X}_{10}(-1) = x_9$
$\hat{X}_{10}(-2) = x_8$
$\vdots$

all of which have already been observed.

> **Key Point:** Equation 6.26 is trivial and you may ask, "Why would we want to 'forecast' something we have already observed?" Good question. The fact is that the iterative equations for computing $\hat{X}_{t_0}(\ell)$ for $\ell > 0$ involve "forecasts" $\hat{X}_{t_0}(m)$ for negative values of m. As defined in (6.26) these "forecasts" are actually values we have already observed. See Example 6.9.

6.2.1.3 Forecasting $a_{t_0+\ell}$ for $\ell > 0$

$$\hat{a}_{t_0}(\ell) = 0 \text{ if } \ell > 0. \tag{6.27}$$

Note that $\hat{a}_{10}(1)$ is the forecast of a_{11} given data to time $t = 10$. Because a_t is random white noise with zero mean, and as of time $t = 10$, a_{11} is unknown and uncorrelated with past values of a_t, it makes intuitive sense to define $\hat{a}_{10}(1) = 0$. So, in general the forecast formula in (6.27) makes sense for all $\ell > 0$.

6.2.2 Forecasting Using an AR(p) Model

We first consider the simplest AR(p) case, that is, forecasting based on an AR(1).

6.2.2.1 Forecasting Using an AR(1) Model

Suppose X_t satisfies an AR(1) model given by $X_t = \phi_1 X_{t-1} + (1 - \phi_1)\mu + a_t$, and that we have observed values for $x_1, x_2, \ldots, x_{10} \left(= x_{t_0} \right)$. As mentioned, we will estimate ϕ_1 with the ML estimate, $\hat{\phi}_1$, and the mean

μ by $\bar{x}$. Using the fitted model, the formula for (unobserved) X_{11} is $X_{11} = \hat{\phi}_1 x_{10} + \left(1 - \hat{\phi}_1\right)\bar{x} + a_{11}$. Because X_{11} and a_{11} have not been observed, we replace all quantities with their forecasts as of time $t = 10$ yielding

$$\hat{X}_{10}(1) = \hat{\phi}_1 \hat{X}_{10}(0) + \bar{x}(1 - \hat{\phi}_1) + \hat{a}_{10}(1).$$

Obviously, by (6.26), $\hat{X}_{10}(0) = x_{10}$ and by (6.27), $\hat{a}_{10}(1) = 0$. This leads to the general formula for calculating forecasts using an AR(1) model.

The **basic formula for forecasting from a fitted AR(1) model** is

$$\hat{X}_{t_0}(\ell) = \hat{\phi}_1 \hat{X}_{t_0}(\ell - 1) + \bar{x}(1 - \hat{\phi}_1). \qquad (6.28)$$

The formula in (6.28) is used recursively. That is, to calculate $\hat{X}_{t_0}(\ell)$ we iteratively calculate

$$\hat{X}_{t_0}(1), \; \hat{X}_{t_0}(2), \; ..., \; \hat{X}_{t_0}(\ell - 1), \; \boldsymbol{\hat{X}_{t_0}(\ell)}.$$

($\hat{X}_{t_0}(\ell)$ is in bold face to emphasize that it was the forecast that was desired.) The other forecasts were involved in the iterative procedure used to find $\hat{X}_{t_0}(\ell)$. For example, to forecast X_{12} based on data to time $t = 10$, the forecast formula in (6.28) says that

$$\hat{X}_{10}(2) = \hat{\phi}_1 \hat{X}_{10}(1) + \bar{x}(1 - \hat{\phi}_1),$$

but we must first calculate $\hat{X}_{10}(1)$ which is given by

$$\hat{X}_{10}(1) = \hat{\phi}_1 \hat{X}_{10}(0) + \bar{x}(1 - \hat{\phi}_1).$$

Based on (6.26), $\hat{X}_{10}(\ell) = X_{t_0 + \ell}$ if $\ell \leq 0$, so it follows that $\hat{X}_{10}(0) = x_{10}$, a *known value*. So, to calculate $\hat{X}_{10}(12)$ we recursively calculate

$$\hat{X}_{10}(1) = \hat{\phi}_1 x_{10} + \bar{x}(1 - \hat{\phi}_1)$$
$$\boldsymbol{\hat{X}_{10}(2)} = \hat{\phi}_1 \hat{X}_{10}(1) + \bar{x}(1 - \hat{\phi}_1).$$

QR 6.5 AR(1)
Forecasting

Example 6.9 AR(1) Forecasting Example
Consider the following *tswge* code:

```
x=c(14.0,14.8,14.0,14.0,13.8,12.4,12.9,11.9,12.0,10.9,10.5,9.2,10.7,11.3,
9.2,8.6,8.2,7.3,7.9,8.8)
aic.x=aic.wge(x,p=0:4,q=0:0) # aic.x$p is the value of p selected by AIC
# AIC picks p=1
est.x=est.ar.wge(x,aic.x$p)
```

```
# the MLE of phi(1)=.92 (est.x$phi)
# white noise variance estimate is .82 (est.x$avar)
mean(x)
# the estimated mean is 11.12
```

After running the above code we see that AIC picks an AR(1) and the fitted AR(1) model is

$$X_t = 0.92X_{t-1} + .89 + a_t, \tag{6.29}$$

where $\hat{\sigma}_a^2 = 0.82$, because $\bar{x}(1-\hat{\phi}_1) = 11.12(1-.92) = .89$. The dataset **x** is plotted in Figure 6.8(a) along with a horizontal line at the mean, $\bar{x} = 11.12$. We see the typical wandering behavior associated with an AR(1) model with positive real root. That is, if the value x_{t_1} is below the mean, then observations in the neighborhood of t_1 will also tend to be below the mean (and similarly for values above the mean).

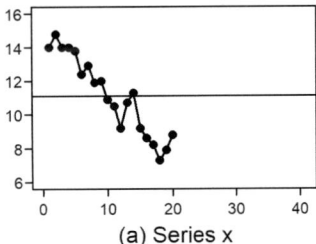
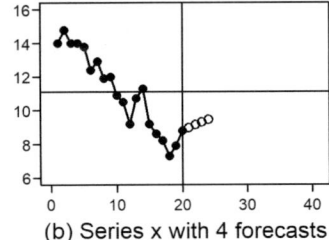
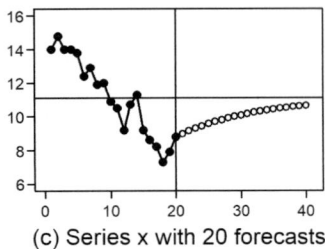

(a) Series x   (b) Series x with 4 forecasts   (c) Series x with 20 forecasts

FIGURE 6.8   (a) Realization x for which the fitted model is (6.29), (b) Realization in (a) along with forecasts for $t = 21, 22, 23$, and 24 calculated by hand, and (c) realization in (a) along with forecasts for $t = 21, 22, \ldots, 40$ calculated using `fore.arma.wge`.

For purposes of calculating forecasts using the formula $\hat{X}_{t_0}(\ell) = \hat{\phi}_1 \hat{X}_{t_0}(\ell-1) + \bar{x}(1-\hat{\phi}_1)$, we note that $\hat{\phi}_1 = .92$ and $x_{20} = 8.8$. The forecast function for this dataset is

$$\hat{X}_{20}(\ell) = .92 \hat{X}_{20}(\ell-1) + .89. \tag{6.30}$$

The forecasts of X_{21} through X_{24} are given by

$$\hat{X}_{20}(1) = .92\hat{X}_{20}(0) + .89 = .92(8.8) + .89 = 8.99$$
$$\hat{X}_{20}(2) = .92\hat{X}_{20}(1) + .89 = .92(8.99) + .89 = 9.16$$
$$\hat{X}_{20}(3) = .92(9.16) + .89 = 9.32$$
$$\hat{X}_{20}(4) = .92(9.32) + .89 = 9.46$$
$$\vdots$$

The realization in **x** and forecasts $\hat{X}_{20}(1), \hat{X}_{20}(2), \hat{X}_{20}(3)$, and $\hat{X}_{20}(4)$ calculated above are shown in Figure 6.8(b). It is important to note that because **x** is from a stationary time series, there is an "attraction" to the mean, in this case $\bar{x} = 11.12$. Thus, it is not surprising that because $x_{20} = 8.8$ is below the mean, forecasts trend upward toward the mean. Forecasts shown in Figure 6.8(c) were obtained using the *tswge* command

```
fore.x=fore.arma.wge(x,est.x$phi,n.ahead=20,limits=FALSE)
```

where **x** is the dataset in Figure 6.8(a), `est.x$phi` are the ML estimates of the AR(1) fit to the data, and `n.ahead=20` instructs the function to calculate and plot the forecasts up to 20 steps ahead, that is,

$\hat{X}_{20}(\ell), \ell = 1,\ldots,20,$ shown in Figure 6.8(c). Also, **limits=FALSE** indicates that we want to omit the prediction limits from the plot (these will be discussed in Section 6.2.5). In Figure 6.8(c) it is clear that the forecasts continue trending upward toward the mean, $\bar{x} = 11.12$. The values for $\hat{X}_{20}(\ell), \ell = 1,\ldots,4$ are given as the first four values in **fore.x\$f** which are

```
fore.x$f 8.980427 9.146823 9.300277 9.441798
```

Note that these vary slightly from the calculations above because the coefficient estimates are rounded to two decimal places.

QR 6.6 AR(1)
Forecasting Real Data

The AR(0) Model: Consider the case of normal noise, a_t. The "AR(0)" model is given by

$$X_t - \mu = a_t, \tag{6.31}$$

where $a_t \sim N\left(0, \sigma_a^2\right)$. If $X_t, t = 1,\ldots,n$ satisfies this model, then $X_1, X_2,\ldots, X_n$ are actually a random sample from a Normal(μ, σ_a^2) distribution, and there is no dependence on time. Consequently, the best estimate of X_{n+1} is $\bar{x} = \dfrac{1}{n}\sum_{t=1}^{n} x_t$. That is, if $t_0 = n$, then $\hat{X}_{t_0}(1) = \bar{x}$. Likewise, because the data are uncorrelated (and not dependent on time), then $\hat{X}_{t_0}(\ell) = \bar{x}$ for any $\ell > 0$. Of course, these are "unexciting" forecasts.

Key Point: Earlier we said that an AR(1) is the simplest of the AR(p) models. Technically, white noise is the simplest of the AR(p) models. Actually, white noise, that is, an AR(0) (and an ARMA(0,0)) is the "simplest model".

QR 6.7 AR(2)
Forecasting

Example 6.10 AR(2) forecasting example
Consider the following *tswge* code.

```
y=c(40.3,36.6,40.1,42.4,40.7,38.5,39.3,42.0,41.5,39.3,37.8,40.4,43.5,41.5,37.
4,37.8,40.6,43.2,40.7,38.71,40.9,40.7,40.2,39.5,39.4)
# y is a time series of length n=25
aic.y=aic.wge(y,p=0:10,q=0:0)
# AIC picks p=2 so we estimate the parameters
est.y=est.ar.wge(y,aic.y$p)
# MLE in $phi: 0.30 -0.90
# white noise variance estimate is $avar=.57
# sample mean is $xbar=40.12
```

The output from the above code shows that AIC selects an AR(2) and the fitted AR(2) model is

$$\left(1 - 0.3B + 0.9B^2\right)\left(X_t - 40.12\right) = a_t, \tag{6.32}$$

where $\hat{\sigma}_a^2 = 0.57.$ The one-line factor table is shown in Table 6.12.

TABLE 6.12   Factor Table for Model: $\left(1 - 0.30B + 0.90B^2\right)\left(X_t - 40.12\right) = a_t.$

FACTOR	ROOTS	$|r|^{-1}$	f_0
$1 - 0.3B + .9B^2$	$.17 \pm .1.04i$	.95	.22

The factor table shows a single 2$^\text{nd}$-order factor whose roots are fairly close to the unit circle ($|r|^{-1} = .95$) and are associated with system frequency $f = .22$ (that is, a cycle length of $1/.22 = 4.5$). The data shown in Figure 6.9(a) have a cyclic behavior with cycle length of about four or five time points. Also, $\bar{x} = 40.12$ is plotted as a horizontal line on the plot.

Key Point: We could have used `aic5.wge` to examine other models, but because the focus of this example is the forecast function, we simply use the model selected by AIC. Although we recommend checking alternatives to the model automatically selected by AIC, we will at times simply use the AIC choice for simplicity of discussion.

The model in (6.32) can be written as

$$X_t = .3X_{t-1} - .9X_{t-2} + 64.19 + a_t \tag{6.33}$$

because $\bar{x}(1 - \hat{\phi}_1 - \hat{\phi}_2) = 40.12(1 - .3 + .9) = 64.19.$ Using (6.33) we have the forecast function

$$\hat{X}_{25}(\ell) = .30\hat{X}_{25}(\ell - 1) - .90\hat{X}_{25}(\ell - 2) + 64.19, \tag{6.34}$$

and because $x_{24} = 39.5$ and $x_{25} = 39.4$, the forecasts of X_{26} through X_{29} are given by

$$\hat{X}_{25}(1) = .30\hat{X}_{25}(0) - .90\hat{X}_{25}(-1) + 64.19 = .30(39.4) - .90(39.5) + 64.19 = 40.46$$

$$\hat{X}_{25}(2) = .30\hat{X}_{25}(1) - .90\hat{X}_{25}(0) + 64.19 = .30(40.46) - .90(39.4) + 64.19 = 40.87$$

$$\hat{X}_{25}(3) = .30(40.87) - .90(40.46) + 64.19 = 40.04$$

$$\hat{X}_{25}(4) = .30(40.04) - .90(40.87) + 64.19 = 39.42$$

Forecasts $\hat{X}_{25}(\ell), \ell = 1, \ldots, 4$ calculated above are shown in Figure 6.9(b), while the forecasts $\hat{X}_{25}(\ell), \ell = 1, \ldots, 20$ shown in Figure 6.9(c) were obtained using the ***tswge*** command

```
fore.y=fore.arma.wge(y,phi=est.y$phi,n.ahead=20,limits=FALSE)
```

where **y** is the dataset specified in the above code and plotted in Figure 6.9(a). The parameters of the fitted model in **est.y$phi** are the ML estimates of the AR(2) fit to the data, and **n.ahead=20** instructs the function to calculate and plot the forecasts up to 20 steps ahead. The use of **limits=FALSE** indicates that we want to omit the prediction limits from the plot (these will be discussed in Section 6.2.5). Note that the 20 forecasts are in vector **fore.y$f**. In Figure 6.9(c) it is clear that the forecasts have a damped cyclic behavior around the mean, $\bar{x} = 40.12.$ and with a cycle length of about 4 to 5. The values for $\hat{X}_{25}(\ell), \ell = 1, \ldots, 4$ are given as the first four values in

fore.x$f: 40.45693 40.86736 40.04601 39.42927

which vary slightly from the calculations above due to the fact that we kept only two decimal places in the coefficient estimates.

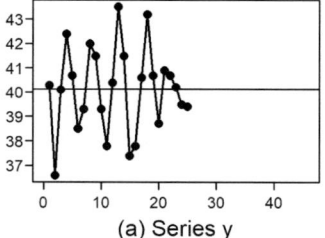
(a) Series y

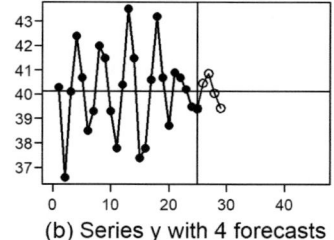
(b) Series y with 4 forecasts

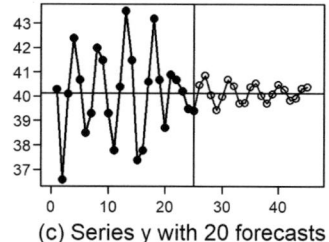
(c) Series y with 20 forecasts

FIGURE 6.9 (a) Realization y for which the fitted model is (6.32), (b) Realization in (a) along with forecasts for $t = 26, 27,\ 28,$ and 29 calculated by hand above, and (c) realization in (a) along with forecasts for $t = 26, 27, \ldots, 45$ calculated using fore.arma.wge.

6.2.3 Basic Formula for Forecasting Using an ARMA(p,q) Model

Equation (6.35) gives the general formula for calculating forecasts from a fitted ARMA(p,q) model.

$$\hat{X}_{t_0}(\ell) = \sum_{j=1}^{p} \hat{\phi}_j \hat{X}_{t_0}(\ell-j) - \sum_{j=\ell}^{q} \hat{\theta}_j \hat{a}_{t_0+\ell-j} + \bar{x}\left[1 - \sum_{j=1}^{p} \hat{\phi}_j\right]. \tag{6.35}$$

Note that for the AR(p) case these forecasts simplify to

$$\hat{X}_{t_0}(\ell) = \sum_{j=1}^{p} \hat{\phi}_j \hat{X}_{t_0}(\ell-j) + \bar{x}\left[1 - \sum_{j=1}^{p} \hat{\phi}_j\right]. \tag{6.36}$$

Thus we see that ARMA(p,q) forecasts for $q > 0$ depend on estimates of the white noise residuals, while these estimates are not involved in AR(p) forecasts. The tswge forecasting routines use the backcasting procedure for computing estimates of the residuals. (See Woodward et al. (2017)).

Key Point: AR forecasts using (6.36) are simply the special case in which $-\sum_{j=\ell}^{q} \hat{\theta}_j \hat{a}_{t_0+\ell-j} = 0$

QR 6.8 ARMA
Forecasting Formula

Example 6.11 Forecasts from an ARMA(2,1) Model

Consider the ARMA(2,1) model

$$\left(1-1.6B+.9B^2\right)(X_t-20)=\left(1-.9B\right)a_t,\tag{6.37}$$

where $\sigma_a^2=1$. Table 6.13 shows that the AR part of the model is associated with a system frequency of $f_0=.09$ (or periods of about 11). The MA part indicates the "removal" of frequency behavior at around $f=0$. Figure 6.10(a) shows a realization of length $n=100$, generated using the command

```
arma21=gen.arma.wge(n=100,phi=c(1.6,-.9),theta=.9,mu=20,sn=5789)
```

TABLE 6.13   Factor Table for ARMA(2,1) Model

AR-FACTOR	ROOTS	$|r|^{-1}$	f_0
$1-1.60B+90B^2$	$.89\pm57i$	.95	.09

MA-FACTOR	ROOTS	$|r|^{-1}$	f_0
$1-.90B$	1.11	.90	.00

The realization shows a cyclic behavior with cycle length about 11 (1/.09). The command

```
aic5.wge(arma21,p=0:12,q=0:4)
```

returns an ARMA(2,1) as the top choice with an ARMA(3,1) being the second choice. The command

```
est21=est.arma.wge(arma21,p=2,q=1)
```

returns the ML fitted model

$$\left(1-1.583B+.899B^2\right)(X_t-20.038)=\left(1-.929B\right)a_t,\tag{6.38}$$

with $\hat{\sigma}_a^2=.84$. Using (6.35) the forecast function is

$$\hat{X}_{100}(\ell)=\sum_{j=1}^{2}\hat{\phi}_j\hat{X}_{100}(\ell-j)-\sum_{j=\ell}^{1}\hat{\theta}_j\hat{a}_{100+\ell-j}+\bar{x}\left[1-\sum_{j=1}^{2}\hat{\phi}_j\right].\tag{6.39}$$

Notice that the term

$$-\sum_{j=\ell}^{1}\hat{\theta}_j\hat{a}_{100+\ell-j}$$

is $-\hat{\theta}_1\hat{a}_{100}$ when $j=1$ and the sum is null for $\ell>1$ because the lower limit (ℓ) is greater than the upper limit (1). So,

$$\hat{X}_{100}(1)=1.583\hat{X}_{100}(0)-.899\hat{X}_{100}(-1)-.929\hat{a}_{100}+20.038(1-1.583+.898)\tag{6.40}$$

The term $20.038(1-1.583+.899)=20.038(.316)=6.332$, and by typing **arma21**, the 100 data values of the realization are displayed. Based on (6.40), in order to compute $\hat{X}_{100}(1)$, the values for $\hat{X}_{100}(0)=x_{100}=19.046$ and $\hat{X}_{100}(-1)=x_{99}=20.596$ are needed. Thus, (6.40) becomes

$$\hat{X}_{100}(1) = 1.583(19.046) - .899(20.596) - .929\hat{a}_{100} + 6.332.$$ (6.41)

However, we are still not in a position to calculate the forecast because we need the value $\hat{a}_{100}$. The residuals are included in the output of **est.arma.wge** in the output vector **est21\$res**. To see the 100 residuals, calculated using backcasting, type **est21\$res**. From this listing of the residuals, it can be seen that $\hat{a}_{100}(100) = -.029$. Consequently,

$$\hat{X}_{100}(1) = 1.583(19.046) - .899(20.596) - .929(-.029) + 6.332 = 17.993.$$

The forecasts for $\hat{X}_{100}(2)$ and $\hat{X}_{100}(3)$, which do not include the moving average term in (6.41), are given by

$$\hat{X}_{100}(2) = 1.583\hat{X}_{100}(1) - .899\hat{X}_{100}(0) + 6.332$$
$$= 1.583(17.993) - .899(19.046) + 6.332 = 17.693$$
$$\hat{X}_{100}(3) = 1.583\hat{X}_{100}(2) - .899\hat{X}_{100}(1) + 6.332$$
$$= 1.583(17.693) - .899(17.993) + 6.332 = 18.164$$

It is obvious that the forecasts calculated above, which required considerable effort, are best calculated using the computer. After running the *tswge* code above to calculate realization **arma21** and estimate the parameters, we can calculate the forecasts using

```
fore.21=fore.arma.wge(arma21,phi=est21$phi,theta=est21$theta,n.ahead=30,
limits=FALSE)
```

The first three forecasts, shown under **fore.21\$f**, are given by **17.99311, 17.696328**,   and **18.16562**. These match closely to those calculated above with slight differences due to rounding. Figure 6.10(b) shows the realization of length $n = 100$ along with the forecasts for the next 30 time points. The forecasts have the appearance of a damping sinusoid centered about the line $\bar{x} = 20.038$. The system frequency associated with $(1 - 1.583B + .899B^2)$ is $f_0 = 0.093$ which is consistent with the fact that the cycle length for the data and the forecasts is about $1/.093 = 10.75$.

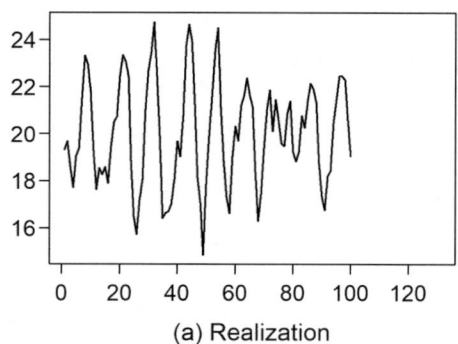
(a) Realization

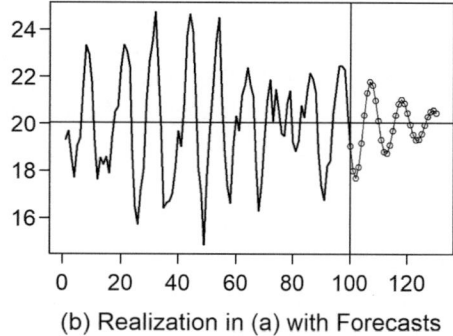
(b) Realization in (a) with Forecasts

FIGURE 6.10   Generated data and forecasts using a fitted ARMA(2,1) model for the next 30 time periods calculated using `fore.arma.wge`.

Example 6.12 Sunspot Numbers with Forecasts

In Example 5.6 we used an AR(9) model for the sunspot data in file **sunspot2.0** that contains sunspot numbers based on the new counting procedure for the years 1700 through 2020. Example 6.8 showed that AIC selects an AR(9) model for these data. The ML estimates are shown in (5.36) and again in the output vector **ss\$phi** obtained in Example 6.8. Table 5.6 shows the factor table for this fitted model. These sunspot numbers are plotted in Figure 6.11 along with forecasts for the 30 years following 2020. There you can see that the forecasts follow a damped sinusoidal pattern with a period of about 10–11 years. This is to be expected based on the factor table in Table 5.6. The horizontal line is at the mean 78.97 and the vertical line is at year 2020. The forecasts were obtained using the *tswge* code

```
data(sunspot2.0)
aic.ss=aic.wge(sunspot2.0,p=0:10,q=0:4)
est.ss=est.ar.wge(sunspot2.0,p=aic.ss$p)  # aic selects p=9
fore.ss=fore.arma.wge(sunspot2.0,phi=est.ss$phi,n.ahead=30,limits=FALSE)
```

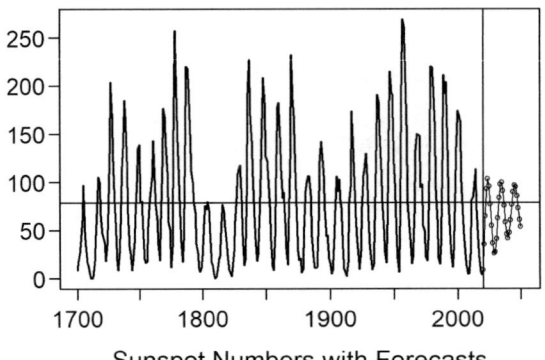

Sunspot Numbers with Forecasts

FIGURE 6.11   Sunspot numbers 1700–2020 with forecasts for the 30 years following 2020 calculated using fore.arma.wge based on the AR(9) fit in Examples 5.6 and 6.8.

Example 6.13 Canadian Lynx Data Revisited
In Example 6.5 we examined the classical Canadian lynx dataset (log base 10) that exhibits an interesting 10-year cycle. In that example we decided that AR(12), ARMA(2,3), and an AR(2) were models worth pursuing further. Examining the factor tables in Table 6.10 we noted that the dominant factor in all three models is a 2nd-order factor associated with a frequency of $f = 0.10$. However, the roots associated with that factor are much closer to the unit circle in the ARMA(2,3) and AR(12) models than they are for the AR(2) model.

Figure 6.12 shows forecasts for 30 years following 1934, and we see that the ARMA(2,3) and AR(12) models detect the 10-year period and predict this pattern to continue while the AR(2) model forecasts quickly damp to the mean. It is interesting to note the similarity between the ARMA(2,3) and AR(12) forecasts. While the ARMA(2,3) and AR(12) produce "more appealing-looking" forecasts than the AR(2) model, this does not indicate they are better forecasts. The assessment of forecast performance will be discussed in Section 6.2.6.

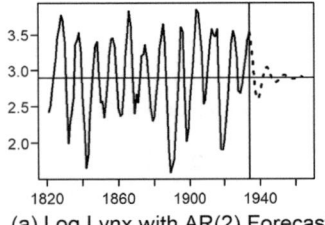
(a) Log Lynx with AR(2) Forecasts

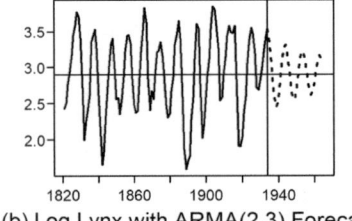
(b) Log Lynx with ARMA(2,3) Forecasts

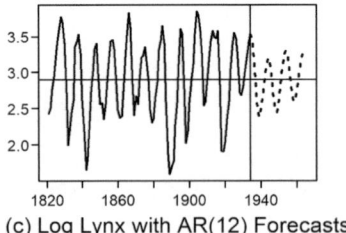
(c) Log Lynx with AR(12) Forecasts

FIGURE 6.12   Log-lynx data showing (a) forecasts using AR(2) fit, (b) forecasts using ARMA(2,3) fit, and (c) forecasts using the AR(12) fit.

6.2.4 Eventual Forecast Functions

In the forecasting examples using AR and ARMA models (which are stationary models) it should be noted that in all cases, the forecasts eventually tend toward the realization mean as the number of steps ahead, ℓ, increases. In Figure 6.13 we replot forecasts from the AR(1) and AR(2) models shown previously in Figure 6.8(c) and Figure 6.9(c). In both cases the forecasts trend toward the sample mean illustrated by the horizontal line, $\bar{x} = 11.12$ and 40.12 in Figures 6.13(a) and 6.13(b), respectively. For stationary AR and ARMA models, the autocorrelations, ρ_k, damp to zero as $k \to \infty$. Thus, the correlations between observed values for X_t up to time t_0 and $X_{t_0+\ell}$ decrease as ℓ increases. Because the observed values give very little information about $X_{t_0+\ell}$ when ℓ is large, it makes sense to simply forecast $\bar{x}$. In both cases in Figure 6.13, the forecasts trend toward the sample mean shown as a horizontal line on the plots.

> **Key Point:** The tendency for the forecasts to eventually tend to the sample mean can be "disappointing" at first glance. Novice forecasters may think that they can "do better" without the use of a model. However, *if the data are stationary*, then it would be a mistake to make "bold" forecasts into the "distant" future because it is essentially uncorrelated with observed data. See Section 6.2.6 Assessing Forecast Performance..

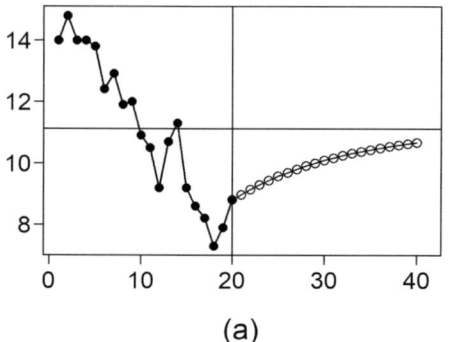
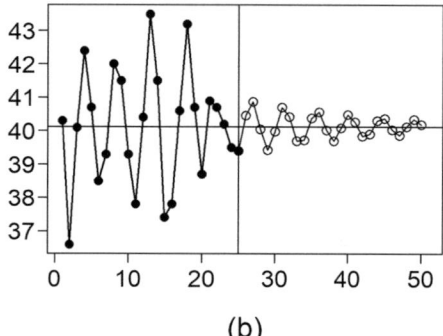

(a)   (b)

FIGURE 6.13   Forecasts previously shown in Figure 6.8(c) and 6.9(c).

6.2.5 Probability Limits for ARMA Forecasts

We have found how to find forecasts of $X_{t_0+\ell}$ based on data to time t_0 using ARMA models. In addition to finding a forecast for $X_{t_0+\ell}$, that is, $\hat{X}_{t_0}(\ell)$, the knowledgeable forecaster will also want to assess the uncertainty associated with the prediction. As an example, we may want to construct limits such that there is a 95% chance that the actual future value of $X_{t_0+\ell}$ falls within these limits. In this section we will learn how to construct such limits.

The difference, $e_{t_0}(\ell) = X_{t_0+\ell} - \hat{X}_{t_0}(\ell)$, is called the *forecast error*. That is, it measures the amount by which the forecast of $X_{t_0+\ell}$ differs from the actual value. The following summarizes the properties of the forecast error.

> **Key Point:** In the following we assume *normal white noise*, a_t, with zero mean and finite variance σ_a^2.

6.2.5.1 Facts about Forecast Errors

First recall that a stationary ARMA model can be expressed in GLP form, $X_t - \mu = \sum_{k=0}^{\infty} \psi_k a_{t-k}$. It follows

that $X_{t_0+\ell}$ satisfies $X_{t_0+\ell} - \mu = \sum_{k=0}^{\infty} \psi_k a_{t_0+\ell-k}$. With this as a foundation, consider the following facts about the

forecast error, $e_{t_0}(\ell)$:

(1) $e_{t_0}(\ell) = \sum_{k=0}^{\ell-1} \psi_k a_{t_0+\ell-k}$

(2) $e_{t_0}(\ell)$ is normally distributed (it is a finite linear combination of normal random variables) with

     (a)   mean $E\left[e_{t_0}(\ell)\right] = 0$ ($E\left[a_t = 0\right]$ for all t)

     (b)   variance $Var\left[e_{t_0}(\ell)\right] = Var\left[\sum_{k=0}^{\ell-1} \psi_k a_{t_0+\ell-k}\right]$

$$= \sigma_a^2 \sum_{k=0}^{\ell-1} \psi_k^2.$$

$1 - \alpha \times 100\%$ Prediction Limits

Using the above facts, it follows that

$$\frac{e_{t_0}(\ell) - 0}{\sigma_a \sqrt{\sum_{k=0}^{\ell-1} \psi_k^2}} \sim N(0,1). \tag{6.42}$$

That is,

$$\frac{X_{t_0+\ell} - \hat{X}_{t_0}(\ell)}{\sigma_a \sqrt{\sum_{k=0}^{\ell-1} \psi_k^2}} \sim N(0,1),$$

from which it follows that

$$Prob\left\{-z_{1-\alpha/2} \leq \frac{X_{t_0+\ell} - \hat{X}_{t_0}(\ell)}{\sigma_a \sqrt{\sum_{k=0}^{\ell-1} \psi_k^2}} \leq z_{1-\alpha/2}\right\} = 1-\alpha, \tag{6.43}$$

where z_β is the $\beta \times 100\%$ percentile of the standard normal distribution. For example, $z_{1-.05/2} = z_{.975} = 1.96$. The theoretical $(1-\alpha) \times 100\%$ prediction interval for $X_{t_0+\ell}$ is

$$\hat{X}_{t_0}(\ell) - z_{1-\alpha/2}\sigma_a\left\{\sum_{k=0}^{\ell-1} \psi_k^2\right\}^{\frac{1}{2}} \leq X_{t_0+\ell} \leq \hat{X}_{t_0}(\ell) + z_{1-\alpha/2}\sigma_a\left\{\sum_{k=0}^{\ell-1} \psi_k^2\right\}^{\frac{1}{2}}$$

or

$$\hat{X}_{t_0}(\ell) \pm z_{1-\alpha/2}\sigma_a\left\{\sum_{k=0}^{\ell-1} \psi_k^2\right\}^{\frac{1}{2}}. \tag{6.44}$$

In practice we will not know σ_a so the white noise variance, σ_a^2, is estimated by $\hat{\sigma}_a^2$ based on the backcasting procedure. Also, the notation ψ_k denotes the ψ-weights of the "true" model, so we denote the ψ-weights of the fitted model by $\hat{\psi}_k$. Consequently, in practice, the $(1-\alpha)\times 100\%$ prediction interval for $X_{t_0+\ell}$ is

$$\hat{X}_{t_0}(\ell)\pm z_{1-\alpha/2}\,\hat{\sigma}_a\left[\sum_{k=0}^{\ell-1}\hat{\psi}_k^2\right]^{\frac{1}{2}}, \tag{6.45}$$

and, for example, a 95% prediction interval is

$$\hat{X}_{t_0}(\ell)\pm 1.96\hat{\sigma}_a\left[\sum_{k=0}^{\ell-1}\hat{\psi}_k^2\right]^{\frac{1}{2}}. \tag{6.46}$$

QR 6.9 ARMA
Forecast Limits

Key Points

1. The $(1-\alpha)\times 100\%$ prediction limits apply to individual forecasts. If 95% prediction limits are used to obtain forecasts $\hat{X}_{t_0}(\ell), \ell = 1,\ldots,m$, we should not be surprised to find that about 5% of the actual values fall outside the prediction limits.
2. As we go from ℓ step-ahead to $\ell+1$ step-ahead forecasts, the forecast limits expand because $\hat{\psi}_\ell^2$ is added to the sum of squared ψ-weights (unless $\hat{\psi}_\ell^2 = 0$).

Example 6.14 (revisited)
Figure 6.14 shows a realization of length $n = 100$ to which we fit the ARMA(2,1) $(1-1.583B+.899B^2)(X_t - 20.038) = (1-.929B)a_t$, where $\hat{\sigma}_a^2 = .84$. Figure 6.14 also shows the forecasts for 30 steps beyond the end of the dataset. If we replace the call to **fore.arma.wge** in Example 6.12 with the following,

```
fore.21=fore.arma.wge(arma21,phi=c(1.583,-.899),theta=.929,n.ahead=30,
limits=TRUE,alpha=.05)
```

then we obtain the forecasts along with the 95% prediction limits shown in Figure 6.14.

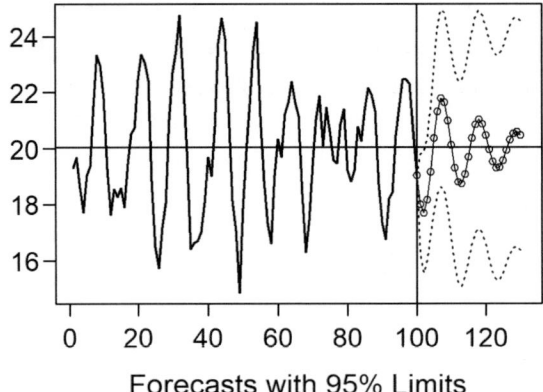

FIGURE 6.14   Data and forecasts in Figure 6.10(b) with 95% forecast limits.

The quantity

$$z_{1-\alpha/2}\, \hat{\sigma}_a \left\{ \sum_{k=0}^{\ell-1} \hat{\psi}_k^2 \right\}^{\frac{1}{2}} \tag{6.47}$$

is referred to as the half-width (or margin of error) of the prediction interval. As mentioned previously the half-widths increase as ℓ increases. Table 6.14 shows the forecasts and half-widths for the 95% prediction limits plotted in Figure 6.14 for a few values of ℓ. Recall that the ψ-weights can be found using *tswge* function `psi.weights.wge.` For the fitted ARMA(2,1) model we use the command

```
psi.weights.wge(phi=c(1.583,-.899),theta=.929,lag.max=4)
```

and find that $\hat{\psi}_k$, $k = 1,\dots,4$ are given by 0.654, 0.136, −0.372, and −0.712, respectively. Of course, $\hat{\psi}_0 = 1$. In Example 6.11 we found that $\hat{\sigma}_a^2 = .843$. Given the above information, the half-width, for $\ell = 1$, is given by

$$\hat{\sigma}_a \sqrt{1} = 1.96\sqrt{.843}\sqrt{1} = 1.96(.918) = 1.800,$$

for $\ell = 2$, is given by

$$1.96\,\hat{\sigma}_a \sqrt{1 + \hat{\psi}_1^2} = 1.96\sqrt{.843}\sqrt{1 + .654^2} = 1.96(.918)(1.1949) = 2.150,$$

for $\ell = 3$, is

$$1.96\hat{\sigma}_a \sqrt{\left(1 + \hat{\psi}_1^2 + \hat{\psi}_2^2\right)} = 1.96(.918)\sqrt{1 + .654^2 + .136^2}$$

$$= 1.96(.918)(1.2026) = 2.164.$$

and for $\ell = 4$, is

$$1.96\,\hat{\sigma}_a \sqrt{\left(1 + \hat{\psi}_1^2 + \hat{\psi}_2^2 + \hat{\psi}_3^2\right)} = 1.96(.918)\sqrt{1 + .654^2 + .136^2 + .372^2}$$

$$= 1.96(.9165)(1.2589) = 2.265.$$

Table 6.14 shows forecasts and half-widths for $\ell = 1,\ldots,4$. Notice that calculation of the half-width for $\ell = 4$ did not require $\hat{\psi}_4$ because the sum of squared ψ-weights only goes to $\ell - 1$. Based on the table we see that the forecast for X_{104} is 19.182, and the 95% forecast limits for X_{104} are $\hat{X}_{100}(4) \pm 2.265$. We expect X_{104} to fall somewhere in the interval 19.182±2.265, that is, between $19.182 - 2.265 = 16.917$ and $19.182 + 2.265 = 21.447$.

Also, Table 6.14 shows forecasts and half-widths for 10, 20, and 50 steps ahead using $\alpha = .05$. Note the following. Theorem 5.4 states that the variance of the GLP $X_t = \sum_{k=0}^{\infty} \psi_k a_{t-k}$ is given by $\sigma_X^2 = \sigma_a^2 \sum_{k=0}^{\infty} \psi_k^2$, which is finite for stationary processes. Consequently, the sum in (6.47) does not increase without bound, and in fact

$$\hat{\sigma}_a \left\{ \sum_{k=0}^{\ell-1} \hat{\psi}_k^2 \right\}^{\frac{1}{2}} \rightarrow \hat{\sigma}_a \left\{ \sum_{k=0}^{\infty} \hat{\psi}_k^2 \right\}^{\frac{1}{2}} = \hat{\sigma}_X, \tag{6.48}$$

so that the half-width in (6.47) converges to $1.96\hat{\sigma}_X$.

For the fitted model $\left(1 - 1.583B + .899B^2\right)\left(X_t - 20.038\right) = \left(1 - .929B\right)a_t$, where $\hat{\sigma}_a^2 = .843$ and using *tswge* command

```
v=true.arma.aut.wge(phi=c(1.583,-.899),theta=.929,vara=.8426)
```

we find that the process variance of the fitted model, $\hat{\sigma}_X^2$, is 4.5242 (**v$acv[1]**) so that $\hat{\sigma}_X = 2.1270$. The right-hand side of (6.48) is $1.96\sqrt{4.5242} = 4.169$. Notice that the half-widths in Table 6.14 seem to converge to 4.169, and the forecasts converge to the sample mean, $\bar{x} = 20.038$ as expected.

TABLE 6.14   Forecasts and 95% Prediction Interval Half-Widths for the ARMA(2,1) Realization plotted in Figure 6.14.

ℓ	FORECAST	HALF-WIDTH
1	17.993	1.800
2	17.693	2.150
3	18.164	2.164
4	19.182	2.265
⋮		
10	20.133	3.313
20	20.480	3.894
50	20.212	4.160
100	20.028	4.169
500	20.038	4.169

Example 6.14 Sunspot Forecasts with 95% Prediction Limits

In Example 6.12 we used the following code to find forecasts for the sunspot data using the AR(9) model in (5.36).

```
data(sunspot2.0)
est.ss=est.ar.wge(sunspot2.0,p=9)
```

In that example we found and plotted forecasts for the next 30 years. We modify the call to

```
forecast.arma.wge
```

so that the plot includes prediction limits.

```
fore.ss=fore.arma.wge(sunspot2.0,phi=est.ss$phi,n.ahead=30,limits=TRUE,
alpha=.05)
```

Figures 6.15(a) and (b) show the forecasts with 95% limits. Notice that the forecast limits are getting a little wider for larger ℓ but the increase is not dramatic.

> **Key Point:** Note that some of the lower limits for the sunspot forecasts are negative. Since the sunspot number cannot be negative, any negative lower limit should be converted to 0.

6.2.5.2 Lack of Symmetry

Although the sunspot data have been analyzed through the years using ARMA models, they are clearly not data from an ARMA model. The troughs are all very similar but there is much variation in peak heights, resulting in the asymmetric appearance in Figure 6.15(a). However, AR and ARMA models fit to the sunspot data, such as the AR(9) model we have examined, do not have realizations with this type of asymmetric appearance. Tong (1990) and others have investigated the use of nonlinear models to model these data. Recall that we encountered a similar problem with the **lynx** dataset and solved the problem by taking logs. However, the annual sunspot number for 1810 is zero so we cannot take logs unless we, for example, add a positive number to each sunspot number. Traditionally, the raw sunspot numbers have been analyzed despite their obvious lack of symmetry.

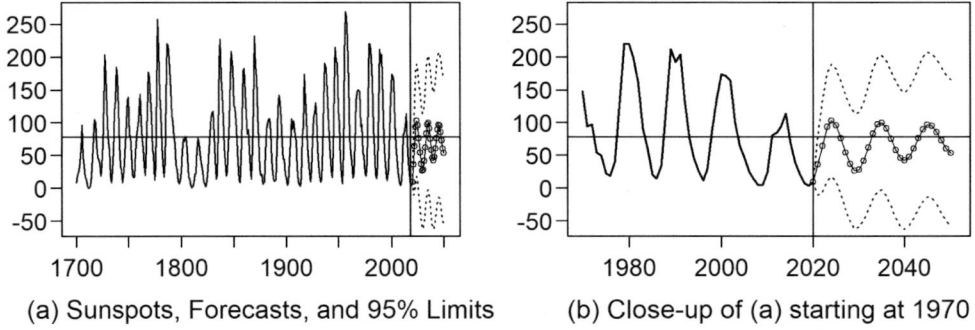

(a) Sunspots, Forecasts, and 95% Limits   (b) Close-up of (a) starting at 1970

FIGURE 6.15   (a) sunspot2.0 with 30 forecasts shown in Figure 6.11 along with 95% forecast limits (b) "close-up" forecasts and limits showing sunspot data only from 1970.

6.2.6 Assessing Forecast Performance

The forecast limits are useful for understanding the reliability of forecasts, that is, for assessing how precise we believe they are. However, the acid test of forecast performance is to compare forecasts with reality. One way to do this is to measure ***how well "actual values" can be predicted***.

6.2.6.1 How "Good" Are the Forecasts?

As an example, given a realization of length 100, you may want to determine how well your model predicts the last 20 values using $t_0 = 80$ as the forecast origin. In order to develop forecast models based on their performance in forecasting known values, we need tools for assessing "how good" the forecasts are. There are many ways to assess the quality, but the two methods we will discuss are the following:

(a) **Check actual values with the forecast limits**. If your model is appropriate and you have obtained, for example, 95% forecast limits, then approximately 95% of the actual values should fall within the forecast limits. If this condition is not satisfied and substantially fewer than 95% of actual values fall within the limits, then this is evidence that your model is not appropriate.

(b) **Quantify the difference between forecasts and actual values**. A useful measure of the quality of the forecasts is the root mean square error. That is, if $x_{t_0+1}, \ldots, x_{t_0+k}$ are the actual values and $f_{t_0+1}, \ldots, f_{t_0+k}$ are the forecasts based on the forecast model at origin t_0, then the *forecast error* at time t_{0+j}, for $1 \leq j \leq k$ is the "forecast value minus actual value", or $f_{t_0+j} - x_{t_0+j}$. A measure of how close the actual values and forecasts are to each other *as a group* is the root mean square error

$$RMSE = \sqrt{\frac{1}{k} \sum_{t=t_0+1}^{t_0+k} (f_t - x_t)^2}. \tag{6.49}$$

The MSE is the average of the squared forecast errors (removing the effect of the signs). The RMSE is the square root of this quantity which reverts back to the units of the problem. The RMSE is a measure of the "typical amount by which forecasts differ from actual values", that is, it measures the typical size of the forecast errors. Use of the RMSE to assess forecast results was introduced in Section 2.2.4.

For example: Given a model and a dataset of length 100, a measure of the RMSE associated with forecasting the last 20 data values using the model is

$$RMSE = \sqrt{\frac{1}{20} \sum_{t=81}^{100} (f_t - x_t)^2}. \tag{6.50}$$

Note: A measure related to the RMSE is the mean absolute deviation (MAD), defined by

$$MAD = \frac{1}{k} \sum_{t=t_0+1}^{t_0+k} |f_t - x_t|.$$

Although forecast performance will probably be *your gold standard* for assessing model appropriateness in practice, in Chapter 9 we will discuss other methods for judging how well your model fits the data.

6.2.6.2 Some Strategies for Using RMSE to Measure Forecast Performance

(1) Suppose we want to forecast, for example, the last 20 data values. We could develop the forecast function based on the first $n-20$ data values, and then forecast values $n-19, n-18, \ldots, n$. Note that if this procedure is repeated for various forecast origins then it requires refitting a new model for *each* new forecast origin using all data up to and including each forecast origin. For example, to forecast the last 20 data values, AIC may choose an ARMA(2,1) model fit to the first $n-20$ data values. However, if forecasting the last 40 data values, AIC may select an AR(4) model fit to the first $n-40$ data values.

(2) A simplification of (1) is to use AIC to choose a model *order*, say AR(4) for the entire dataset, and then refit models for each forecast origin, t_0, using an AR(4) model fit to the realization $x_1, x_2, \ldots, x_{t_0}$.

(3) A further simplification is to fit a model using the entire dataset, and then use the associated forecast function to forecast from each forecast origin of interest.

> **Key Point:** All of these strategies are designed to help answer the fundamental question at hand. That is, ***"How well will the model forecast future values when they become available?"***

The following rather lengthy example illustrates the ideas discussed above.

Example 6.15 Forecasting the sunspot data

In Example 6.8 we fit an AR(9) model to the data in file **sunspot2.0** which contains sunspot data from 1700–2020 using the new numbering system (Clette et al., 2016). Figures 6.16(a)–(d) show forecasts with forecast origins, $t_0 = 2010, 2000, 1990,$ and 1980, respectively. We have used strategy (2) above and we fit an AR(9) model to the available data up to and including the forecast origin. Notice that 2010=1700+310, ..., 1980=1700+280. Because **sunspot2.0** is sequenced on the integers 1 through 319, the four models can be obtained using the *tswge* commands

```
data(sunspot2.0)
a=est.arma.wge(sunspot2.0[1:311],p=9)
b=est.arma.wge(sunspot2.0[1:301],p=9)
c=est.arma.wge(sunspot2.0[1:291],p=9)
d=est.arma.wge(sunspot2.0[1:281],p=9)
```

We leave it to the reader to check, but the commands above produce four very similar models and factor tables. From all forecast origins it is seen that the forecasts have a damping sinusoidal appearance with a period length of about 10.5 years.

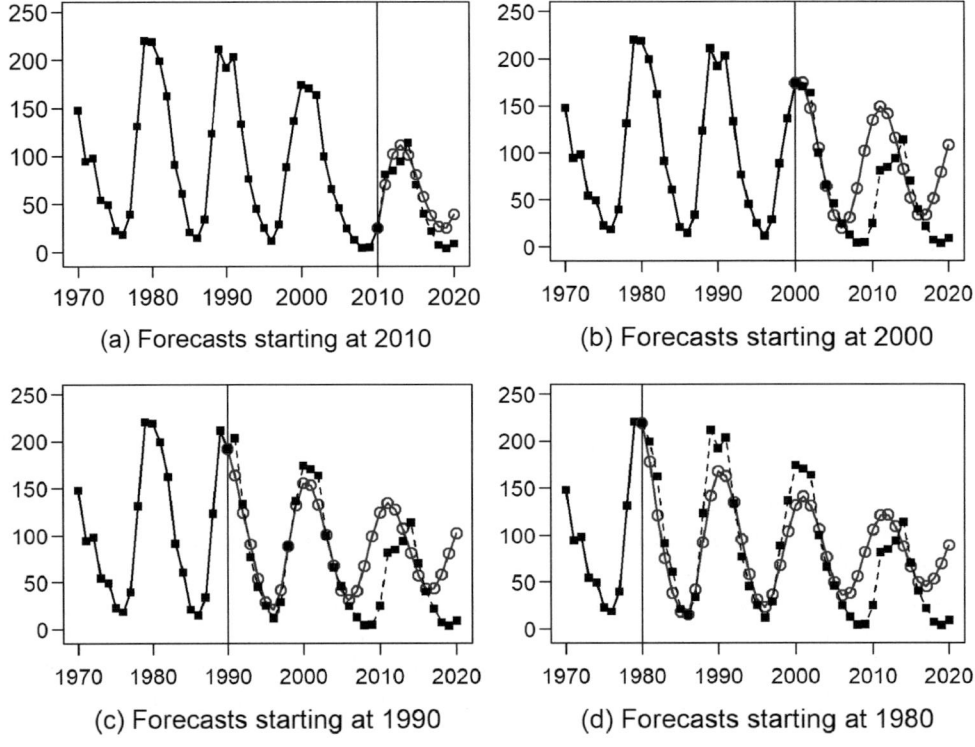

(a) Forecasts starting at 2010

(b) Forecasts starting at 2000

(c) Forecasts starting at 1990

(d) Forecasts starting at 1980

FIGURE 6.16 Comparing forecasts (open circles) with actual values (solid squares) for forecast origins (a) 2010, (b) 2000, (c) 1990, and (d) 1980 using the sunspot data in sunspot2.0.

Figure 6.16 illustrates an important issue related to the sunspot data that was discussed in Chapter 1. That is, while on the average, cycle lengths tend to be about 11 years, in reality each actual sunspot cycle is not of exactly the same length. For example, the cycle starting at the peak at 1980 is of length close to 11 years, while the cycle beginning at 2000 is of length about 12 years. Forecasts starting at origin 2010 do an excellent job of tracking the actual behavior while forecasts from 2000 get "off cycle" at about 2007 and as a whole are disappointing. Forecasts from origin 1990 are quite good for the years 1991–2006, but again get off cycle at about 2007. Finally, forecasts from 1980 stay "in cycle" for the 26 years from 1981 –2006. However, these forecasts tend to underestimate the cycle peaks. In the following we will quantify the forecast performance described above.

(1) Do the Actual Values Tend to Stay with the Prediction Limits?

We first evaluate the performance of these forecasts by determining whether the actual values stay within the prediction limits. That is, for the 95% limits, we should be concerned if fewer than 95% of the actual values stay within the limits for a given forecast origin. Figure 6.17(a)–(d) shows the forecasts in Figure 6.16 along with 95% prediction limits. It can be seen that in all cases, the actual values are within the prediction limits. Note that for forecasts with forecast origin 1980, the actual value is close to the upper limit at 1989. However, in this case the actual value is 211.1 and the upper limit is 231.85. Again note that many of the lower limits were negative, and in these cases it would have been appropriate to convert negative lower limits to zero. Based on the forecast limits we see no reason for concern.

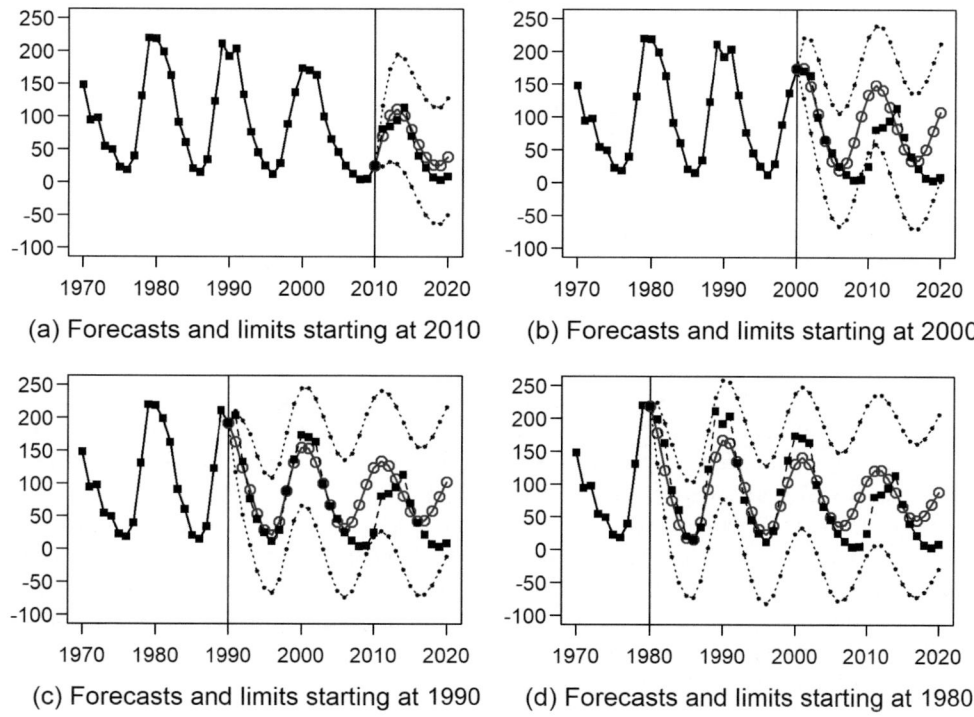

(a) Forecasts and limits starting at 2010

(b) Forecasts and limits starting at 2000

(c) Forecasts and limits starting at 1990

(d) Forecasts and limits starting at 1980

FIGURE 6.17 Forecasts in Figure 6.16 showing associated 95% prediction limits.

(2) Quantifying Forecast Performance

The fact that actual values stay within the 95% limits is not necessarily an indication that the forecasts are "good" or the best we can do. For example, the forecasts for 2007–2020 beginning at forecast origins 1980, 1990, and 2000 were "off cycle". This is still a cause for concern even though the forecasts stay within the limits. Table 6.15 shows the actual values and associated forecasts for years 2011–2020 based on forecast origin 2010. The last column shows the forecast errors. Recall that the forecasts in Figures 6.16 were using AR(9) models fit to data up to and including the forecast origin.

TABLE 6.15   Actual Values and Forecasts from Forecast Origin 2010

FORECAST YEAR	ACTUAL	FORECAST	FORECAST ERROR
2011	80.8	70.3	−10.5
2012	84.5	102.0	17.5
2013	94.0	111.2	17.2
2014	113.3	101.0	12.3
2015	69.8	80.5	10.7
2016	39.8	57.5	17.7
2017	21.7	38.1	16.4
2018	7.0	26.5	19.5

In order to quantify the "typical size of the forecast errors ignoring signs" from the four forecast origins, we calculate the associated forecast RMSEs. For the above data the RMSE is 15.6 and the MAD is 15.2, and it is clear that both measures give a "typical" size of the forecast error.

In Table 6.16 we show the RMSEs associated with forecasts from forecast origins 2010, 2000, 1990, and 1980. The RMSE for origin 2010 is 15.6 as mentioned above.

TABLE 6.16   RMSE for Forecasts in Figures 6.15 and 6.16

FORECAST ORIGIN	RMSE
2010	15.6
2000	45.6
1990	36.2
1980	32.3

As observed earlier, the forecasts as a group from origin 2010 were quite good and the RMSE is the smallest of those tabled above. The "disappointing" forecasts from origin 2000 resulted in the largest RMSE, 45.6. Forecasts from origins 1990 and 1980 were satisfactory for early lags before getting off cycle at 2007, and their RMSEs are smaller than those for origin 2000.

As mentioned, the RMSE values give us a measurement of the size of the forecast errors and are useful for informing us concerning whether the forecasts are "doing well enough for our purposes". Another use of the RMSE is to compare forecasts from two or more competing models. In our case AIC measures were consistent in selecting an AR(9) model for forecasting from the origins used in the above analyses. Box, Jenkins, and Reinsel (2008) analyze a shorter sunspot dataset based on the previous counting measure. They chose an AR(2) model based on the partial autocorrelations.

Using an AR(2) Model for Forecasting
In Examples 6.12 and 6.14 we examined forecasts from the AR(9) model fit to the **sunspot2.0** dataset. At this point we will consider using AR(2) fits for forecasting. In Figure 6.15 we showed forecasts for the next 30 years obtained using the AR(9) model along with 95% prediction limits. Figure 6.18 is the analogous figure based on an AR(2) fit.

The following code fits an AR(2) model to the **sunspot2.0** data.

```
data(sunspot2.0)
est.ss=est.ar.wge(sunspot2.0,p=2)
```

The resulting model is

$$\left(1-1.38B+.69B^2\right)\left(X_t-78.97\right)=a_t,\tag{6.51}$$

with $\hat{\sigma}_a^2=659.39$. The single row factor table is given in Table 6.17 where it can be seen that there is a single factor, and, not surprisingly, an associated system frequency (.094) that corresponds to a cycle length of about 10.5 years. However, the associated roots are not as close to the unit circle as they are in the AR(9) model.

TABLE 6.17 Factor Table AR(2) Fit to sunspot2.0 Data

AR-FACTOR	ROOTS	$|r|^{-1}$	f_0
$1-1.38B+.69B^2$	$1.0\pm67i$	.83	.094

We use the following code to find the forecasts and prediction limits using an AR(2) model.

```
fore.ss=fore.arma.wge(sunspot2.0,phi=est.ss$phi,n.ahead=30,limits=TRUE)
```

The primary difference between forecasts from the AR(9) and AR(2) models is that the AR(9) model picks up the 10.5-year cycle and seemingly does a better job forecasting it to continue into the future. On the other hand, after about 10 steps ahead, the AR(2) model essentially forecasts the series mean.

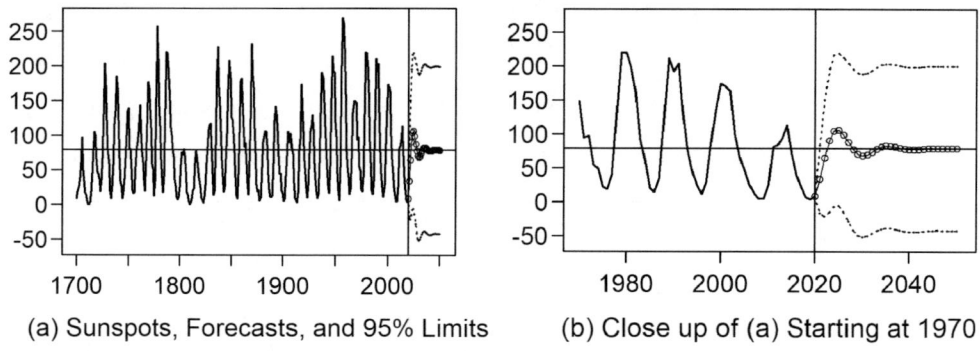

(a) Sunspots, Forecasts, and 95% Limits   (b) Close up of (a) Starting at 1970

FIGURE 6.18 (a) sunspot2.0 with 30 forecasts from AR(2) model along with 95% forecast limits (b) "close-up" forecasts and limits showing sunspot data only from 1970.

Figure 6.19 shows the AR(9) forecasts in Figure 6.16 as open circles, AR(2) forecasts with "×," and actual values with solid squares. In these figures we see that again that the AR(2) forecasts are rather "boring" in that they tend quickly to the mean and seem to minimize the prominent 10−11 year cycle. Table 6.18 shows the RMSE for the two models based on four forecast origins. Surprisingly, at forecast origin 2000, the RMSE for the AR(2) forecasts is smaller than for the AR(9). The lesson is that the AR(9) forecasts "boldly" predicted a cyclic behavior that was off cycle, suggesting that in this particular case, it was better to be very conservative. Even though the AR(9) forecasts also got "off-cycle" at 2007 for forecast origins 1990 and 1980, the fact that the AR(9) forecasts tracked well until 2007 caused the AR(9) forecasts to be preferable.

Key Points

1. As discussed in Chapter 1, the approximate 11-year cycle in the sunspot data is not well under-stood. For example, scientists do not know when a future cycle may be shorter or longer than is typical. In the presence of this type of cyclic behavior, it may be a mistake to be too confi-dent about predictions based on a certain cycle length to continue. Notice that even the AR(9) forecasts become more "conservative" at longer cycle lengths (and will eventually tend to the mean) providing protection from "overconfidence".

2. We mentioned in Chapter 1 that some cycles, such as monthly temperature patterns at a certain location, are explainable. In this case the cycle length is based on the Earth's rotation around the sun. In such cases, these are referred to as "cyclic data" with fixed cycle length. If the data relate to the calendar year, then they are referred to as "seasonal" data. For fixed cycle length data, forecasts can be made more "boldly" into the future. In fact, this type of cyclic behavior may be indicative of the fact that a signal-plus-noise model is appropriate. See Chapter 8.

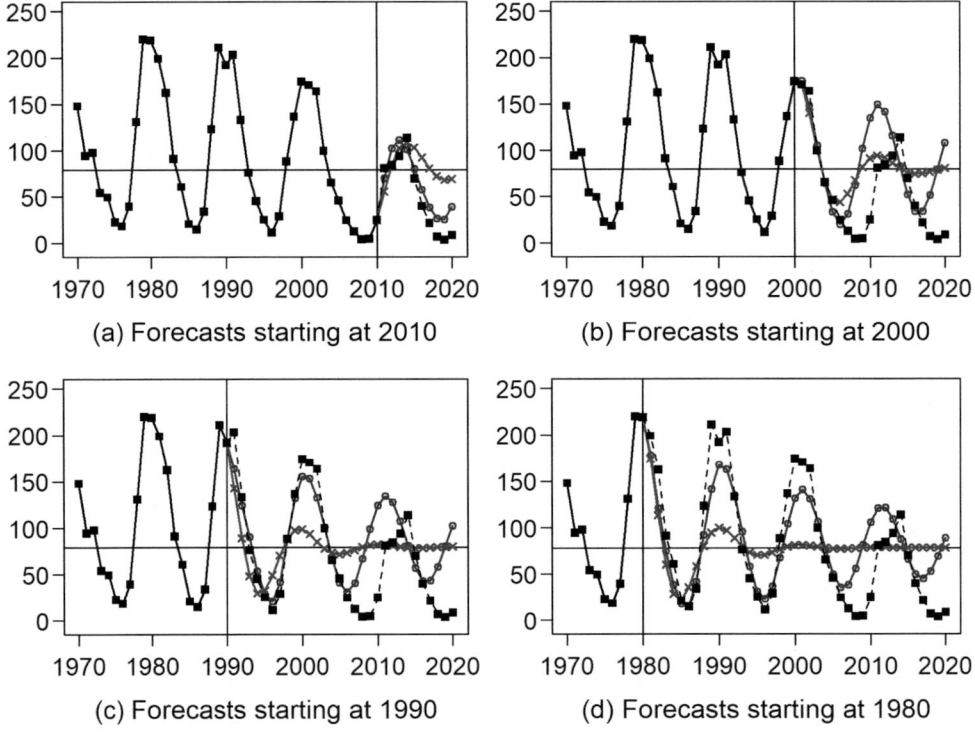

FIGURE 6.19   Comparing AR(9) forecasts (open circles) and AR(2) forecasts (×) with actual values (solid squares) for forecast origins (a) 2010, (b) 2000, (c) 1990, and (d) 1980 using the sunspot data in `sunspot2.0`.

TABLE 6.18   RMSE for Forecasts Based on AR(9) and AR(2) Fits

FORECAST ORIGIN	RMSE AR(9)	RMSE AR(2)
2010	18.3	45.0
2000	51.4	43.7
1990	41.4	49.6
1980	35.5	55.3

A Rolling Window RMSE Approach

In the previous example, the AR(9) with a forecast origin of 2010 (and time horizon of 10) was favored over the AR(2) model based on the lower RMSE. It is important to note that this assessment is based on only the last 10 observations of the **sunspot2.0** series. What if this model were evaluated on a different 10-year window of sunspot data? Would the AR(9) still be preferred? To investigate this question, we consider the performance of each of these models during a 10-year window near the *beginning* of the series.

Question: Could we evaluate the models on the first 10 years of recorded sunspot data, that is, 1700–1709?

Not quite, right? To forecast sunspots for the year 1700, the AR(2) model would require knowing the sunspot data for 1698 and 1699, and the AR(9) would require the data from 1691−1699. However, these data are unavailable in **sunspot2.0**. Consequently, given this particular dataset, the first year that the AR(2) model can forecast is 1702, while the first year that the AR(9) model can forecast is 1709. Furthermore, the *backcasting* procedure used to obtain better estimates of the residuals requires one additional observation to calculate forecasts from these models. For this reason, the earliest 10-year window for which these models can provide forecasts is 1703–1712 and 1710–1719 for the AR(2) and AR(9), respectively. For the purpose of comparison, consider the AR(2) and AR(9) forecasts of sunspots from 1710–1719.

Figure 6.20 (a) and (b) below displays the performance of these forecasts with their corresponding RMSEs.

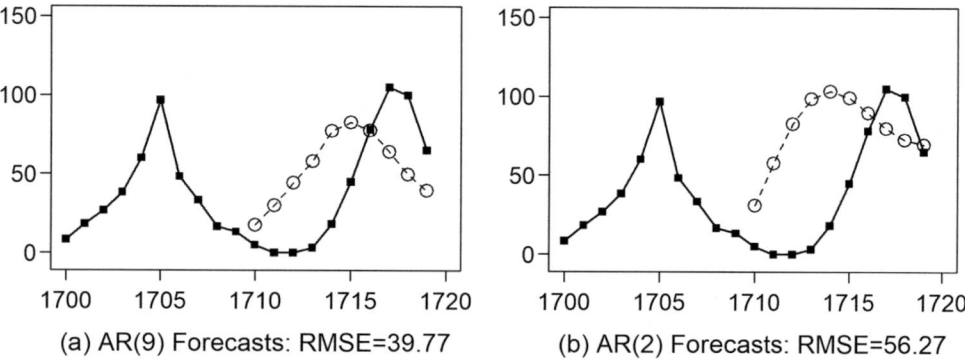

(a) AR(9) Forecasts: RMSE=39.77   (b) AR(2) Forecasts: RMSE=56.27

FIGURE 6.20   Horizon 10 forecasts from 1710–1719 using (a) an AR(9) and (b) an AR(2) Model

The AR(2) forecasts produce an RMSE of 56.27 compared to an RMSE of 39.77 for the AR(9) forecasts. As was the case for the forecast origin beginning in 2010, the AR(9) again outperforms the AR(2) with a forecast origin beginning in 1710.

A natural question is, "What about a third 10-year window?" which would then lead to, "What about a fourth 10-year window?", and so on. Table 6.19 displays the two 10-year windows we have already assessed (in italics), along with three additional randomly selected 10-year windows from the series with their corresponding RMSEs (ordered chronologically):

TABLE 6.19   AR(9) and AR(2) RMSEs for a horizon of 10 for five different "windows" in the sunpot2.0 series.

WINDOW	RMSE AR(9)	RMSE AR(2)
1710 – 1719	*39.77*	*56.27*
1790 – 1799	86.97	54.73
1845 – 1854	32.11	54.33
1986 – 1995	35.05	68.97
2011 – 2020	*18.3*[1]	*45.01*

[1] Note that the RMSEs for the last window in Table 6.19 for the AR(9) and AR(2) are calculated using the mean of the **sunspot2.0** data only up to year 2010 to match the output

TABLE 6.19 (*continued*)

seen in Table 6.18. The other RMSEs are calculated using the mean of the full data so that forecasts will approach the same sample mean. This doesn't make a big difference for the window starting at 2011; (values using the full mean are 17.65 and 44.53 respectively). However, it has the potential to make a very impactful difference as the data before the current window becomes sparse. For example, the sample mean for the first window in Table 6.19 would have been calculated using only 10 of the 321 values (1700–1709) in the **sunspot2.0** series.

We note that the AR(9) had a lower RMSE in all but one of the windows in Table 6.19. The obvious question is, "What if the models were compared only using the window from 1790–1799?" This table shows that the choice of model, with respect to the RMSE, is largely dependent on the choice of the window.

To address the dependency on the window, one might choose to use the average of the five windows for each model to obtain a more stable assessment. This would yield an average RMSE for the AR(9) model of 42.44 compared to 55.86 for the AR(2) model. The AR(9) model would again be favored. However, this time the comparison would be using information from all five windows thus reducing the impact from any single window.

The natural progression of this idea is to then include as many windows as the series will allow. For the AR(9) and the **sunspot2.0** data, this would entail starting with the first possible window (1710–1719) and calculating the RMSE and then shifting ("rolling") the window ahead one year and calculating the RMSE for the second window (1711–1720). This process of rolling the window ahead one year and calculating the RMSE continues until the RMSE is calculated on the final window (2011–2020). This is the strategy employed by the "*Rolling Window RMSE*". The rolling window approach will result in the comparison of 302 windows (and thus 302 RMSEs) using the AR(9) model on the **sunspot2.0** data.

Using *tswge* function **roll.win.rmse.wge**, the rolling window RMSE (rwRMSE) is calculated by taking the average of these 302 RMSEs, which is found to be 37.67. A tabular illustration of this process for the AR(9) is displayed in the left-hand side of Table 6.20.

TABLE 6.20   Illustration of the Windows in the Rolling Window RMSE Calculation for a Horizon of 10 for the AR(9) and AR(2) models. The Average of the RMSEs is in the Last Row and is the "Rolling Window RMSE" (rwRMSE)

	AR(9)			AR(2)	
WINDOW #	WINDOW	RMSE	WINDOW #	WINDOW	RMSE
1	1710 – 1719	39.77	1	1703 – 1719	56.27
2	1711 – 1720	36.28	2	1704 – 1720	45.52
3	1712 – 1721	30.81	3	1705 – 1721	35.68
	...	...		...	...
166	1875 – 1884	50.52	166	1875 – 1884	33.88
167	1876 – 1885	51.85	167	1876 – 1885	27.84
168	1877 – 1886	53.23	168	1877 – 1886	32.12
	...	...		...	...
300	2009 – 2018	35.15	306	2009 – 2018	31.78
301	2010 – 2019	19.28	307	2010 – 2019	39.37
302	2011 – 2020	17.65[1]	308	2011 – 2020	44.53
	rwRMSE	**37.67**			**48.81**

[1] Differs from value in Table 6.19 as per the discussion in footnote 11.

Note in the right half of Table 6.20, an analogous process is implemented for the AR(2) in which, because of the smaller order ($p=2$), there are now 308 possible windows across the series ranging from starting points of 1703 to 2011. The rwRMSE for the AR(2) is 48.61.

It can again be seen that at different windows in the series, the models may trade in achieving the smallest RMSE. This dependence is now *eliminated* by averaging across *all* possible windows. The AR(9) is ultimately found to be favored with respect to this measure because the AR(9) rwRMSE of 37.67 is smaller than the 48.81 produced by the AR(2).

The function **roll.win.rmse.wge** in *tswge* will calculate the rwRMSE given the data, the model, and the horizon. In the code and output below, the AR(9) model is first estimated for the **sunspot2.0** series as it was in Example 6.12 and the subsequent model is then used in the call to **roll.win.rmse.wge**.

```
est9 = est.arma.wge(sunspot2.0, p = 9, factor = FALSE)
roll.win.rmse.wge(sunspot2.0, phi = est9$phi, h = 10)
```

Output:

```
"Please Hold For a Moment, TSWGE is processing the Rolling Window RMSE with 302
windows."
"The Summary Statistics for the Rolling Window RMSE Are:"
Min. 1st Qu. Median  Mean 3rd Qu.  Max.
8.399 23.274 35.040 37.666 47.226 96.858
"The Rolling Window RMSE is: 37.666"
```

Examining the output above, we see that the first line shows that 302 windows were used in the calculation of the rwRMSE, which in the last line is shown to be 37.67 (as reported in Table 6.20). Since the rwRMSE in this case is calculated from 302 RMSEs, it may also be useful to know the spread of the RMSEs for the individual windows. The Tukey five-number summary is provided by **roll.win.rmse.wge** to reflect the spread and distribution of these RMSEs. The **roll.win.rmse.wge** function also outputs a histogram of the RMSEs shown in Table 6.21. Calculating the rwRMSE for the AR(2) is left as an exercise (Problem 6.11).

QR 6.10 Forecast
Assessment Methods

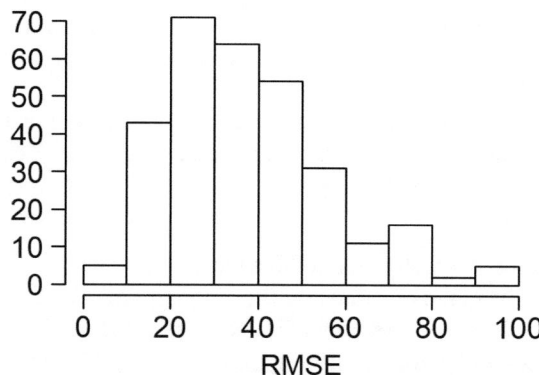

FIGURE 6.21 Histogram of RMSEs for AR(9) fits to all 302 horizon 10 windows.

6.3 CONCLUDING REMARKS

In the previous chapter, the ARMA(p,q) model was defined and various properties of AR(p), MA(q), and ARMA(p,q) models were established. It was shown that, for example, realizations from AR(1) models exhibit noticeably different behavior than realizations from AR(2) models. Therefore, given the importance of the order of a time series model, a key objective of this chapter was to learn how to identify the orders p and q of an ARMA(p,q) model.

The issues of ARMA model identification (deciding on the orders p and q of your model) and estimating the parameters were discussed. The main tools presented for model identification were AIC and it's variations. Maximum likelihood estimation of ARMA parameters was presented as the gold standard. Burg estimation and Yule-Walker estimation were discussed for the special case of AR(p) models. A backcasting method for estimating the residuals and the white noise variance was discussed..

Finally, the methodology of calculating forecasts from ARMA(p,q) models was discussed in detail and examples were given to illustrate the process. The assessment of forecasts can be measured by root mean square error (RMSE) or the rolling window RMSE (rwRMSE).

APPENDIX 6A

TSWGE FUNCTIONS

(a) `aic.wge(x,p=pmin:P,q=qmin:Q,type='aic')` is a function that finds the best ARMA(p,q) model using ML estimation within the p and q ranges specified in the call statement. The function can base the model identification on AIC, AICC, or BIC.

   x is a vector containing the realization to be analyzed.
   p=pmin:P specifies the range of values for the autoregressive order p to be used in the search for the best model. (default is **p=0:5** and it is common for **pmin** to be 0)
   type specifies the identification criterion type to be used. Choices are type= **'aic'**, **'bic'**, or **'aicc'**. (default is **'aic'**)

(b) `aic.ar.wge(x,p=pmin:P,type='aic',method='mle')` is a function that finds the best AR(p) model within the **p** range specified in the call statement. The function can base the model identification on AIC, AICC, or BIC and can be used with estimation methods MLE, Burg, and Yule-Walker.

   x is a vector containing the realization to be analyzed.
   p=pmin:P specifies the range of values for the autoregressive order p to be used in the search for the best model. (default is **p=0:5** and it is common for **pmin** to be 0)
   type specifies the identification criterion type to be used. Choices are type= **'aic'**, **'bic'**, or **'aicc'**. (default is **'aic'**)
   method specifies the estimation method used. Choices are **method= 'mle'**, **'burg'**, or **'yw'**. (default is **'mle'**)

(c) `aic5.wge(x,p=pmin:pmax,q=qmin:qmax,type)` finds the five best ARMA(p,q) models within the **p** and **q** range in the call statement using the selection criterion specified by **type**.

x is a vector containing the realization to be analyzed.

p=pmin:P specifies the range of values for the autoregressive order p to be used in the search for the best model. (default is **p=0:5** and it is common for **pmin** to be 0)

type specifies the identification criterion type to be used. Choices are **type= 'aic'**,**'bic'**, or **'aicc'**. (default is **'aic'**)

method specifies the estimation method used. Choices are **method= 'mle'**, **'burg'**, or **'yw'**. (default is **'mle'**)

Note: The program outputs the model orders of the top five models (and associated criterion values) within the range of p and q specified in the function statement.

(d) **aic5.ar.wge(x,p=pmin:P,type='aic',method='mle')** finds the five best AR(p) models within the **p** range in the call statement using the selection criterion specified by **type** and estimation criterion indicated by **method**.

x is a vector containing the realization to be analyzed

p=pmin:P specifies the range of values for the autoregressive order p to be used in the search for the best model. (default is **p=0:5** and it is common for **pmin** to be 0)

type specifies the identification criterion type to be used. Choices are **type= 'aic'**,**'bic'**, or **'aicc'**. (default is **'aic'**)

method specifies the estimation method used. Choices are **method= 'mle'**,**'burg'**, or **'yw'**. (default is **'mle'**)

Note: The program outputs the model orders of the top five models (and associated criterion values) within the range of p specified in the function statement.

(e) **est.arma.wge(x, p=0,q=0, factor=TRUE,method='mle')** obtains parameter estimates of a stationary ARMA(p,q) that is fit to the data in the vector **x**.

x is a vector containing the realization to be analyzed

p is the order of the AR part

q is the order of the MA part

factor is a logical variable, and if **factor=TRUE** (default), a factor table is printed for the estimated model.

(f) **est.ar.wge(x, p=2,factor=TRUE,method='mle')** obtains parameter estimates of a stationary autoregressive model of order **p** that is fit to the data in the vector **x**.

x is a vector containing the realization to be analyzed

p is the order of the AR model (default is **p=2**)

factor is a logical variable, and if **factor=TRUE** (default), a factor table is printed for the estimated model.

method specifies the method used to estimate the parameters: either **'mle'** (default), **'burg'**, or **'yw'**

(g) **fore.arma.wge(x,phi,theta,n.ahead,lastn,plot,alpha,limits)** forecasts **n.ahead** steps ahead for stationary ARMA models (of which AR models are a special case with **theta=0**). The forecasts and forecast limits can be optionally plotted. You can calculate and plot forecasts **n.ahead** steps beyond the end of the series, or you can forecast the last **n.ahead** data values (if **lastn=TRUE**).

Note: The components of **phi** and/or **theta** can be obtained by first using **mult.wge** if the models are given in factored form.

Note: The ARMA model can be a given/known model, but the most common usage will be to fit an ARMA(p,q) model to the data and base the forecasts on the fitted model.

x is a vector containing the time series realization $x_t, t = 1,\ldots,n$

phi is a vector of autoregressive coefficients (default=0)

theta is a vector of moving average coefficients (default=0)

n.ahead specifies the number of steps ahead you want to forecast (default=2)

lastn is a logical variable (default=**FALSE**).

 If **lastn=FALSE** (default) then the forecasts are for $x(n+1), x(n+2), \ldots, x(n+n.ahead)$ where n is the length of the realization.

 If **lastn=TRUE**, then the program forecasts the last **n.ahead** values in the realization.

plot is a logical variable. (default=**TRUE**)

 If **plot=TRUE** then the forecasts will be plotted along with the realization.

 If **plot=FALSE** then no plots are output.

alpha specifies the significance levels of the prediction limits. (default=.05)

limits is a logical variable. (default=**TRUE**)

 If **limits=TRUE** then the (1–**alpha**)×100% prediction limits will be plotted along with the forecasts.

(h) **pacf.wge(x,   lag.max=20,plot=TRUE,method)** calculates and optionally plots the estimated partial autocorrelations

 x is a vector containing the realization to be analyzed

 lag.max is the maximum lag at which the PACF is calculated. (default is **p=20**)

 method specifies the method used to estimate the parameters: either **'mle'**, **'burg'**, or **'yw'**

 plot is a logical variable. (default=**TRUE**)

 If **plot=TRUE** then the PACFs will be plotted along with 95% limit lines.

 If **plot=FALSE** then no plots are output but the PACFs are contained in the output vector **$pacf**.

(i) **roll.win.rmse.wge(series,horizon=1,s=0,d=0,phi=0,theta=0)** is a function that creates as many "windows" as is possible with the data vector **series** and calculates an RMSE for each window. The resulting "rolling window RMSE" is the average of the individual RMSEs from each window.

 series is the data

 horizon is the number of steps ahead to be forecasted.

 s is the order of the seasonal difference, default=1

 d is the order of the difference

 phi is the vector of AR coefficients

 theta is the vector of MA coefficients

 Output

 rwRMSE is the average of the individual RMSEs of each window

 numwindows is the number of windows

 horizon is the number of observations ahead to be forecasted.

 s is the order of the seasonal difference, default=1

 d is the order of the difference

 phi is the vector of AR coefficients

 theta is the vector of MA coefficients

 RMSEs is an array of the calculated RMSEs for each window.

PROBLEMS

6.1 (a) Factor the Yule-Walker and the Burg estimates fit to the dataset **A** shown in Figure 5.19(a) and compare them with the factors of those of the true model.

(b) Generate three realizations from the ML, Burg, and YW models *fit* to dataset **A**. Plot the realizations, sample autocorrelations, and Parzen spectral density estimates for the three realizations. Compare and contrast these with the corresponding plots in Figures 5.19(a)–(c) for the original data generated from the "true model".

6.2 The second choice of a model for the **sunspot2.0** data using **aic5.ar.wge** is an AR(10) model. Use factor tables to compare the AR(9) and AR(10) models and explain which model you would choose and why.

6.3 Generate a realization using the command

```
x=gen.arma.wge(n=150,phi=c(2.6,-3.34,2.46,-.9024),sn=3233)
```

Use **aic.ar.wge(x,p=0:10,type= 'a',method= 'b')** to identify the order of the selected model letting **a=aic, bic,** and **aicc** and **b=mle, burg,** and **yw**.

(a) Use all nine combinations of **type** and **method**.

(b) If an AR(4) model is not selected, examine the factor table of the estimated model and compare it with the factor table for the model from which the realization is generated, that is, **phi=c(2.6,-3.34,2.46,.9024)**

6.4 Consider the dataset in Example 6.3 for which the Yule-Walker estimates are poor.

```
ar4=gen.arma.wge(n=100,phi=c(2.76,-3.76,2.6,-.89),sn=468)
```

Use **aic5.ar.wge** using the Burg and Yule-Walker methods with p ranging from 0 to 10.

(a) What is the top model selected by Burg?

(b) What is the top model selected by Yule-Walker?

(c) Factor each of the models obtained in (a) and (b) and compare the factors with those of the actual AR(4) model.

6.5 Use **aic5.ar.wge** using AIC and ML estimates to find the top model for the following *tswge* datasets:

(a) **dfw.mon**

(b) **dfw.yr**

(c) **wtcrude2020**

6.6 Use Base R function **pacf** to identify the order of the AR model you would fit to the datasets in Problem 6.5. Additionally, use pacf on the:

(d) **x500=gen.arma.wge(n=500,phi=c(.13,1.4414,-.0326,-.8865),sn=9310)**

(e) **ar3** in Problem 6.4

6.7 Calculate the data **logss**=log(**sunspot2.0**+10)

(a) Use **plotts.sample.wge** on **sunspot2.0** and **logss**. Compare and contrast the results. What has been the effect of taking the logarithm?

(b) Why didn't we calculate log(**sunspot2.0**)?

(c) Use AIC and BIC to select model orders for the **logss** data letting **p=0:10** and **q=0:4**. How do these model orders compare with AIC and BIC fits to the raw **sunspot2.0** data?

(d) Find ML estimates for the models selected by AIC for the raw sunspot 2.0 and **logss** data.

(e) Use the ML model fit to **logss** to forecast the last 10 and 20 years of **logss** data. Plot the forecasts.

(f) Find the RMSE and MAD for the forecasts in (e).

6.8 Consider the model: $(1 - B + .9B^2)(1 - .95B)(1 + .9B)(X_t - 100) = a_t$ where $\sigma_a^2 = 1$.

(a) Use **mult.wge** (see Appendix 5A) to find the coefficients of the resulting 4th order model.

(b) Generate a realization of length $n = 200$ from this model.

(c) Use **aic.ar.wge** or **aic5.ar.wge** (based on AIC and Burg estimates) to fit a model to your realization. Did AIC select $p = 4$? Factor your model and discuss how its factors compare with those of the true model.

(d) Use your fitted model to
  (i) Forecast 20 time periods beyond the end of the realization and show the 80% prediction limits.
  (ii) Forecast the last 20 time periods of the data and find the RMSE and MAD for these forecasts.

6.9   Show that the commands below produce four similar models and factor tables.

```
data(sunspot2.0)
par(mfrow=c(2,2),mar=c(4,2,1,1))
a=est.ar.wge(sunspot2.0[1:311],p=9)
b=est.ar.wge(sunspot2.0[1:301],p=9)
c=est.ar.wge(sunspot2.0[1:291],p=9)
d=est.ar.wge(sunspot2.0[1:281],p=9)
```

6.10   Verify the rwRMSE shown in Table 6.20 for the AR(2) model.

6.11   In Table 6.18 it was shown that the AR(9) achieved a lower RMSE for all but one of the tested forecast origins. Calculate a rolling window RMSE for each model and compare to the results of Table 6.18.

6.12   Use **pacf.wge** to select the "best" AR model for the following datasets:

(a) **sunspot2.0**
(b) log(base 10) **lynx**
(c) Example 6.3 data vector **ar4**
(d) **dfw.mon**
(e) **wtcrude2020**

For each dataset plot the PACFs and "assess model order" using ML, Burg, and YW estimates.

ARIMA and Seasonal Models

7

Chapters 5 and 6 focused on ARMA models (along with their special cases, AR and MA models) and how these models are used for modeling and forecasting *stationary time series data*. The assertion was made that many (or most) time series do not satisfy the conditions of stationarity given in Section 3.3, and consequently there is a need for time series models that focus on modeling nonstationary behavior in time series. There are several ways that a time series can be nonstationary. Among these violations of stationarity are:

1. *The mean changes with time:* This often occurs when there is a deterministic component such as a linear trend. This is the topic of Section 8.1.
2. *The variance changes with time:* The **AirPassengers** data first shown in Figure 1.5 are an example of a time series for which the variance seems to be increasing in time.
3. *The correlation structure depends on time*: An example of this is the bat echolocation data and the seismic data in Figure 3.15(a) and (b), respectively.[1]
4. *The AR-characteristic equation has one or more roots on the unit circle:* Recall that Theorem 5.7 states that an ARMA(p,q) process is stationary if and only if all the roots of the AR-characteristic equation fall outside the unit circle.

In this chapter we will consider models which are nonstationary due to "Violation 4" with some discussion of "Violation 2". Chapter 8: Time Series Regression will discuss models for time series whose mean value changes with time.

7.1 ARIMA(*p, d, q*) MODELS

The ARIMA(p,d,q) models have proven to be very useful in practice for modeling data that have a wandering behavior but do not seem to have an attraction to a process mean. This model is very popular in the modeling of economic time series, but the applications of these models are extensive. In Section 7.1.1 we define the ARIMA(p,d,q) model and discuss its properties while in Section 7.1.2 we describe how to fit an ARIMA model to a set of time series data.

---

1 A broad class of time series which have autocorrelations that depend on time are the G-stationary time series discussed in Chapter 13 of Woodward et al. (2017). These models will not be discussed in this text, but readers who want more information on G-stationary processes should refer to Woodward et al. (2017) and research articles referenced therein.

7.1.1 Properties of the ARIMA(*p,d,q*) Model

The process, X_t, is an ARIMA(p,d,q) process if $Y_t = (1-B)^d X_t$ satisfies a stationary ARMA(p,q) model. An ARIMA(p,d,q) model is referred to as an *autoregressive integrated moving average process of orders* $p, d,$ *and* q. The operator $(1-B)^d$ is defined in Definition 7.1.

Definition 7.1: $(1-B)^d$ The operator $(1-B)^d$ is the *d*th difference operator where *d* is a non-negative integer.

(a) If $d = 1$ then $(1-B)^d = (1-B)$ and $(1-B)X_t = X_t - X_{t-1}$.
(b) If $d = 2$ then $(1-B)^d = (1-B)^2 = (1-B)(1-B) = 1 - 2B + B^2$ so that
$$(1-B)^2 X_t = (1 - 2B + B^2)X_t = X_t - 2X_{t-1} + X_{t-2}.$$
(c) If $d = 3, 4, \ldots$, then $(1-B)^d$ is defined analogously.
(d) If $d = 0$, then $(1-B)^0 X_t = X_t$.

Definition 7.2: The ARIMA(p,d,q) model is defined by

$$\phi(B)(1-B)^d X_t = \theta(B)a_t, \tag{7.1}$$

where $\phi(B)$ and $\theta(B)$ are "stationary and invertible" operators for which all roots of the characteristic equations $\phi(z) = 0$ and $\theta(z) = 0$ lie outside the unit circle. Note that the "autoregressive part" of this model is $\phi(B)(1-B)^d$.

7.1.1.1 Some ARIMA(*p,d,q*) Models

The following are examples of ARIMA(p,d,q) models which we will discuss in more detail in the following and will be referred to as Models (a), (b), and (c), respectively.

(a) $(1-B)X_t = a_t$: This is an ARIMA(0,1,0), the simplest of all nonstationary ARIMA(p,d,q) models.
(b) $(1 - 1.3B + .65B^2)(1-B)X_t = a_t$: An ARIMA(2,1,0) model with $\phi(B) = 1 - 1.3B + .65B^2$.
(c) $(1 - 1.3B + .65B^2)(1-B)^2 X_t = a_t$: An ARIMA(2,2,0) model with $\phi(B) = 1 - 1.3B + .65B^2$.

Key Points:

1. ARIMA(p,d,q) processes with $d > 0$ have the property that one or more roots of the autoregressive characteristic equation are ***on the unit circle***. We say that these models have ***unit roots***.
2. Because of the unit root(s), these models do not satisfy the property of stationary ARMA models which require all roots of the autoregressive characteristic equation to be outside the unit circle. However, these processes are on the "border" of the stationary region.

7.1.1.2 Characteristic Equations for Models (a)–(c)

(a) The ARIMA$(0,1,0)$ model $(1-B)X_t = a_t$ has the associated characteristic equation $1-z=0$, which has a single root $r=1$. Because this root is *on* the unit circle, the ARIMA$(0,1,0)$ model is not a stationary AR(1) model.

(b) The autoregressive part of the model in (b) is $(1-1.3B+.65B^2)(1-B)$ so the autoregressive characteristic equation is $(1-1.3z+.65z^2)(1-z)=0$. Based on this factorization of the characteristic equation we see that $z=1, 1+0.73i$, and $1-0.73i$ are solutions. Consequently, the process is nonstationary with a root of $r=1$ (which is on the unit circle) and a complex conjugate pair that are outside the unit circle.

(c) The autoregressive component in (c) is $(1-1.3B+.65B^2)(1-B)^2$ so the autoregressive characteristic equation is $(1-1.3z+.65z^2)(1-z)^2=0$ which has two roots of $+1$ on the unit circle.

(1) Model (a) ARIMA(0,1,0): $(1-B)X_t = a_t$

We now consider the simple ARIMA$(0,1,0)$ model

$$(1-B)X_t = a_t. \tag{7.2}$$

This has the appearance of an AR(1) model with $\phi_1 = 1$. However, because the characteristic equation $1-z=0$ has a root $r=1$, which is *on* the unit circle, the ARIMA$(0,1,0)$ model is nonstationary as discussed above. However, every AR(1) model for which ϕ_1 is less than 1 in absolute value corresponds to a stationary model, no matter how close ϕ_1 is to 1. For example, $(1-.999B)X_t = a_t$ is stationary while $(1-B)X_t = a_t$ is not (see Problem 7.1).

7.1.1.3 Limiting Autocorrelations

From Equation 5.6 we know that for the stationary AR(1) process $(1-\phi_1 B)(X_t - \mu) = a_t$, with $|\phi_1| < 1$, the variance $\sigma_X^2 (= \gamma_0)$ is given by $\gamma_0 = \dfrac{\sigma_a^2}{1-\phi_1^2}$. Thus, it follows that the process variance $\sigma_X^2 \to \infty$ as $\phi_1 \to 1$.

The autocovariance, γ_k for $k>0$, of a stationary AR(1) process is given by $\gamma_k = \gamma_0 \phi_1^k$ which also approaches ∞ as $\phi_1 \to 1$. Because the autocorrelation of an AR(1)-process is $\rho_k = \dfrac{\gamma_k}{\gamma_0} = \dfrac{\gamma_0 \phi_1^k}{\gamma_0} \cdot \dfrac{\infty}{\infty}$ if $\phi_1^k = 1$, we clearly need another definition for the autocorrelation function when $\phi_1 = 1$. Note that if $|\phi_1| < 1$ then $\gamma_0 = \sigma_X^2$ is finite.

So, $\rho_k = \dfrac{\gamma_k}{\gamma_0} = \dfrac{\gamma_0 \phi_1^k}{\gamma_0} = \phi_1^k$ for each $|\phi_1| < 1$ as $\phi_1 \to 1$. Consequently, it is reasonable to define the autocorrelation function for the limiting case, that is, $\phi_1 = 1$, as the limit of $\rho_k = \phi_1^k$ as $\phi_1 \to 1$. That is, the "autocorrelation" for $\phi_1 = 1$ is defined to be $1^k = 1$ for each k.

> **Key Point:** In general, for ARIMA and seasonal models, *the autocorrelation is defined as a limit.* These limiting autocorrelations are also referred to as "extended autocorrelations" and are denoted by ρ_k^* in Chapter 5 of Woodward et al., 2017. In the current book we will continue to simply refer to them as autocorrelations and denote them by ρ_k.

Also, note that the spectral density of the ARIMA(0,1,0) has a peak at zero that is infinite. Figure 7.1(a) shows the (limiting) autocorrelations which are $\rho_k = 1$ for all integer k. Figure 7.1(b) is a plot of the spectral density for this model, for which the peak in the spectral density is technically infinite (which we cannot plot). The *tswge* package doesn't have a routine for plotting the autocorrelations and spectral densities for nonstationary processes due to the unbounded behavior associated with them. Consequently, the plots in Figure 7.1 were obtained by using a very nearly nonstationary model.[2]

```
plotts.true.wge(phi=.99999,plot.data=FALSE)
```

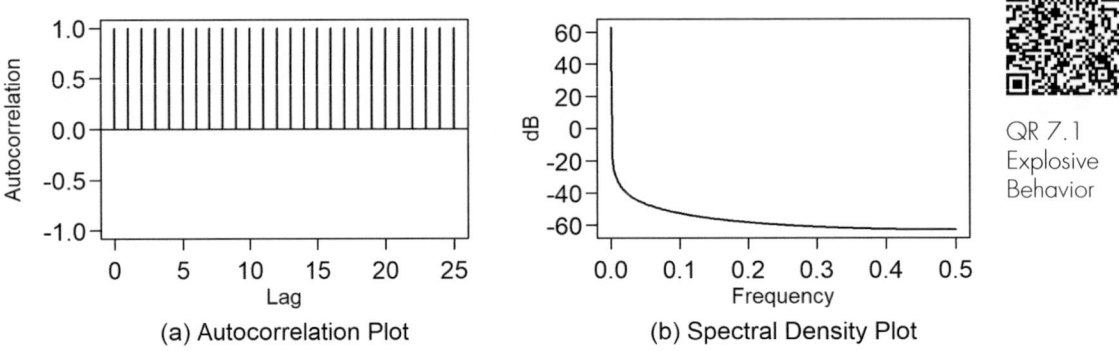

QR 7.1
Explosive
Behavior

(a) Autocorrelation Plot   (b) Spectral Density Plot

FIGURE 7.1   Model based (a) autocorrelations and (b) spectral density for $(1-B)X_t = a_t$.

Figure 7.2(a) is a plot of a realization of length $n = 200$ from the ARIMA(0,1,0) model along with the sample autocorrelations and Parzen spectral density estimate. The realization has the wandering behavior typical for an AR(1) model with ϕ_1 close to one. Recall that Figure 5.6(a) is a plot of a realization from the nonstationary model in (7.2). The realizations in Figure 5.6(a) and Figure 7.2(a) are similar in that the data seem to be wandering. The data in Figure 5.6(a) show an overall "upward" wandering while Figure 7.2(a) shows a downward trending behavior. In Figure 7.2(b) we show the sample autocorrelations, and note that they are not identically equal to one for all lags. However, the sample autocorrelations for small lags are quite close to one and the damping is relatively slow. The Parzen spectral density estimate has a sharp peak at $f = 0$, which is finite. The realization in Figure 7.2(a) was generated using the *tswge* command **gen.arima.wge** and the plots below were obtained using the commands[3]. Note that the default value is **mu=0**. The effect of specifying **mu=50** is to add 50 to each generated data value compared to the use of the default value **mu=0**.

```
x010=gen.arima.wge(n=100,d=1,sn=409)
plotts.sample.wge(x010)
```

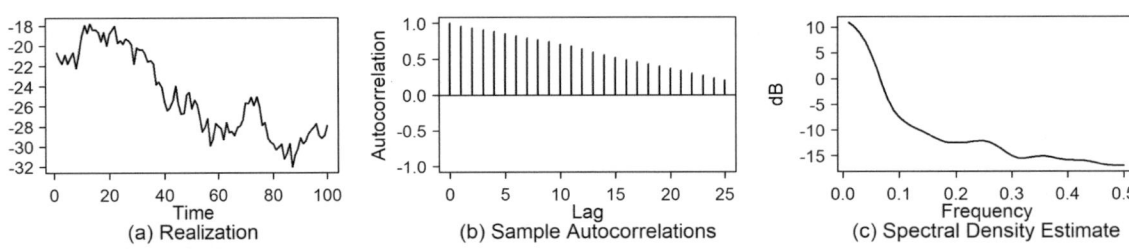

(a) Realization   (b) Sample Autocorrelations   (c) Spectral Density Estimate

FIGURE 7.2   (a) Realization, (b) sample autocorrelations, and (c) Parzen spectral density estimate for a realization of length $n = 100$ from an ARIMA(0,1,0) model.

---

2 Note that for the stationary model $(1-.99999B)X_t = a_t$ it follows that $\rho_{25} = \phi_1^{25} = .99999^{25} = .99975$. So, although the true autocorrelations for an ARIMA(0,1,0) model are all equal to one, the autocorrelations above are visually indistinguishable from one. Similarly, the spectral density (dB) at $f = 0$ is quite large but not infinite.

3 The *tswge* function **gen.arima.wge** generates a realization from an ARIMA model. It is described in Appendix 7A.

Key Point: A realization with wandering behavior, slowly damping sample autocorrelations, and a peak in the Parzen spectral density estimate at $f = 0$ "suggests" that a $(1 - B)$ factor may be present in the data. Note the quotes around "suggests". We will return to this issue in Section 7.1.2 when we discuss model identification.

7.1.1.4 Lack of Attraction to a Mean

Recall that in the discussion of stationary ARMA models, it was noted that one of their features was that realizations tend to show an "attraction to the mean". That is, if $\mu = 50$ is the theoretical mean of a stationary ARMA model, then realizations from that model that drift very far from the mean (50) will eventually "be attracted back toward the mean". In fact, for sufficiently long realizations from a stationary model, the sample mean, $\bar{X} = \frac{1}{n}\sum_{t=1}^{n} X_t$, will tend to be a good estimator of the true mean.

You may have noticed that the ARIMA(p,d,q) processes in (7.1) and (7.2) did not contain a mean term which are in the models for stationary models. To understand why, recall that the AR(1) model, $(1 - \phi_1 B)(X_t - \mu) = a_t$, is sometimes written as $X_t - \phi_1 X_{t-1} = (1 - \phi_1)\mu + a_t$ where $(1 - \phi_1)\mu$ is referred to as the moving average constant. Consider the ARIMA(0,1,0) model written to include a mean, that is $(1 - B)(X_t - \mu) = a_t$. The moving average constant is $(1 - \phi_1)\mu = (1 - 1)\mu = 0$, regardless of μ. Consequently, the model $(1 - B)(X_t - \mu) = a_t$ doesn't actually depend on μ. Consequently, the models $(1 - B)(X_t - 50) = a_t$, $(1 - B)(X_t + 600) = a_t$, and $(1 - B)X_t = a_t$ are all the same. Thus, it is common to write the ARIMA model without the mean term as we have done in (7.1) and (7.2).

Figure 7.3(a) shows a realization of length $n = 10,000$ from the stationary (but nearly nonstationary) model $(1 - .999B)(X_t - 50) = a_t$, generated using the command

```
x9=gen.arma.wge(n=10000,phi=.999,mu=50,sn=509)
```

Note that whenever values of the realization tend to stray very far from the mean of 50 (shown on the graph as a horizontal line), there will eventually be a tendency (especially visible in longer realizations) for the values to return toward 50. Figure 7.3(b) is a realization of length $n = 10,000$ from the nonstationary model $(1 - B)(X_t - 50) = a_t$, generated using the command

```
x1=gen.arima.wge(n=10000,d=1,mu=50,sn=340)
```

where of course 50 is the "mean in name only", and it is reasonable to question the meaning of **mu** in the call statement. Function **gen.arma.wge** generates realizations by setting the first p data values equal to **mu**. Similarly, in **gen.arima.wge**, the first $p + d$ observations are set equal to the value of **mu**. The functions then generate 2000 data values using the recursive relationship associated with the model. The final series of length n returned from the functions are the data values $2001, 2002, \ldots, 2000 + n$. Using this procedure, the stationary realization in Figure 7.3(a) did indeed tend to wander around 50 as previously mentioned, but the nonstationary realization Figure 7.3(b) shows no attraction to the "mean" (50) and in fact 50 is not even on the vertical scale.

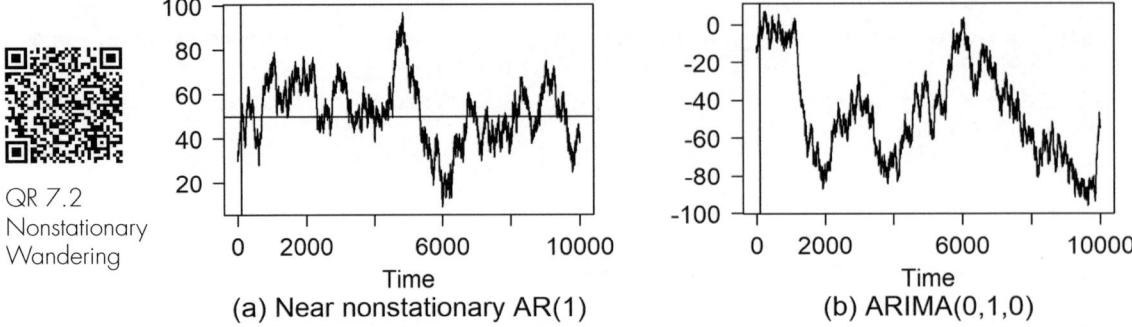

QR 7.2
Nonstationary
Wandering

(a) Near nonstationary AR(1)

(b) ARIMA(0,1,0)

FIGURE 7.3 Realizations of length $n = 10,000$ from (a) $(1-.999B)(X_t - 50) = a_t$ and (b) $(1-B)(X_t - 50) = a_t$.

7.1.1.5 Random Trends

Figures 7.4(a) and (b) show the first 100 observations from the corresponding plots in Figure 7.3. Because of the length of the realizations plotted in Figure 7.3, it is difficult to "get a good look" at the first 100 observations. The plots in Figure 7.4 are the observations to the left of the vertical lines at $t = 100$ in the plots in Figure 7.3 (trust us!). The point is that ARIMA(0,1,0) models and nearly nonstationary AR(1) models produce *random trends*. The longer realizations in Figure 7.3 show that these random trends may result in significant time intervals in which the data are "linearly trending up" or "linearly trending down". There also seems to be no "warning" that a turn in direction is imminent. The longer realizations in Figures 7.3(a) and (b) make it clear that it would have been a mistake to predict the trends visible in the first one hundred time values to continue. In fact, if the realizations in Figures 7.4(a) and (b) were all that were available to the analyst, it would be difficult (or maybe impossible) to determine whether the visible trend is due to an actual deterministic trend component in the data (that we would be tempted to predict to continue into the future). The decision whether to predict an observable trend to continue into the future must often be made by the analyst based on domain knowledge. We will discuss this further in the context of forecasting in Section 7.1.3. The "partial realizations" in Figure 7.4 were obtained using the commands

```
plotts.wge(x9[1:100])
plotts.wge(x1[1:100])
```

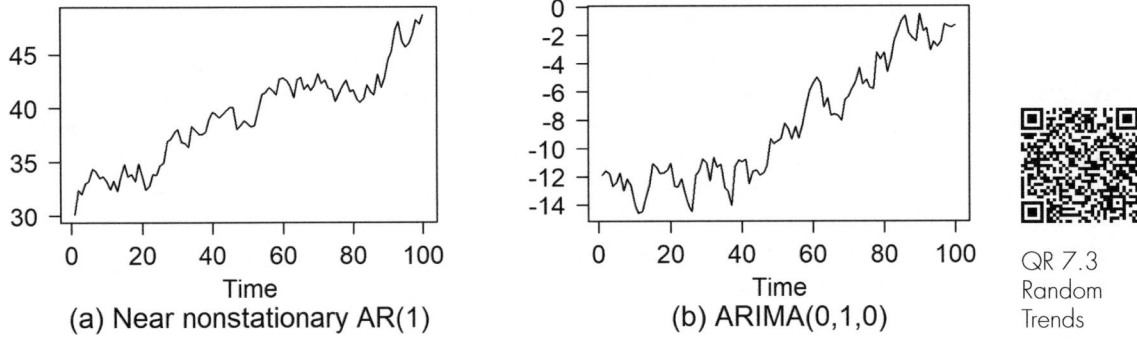

(a) Near nonstationary AR(1)

(b) ARIMA(0,1,0)

QR 7.3
Random
Trends

FIGURE 7.4 Realizations of length $n = 100$ from (a) $(1-.999B)(X_t - 50) = a_t$ and (b) $(1-B)(X_t - 50) = a_t$.

Key Points:

1. There is no attraction to a mean for realizations from the ARIMA(0,1,0) model in (7.2). This is a common characteristic of all ARIMA(p,d,q) models.
2. "Shorter" realization lengths from an ARIMA(0,1,0) (and an AR(1) with ϕ_1 close to one) can show trending behavior, but these are "random trends" and should not be predicted to continue.
3. Figure 7.3(b) shows that due to the lack of an attraction to the mean, the random trending behavior in an ARIMA(0,1,0) model can be quite lengthy. Specifically, the realization has a long downward trend starting at about $t = 6,000$. However, this trend reverses dramatically at about $t = 9,900$.
4. The effect of including `mu=50`, say, in the `gen.arima.wge` command is to add 50 to each generated value. However, as discussed, 50 is not the mean of the process.

Try This: One of the best ways to get a feel for the idea of random trends is to try it out! (This is a relevant recommendation for topics throughout the book.) The following code could be used to generate Figures 7.4(a) and (b):

```
x9.100=gen.arma.wge(n=100,phi=.999,mu=50,sn=509)
x1.100=gen.arima.wge(n=100,d=1,mu=50,sn=340)
```

Take a few minutes and run these commands using different `sn` values (or removing the `sn=` definition to produce randomly selected realizations). Examine the realizations that the models produce, and decide whether many of these realizations (that only have random trends) would have led you to believe there was an underlying linear (deterministic) component in the generating model.

7.1.1.6 Differencing an ARIMA(0,1,0) Model

Suppose we "difference" a realization from the ARIMA(0,1,0) model $(1-B)X_t = a_t$. That is, we calculate $Y_t = X_t - X_{t-1}$. Note that the ARIMA(0,1,0) can be written as $(1-B)X_t = X_t - X_{t-1} = a_t$, and recalling that because $Y_t = X_t - X_{t-1}$, it follows that $Y_t = a_t$. That is, if you difference data that are properly modeled as an ARIMA(0,1,0) model, the differenced data should have the appearance of white noise.

Note that for a realization $x_1, x_2, \ldots, x_n$, the "differenced" realization $y_t = x_t - x_{t-1}$ has only $n-1$ observations because $y_1 = x_1 - x_0$ cannot be computed (x_0 is not an observed data value). We often "violate" notation by referring to the differenced data as $y_1 = x_2 - x_1$, $y_2 = x_3 - x_2, \ldots, y_{n-1} = x_n - x_{n-1}$.[4] The command `artrans.wge` in *tswge* performs "autoregressive transformations". These are transformations of the form

$$Y_t = X_t - \phi_1 X_{t-1} - \cdots - \phi_p X_{t-p}.$$

As an example, in order to apply the transformation $Y_t = X_t + .2X_{t-1} - .48X_{t-2}$ to a dataset **x** in R, the command would be

```
y=artrans.wge(x,phi.tr=c(-.2,.48))
```

This command takes the data in realization **x**, applies the transformation $Y_t = X_t + .2X_{t-1} - .48X_{t-2}$ to the data and saves the transformed data in **y**. For the simple difference transformation, $Y_t = X_t - X_{t-1}$, it follows that `phi.tr=1`.

---

4 Actually, by definition, $y_1 = x_1 - x_0$, $y_2 = x_2 - x_1, \ldots, y_n = x_n - x_{n-1}$. The first value, $y_1 = x_1 - x_0$, cannot be calculated, leaving us with the realization, $y_2, y_3, \ldots, y_n$. Our "violation" of notation is, for notational convenience, to refer to the y_t realization as $y_1, y_2, \ldots, y_{n-1}$.

The differences of the data in Figure 7.2(a) are obtained using the command

```
x010=gen.arima.wge(n=100,d=1,sn=409)
d1=artrans.wge(x010,phi.tr=1)
```

Figure 7.5 shows a plot of the differences (**d1**) of the ARIMA(0,1,0) data in Figure 7.2(a) along with the sample autocorrelations and Parzen spectral density estimate of the differenced data. The data appear uncorrelated and the sample autocorrelation plot in Figure 7.5(b) includes the 95% limit lines as a reference. All of the sample autocorrelations are within the 95% limit lines although $\hat{\rho}_1$ and $\hat{\rho}_{13}$ are close to the boundaries. The Parzen spectral density estimate in Figure 7.5(c) is fairly flat. The plots in Figure 7.5 are consistent with white noise. Figure 7.5 is obtained using the code

```
x010=gen.arima.wge(n=100,d=1,sn=409,plot=FALSE)
d1=artrans.wge(x010,phi.tr=1)
plotts.sample.wge(d1, arlimits=TRUE,speclimits=c(-20,10))
```

> **Key Point:** If you difference a realization of length n from an ARIMA(0,1,0) model, the differenced data should appear to be a white noise realization of length $n-1$.

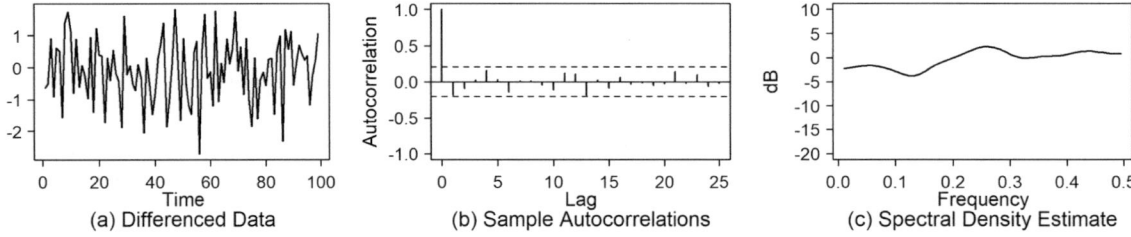

FIGURE 7.5   (a) Data in Figure 7.2(a) differenced, (b) sample autocorrelations, and (c) Parzen spectral density estimate for the differenced data in (a).

7.1.1.7 ARIMA Models with Stationary and Nonstationary Components

To this point we have considered the ARIMA(0,1,0) model in some detail. In this section we will consider ARIMA(p,d,q) models that have both stationary and nonstationary components. We note that Models (b) and (c) are both ARIMA(2,d,0) models containing the stationary AR component $1-1.3B+.65B^2$. Before examining Models (b) and (c) we first consider the stationary AR(2) model $\left(1-1.3B+.65B^2\right)X_t = a_t$.

7.1.1.8 The Stationary AR(2) Model: $(1-1.3B+.65B^2)X_t = a_t$

The factor table in Table 7.1 shows that this AR(2) model is associated with a system frequency of about $f = .10$, and that the roots of the characteristic equation are outside, but not very close to, the unit circle. Specifically, $|r|^{-1} = .81$.

TABLE 7.1   Factor Table for $\left(1-1.3B+.65B^2\right)X_t = a_t$

AR-FACTOR	ROOTS	$|r|^{-1}$	f_0
$1-1.3B+.65B^2$	$1.0\pm.73i$	.81	.10

Figure 7.6 shows a realization of length $n = 200$ along with the associated sample autocorrelations and Parzen spectral density estimate. The cyclic realization of length $n = 200$ with about 20 periods, the sample autocorrelations, and especially the Parzen spectral density estimate all indicate that the system frequency is about $f = .10$. We will refer back to Figure 7.6 and Table 7.1 as we analyze Models (b) and (c) and a seasonal model in Section 7.2. The plots below were obtained using the code

```
ar2=gen.arma.wge(n=200,phi=c(1.3,-.65),sn=637,plot=FALSE)
plotts.sample.wge(ar2)
```

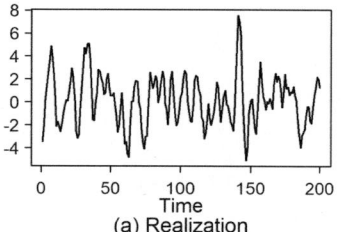 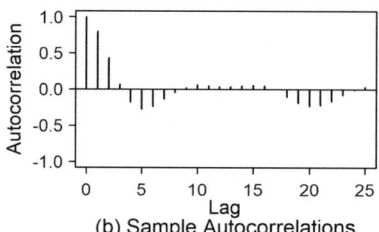 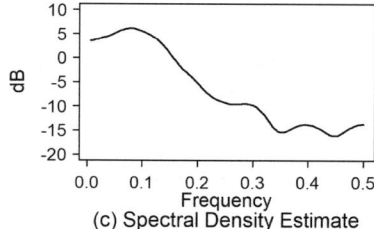

(a) Realization   (b) Sample Autocorrelations   (c) Spectral Density Estimate

FIGURE 7.6 (a) Realization of length $n = 200$, (b) sample autocorrelations, and (c) Parzen spectral density estimate for the stationary AR(2) model $\left(1 - 1.3B + .65B^2\right)X_t = a_t$.

Model (b) ARIMA(2,1,0): $(1 - 1.3B + .65B^2)(1 - B)X_t = a_t$
We next consider the ARIMA(2,1,0) model

$$\left(1 - 1.3B + .65B^2\right)\left(1 - B\right)X_t = a_t,$$

which we have referred to as Model (b). Table 7.2 shows the factor table for the nonstationary and stationary autoregressive components of this model.

TABLE 7.2 Factor Table for Model (b): $\left(1 - 1.3B + .65B^2\right)(1 - B)X_t = a_t$

AR-FACTOR	ROOTS	$|r|^{-1}$	f_0
$1 - B$	1	1	0
$1 - 1.3B + .65B^2$	$1.0 \pm .73i$	.81	.10

The factor table for the stationary AR(2) component of the model is shown in Table 7.1. Table 7.2 shows the additional factor of $1 - B$ which is on the unit circle and has a system frequency of $f = 0$.

Figure 7.7(a) shows a realization of length $n = 200$ from Model (b), and it is clear that the nonstationary part of the model, $(1 - B)$, dominates the behavior of the ARIMA(2,1,0) realization, and obscures the properties of the stationary AR(2) components of the model, sometimes to the point that they are not visible at all. Figures 7.7(b) and 7.7(c) show the sample autocorrelations and Parzen spectral density estimate, respectively, of the realization in Figure 7.7(a). The basic behaviors of the realization, sample autocorrelations, and spectral density are very similar to those in Figure 7.2, which were associated with a realization from an ARIMA(0,1,0). That is, the sample autocorrelations are slowly damping, and the spectral density has a strong peak at zero, which are the behaviors associated with the $(1 - B)$ component in the model. The sample autocorrelations and Parzen spectral density estimate fail to indicate the presence of the AR(2) component.

We next difference the data in Figure 7.7(a). Recall that in the ARIMA(0,1,0) case, the differenced data were white noise. We issue the following commands to generate the data in Figure 7.7(a) and difference the data using **artrans.wge**.

```
x210=gen.arima.wge(n=200,phi= c(1.3,-.65),d=1,sn=4855)
d1=artrans.wge(x210,phi.tr=1)
# plot the data (first row) and the differenced data (second row)
plotts.sample.wge(x210)
plotts.sample.wge(d1)
```

QR 7.4
Differencing

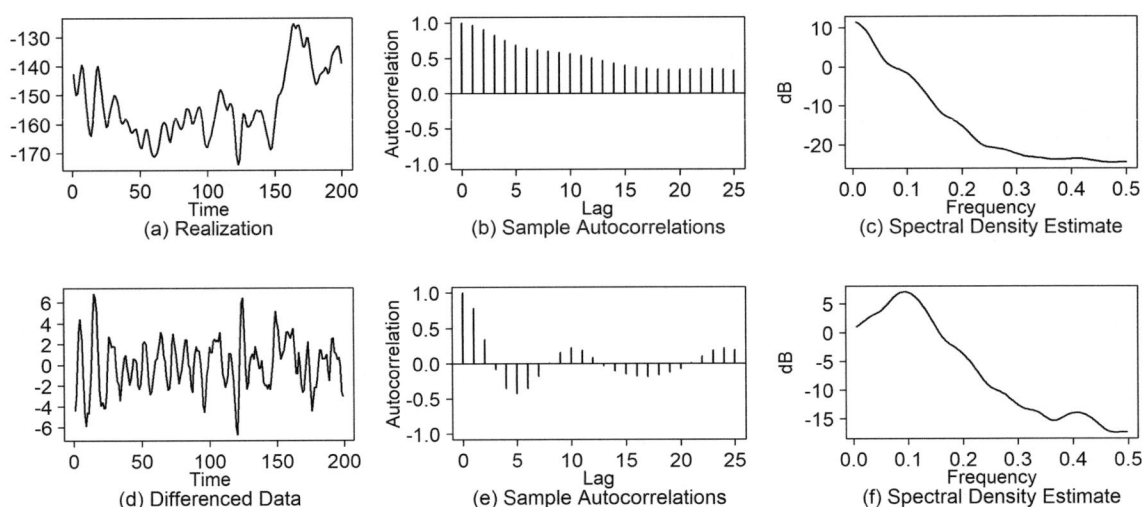

FIGURE 7.7   (a) Realization, (b) sample autocorrelations, and (c) Parzen spectral density estimate for a realization of length $n = 200$ from ARIMA(2,1,0) Model (b), (d) realization in (a) after differencing, (e) sample autocorrelation, and (f) Parzen spectral density of differenced data in (d).

The differenced dataset, **d1**, is a realization from the differenced process $Y_t = (1-B)X_t$. Because $(1-1.3B+.65B^2)(1-B)X_t = a_t$, it follows that Y_t satisfies the stationary model, $(1-1.3B+.65B^2)Y_t = a_t$. Figure 7.7(d) shows a plot of the differenced dataset, **d1**, along with the sample autocorrelations in Figure 7.7(e), and spectral density estimate in Figure 7.7(f). It is clear that the data are not white noise, the realization seems to be cyclic, the sample autocorrelations have a damped sinusoidal behavior, and the spectral density has a peak at about $f = .10$. This behavior is consistent with the factor table in Table 7.1 for the stationary AR(2) component of the model, and the plots in Figures 7.7(d)−(f) are quite similar to those in Figure 7.6 for a realization from the corresponding stationary AR(2) model. That is, differencing the data has "revealed" the AR(2) behavior in the data that was not visible before the difference was taken.

Key Points:

1. Differencing the data automatically adds a $1-B$ factor in the final model.
2. Sample autocorrelations that have a slowly damping exponential appearance are an indication that a factor, $(1-B)$, should be considered in the model. Further discussion of ARIMA model identification will be given in Section 7.1.2.
3. The stationary and invertible AR and MA components of an ARIMA$(p,1,q)$ model may not be apparent until the data are differenced, which removes the dominating nonstationary factor, $(1-B)$.
4. In the end, the final model will include all the factors (both stationary and nonstationary) that were revealed/discovered.

Model (c): ARIMA(2,2,0) $(1 - 1.3B + .65B^2)(1 - B)^2 X_t = a_t$

Consider the ARIMA(2,2,0) model

$$(1 - 1.3B + .65B^2)(1 - B)^2 X_t = a_t,$$

which we refer to as Model (c). This model differs from the ARIMA(2,1,0) Model (a) in that there are now *two* factors of $(1 - B)$. That is, the model has two unit roots. Because the single unit root in the ARIMA(2,1,0) model in Model (b) dominated the behavior of the realization and its characteristics, it is reasonable to expect that a similar model, but with two unit roots, will *really* be dominated by the unit roots. If this is what you expected, you would be correct! Figure 7.8(a–c) shows a realization from this ARIMA(2,2,0) model along with sample autocorrelations and Parzen spectral density estimate.

The realization in Figure 7.8(a) has the appearance of a smooth curve with sections of near linear behavior and a large variance as can be seen by values of the realization. The $(1 - B)^2$ behavior *totally* dominates the realization and there is absolutely no indication of the stationary AR(2) components of the model. The sample autocorrelations in Figure 7.8(b) damp very slowly, which is consistent with the behavior of the data.[5] As expected, the Parzen spectral density estimate in Figure 7.8(c) has a peak at $f = 0$. The data in Figure 7.8(a) were generated using the *tswge* command

```
x220=gen.arima.wge(n=200,phi=c(1.3,-.65),d=2,sn=450)
```

Based on our strategy for "stationarizing" the ARIMA(2,1,0) model, it seems reasonable that the data will need to be differenced two times to "uncover" the stationary components of the model. Because $(1 - B)^2 = (1 - 2B + B^2)$ there are two ways to difference the data "two times". We could use the commands

```
d1=artrans.wge(x220,phi.tr=1)
d2=artrans.wge(d1,phi.tr=1)
```

or

```
d2=artrans.wge(x220,phi.tr=c(2,-1))
```

It is instructive to use the first approach and apply the 1st-order differences sequentially. Figures 7.8(d–f) show the differenced data **(d1)**, the sample autocorrelations of the differenced data, and the Parzen spectral density estimate, respectively. The differenced data in Figure 7.8(d) do not have the extremely smooth appearance and high variance behavior of the data in Figure 7.8(a) (note the difference in the vertical axis values). However, the data do not appear to be attracted to a mean and the sample autocorrelations have much the same behavior as those in Figure 7.8(b). The twice differenced data in Figure 7.6(g), that is, $Y_t = (1 - B)(1 - B)X_t = (1 - B)^2 X_t$, follow the model for $Y_t = (1 - B)^2 X_t$ which is $(1 - 1.3B + .65B^2)Y_t = a_t$. Twice differencing the data in Figure 7.8(g) removes the influence of the dominant nonstationary components and reveals any stationary behavior that may remain (which in this case we know satisfies $(1 - 1.3B + .65B^2)Y_t = a_t$). Remarkably (but we really should not be surprised), the realization, sample autocorrelations, and spectral density in Figures 7.8(g)–(i) are similar to those shown in Figure 7.6. Note the similarities which include the once again visible peak in the spectral density at $f = .10$ in Figure 7.8(i).

---

5 The true/model based autocorrelations and spectral density from any $\text{ARMA}(p,d,q)$ model with $d \geq 1$ are the same as those shown in Figure 7.1 for the ARIMA(0,1,0) model (see Woodward, et al., 2017).

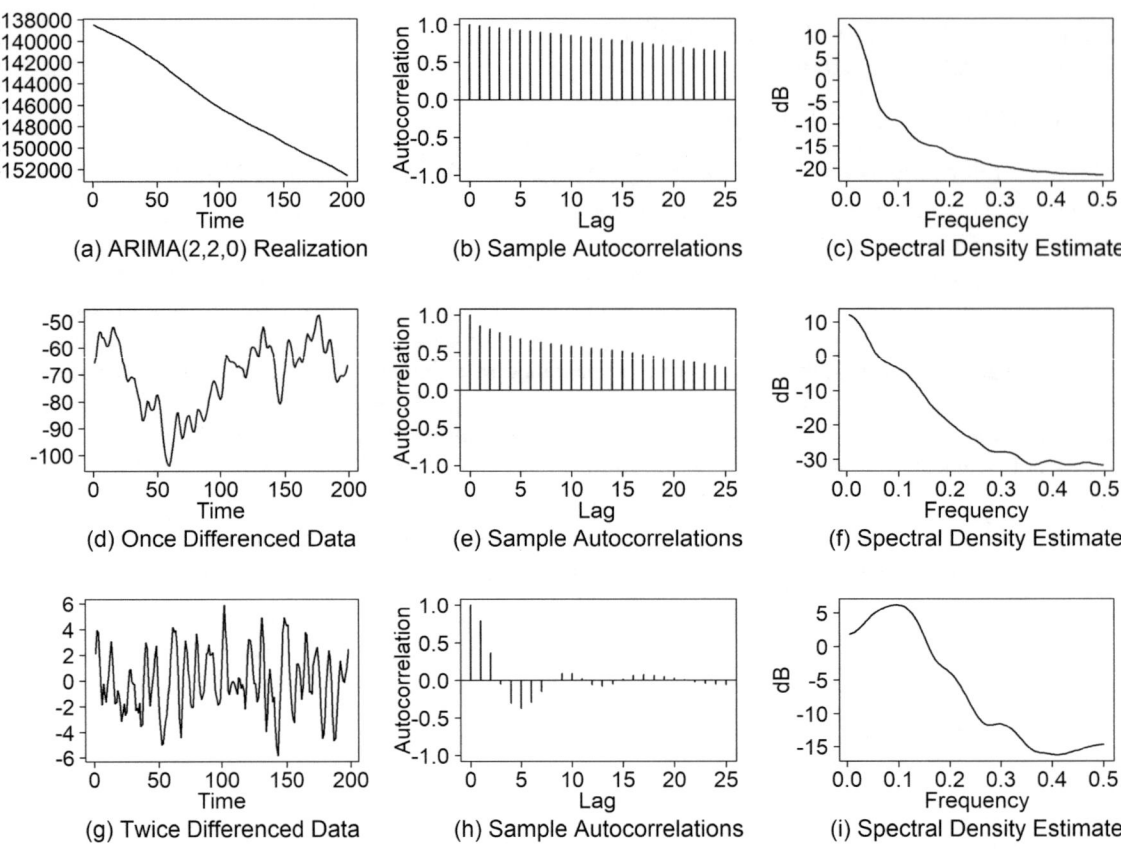

FIGURE 7.8   (a) Realization, (b) sample autocorrelations, and (c) Parzen spectral density estimate for ARIMA(2,2,0) Model (c); (d) data in (a) differenced, (e) sample autocorrelation, and (f) Parzen spectrum of differenced data in (d); (g) data in (d) differenced, (h) sample autocorrelations, and (i) Parzen spectrum of data in (g).

The plots in Figure 7.8 are created using the commands

```
x220=gen.arima.wge(n=200,phi=c(1.3,-.65),d=2,sn=450)
d1=artrans.wge(x220,phi.tr=1,plot=FALSE)
d2=artrans.wge(d1,phi.tr=1,plot=FALSE)
plotts.sample.wge(x220)
plotts.sample.wge(d1)
plotts.sample.wge(d2)
```

Key Points:

1. Data with strong positive correlations resulting in "smooth" realizations such as seen in Figure 7.8(a) may require two differences to recover the underlying, more subtle stationary components of the data.
2. In the case of $d = 2$, notice that there is absolutely no indication of the stationary components in the original data.

7.1.2 Model Identification and Parameter Estimation of ARIMA(*p,d,q*) Models

In this section we discuss the issues involved with fitting the ARIMA(p,d,q) model in (7.1) to a time series realization. As noted in the examination of ARIMA models (a)–(c) in Section 7.1, the behavior of the data from an ARIMA(p,d,q) model with $d \geq 1$ is dominated by the $(1-B)^d$ factor. Realizations from the ARIMA models tend to wander aimlessly and are not attracted to an overall mean level while sample autocorrelations have the appearance of a slowly damping exponential. The stationary components in realizations, autocorrelations, and spectral densities are typically only visible after the data have been differenced *d* times.

7.1.2.1 Deciding Whether to Include One or More 1–B Factors (That Is, Unit Roots) in the Model

Consider the realization, sample autocorrelations, and Parzen spectral density estimate in Figure 7.2. These data are aimlessly wandering and the sample autocorrelations have a slowly damping exponential behavior. The question at hand is the following:

Should we fit

(a) an ARIMA model with $d \geq 1$ (that is, include a unit root)

or

(b) a stationary ARMA model that has a root of the AR characteristic equation close to one?

In the following we discuss possible approaches to answering this question.

(a) A naive approach: An obvious approach is to fit a model to the dataset (say using ML estimates) and check the factor table for one or more factors of $1-B$.

First we simplify things by assuming a 1st-order model. For example, consider realizations from the ARIMA(0,1,0) model $(1-B)X_t = a_t$. It would be ideal if we could simply use ML estimates to fit a 1st-order AR model, that is, $(1-\phi_1 B)X_t = a_t$, to the data and check to see whether

(a) $\hat{\phi}_1 = 1$ (in which case we identify the underlying model as an ARIMA(0,1,0))

or

(b) $|\hat{\phi}_1| < 1$ (that is, we would identify the model as a stationary AR(1)).

Oh, if it were only that simple!

To check the viability of this procedure, we generated 100 realizations of length $n = 100$ from the ARIMA(0,1,0) model. For each realization we found the ML estimate, $\hat{\phi}_1$, associated with a 1st-order model. The average of the ML estimates was obtained along with the maximum and minimum from the 100 replications. The results are shown in the first column of Table 7.3 where it can be seen that while the estimate did equal the "hoped for" $\hat{\phi}_1 = 1$, this actually occurred (rounded to 3 decimal places) in *only four realizations out of 100*. In fact, the average $\hat{\phi}_1$ is .958, and in one realization, the ML estimate was as small as $\hat{\phi}_1 = .797$. Consequently, the above strategy ***absolutely does not work***!

The simulation procedure was repeated using realizations from the stationary (but nearly nonstationary) models $(1-.99B)X_t = a_t$ and $(1-.975B)X_t = a_t$. The results for these models, shown in the second and third columns, respectively, of Table 7.3 are very similar to those for the actual ARIMA(0,1,0) model. Also, some of the realizations from the stationary models yielded $\hat{\phi}_1 = 1$ to three decimal places. Specifically, the results in Table 7.3 illustrate the difficulty of deciding whether to include a factor of $1-B$ in the model.

TABLE 7.3   Simulation Results Comparing 100 Realizations of Length $n = 100$ from the Models $(1-B)X_t = a_t$, $(1-.99B)X_t = a_t$, and $(1-.975)X_t = a_t$

	$(1-B)X_t = a_t$	$(1-.99B)X_t = a_t$	$(1-.975B)X_t = a_t$
Average $\hat{\phi}_1$	0.958	0.950	0.928
Maximum $\hat{\phi}_1$	1	1	1
Minimum $\hat{\phi}_1$	0.797	0.850	0.786

In fact, any real dataset is almost certainly neither perfectly fit by an ARMA nor an ARIMA model. Recall the opening of Chapter 5 in which we mentioned that real data are almost never perfectly fit by an ARMA model. This is true of ARIMA models and essentially any mathematical model fit to a set of data. Such models are used because they are useful in the analysis of data. To this end, we cannot overstate George Box's famous quote

George E. P. Box:
"All models are wrong , but some are useful."

This is the spirit in which the model identification and parameter estimation in Chapter 6 and the remainder of this book should be viewed. That is, in Chapter 6, given a set of data which seemed to satisfy the basic properties of stationarity, AIC-type measures were used to find an ARMA model that was helpful in describing the behavior of the data and was used to provide a method for forecasting future values. In the current section we introduce another level of complication in the attempt to find an imperfect but useful model. Specifically, we need to decide whether a model should include a unit root, that is, a factor, $1-B$, and if so, how many, *before* we use AIC-type measures to select model orders and ultimately estimate the parameters of the stationary components of the model.

Key Points:

1. Fitting a model to data from an ARIMA model using ML estimates will nearly always produce a stationary model.
2. Consequently, unit roots cannot be identified by simply fitting a model to the data and checking the factor table for unit roots.
3. Points 1 and 2 and the simulations reported in Table 7.3 illustrate the fact that the decision whether to include a unit root in the model is a difficult one.
4. As will be discussed in the following, the decision whether to include a unit root in the model is one the investigator will need to make, based to some extent on physical considerations/domain knowledge.

(1) Unit Root Tests

Unit root tests are more sophisticated procedures (than the naïve approach in (a)) that give the investigator a "yes or no" answer concerning whether to include a unit root. The unit root tests are appealing because they "make the decision for you". We will not discuss the tests in detail but will show how to apply the tests to data and discuss their performance. The tests we will consider are:

(a) ADF (Augmented Dickey-Fuller) tests
(b) KPSS (Kwiatkowski-Phillips-Schmidt-Shin) tests

(a) The Augmented Dickey-Fuller Test

This test has been in use for many years to test the hypotheses:

H_0 : the model contains a unit root

H_a : the model does not contain a unit root (and is thus stationary)

Test statistic: τ

Rejection region: Reject H_0 if $\tau < d_\alpha$ where d_α is the α-level critical value.

Dickey (1976) obtains the (complicated) limiting distribution of the test statistic. If $\tau \geq d_{.05}$ then H_0 is not rejected and the Dickey-Fuller test "detects" a unit root. Note that rejection of the null hypothesis leads to a conclusion that the process is stationary. Thus, the conclusion of a unit root is based on *not rejecting the null hypothesis*. This is in a sense "backward" because as data scientists we know that failure to reject the null hypothesis does not imply belief that the null hypothesis is true, but simply that there was not sufficient evidence to reject it. To conduct unit root tests, we will use the function **ur.df** in the R package *urca*. There are a variety of options, and we use the following command that includes a constant but not trend in the model and uses AIC to select the number of lags. See Dickey and Fuller (1979) or Fuller (1996). The following command uses the Dickey-Fuller test for a unit root given a realization **x**.

```
h.df=ur.df(x,type='drift',selectlags='AIC')
```

The data in Figure 7.2(a) are a realization from an ARIMA(0,1,0) model (the model has a *unit root*). The data are generated using the command

```
x010=gen.arima.wge(n=100,d=1,sn=409)
```

After running the **ur.df** command using the **x010** data, the output, **h**, contains several items. For our purposes the test statistic, τ, is found in **h.df@teststat[1]** and $\tau = -1.076$. The 5% critical value, $d_{.05}$, is found in the output value **d05=h.df@cval[1,2]**, where we see that $d_{.05} = -2.89$. The null hypothesis of a unit root is rejected at the $\alpha = .05$ level if $\tau < d_{.05}$. Because $-1.076 \geq -2.89$ we *do not reject* the null hypothesis at the $\alpha = .05$ level, and correctly *conclude that the model contains a unit root*.

(b) The KPSS Test

The KPSS test (see Kwiatkowski, Phillips, Schmidt, and Shin (1992)) has the desirable feature that it is designed to test the hypotheses

H_0 : the model is stationary

H_a : the model contains a unit root (and is thus non stationary)

Test statistic: *kpss*

Reject region: Reject H_0 if $kpss > k_\alpha$ where k_α is the α-level critical value.

This is a more desirable setting for "concluding" a model has a unit root because the probability of rejecting the null hypothesis (concluding a unit root) when the null hypothesis is true (the model is stationary) can be made small (say, for example, less than .05). The KPSS test can be implemented using the function **ur.kpss** which is also in the *urca* package. The command below states that a trend term is not included in the model (**type='mu'**). Also, **lags='short'** specifies the formula to use for calculating the number of lags. (See *urca* documentation.) To apply the KPSS test to the data in **x010**, we use the command

```
h.kpss=ur.kpss(x010,type='mu',lags='short')
```

The output, **h.kpss**, contains several elements. The KPSS test statistic, *kpss*, is found in **h.kpss@ teststat,** and in this case $kpss = 1.795$. The critical value, $k_{.05}$, is found in **h.kpss@cval[2]**. In this case, $k_{.05} = .463$, and because $1.795 > .463$, the null hypothesis of stationarity is correctly rejected at the .05 level. The conclusion is that the model contains a unit root.

(2) Further Analysis of the Performance of Unit Root tests
We have seen that the ADF and KPSS tests gave the "correct" answers on the ARIMA(0,1,0) data in Figure 7.2(a). However, before recommending the tests for deciding whether a model has unit roots, we further investigate their performance using simulations.

(i) Ability of the tests to correctly detect unit roots when they are in the model (that is, when the model is nonstationary)
A useful test for unit roots should detect unit roots when the model in fact contains unit roots. To examine the ADF and KPSS tests in this regard, we generated 100 realizations from the ARIMA(0,1,0) model. We then calculated the percentage of times that the test found a unit root. Realization lengths $n = 100, 200, 500$, and 10000 were used. These results are given in Table 7.4, and note that for a good test, the percentages of "detection" in the table should be "high". The results in the table show that both tests detect a unit root a high percentage of the time, and that the KPSS test seems to perform better with increasing realization length while the ADF test does not.

TABLE 7.4 Simulation Results Showing Percentage of 100 Realizations from an ARIMA(0,1,0) Model for which the Test *Correctly* Detected a Unit Root

	$n = 100$	$n = 200$	$n = 500$	$n = 10000$
ADF Test	91%	95%	96%	95%
KPSS Test	90%	97%	100%	100%

QR 7.5 ADF and KPSS Performance

(ii) Ability of the tests to correctly not detect unit roots when models are stationary
Table 7.4 shows that if a model has a unit root, the ADF and KPSS tests are "good" about detecting it. In addition, if a test is designed to tell us whether a model contains a unit root, then *when there is not a unit root in the model, it should not detect one*. We state the obvious: All *stationary* AR(1) models do not have a unit root. But of course, there are many "stationary" AR(1) models. Table 7.5 shows the extent to which these tests can distinguish between ARIMA(0,1,0) models with unit roots and stationary AR(1) models with roots "close to" the unit circle. Table 7.5 is based on realizations of length $n = 200$ from the stationary AR(1) models with $\phi_1 = .9, .95, .99$, and .999. The percentage of times that the test (incorrectly) detected a unit root are shown in Table 7.5. Because these models do not have a unit root, we would hope that the percentages are small.

TABLE 7.5 Simulation Results Showing Percentage of 100 Realizations of length $n = 200$ from a Stationary AR(1) Model for which the Test *Incorrectly* Detected a Unit Root

	$\phi_1 = .999$	$\phi_1 = .99$	$\phi_1 = .95$	$\phi_1 = .9$
ADF Test	95%	95%	72%	17%
KPSS Test	89%	90%	76%	43%

QR 7.6 ADF and KPSS Type II Error

The results of Table 7.5 show that for stationary AR(1) models with $\phi_1 \geq .95$, the ADF and KPSS tests are very likely to *incorrectly* detect a unit root. Even for $\phi_1 = .9$ the tests detect a unit root substantially more than 5% of the time. Other results not tabled show that for $\phi_1 = .8$ the ADF test *never* detected a unit root, but the KPSS detected one 28% of the time. In summary, if these tests detect a unit root, the investigator cannot feel confident that the model actually contains a unit root!

Key Point:

1. Whenever a unit root test "detects" a unit root, the correct interpretation is that a model with unit roots is "plausible".
2. We do not recommend using the ADF, KPSS, or other similar tests as the *definitive* basis of a yes-or-no decision to include a unit root.
3. We emphasize again that the final decision should include physical considerations/domain knowledge (when it exists). See Example 7.1(a).

The Key Points above are disappointing, because it would have been ideal if the unit root tests would provide investigators with a decisive yes-or-no conclusion concerning whether to include a unit root in your model. *However, they do not!*

7.1.2.2 General Procedure for Fitting an ARIMA(p,d,q) Model to a Set of Time Series Data

Once the decision to include a unit root is made, the general procedure for fitting an $\text{ARIMA}(p,d,q)$ model is as follows:

1. *Difference the data:* Given that the decision has been made to include a unit root, difference the data. Suppose the data are in **x**. Then difference the data using the command

   ```
   d1=artrans.wge(x,phi.tr=1)
   ```

2. *Examine the differenced data:*
   If the model for the differenced data should include a unit root
       –   difference the data again
   If the differenced data appear to be stationary,
       –   then set $d = 1$ and include the factor $1 - B$ in the model
       –   go to step 4
3. *Continue to difference the data enough times to produce stationarity:*
       –   the number of differences required to "stationarize" the data is the estimated value of d.
       –   the factor $(1 - B)^d$ should be included in the final model.
4. *Fit an ARMA(p,q) model to the stationary data.*

Key Points:

1. Don't "overdo" the differencing.
2. These authors have never encountered real data for which $d = 3$ or more was appropriate.

Obtaining a Final ARIMA (p,d,q) Model

In this section we use the above general procedure for obtaining final ARIMA models for the datasets from Models (a), (b), and (c) in Section 7.1.1. In all three cases the issue of deciding whether borderline situations lead to stationary or nonstationary models will not be addressed. Nonstationary ARIMA models will be used in cases where they are suggested as *plausible* using such tools as the unit root tests and the damping behavior of the sample autocorrelations. Because these are simulated datasets, there are no physical or domain knowledge factors to consider. In Example 7.1(a) we consider these factors in our modeling of Dow Jones data.

Model (a): Final Model Fit to ARIMA(0,1,0) Data

Recall that the data in Figure 7.2(a) were generated using the command

```
x010=gen.arima.wge(n=100,d=1,sn=409)
```

The following analysis proceeds as if the true model is unknown. As previously noted, the data are wandering and the sample autocorrelations have a slowly damping exponential behavior. The dataset **x010** is the dataset we previously used to illustrate the ADF and KPSS tests. In both cases, the tests found a unit root. Assuming that the decision has been made to include a unit root in the model, the procedure is as follows:

1. ***Difference the data:***
   ```
   d1=artrans.wge(x010,phi.tr=1)
   ```
2. ***Examine the differenced data:*** The differenced data, sample autocorrelations, and Parzen spectral density estimate were shown in Figure 7.5 (a)–(c). It was previously noted that the differenced data appear to be stationary (and in fact looked like white noise). Therefore, we select $d = 1$, and thus the model contains a factor $1 - B$. Go to Step 4.
4. ***Fit an ARMA (p,q) model to the differenced data:*** As noted in Step 2, the differenced data appear to be uncorrelated, the sample autocorrelations are very close to zero and tend to stay within the 95% limit lines, and the Parzen spectral density estimate is fairly flat. These are characteristics of white noise data and indicate that no further analysis is needed (that is, we fit an ARMA(0,0) model to the data). The white noise variance is estimated by the variance of the differenced data using the base R command **var(d1)** from which we find that the variance of **d1** is 1.035. Consequently, because $d = 1$, we include a first difference, $1 - B$, term to the model, and the final model is

$$(1 - B)X_t = a_t, \text{ with } \hat{\sigma}_a^2 = 1.035.$$

Model (b): Final Model Fit to ARIMA(2,1,0) Data

The data in Figure 7.7(a) were generated using the command

```
x210=gen.arima.wge(n=200,phi=c(1.3,-.65),d=1,sn=4855)
```

The data are again analyzed as if the true model is unknown. As noted previously, the data and sample autocorrelations suggest the possibility of a unit root. Applying the ADF test yields $\tau = -2.56$ and $d_{.05} = -2.88$, so because $-2.56 \geq -2.88$, the ADF test does not reject the null hypothesis of a unit root. For the KPSS test, $kpss = 3.69$ and $k_{.05} = .463$. So, because $3.69 > .463$ the KPSS test detects a unit root. Assuming that we have decided to include a unit root in the model based on the available knowledge, we proceed as follows.

1. ***Difference the data:***
   ```
   d1=artrans.wge(x210,phi.tr=1)
   ```

2. ***Examine the differenced data:*** The differenced data, shown in Figure 7.7(d), appear to be stationary (but not white noise). (Also, the ADF and KPSS tests fail to detect a unit root.) Therefore, we select $d = 1$. Thus, the model contains a factor $1 - B$ Go to Step 4.

4. ***Fit an ARMA (p,q) model to the differenced data:*** The following command uses AIC to select ARMA model orders for the differenced data (**d1**) letting p range from 0 to 10 and q from 0 to 2.

```
aic.wge(d1,p=0:10,q=0:2)
```

AIC selects $p = 2$ and $q = 0$, so we estimate the parameters of the AR(2) model fit to the stationary differenced data using the command

```
est.ar.wge(d1,p=2)
```

The ML estimates for the differenced data are $\hat{\phi}_1 = 1.37$, $\hat{\phi}_2 = -.73$, and $\hat{\sigma}_a^2 = 1.036$. The final model is

$$(1 - 1.37B + .73B^2)(1 - B)X_t = a_t, \tag{7.3}$$

with $\hat{\sigma}_a^2 = 1.036$. This model is "similar" to the true model (Model (b)), and the factor table (not shown here) shows that the fitted model has system frequency $f = .10$ which is consistent with the features of the true AR(2) factor (see Table 7.1). Notice that because of the factor $1 - B$ in the model, we do not estimate the mean.

(4) Model (c): Final Model Fit to ARIMA(2,2,0) Data

The data in Figure 7.8(a) were generated using the following command, and again the data are analyzed as if the true model is unknown.

```
x220=gen.arima.wge(n=200,phi=c(1.3,-.65),d=2,sn=450)
```

The data in Figure 7.8(a) have a *dramatic* wandering behavior along with sample autocorrelations that damp slowly. This strongly suggests a unit root. Also, $\tau = -1.167$ and $d_{.05} = -2.88$, and since $-1.167 \geq -2.88$, the ADF test does not reject a unit root. Also, $kpss = 4.097$ and $k_{.05} = .463$, and because $4.097 > .463$, the KPSS test detects a unit root. Evidence suggests a unit root, so we proceed as follows.

1. ***Difference the data:***

```
d1=artrans.wge(x220,phi.tr=1)
```

2. ***Examine the differenced data:*** The differenced data, shown in Figure 7.8(d), have a wandering appearance and associated sample autocorrelations that are slowly damping. For the differenced data, $\tau = -3.878$ is less than $d_{.05} = -2.88$, so the ADF test rejects the null hypothesis of a unit root. However, $kpss = 1.060$ is greater than $k_{.05} = .463$ so the KPSS test rejects the null hypothesis of stationarity and concludes that the differenced data has a unit root. Despite the conflicting results, we decide to difference the data again.

3. ***Continue to difference the data enough times to produce stationarity:*** We difference the data again using the command

```
d2=artrans.wge(d1,phi.tr=1)
```

The twice differenced data are plotted in Figure 7.8(g) along with sample autocorrelations and spectral density in Figures 7.8 (h) and (i), respectively. These data appear to be stationary but not white noise. (The ADF and KPSS tests do not detect another unit root.) We conclude that $d = 2$, and thus that the model contains a factor $(1 - B)^2$.

4. **_Fit an ARMA (p,q) model to the twice differenced data:_** The following command uses AIC to select ARMA model orders for the differenced data (in **d2**) letting p range from 0 to 10 and q from 0 to 2.

```
aic.wge(d2,p=0:10,q=0:2)
```

AIC selects $p = 2$ and $q = 0$, so we use the following command to estimate the parameters of the AR(2) model fit to the stationary differenced data.

```
est.ar.wge(d2,p=2)
```

The ML estimates for the differenced data are $\hat{\phi}_1 = 1.33$, $\hat{\phi}_2 = -.69$, and $\hat{\sigma}_a^2 = 1.06$. The final model is

$$(1 - 1.33B + +.69B^2)(1 - B)^2 X_t = a_t,$$ (7.4)

with $\hat{\sigma}_a^2 = 1.06$. This model is very similar to the actual ARMA(2,2,0) model, and again, because the factor $(1 - B)^2$ is in the model, we do not estimate the mean.

Key Points:

1. If the original data appear to be stationary, then we do not difference the data and fit an ARMA(p,q) model to the data.
2. Each time we difference the data, we automatically add a $1 - B$ factor to the model.
3. For the data from Models (a)–(c), we "know" that the true models are ARIMA. Had these been real datasets for which we did not know a true model, then the most we could say is that ARIMA models with $d = 1,1$, and 2 are *plausible* models for datasets **x010**, **x210**, and **x220**, respectively.
4. As is the case throughout the book, running the code in these examples and obtaining the results builds confidence and momentum.
   – **_Run the code!_**

QR 7.7 Building an ARIMA Model

Example 7.1(a) DOW Jones Data

In Example 5.1 we considered dataset **dow1985** in *tswge* which is a record of the DOW Jones closing averages by month from March 1985 through December 2020. (Source: Yahoo! Finance). The **dow1985** data, sample autocorrelations, and Parzen spectral density estimate are shown in Figure 7.9(a), (b), and (c), respectively. The data have an upward wandering behavior and the sample autocorrelations have a slowly damping exponential appearance. The ADF and KPSS tests both suggest a unit root. The plots in Figure 7.9 can be obtained using the *tswge* commands

```
data(dow1985)
plotts.sample.wge(dow1985)
ddow=artrans.wge(dow1985,phi.tr=1,plot=FALSE)
plotts.sample.wge(ddow,arlimits=TRUE,speclimits=c(-10,10))
```

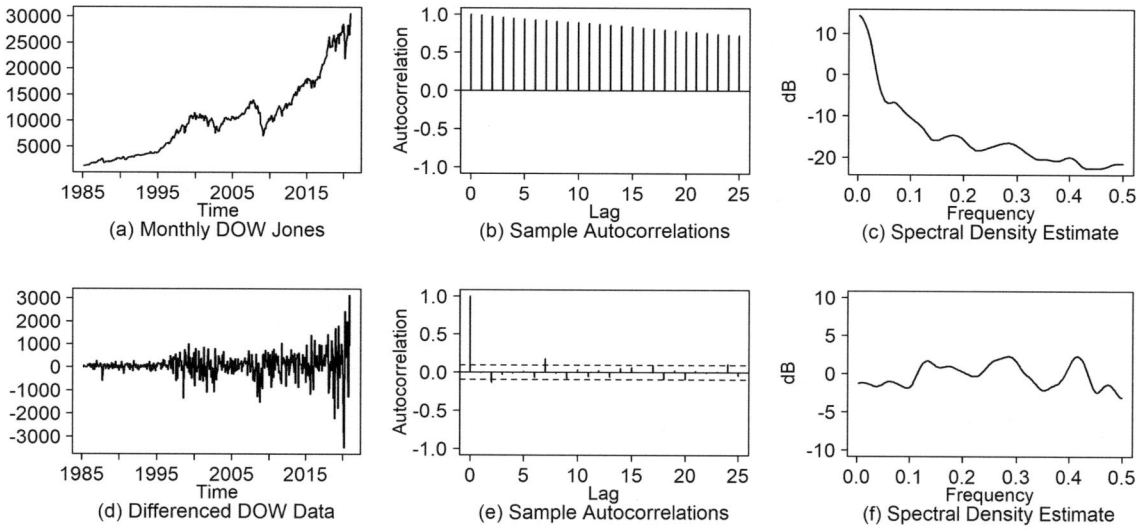

FIGURE 7.9 (a) DOW Jones closing average by month from March 1985 through December 2020 with (b) sample autocorrelations and (c) Parzen spectral density estimate for the data in (a), (d) DOW data in (a) differenced along with (e) sample autocorrelations and (f) Parzen spectral density estimate of the differenced data in (d).

Should We Use a Stationary or Nonstationary Model for DOW Data?

The Dow Jones data provide a good example concerning the decision whether to use a nonstationary or stationary model. The mean of the DOW data from 1985 through 2020 is $10,352. If we decided to fit a stationary model by simply estimating the parameters of an AR(1) fit to the data, the model would be $\left(1-.999B\right)\left(X_t - 10352\right) = a_t$. Notice that the mean term was included in the model because it is stationary. Use of the stationary model implies that the DOW will have some attraction to a mean value which is estimated by $\bar{X}=\$10,352$.

Example of Domain Knowledge:

Although there are dips, for example, around 2009, historically the stock market increases in the long term. Figure 7.10 is a plot of *tswge* dataset **dow.annual** which shows the historical annual closing averages from 1915–2020. There certainly seems to be an upward trend with no tendency to be attracted back to an overall mean. Consequently, for the 1985–2020 data modeled in this example, we select the nonstationary model, believing that there is no expectation for the future Dow Jones averages to be attracted back to earlier mean levels.[6]

---

6 The authors make this claim even though as we write this footnote on Friday, February 28, 2020, the DOW Jones average has dropped by about 3,500 points for the week over concerns about the coronavirus. We will keep this footnote unchanged, and by the time you read the footnote you will know whether the DOW continued to wander (mostly upward) or whether it turned downward and tended to be attracted to the mean (which is estimated in this dataset by $\bar{X} = \$10,352$). As investors, we hope the former is the case!

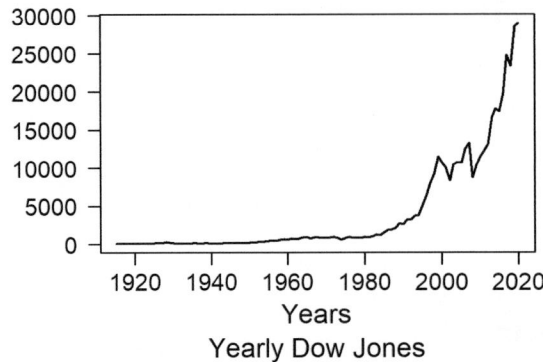

FIGURE 7.10   DOW Jones year-end averages for 1915–2020.

Back to the analysis of the **dow1985** data, we follow the steps used before:

1. ***Difference the data:***

   ```
   d1=artrans.wge(dow1985,phi.tr=1)
   ```

2. ***Examine the differenced data:*** The differenced data (**d1**) appear to be random but with an increasing variance in time. The sample autocorrelations in Figure 7.9(e) are all small (95% limit lines are added as a reference), and the spectral density in Figure 7.9(f) is fairly "flat". These suggest white noise except for the increased variability in time. At this point we will model the differenced data as white noise and further analysis is not needed. We conclude that $d = 1$ so that the model contains a factor of $1 - B$. Go to Step 4.

3. ***Fit an ARMA (p,q) model to the differenced data:*** The white noise variance is estimated by the variance of the differenced data. Using the Base R command **var(d1)** we find that the variance of the differenced data is 294,793. Consequently, because the first difference added a $1 - B$ term to the model, the final model is

$$(1 - B)X_t = a_t, \text{with } \hat{\sigma}_a^2 = 294,793.$$

Note: The increased variability is concerning. Consequently, it might be beneficial to take the log of the DOW data before analysis. See Problem 7.4(b).

Example 7.1(b) US Population Data

Figure 7.11(a) shows the estimated US population by year from 1900 through 2020 (shown here in millions of people). (Source: usafacts.org). These data are in *tswge* file **uspop** and have the appearance of nonstationary data, similar to the ARIMA(2,2,0) data in Figure 7.8(a) except for the fact that the population is increasing in Figure 7.11(a) while the data in Figure 7.8(a) are decreasing.

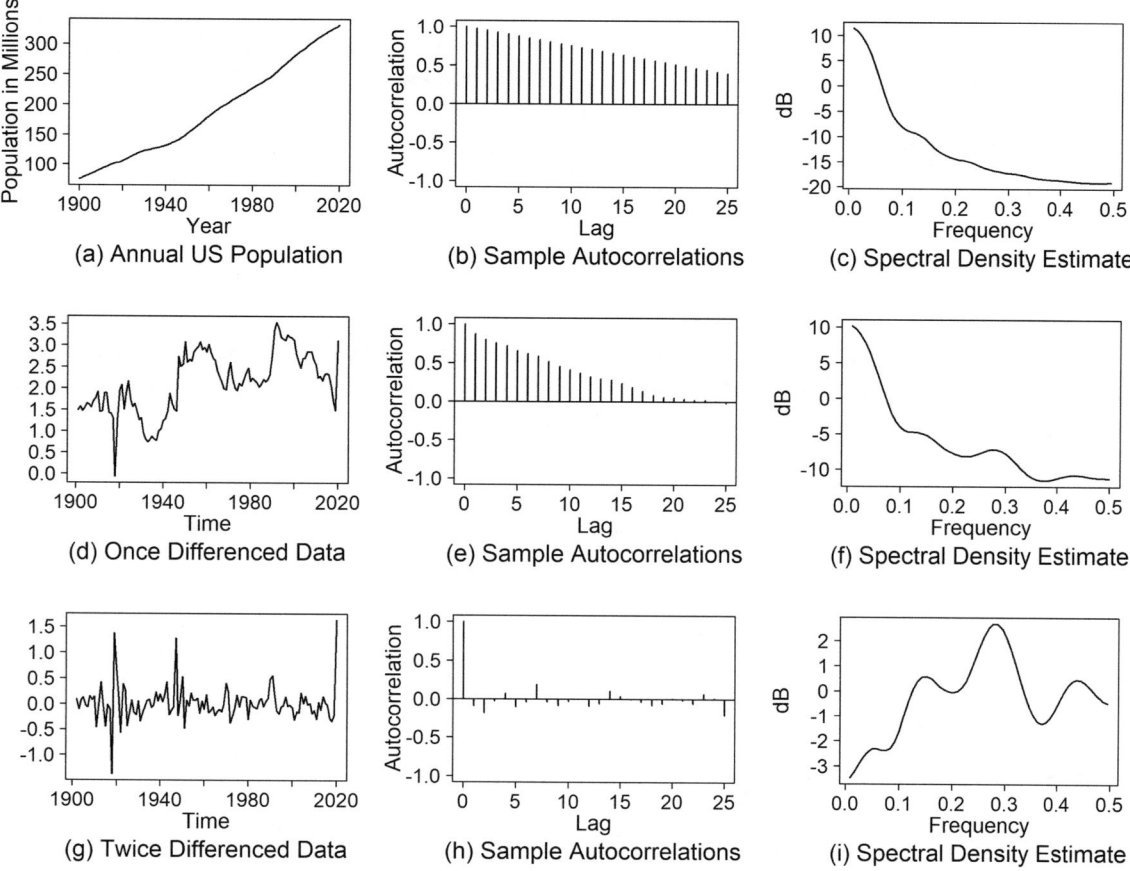

FIGURE 7.11 (a) US annual population data in millions, (b) sample autocorrelations and (c) Parzen spectral density estimate of data in (a); (d) data in (a) differenced, (e) sample autocorrelation, and (f) Parzen spectral density estimate of differenced data in (d); (g) data in (d) differenced, (h) sample autocorrelations, and (i) Parzen spectral density estimate of data in (g).

The slowly damping sample autocorrelations in Figure 7.11(b) suggest the need to difference the data. Also, $\tau = 1.956$ and $d_{.05} = -2.88$, and since $1.956 \geq -2.88$, the ADF test does not reject a unit root. Also, $k_{.05} = 2.496$ and $k_{.05} = .463$, and because $2.496 > .463$, the KPSS test detects a unit root. Evidence suggests a unit root, so we proceed as follows.

1. ***Difference the data:***

   ```
   d1.pop=artrans.wge(uspop,phi.tr=1)
   ```

2. ***Examine the differenced data:*** The differenced data are shown in Figure 7.11(d) along with sample autocorrelations and spectral density in Figures 7.11 (e) and (f), respectively. The differenced data have a wandering appearance and associated sample autocorrelations that are slowly damping. For the differenced data, $\tau = -2.324$ and $d_{.05} = -2.88$, so because $-2.324 \geq -2.88$ the ADF test fails to reject the null hypothesis of a unit root. Also, $kpss = 1.2759$ is greater than $k_{.05} = .463$, so the KPSS test rejects the null hypothesis of stationarity and concludes that the differenced data has a unit root. Consequently, we decide to difference the data again.

3. ***Continue to difference the data enough times to produce stationarity:*** We difference the data again using the command

```
d2.pop=artrans.wge(d1.pop,phi.tr=1)
```

The twice differenced data are plotted in Figure 7.11(g) along with sample autocorrelations and spectral density in Figures 7.11 (h) and (i), respectively. These data appear to be stationary and possibly white noise. We conclude that $d = 2$, and thus that the model contains a factor $(1 - B)^2$.

4. **Fit an ARMA (p,q) model to the differenced data:** Using the commands

```
aic.wge(d2.pop,p=0:10,q=0:4)
aic.wge(d2.pop,p=0:10,q=0:4,type='bic')
```

AIC selects $p = 0$ and $q = 2$ and BIC selects white noise ($p = 0, q = 0$). Based on the BIC and the fact that the sample autocorrelations tend to stay within the 95% limits, we decide to consider **d2.pop** to be white noise, and the final model is

$$(1 - B)^2 X_t = a_t,$$

with $\hat{\sigma}_a^2 = .113$. Again, because the factor $(1 - B)^2$ is in the model, we do not estimate the mean.

Note: The **d2.pop** data, having small sample autocorrelations, seems to have more variability early in the 20th century. The "rate of change" was greater with the smaller population. This is not unusual for "real" datasets for which we do not know the true model, but hopefully we have found a useful one. See Problem 7.4 for an alternative approach to analyzing the **uspop** data.

7.1.3 Forecasting with ARIMA Models

Because ARIMA models are basically limiting versions of ARMA models where roots of the characteristic equation are allowed to be on the unit circle, we quickly review forecasts using ARMA models.

7.1.3.1 ARMA Forecast Formula

The basic formula for forecasts from an $\text{ARMA}(p,q)$ model is

$$\hat{X}_{t_0}(\ell) = \sum_{j=1}^{p} \hat{\phi}_j \hat{X}_{t_0}(\ell - j) - \sum_{j=\ell}^{q} \hat{\theta}_j \hat{a}_{t_0+\ell-j} + \overline{x}\left[1 - \sum_{j=1}^{p}\hat{\phi}_j\right].$$

ARIMA Forecast Formula

For a fitted $\text{ARIMA}(p,d,q)$ model, $\hat{\phi}(B)(1 - B)^d X_t = \hat{\theta}(B)a_t$, the "AR" part is the $p + d$ order operator $\hat{\phi}^*(B) = \hat{\phi}(B)(1 - B)^d$ or in expanded form $\hat{\phi}^*(B) = 1 - \sum_{j=1}^{p+d}\hat{\phi}_j^* B^j$,

where $\hat{\phi}_j^*$ denotes the jth coefficient. For example, consider the ARIMA(2,1,0) model fit to the data in Figure 7.7(a). The model, given in (7.3) is $(1 - 1.25B + .64B^2)(1 - B)X_t = a_t$. In this case, $\hat{\phi}^*(B) = (1 - 1.25B + .64B^2)(1 - B)$, so by multiplying operators it follows that $\hat{\phi}^*(B) = 1 - 2.25B + 1.89B^2 - .64B^3$, and $\hat{\phi}_1^* = 2.25$, $\hat{\phi}_2^* = -1.89$, and $\hat{\phi}_3^* = .64$.

For the $\text{ARIMA}(p,d,q)$ case, the forecast formula analogous to the ARMA formula above is

$$\hat{X}_{t_0}(\ell) = \sum_{j=1}^{p+d} \hat{\phi}_j^* \hat{X}_{t_0}(\ell - j) - \sum_{j=\ell}^{q} \hat{\theta}_j \hat{a}_{t_0+\ell-j} + \overline{x}\left[1 - \sum_{j=1}^{p+d}\hat{\phi}_j^*\right] \tag{7.5}$$

Key Points:

1. The characteristic polynomial associated with the $p + d$ order operator can be written as:

    (a) $\hat{\phi}^*(z) = \hat{\phi}(z)(1-z)^d$

    (b) $\hat{\phi}^*(z) = 1 - \sum_{j=1}^{p+d} \hat{\phi}_j^* z^j$

2. Formula (a) shows that $\hat{\phi}^*(1) = \hat{\phi}(z)(1-1)^d = 0.$

3. From (2) and Formula(b) it follows that $\hat{\phi}^*(1) = 1 - \sum_{j=1}^{p+d} \hat{\phi}_j^* = 0.$

4. From (3) it follows that the last term in (7.5) is always zero if $d > 0$.

Based on Key Point (4) it follows that the last term in (7.5) is always zero if $d > 0$. So, the ARIMA(p,d,q) forecast formula in (7.5) can be simplified to

$$\hat{X}_{t_0}(\ell) = \sum_{j=1}^{p+d} \hat{\phi}_j^* \hat{X}_{t_0}(\ell - j) - \sum_{j=\ell}^{q} \hat{\theta}_j \hat{a}_{t_0 + \ell - j}. \tag{7.6}$$

For the model in (7.3) the forecasts become

$$\hat{X}_{t_0}(\ell) = 2.25 \hat{X}_{t_0}(\ell - 1) - 1.89 \hat{X}_{t_0}(\ell - 2) + .64 \hat{X}_{t_0}(\ell - 3)$$

Key Points:

1. From (7.6) it follows that the sample mean does not play a part, not even in the forecasts.
2. This is consistent with our expression of ARIMA models (which do not include a mean).

We continue our discussion of ARIMA forecasts with the ARIMA(0,1,0) model.

Forecasts for an ARIMA(0,1,0) Model

Figure 7.2(a) showed a realization of length $n = 100$ from an ARIMA(0,1,0) model with $\sigma_a^2 = 1$. The data in Figure 7.2(a) seem to be linearly decreasing, and because of this, a time series regression fit to the data would probably be assessed to have a negative, statistically significant slope. Based on examination of the data, our heuristic forecasts might be for the future values to continue to decrease. For the ARIMA(0,1,0) model, $p = 0$, $d = 1$, and $\hat{\phi}^*(B) = \hat{\phi}(B)(1-B)^d = 1 - B$. That is, $\hat{\phi}_1^* = 1$, $\hat{\phi}_j^* = 0$ for $j > 1$, and from (7.6) it follows that $\hat{X}_{t_0}(\ell) = \hat{X}_{t_0}(\ell - 1)$. Using this expression, $\hat{X}_{t_0}(1) = \hat{X}_{t_0}(0) = X_{t_0}$, $\hat{X}_{t_0}(2) = \hat{X}_{t_0}(1) = X_{t_0}$, and so forth.

The data from Figure 7.2(a) (and shown again in Figure 7.12(a)) and the forecasts shown in Figure 7.12(b) are obtained using the commands

```
x010=gen.arima.wge(n=150,d=1,sn=409)
fore.010=fore.arima.wge(x010[1:100],d=1,n.ahead=50,limits=FALSE)
```

Figure 7.12(b) forecasts show no tendency to move downward or toward some mean level. Instead, the best forecast of future values is the last value observed. These are very "boring" and "disappointing" forecasts, and they seem to not be "as good as we could do" without using a model. However, because the model $(1-B)X_t = a_t$ can be written as $X_t = X_{t-1} + a_t$, where a_t is zero-mean white noise, it follows that X_{t_0+1} is as likely to be greater than X_{t_0} as it is to be less than X_{t_0}. Consequently, there is no reason to predict that, for example, X_{t_0+1} should be above or below X_{t_0}. That is, $\hat{X}_{t_0}(1) = X_{t_0}$ is a reasonable predictor. Figure 7.12(b) also shows the 95% prediction limits obtained using the formula

$$\hat{X}_{t_0}(\ell) \pm 1.96 \hat{\sigma}_a \left\{ \sum_{k=0}^{\ell-1} \hat{\psi}_k^2 \right\}^{\frac{1}{2}},$$

where the ψ-weights are based on the coefficients from the "full" AR part. For the ARIMA(0,1,0) model, $\hat{\phi}^*(B) = 1 - B$, so the ψ-weights can be found using the command

```
psi.weights.wge(phi=1,lag.max=50)
```

The resulting ψ-weights are $\hat{\psi}_j = 1$ for each $j > 1$. Consequently, the series $\sum_{k=0}^{\ell-1} \hat{\psi}_k^2 = \sum_{k=0}^{\ell-1} 1 = \ell$, does not converge as $\ell \to \infty$, so the limits increase without bound. The 95% prediction limits in Figure 7.12(b) indicate that there is substantial uncertainty regarding future values.

The first 100 data values in **x010** were used to obtain the model, and to further examine the forecasts in Figure 7.12(b) we consider forecasts for X_{101} through X_{150}. The actual data values x_{101} through x_{150} are available in the full **x010** dataset. Figure 7.12(c) is a plot of Figure 7.12(b) along with the points **x010[101:150]**, and we see that in fact, the trend did not continue, and the forecasts were about as good as you could expect. The wide prediction limits indicate that forecasts very far into the future may not be very good. The trend seen in Figure 7.12(a) is a random trend, and cannot be "trusted" to continue into the future.

The plots in Figure 7.12 were obtained using the *tswge* commands

```
par(mfrow=c(1,3))
t=1:150
plot(t[1:100],x010[1:100],xlim=c(0,150), type='l')
fore.arima.wge(x010[1:100],d=1,n.ahead=50)
points(t[101:150],x010[101:150], type='l')
```

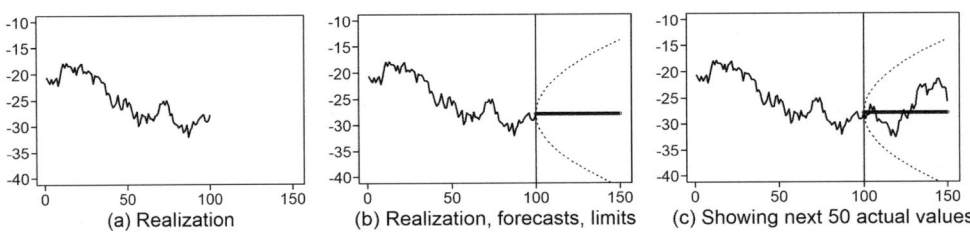

FIGURE 7.12 (a) Figure 7.2(a) showing realization of length $n = 100$ from an ARIMA(0, 1, 0) model, (b) realization in (a) along with first 50 step ahead forecasts and 95% limits, and (c) same as (b) showing the next 50 actual data values in the series x010.

Forecasts from an AR(1) Model Fit to x010

Although the **x010** dataset is a realization from the nonstationary ARIMA(0,1,0) model, in practice we will not "know" this. Referring back to the discussion about whether to include a unit root, suppose the decision was made to fit a stationary AR(1) model to the **x010** data shown again in Figure 7.12(a). Using the command

```
est.ar.wge(x010,p=1)
```

the estimated stationary model is

$$(1-.965B)(X_t + 25.56) = a_t, \tag{7.7}$$

with $\hat{\sigma}_a^2 = 1.02$. Because this is a stationary model, forecasts trend "upward" toward the sample mean, $\bar{x} = -25.56$. This tendency is visible in Figure 7.13(b). Figure 7.13(b) also shows the 95% prediction limits which are "tighter" than those in Figure 7.12(b). That is, if we believe the model is stationary, we are not as concerned that data will trend far away from the mean. Figure 7.13(c) adds the actual values as shown in Figure 7.12(c). Both sets of forecasts (those in 7.12 and 7.13) were "good". One measure of the forecast quality is that in both cases, all forecasts stayed within the 95% limits.

The plots in Figure 7.13 were obtained using the following code:

```
t=1:150
plot(t[1:100],x010[1:100],xlim=c(0,150), type='l')
fore.arma.wge(x010[1:100],phi=.965,n.ahead=50)
points(t[101:150],x010[101:150], type= 'l')
```

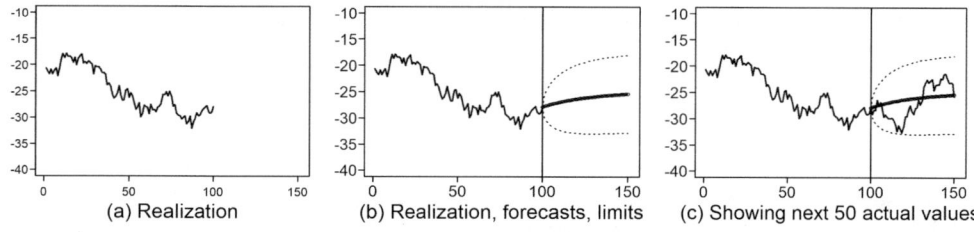

(a) Realization (b) Realization, forecasts, limits (c) Showing next 50 actual values

FIGURE 7.13 (a) Figure 7.12(a) showing realization of length $n = 100$ from an ARIMA(0,1,0) model, (b) realization in (a) along with first 50-step ahead forecasts and 95% limits based on fitting the stationary model in (7.7) to the data, and (c) same as (b) showing the next 50 actual data values in the series x010.

Comparing Forecasts from an AR(1) model fit to x010

The AR(1) model and the ARIMA(0,1,0) model discussed previously can be compared using the rolling window RMSE method introduced in Chapter 6. Below, the rolling window RMSE is calculated for both models again with a horizon of 50:

```
roll.win.rmse.wge(x010,horizon = 50,d = 1) #ARIMA(0,1,0)
```

Output: "Please Hold For a Moment, TSWGE is processing the Rolling Window RMSE with 97 windows."
"The Summary Statistics for the Rolling Window RMSE Are:"
Min. 1st Qu. Median Mean 3rd Qu. Max.
1.414 2.329 3.446 3.844 5.743 7.632
"The Rolling Window RMSE is: 3.844"

```
roll.win.rmse.wge(x010,horizon = 50, phi = .965) #AR(1)
```

Output: "Please Hold For a Moment, TSWGE is processing the Rolling Window RMSE with 99 windows."
"The Summary Statistics for the Rolling Window RMSE Are:"
Min. 1st Qu. Median Mean 3rd Qu. Max.
2.156   2.577   2.873   2.961   3.287   4.328
"The Rolling Window RMSE is: 2.961"

Interestingly, although the ARIMA(0,1,0) is the "right" model, the rolling window RMSE is lower for the AR(1) is 2.961 as compared to 3.844 for the ARIMA(0,1,0) model.

Key Points:

1. Trends seen in ARIMA realizations with $d = 1$ are random trends and a given realization may show time intervals with upward trending behavior along with other intervals of downward trending behavior.
   - Realizations can "turn on a dime."
2. If an ARIMA model with $d = 1$ is an appropriate model for a set of data, any observable trending behavior should not be "boldly" predicted to continue.
3. The forecasts depend *heavily* on the type of model selected.
   - Fitting an ARIMA model with $d = 1$ produces flat forecasts suggesting the trending might or might not continue.
   - Fitting a stationary ARMA model produces forecasts that move toward the sample mean.
   - Fitting a line-plus-noise model (Chapter 8) produces forecasts that that predict the existing trend to continue.

Forecasts for an ARIMA(2,1,0) Model

Figure 7.7(a) showed a realization of length $n = 200$ from the ARIMA(2,1,0) model with $\sigma_a^2 = 1$, referred to as Model (b) in Section 7.1. The data in Figure 7.7(a) from the ARIMA(2,1,0) Model (b) (shown again in Figure 7.14(a)) and the forecasts in Figure 7.14(b) (based on the fitted model in (7.3)) were found using the commands

```
x210=gen.arima.wge(n=250,phi= c(1.3,-.65),d=1,sn=4855)
fore.210=fore.arima.wge(x210,phi=c(1.25,-.64),d=1,n.ahead=50)
```

In Section 7.1.2 we fit the ARIMA(2,1,0) model $(1-1.25B+.64B^2)(1-B)X_t = a_t$, with $\hat{\sigma}_a^2 = 1.03$ to this dataset. Figure 7.14(b) shows the data in Figure 7.14(a) along with forecasts for the next 50 steps ahead. Close examination shows that the stationary AR(2) part of the model influences the first few forecasts, but that the forecasts quickly level out, in this case to -89.3. The last data value (that is, at $t = 200$) is -88.9. Consequently, the forecasts level off at a value that is neither the last value nor the sample mean of the first 200 values, $\bar{x} = -103.3$. Figure 7.14(c) shows the 50 forecasts along with forecast limits, and the next 50 actual data values in the simulated series. In this case the actual values decline rapidly while the forecasts are all close to the value at $t = 200$. The forecasts are not "good", but the data values stay within the 95% prediction limits.

Key Points:

1. Forecasts from am ARIMA(0,1,0) are simply a repetition of the last value
2. Short lead-time forecasts from an ARIMA($p,1,q$) model with either $p > 0$ or $q > 0$ adjust for the underlying ARMA(p,q) behavior but eventually level out to a "horizontal" forecast line that tends to be different from the last observed data value.

The plots in Figure 7.14 were produced with the following code

```
x210=gen.arima.wge(n=250,phi=c(1.3,-.65),d=1,sn=4855,plot=FALSE)
t=1:250
plot(t[1:200],x210[1:200],xlim=c(0,250),type='l',xlab='Time')
fore.210=fore.arima.wge(x210[1:200],phi=c(1.25,-.64),d=1,n.ahead=50)
fore.210=fore.arima.wge(x210[1:200],phi=c(1.25,-.64),d=1,n.ahead=50)
points(t[201:250],x210[201:250],type='l')
```

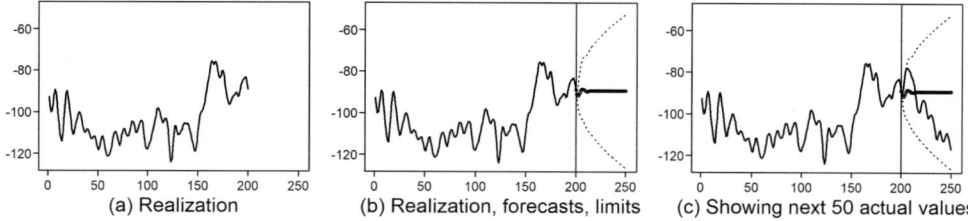

(a) Realization   (b) Realization, forecasts, limits   (c) Showing next 50 actual values

FIGURE 7.14   (a) Realization of length $n = 200$ from ARIMA(2,1,0) Model (b), (b) realization in (a) along with first 50 step ahead forecasts and 95% limits, and (c) same as (b) showing the next 50 actual data values in the series **x210**.

Forecasts for ARIMA (p,d,q) Models with d = 2.

We now return to the ARIMA(2,2,0) realization in Figure 7.8(a). These data are generated using the command

```
x220=gen.arima.wge(n=200,phi=c(1.3,-.65),d=2,sn=450)
```

The data follow a smooth curve which in the case of Figure 7.8(a) is almost linear. In contrast to the ARIMA models with $d = 1$, forecasts from an ARIMA model with $d = 2$ *do predict a current trend to continue.* It can be shown that the forecasts from an ARIMA(0,2,0) model follow a straight line that passes through the last two points in the realization. (See Woodward et al. (2017.)

The model (shown in Equation 7.4) fit to the data in Figure 7.8(a) is $(1-1.33B+.69B^2)(1-B)^2 X_t = a_t$, with $\hat{\sigma}_a^2 = 1.06$. The data and forecasts up to 50 steps ahead are shown in Figure 7.15(b) along with 95% prediction limits. Figure 7.15(c) shows the realization in Figure 7.15(b) along with the actual data values for times $t=201-250$. As can be seen, the forecasts are quite good.

To further illustrate the nature of realizations from the ARIMA(2,2,0) model, Figure 7.15(d) shows a realization of length $n = 2,000$. The data values to the left of the vertical line at $n = 200$ are those plotted in Figure 7.15(a). As can be seen from Figure 7.15(d), realizations from ARIMA models with $d = 2$ tend to wander, but "change course" much more slowly than in the case of $d = 1$. Figure 7.15(d) shows that for fairly short-term forecasts, the decision to predict the existing trend to continue is reasonable. Plots in Figure 7.15 are produced using the following *tswge* code

```
par(mfrow=c(2,2))
x220=gen.arima.wge(n=2000,phi=c(1.3,-.65),d=2,sn=450,plot=FALSE)
t=1:2000
plot(t[1:200],x220[1:200],xlim=c(0,250),type='l',xlab='Time')
fore.220=fore.arima.wge(x220[1:200],phi=c(1.33,-.69),d=2,n.ahead=50)
abline(v=200)
fore.220=fore.arima.wge(x220[1:200],phi=c(1.33,-.69),d=2,n.ahead=50)
points(t[201:250],x220[201:250],type= 'l',lwd=2,col='black')
```

```
abline(v=200)
plot(t,x220, type='l')
abline(v=200)
```

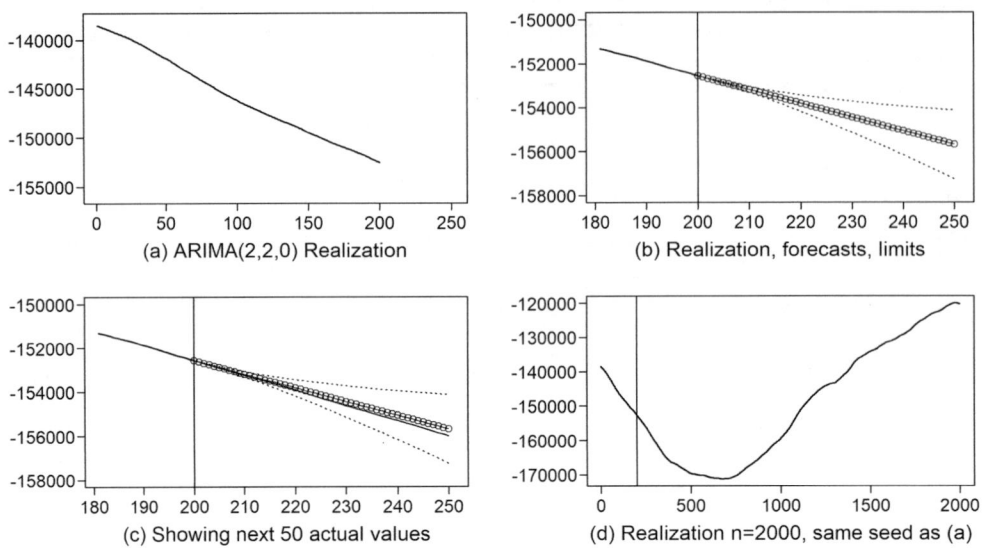

(a) ARIMA(2,2,0) Realization

(b) Realization, forecasts, limits

(c) Showing next 50 actual values

(d) Realization n=2000, same seed as (a)

FIGURE 7.15 (a) Figure 7.8(a) showing realization of length $n = 200$ from ARIMA(2,2,0) Model (c); (b) last 20 data values in (a) along with forecasts and forecast limits for times $t = 201$-250; (c) Same as (b) including true values (solid line) slightly below the forecasts; and (d) Realization of length $n = 2,000$ for which the first 200 points are plotted in Figure 7.15(a)

7.2 SEASONAL MODELS

Seasonal models are nonstationary models that are useful for modeling data such as monthly or quarterly sales data that exhibit repetitive behavior. For example, each year December sales may tend to be high while sales in certain other months are traditionally low. In this section we will discuss seasonal ARIMA models which are defined in Definition 7.3.

Definition 7.3: A seasonal ARIMA model is a model of the form

$$\phi(B)(1 - B^s)X_t = \theta(B)a_t,$$ (7.8)

where s is a non-negative integer, and where $\phi(B)$ and $\theta(B)$ are stationary and invertible operators of orders p and q, respectively (for which all roots of the characteristic equations $\phi(z) = 0$ and $\theta(z) = 0$ lie outside the unit circle).

Section 7.2.1 discusses the properties of the seasonal models, Section 7.2.2 illustrates how to fit these models to data, and Section 7.2.3 discusses forecasting using seasonal models.

7.2.1 Properties of Seasonal Models

The variable s is the length of the season. For example, for quarterly sales data which exhibit seasonal behavior, $s = 4$, while seasonal monthly data would be associated with $s = 12$. For the model in (7.8) to be a seasonal model, we require that $s > 1$. For example, if $s = 0$ or $s = 1$, then the model in (7.8) simplifies to an $\text{ARMA}(p,q)$ and an $\text{ARIMA}(p,1,q)$ model, respectively, neither of which is a seasonal model.

7.2.1.1 Some Seasonal Models

In the following examples we will examine the properties of the seasonal models below which will be referred to as Models (A) and (B).

(A) $\left(1 - B^s\right)X_t = a_t$: we consider the cases $s = 4$ and $s = 12$.
(B) $\left(1 - 1.3B + .65B^2\right)\left(1 - B^4\right)X_t = a_t$

(1) Seasonal Model (A): $(1 - B^s)X_t = a_t$

Model (A) is a "purely" seasonal model because it has no other autoregressive or moving average components. In this example the focus will be on $s = 4$ and $s = 12$.

Models with $s = 4$ will be referred to as models for quarterly data. For such data, quarters one through four correspond to year one, quarters five through eight are year two, and so forth.[7] The model $\left(1 - B^4\right)X_t = a_t$ can be written as $X_t - X_{t-4} = a_t$, or in the more instructive form

$$X_t = X_{t-4} + a_t. \tag{7.9}$$

Equation 7.9 shows that the value of the process at time t is equal to the value at time $t - 4$ plus a random noise component. A realization of length $n = 72$ from (7.9) was generated using the ***tswge*** command

```
xA4=gen.arima.wge(n=72,s=4,mu=50,sn=52)
```

The first 24 data values in this realization are shown in Figure 7.16(a). The dotted vertical lines separate the data into years. For convenience of discussion we will refer to the data as some measure of sales in thousands of dollars. Notice that in the first year, the first quarter ($t = 1$) sales tend to be about 18 ($18k) with an increase in quarter two to about $42k, followed by slight drops in the third and fourth quarters to sales in the range of $36k each. The fifth quarter (that is, the first quarter in year two) drops to about $18k as in the first year, and the general behavior is repeated for the following five years of data in the plot.

Figure 7.16(b) shows the first two years of a different set of monthly sales data using

```
xA12=gen.arima.wge(n=72,s=12,mu=50,sn=100)
```

From the plot it is seen that monthly sales tend to be highest in January, February, June, and November with the lowest sales in April, May, and July. (Strange sales pattern!!) The plots in Figure 7.16 plots were obtained using the commands

```
xA4=gen.arima.wge(n=72,s=4,mu=50,sn=52)
xA12=gen.arima.wge(n=72,s=12,mu=50,sn=100)
plot(xA4[1:24],type='l',xlab='Time')
abline(v=4.5,lty=2);abline(v=8.5,lty=2);abline(v=12.5,lty=2)
```

---

7 While quarterly data based on "3-month quarters" is a very common application for seasonal data with $s = 4$, a "quarter" need not have this interpretation and could be appropriate for data recorded at other time intervals.

```
abline(v=16.5,lty=2);abline(v=20.5,lty=2)
plot(xA12[1:24],type='l',xlab='Time')
abline(v=12.5,lty=2)
```

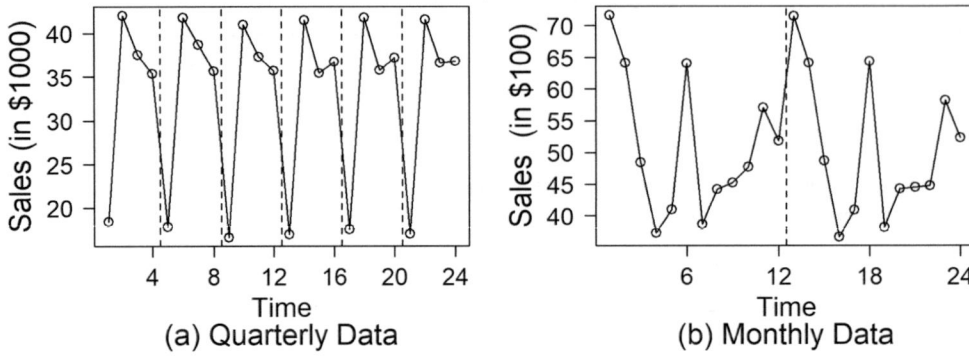

FIGURE 7.16   Realizations of length 24 from the seasonal Model (A) (a) with $s = 4$, and (b) with $s = 12$.

Key Point: For both $s = 4$ and $s = 12$, the seasonal nature within a year is not sinusoidal nor does it seem to follow any such mathematical pattern. However, the sales within a year do follow a particular, possibly irregular pattern, and the "seasonal" aspect of the data is the fact that this pattern is approximately repeated from "year to year".

(a) Factor Tables for Purely Seasonal Models with $s = 4$ and $s = 12$

In this section we show the factor tables for the purely seasonal models of order $s = 4$ and $s = 12$. Before showing the factor table for $s = 4$ we note that the associated characteristic polynomial is $1 - z^4$. From algebra, recall that $1 - z^4$ can be factored as $(1 - z^2)(1 - z^2) = (1 - z)(1 - z)(1 - z^2)$. Consequently, $(1 - B^4)X_t = a_t$ can be factored as

$$\left(1 - B^4\right)X_t = \left(1 - B\right)\left(1 + B\right)\left(1 + B^2\right)X_t = a_t. \tag{7.10}$$

The factors in (7.10) are the three factors in the associated factor table. To find the factor table, we issue the *tswge* command

```
factor.wge(phi=c(0,0,0,1))
```

or

```
factor.wge(phi=c(rep(0,3),1))
```

because the polynomial $1 - z^4$ has zero coefficients for z, z^2, and z^3 and the coefficient on z^4 is $+1$. (Remember the treatment of signs for autoregressive and moving average coefficients.) The resulting factor table is given in Table 7.6, where it can be seen that $(1 - B^4)X_t = a_t$ is associated with system frequencies of 0, .25, and .5 and also that all of the roots are on the unit circle.

TABLE 7.6   Factor Table for $\left(1 - B^4\right)X_t = a_t$

AR-FACTOR	ROOTS	$|r|^{-1}$	f_0
$1 - B$	1	1	0

AR-FACTOR	ROOTS	$|r|^{-1}$	f_0
$1 + B^2$	$\pm i$	1	.25
$1 + B$	-1	1	.50

Factoring the characteristic polynomial $1 - z^{12}$ is more complicated and we use the **factor.wge** command for this purpose. The factor table for the purely seasonal model $(1 - B^{12})X_t = a_t$ is obtained using the command

```
factor.wge(phi=c(0,0,0,0,0,0,0,0,0,0,0,1))
```

or

```
factor.wge(phi=c(rep(0,11),1))
```

and is given in Table 7.7.

QR 7.8 Factor Table and Spectral Density of Model with $1 - B^s$

TABLE 7.7   Factor Table for $(1 - B^{12})X_t = a_t$

AR-FACTOR	ROOTS	$|r|^{-1}$	f_0
$1 - B$	1	1	.000
$1 - 1.732B + B^2$	$.866 \pm .5i$	1	.083
$1 - B + B^2$	$.5 \pm .866i$	1	.167
$1 + B^2$	$\pm i$	1	.250
$1 + B + B^2$	$-.5 \pm .866i$	1	.333
$1 + 1.732B + B^2$	$-.866 \pm .5i$	1	.417
$1 + B$	-1	1	.500

These two factor tables will be very useful as we model data and decide whether to include a seasonal factor in the model.

Key Points:

1. The factor tables for $(1 - B^4)X_t = a_t$ and $(1 - B^{12})X_t = a_t$ show that all roots of the associated characteristic equations are on the unit circle. Consequently, these models are nonstationary. This is true for $(1 - B^s)X_t = a_t$ for any positive integer s.
2. Quarterly data is not the same as $s = 4$ and monthly data does not necessarily indicate that the use of a model with $s = 12$ is warranted. See (3) below.
3. Factor Tables 7.6 and 7.7 show the factors associated with the seasonal models $(1 - B^4)X_t = a_t$ and $(1 - B^{12})X_t = a_t$, respectively. If we have quarterly or monthly data, then the factor table associated with a model fit to the data *should have factors similar to those in the above corresponding factor tables in order to be justified to include the seasonal factor* $1 - B^4$ or $1 - B^{12}$ *in the model*. We will return to this in Section 7.2.2 where we discuss model identification.(4)

(b) Autocorrelations and Spectral Density for $(1 - B^4)X_t = a_t$ *and* $(1 - B^{12})X_t = a_t$

Consider the quarterly seasonal model $(1 - B^4)X_t = a_t$ and note again that the realization in Figure 7.16(a) shows a very strong correlation between data values that are four quarters apart. That is, $\hat{\rho}_4$ should be strongly positive. Also, sales for the same quarter two years apart should also be positively correlated. That is, we should also expect $\hat{\rho}_8$ to be strongly positive. In general, $\hat{\rho}_{4m}$ will tend to be positive for integers m. Also note that when the seasonal model $(1 - B^4)X_t = a_t$ is written in the form $X_t = X_{t-4} + a_t$, it is clear that the model does not "say anything" about the relationship between, for example, X_t and X_{t+1}. Consequently, the sample autocorrelations, $\hat{\rho}_k$, for integers k that are not multiples of four would be expected to be smaller in magnitude than those for which k is a multiple of four. Using the same logic, for the monthly seasonal model $(1 - B^{12})X_t = a_t$, sample autocorrelations $\hat{\rho}_k$ where k is an integer multiple of 12 will tend to be positive and larger in magnitude than the other sample autocorrelations.[8]

Figure 7.17(a) is a plot of a realization of length $n = 72$ from a purely seasonal model with $s = 4$. Note that the data in Figure 7.16(a) are simply the first 24 data values of the realization in Figure 7.17(a). The sample autocorrelations and Parzen spectral density estimate for the data in Figure 7.17(a) are shown in Figures 7.17(b) and (c), respectively. The quarterly behavior can be seen in Figure 7.17(a), but it is not as obvious as it was in the first 24 values shown in Figure 7.16(a). The sample autocorrelations at lags 4, 8, 12, ... are "large" and positive which is consistent with the above discussion. Also, the spectral density has peaks at the system frequencies (see Table 7.6), $f = 0, .25,$ and $.5$.

Figure 7.17(d) is a realization of length $n = 72$ from a purely seasonal model with $s = 12$, and for which the first 24 values are shown in Figure 7.16(b). As would be expected, the sample autocorrelations in Figure 7.17(e) at lags $k = 12, 24$, and so forth are large and positive while the peaks in the Parzen spectral density estimate in Figure 7.17(f) correspond to the system frequencies in Table 7.7 (although the peaks at $f = 0$ and $.5$ are weak). The plots in Figure 7.16 can be obtained using the commands

```
xA4=gen.arima.wge(n=72,s=4,mu=50,sn=52)
xA12=gen.arima.wge(n=72,s=12,mu=50,sn=100)
plotts.sample.wge(xA4)
plotts.sample.wge(xA12,trunc=32)
```

> **Key Point:** In practice, the main feature of sample autocorrelations for seasonal data is that autocorrelations at multiples of 4 for $s = 4$ (or of 12 for $s = 12$) tend to be positive and larger in magnitude than sample autocorrelations at lags that are not multiples of s.

---

8 The seasonal model $(1 - B^s)X_t = a_t$ is a nonstationary model, and as in the ARIMA case, the autocorrelations are defined as "limiting autocorrelations". It can be shown that the limiting autocorrelations for $(1 - B^s)X_t = a_t$ are $\rho_0 = \rho_s = \rho_{2s} = \cdots = 1$; that is, $\rho_{ks} = 1$ for all integer k, positive or negative. The model-based autocorrelations for lags that are not integer multiples of s are equal to zero.

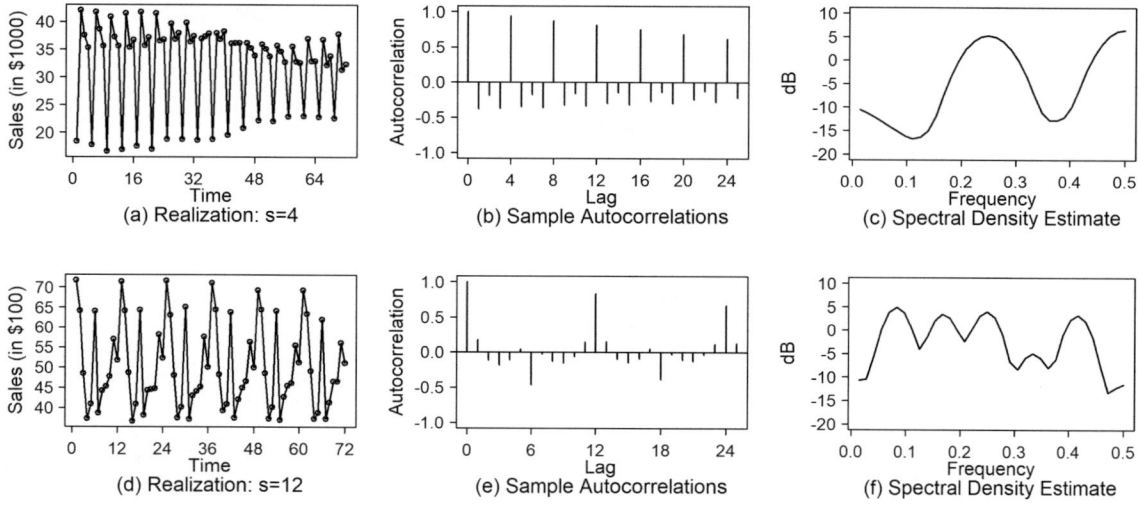

FIGURE 7.17 (a) Realization of length n = 72 from the model $(1 - B^4)X_t = a_t$; (b) sample autocorrelations; (c) Parzen spectral density estimate; (d) Realization of length n=72 from the model $(1 - B^{12})X_t = a_t$; (e) sample autocorrelations; (f) Parzen spectral density estimate with M=32.

Seasonal Model (B): $(1 - 1.3B + .65B^2)(1 - B^4)X_t = a_t$

Seasonal Model (B) not only has the seasonal factor $(1 - B^4)$ but also contains the stationary autoregressive factor, $(1 - 1.3B + .65B^2)$, which is the same AR(2) component used in the previous ARIMA examples. Recall that in the ARIMA case, the factors $(1 - B)^d$ dominated the behavior associated with the stationary factors. In the case of seasonal Model (B), the nonstationary factor is $(1 - B^4)$. The command

```
xB=gen.arima.wge(n=100,s=4,phi= c(1.3,-.65),sn=290)
```

generates a realization from Model (B). Figures 7.18(a−c) display the resulting realization, sample autocorrelations, and Parzen spectral density estimate. Again, for convenience of discussion, we assume the simulated data are sales (in $1,000). The realization in Figure 7.18(a) is of length $n = 100$ (that is, 25 years). The quarterly behavior of the data is not very obvious, although throughout the realization the first quarter sales tend to be much lower than those in the other three quarters. The main thing to notice about the sample autocorrelations is that they are strongly positive at lags that are multiples of four. It will often be the case that sample autocorrelations at lags say, $k = 2,6,\ldots$ may not be very close to zero but will tend to be smaller in magnitude than those at multiples of four.

Recall that in order to reveal the effect of the stationary factors in the ARIMA data we needed to difference the data. In the seasonal case, we apply a *seasonal difference* $Y_t = X_t - X_{t-s}$, in order to remove the dominating nonstationary behavior. In the $s = 4$ case, seasonally differencing the data in Figure 7.18(a) yields $Y_t = X_t - X_{t-4} = (1 - B^4)X_t$. Notice that because $(1 - 1.3B + .65B^2)(1 - B^4)X_t = a_t$, then $Y_t = (1 - B^4)X_t$ satisfies the stationary AR(2) model $(1 - 1.3B + .65B^2)Y_t = a_t$. The seasonal transformation is accomplished using the *tswge* command

```
sB=artrans.wge(xB,phi.tr=c(0,0,0,1))
```

The seasonally transformed data are shown in Figure 7.18(d) along with the associated sample autocorrelations and Parzen spectral density estimates in Figures 7.18 (e) and (f), respectively. The cyclic behavior has a cycle length of about 10 time units (quarters in our case). The transformed data and their characteristics appear similar to those shown in Figure 7.6 for the associated stationary AR(2) model, but these features were not apparent in the original dataset, **xB**.

The plots in Figure 7.17 were obtained using the following code

```
xB=gen.arima.wge(n=100,s=4,phi= c(1.3,-.65),sn=290)
sB=artrans.wge(xB,phi.tr=c(0,0,0,1),plot=FALSE)
plotts.sample.wge(xB)
plotts.sample.wge(sB)
```

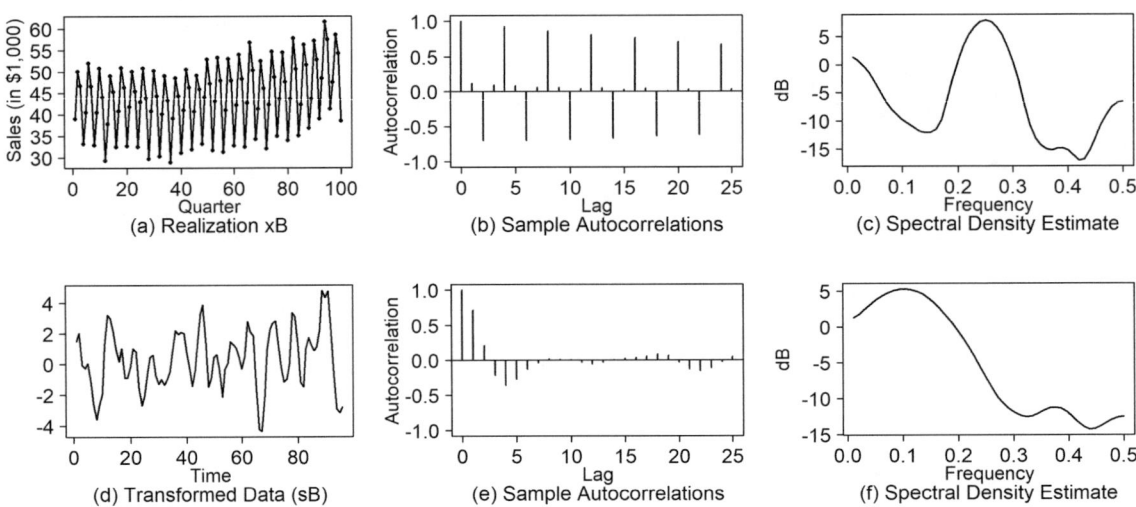

FIGURE 7.18   (a) Realization (xB) of length n=100, (b) sample autocorrelations, and (c) Parzen spectral density estimate from Model B: $\left(1-1.3B+.65B^2\right)\left(1-B^4\right)X_t = a_t$, (d) data in (a) transformed by $\left(1-B^4\right)$, (e) sample autocorrelations, and (f) Parzen spectral density estimate of the data in (d).

The Airline Model

We return to the Base R **AirPassengers** dataset which was discussed at length in Chapters 1 and 2. In Chapter 2 we produced Holt-Winters forecasts for the **AirPassengers** data and for the logarithms of the **AirPassengers** data. Recall that the two datasets differ in that the within-year variability increased in the raw **AirPassengers** data while this variability was fairly constant in the log data. See Figure 2.7. The log data are shown in Figure 7.19(a) along with the associated sample autocorrelations in Figure 7.19(b) and the Parzen spectral density estimate in Figure 7.19(c). The plots in Figure 7.19 are obtained using the commands

```
data(AirPassengers)
logAP=log(AirPassengers)
plotts.sample.wge(logAP)
```

Key Points:

1. In the introduction to this chapter it was noted that a violation of stationarity occurs if the process variance changes with time.
2. The logarithm of the **AirPassengers** data is taken so that the variance doesn't change with time.

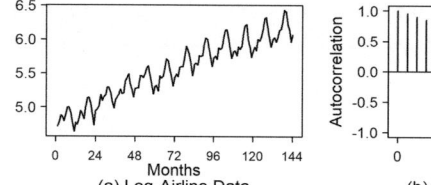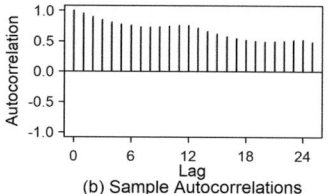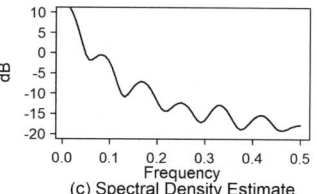

FIGURE 7.19   (a) Log Air Passengers data, (b) sample autocorrelations of the data in (a), and (c) Parzen spectral density estimate.

Notice that the **AirPassengers** data are like the seasonal dataset in Figure 7.17(d) but have the additional feature that there is a strong trending behavior. Also, the sample autocorrelations have slight "bumps" at lags 12 and 24, but they are dominated by the appearance of a slowly damping exponential behavior. The Parzen spectral density estimate has peaks at the system frequencies in Table 7.7 but the peaks at and near $f = 0$ are accentuated. Many analysts have fit models to the log **AirPassengers** data of the form:

$$\phi(B)(1-B)\left(1-B^{12}\right)X_t = \theta(B)a_t. \tag{7.11}$$

In fact, models containing $(1-B)\left(1-B^{12}\right)$ are sometimes referred to as "Airline Models". Note that the system frequencies for $1-B^{12}$ shown in Table 7.7 contain a factor, $1-B$, so that model (7.11) has two factors of $1-B$. Consequently, the behavior of data from the Airline Model would be expected to have seasonal behavior along with trending behavior such as was seen in the realization in Figure 7.8(a) from a model containing $(1-B)^2$. Models containing the factors $(1-B)\left(1-B^{12}\right)$ will be discussed in more detail when we discuss forecasting using nonstationary models.

7.2.2 Fitting Seasonal Models to Data

In this section we discuss the issues involved with fitting a seasonal model to a time series realization. As mentioned previously, in seasonal datasets the nonstationary component is the seasonal factor, $1-B^s$. The procedure for finding models for seasonal data is a generalization of the technique used for ARIMA models:

(1) *Select the value of s:* That is, determine whether a seasonal factor is appropriate for the data, and if so select the value of s.

(2) *Seasonally difference the data:* For example, if $s = 4$, then seasonally difference the data to obtain $Y_t = \left(1-B^4\right)X_t = X_t - X_{t-4}$. Letting **x** denote the original data, then the seasonally differenced data can be obtained using the command

```
y=artrans.wge(x,phi.tr=c(0,0,0,1))
```

(3) *Fit an ARMA(p,q) model to the seasonally differenced data:* Note that for data satisfying the Airline Model, the seasonally differenced data, Y_t, may show evidence of a unit root, in which case the Y_t (the seasonally differenced data) would need to be differenced.

Step 1, that is, determining whether a seasonal factor is required and if so, identifying the seasonal order, s, is the most difficult step in this procedure. We highly recommend the overfitting procedure described below for this purpose.

7.2.2.1 Overfitting

As mentioned previously, simply because data are collected monthly does not imply that a seasonal factor, $1 - B^{12}$, should be included in the model. Recall that the factor tables in Tables 7.6 and 7.7 show the underlying 1st- and 2nd-order factors of the polynomials $1 - B^4$ and $1 - B^{12}$, respectively. Although $s = 4$ and $s = 12$ are the most common seasonal factors, similar factor tables could be obtained for any positive integer value of s.

The strategy of overfitting is based on a result due to Tiao and Tsay (1983) that loosely says that if a high order AR(p) model is fit to a realization from a nonstationary model with roots on the unit circle, then factors associated with "nonstationary factors" will "show up" in the AR factor tables for autoregressive fits to the data for a wide range of autoregressive orders. Specifically, overfitting involves fitting several "high" order AR models to a set of data to identify possible nonstationarities. For a seasonal model with $s = 4$ there are three factors associated with roots on the unit circle: $1 - B$, $1 + B$, and $1 + B^2$. The Tiao-Tsay result says that high order AR fits to data from the model $\phi(B)(1 - B^4)X_t = \theta(B)a_t$ should consistently tend to show factors close to these three factors. Woodward et al. (2017) recommend fitting "high order AR models" using Burg estimates. Recall that Burg estimates always produce a stationary model, so we will be looking for factors "close" to the three seasonal factors in the $s = 4$ case. Our strategy will be to use Burg estimates for overfitting.[9] The overfitting procedure will be illustrated as we fit seasonal models below.

Model (A): Final Model for Data in Figure 7.17(d)
Consider again the data in Figure 7.17(d) which were generated from the seasonal model

$$\left(1 - B^{12}\right)X_t = a_t. \tag{7.12}$$

where $\sigma_a^2 = 1$. These data were generated using the command

```
xA12=gen.arima.wge(n=72,s=12,mu=50,sn=100)
```

(1) ***Select the value for s:*** The overfitting procedure involves fitting a few high order *autoregressive models* to the data and examining the factor tables. Table 7.8 shows the factor table for an AR(16) fit to the data. Note that if the data are monthly, and because we suspect that a 12th-order seasonal factor is appropriate, then the AR order of the fit to the data should be larger than 12. The command for obtaining the overfit factor table in Table 7.8 is[10]

```
est.ar.wge(xA12,p=16,type='burg')
```

---

9 ML estimates could also be used, but there are cases in which the ML estimation procedure does not converge.

10 Note that the factor table is returned in the output from **est.ar.wge** so we did not need to use **factor.wge**.

TABLE 7.8 $p = 16$ Overfit Factor Tables for Realization from $(1 - B^{12})X_t = a_t$ AR(16) Overfit

AR-FACTOR	ROOTS	$\lvert r \rvert^{-1}$	f_0
$1 - 1.736B + .9996B^2$	$.868 \pm .497i$	.9998	.083
$1 + 1.723B + .9992B^2$	$-.862 \pm .507i$	.9996	.415
$1 - .012B + .999B^2$	$.006 \pm 1.001i$	.9995	.249
$1 + .976B + .993B^2$	$-.491 \pm .875i$	.997	.331
$1 - .964B + .993B^2$	$.485 \pm .878i$	.997	.167
$1 - .991B$	-1.009	.991	.500
$1 - .981B$	1.019	.981	.000
$1 - .906B + .736B^2$	$.616 \pm .990i$	.858	.161
$1 + .816B + .333B^2$	$-1.225 \pm 1.225i$	.577	.375

Take a few minutes to compare the first seven rows of Table 7.8 (shown in bold type) with Table 7.7. Note that Table 7.7 shows the actual factored version of $1 - B^{12}$ while Table 7.8 is the factor table associated with an AR(16) fit to the data in Figure 7.17(d). Notice how the first seven lines of Table 7.8 closely approximate Table 7.7 (the factors are in different orders in the two tables). Notice also that the other two factors in Table 7.8 are much further from the unit circle than are the "seasonal" factors in the first seven lines of the table. If you have actual monthly data, and if the associated overfit factor table is similar to the one in Table 7.7, then a seasonal model with $s = 12$ is a reasonable model. Overfitting in this setting usually involves looking at factor tables for a variety of values for $p > 12$. For example, factor tables for $p = 14, 15,$ and 17 (not shown – but you should check them out) exhibit similar patterns.[11]

(2) ***Seasonally difference the data:*** Based on the previous discussion, we select $s = 12$ and seasonally difference the data, which adds a factor of $1 - B^{12}$ to the model. The command to seasonally difference the data in **xA12** is

```
sA12=artrans.wge(xA12,phi.tr=c(rep(0,11),1))
```

(3) ***Fit an ARMA(p,q) model to the seasonally differenced data:*** For convenience, Figure 7.17(d–f) is repeated as Figure 7.20(a–c). These show the data in **xA12**, along with the corresponding sample autocorrelations and Parzen spectral density estimate. The seasonally differenced data, **sA12**, are shown in Figure 7.20(d) along with the sample autocorrelations and Parzen spectral density estimate in Figures 7.20(e) and (f), respectively. The seasonally differenced data appear uncorrelated, the sample autocorrelations are very close to zero and tend to stay within the 95% limit lines,and the Parzen spectral density estimate is fairly flat. These are characteristics of white noise data which indicate that no further analysis is needed. This is consistent with the fact that the true model is $(1 - B^{12})X_t = a_t$, in which case the seasonally differenced data, $Y_t = (1 - B^{12})X_t$ follows the model $Y_t = a_t$. That is, the seasonally differenced data are from a white noise model. The white noise variance is estimated by the variance of the differenced data using the Base R command **var(sA12)** from which we find that the sample variance of **sA12** is .97. Consequently, because the seasonal difference added a $1 - B^{12}$ term to the model, the final model is

$$(1 - B^{12})X_t = a_t,$$

with $\hat{\sigma}_a^2 = .97$, which is very similar to the true model. The plots in Figure 7.20 can be obtained using the code

---

11 Factor tables of excessively high order (say 25 or more) will begin to bring several roots closer to the unit circle.

```
xA12=gen.arima.wge(n=72,s=12,mu=50,sn=100)
sA12=artrans.wge(xA12,phi.tr=c(rep(0,11),1),plot=FALSE)
plotts.sample.wge(xA12, arlimits=TRUE,speclimits=c(-20,10),trunc=30)
plotts.sample.wge(sA12, arlimits=TRUE,speclimits=c(-20,10),trunc=30)
```

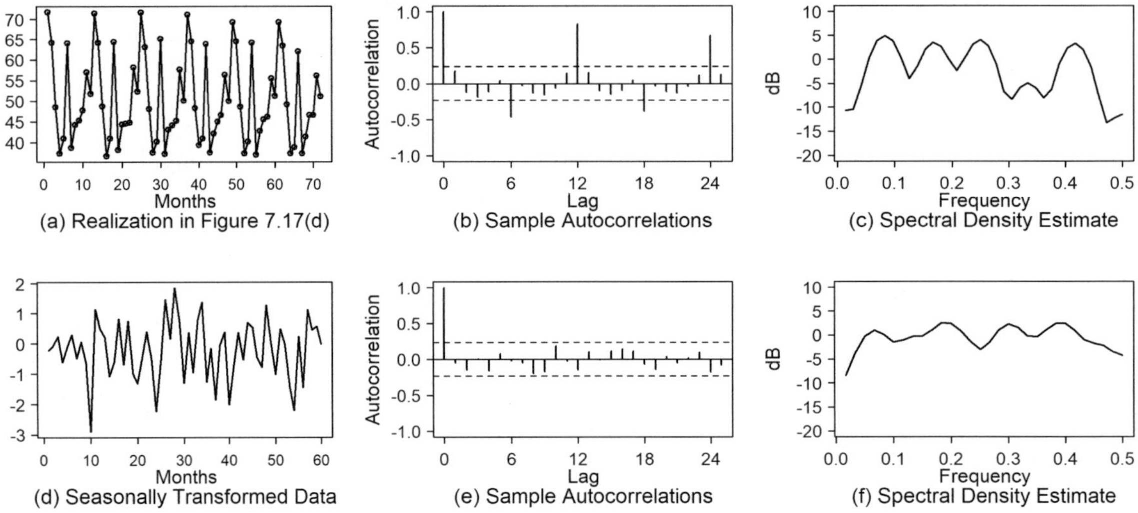

FIGURE 7.20 (a) Data in Figure 7.17(d), (b) sample autocorrelations of data in (a), (c) Parzen spectral density of data in (a) with M=32, (d) data in (a) seasonally transformed with $s = 12$, (e) autocorrelations of the data in (d), and (f) Parzen spectral density estimate of data in (d) with M=32.

Model (B): Final Model for Data in Figure 7.18(a)

Consider again the data in Figure 7.18(a) which were generated from the seasonal model

$$\left(1-1.3B+.65B^2\right)\left(1-B^4\right)X_t = a_t.$$

where $\sigma_a^2 = 1$. The realization in Figure 7.18(a) was generated using the command

```
xB=gen.arima.wge(n=100,phi=c(1.3,-.65),s=4,sn=290)
```

(1) **Select the value for s:** We again use the overfitting procedure. Because we are assuming that this is quarterly data we use the overfit procedure with values of p such as $p = 8, 10$, and 12 (that is, we do not have to assure that the overfit order is greater than 12 as we did when we assumed a 12th-order seasonality). However, overfitting with $p = 14$ gives similar results to those used below. The overfit factor table is shown for $p = 10$ in Table 7.9. We examined other values of p and obtained similar results. The command for obtaining the overfit factor table in Table 7.9 is

```
est.ar.wge(xB, p=10,type='burg')
```

Comparing the first 3 lines of Table 7.9 (shown in bold) with Table 7.6 we see that the factors in Table 7.9 with roots closest to the unit circle are very similar to the factors of $1-B^4$, that is, $1-B, 1+B^2$, and $1+B$. The other factors in Table 7.9 are associated with roots much further from the unit circle. Note that $1-.974B$ is not as close to the unit circle as the other two. However, its presence in the factor table along with the other two factors that are very close to $1 + B^2$ and $1 + B$ suggests that a seasonal factor of $1 - B^4$ is a reasonable choice, especially if the data have a natural 4th-order seasonal behavior, such as quarterly sales data. Therefore, we decide to fit a seasonal model with $s = 4$.

TABLE 7.9 $p = 10$ Overfit Factor Tables for Realization from $\left(1 - 1.3B + .65B^2\right)\left(1 - B^4\right)X_t = a_t$ AR(10) Overfit

AR-FACTOR	ROOTS	$|r|^{-1}$	f_0
$1 - .0013B + .9990B^2$	$.000 \pm 1.0005i$	.9995	.250
$1 + .995B$	-1.005	.995	.500
$1 - .974B$	1.027	.974	.000
$1 - .826B + .633B^2$	$.653 \pm 1.075i$	.795	.163
$1 - 1.140B + .612B^2$	$1.139 \pm .579i$	.783	.075
$1 + B + .366B^2$	$-1.365 \pm .931i$	.605	.405

(2) *Seasonally difference the data:* Based on the previous discussion, we select $s = 4$, and seasonally difference the data using

```
sB4=artrans.wge(xB,phi.tr=c(0,0,0,1))
```

The seasonal 4th-order seasonal difference adds a factor of $1 - B^4$ to the model.

(3) *Fit an ARMA(p,q) model to the seasonally differenced data:* The seasonally differenced data in **sB4** are shown in Figure 7.18(d) along with the sample autocorrelations and Parzen spectral density estimate in Figures 7.18(e) and (f). Again, we see cyclic behavior and quickly damping sample autocorrelations only after seasonally differencing the data. Specifically, the data appear to be stationary but not white noise. The following command uses AIC to select ARMA model orders for the seasonally differenced data (in **sB4**) letting p range from 0 to 10 and q from 0 to 2.

```
aic.wge(sB4,p=0:10,q=0:2)
```

AIC selects $p = 2$ and $q = 0$, so the following command is used to estimate the parameters of the AR(2) model fit to the stationary, seasonally differenced data.

```
est.ar.wge(sB4,p=2)
```

The ML estimates for the differenced data are $\hat{\phi}_1 = 1.19$, $\hat{\phi}_2 = -.64$, and $\hat{\sigma}_a^2 = 1.04$, so the final model is

$$\left(1 - 1.19B + .64B^2\right)\left(1 - B^4\right)X_t = a_t, \tag{7.13}$$

with $\hat{\sigma}_a^2 = 1.04$. This model is "similar" to the true Model (B), and the factor table (check it out!) shows that the fitted model has a system frequency close to $f = .10$, which is consistent with the features of the true AR(2) model (see Table 7.1). Recall that because the factor $1 - B$ is a factor of $1 - B^4$, we do not estimate the mean.

Example 7.2 Air Passengers Data

In Figure 7.19(a) we showed the log of the monthly number of airline passengers (in thousands) for the 12 years, 1949−1960. We noted that the **AirPassengers** data are similar in appearance to the seasonal dataset in Figure 7.17(d) with the additional feature that there is a strong trending behavior. We also noted that the sample autocorrelations (Figure 7.19(b)) have slight "bumps" at lags 12 and 24 but are dominated by the appearance of a slowly damping exponential behavior. The spectral density (Figure 7.19(c)) has peaks at the system frequencies in Table 7.7, but the peaks at and near $f = 0$ are accentuated. Figures 7.19(a–c) are obtained using the commands

```
data(AirPassengers)
logAP=log(AirPassengers)
plotts.sample.wge(logAP)
```

(1) ***Select the value for s:*** Because these are monthly data, we use the overfit procedure to determine whether the factors associated with $1 - B^{12}$ are in the model. For this purpose we fit an AR(15) using Burg estimates

```
est.ar.wge(logAP,p=15,type='burg')
```

The resulting factor table is shown in Table 7.10, and in Table 7.11 we compare the most dominant factors of the AR(15) overfit (shown in bold) with those of $1 - B^{12}$.

TABLE 7.10 $p = 15$ Overfit Factor Table with $p = 15$ for Log `AirPassengers` Data AR(15) Overfit

AR-FACTOR	ROOTS	$|r|^{-1}$	f_0
$1-1.723B+.995B^2$	$.867\pm.504i$	.998	.084
$1-.986B+.995B^2$	$.496\pm.872i$	.998	.168
$1+.987B+.985B^2$	$-.501\pm874i$	.993	.333
$1-.038B+.979B^2$	$0.19\pm.1.011i$	.989	.247
$1-1.9697B+.9704B^2$	$1.015\pm.022i$	.985	.003
$1-1.689B+.950B^2$	$-.889\pm.512i$	.975	.417
$1+.854B$	-1.171	.854	.500
$1+.600B$	-1.667	.600	.500
$1-.111B$	9.007	.111	.000

TABLE 7.11 Comparison of the Factors of $1 - B^{12}$ with Those Shown in Table 7.10

FACTOR IN AR(15) OVERFIT	ASSOCIATED FACTOR OF $1-B^{12}$
$1-1.723B+.995B^2$	$1-1.732B+B^2$
$1-.986B+.995B^2$	$1-B+B^2$
$1+.987B+.985B^2$	$1+B+B^2$
$1-.038B+.979B^2$	$1+B^2$
$1-1.9697B+.9704B^2$	See discussion below
$1-1.689B+.950B^2$	$1-1.732B+B^2$
$1+.854B$	$1+B$

From Table 7.11 we see that the factor $1 - B$ does not seem to be represented in the AR(15) factor table. However, the factor $1 - 1.9697B + .9704B^2$ is close to the factor $1 - 2B + B^2 = (1 - B)(1 - B)$, so that the AR(15) factor table suggests two factors of $1 - B$. This is consistent with the Airline Model in which $(1 - B)(1 - B^{12})$ contains two $1 - B$ factors. Notice also that the factor $1 + .854B$ is not very close to the unit circle. However, because there is such a close approximation to the other factors of $(1-B)(1-B^{12})$, we believe that there is sufficient evidence to suggest fitting a model with these two "Airline Model" factors.[12]

(2) *Transform the data using $(1-B)(1-B^{12})$*: To perform this transformation on `logAP`, we first apply a seasonal transformation and then difference the transformed data. We use the commands

```
s12=artrans.wge(logAP,phi.tr=c(rep(0,11),1))
s12d1=artrans.wge(s12,phi.tr=1)
```

(3) *Fit an ARMA(p,q) model to the transformed data:* The log `AirPassengers` data along with the sample autocorrelations and Parzen spectral density estimate are shown in the first row of Figure 7.21. The log `AirPassengers` data transformed by $(1-B)(1-B^{12})$ are shown in Figure 7.21(d) along with sample autocorrelations and spectral density in Figures 7.21(e)–(f).

The sample autocorrelations at lags $k=1$ and $k=12$, shown in Figure 7.21(e), are outside the limit lines leading us to believe that there may still be some 12th-order seasonal behavior to be modeled. For this reason, we issue the following command to let AIC select the top five ARMA model orders for the transformed data (in `s12d1`) letting p range from 0 to 15 and q from 0 to 2. (We let p extend beyond 12 to capture 12th-order seasonal behavior that might still be present.)

```
aic5.wge(s12d1,p=0:15,q=0:2)
```

AIC selects an ARMA(12,1) followed by an AR(13) as the top two models. We fit the data with an AR(13) model and use Burg estimates to be consistent with the model selected by Gray and Woodward (1981). The command

```
est.ar.wge(s12d1,p=13,type='burg')
```

produces the model

$$\phi_{13}(B)(1-B)(1-B^{12})X_t = a_t, \qquad (7.14)$$

where

$$\phi_{13}(B) = 1+.41B+.06B^2+.13B^3+.09B^4-.05B^5-.09B^6+.01B^7-.03B^8$$
$$-.15B^9-.01B^{10}+.11B^{11}+.44B^{12}+.14B^{13}, \qquad (7.15)$$

and $\hat{\sigma}_a^2 = 0.0013$. These commands produce the plots in Figure 7.21.

```
data(AirPassengers)
logAP=log(AirPassengers)
s12=artrans.wge(logAP,phi.tr=c(rep(0,11),1),plot=FALSE)
s12d1=artrans.wge(s12,phi.tr=1,plot=FALSE)
plotts.sample.wge(logAP)
plotts.sample.wge(s12,arlimits=TRUE, speclimits=c(-20,10))
plotts.sample.wge(s12d1,arlimits=TRUE, speclimits=c(-20,10))
```

QR 7.9 Air Passenger Example

---

12 Note that the factor immediately following $1+.854B$ in the AR(15) overfit is $1+.600B$ which "enhances" the effect of the factor associated with $f=.5$.

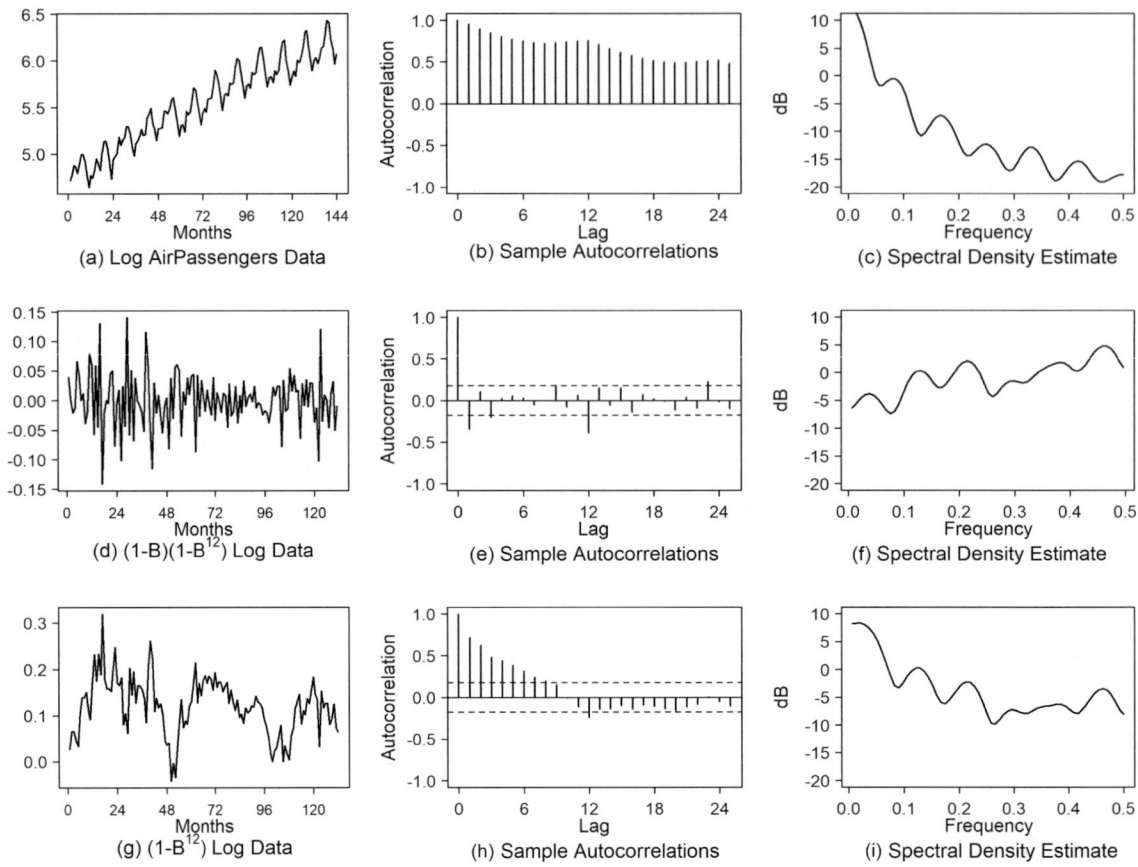

FIGURE 7.21 (a) log AirPassengers data in Figure 7.19, (b) sample autocorrelations, and (c) Parzen spectral density estimate, (d) log AirPassengers data transformed by $(1-B)(1-B^{12})$, (e) sample autocorrelations of the data in (d) showing limit lines, and (f) Parzen spectral density estimate, (g) log AirPassengers data transformed by $(1-B^{12})$, (h) sample autocorrelations of the data in (i) showing limit lines, and (j) Parzen spectral density estimate.

An Alternative Model for the Log AirPassengers Data

Another strategy might be to examine the factors in Table 7.10 and decide to transform the data by $1-B^{12}$ (instead of $(1-B)(1-B^{12})$) and then model the transformed data. To this end, Figures 7.21(g)–(i) show the log **AirPassengers** data transformed by $1-B^{12}$ along with the associated sample autocorrelations and spectral density. The data are wandering, the sample autocorrelations damp (fairly quickly), and the spectral density has a peak at $f=0$. Table 7.12 is the overfit factor table using $p=15$ for log **AirPassengers** data transformed only by $1-B^{12}$.

TABLE 7.12 $p=15$ Overfit Factor Table for the Log AirPassengers Data Transformed by $1-B^{12}$

AR-FACTOR	ROOTS	$|r|^{-1}$	f_0
$1-1.323B+.912B^2$	$.726\pm.755i$	.955	.128
$1-.453B+.910B^2$	$.249\pm1.018i$	.954	.212
$1+1.821B+.885B^2$	$-1.029\pm.268i$	.941	.460
$1-1.828B+.876B^2$	$1.044\pm.230i$	.936	.035
$1+1.289B+.820B^2$	$-.786\pm.776i$	.906	.376

TABLE 7.12   Continued

AR-FACTOR	ROOTS	$|r|^{-1}$	f_0
$1+.545B+.753B^2$	$-.362\pm1.094i$	.868	.301
$1-.837B$	1.195	.837	.000
$1+.274B+.095B^2$	$-1.445\pm2.907i$	.308	.323

We would have expected to see a factor close to $1-B$ to show up in the factor table after transforming only by $1-B^{12}$. However, the overfit table is not suggestive of a nearly nonstationary factor close to $1-B$. The related factor is $1-.837B$ which is the next to last factor in Table 7.12. This suggests modeling the $1-B^{12}$ transformed data with a stationary model. Using this strategy in obtaining the model (7.15), we use AIC to find the "best" AR model using Burg estimates for the seasonally differenced data in **s12**. Because we are dealing with data having 12th -order seasonality, we let p range from 0 to 15 using the commands

```
s12=artrans.wge(airlog,phi.tr=c(rep(0,11),1))
aic5.wge(s12,p=0:15,q=0:0)
# AIC selects an AR(13)
est.ar.wge(s12,p=13)
```

AIC selects an AR(13), and the final model is

$$\phi_{s13}(B)(1-B^{12})X_t = a_t, \tag{7.16}$$

where

$$\phi_{s13}(B)=1-.54B-.28B^2+.08B^3-.02B^4-.14B^5-.05B^6+.10B^7$$
$$-.04B^8-.14B^9+.14B^{10}+.14B^{11}+.32B^{12}-.30B^{13}. \tag{7.17}$$

and $\hat{\sigma}_a^2 = 0.0013$. We will return to models (7.14) and (7.16) for the log **AirPassengers** data when we compare their forecast performances.

Example 7.3 DFW Temperature Data
Dataset **dfw.mon**, which is a *tswge ts* file, consists of monthly average temperatures in degrees Fahrenheit for the Dallas-Ft.Worth area from January 1900 through December 2020. This dataset was plotted in Figure 1.20 where the data are so dense that it is difficult to detect seasonal patterns. Figure 1.4 shows a subset of the DFW data from 1979 through 1986 and focuses on the seasonal pattern. It is noted there that the data have a smooth progression from summer to winter and again from winter to summer with the average temperatures in the Fall and Spring being similar. The resulting overall pattern is "sort of" sinusoidal, or pseudo-sinusoidal.

In this example we will use the temperature data from January 2000 through December 2020. The following commands create the desired dataset and plots in Figures 7.22(a)–(c).

```
data(dfw.mon)
DFW.2000=window(dfw.mon,start=c(2000,1))
plotts.sample.wge(DFW.2000)
```

The seasonal pattern can be seen in the data, the sample autocorrelations have a slowly damping sinusoidal behavior with cycle length 12, and the dominant feature of the spectral density estimate is a peak at about $f = .08 = 1/12$.

In order to verify that seasonal factors approximating those in Table 7.7 for monthly seasonal datasets are present in overfit tables, we fit AR(15), AR(16), and AR(17) models using Burg estimates for the data in **DFW.2000**. These tables are similar to each other, and the factor table for the AR(16) fit is shown in Table 7.13. Examination of Table 7.13 shows that the overfit factor table is not what was expected, and differs dramatically from those shown in Table 7.7 for monthly seasonal datasets. In the case of dataset **DFW.2000**, the factor $1 - 1.728B + .998B^2$ totally dominates the model fit and approximates the factor $1 - 1.732B + B^2$ very closely. Note that some other factors in Table 7.13 approximate factors of $1 - B^{12}$. For example, the second factor, $1 - .958B + .910B^2$, is an approximation of $1 - B + B^2$. Recall, however, that in Tables 7.10 and 7.11 the estimates of the factors of $1 - B^{12}$ in the AR overfit table for the log **AirPassengers** data are each about the same distance from the unit circle.[13] The factor $1 - 1.728B + .998B^2$ is associated with a frequency of $f = .084$, that is, a period of $1 / .084 = 12$ months. Consequently, the data are not "seasonal" in the sense of Table 7.7 but instead have a cyclic nature with a period of 12. This brings to mind the comment earlier that the data have a pseudo-sinusoidal behavior. This contrasts with the **AirPassengers** data in which the "seasonal" pattern within a year was not sinusoidal but followed a fairly irregular pattern that persisted from year to year. Examination of Figures 7.22(b) and (c) show that the data behave like that of an AR(2) model with system of frequency about $f = 1 / 12$ and with roots very close to the unit circle

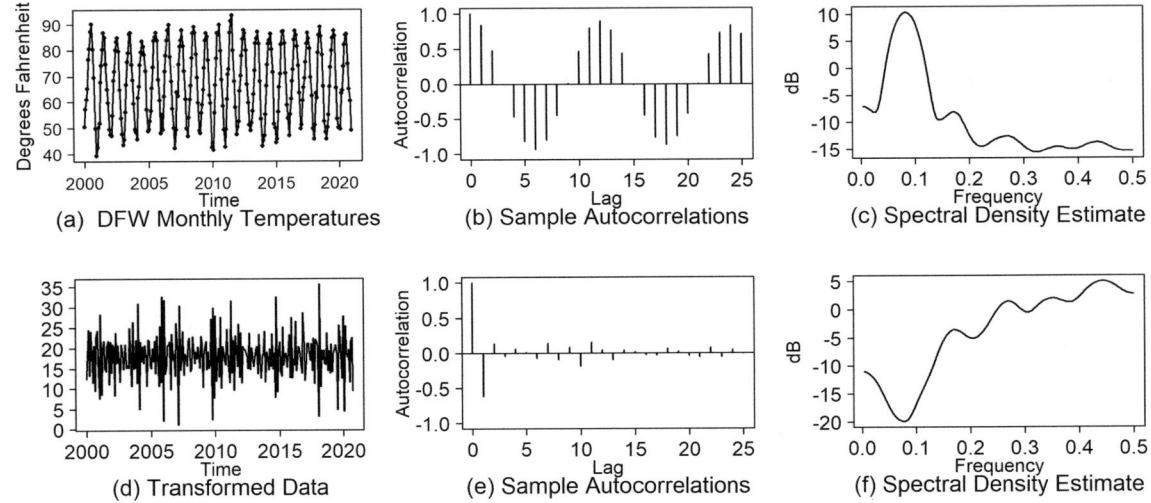

FIGURE 7.22 (a) DFW average monthly temperatures 2000–2020; (b) sample autocorrelations; (c) Parzen spectral density estimate for data in (a), (d) data in (a) transformed by $Y_t = (1 - 1.732B + B^2)X_t$; (e) sample autocorrelations; (f) Parzen spectral density estimate of the data in (d).

The above discussion suggests a model containing the nonstationary factor $1 - 1.732B + B^2$, or a nearly nonstationary factor close to it. We know from previous experience that nonstationary factors (in ARIMA and seasonal models) obscure underlying stationary behavior. Consistent with our analysis of ARIMA and stationary models, we transform the data by $Y_t = (1 - 1.732B + B^2)X_t$ to reveal the stationary aspects of the data.

The data are transformed and the transformed data, sample autocorrelations, and Parzen spectral density estimate are plotted in Figures 7.22(d)–(f), respectively, using the commands

```
DFW.tr12=artrans.wge(DFW.2000,phi.tr=c(1.732,-1),plot=FALSE)
plotts.sample.wge(DFW.tr12)
```

---

13 In Table 7.10 the factor $1 + .854B$ was further from the unit circle than the other factors. We used the factoring $(1 - B)(1 - B^{12})$ because the other factors were consistently close to factors of $(1 - B)(1 - B^{12})$.

TABLE 7.13 $p = 15$ Overfit Factor Table for DFW Monthly Temperature Data in `DFW.2000` AR(15) Overfit

AR-FACTOR	ROOTS	$|r|^{-1}$	f_0
$1 - 1.728B + .998B^2$	$.866 \pm .502i$	.999	.084
$1 - .958B + .910B^2$	$.527 \pm .907i$	.954	.166
$1 - .914B$	1.094	.914	.000
$1 + 1.648B + .804B^2$	$-1.025 \pm .440i$	.897	.436
$1 - .138B + .778B^2$	$-.089 \pm 1.130i$	.882	.262
$1 + 1.063B + .760B^2$	$-.699 \pm .909i$	.872	.354
$1 + .816B$	-1.226	.816	.500
$1 + .163B + .449B^2$	$-.181 \pm 1.482i$	.670	.269
$1 - .660B$	1.514	.660	.000

Although the transformed data appear to be white noise, the large negative autocorrelation at lag 1 and the dip in the spectral density at about $f = .08$ suggest that modeling should continue. Using the command

`aic5.wge(DFW.tr12,p=0:15,q=0:2)`

AIC selects an ARMA(11,1). AICC also selects an ARMA(11,1) while BIC selects an AR(10). Because of the consistency of choices, we fit an ARMA(11,1) model for Y_t. This model is $\hat\phi_{11}(B)Y_t = (1 - .95B)a_t$, where

$$\hat\phi_{11}(B) = 1 + .38B - .02B^2 - .30B^3 - .52B^4 - .56B^5 - .46B^6 - .36B^7 + .01B^8 + .30B^9 + .51B^{10} + .23B^{11}.$$

Because $Y_t = (1 - 1.732B + B^2)X_t$ and $\bar{x} = 67.36$, it follows that the final model for the temperature data is

$$\hat\phi_{11}(B)(1 - 1.732B + B^2)(X_t - 67.36) = (1 - .95B)a_t, \qquad (7.18)$$

and $\hat\sigma_a^2 = 9.41.$ [14]

Key Points:

1. **Do not** include $1 - B^{12}$, for example, in the model simply because there is a 12th-order nonstationarity, or because the data show a period of 12, or just because you have monthly data. – **Check factor tables!**
2. The factor $1 - 1.732B + B^2$ produces sinusoidal-type behavior of period 12. For the **DFW.2000** data one might also consider a cosine+noise model discussed in Section 8.2.

7.2.3 Forecasting Using Seasonal Models

In this section we discuss forecasting using the seasonal model $\phi(B)(1 - B^s)X_t = \theta(B)a_t$. For this seasonal model, the "AR part" is the $p + s$ order factor $\phi(B)(1 - B^s)$, and, analogous to the ARIMA case, we use the notation $\hat\phi^*(B) = \hat\phi(B)(1 - B^s)$ to denote the AR part of the fitted seasonal model. Forecasts for seasonal models use the following formula which is analogous to (7.6) for ARIMA models:

---

14 The mean, 67.36, is included in model (7.18) because $1 - 1.732 + 1 \neq 0$, and so the mean plays a role in the model.

$$\hat{X}_{t_0}(\ell) = \sum_{j=1}^{p+s} \hat{\phi}_j^* \hat{X}_{t_0}(\ell-j) - \sum_{j=\ell}^{q} \hat{\theta}_j \hat{a}_{t_0+\ell-j}. \qquad (7.19)$$

The constant term is zero in seasonal models because $1-B^s$ has a factor of $1-B$ for all $s > 0$.[15] We begin our discussion of forecasting from seasonal models by considering forecasts from the model $(1-B^4)X_t = a_t$.

Model (A): Forecasts for the Purely Seasonal Model $(1-B^4)X_t = a_t$

Figure 7.16(a) showed the first 24 data values from a realization of length $n = 72$ from the purely seasonal model with $s = 4$. The full dataset was generated using the command

```
xA4=gen.arima.wge(n=72,s=4,mu=50,sn=52)
```

Figure 7.23(a) is a plot of the data in Figure 7.16(a) along with forecasts for the next 12 quarters (three years). By carefully observing forecasts for $t = 25, 26, 27$, and 28 it can be seen that these are exact replicas of the data values for the last four quarters observed, that is, $t = 21, 22, 23$ and 24. Notice that Figure 7.23(a) shows forecasts for the next three years, and for each of these years, the forecasts are the values observed for $t = 21, 22, 23$, and 24. This is analogous to the ARIMA(0,1,0) model for which forecasts are simply replications of the last data value.

In order to understand these forecasts, we write $(1-B^4)X_t = a_t$ as $X_t = X_{t-4} + a_t$. Using this expression, it can be seen that, for example, the value for $t = 25$ is the value for $t = 21$ plus a zero-mean white noise value that is as likely to be positive as it is to be negative. Consequently, it makes intuitive sense that $\hat{X}_{24}(1) = x_{21}$, that is, x_{21} is the "best" forecast for X_{25}.

Figure 7.23(b) shows the purely seasonal (monthly) data with $s = 12$, shown previously in Figure 7.16(b) along with forecasts for the next year (that is, 12 months). Close examination shows that these forecasts are replicas of the sales for the last year observed.

```
xA4=gen.arima.wge(n=72,s=4,mu=50,sn=52)
xA12=gen.arima.wge(n=72,s=12,mu=50,sn=100)
par(mfrow=c(1,2))
fore.arima.wge(xA4[1:24],s=4,n.ahead=12,limits=FALSE)
abline(v=24.5,lwd=2)
fore.arima.wge(xA12[1:24],s=12,n.ahead=12,limits=FALSE)
abline(v=24.5,lwd=2)
```

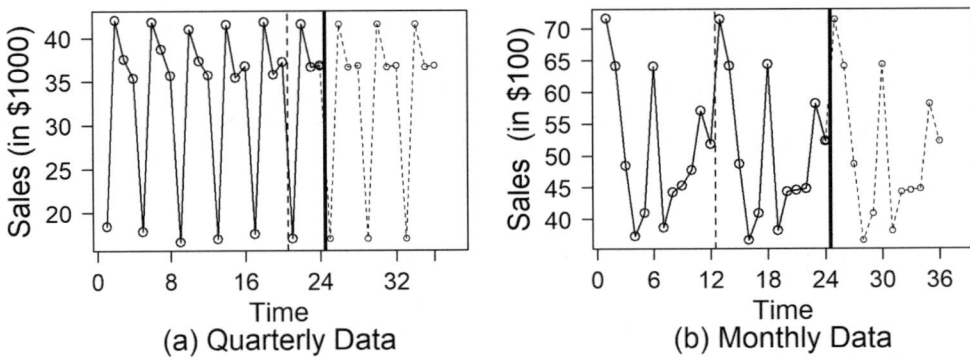

(a) Quarterly Data　　(b) Monthly Data

FIGURE 7.23　(a) Realization of length 24 from a seasonal model with $s = 4$ along with forecasts for the next 12 quarters and (b) realization of length 24 from a seasonal model with $s = 12$ along with forecasts for the next 12 months.

15　The characteristic equation $1-z^s=0$, that is $z^s=1$, always has a solution $z=1$. That is, $1-z$ is a factor of $1-z^s$.

Model (B): Forecasts Using Data from Seasonal Model $(1-1.3B+.65B^2)(1-B^4)X_t = a_t$

Figure 7.18(a) showed a realization of length $n = 100$ from Model (B), $(1-1.3B+.65B^2)(1-B^4)X_t = a_t$, with $\sigma_a^2 = 1$. The data in Figure 7.18(a) were generated using the command

```
xB=gen.arima.wge(n=100,phi=c(1.3,-.65),s=4,sn=290)
```

In Section 7.2.2 the model $(1-1.19B+.64B^2)(1-B^4)X_t = a_t$, with $\hat{\sigma}_a^2 = 1.04$ was fit to the data. The forecasts for the next 12 quarters (three years) are calculated using the command

```
fore.xB=fore.arima.wge(xB,phi=c(1.19,-.64),s=4,n.ahead=12)
```

Figure 7.24(a) shows the realization previously plotted in Figure 7.18(a). Figure 7.24(b) shows a close-up showing the last 12 data values, the first 12 forecasts, and the 95% limits. The forecasts look like exact replicas of the last four data values. In this case, however, the last four data values are 58.5, 74.6, 21.8, and 17.7, while the first four forecasts are 55.1, 73.4, 20.1, and 16.1, respectively. Thus, the AR component has modified the forecasts slightly. Also, note that the prediction limits are very "tight". Figure 7.24(c) shows Figure 7.24(b) along with the next 12 data values in the simulated series. The true values are indicated by solid squares. Notice that it is difficult to see the forecasts in this plot because they are very close to the true values. Additionally, the true values fall within the prediction limits. That is, the forecasts are quite good. The plots in Figure 7.24 were generated using the code

```
xB=gen.arima.wge(n=112,phi=c(1.3,-.65),s=4,sn=290,plot=FALSE)
t=1:112
plot(t[1:100],xB[1:100],type='l',xlim=c(0,120))
fore.xB=fore.arima.wge(xB[89:100],phi=c(1.19,-.64),s=4,n.ahead=12)
fore.xB=fore.arima.wge(xB[89:100],phi=c(1.19,-.64),s=4,n.ahead=12,limits=
FALSE)
points(t[13:24],xB[101:112],type='o')
```

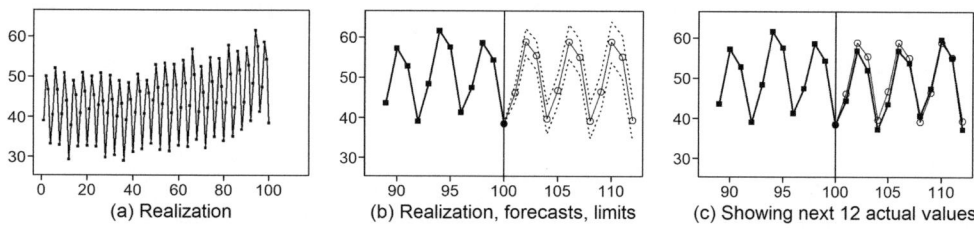

FIGURE 7.24   (a) Realization of length $n = 100$ from Model (B), (b) close-up of last   12 data values in (a) along with first 12 forecasts and 95% limit lines, and (c) plot in (b) along with actual data values for $t = 101-112$.

Example 7.4 Forecasting the `AirPassengers` Data

(a) Using the "Airline Model" in (7.14): In Example 7.2 the "Airline Model", $\phi_{13}(B)(1-B)(1-B^{12})X_t = a_t$, where $\phi_{13}(B)$ as defined in (7.15) was fit to the log `AirPassengers` data. To simplify notation, the data will be numbered in months, 1:144. In this example forecasts will be compared with actual values of the log `AirPassengers` data using several forecast origins:

(a) Forecast origin $t_0 = 132$, forecast last 12 months
(b) Forecast origin $t_0 = 120$, forecast last 24 months
(c) Forecast origin $t_0 = 108$, forecast last 36 months
(d) Forecast origin $t_0 = 96$, forecast last 48 months

Since we fit an AR(13) model to $Y_t = (1-B)(1-B^{12})X_t$ when analyzing the entire dataset in Example 7.2, we will fit an AR(13) using data up to and including the forecast origin for each set of forecasts. The forecasts are shown in Figure 7.25, where it can be seen that for all forecast origins, the forecasts are quite good. Specifically, the seasonal behavior is forecast well and the trend in the data is forecast to continue. This is due to the fact that the Airline Model operator, $(1-B)(1-B^{12})$, has two factors of $1-B$. Thus, the trend is forecast to continue analogously to the situation for ARIMA models with $d = 2$.

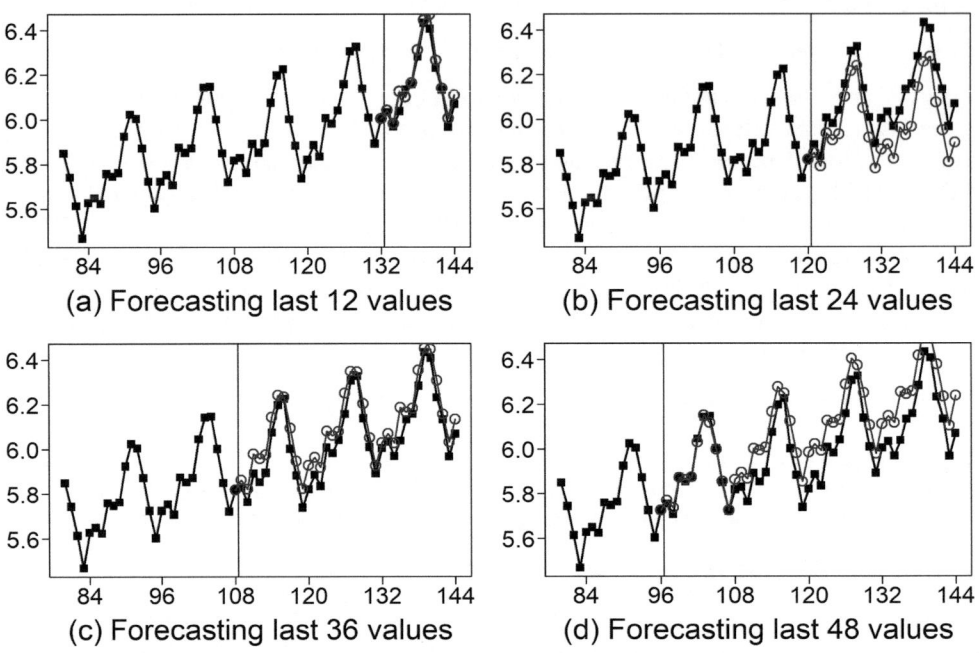

FIGURE 7.25   Comparing forecasts (open circles) with actual values (solid squares) for the last 12, 24, 36, and 48 months, respectively, for the log AirPassengers data using the Airline Model in (7.14).

(b) Using the Alternative Model in (7.16) for AirPassengers data:

We next consider model (7.16), which has a seasonal factor but not a second factor of $1-B$. In Figure 7.26 we compare forecasts of the last 12, 24, 36, and 48 months with actual values using the procedure described above for forecasts shown in Figure 7.25. Clearly, the upward trend is not forecast to continue, and in Figures 7.26(b)–(d) the forecasts are quite poor.

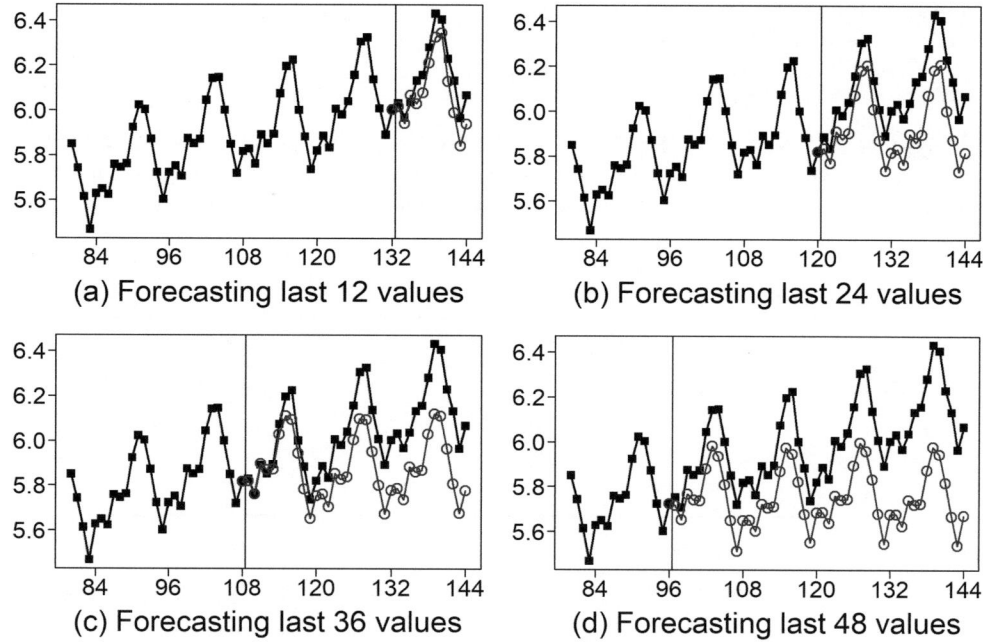

FIGURE 7.26 Comparing forecasts (open circles) with actual values (solid squares) for forecasting the last 12, 24, 36, and 48 months, respectively, for the Log `AirPassengers` data using model (7.16).

(c) Comparing the "Airline Model" to the ARIMA(13,1,0) in (7.16)

In order to compare these two models with horizon 12, the rolling window RMSE was calculated for each model below:

```
data(AirPassengers)
logAP=log(AirPassengers)
s12=artrans.wge(logAP,phi.tr=c(rep(0,11),1),plot=FALSE)
s12d1=artrans.wge(s12,phi.tr=1,plot=FALSE)
estAIRLINE = est.ar.wge(s12d1,p=13,method='burg')

roll.win.rmse.wge(logAP, s = 12, d = 1, phi = estAIRLINE$phi, horizon = 12)
```

Output:
```
"Please Hold For a Moment, TSWGE is processing the Rolling Window RMSE with 105
windows."
"The Summary Statistics for the Rolling Window RMSE Are:"
Min. 1st Qu. Median  Mean 3rd Qu.  Max.
0.01750 0.03663 0.04530 0.05466 0.07067 0.14778
"The Rolling Window RMSE is: 0.055"
```

```
aic5.wge(s12,p=0:15,q=0:0)
# AIC selects an AR(13)
estALT = est.ar.wge(s12,p=13)

roll.win.rmse.wge(airlog, s = 12, d = 0, phi = estALT$phi,
horizon = 12)
```

QR 7.10
Rolling Window

Output:
```
"Please Hold For a Moment, TSWGE is processing the Rolling Window
RMSE with 106 windows."
```

```
"The Summary Statistics for the Rolling Window RMSE Are:"
Min. 1st Qu. Median  Mean 3rd Qu.  Max.
0.0381 0.0984 0.1165 0.1147 0.1429 0.1844
"The Rolling Window RMSE is: 0.115"
```

The "Airline Model" achieves the lower rolling window RMSE (.055) for a horizon of 12 when compared to the alternative model (.115). Table 7.14 below shows that the "Airline Model" achieves the lower rolling window RMSE for horizons 24, 26, and 48 as well, thus clearly providing evidence that it is the more useful model.

TABLE 7.14   Rolling Window RMSEs for "Airline Model" and its Alternative

HORIZON	"AIRLINE MODEL"	ALTERNATIVE MODEL
12	.055	.155
24	.085	.189
36	.113	.253
48	.134	.328

Key Point:

1. Forecasts from an "Airline Model" such as (7.14) containing the factors $(1-B)$ and $(1-B^{12})$ forecast any existing trend to continue because the model has two unit roots.
2. Model (7.16) does not forecast the trend in the **AirPassengers** data to continue because the model contains only one unit root.

Example 7.5 Forecasting the DFW Monthly Temperature data

In Example 7.3 the nonstationary "seasonal" model

$$\hat{\phi}_{11}(B)(1-1.732B+B^2)(X_t-67.36)=(1-.95B)a_t, \tag{7.20}$$

was fit to the **DFW.2000** data, where $\hat{\phi}_{11}(B)$ is given in Example 7.3 and $\hat{\sigma}_a^2=9.41$. Seasonal is in quotes because in this case, even though the data are monthly, the "seasonal" factor is $1-1.732B+B^2$ instead of $1-B^{12}$. Recall that we transformed the **DFW.2000** data to obtain $Y_t=(1-1.732B+B^2)X_t$ using the command

```
DFW.tr2=artrans.wge(DFW.2000,phi.tr=c(1.732,-1),plot=FALSE)
```

and then used AIC to select the model for the transformed data in **DFW.tr2**. AIC chose an ARMA(11,1) whose parameters are found using the command

```
est.tr2=est.arma.wge(DFW.tr2,p=11,q=1)
```

The vector **est.tr2$phi** contains the 11 AR coefficients of the ARMA(11,1) fit. In order to forecast using **fore.arma.wge** we must obtain the 13th-order AR operator obtained by multiplying the AR coefficients in **est.tr2** with **seas.12=c(1.732,-1)** using the commands

```
seas.12=c(1.732,-1)
DFW.13=mult.wge(fac1=est.tr2$phi,fac2=seas.12)
```

The resulting AR coefficients are in the vector **DFW.13$model.coef** while the MA coefficient is in **est. tr2$theta**. Figure 7.27 shows forecasts of the last 12, 36, 84, and 132 months.[16] These forecasts were obtained using the code

```
DFW.full=DFW.13$model.coef
f12=fore.arma.wge(DFW.2000,phi=DFW.13$model.coef,theta=est.tr2$theta,
   lastn=TRUE,n.ahead=12,limits=FALSE)
f36=fore.arma.wge(DFW.2000,phi=DFW.13$model.coef,theta=est.tr2$theta,
   lastn=TRUE,n.ahead=36,limits=FALSE)
f84=fore.arma.wge(DFW.2000,phi=DFW.13$model.coef,theta=est.tr2$theta,
   lastn=TRUE,n.ahead=84,limits=FALSE)
f132=fore.arma.wge(DFW.2000,phi=DFW.13$model.coef,theta=est.tr2$theta,
   lastn=TRUE,n.ahead=132,limits=FALSE)
```

Examination of Figure 7.27 shows that the forecasts from forecast origins considered are very good, staying in sync with the cyclic behavior and predicting peaks and troughs quite well.

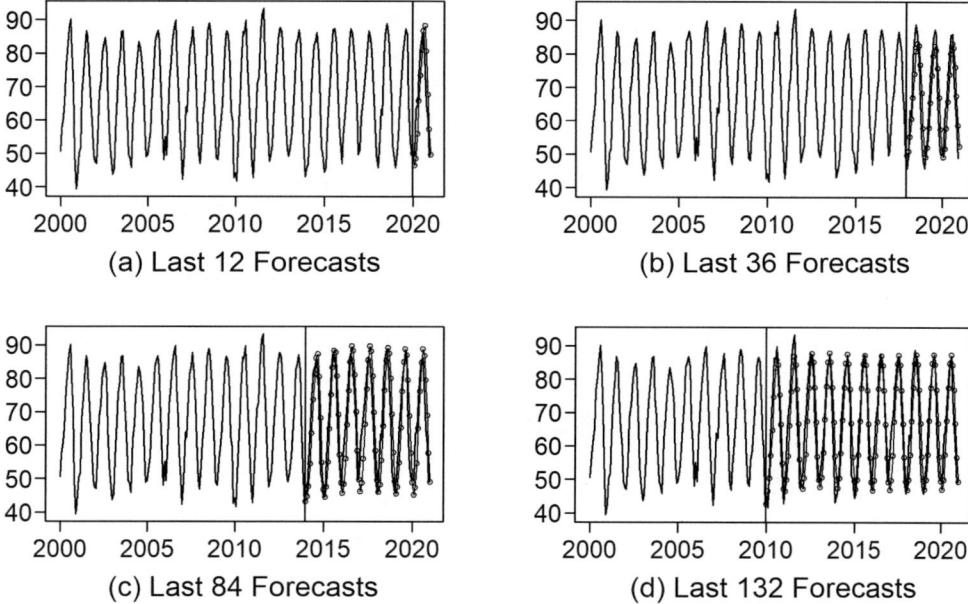

FIGURE 7.27   DFW monthly temperature with forecasts for last 12, 36, 84, and 132 months.

7.3 ARCH AND GARCH MODELS

Figure 7.28 is a plot of the Dow Jones daily rates of return from 1971 through 2020.[17] This is an interesting set of data. The rates of return clearly look more variable (volatile) during some time periods than they do in others. Isolated extremes on Black Monday (October 19, 1987) and the early stages of COVID pandemic awareness in early March, 2020 are visible. Also, the Great Recession from 2007–2009 was a period of continuing volatility.

---

16 It can be seen from factor tables that **DFW.13$model.coef** has a complex pair of roots on the unit circle. Even though this is not a stationary model, the function **fore.arma.wge** produces the correct forecasts.

17 We calculated rate of return as 100*(current value-previous value)/previous value.

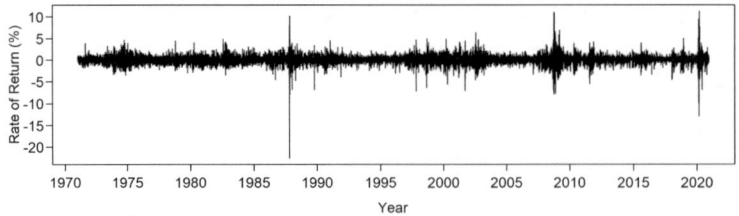

FIGURE 7.28   Daily Dow Jones rates of return from 1971 through 2020.

A key feature of the data is that the variance changes over time. The variance doesn't change gradually in time but rather experiences sudden periods of increased and decreased variability (or volatility). This type of behavior is called *conditional heteroskedasticity*. It is important to note that because the variance changes with time, this type of data should be modeled as a *nonstationary process*.

Taking a "closer look" at the data, in Figure 7.29(a) we plot the daily rates of return for the first 1,000 trading days beginning in July 1999. A unique feature of these data is that the sample autocorrelations in Figure 7.29(b) behave like sample autocorrelations of white noise. So, this is a type of "noise" that is uncorrelated yet nonstationary (the variance changes with time). Economists have encountered data such as these over the years, yet standard models such as ARMA and ARIMA are not designed to model this type of behavior. Figure 7.29(c) is a plot of the squared rates of return, and this plot really accentuates the spikes. Interestingly, the sample autocorrelations of the squared data in Figure 7.29(d) behave somewhat like the autocorrelations of an AR process. That is, the squared data are *correlated*. In a seminal paper, Engle (1982) introduced the *autoregressive conditional heteroskedastic* (ARCH) model to explain this behavior.[18] Based on data such as those in Figure 7.28, Engle's idea was to model the squared data as an AR process.

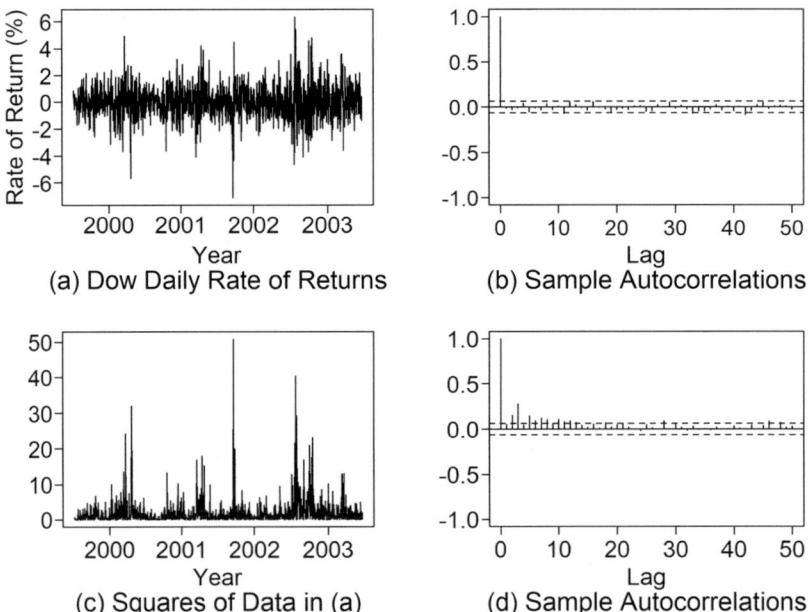

FIGURE 7.29   (a) Dow daily rates of return for first 1,000 days following July 1, 1999, (b) sample autocorrelations of the data in (a), (c) Data in (a) squared, and (d) sample autocorrelations of the squared data in (c).

---

18 The ARCH model idea was such a breakthrough that Robert Engle and his research partner Clive Granger were awarded the Nobel Prize in Economics in 2003.

7.3.1 ARCH(1) Model

The simplest ARCH model is an ARCH(1) model. Specifically, a time series u_t is said to satisfy a 1st-order ARCH model if

$$u_t = \varepsilon_t \sqrt{\alpha_0 + \alpha_1 u_{t-1}^2},$$ (7.21)

where α_0 and α_1 are constants, ε_t is zero-mean normally distributed white noise with unit variance, and ε_t is uncorrelated with u_{t-j}^2, $j > 0$.

Note that because ε_t has mean zero, so does u_t. Squaring both sides of (7.21) gives

$$\begin{aligned} u_t^2 &= \varepsilon_t^2 \left(\alpha_0 + \alpha_1 u_{t-1}^2 \right) \\ &= \alpha_0 + \alpha_1 u_{t-1}^2 + w_t, \end{aligned}$$ (7.22)

where w_t is a zero-mean, uncorrelated noise process.[19] See Appendix 7B. Thus, (7.22) shows that u_t^2 satisfies an AR(1) type model.

Notes:

(1) The behavior of the rates of return data is that there are dramatic changes in variability of the data, u_t, in different time periods. In Figure 7.29(d) we see the autocorrelations of the squared residuals that Engle modeled using an AR model. However, the AR model in (7.22) is an AR model for u_t^2, not for the *variance*. The two concepts are intimately related. Note that the mean of $u_t = 0$, and consequently

$$\begin{aligned} \text{Var}(u_t) &= E\left[\left(u_t - \mu_{u_t} \right)^2 \right] \text{(definition of variance)} \\ &= E\left[u_t^2 \right]. \end{aligned}$$

That is, the *variance of u_t is the expected value of u_t^2*.

(2) In the ARCH setting, the quantity under the radical in (7.21), $\alpha_0 + \alpha_1 u_{t-1}^2$, is the conditional variance of u_t given u_{t-1}. That is, standing at time $t-1$, and having already observed u_{t-1}, the conditional variance of u_t at time t is given by

$$\sigma_{t|t-1}^2 = \alpha_0 + \alpha_1 u_{t-1}^2.$$ (7.23)

(3) *Interpretation of (7.23) using the rates of return data as an example:* Regions of high volatility are periods in which the behavior of the rates of return from day to day is less predictable (more uncertain) than it is in low volatility periods. In the ARCH(1) model, $\alpha_1 \geq 0$, so if the value u_{t-1} is "large", then $\sigma_{t|t-1}^2 = \alpha_0 + \alpha_1 u_{t-1}^2$ is "large". This implies that, at time $t-1$, there is a "high" level of uncertainty about the rates of return on day t.

---

19 Note that we haven't used the notation a_t in these definitions. For example, ε_t is restricted to have unit variance and to be normally distributed. Neither of these conditions are imposed on the white noise in this book that we have denoted by a_t. Also, the variance of w_t depends on time, but a_t has constant variance. See Definition 3.14.

Key Points:

1. If u_{t-1} is "larger than usual" then there is increased *uncertainty* about the upcoming value of u_t.
2. Point (1) does not indicate that u_t itself will be large. It may be large, small, or "in the middle".
3. Equation (7.23) says that the *uncertainty* about the upcoming value, u_t, is large.[20]

Generating realizations from an ARCH(1) process. The *tswge* function `gen.arch.wge` *generates* realizations from an ARCH model. The following commands generate the ARCH(1) realizations in Figure 7.30. The sample autocorrelations can be found using standard techniques.

```
set.seed(11)
par(mfrow=c(4,1), mar(2,1,1,1))
u1=gen.arch.wge(500,alpha0= 1,alpha=0)
u2=gen.arch.wge(500,alpha0=.6,alpha=.4)
u3=gen.arch.wge(500,alpha0=.3,alpha=.7)
u4=gen.arch.wge(500,alpha0=.1,alpha=.9)
```

When $\alpha_1 = 0$, then u_t^2 in (7.22) is a "regular" white noise process. This is consistent with Figure 7.30(a) which is a realization of u_t^2 in which $\alpha_1 = 0$. Note that the volatility increases as α_1 goes from .4 to .7 to .9 in Figures 7.30(c), (e), and (g), respectively. Note also that the sample autocorrelations in all cases behave like sample autocorrelations of white noise.

---

20 Suppose we have two random variables, X and Y, both with mean $\mu = 50$, but $\sigma_X^2 = 1$ and $\sigma_Y^2 = 100$. Suppose for this example we assume that they are both normally distributed and that we are about to observe a new observation, x and y, from each of the random variables. We know there is about a 95% chance that the new observation in each case will be within two standard deviations of the mean. That is, for random variable X, we are about 95% confident that the new observation will be between 48 and 52. However, for random variable Y, we are 95% confident that the new observation will be between $50 \pm 2(10)$, that is, between 30 and 70. *Obvious lesson:* There is more uncertainty about the *yet to be observed value* of y than there is about the upcoming value of x. So, a large value of u_{t-1} suggests that, at time $t-1$, there is a lot of uncertainty concerning the value of u_t.

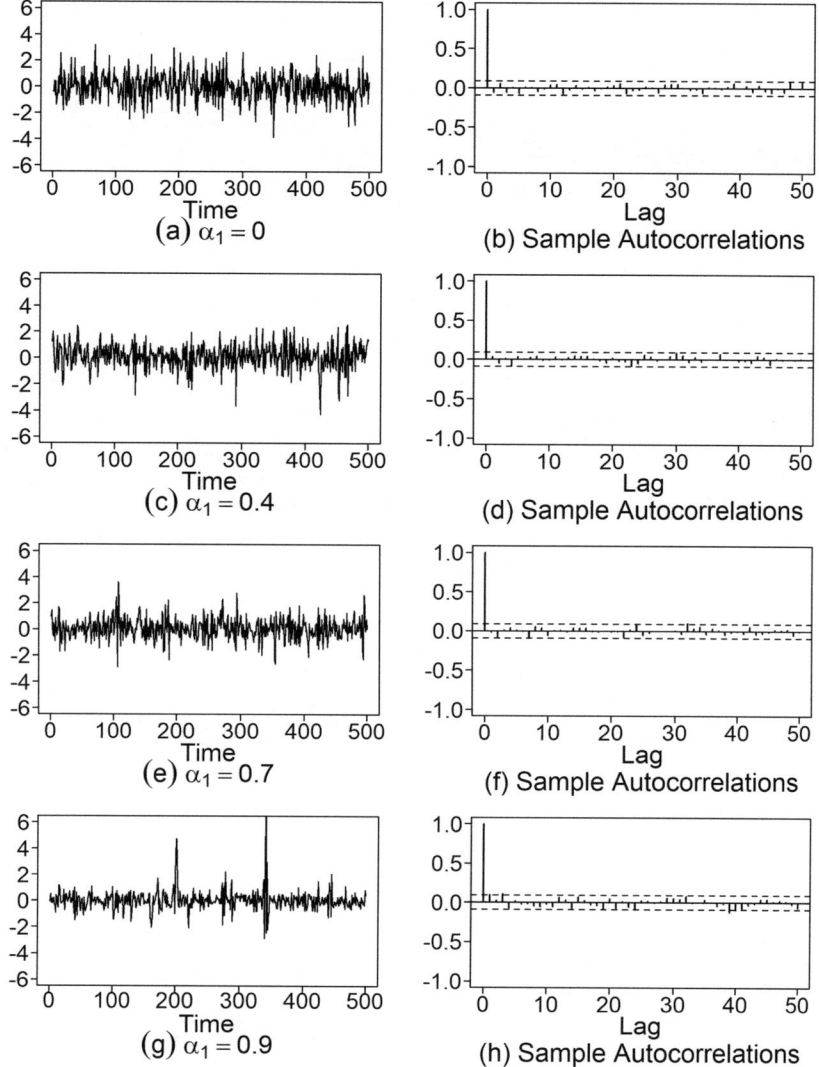

FIGURE 7.30   Realizations and associated sample autocorrelations from ARCH(1) models for various values of α_1.

Figures 7.31 (a), (c), (e), and (g) show the squares of the values in Figures 7.30 (a), (c), (e), and (g), respectively. In these plots we again see that the volatile behavior increases as α_1 increases. Also note in Figure 7.31(b) that the sample autocorrelations of the squared data for the $\alpha_1 = 0$ case (true white noise) are uncorrelated. However, as α_1 increases, the sample autocorrelations exhibit AR(1)-like autocorrelation behavior. The ARCH realizations have the appearance and the behavior associated with the DOW rates of return data, implying that the ARCH model seems to be appropriate for modeling these types of data.

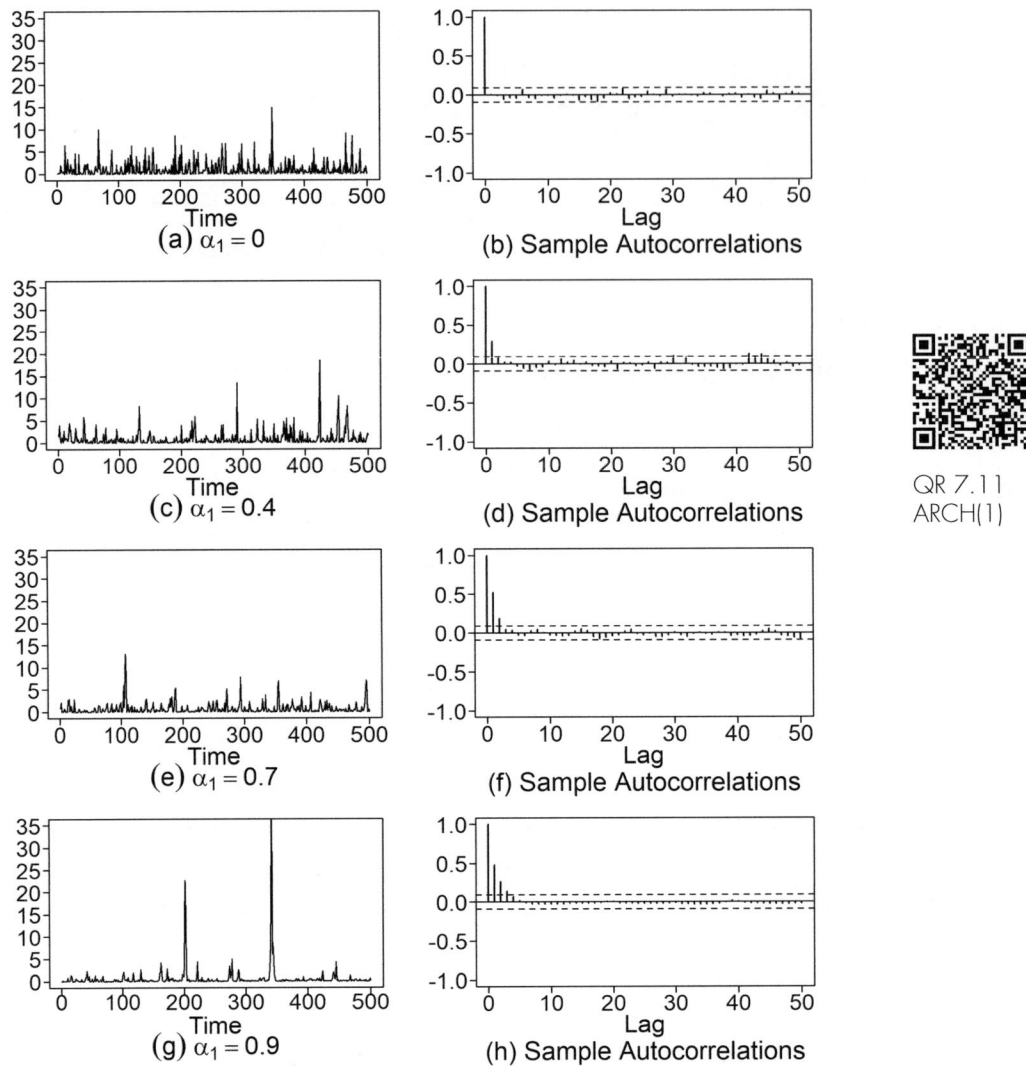

FIGURE 7.31 Squared values of the realizations in Figure 7.30 and associated sample autocorrelations.

QR 7.11
ARCH(1)

7.3.2 The ARCH(p) and GARCH(p,q) Processes

Notice that the periods of high volatility in the ARCH(1) realizations tended to be "spiked" in nature. The ARCH(1) model says how the size of the data value on one day predicts the volatility the next day. However, the Great Recession and the early 2000s (Gulf War early stages?) were prolonged periods of increased volatility. The ARCH(p) model is a generalization of the ARCH(1) that models longer stretches of volatility and is defined by

$$u_t = \varepsilon_t \sqrt{\alpha_0 + \alpha_1 u_{t-1}^2 + \cdots + \alpha_p u_{t-p}^2},$$

(7.24)

where α_0, $\alpha_1, \ldots, \alpha_p$ are constants, ε_t is zero-mean normally distributed white noise with unit variance, and ε_t is uncorrelated with u_{t-j}^2, $j > 0$. Using the logic for the ARCH(1) model, it can be shown that

$$u_t^2 = \alpha_0 + \alpha_1 u_{t-1}^2 + \cdots + \alpha_p u_{t-p}^2 + w_t, \qquad (7.25)$$

where w_t is a zero-mean white noise process, and thus u_t^2 has an AR-type behavior. Analogous to (7.23) the conditional variance of u_t given data at times up to and including $t-1$, is given by[21]

$$\sigma_{t|t-1}^2 = \alpha_0 + \alpha_1 u_{t-1}^2 + \cdots + \alpha_p u_{t-p}^2. \qquad (7.26)$$

In this case, "large" values of $\alpha_0 + \alpha_1 u_{t-1}^2 + \cdots + \alpha_p u_{t-p}^2$ imply that "standing at time $t-1$ there is "high" uncertainty about the value of u_t at time t.

An important generalization (analogous to the ARMA generalization of an AR model) is the GARCH(p,q) process (see Bollerslev, 1986) which is defined by

$$u_t = \varepsilon_t \sqrt{\alpha_0 + \alpha_1 u_{t-1}^2 + \cdots + \alpha_p u_{t-p}^2 + \beta_1 \sigma_{t-1|t-2}^2 + \cdots + \beta_q \sigma_{t-q|t-q-1}^2}. \qquad (7.27)$$

Analogous to (7.26), it follows that

$$\sigma_{t|t-1}^2 = \alpha_0 + \alpha_1 u_{t-1}^2 + \cdots + \alpha_p u_{t-p}^2 + \beta_1 \sigma_{t-1|t-2}^2 + \cdots + \beta_q \sigma_{t-q|t-q-1}^2, \qquad (7.28)$$

where ε_t is a zero-mean normal white noise process uncorrelated with u_{t-j}^2 for $j > 0$. That is, $\sigma_{t|t-1}^2$ is a linear combination of previous observed values u_t^2 and previous conditional variances. As with the ARMA model, a GARCH model can often be obtained that has fewer parameters than an appropriately fitting ARCH model.

Key Points:

1. It is possible for the residuals from an ARMA or ARIMA fit to appear to be white and pass the standard diagnostic tests, but for the squared residuals to not be white.
2. It is good practice to check the squared residuals, especially if conditional heteroskedasticity is suspected.

Generating Realizations from ARCH/GARCH Models

Both the ARCH(p) with $p > 1$ and the GARCH(p,q) can model volatility that continues over a period of time. To better understand the ARCH(p), $p > 1$ and GARCH(p,q) behaviors, in Figure 7.32(a–c) we show realizations from ARCH(2), ARCH(3), and GARCH(1,1) models, respectively. The models are:

---

21 For the case of an ARCH(1), the interpretation we gave of $\sigma_{t|t-1}^2$ was that it was the conditional variance of u_t given data at time $t-1$. We continue to use the same notation in the ARCH(p) case, but note that it is more accurately defined as the conditional variance *given data up to and including time $t-1$*.

(a)   ARCH(2) with $\alpha_0 = .1, \alpha_1 = .5,\ \alpha_2 = .4$
(b)   ARCH(3) with $\alpha_0 = .1$ and $\alpha_1, \alpha_2, \alpha_3$ given by .4,.3, and .2, respectively.
(c)   GARCH(1,1) with $\alpha_0 = .1, \alpha_1 = .4$, and $\beta_1 = .5$

In each case (especially for the ARCH(3) and GARCH(1,1) cases) the realizations show periods of more sustained volatility than did the ARCH(1) realizations in Figure 7.30. The realizations in Figure 7.32 were generated using the commands

```
u1=gen.arch.wge(1000,alpha0=.1,alpha=c(.5,.4),sn=17)
u2=gen.arch.wge(1000,alpha0=.1,alpha=c(.4,.3,.2),sn=32)
u3=gen.garch.wge(1000,alpha0=.1,alpha=.4,beta=.5,sn=141)
```

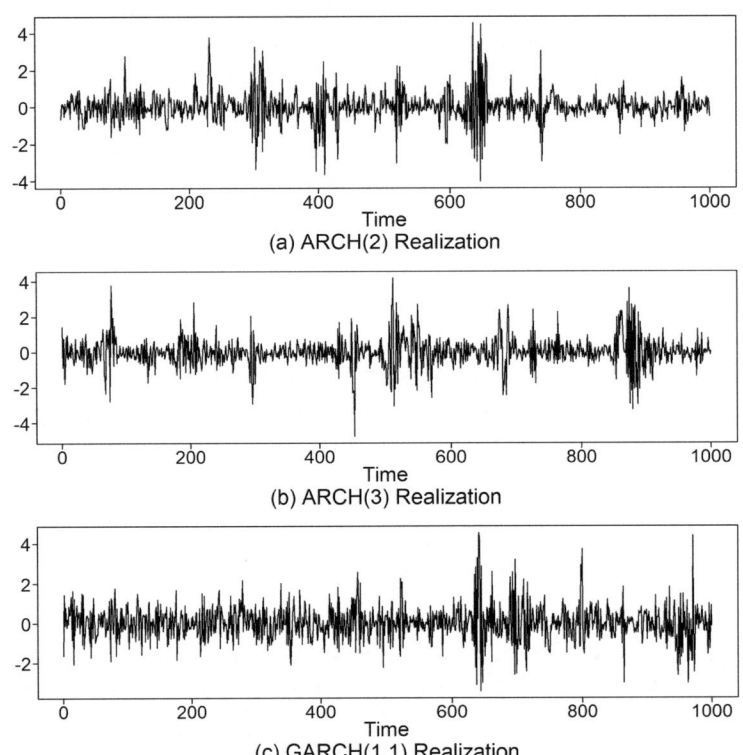

FIGURE 7.32   Realizations from an ARCH(2), ARCH(3), and GARCH(1,1) model, respectively.

7.3.3 Assessing the Appropriateness of an ARCH/GARCH Fit to a Set of Data

The question arises concerning how to evaluate the suitability of an ARCH/GARCH fit to a set of data. The most common way is to examine the residuals. What are the residuals? The answer to this question is not as obvious as it is for, say, an ARMA model. In the following we discuss the calculation and assessment of residuals from a GARCH fit to a set of data. Appropriate simplifications are in order if the model is ARCH.

Consider the defining relation (7.27) for the GARCH model. After a GARCH model has been fit, then

$$\hat{\varepsilon}_t \sqrt{\hat{\alpha}_0 + \hat{\alpha}_1 u_{t-1}^2 + \cdots + \hat{\alpha}_p u_{t-p}^2 + \hat{\beta}_1 \sigma_{t-1|t-2}^2 + \cdots + \hat{\beta}_q \sigma_{t-q|t-q-1}^2}$$

is the model-based estimate of u_t. Consequently, the residual associated with the fit at time t is

$$\hat{\varepsilon}_t = u_t \Big/ \sqrt{\hat{\alpha}_0 + \hat{\alpha}_1 u_{t-1}^2 + \cdots + \hat{\alpha}_p u_{t-p}^2 + \hat{\beta}_1 \sigma_{t-1|t-2}^2 + \cdots + \hat{\beta}_q \sigma_{t-q|t-q-1}^2}.$$

Under the assumptions of the model, ε_t is zero-mean normally distributed white noise with unit variance. Thus, a check for model appropriateness is to check the residuals, $\hat{\varepsilon}_t$, to see if they appear to be white noise and to check for normality. This can be done by checking the plot and sample autocorrelations of the residuals versus their 95% limit lines.

The assumption of approximate normality can be assessed using measures such as the Shapiro-Wilk test (see Shapiro and Wilk, 1965) which tests the null hypothesis that the data are normally distributed. We will return to the use of this test in Chapter 9. Note that if the residuals, $\hat{\varepsilon}_t$, are approximately uncorrelated normal random variables then the squared residuals should also be uncorrelated.[22] We will illustrate the concepts discussed above in the following sections.

7.3.4 Fitting ARCH/GARCH Models to Simulated Data

The CRAN package *tseries* has functions designed to analyze data using ARCH and GARCH models. Specifically, the *tseries* function `garch` can be used to fit an ARCH/GARCH model to a set of data. In the following we use the `garch` command to estimate the coefficients of the realizations `u1`, `u2`, and `u3` above.

(a) **`u1`: ARCH(2)** model in which the parameters $\alpha_0, \alpha_1,$ and α_2 are .1, .5, and .4, respectively.

   tseries commands:[23]

```
u1.out=garch(u1,order=c(0,2))
summary(u1.out)
```

Output from `garch`:

```
Coefficient(s):
      Estimate   Std. Error   t value   Pr(>|t|)
a0    0.10336    0.01064      9.711     <2e-16     ***
a1    0.47753    0.06412      7.448     9.48e-14   ***
a2    0.41433    0.06508      6.367     1.93e-10   ***


Diagnostic Tests:
        Jarque Bera Test
data: Residuals
X-squared = 1.7844, df = 2, p-value = 0.4097


        Box-Ljung test
data: Squared.Residuals
X-squared = 0.12406, df = 1, p-value = 0.7247
```

QR 7.12
ARCH(p) and
GARCH(p,q)

The output shows that the estimates of all parameters were highly significant and close to the true values. The Jarque-Bera test (J. Bera, 1980, 1987) is a test to check the residuals for skewness and kurtosis consistent with normality and in this case failed to reject normality. Additionally, the

---

22 Three results from mathematical statistics are being used here: (1) If random variables are independent, then they are uncorrelated. (2) If normal random variables are uncorrelated, then they are independent. (3) Functions of independent random variables are independent (and thus uncorrelated). See Wackerly, Mendenhall, and Scheaffer (2008).

23 Using the notation adopted in this book, the **order** parameter in the **garch** function is the ordered pair (q, p). Be careful!

Ljung-Box test (or sometimes called the Box-Ljung test) is designed to test the null hypothesis of white noise. This test will be introduced in Section 9.1. In the output above, the Ljung-Box test on the squared residuals has a *p*-value of .536, thus not rejecting white noise for the squared residuals. This is consistent with normal residuals in which case if normally distributed residuals are uncorrelated then so are the squared residuals. (See Footnote 22.)

We additionally ran the Base R **shapiro.test** for normality on the residuals and obtained a *p*-value of .957. (See Section 9.1.2 for a discussion of tests for normality.) The residuals are contained in the vector **u1.out$res** but care must be taken because the first two values are NA.

(b) **u2**: ARCH(3) model in which the parameters $\alpha_0, \alpha_1, \alpha_2$, and α_3 are .1, .4, .3, and .2, respectively.

tseries commands:

```
u2.out=garch(u2,order=c(0,3)) # recall, order=c(q,p)
summary(u2.out)
```

Output from **garch**:

```
Coefficient(s):
      Estimate  Std. Error  t value  Pr(>|t|)
a0    0.08632   0.01085     7.955    1.78e-15  ***
a1    0.41018   0.05893     6.960    3.40e-12  ***
a2    0.35192   0.05996     5.869    4.37e-09  ***
a3    0.19469   0.04684     4.157    3.23e-05  ***


Diagnostic Tests:
        Jarque Bera Test
data: Residuals
X-squared = 0.70133, df = 2, p-value = 0.7042


        Box-Ljung test
data: Squared.Residuals
X-squared = 0.021352, df = 1, p-value = 0.8838
```

The output shows that the estimates of all parameters were highly significant and close to the true values. The Jarque-Bera test failed to reject normality of the residuals. The Ljung-Box test on the squared residuals does not reject white noise. The Shapiro-Wilk test has a *p*-value of .587 and thus does not reject white noise.

(c) **u3**: GARCH(1,1) model in which the parameters α_0, α_1, and β_1 are .1, .4, and .5, respectively.

tseries commands:

```
u3.out=garch(u3,order=c(1,1))
summary(u3.out)
```

Output:

```
Coefficient(s):
      Estimate  Std.Error  t value  Pr(>|t|)
a0    0.13799   0.02431    5.676    1.38e-08  ***
a1    0.43508   0.05764    7.548    4.42e-14  ***
b1    0.39874   0.05949    6.703    2.04e-11  ***


Diagnostic Tests:
        Jarque Bera Test
data: Residuals
X-squared = 0.50614, df = 2, p-value = 0.7764
```

```
            Box-Ljung test
data: Squared.Residuals
X-squared = 0.0038326, df = 1, p-value = 0.9506
```

The output shows that the estimates of all parameters were highly significant and close to the true values. The Jarque Bera test fails to reject normality of the residuals. The Ljung-Box test on the squared residuals does not reject white noise. The Shapiro-Wilk p-value was .941 which does not reject normality. Again, the fit seems to be good.

Summary:

The model fits all seem to be good with diagnostics "checking out". However, in these cases we did know the true model orders and the ε_ts were normally distributed in the simulations. We close out the discussion of ARCH/GARCH by fitting a GARCH model on the DOW daily rates of return data in Figure 7.29.

7.3.5 Modeling Daily Rates of Return Data

The Dow daily rates of return data in Figure 7.28 shows the volatile behavior for which the ARCH or GARCH models were designed to model. The *tswge* dataset **rate** contains daily rates of return from 1971 through 2020. The data in Figure 7.29 were for the first 1,000 trading days beginning in July 1999. The commands below retrieve and analyze the rate data for these 1,000 days. We examined several models and the two that performed best in terms of producing residuals and squared residuals that were approximately white noise were GARCH(1,1) and GARCH(2,1) with a slight edge given to GARCH(2,1). The following are the results for the GARCH(2,1) fit.

```
data(rate)
rate.2000= rate[7200:8199]
rate.2000.out=garch(rate.2000,order=c(1,2) # recall, order=c(q,p))
summary(rate.2000.out)
```

Output:

```
Coefficient(s):

      Estimate  Std. Error  t value  Pr(>|t|)
a0    0.08490   0.02628     3.230    0.001237   **
a1    0.01565   0.02617     0.598    0.549917
a2    0.09371   0.02666     3.516    0.000439   ***
b1    0.84668   0.02457     34.461   < 2e-16    ***

Diagnostic Tests:
        Jarque Bera Test
data: Residuals
X-squared = 58.251, df = 2, p-value = 2.244e-13

        Box-Ljung test
data: Squared.Residuals
X-squared = 0.0015587, df = 1, p-value = 0.9685
```

An interesting feature of the GARCH(2,1) fit is that the data value two days ahead makes the most difference concerning the rate on day t. However, the main factor predicting volatility is the conditional variance $\sigma^2_{t-1|t-2}$. The Jarque-Bera test "doesn't like" the skewness and kurtosis of the residuals *at all* and strongly rejects normality. The residuals and their sample autocorrelations are shown in Figures 7.33(a) and (b). The Ljung-Box test on the squared residuals did not reject white noise (which is an indicator of normally distributed noise). See Figures 7.33(c) and (d). However, the Shapiro-Wilk test strongly rejected normality with a p-value less than .0001. All things considered, the GARCH(2,1) model seems to be our best (but not perfect) choice. Recall Box's quote about useful models.

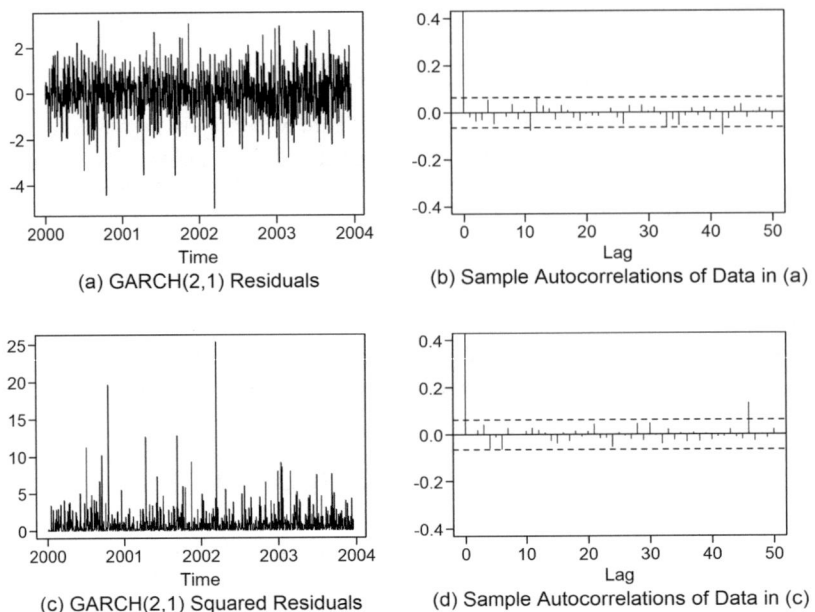

FIGURE 7.33   (a) Residuals from the GARCH(2,1) fit to the Dow rates of return data for 1,000 trading days beginning in July 1999 and (b) sample autocorrelations of the data in (a), (c) squares of the residuals in (a), and (d) sample autocorrelations of the data in (c).

7.4 CONCLUDING REMARKS

In Chapter 6, we proposed techniques for analyzing stationary time series data. However, not all data satisfy the conditions of stationarity. In this chapter we analyzed data for which the roots of the fitted autoregressive characteristic equation are on the unit circle. These data violate stationarity (see Violation 4 in the chapter introduction). We introduced the ARIMA and seasonal ARIMA models to analyze such data. We then discussed the ARCH/GARCH models for analyzing a class of time series data for which the variance changes erratically over time (another violation of stationarity).

Many economic datasets have a type of "wandering" behavior not consistent with stationarity. The autoregressive integrated moving average (ARIMA(p,d,q)) model is often used to analyze this type of nonstationary data. If $d = 0$ then an ARIMA(p,d,q) model simplifies to a stationary ARMA(p,q) model. If $d > 0$, then the ARIMA(p,d,q) model is said to contain one or more "unit roots". We discussed the assessment of whether a model containing a unit root should be used. We showed how to fit ARIMA(p,d,q) to data and how to use these models for forecasting future values. Model identification, model estimation, and forecasting techniques were illustrated using examples.

Seasonal models were introduced for the modeling of nonstationary data that contain repetitive seasonal (for example, monthly or quarterly) patterns. We introduced the seasonal ARIMA models and presented the method of overfitting. Overfitting helps identify seasonal patterns by fitting "high-order" AR models to the data and examining the resulting factor tables. We discussed methods for estimating the parameters of a seasonal ARIMA model and forecasting future values. These techniques were illustrated using examples.

Finally, ARCH/GARCH models were presented as effective models for data which exhibit suddenly erratic change in variance behavior, as is frequently seen in important economic and financial data. We discussed the properties of these models and presented techniques for fitting ARCH/GARCH models to these types of data. Again, these techniques were illustrated using examples.

In the next chapter, we will discuss nonstationary models that contain a deterministic component.

APPENDIX 7A

TSWGE FUNCTIONS

(a) **artrans.wge(x,phi.tr,lag.max,plottr)** performs a transformation $y_t = \phi(B)x_t$ where the coefficients of $\phi(B)$ are given in **phi.tr**.

x = original realization.
phi.tr = vector of transforming coefficients
lag.max = maximum lag for calculated sample autocorrelations (for original and transformed data) (default = 25)
    plottr = logical variable.
    If **plottr = TRUE** (default) then the function plots the original and transformed realizations and associated sample autocorrelations.

(b) **fore.arima.wge(x,phi,theta,d,s,n.ahead,lastn,plot,alpha,limits)** forecasts **n.ahead** steps ahead for stationary ARIMA and seasonal models. The forecasts and forecast limits can optionally be plotted. Forecasts can be calculated and plotted **n.ahead** steps beyond the end of the series, or you can forecast the last **n.ahead** data values (if **lastn=TRUE**). Note: The components of **phi** and/or **theta** may be obtained using **mult.wge** if the models are given in factored form.

Note: The ARIMA model can be a given/known model, but the more common usage will be to fit an ARMA(p,q) model to the data and base the forecasts on the fitted model

x is a vector containing the time series realization $x_t, t = 1, \ldots, n$
phi is a vector of autoregressive coefficients (default=0)
theta is a vector of moving average coefficients (default=0)
d is the order of the difference operator, $(1 - B)^d$, in the model (default = 0)
s is the seasonal order (default = 0)
n.ahead specifies the number of steps ahead you want to forecast (default=2)
lastn is a logical variable (default=**FALSE**)
    If **lastn=FALSE** (default) then the forecasts are for $x(n+1), x(n+2), \ldots, x(n + n.ahead)$ where n is the length of the realization.
    If **lastn=TRUE**, then the program forecasts the last **n.ahead** values in the realization.
plot is a logical variable. (default=**TRUE**)
    If **plot=TRUE** then the forecasts will be plotted along with the realization.
    If **plot=FALSE** then no plots are output.
alpha specifies the significance levels of the prediction limits. (default=.05)
limits is a logical variable. (default= **TRUE**)
    If **limits=TRUE** then the (1-**alpha**)×100% prediction limits will be plotted along with the forecasts.

(d) **gen.arima.wge(n,phi,theta,d,s,lambda,vara,sn)** is a function that generates a realization of length **n** from a given AR, MA, ARMA, ARIMA, or seasonal model.
n is the realization length
phi is a vector of AR parameters (of the stationary part of the model) (default = 0)
theta is a vector of MA parameters (using signs as in this text) (default = 0)
d specifies the number of factors of $(1 - B)$ in the model.
s specifies the order of the seasonal difference, $1 - B^s$.
vara is the white noise variance (default = 1)

plot = TRUE (default) produces a plot of the realization generated

sn determines the seed used in the simulation.

Notes:

(1) This function uses a call to the Base R function **arima.sim** which uses the same signs as this text for the AR parameters but opposite signs for MA parameters. The appropriate adjustments are made within **gen.arima.sim** so that input vectors **phi** and **theta** for **gen.arima.wge** contain parameters using the signs as in this text. However, if you use **arima.sim** directly (which has options not employed in **gen.arima.sim**), then you must remember that the signs needed for the MA parameters are opposite to those in **gen.arima.wge**.

(2) **sn = 0** (default) produces results based on a randomly selected seed. If you want to reproduce the same realization on subsequent runs of **gen.arima.wge** then set **sn** to the same positive integer value on each run.

Example: The command

```
x = gen.arima.wge(n = 200,phi = c(1.6,-.9),theta = .9,d = 2, vara = 1)
```

generates and plots a realization from the model $(1-1.6B+0.9B^2)(1-B)^2 Xt = (1-0.9B)a_t$ and stores it in the vector **x**.

APPENDIX 7B

The steps to obtain the equality in (7.22) are as follows:

$$u_t^2 = \varepsilon_t^2 \left(\alpha_0 + \alpha_1 u_{t-1}^2 \right)$$
$$= \varepsilon_t^2 \left(\alpha_0 + \alpha_1 u_{t-1}^2 \right) + \left(\alpha_0 + \alpha_1 u_{t-1}^2 \right) - \left(\alpha_0 + \alpha_1 u_{t-1}^2 \right)$$
$$= \alpha_0 + \alpha_1 u_{t-1}^2 + \left(\varepsilon_t^2 - 1 \right) \left(\alpha_0 + \alpha_1 u_{t-1}^2 \right)$$
$$= \alpha_0 + \alpha_1 u_{t-1}^2 + w_t$$

Recall that $E(\varepsilon_t) = \mu_{\varepsilon_t} = 0$ and $Var(\varepsilon_t) = 1$. Thus we have

$$Var(\varepsilon_t) = E\left[\left(\varepsilon_t - \mu_{\varepsilon_t} \right)^2 \right]$$
$$= E\left[\varepsilon_t^2 \right]$$
$$= 1,$$

so, $E\left[\left(\varepsilon_t^2 - 1 \right) \right] = 0$. Also, by assumption, ε_t and u_{t-1} are uncorrelated, so

$$E\left[w_t \right] = E\left[\left(\varepsilon_t^2 - 1 \right) \left(\alpha_0 + \alpha_1 u_{t-1}^2 \right) \right]$$
$$= E\left[\left(\varepsilon_t^2 - 1 \right) \right] E\left(\alpha_0 + \alpha_1 u_{t-1}^2 \right)]$$
$$= 0.$$

So, $u_t^2 = \alpha_0 + \alpha_1 u_{t-1}^2 + w_t$,

where w_t, is a zero-mean process.

TSWGE DATASETS RELATED TO THIS CHAPTER

uspop – estimated US population by year from 1900 through 2020

patemp – average monthly temperatutre (in Fahrenheit) for Pennsylvania from Januaeary 990 through December 2004.

freeze – minimum temperature over 10-day periods at a location in South America for 500 consecutive 10-day periods.

rate – DOW daily rates of return from 1971-2020

cement - Quarterly usage of metric tons (in thousands) of Portland cement used from the first quarter of 1973 through the fourth quarter of 1993 in Australia

freight - 9 years of monthly freight shipment data

lavon – Lake level data given in feet above sea level. Quarterly data, 1982-2009.

PROBLEMS

7.1 Generate a realization of length $n = 1000$ from each of the following models:
   (a) $(1-B)(X_t - 300) = a_t$.
   (b) $(1-.999B)(X_t - 300) = a_t$.
   Explain the role of $\mu = 300$ in each realization.

7.2 Generate realizations of length $n = 200$ from each of the following models:
   (a) $(1+1.2B+.9B^2)(1-B)X_t = a_t$.
   (b) $(1+1.2B+.9B^2)(1-B)^2 X_t = a_t$.
   (c) $(1+1.2B+.9B^2)(1-B^2)X_t = a_t$.
   (d) $(1+1.2B+.9B^2)(1-B^{12})X_t = a_t$.
   Use the techniques described in this chapter to fit models to the datasets and specify your final models.

7.3 Using the monthly **cement** sales data in **tswge**, overfit a 13th, 15th, and 17th-order AR using Burg estimates.
   (a) Explain whether you believe a $1 - B^{12}$ factor should be included in your model.
   (b) Should $(1-B)(1-B^{12})$ be included in the model? Explain.
   (c) Whether you believe the model should include the seasonal factors $1-B^{12}$ or $(1-B)(1-B^{12})$ go ahead and model the data using the seasonal factor you like best.

7.4 (a) Repeat the analysis in Example 7.1(b) using the logarithm of the US population data in *tswge* ts object **uspop**. How do the residuals of the "log" model compare with those in Example 7.1(b).
   (b) As in (a) analyze the logarithms of the **dow1985** data.

7.5 In this problem you will analyze the average monthly temperature in degrees Fahrenheit for Pennsylvania from January 1990 through December 2004. The data are in *tswge* file **patemp**.
   (a) Plot the data, sample autocorrelations, and Parzen spectral density estimate.
   (b) Use Burg estimates to overfit $p = 14, 16$, and 18.
   (c) Do you see evidence of a $1 - B^{12}$? Explain.
   (d) Fit a final model for these data, factor your model, and explain your findings.
   (e) Use this model to forecast the temperatures for the next 5 years.
   (f) Assess the accuracy of forecasts by finding the RMSE of your forecasts for the last 5 years of data.

7.6 The Tesla data in *tswge* dataset **tesla** contains daily data from January 1, 2020 through April 30, 2021.
   (a) Plot the data, sample autocorrelations, and spectral density.

(b) Fit a model to the data. Does your model include a unit root? Explain why or why not.

(c) Forecast the Tesla stock data through the end of 2021.

(d) Find the actual data values for this time frame.

(e) Compare your forecasts with the actual values.

(f) Find the RMSE of your forecasts and discuss your results.

7.7   Using the **dow1985** dataset:

(a) Difference the data using **artrans.wge** and plot the differenced data.

(b) Compute the monthly rates of return for the **dow1985** data and plot these rates of return.

(c) Discuss the relationship between the plots in (a) and (b).

7.8   The *tswge* dataset **rate** contains the DOW daily rates of return for the year's daily rates of returns for the years 1971 through 2020. In Section 7.3.5 we fit a GARCH model to the first 1,000 days beginning in July 1999. Repeat the analysis used in that section to analyze:

(a) the first 1,000 days beginning in 1971.

(b) the first 500 days beginning in 1990.

(c) the 1500 days beginning in 1980.

7.9   In Example 7.4 an "Airline Model" (ARIMA(13,1,0), s = 12) was compared to an alternative model (ARIMA(13,0,0), s= 12). The "Airline Model" was fit by taking the first difference and then taking a seasonal difference with s = 12. At that point, we used **aic5.wge** to select the model orders for the stationary portion of the model. AIC selected $p = 13$.

(a) Fit a third model using **aic5.wge** to select the model orders for the stationary portion of the model and then use the p and q that have the second smallest AIC.

(b) Evaluate this model against the other two using the rolling window RMSE for a horizon of 12.

(c) Evaluate this model using the rolling window RMSE for horizons of 24, 36, and 48 as well.

7.10   The *tswge* ts object **us.retail** is a record of the quarterly US retail sales (in $millions) from the fourth quarter of 1999 through the second quarter of 2021.

(a) Plot the data, sample autocorrelations, and Parzen spectral density estimate

(b) Use Burg estimates to overfit using $p = 8$, 10, *and* 12.

(c) Do you see evidence of a $1 - B^4$? Explain.

(d) Fit a final model for these data and explain your findings.

(e) Use this model to forecast the temperatures for the next 5 years.

7.11   Find appropriate models for the following data sets using the procedures discussed in this chapter. Discuss features of your model including seasonality, etc. Use the overfitting procedure outlined in Problem 8.5(a).

(a) **freight:** 9 years of monthly freight shipment data

(b) **freeze:** Each data value represents the minimum temperature over a 10-day period at a location in South America.

(c) **lavon:** Gives lake levels of Lake Lavon (a reservoir northeast of Dallas) from 1982 through 2009. The data values are feet above sea level and are obtained at the end of each quarter (i.e., March 31, June 30, September 30, and December 31).

(d) **MedDays:** Median days a house stayed on the market for the months between July 2016 and April 2021.

(e) **HSN1F:** Monthly data containing the number of new houses sold in US between 1965–2020.

(f) **NAICS:** US total sales in retail trade and food services in billions of dollars from Jan 2000 through December 2019 (not seasonally adjusted)

(g) **NAICS.adj:** US total sales in retail trade and food services in billions of dollars from Jan 2000 through December 2019 (seasonally adjusted) (See Problem 2.4 (f))

(h) **NSA:** Monthly vehicle sales in the United States from January 1976 to July 2021

Time Series Regression

8

In Chapter 7 we discussed the nonstationary ARIMA and seasonal models. These models are nonstationary because the AR-characteristic equation has one or more roots on the unit circle. In the current chapter we focus on the nonstationary process for which the mean changes with time. One such time series model is

$$X_t = b_0 + b_1 t + Z_t, \tag{8.1}$$

where Z_t is a zero-mean stationary process. If Z_t were white noise, then (8.1) would be a simple linear regression (SLR) model with independent variable time t However, the assumption regarding the Z_t process is that it is stationary, which allows for correlated errors. Model (8.1) is sometimes called a *regression model with correlated errors*. Model (8.1) is a special case of the model

$$X_t = s_t + Z_t, \tag{8.2}$$

where s_t is a deterministic (non-random) *signal* and Z_t is a zero-mean stationary process. Model (8.2) is often referred to as a signal+noise model. Model (8.1) is the special case in which $s_t = b_0 + b_1 t$, but other possible signals are $s_t = b_0 + b_1 t + b_2 t^2$, $s_t = \log(t), X s = \cos(2\pi f t)$ and so forth. For the signal+noise model in (8.2), $\mu_t = E[X_t] = E[s_t] + E[Z_t] = s_t$, because s_t is nonrandom and $E[Z_t] = 0$. Unless s_t is a constant function, then the mean, μ_t, changes with time t, and consequently X_t in (8.2) is *nonstationary*. In this chapter we will direct our attention to cases in which s_t is linear or a sinusoidal curve.

> **Key Points:**
>
> 1. We will model the stationary error process Z_t using an $\mathrm{AR}(p)$ model.
> 2. Z_t could be modeled as an $\mathrm{ARMA}(p, q)$ process, but it is common to use an $\mathrm{AR}(p)$ model in this case.

8.1 LINE+NOISE MODELS

In this section we consider the line+noise model in (8.1). While this appears to be an easy model to fit to data and to interpret, there are issues that need to be considered. In Section 5.1.1.6 and 7.1.1.5, it was noted that data from models, $(1 - \phi_1 B) Z_t = a_t$, can produce realizations with trending behavior when ϕ_1 is slightly less than or equal to one (see Figure 7.3(a)). This "random trending" is amplified in the presence of one or more unit roots, but in this chapter we will focus on the situation in which Z_t is stationary.

Question: Given a realization with trending behavior, how can we tell whether the trending is due to random trending or due to an underlying regression line?

Model (8.1) seems to be the ideal model for helping answer this question. That is, using (8.1), it seems that we can let the "model decide" the source of the trending behavior. For example, the model has the option of attributing any trending either to (a) a deterministic signal ($b_1 \neq 0$) or (b) a random trend in Z_t ($b_1 = 0$). In this section we will discuss techniques for answering the above question.

Before addressing the problem of deciding whether there is an underlying "linear signal" in the data, it is important to understand the following Key Points.

Key Points:

1. The finding of a significant slope implies that the process is nonstationary because the mean of the process, μ_t, is non-constant and depends on t.
    -   If $\hat{b}_0$ and $\hat{b}_1$ denote the estimates of b_0 and b_1 (and it is determined that $\hat{b}_1$ is significantly different from zero), then μ_t is estimated by $\hat{\mu}_t = \hat{b}_0 + \hat{b}_1 t$.
2. Given that conditions do not change, the finding of a significant slope suggests that forecasts would predict the existing trend line to continue.
3. Forecasting a functional form (such as a line) to continue into the future is an example of **extrapolation.**
    -   Good data scientists know *to be careful* when extrapolating beyond the range of the observed independent variables (in this case, time).

Before dealing with the issue of forecasting, we address the problems related with testing for the presence of a significant linear signal.

8.1.1 Testing for Linear Trend

In the following we discuss possible techniques the analyst might use to decide whether an apparent linear trend in a set of data implies that a (deterministic) linear trend term should be included in the model.

8.1.1.1 Testing for Trend Using Simple Linear Regression

We begin by using "what we already know". That is, for a simple linear regression (SLR), the standard procedure is to test for significance of the slope. In that setting, the hypothesis $H_0 : b_1 = 0$, is tested using the statistic $\hat{b}_1 / SE(\hat{b}_1)$, which has a t-distribution when the errors are uncorrelated and normally distributed. See for example, Moore, McCabe, and Craig (2021). In our setting, the independent variable is time t, and the realization is a sample from the dependent variable X_t.[1] In this section we examine the validity of using the standard simple linear regression t-test to test the significance of the slope in (8.1).

---

1 Note the possibility of notational confusion because in usual simple linear regression, the dependent variable is typically denoted by Y and the independent variable is called X. Here, X_t denotes the dependent variable.

Key Point: We use the simple linear regression approach here although we acknowledge the fact that the *assumption of uncorrelated errors is violated* (unless the stationary errors, Z_t, are white noise).

The *tswge* function `slr.wge` performs this t-test and outputs the least squares estimates of b_0 and b_1. Using the SLR testing approach, we would conclude that there is a "deterministic" linear trend if H_0 is rejected. As an example, consider the model

$$X_t = 10 + .08t + Z_t,\tag{8.3}$$

where Z_t is a realization from the AR(1) model $(1-.95B)Z_t = a_t$ with $\sigma_a^2 = 1$. The realization in Figure 8.1(a) was obtained using the R code

```
z=gen.arma.wge(n=100,phi=.95,sn=447,plot=FALSE)
t=1:100
line=10+.08*t
x=line+z
plotts.wge(x)
```

Figure 8.1(b) shows the realization in Figure 8.1(a) along with the least squares regression line whose coefficients are found using the simple linear regression function `slr.wge` in *tswge*. Using the following command

```
reg=slr.wge(x)
```

we obtain from the output

```
reg$b0hat (Intercept) 11.94136
t_test$b1hat Time 0.09125542
```

The p-value for the test, $H_0 : b_1 = 0$, is given by

```
t_test$pvalue [1] 7.434276e-23
```

In this case obviously $p < .001$. Figure 8.1(b) is obtained using the commands

```
plotts.wge(x)
fit=reg$b0hat+t*reg$b1hat
points(fit,type='l')
```

Consequently, the regression line has a highly significant slope based on the usual SLR test statistic. Figure 8.1(c) shows the residuals $x(t) - 11.941 - .091t$. These residuals should cause concern because they are clearly not uncorrelated, which, as noted, is a key assumption in the simple linear regression t-test for the significance of $\hat{b}_1$. Figure 8.1(c) can be produced using the commands

```
resid=x-fit
plot(resid,type='l')
abline(h=0)
```

Key Points:

1. The p-value output from `slr.wge` is based on an assumption of uncorrelated residuals which is often violated in time series data.
2. For this reason, p-values obtained using SLR methods should be viewed skeptically.

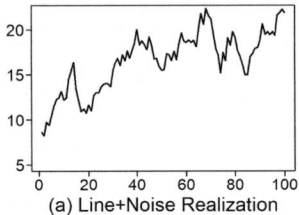 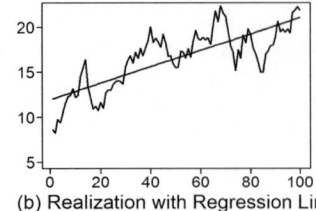 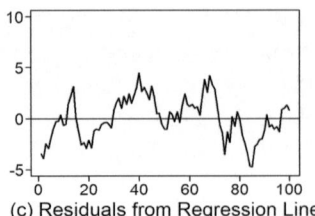

FIGURE 8.1   (a) Realization from line+noise model in (8.3), (b) realization in (a) with fitted regression line, and (c) residuals from fitted regression line.

It is reasonable to ask the following question.

> **Question:** "Does the lack of uncorrelated residuals really make a difference or is that assumption just something that statisticians worry about?"[2]

Example 8.1 Analysis of realizations from a line+noise model and an AR(1) model

To assist in examining the impact of correlated residuals, we consider the two models:

Model (a): $X_t = 10 + .1t + Z_{1t}$
Model (b): $X_t = 10 + Z_{2t}$

where Z_{1t} and Z_{2t} are the AR(1) processes $(1-.9B)Z_{1t} = a_t$ and $(1-.95B)Z_{2t} = a_t$, respectively, both with $\sigma_a^2 = 1$. Clearly, Model (a) has a deterministic linear signal while Model (b) does not. Models (a) and (b) and the *tswge* code to generate the realizations in Figures 8.2(a) and (b) are given below.

Model (a): $X_{1t} = 10 + .1t + Z_{1t}$ where z_{1t} is the AR(1) realization generated by the command

```
z1=gen.arma.wge(n=100,phi=.90,sn=65987)
time=1:100
x1=10+.1*time+z1
```

Alternatively, the *tswge* command

```
x1=gen.sigplusnoise.wge(n=100,b0=10,b1=.1,phi=.9,sn=65987)
```

generates the same realization, **x1**.

Model (b): $X_{2t} = 10 + Z_{2t}$ where z_{2t} is the AR(1) realization generated by the *tswge* command

```
z2=gen.arma.wge(n=100,phi=.95,sn=6587)
time=1:100
x2=10+z2
```

---

2 Consider the one-sample *t*-test. The mathematical validity of the *t*-test requires the data to be from a normal distribution. However, because of the central limit theorem, the normality assumption is not critical for validity in the case of "large" sample sizes (often taken to be $n > 30$). The normality assumption is still required for the test statistic to *truly* follow a *t* distribution, but the *t*-test performs well for large sample sizes in the case of non-normal data. The question here is whether the assumption of uncorrelated residuals is similarly a condition that only has underlying mathematical implications.

Figures 8.2(a.1) and (a.2) show a decomposition of x_{1t}. That is, Figure 8.2(a) is the sum of the deterministic linear signal, $s_t = 10 + .1t$, plotted in Figure 8.2(a.1) and the AR(1) realization, z_{1t}, in Figure 8.2(a.2). The AR(1) realization z_{1t} has short-term, random wandering behavior and is positively correlated data. Because the mean, $\mu_t = s_t = 10 + .1t$ depends on time, Model (a) is nonstationary.

Model (b) has a horizontal signal, $s_t = 10$, plotted in Figure 8.2(b.1). The realization in Figure 8.2(b) is obtained by adding 10 to each value in the zero-mean error series in Figure 8.2(b.2). For Model (b), $\mu_t = 10$ does not depend on time, and thus x_{2t} is a realization from a *stationary* process (and $b_1 = 0$).

Applying the commands

```
slr.wge(x1)
slr.wge(x2)
```

we obtain $p < .001$ in each case. The decision to reject $H_0 : b_1 = 0$ is correct for realization **x1**. However, in the case of realization **x2** there is no deterministic trend, and the random upward trending behavior has "tricked" the *t*-test into rejecting $H_0 : b_1 = 0$, even though it is true.

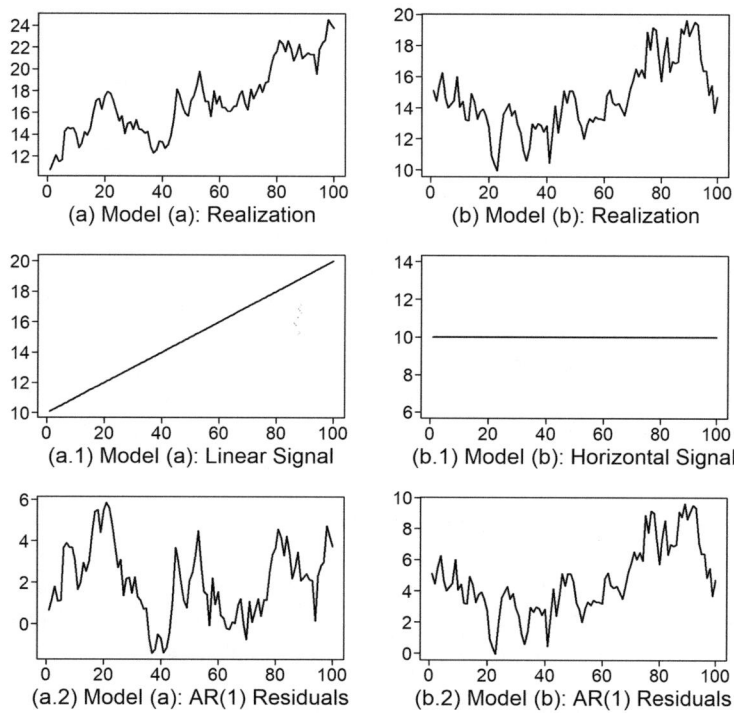

FIGURE 8.2   (a) Realization from x_{1t} from Model (a), (b) realization x_{2t} from Model (b), (a.1) the line $s_t = 10 + .1t$ and (b.1) the horizontal line $s_t = 10$; (a.2) AR(1) realization z_{1t} and (b.2) AR(1) realization z_{2t}.

Key Points:

1. Because positively correlated stationary time series can show trending behavior in the absence of an actual deterministic trend (as was the situation in Model (b)), it is difficult to correctly detect trend in the presence of positively correlated residuals.

> 2. The model selected has a major impact on forecasts!
>
> − Forecasts from a stationary model tend toward the sample mean (see Section 6.2.5)
> − We will see that forecasts from a line+noise model such as (8.1) will tend toward the fitted regression line and will tend to move upward or downward depending on the estimated slope.

8.1.1.2 A t-test Simulation

The situation in Model (b) brings up the following questions:

> **Questions:**
>
> (1) How often does the standard t-test detect a trend in stationary AR(1) realizations (which do not have trend components)?
> (2) How important is the "uncorrelated" assumption when using the standard SLR t-test on time series data?

We use a simulation study to investigat these questions. Realizations were generated from the model $X_t = 10 + Z_t$ where Z_t is a zero mean AR(1) model with $\sigma_a^2 = 1$ and where ϕ_1 takes on the values .8, .9, .95, and .99. These AR(1) models, with positive ϕ_1, have realizations that are positively autocorrelated. In the simulation study, realization lengths $n = 100$, 200, and 500 were used, and 1,000 realizations were generated from each of the twelve "ϕ_1 and realization length" combinations. For each realization the hypothesis $H_0 : b_1 = 0$ was tested at the $\alpha = .05$ level using SLR methods. Notice that the models are all stationary with mean 10 (that is, *the null hypothesis is true in all cases*), and whenever a trend was detected by obtaining a significant slope (rejecting $H_0 : b_1 = 0$), a Type I error was made. That is, the null hypothesis has been rejected when it is true.

For a given "ϕ_1 /realization length" combination, the value in Table 8.1 is the percentage of realizations out of 1,000 for which the standard t-test rejected the null and found a trend (significant slope). Because these are designed to be $\alpha = .05$ level tests and the null hypothesis is true, a trend should be detected about 5% of the time. If positive correlation in the residuals is not a "big deal", then we would expect each of the tabled values to be about 5%. A quick glance at the table says that something is very wrong!

TABLE 8.1 Percentage of times out of 1,000 replications that the t-test incorrectly found a significant slope for the given stationary AR(1) models and tabled realization lengths. The tests were all conducted at the $\alpha = .05$ level

		ϕ_1			
		.8	.9	.95	.99
	100	50.6%	66.1%	74.7%	84.3%
n	200	50.8%	63.5%	76.3%	87.7%
	500	53.1%	65.9%	75.9%	90.7%

Note that the SE of tabled values is at most 1.58%.[a]
[a] A conservative SE of a tabled value is $\sqrt{(.5 * .5)/1000} \times 100\% = 1.58\%$. See Moore et al. (2017).

The realizations in the simulation were generated from the model $X_{2t} = 10 + Z_{2t}$. Recall that Figure 8.2(b) shows a particular realization for the case in which Z_{2t} is an AR(1) model with $\phi_1 = .95$. That is, it is a stationary realization (the slope is zero), but for which correlation structure produced a "random trending behavior" that "tricked" the t-test into finding a significant slope. The questions above had to do with how often this might happen. For example, the tabled value for $\phi_1 = .95$ and $n = 100$ is 66.1% (Wow!) That is, about two-thirds of realizations of length $n = 100$ from the stationary model, $(1 - .95B)X_t = a_t$, will have enough "trending behavior" for the t-test to (incorrectly) find a significant slope. Thus, we see that incorrectly finding a significant slope occurs *much* more often than the nominal 5%, and consequently, the usual t-test is failing to perform properly in this case.

Key Points:

1. The percentage of cases in which a hypothesis test actually rejects the null hypothesis when it is true is called the "observed significance level".
2. The values in Table 8.1 are observed significance levels of the t-test under each "ϕ_1 and realization length" scenario.

Examination of all tabled values in Table 8.1 shows that the results of the t-test are uniformly ***awful***. In the terminology of the Key Points above, the "observed significance levels" are *unacceptably high*. At the extreme in the table, for realizations of length $n = 500$ and $\phi_1 = .99$, the SLR test (incorrectly) detected a significant trend about 90% of the time (instead of the hoped for 5%)!

Key Point: The finding of a significant slope when using the t-test on time series data "cannot be trusted."

QR 8.1 A t-test
Simulation

Important Note:

Before continuing, it is critically important to understand the implications of the information in Table 8.1. Hypothesis testing is designed so that if the null hypothesis is rejected, say at the $\alpha = .05$ level, then the analyst can feel "confident" that the null hypothesis is not true. In the current case, rejection of the null hypothesis should provide confidence that there is a real deterministic linear trend signal in the data. However, in each simulation scenario, even though the null hypothesis, $H_0 : b_1 = 0$, is true, there is *at least a 50% chance of rejecting the null hypothesis*. That is, if the standard SLR t-test for slope is used in this "correlated errors" setting and the result of the standard t-test "says" to reject the null hypothesis, the analyst should have *no confidence* that there really is a deterministic linear signal driving the data.

Key Points:

1. The requirement of uncorrelated residuals associated with the usual t-test for slope is a Big Deal!

2. ***Don't use the usual t-test for detecting a trend in time series data unless the residuals from the regression line appear to be uncorrelated.***
3. What is needed are tests of $H_0 : b_1 = 0$ that take into account or adjust for the correlation in the residuals.

8.1.1.3 Cochrane-Orcutt Test for Trend

The simulation study in Section 8.1.1.2 shows that the standard t-test for slope is not appropriate for the time series setting of Equation 8.1. This is a well-known fact, but we used it here because using the SLR is the first thing someone with statistical (but not time series) training would think of doing. Obviously, there is a need for better tests that take into consideration the correlation in the residuals. The most common such test is the Cochrane-Orcutt (1949) test. We consider again the line+noise model, $X_t = b_0 + b_1 t + Z_t$, given previously in (8.1). The outline below is for the case in which Z_t satisfies the AR(1) model $(1 - \phi_1 B)Z_t = a_t$. The AR(1) is used in the outline for simplicity, but the restriction to AR(1) errors is not imposed in the analyses we will discuss. The version of the Cochrane-Orcutt (CO) procedure using an AR(1) to model the residuals is as follows:

(1) Fit a linear regression line to the original data by calculating the least squares estimates $\hat{b}_0$ and $\hat{b}_1$.
(2) Find $\hat{Z}_t = X_t - \hat{b}_0 - \hat{b}_1 t$ (which we will assume here to be approximately distributed as AR(1)).
(3) Calculate $\hat{\phi}_1$ by fitting an AR(1) model to $\hat{Z}_t$.
(4) Compute $\hat{Y}_t = (1 - \hat{\phi}_1 B)X_t$, $t = 2,\ldots,n$.

Note that $\hat{Y}_t = (1 - \hat{\phi}_1 B)X_t = (1 - \hat{\phi}_1 B)(b_0 + b_1 t + Z_t)$

$$= (1 - \hat{\phi}_1 B)(b_0 + b_1 t) + (1 - \hat{\phi}_1 B)Z_t$$

$$= b_0 + b_1 t - \hat{\phi}_1 b_0 - \hat{\phi}_1 b_1 (t - 1) + (1 - \hat{\phi}_1 B)Z_t,$$

using the facts that $Bb_0 = b_0$ and $Bt = t - 1$. Rearranging terms

$$\hat{Y}_t = b_0(1 - \hat{\phi}_1) + b_1(t - \hat{\phi}_1(t - 1)) + (1 - \hat{\phi}_1 B)Z_t$$
$$= c + b_1 t_{\hat{\phi}_1} + g_t,$$

(8.4)

where $c = b_0(1 - \hat{\phi}_1)$, $t_{\hat{\phi}_1} = t - \hat{\phi}_1(t - 1)$, and $g_t = (1 - \hat{\phi}_1 B)Z_t$. [3]

QR 8.2 Cochrane-Orcutt Method

---

[3] The AR(1) results above easily extend to the general case of an AR(p) fit to Z_t. If $\hat{\phi}(B) = 1 - \hat{\phi}_1 B - \cdots - \hat{\phi}_p B^p$, then $\hat{Y}_t = c + b_1 t_{\hat{\phi}} + g_t$, where $c = b_0(1 - \hat{\phi}_1 - \cdots - \hat{\phi}_p)$, $t_{\hat{\phi}_1} = t - \hat{\phi}_1(t - 1) - \cdots - \hat{\phi}_p(t - p)$, $g_t = \hat{\phi}(B)Z_t$.

Some things to notice regarding (8.4) are the following:

(a) This is a new line+noise model with new constant, c, but with *the same slope b_1*
(b) The new "time" variable is $t_{\hat{\phi}_1}$
(c) The new residuals are denoted as g_t: Because $(1 - \phi_1 B)Z_t = a_t$ can be inverted to obtain $Z_t = (1 - \phi_1 B)^{-1} a_t$, then g_t can be expressed as

$$g_t = \left(1 - \hat{\phi}_1 B\right) Z_t = \left(1 - \hat{\phi}_1 B\right)\left(1 - \phi_1 B\right)^{-1} a_t.$$

Because the operators $\left(1 - \hat{\phi}_1 B\right)$ and $\left(1 - \phi_1 B\right)^{-1}$ *nearly cancel* each other, it follows that g_t is similar to a_t and thus g_t should be "nearly uncorrelated".

In summary, the CO method uses a transformation that results in a new line+noise model with the same slope, b_1, and nearly uncorrelated residuals. Thus, to test $H_0 : b_1 = 0$, the CO method uses the usual regression t-test on this "new" data because the assumption of uncorrelated errors is reasonable. Using this transformed data, $H_0 : b_1 = 0$ is tested using the test statistic

$$t_{CO} = \frac{\hat{b}_1^{(CO)}}{SE(\hat{b}_1^{(CO)})}, \tag{8.5}$$

where $\hat{b}_1^{(CO)}$ is the least squares (regression) estimate of b_1 in the "new" model, and $SE\left(\hat{b}_1^{(CO)}\right)$ is its usual standard error.

Key Points:

1. The CO method employs a transformation to produce almost uncorrelated residuals.
2. *tswge* command `co.wge` implements the CO procedure by
    - finding the least squares regression line
    - computing the residuals, $\hat{Z}_t$, from the regression line
    - fitting the residuals using an $AR(p)$ model with operator $\hat{\phi}(B)$ (based on AIC-type measures to identify p. Burg estimates are used because they always obtain a stationary solution.)
    - transforming the data to obtain $\hat{Y}_t = \hat{\phi}(B)X_t = c + b_1 t_{\hat{\phi}_1} + g_t$.
    - using (8.5) to test for the significance of the slope in this transformed model under the assumption that g_t is "nearly" uncorrelated data.

Table 8.2 shows simulation results analogous to those in Table 8.1. That is, 1,000 replications were generated from each "ϕ_1 and realization length" combination, and the value in the table is the percentage of realizations out of 1,000 for which the CO test found a significant slope while testing at the $\alpha = .05$ level. Note again that for each "ϕ_1 and realization length" combination, the null hypothesis is true. That is, if the CO test appropriately corrected for the correlation in the residuals, the tabled values should each be close to 5%.

Table 8.2 results provide some "good news" and some "bad news". First the good news. The values in Table 8.2 are much closer to 5% than are those in Table 8.1. However, the bad news is that, although the CO method clearly made an adjustment for the correlation structure, the adjustment was not sufficient to produce a valid 5% level test. The only value in the table that is within two SE's of 5% is 7.5% for

"$n = 500 / \phi_1 = .8$". Most of the other values are unacceptably high. For example, given a "stationary" realization of length $n = 100$ with $\phi_1 = .95$, there is a 32.7% chance that the CO test will (incorrectly) "detect" a significant trend. The implication is that if the CO test detects a significant linear trend, there is still a lack of "confidence" that the trend is not just a random trend in a realization from a stationary model (and should not be predicted to continue).

TABLE 8.2   Percentage of times out of 1,000 replications that the CO test incorrectly found a significant slope for the given AR(1) models and tabled realization lengths. The tests were all conducted at the $\alpha = .05$ level

		.8	.9	.95	.99
	100	15.0%	21.4%	32.7%	48.0%
n	200	10.8%	12.9%	20.0%	42.3%
	500	7.5%	8.8%	11.4%	29.5%

Note that the SE of tabled values is at most 1.58%.

Example 8.1 (revisited)

The *tswge* command `co.wge` can be used to apply the CO test to test for linear trend in realizations **x1** and **x2** in Example 8.1. Recall that **x1** is a realization from Model (a) (with trend) and **x2** is from Model (b) (without trend). Applying the commands

```
co.wge(x1)
co.wge(x2)
```

we obtain $p < .001$ for **x1** from Model (a). That is, CO correctly identifies the underlying trend. For realization **x2** from Model (b), the p-value for the CO test is .016, implying that CO detects a trend (but the model has no trend). Again, note that in Table 8.2 we see that incorrectly detecting a trend for the $n = 100$ and $\phi_1 = .95$ case tends to occur about 32.7% of the time.

8.1.1.4 Bootstrap-Based Test for Trend

Although the Cochrane-Orcutt method is widely used to test for trend in time series data, the results in Table 8.2 clearly show it can have excessively high "observed significance levels". Again, the problem with observed significance levels is well known (see Woodward and Gray (1993), Vogelsang (1998), and Sun and Pantula (1999), to name a few). Other tests, discussed in Woodward and Gray (1993) are available, but they all have a similar problem. Woodward, Bottone, and Gray (1997) addressed the problem of testing for trend using a bootstrap-based approach. This is denoted as the WBG test. The WBG testing approach, implemented in *tswge* function `wbg.boot.wge`, can be summarized as follows:

- The t_{CO} statistic is calculated as described previously on the original (observed) dataset.
- An autoregressive model is fit to the original data. Note that the data may or may not actually contain a deterministic trend. Fitting an AR model implicitly assumes that any trending behavior in the series is due to the correlation structure alone. That is, the assumption is being made that the null hypothesis ($H_0 : b_1 = 0$) is true.
- B realizations are generated from the autoregressive model fit to the original data. A parametric bootstrap procedure is used which assumes normal errors and generates the a_ts in the generated realizations as random normal deviates.
    - If normality of the errors is questionable, then a nonparametric bootstrap could be used.
- For the bth realization, $b = 1, \ldots, B$, we calculate $t_{CO}^{(b)}$ using the CO procedure. The procedure is as follows:
    - Find the least squares regression line
    - Compute the residuals, $\hat{Z}_t$, from the regression line

- Fit the residuals with an AR(p) model with pth order operator $\hat{\phi}(B)$ (using AIC-type measures based on Burg estimates to identify p)
- Transform the data to obtain $\hat{Y}_t = \hat{\phi}(B)X_t$
- Calculate $t_{CO}^{(b)}$ as in (8.5) for each of the B bootstrap realizations, $b = 1, \ldots, B$.
- Recalling that t_{CO} is the test statistic calculated on the original dataset, then for a two-sided test, $H_0: b_1 = 0$ is rejected at significance level α if $|t_{CO}| > t_{1-\alpha/2}^*$, where t_β^* is the βth empirical quantile of the collection $t_{CO}^{(b)}$, $b = 1, \ldots, B$ with standard adjustments made for one-sided tests.

Key Point: We use **_Burg estimates_** in the WBG method.

1. For each of the B bootstrap realizations, AIC-type methods are used for identifying p. Using the Burg estimates speeds up the process which will involve using an AIC-type procedure on the original data and all B realizations (where B is typically 199, 399, …).
2. The Burg estimates always find a stationary solution. ML estimates may sometimes fail to converge or may find an explosively nonstationary solution.
3. The computational speed with which Burg estimates can be computed makes them ideal for the WBG bootstrap approach because of the hundreds of realizations that are analyzed.

The simulation scenarios in Table 8.1 and Table 8.2 were repeated using the WBG method. Table 8.3 shows the results of these simulations. In the table it can be seen that all observed significance levels are within two standard errors in all except two cases: for "$n = 100 / \phi_1 = .99$" and for "$n = 200 / \phi_1 = .99$".

Note that while these two observed significance levels using the WBG bootstrap were about 11% and certainly larger than the nominal 5% level, the corresponding observed significance levels in Table 8.2 using the CO test were about 45%.

QR 8.3 Bootstrap
Test for Trend

TABLE 8.3   Percentage of times out of 1,000 replications and using $B = 199$ that the WBG test incorrectly found a significant slope for the given AR(1) models and tabled realization lengths. The tests were all conducted at the $\alpha = .05$ level

		.8	.9	.95	.99
	100	4.7%	5.2%	7.4%	11.9%
n	200	4.2%	5.2%	5.5%	10.2%
	500	4.6%	5.3%	5.5%	8.0%

Note that the SE of tabled values is at most 1.58%.

Example 8.1 (revisited again) The *tswge* command `wbg.boot.wge` can be used to apply the WBG bootstrap test for trend in realizations `x1` and `x2` in Example 8.1. We use the commands

```
wbg.boot.wge(x1)
wbg.boot.wge(x2)
```

and obtain $p = .008$ for **x1** from Model (a). That is, the WBG test correctly identifies the underlying trend. For realization, **x2**, from Model (b), $p = .120$, and the WBG test correctly fails to reject the null hypothesis of no trend. Recall that both the SLR and CO tests detected trend for Models (a) and (b) although Model (b) did not have a trend component. From Table 8.3 we see that we only reject the null hypothesis under this scenario ($n = 100$ and $\phi_1 = .95$) about 7.4% of the time (as compared to 32.7% for CO) .

It should be noted that, because of the fact that the bootstrap randomly generates 399 (default) realizations from the fitted AR model, the results of subsequent **wbg.boot.wge** commands will differ somewhat. For example, we issued the above commands four more times. The p-values for the four issuances of each command are as follows:

x1 ($b_1 \neq 0$)	x2 ($b_1 = 0$)
.013	.148
.020	.113
.030	.143
.018	.092

In each case the results are consistent with those given above. In order to be able to replicate a particular bootstrap result you can set a value for seed **sn**. For example, issuing the commands

```
wbg.boot.wge(x1,sn=234)
wbg.boot.wge(x2,sn=183)
```

you will always obtain p-values .020 and .113, respectively.

Don't Miss This Point about Power

Simulation results (not shown) based on generating 1,000 realizations from line+noise Model (a) show that the SLR, CO, and WBG test (correctly) found a significant trend in 99.8%, 91%, and 50.7% of the realizations, respectively.[4] In these simulations the SLR and CO tests have impressively high powers, but they come at a cost. Consider again Tables 8.1 and 8.2 for the case $b_1 = 0, n = 100$ and $\phi_1 = .9$, (for which there is no deterministic linear term). In this setting the SLR and CO tests (incorrectly) detected a significant deterministic linear trend 66.1% and 21.4% of the time, respectively. On the other hand, Table 8.3 shows that the WBG test (incorrectly) rejected $H_0 : b_1 = 0$ 5.2% of the time, which is consistent with what would be expected for an $\alpha = .05$ level test. That is, if $H_0 : b_1 = 0$ is rejected, the analyst can be "confident" that a deterministic linear trend component does exist. However, the SLR and CO tests do not provide similar protection. That is, they "cry wolf" (reject H_0) too many times when there is no wolf (H_0 is true).

Key Points:

1. The WBG bootstrap-based test controls the observed significance level much better than the SLR and CO tests in the presence of correlated residuals.
2. The SLR and CO tests have higher power than WBG, but the higher powers are artificial because they are not controlling the observed significance level.

---

4 The 99.8%, 91%, and 50.7% quoted here are measures of the power of the tests under the particular alternative $b_1 = .1$, n = 100, 2 and $\phi_1 = .9$.

8.1.1.5 Other Methods for Testing for Trend in Time Series Data

Many other tests are available for testing for trend in the presence of correlated residuals. Among these, Beach and MacKinnon (1978) developed an ML-based method, and Cowpertwaite and Metcalfe (2009) recommend a generalized least squares (GLS) approach in the presence of AR(1) residuals. Consider the previous case in which the true model is $(1 - 0.9B)X_t = a_t$, and suppose tests are run at the 5% level of significance. Table 8.2 shows that the CO method incorrectly detects a trend about 21% of the time. Woodward et al. (1997) showed that the ML approach of Beach and MacKinnon (1978) detects a trend about 17% of the time, and further simulations, not shown here, indicate that the GLS approach detects a trend about 12% of the time. The simulations discussed in Woodward, et al. (1997) show that when X_t is a stationary AR(1) with $\phi_1 = .9$ or higher, realizations of length $n = 1000$ or more may be required in order for the CO or ML tests to have observed significance levels close to the nominal levels. Further discussion of these inflated significance level issues can be found in Park and Mitchell (1980), Woodward and Gray (1993), Woodward et al. (1997), and Woodward (2013).

> **Key Point:** We recommend using the WBG approach because it controls the observed significance levels at near the nominal level and is easily implemented using *tswge* function `wbg.boot.wge`.

8.1.2 Fitting Line+Noise Models to Data

Section 8.1.1 discussed the issue of determining whether a deterministic linear signal should be included in the model for a given time series realization. After this decision has been made, a final model will need to be fit to the data. Consider the model

$$X_t = b_0 + b_1 t + Z_t,$$

previously given in (8.1) where Z_t is a zero-mean stationary mode that we will model using an AR(p). When analyzing a set of time series data for which we believe the appropriate model could include a deterministic linear component, we recommend using the following procedure, which follows the strategy in Section 8.1.1.

Procedure: First check for the existence of a deterministic linear trend. That is, test $H_0 : b_1 = 0$ (we recommend using the WBG method).

   (a) ***If $H_0 : b_1 = 0$ is rejected:*** Then fit a line+noise model to the data using the following steps:

      (i) Fit a least squares line to the data $\hat{s}_t = \hat{b}_0 + \hat{b}_1 t$ using the command

```
xa=slr.wge(x1)
# among the output are xa$b0hat, xa$b1hat # SLR estimates b̂₀, b̂₁
```

      (ii) Find the residuals $\hat{Z}_t = X_t - \hat{b}_0 - \hat{b}_1 t$. The output from `slr.wge` contains these residuals in
```
xa$res
```

     (iii) Fit an AR model to the residuals using Burg estimates
```
xa.aic=aic.burg.wge(xa$res,p=1:5)
```

     (iv) The final fitted model is $X_t = \hat{b}_0 + \hat{b}_1 t + \hat{Z}_t$, where $\hat{Z}_t$ is the maximum likelihood AR fit to the residuals and $\hat{\sigma}_a^2$ is the white noise variance estimate.

   (b) ***If $H_0 : b_1 = 0$ is not rejected:*** Then the evidence suggests that the fitted model does not contain a linear signal, and we fit an ARMA(p,q) or ARIMA(p,d,q) model to X_t as discussed in Chapters 6 and 7.

Key Points:

1. The SLR, CO, and WBG tests are designed to test whether $b_1 = 0$.
2. If it is concluded that $b_1 \neq 0$, then regardless of the testing method used to arrive at that decision, the final model is obtained by
   - Using `slr.wge` to fit a regression line to the data
   - Finding the residuals, $\hat{Z}_t$, from the regression line and fitting an AR(p) model to them.
   - Although we have used Burg estimates up until this point, it is common practice to use ML estimates for the final model (but Burg estimates would also be a good choice).

Example 8.2 Fitting models to datasets x1 and x2 in Figure 8.2(a) and (b), respectively.

Model (a) *Realization* **x1** *from* $X_{1t} = 10 + .1t + Z_{1t}$ *where* Z_{1t} *is the AR(1) process* $(1 - .9B)Z_t = a_t$.
Recall that the realization, **x1**, is generated by the command

```
x1=gen.sigplusnoise.wge(n=100,b0=10,b1=.1,phi=.9,sn=65987)
```

Note: We know that this realization came from a model with a linear signal, but we analyze the data as if we do not know this.

Procedure: First check for the existence of a deterministic linear signal. That is, test $H_0 : b_1 = 0$ (we recommend using the WBG method) using the command

```
wbg.boot.wge(x1)
# the resulting p-value is .030
# repeated issuance of the command produces similar p-values #try it!
```

Conclusion: We reject $H_0 : b_1 = 0$ (at the $\alpha = .05$ level) and decide to fit a model containing a linear trend. Fit the model using the following steps:

(1) **Fit a least squares line to the data** $\hat{s}_t = \hat{b}_0 + \hat{b}_1 t$ using the command

```
xa=slr.wge(x1)
# among the output are xa$b0hat, xa$b1hat # SLR estimates b̂₀,b̂₁
# xa$b0hat=12.028 xa$b1hat=.059
```

(2) **Find the residuals** $\hat{Z}_t = X_t - \hat{b}_0 - \hat{b}_1 t$ **:** For the current dataset $\hat{Z}_t = X_t - 12.028 - .059t$.

   The output from `slr.wge` contains these residuals in `xa$res`.

(3) **Fit an AR model to the residuals** (using ML estimates for the final model)

```
xa.aic=aic.wge(xa$res,p=0:5,q=0:0)
```

   AIC selects the AR(3) model below for the residuals (even though the original model was an AR(1))

$$\left(1 - .98B + .08B^2 + .14B^3\right)\hat{Z}_t = a_t$$

   where $\hat{\sigma}_a^2 = .971$.

(4) **The final fitted model is**

$$X_t = 12.028 + .059t + \hat{Z}_t,$$

where $\left(1 - .98B + .08B^2 + .14B^3\right)\hat{Z}_t = a_t$ with $\hat{\sigma}_a^2 = .971$.

Note: The above sequence of commands was used to illustrate the steps involved in fitting line+noise models to data. After deciding to include a line in the final model, the *tswge* command

```
fit.sig=fore.sigplusnoise.wge(x1,linear=TRUE)
```

can be used to fit a line+noise model to the data using AIC to select p, and MLE to estimate the parameters of the pth order AR model fit to $\hat{Z}_t$. As the name of the command suggests, **fore. sigplusnoise.wge** also computes forecasts which will be discussed in Section 8.1.3. For now, among the output from the above command are

```
fit.sig$b0hat 12.028
fit.sig$b1hat .059
fit.sig$phi.z 0.98 -0.08 -0.14
fit.sig$wnv .971
```

which are consistent with the final model in (iv) above.

Model (b) Realization **x2** from $X_{2t} = 10 + Z_{2t}$ where Z_{2t} is the AR(1) process $\left(1 - .95B\right)Z_t = a_t$. Recall that the realization, **x2**, is generated using the commands

```
z2=gen.arma.wge(n=100,phi=.95,sn=6587)
x2=10+z2
```

Note: We know that this realization came from a stationary model without a linear signal, but the analysis will proceed as if this information is unknown.

Procedure: *Check for the existence of a deterministic linear signal.* That is, test $H_0 : b_1 = 0$ (again, we recommend using the WBG method). We issue the commands

```
wbg.boot.wge(x2)
# the resulting p-value is .12
# repeated issuance of the command produces similar p-values
```

Conclusion: We do not reject $H_0 : b_1 = 0$ (at $\alpha = .05$) and decide to model the data using techniques for modeling ARMA and ARIMA models. The data, sample autocorrelations, and Parzen spectral density estimate obtained using **plotts.sample.wge(x2)** are shown in Figures 8.3(a)–(c).

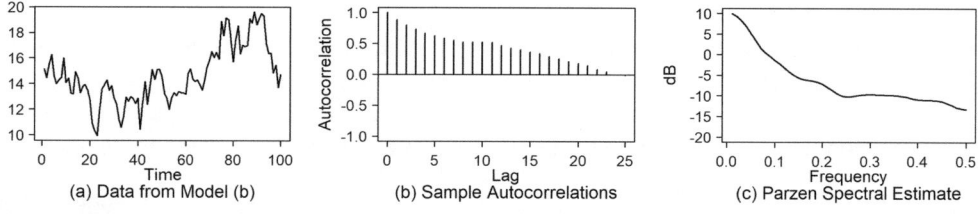

(a) Data from Model (b)   (b) Sample Autocorrelations   (c) Parzen Spectral Estimate

FIGURE 8.3 (a) Data from Model (b), (b) sample autocorrelations of the data in (a), and (c) Parzen spectral density estimate of the data in (a).

The slowly damping sample autocorrelations are suggestive of a possible unit root, but because this dataset is simulated data and we have no physical reason to include a unit root, we simply fit an ARMA(p,q) model to the data. Using the command

`aic.wge(x2,p=0:8,q=0:2)`

AIC selects the AR(1) model $(1-.878B)(X_t - 14.656) = a_t$ where $\hat{\sigma}_a^2 = 1.073$.

QR 8.4 Line+Noise
Model Fitting

A final note: In Section 8.1.3 we will discuss forecasting with line+noise models. We again note that these forecasts "look, smell, and act" like *extrapolation* in a simple linear regression setting. Be careful when projecting a line (or other deterministic function) beyond the range of the independent variables (in this case, "time").

8.1.3 Forecasting Using Line+Noise Models

In this section we discuss forecasting future values using a fitted line+noise model. Assume that, using the techniques of Section 8.1.2, a time series, X_t, has been modeled using a line+noise model

$$X_t = \hat{b}_0 + \hat{b}_1 t + \hat{Z}_t, \tag{8.6}$$

where $\hat{Z}_t$ is modeled as a zero-mean AR(p) process $\hat{\phi}(B)\hat{Z}_t = \hat{a}_t$. Suppose data are observed for times $t = 1,2,\ldots,t_0$ and forecasts $\hat{X}_{t_0}(\ell)$ are desired for $\ell = 1,\ldots,k$. Forecasts are obtained as follows:

(1) ***Forecast the residual series, $\hat{Z}_t$:***
   Because $\hat{Z}_t = X_t - \hat{b}_0 - \hat{b}_1 t$ is modeled as a stationary AR(p) process, forecasts $\hat{Z}_{t_0}(\ell)$, $\ell = 1,\ldots,k$ can be found using `fore.arma.wge` based on its stationary model fit using techniques of Section 6.2.

(2) ***Convert the forecasts in (1) to forecasts, $\hat{X}_{t_0}(\ell)$, for $X_{t_0+\ell}$:***
   The forecasts, $\hat{Z}_{t_0}(\ell)$, can be converted to forecasts, $\hat{X}_{t_0}(\ell)$, for $X_{t_0} + \ell$ using the formula

$$\hat{X}_{t_0}(\ell) = \hat{b}_0 + \hat{b}_1 t + \hat{Z}_{t_0}(\ell). \tag{8.7}$$

Notes:

(1) If a line+noise model has been fit to a set of data, then the tacit assumption is that this fitted line can be appropriately "extrapolated" into the future.
(2) Because $\hat{Z}_t$ is modeled as a stationary process with zero mean, the eventual forecasts, $\hat{Z}_{t_0}(\ell)$' will tend toward zero. Consequently, the eventual forecasts, $\hat{X}_{t_0}(\ell)$, will tend toward the fitted line.

QR 8.5 Line+
Noise Forecasting

Example 8.3 Forecasting realization `x1` using the model in (8.7).

Assume that a decision has been made to fit a line+noise model to the **x1** data and we want to forecast 25 steps beyond the end of the realization. The estimates of the fitted line from Example 8.2 Model (a) are found by issuing the command

```
xa=slr.wge(x1)
```

From the output we obtain

$\hat{b}_0$ is 12.028 (**xa$b0hat**)
$\hat{b}_1$ is .059 (**xa$b1hat**)
$\hat{z}_t$, the residual series (**xa$res**)

AIC selected an AR(3) model for the residual series using

```
xa.aic=aic.wge(xa$res,p=0:5)
```

The output includes:

AIC selection: 3 (**xa.aic$aic**)
$\hat{\phi}_1, \hat{\phi}_2$, and $\hat{\phi}_3$: 0.98, −0.08, and −0.14 (**xa.aic$phi**)
    $\hat{\sigma}_a^2$: .971 (**xa.aic$vara**)

Forecasts are obtained as follows:

(1) Forecast the residual series: Forecast the residual series using the command

```
f.z1=fore.arma.wge(xa$res,xa.aic$phi,n.ahead=25)
```

The forecasts for the residual series along with 95% limit lines are shown in Figure 8.4(a).

(2) Convert the forecasts in (1) to forecasts, $\hat{X}_{t_0}(\ell)$:

The forecasts for 25 steps ahead of the line+noise dataset **x1**, can be calculated using the commands below.

```
tf=101:125
f.x1=f.z1$f+xa$b0hat+xa$b1hat*tf
```

The vector **f.x1** contains the forecasts $\hat{X}_{t_0}(\ell), \ell =1, \ldots, 25$.

Notes:

(1) The commands above were provided to show the forecasting procedure step-by-step.
(2) In practice the forecasts for the line+noise fit can be easily obtained using the command

```
fore.sigplusnoise.wge(x1,n.ahead=25,limits=TRUE)
```

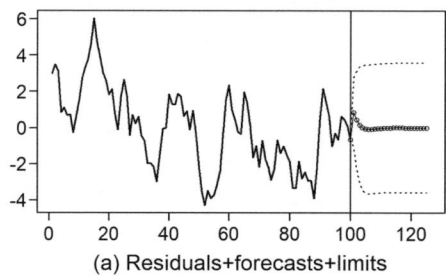

(a) Residuals+forecasts+limits

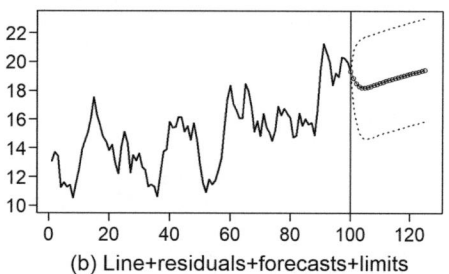

(b) Line+residuals+forecasts+limits

FIGURE 8.4 (a) Forecasts up to 25 steps ahead for residual series $\hat{Z}_t$ along with 95% prediction limits (b) corresponding forecasts and 95% prediction limits for the line+noise signal in x1.

Key Points:

1. As mentioned previously, the fact that we *have decided* to include a deterministic linear component in our model tacitly implies that the eventual forecasts will follow the fitted line.
   - This is extrapolation!
2. Before including a line in a model to be used for forecasting, you should be certain that "it belongs there" and that you have good reason to believe that current forces impacting the data will continue.

8.2 COSINE SIGNAL+NOISE MODELS

In this section, we address the case in which the signal, s_t, is a cosine function. The signal+noise model in this case is

$$X_t = C_0 + C_1 \cos(2\pi ft + U) + Z_t \tag{8.8}$$

where Z_t is a zero-mean stationary process which we will model using an autoregressive model.[5]
    Figure 8.5(a) shows a realization from the model

$$X_t = 5\cos(2\pi(.1)t + \pi/3) + Z_t, \tag{8.9}$$

where $(1 - .75B)Z_t = a_t$, and $\sigma_a^2 = 1.5$. Note that $C_0 = 0$. The realization was created using the *tswge* commands

```
z=gen.arma.wge(n=100,phi=.75,sn=921,vara=1.5)
time=1:100
sig.cos=5*cos(2*pi*.1*time+pi/3)
x.cos=sig.cos+z
```

QR 8.6 Cosine+ Noise Model

Figure 8.5(a) shows the cosine-based cyclic behavior and shows that the amplitudes of the peaks and troughs vary from cycle to cycle. Figure 8.5(b) shows the underlying cosine signal while Figure 8.5(c) is a plot of the noise realization, z_t. The plot in Figure 8.5(a) is simply the sum of the signals plotted in Figures 8.5(b) and (c).

---

5 Woodward et al. (2017) refer to the model in (8.8) as a harmonic component model. Also, because $\sin(\theta) = \cos\left(\frac{\pi}{2} - \theta\right)$ the "cosine signal+noise models" could just as easily be called "sine signal+noise" models or "sinusoidal" signal+noise models.

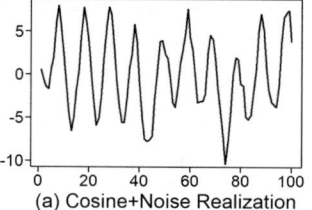 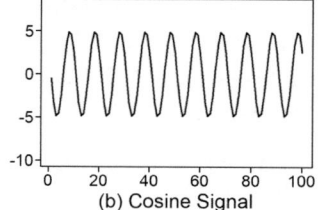 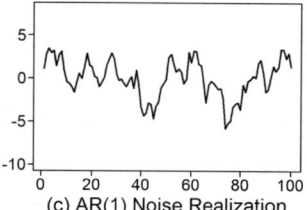

(a) Cosine+Noise Realization   (b) Cosine Signal   (c) AR(1) Noise Realization

FIGURE 8.5 (a) Realization from cosine signal+noise model in (8.9), (b) plot of the deterministic curve $5\cos\left(2\pi(.1)t + \pi/3\right)$, and (c) realization from AR(1) model $(1-.75B)Z_t = a_t$ with $\sigma_a^2 = 1.5$.

Cosine signal+noise models could also contain cosines at several different frequencies. We restrict our analyses to a cosine signal at a single frequency.

> **Key Point:**
>
> 1. As was the case with the line+noise models, a major step in the modeling process is deciding whether to include a (deterministic) cosine signal in the model
> 2. We will address this important "decision" issue in Section 8.2.3.

8.2.1 Fitting a Cosine Signal+Noise Model to Data

Before fitting a model of the form (8.8) to a set of data, it is common to re-write (8.8) in the form

$$X_t = C_0 + A\cos\left(2\pi ft\right) + B\sin\left(2\pi ft\right) + Z_t, \tag{8.10}$$

where $A = \cos(U)$ and $B = -\sin(U)$.[6] If the *decision* has been made to fit a model of the form (8.10) to a dataset **x**, then we recommend the following procedure:

(1) Estimate the parameters of $C_0 + A\cos(2\pi ft) + B\sin(2\pi ft)$:

That is, C_0, A, B, and f need to be estimated. The presence of the parameter, f, causes this to be a nonlinear estimation problem.

(a) Estimate f : Our experience is that if the sinusoidal term is sufficiently strong in the signal, then estimating f can be accomplished by examining the data, finding the location of the dominant peak in the spectral density, and/or using the factor table and domain knowledge.

(b) Estimate C_0, A and B : Notice that if f is given (or an estimated value is selected for use) then (8.10) is no longer nonlinear, and can be viewed as a multiple linear regression,

$$X_t = C_0 + AH_1\left(t\right) + BH_2\left(t\right) + Z_t. \tag{8.11}$$

---

6 Use the trigonometric identity $\cos(F+G) = \cos(F)\cos(G) - \sin(F)\sin(G)$. In our case, $F = 2\pi ft$ and $G = U$ in (8.8). Consequently, in (8.10), $A = \cos(U)$ and $B = -\sin(U)$. Given the above identities, then $C_1\cos(2\pi ft + U)$ in (8.8) can be written as $A\cos(2\pi ft) + B\sin(2\pi ft)$ where $A = C_1\cos(U)$ and $B = -C_1\sin(U)$.

where $H_1(t) = \cos(2\pi ft)$ and $H_2(t) = \sin(2\pi ft)$. Letting $\hat{f}$ denote the estimate for f, the estimated independent variables are $\hat{H}_1(t) = \cos(2\pi \hat{f}t)$ and $\hat{H}_2(t) = \cos(2\pi \hat{f}t)$. Multiple linear regression techniques are then applied to find estimates $\hat{C}_0$, $\hat{A}$ and $\hat{B}$

(2) **Calculate the residuals, $\hat{Z}_t$:** The residuals are $\hat{Z}_t = X_t - \hat{C}_0 - \hat{A}\cos(2\pi \hat{f}t) - \hat{B}\sin(2\pi \hat{f}t)$.

(3) **Fit an AR model to the residuals:** We will use AIC to identify the AR order and find the ML estimates of the parameters of the AR model.

(4) **Specify the final model:** The final fitted model is

$$X_t = \hat{C}_0 + \hat{A}\cos(2\pi \hat{f}t) + \hat{B}\sin(2\pi \hat{f}t) + \hat{Z}_t. \tag{8.12}$$

The final model should specify the order and parameter estimates of the AR model fit to the data, and should include the estimated white noise variance.

Example 8.4 Fitting a Cosine Signal+Noise Model to the Data in Figure 8.5(a)

We use the steps listed above to fit a cosine signal+noise model to the data in Figure 8.5(a) which were generated from the model in (8.9). The code for generating the data is given following equation (8.9) and the data to be analyzed are in **x.cos**. As discussed above, before estimating the parameters we express the model in (8.9) in the form given in (8.10). Using the footnote, $A = 5\cos(\pi/3) = 5(.5) = 2.5$ and $B = -5\sin\left(\dfrac{\pi}{3}\right) = -5(.866) = -4.33$. In the form of (8.10), model (8.9) becomes

$$X_t = 2.5\cos(2\pi ft) - 4.33\sin(2\pi ft) + Z_t, \tag{8.13}$$

where $(1 - .75B)Z_t = a_t$ with $\sigma_a^2 = 1.5$.

Figure 8.6(a) is a replot of the signal+noise realization in Figure 8.5(a) along with the sample autocorrelations and Parzen spectral density estimate.

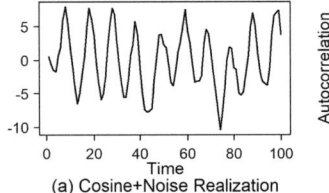 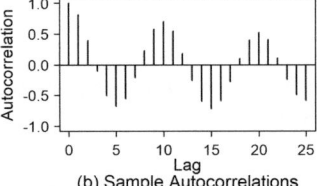 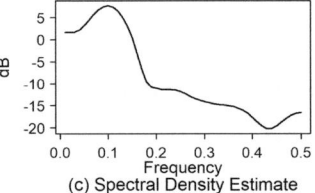

(a) Cosine+Noise Realization   (b) Sample Autocorrelations   (c) Spectral Density Estimate

FIGURE 8.6   (a) Cosine signal+noise realization in Figure 8.5(a), (b) sample autocorrelations of the data in (a), and (c) Parzen spectral density estimate of the data in (a).

(1) **Estimate the parameters of $C_0 + A\cos(2\pi ft) + B\sin(2\pi ft)$:**

The parameter estimation procedure involves the following two steps:

(a) **Estimate f:** By examining the data it can be seen that the data pass through about 10 cycles in the realization of length $n = 100$, suggesting a frequency of $f = .1$. The spectral density provides supporting information in that it has a peak at about $f = .1$. The sample autocorrelations, which do not always provide informative frequency information, have a damped sinusoidal behavior with a period of about 10, again suggesting a frequency of about

$f = .1$. Table 8.4 shows the factor table for an AR(10) fit to the data in Figure 8.5(a) and 8.6(a) using Burg estimates. Overfit models with $p = 8, 12,$ and 14 all had a primary system frequency of $f = .099$. Consequently, we choose to use the estimate, $\hat{f} = .1$. In the R code below we use the notation `f.est=.1`.

TABLE 8.4   Factor Table for AR(10) Overfit to Cosine Signal+Noise Data

AR-FACTOR	ROOTS	$|r|^{-1}$	f_0
$1-1.61B+.97B^2$	$.83\pm.59i$	.99	.099
$1-.92B$	1.08	.92	0
$1+1.54B+.66B^2$	$-1.16\pm.39i$	.81	.448
$1+.76B+.59B^2$	$-.64\pm1.13i$	.77	.332
$1-.26B+.53B^2$	$.25\pm1.35i$	.73	.221
$1-.33B$	3.06	.33	0

(b) Estimate $C_0, A,$ and B:

$\hat{H}_1(t) = \cos\left(2\pi\hat{f}t\right)$ and $\hat{H}_2(t) = \cos\left(2\pi\hat{f}t\right)$ are computed using the following code.
(Recall that the data are in **x.cos**.)

```
n=length(x.cos)
h1=rep(0,n)
h2=rep(0,n)
f.est=.1
for(t in 1:n)  h1[t]=cos(2*pi*f.est*t)
for(t in 1:n)  h2[t]=sin(2*pi*f.est*t)
```

The parameters $C_0, A,$ and B are estimated using the R command

```
h.est=lm(x.cos~h1+h2)
```

The estimates for $C_0, A,$ and B are found in **h.est$coefficients[k]**, `k=1,2`, and `3`, respectively. Specifically, we obtain the following

```
h.est$coefficients[1]  # -.0339
h.est$coefficients[2]  # 2.7590
h.est$coefficients[3]  # -4.6649
```

(2) Calculate the residuals, $\hat{Z}_t$: The residuals, $\hat{Z}_t = X_t - \hat{C}_0 - \hat{A}\cos\left(2\pi\hat{f}t\right) - \hat{B}\sin\left(2\pi\hat{f}t\right)$, are calculated using the commands:

```
z.cos=rep(0,n)
C0=h.est$coefficients[1]
A=h.est$coefficients[2]
B=h.est$coefficients[3]
for(t in 1:n)  z.cos[t]=x.cos[t]-C0-A*h1[t]-B*h2[t]
```

(3) Fit an AR model to the residuals: The command

```
z.cos.aic=aic.wge(z.cos,p=1:5)
```

selects an AR(1) model. The following command estimates the parameters of the AR(1) model, computes the residuals, $\hat{a}_t$, and returns the parameter estimates for the selected model.

```
ar.cos=est.ar.wge(z.cos,p=z.cos.aic$p)
```

Among the output are the following:

```
ar.cos$phi [1]  0. 8315
ar.cos$avar [1]  1.365834
```

The model fit to the residuals is

$$(1-.83B)\hat{Z}_t = a_t, \tag{8.14}$$

where $\hat{\sigma}_a^2 = 1.37$ which is close to the true value, $\sigma_a^2 = 1.5$.[7] The residuals, $\hat{a}_t$, are contained in **ar.cos$res**. These are not plotted here, but they appear to be white noise.

(4) Specify the final model: The final fitted model is

$$X_t = -.03 + 2.76\cos\left(2\pi(.1)t\right) - 4.66\sin\left(2\pi(.1)t\right) + \hat{Z}_t. \tag{8.15}$$

where $\hat{Z}_t$ satisfies the AR(1) model in (8.14). Examination of the fitted model in (8.15) shows that it is quite similar to the actual model in (8.13).

8.2.2 Forecasting Using Cosine Signal+Noise Models

Consider the fitted cosine signal+noise model

$$X_t = \hat{C}_0 + \hat{A}\cos\left(2\pi\hat{f}t\right) + \hat{B}\sin\left(2\pi\hat{f}t\right) + \hat{Z}_t. \tag{8.16}$$

where $\hat{Z}_t$ is modeled as a zero-mean AR(p) model, $\hat{\phi}(B)\hat{Z}_t = a_t$, using the techniques discussed in Chapter 6. In this section we develop forecasts based on the fitted model in (8.16). The forecast procedure for cosine signal+noise models is similar to that used for the line+noise models in Section 8.1.3. There are basically two steps:

(1) Forecast the residual series, $\hat{Z}_t$:

Because $\hat{Z}_t = X_t - \hat{C}_0 - \hat{A}\cos\left(2\pi\hat{f}t\right) - \hat{B}\sin\left(2\pi\hat{f}t\right)$ is modeled as a stationary AR(p) process, forecasts $\hat{Z}_{t_0}(\ell)$, $\ell = 1,...,k$, can found using **fore.arma.wge** based on the stationary model fit to $\hat{Z}_t$ using methods discussed in Section 6.2.

(2) Convert the forecasts in (1) to forecasts, $\hat{X}_{t_0}(\ell)$, for $X_{t_0+\ell}$:

The forecasts, $\hat{Z}_{t_0}(\ell)$, can be converted to forecasts $\hat{X}_{t_0}(\ell)$ for $X_{t_0+\ell}$ using the formula

$$\hat{X}_{t_0}(\ell) = \hat{C}_0 + \hat{A}\cos\left(2\pi\hat{f}t\right) + \hat{B}\sin\left(2\pi\hat{f}t\right) + \hat{Z}_{t_0}(\ell). \tag{8.17}$$

---

7 The sample mean of the **ar.cos$res** data was zero to over 4 decimal places which is also very close to the true value.

Notes:

(a) If a cosine signal+noise model has been fit to a set of data, then the tacit assumption is that this fitted signal can be appropriately "extrapolated" into the future.

(b) Because $\hat{Z}_t$ is modeled as a stationary process with zero mean, the eventual forecasts, $\hat{Z}_{t_0}(\ell)$, will tend toward zero. Consequently, the eventual forecasts, $\hat{X}_{t_0}(\ell)$, will tend toward the fitted cosine curve,

$$\hat{C}_0 + \hat{A}\cos\left(2\pi\hat{f}t\right) + \hat{B}\sin\left(2\pi\hat{f}t\right)$$

as ℓ gets large.

Key Points:

1. The *tacit assumption* that this cosine signal can be appropriately "extrapolated" into the future is not one to be made lightly.

2 The important decision whether to use a cosine signal+noise model will be discussed in Section 8.2.3.

Example 8.5 Forecasting realization x.cos (shown in Figure 8.5(a) and 8.6(a)) using the fitted model in (8.15). In this example we will produce two types of forecasts:

(A) Forecasts 50 steps beyond the end of the series with prediction limits.

(B) Forecasts of the last k values in the dataset for comparison of actual data values.

The forecasts will be made under the assumption that the cosine signal+noise model (8.15) has been fit to the data in **x.cos**.

(A) The forecasts are based on the two-step procedure below:

(1) Forecast the residual series, $\hat{Z}_t$:

The forecasts, $\hat{Z}_{t_0}(\ell), \ell = 1,\ldots,50$ along with 95% forecast limits are plotted in Figure 8.7(a). Recall that the residuals are in the file **z.cos$res** and that the parameters of the AR(1) model fit to the data in **z.cos$res** are contained in **z.cos$phi**. The following command forecasts residuals up to 50 steps ahead.

```
f.ar.cos=fore.arma.wge(ar.cos$res,ar.cos$phi,n.ahead=50)
```

Because $\hat{Z}_{100} = -1.02$ is below the sample mean (zero), the forecasts trend upward toward zero. Specifically, $\hat{Z}_{100}(1) = -.085$.

(2) Convert the forecasts in (1) to forecasts, $\hat{X}_{t_0}(\ell)$, for $X_{t_0+\ell}$: The forecasts up to 50 steps ahead for the cosine signal+noise dataset, **x.cos**, can be calculated using the commands below.

```
f.x.cos=rep(0,length(x.cos))
for (t in 1:length(x.cos))
f.x.cos[t]=f.ar.cos$f[t]+C0+A*h1[t]+B*h2[t]
```

Note that the forecasts in Figure 8.7(b) are simply the forecasts *of the residuals* in Figure 8.7(a) plus the deterministic function, $H(t) = -.03 + 2.76\cos(2\pi(.1)t) - 4.66\sin(2\pi(.1)t)$. That is, the forecast $\hat{X}_{100}(1) = H(t) + \hat{Z}_{100}(1)$

$$= -.03 + 2.76\cos(2\pi(.1)t) - 4.66\sin(2\pi(.1)t) + \hat{Z}_{100}(1),$$

so that

$$\hat{X}_{100}(1) = -.03 + 2.76\cos(2\pi(.1)101) - 4.66\sin(2\pi(.1)101) + \hat{Z}_{t_0}(1) = -1.39.$$

The forecasts tend toward the sinusoidal function, $H(t)$, as t gets large.

8.2.2.1 Using fore.sigplusnoise.wge:

While the above sequence of code illustrates the *procedure* for finding the cosine signal+noise forecasts, the forecasts in Figure 8.7(b) can be more be easily obtained using the *tswge* command

```
fore.sigplusnoise.wge(x.cos,linear=FALSE,freq=.1,n.ahead=50,limits=TRUE)
```

Note:
The command **fore.sigplusnoise.wge** is designed to provide forecasts using either line+ noise or cosine signal+noise models. The line+noise model is the default option. See Appendix 8A for more details.

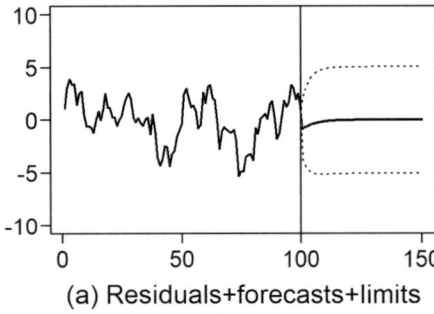
(a) Residuals+forecasts+limits

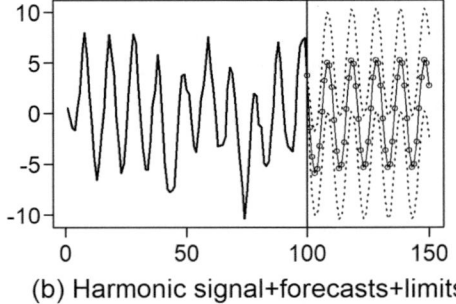
(b) Harmonic signal+forecasts+limits

FIGURE 8.7 (a) Forecasts up to 50 steps ahead for residual series z.cos along with 95% prediction limits (b) corresponding forecasts and prediction limits for the signal+noise signal x.cos.

(B) The following code uses the fitted model to calculate the "last k" forecasts for $k = 10,43,67$, and 84. In this case, we used the entire dataset to obtain model (8.15) and based all forecasts on this model.

```
fore.sigplusnoise.wge(x.cos,linear=FALSE,freq=.1,lastn=TRUE,n.ahead=
10,limits=FALSE)
fore.sigplusnoise.wge(x.cos,linear=FALSE,freq=.1,lastn=TRUE,n.ahead=
43,limits=FALSE)
fore.sigplusnoise.wge(x.cos,linear=FALSE,freq=.1,lastn=TRUE,n.ahead=
67,limits=FALSE)
fore.sigplusnoise.wge(x.cos,linear=FALSE,freq=.1,lastn=TRUE,n.ahead=
84,limits=FALSE)
```

Figure 8.8 shows the forecasts (open circles connected by dotted lines) for the last 10, 43, 67, and 84 values. In each case the forecast cycles are "in sync" with the actual cycles, even for the forecasts up to 84 steps ahead.

> **Key Point:** The cosine signal+noise model produces realizations that have cyclic behavior with *fixed cycle length* in the terminology of Section 1.2.1.

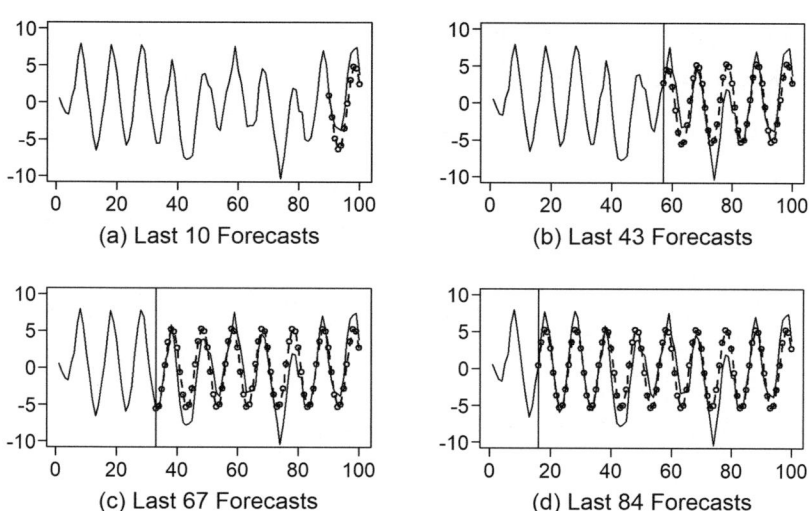

FIGURE 8.8   Forecasts for last *k* values in x.cos dataset for *k* = 10, 43, 67, and 84, respectively.

8.2.3 Deciding Whether to Fit a Cosine Signal+Noise Model to a Set of Data

In this section we address issues similar to those in Section 8.1.1. In that section we discussed the decision whether to include a line (deterministic linear signal) when modeling a dataset that has trending behavior. It was emphasized that this is a difficult decision that is critically important, especially when using the fitted model to forecast into the future. For example, we showed that an AR(1) model with ϕ_1 close to or equal to +1 (but not greater than +1) often has realizations with random trends that have the appearance of (deterministic) linear trending behavior.[8] Simulation studies in Section 8.1.1 show that standard tests for linear trend have difficulty distinguishing between "random trends" in AR models and deterministic trends in line+noise models. We recommend the use of the WBG bootstrap-based test for distinguishing between the two types of trend. However, we noted that the decision should not be based solely on any testing procedure.

---

8 Actually, realizations from (stationary) AR (*p*) and ARMA (*p*,*q*) models with a factor close to +1 can also exhibit the random trending behavior observed for the AR(1) models discussed in Section 8.1.1.

In the cosine setting there is a similar uncertainty:

(a) Cosine signal+noise models produce realizations with cyclic-type behavior.

(b) AR(2) models associated with a pair of complex roots fairly close to the unit circle produce realizations with cyclic-type behavior based on the associated system frequency.[9]

See Section 5.1.2.4.

As an illustration of point (b) we have generated a realization from the AR(2) model,

$$\left(1-1.55B+.925B^2\right)X_t = a_t,$$ (8.18)

with $\sigma_a^2 = .8$. This realization is generated using the *tswge* command

```
x.ar2=gen.arma.wge(n=100,phi=c(1.55,-.925),vara=.8,sn=281)
```

The cosine signal+noise realization, `x.cos`, is plotted in Figure 8.9(a) along with the AR(2) realization, `x.ar2`, in Figure 8.9(b).

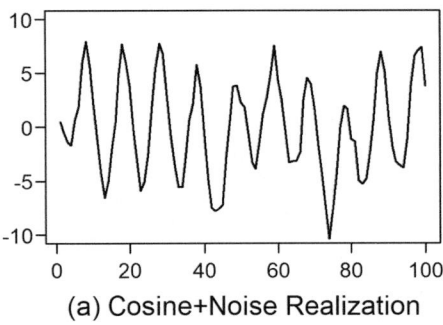

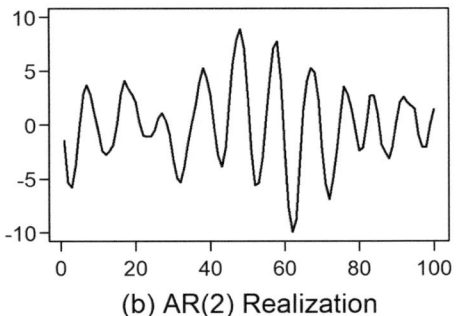

(a) Cosine+Noise Realization   (b) AR(2) Realization

FIGURE 8.9   (a) Cosine signal+noise realization, x.cos, and (b) AR(2) realization, x.ar2.

The realizations are quite similar. The system frequency associated with (8.18) is

$$f_0 = (1/2\pi)\cos^{-1}\left(\frac{1.55}{2\sqrt{.925}}\right) = .1,$$

which is the frequency associated with the cosine signal+noise model in Equation 8.9. We saw in Section 8.1.4 that the decision to include a deterministic linear trend component implies that forecasts of the future will eventually fall along the fitted line. In a similar manner, the decision to include a cosine signal in a time series model implies that forecasts will eventually follow the fitted cosine curve.

Assume that we do not know the source of the realization in Figure 8.9(b), and that because of its cyclic behavior, we decide to fit a cosine signal+noise model of the form (8.12). We use the *tswge* command

```
x.ar2.h=fore.sigplusnoise.wge(x.ar2,linear=FALSE, f=.1)
```

and obtain the following:

```
x.ar2.h$b[1]   0.0042891 (C₀)
x.ar2.h$b[2]  -0.4445196 (A)
x.ar2.h$b[3]  -3.4193501 (B)
```

---

9 AR (*p*) and ARMA (*p,q*) models with an irreducible 2nd-order factor associated with roots "close" to the unit circle will also tend to show cyclic-type behavior.

so that the fitted model is

$$X_t = .00 - .44\cos\left(2\pi(.1)t\right) - 3.42\sin\left(2\pi(.1)t\right) + \hat{Z}_t, \tag{8.19}$$

where AIC selects an AR(3) model for $\hat{Z}_t$. We then use the cosine signal+noise model for forecasting the realization **x.ar2**. Figures 8.10(a)–(d) show forecasts from four different forecast origins and are obtained using the commands

```
fore.sigplusnoise.wge(x.ar2,linear=FALSE,freq=.1,lastn=TRUE,n.ahead=
10,limits=FALSE)
fore.sigplusnoise.wge(x.ar2,linear=FALSE,freq=.1,lastn=TRUE,n.ahead=
43,limits=FALSE)
fore.sigplusnoise.wge(x.ar2,linear=FALSE,freq=.1,lastn=TRUE,n.ahead=
67,limits=FALSE)
fore.sigplusnoise.wge(x.ar2,linear=FALSE,freq=.1,lastn=TRUE,n.ahead=
84,limits=FALSE)
```

The forecasts in Figure 8.10 show that although the system frequency of the generating AR(2) model is $f_0 = .1$, forecasts based on a deterministic cosine signal+noise model are quite poor. For example, the forecasts in Figure 8.10(b) tend to be "backward". That is, they forecast "up" when the data are "down" and vice versa. We noted this type of behavior in Example 6.15 when forecasting the sunspot data. Also note that forecasts of the last 43, 67, and 84 points stay "in sync" with the actual data until about the last two cycles.

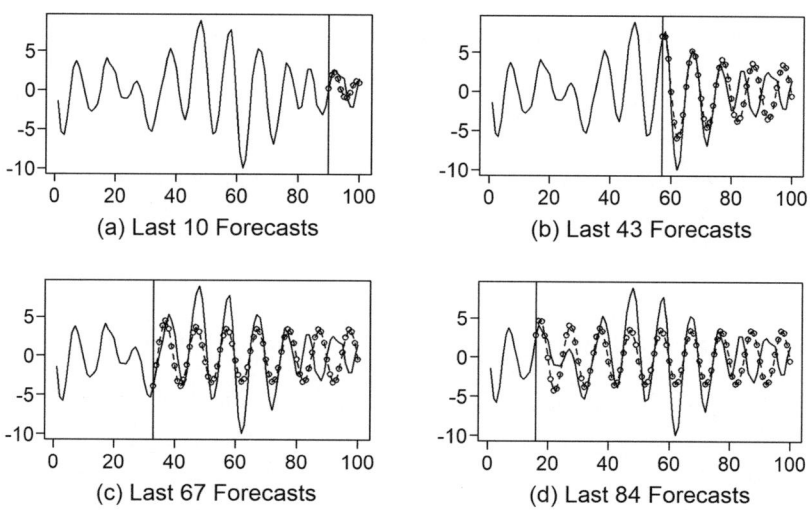

FIGURE 8.10   Forecasts for last k values in x.ar2 using a cosine signal+noise model fit to the data. Forecasts are shown for $k = 10, 43, 67,$ and 84, respectively.

8.2.3.1 A Closer Look at the Cyclic Behavior

Realizations **x.cos** in Figure 8.9(a) and **x.ar2** in Figure 8.9(b) are both described as "cyclic". That is, they show a cyclic behavior but are not "perfectly" sinusoidal. The major difference between the two realizations is that, in the language of Section 1.2.1, **x.cos** has fixed cycle length and **x.ar2** does not. Figure 8.11 is similar to Figures 1.2 and 1.4 for the sunspot data and DFW monthly temperatures, respectively. Figures 8.11(a) and (b) show cycle lengths for the nine full cycles in **x.cos** and in **x.ar2**, respectively. There is very little variation in cycle lengths for the **x.cos** data. Nearly all were length 10 and while cycle four was of length 11 it was "adjusted for" in cycle six, which is of length 9. The first six cycles in **x.ar2** are just as consistently close to 10 as those in **x.cos**. However, cycle seven is of length 7 and cycles six to nine are all 9 or less and there seems to be no adjustment in cycle lengths to get back in "sync". This variability in cycle length is the cause of the poor forecasts toward the end of the dataset and is not "adjusted for". This behavior is a characteristic of cyclic autoregressive data of this type which do not have fixed cycle lengths.

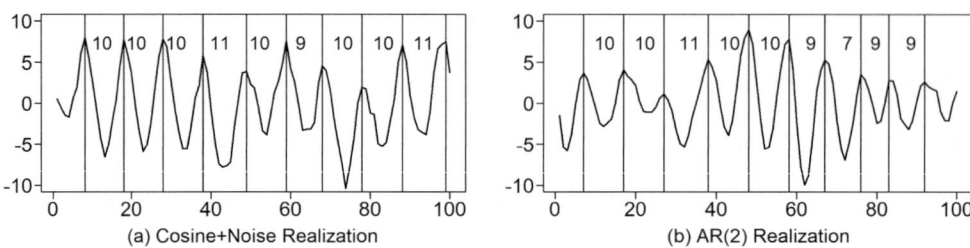

(a) Cosine+Noise Realization   (b) AR(2) Realization

FIGURE 8.11 (a) Realization x.cos and (b) x.ar2 showing cycles and cycle lengths for each of the nine full cycles in each dataset.

Key Point: The amount of variability among cycle lengths provides useful information when deciding whether a cosine signal+noise model should be fit to a given set of data.

We next fit an AR model to the data in **x.ar2**. AIC selects the AR(3) model, in unfactored and factored form,

$$\left(1-.728B+1.230B^2-.193B^3\right)\hat{X}_t=\left(1-1.152B+.908B^2\right)\left(1+.213B\right)\hat{Z}_t=a_t,$$

where the estimated white noise variance is $\hat{\sigma}_a^2=.622$. Thus the dominant 2nd-order factor is very close to the factor in the AR(2) model in (8.18). Figure 8.12 displays the AR-based forecasts analogous to those shown in Figure 8.10 from the cosine signal+noise fit.

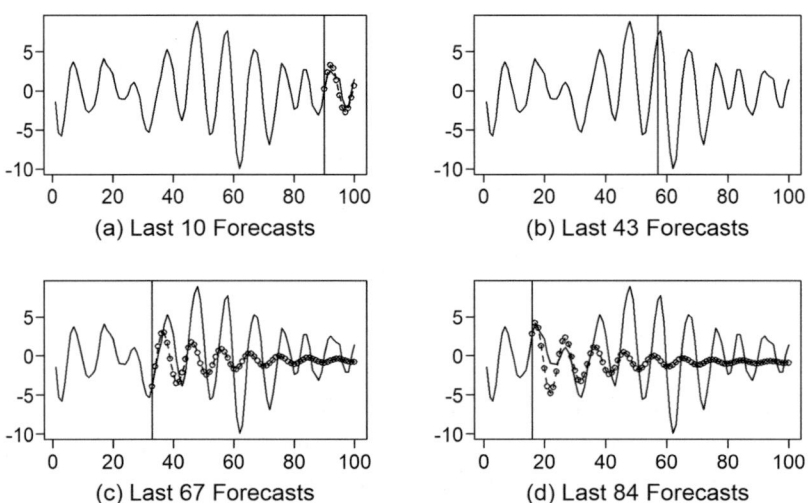

(a) Last 10 Forecasts   (b) Last 43 Forecasts

(c) Last 67 Forecasts   (d) Last 84 Forecasts

FIGURE 8.12 Forecasts for last k values in x.ar2 using the AR(3) model fit to the data. Forecasts are shown for $k=10,43,67$ and 84, respectively.

These forecasts, based on an AR(p) fit, "understand" that cycle-length variability may occur. Since the farther we wish to forecast into the future the less confident we are of the timings of peaks and troughs, it is reasonable that forecasts at longer steps-ahead are close to the mean.

> **Key Point:** Before using a cosine signal+noise model to forecast cyclic behavior very far into the future, you should have physical or empirical evidence concerning the permanence of the cyclic behavior.

Example 8.6 Re-examination of the DFW monthly temperature data, sunspot data, and lynx data
Each of these datasets has cyclic behavior. We will examine each with respect to whether a cosine signal+noise model should be fit to the data.

(a) Dallas Ft. Worth Monthly Temperature Data

In Example 7.3 we fit a nonstationary ARIMA model to the DFW monthly temperature data in *tswge* file **dfw.mon**. This dataset consists of the average monthly temperatures (in degrees Fahrenheit) for the DFW area from January 1900 through December 2020. Consistent with Example 7.3 we will use the data from January 2000 through December 2020 which is in the *ts* file created by the following.

```
dfw.2000=window(dfw.mon,start=c(2000,1))
```

The data have a cyclic behavior with a cycle length of 12 and are plotted in Figure 8.13(a). In Example 7.3 it was noted that the "seasonal" component in **dfw.2000** could be captured using the single factor $1-1.732B+B^2$ (which is associated with a system frequency of $f=1/12$), in contrast to data such as the **AirPassengers** data for which we used a factor $1-B^{12}$.[10] The reason for the difference is that the "seasonal" component of **dfw.2000** has a cyclic behavior that is sinusoidal in nature. This leads us to consider a cosine signal+noise model for the **dfw.2000** data. Figure 8.13(a) shows the **dfw.2000** data and Figure 8.13(b) displays the cycle lengths. Not surprisingly, because of the stability of seasonal data the cycle lengths were 11, 12, or 13 for every cycle. See Figure 1.4 and the associated discussion about temperature patterns in the DFW area. Monthly temperature data such as **dfw.2000** seem ideal for cosine signal+noise models.

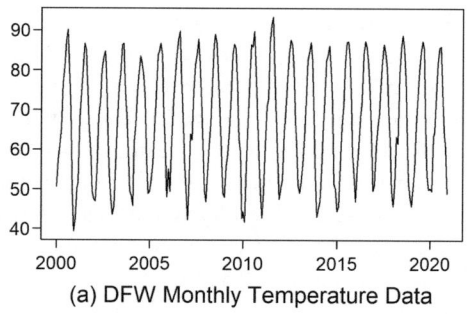
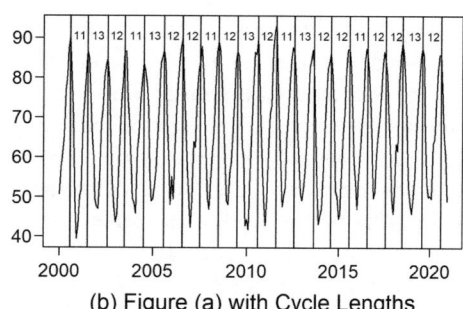

FIGURE 8.13   (a) DFW temperature data in DFW.2000 and (b) Figure (a) showing cycle lengths.

The command

```
fore.sigplusnoise.wge(dfw.2000,linear=FALSE,f=.0833)
```

---

10 Recall that $1-1.732B+B_2$ is one of the factors of $1-B^{12}$.

produces the cosine signal+noise model

$$X_t = 67.35 - 16.97\cos\big(2\pi(.0833)t\big) - 10.26\sin\big(2\pi(.0833)t\big) + \hat{Z}_t, \qquad (8.20)$$

where AIC selects the AR(1) model $\big(1 - 0.34B\big)\hat{Z}_t = a_t$, with $\hat{\sigma}_a^2 = 9.91$.[11]

(a.1) Forecasts of dfw.2000 *data using cosine signal+noise model (8.20)*

Figure 8.14 shows forecasts for the last 12, 36, 84, and 132 months using the cosine signal+noise model in (8.20). The forecasts stay "in sync" with the actual data, and they estimate the peaks and troughs well.

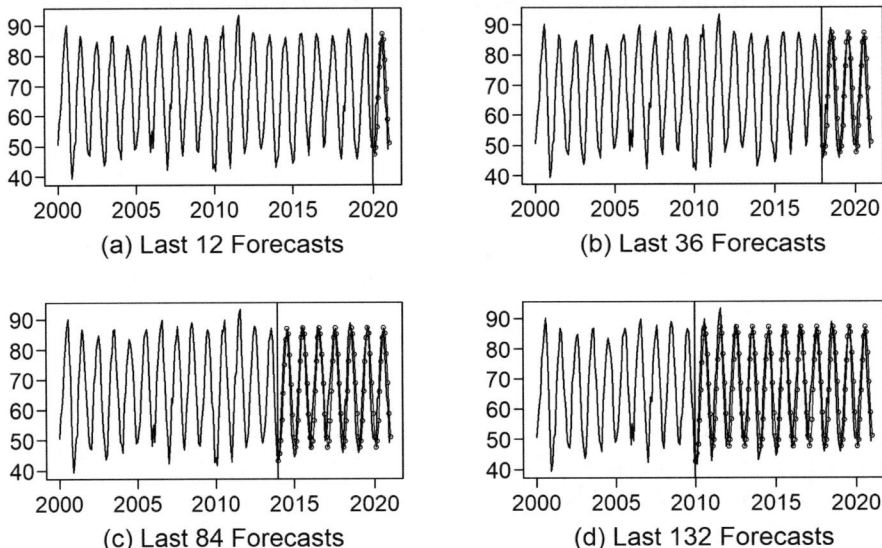

FIGURE 8.14   DFW monthly temperature with forecasts for last 12, 36, 84, and 132 months.

(b) Sunspot Data

The sunspot data have been analyzed in various places in this book. They are introduced in Section 1.2.1 and in Example 6.12 forecasts were obtained although it was noted that these data have more variability in the peaks than in the troughs, so that an $\text{ARMA}(p,q)$ model will not be able to explain this nonlinear behavior. We encountered a similar situation with the lynx data, and it is common to analyze the "log lynx" data (which we did). The logs of the sunspot data are not typically analyzed because there are two instances of zero sunspot activity. In this example we will examine the idea of fitting a cosine signal+ noise model to the sunspot data so it will be helpful to remove the asymmetry from the data prior to any analysis. In order to remove the asymmetry, we will go against tradition and analyze $\log(x_t + 10)$ where x_t is the sunspot number at year t. In the following we will refer to $\log(x_t + 10)$ as the "log sunspot data".

Figure 8.15(a) shows the sunspot data in **sunspot2.0**, while Figures 8.15(b) and (c) are plots of the associated sample autocorrelations and Parzen spectral density estimate, respectively. Figure 8.15(d) is a plot of $\log(x_t + 10)$ (**sunspot2.0+10**) along with sample autocorrelations and Parzen spectral density estimate. Figure 8.15(d) has a more symmetric ARMA-like appearance, and it also has the appearance of a cosine signal+noise realization. Interestingly, the sample autocorrelations and spectral density estimate are very similar in appearance to those associated with the "raw" sunspot data.

---

11 The sample mean is close to zero (within 4 decimal places) as expected

AIC selects an AR(9) model for the log sunspot data. A factor table for this model (not shown) indicates a 2nd-order factor associated with frequency $f = .092$ (period $1/.092 = 10.9$) and a 1st-order factor associated with frequency $f = 0$. This is consistent with the factor table shown in Table 5.6 for the AR(9) fit to the raw sunspot data.

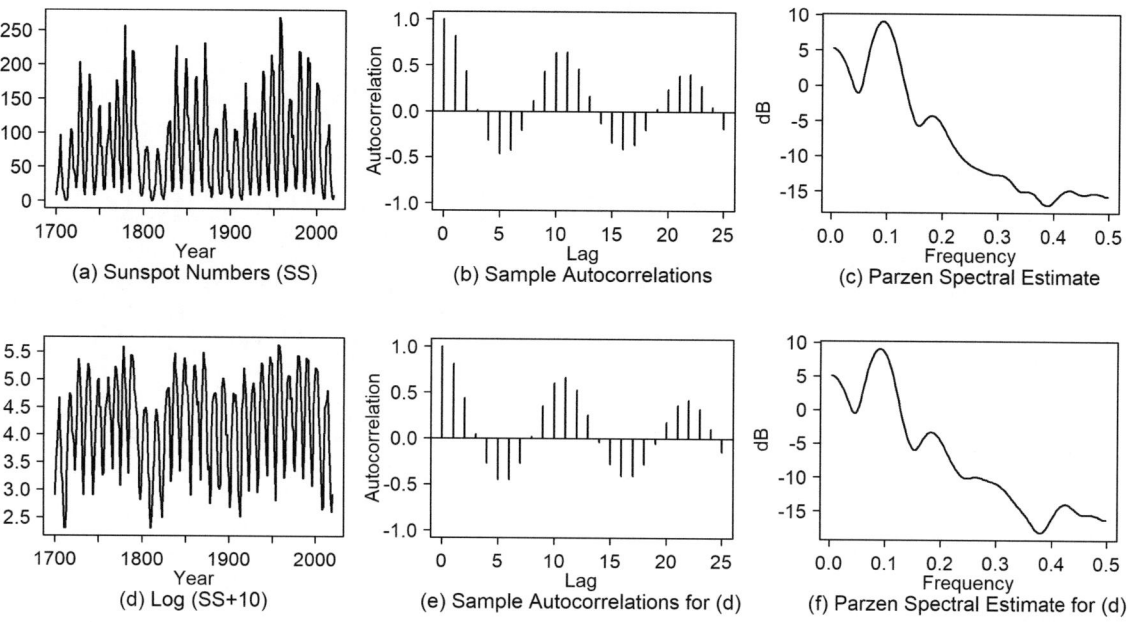

FIGURE 8.15   (a–c) sunspot data in `sunspot2.0`, sample autocorrelations, and Parzen spectral density estimate (d–f) log sunspot, sample autocorrelations, and Parzen spectral density estimate.

We next consider whether a cosine signal+noise model is appropriate for the "log sunspot" data. Figure 8.16 is a plot of the log sunspot data along with cycle lengths. The plot shows quite a lot of variability in cycle lengths, which go from a low of 7 to a high of 17 making a viable cosine signal+noise model unlikely. Any cosine fit to the data will produce forecasts that become out of sync with the data and will very likely at times forecast a peak when the data are at a trough by the fact that the cyclic behavior that is forecast has weak amplitudes.

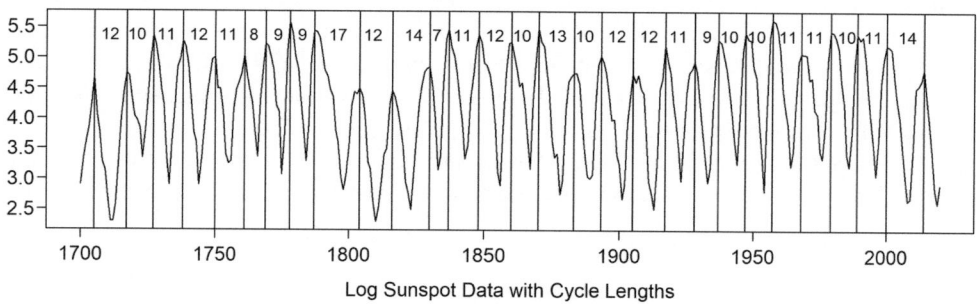

FIGURE 8.16   Log sunspot data showing cycle lengths for each sunspot cycle.

The cosine signal+noise model fit to the log sunspot data is interesting in that the cosine component is weak and most of the cyclic behavior in the model is accounted for by Z_t. The forecasts shown in Figure 8.17 are consistent with this interpretation.

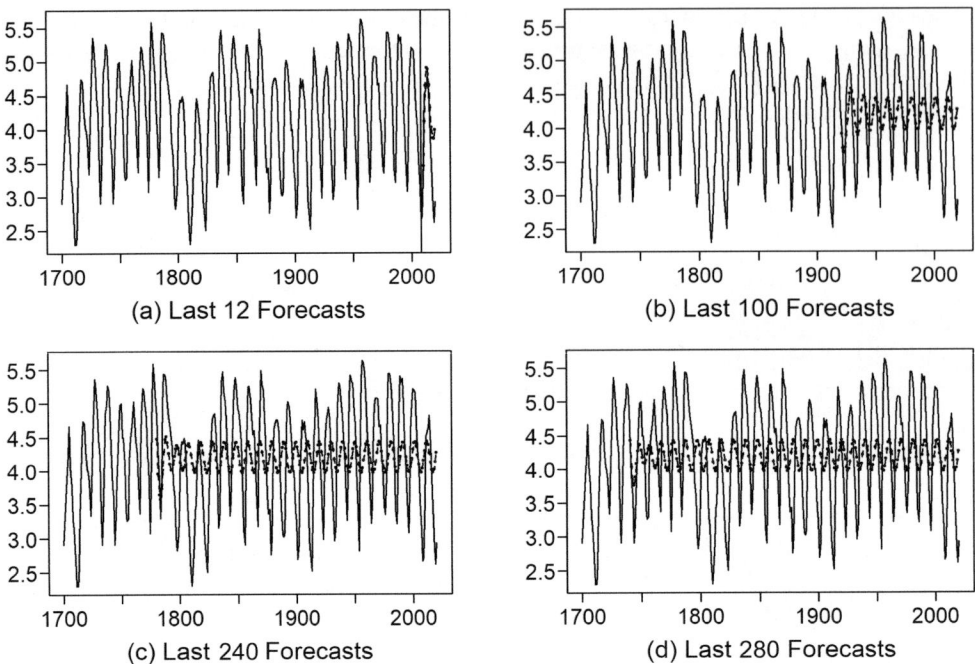

FIGURE 8.17 Log sunspot data with forecasts for last 12, 100, 240, and 280 years using a cosine signal+ noise model fit to the data.

(c) Lynx Data

We return to the lynx dataset examined in Examples 5.3, 6.5, and 6.13. The feature of the data that has caused it to be of interest is the regular cyclic behavior of the data with cycle lengths of about 10 years. In Example 5.3 it was noted that because of the asymmetry in the actual lynx data, researchers have historically analyzed the log (base 10) of the number of lynx trapped. The lynx data are contained in Base R dataset **lynx**. The log-lynx data are shown in Figure 8.18(a).

Researchers have debated whether an AR model or a cosine signal+noise model is appropriate for modeling the log-lynx data. Campbell and Walker (1977) fit a cosine signal+noise model of the form (8.12) to the data while Tong (1977) fit an AR(11) model which is similar to the AR(12) model we fit in Example 6.5.[12] Priestley (1962) proposed the $P(\lambda)$ test, which was designed to determine whether a cosine signal should be included in a model. Bhansali (1979) used the $P(\lambda)$ test and concluded that the log-lynx data contained a cosine signal. The factored form of Tong's model has a 2nd-order factor which has roots very close to the unit circle (the absolute reciprocal is 0.98) associated with frequency $0.104(=1/9.6)$. Woodward and Gray (1983) generated realizations from Tong's model and showed that the $P(\lambda)$ test usually incorrectly "detected" a cosine signal in these AR(11) realizations. Consequently the fact that the $P(\lambda)$ test detected a cosine component does not imply the presence of one.

Figure 8.18(a) shows a plot of the log-lynx data and Figure 8.18(b) shows the cycle lengths in the log-lynx data. There it can be seen that the cycle lengths are consistently 9 or 10 with the exception of the last cycle of 12.

---

12 Tong's model is $(1-1.13B+.51B^2-.23B^3+.29B^4-.14B^5+.14B^6-.08B^7+.04B^8-.13B^9-.19B^{10}+.31B^{11})(X_t-2.90)=a_t$. This model is very close to the YW model obtained using **est.ar.wge**.

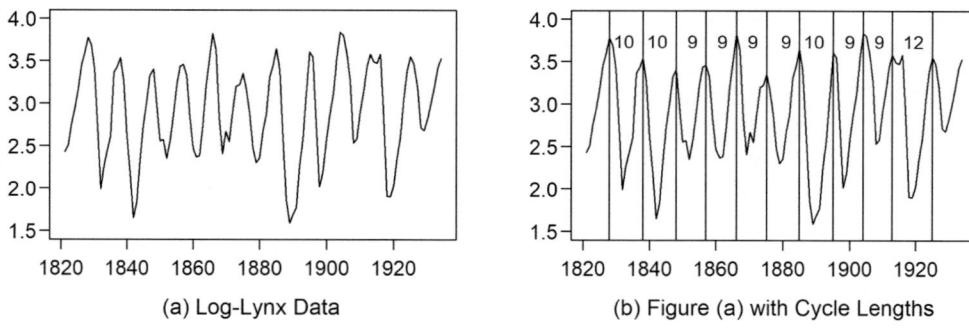

FIGURE 8.18   (a) Log-lynx data in **llynx** and (b) Figure (a) showing cycle lengths.

Using *tswge* command **fore.sigplusnoise.wge**, the resulting cosine signal+noise model is

$$X_t = 2.91 - .093\cos\left(2\pi(.103)t\right) - .607\sin\left(2\pi(.103)t\right) + \hat{Z}_t, \tag{8.21}$$

where $\left(1 - 1.065B + 0.376B^2\right)\hat{Z}_t = a_t$, with $\hat{\sigma}_a^2 = .04$. The last 12, 36, 60 and 84 step-ahead forecasts using (7.41) are shown in in Figure 8.19. There we see that the forecasts stay "in sync". Biologists suggest a possible explanation for the cyclic behavior of the lynx data is that it is related to the corresponding cycle for the horseshoe hare, which is a dietary staple for the lynx. However, the explanation for cyclic behavior is not as explainable as that for the DFW monthly temperature data. Given the information above, it is clear that the cosine signal+noise model and the ARMA(2,3) (or AR(11)) provide reasonable fits to the log-lynx data. We re-emphasize the following Key Points.

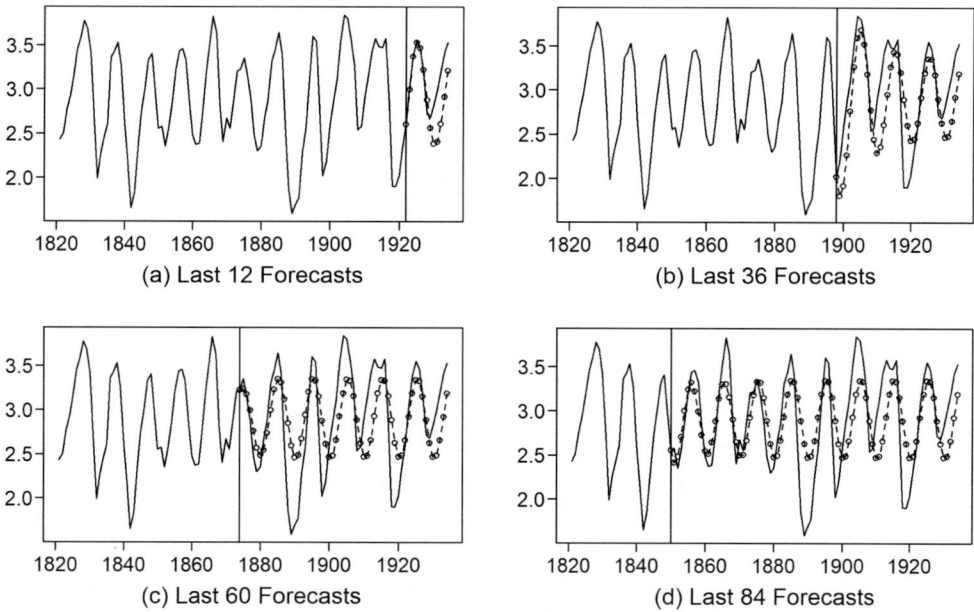

FIGURE 8.19   Log-lynx data with cosine signal+noise forecasts for last 12, 36, 60, and 84 years.

Key Points:

1.   Before using a cosine signal+noise model to forecast cyclic behavior very far into the future, you should have physical or empirical evidence concerning the permanence of the cyclic behavior.
2.   Tong (1977) argued that it was more reasonable to use ARMA-type models (which are capable of describing random phase shifts) than a fixed cycle-length model with an unusual cycle length such as $1/.104=9.6$ years. Tong's reasoning is compelling, but "All models are wrong, but some are useful." (Maybe we've already said that?) That is, neither model is correct and both may be helpful.

Figure 8.20 shows a plot analogous to Figure 8.19 for a realization generated from the ARMA(2,3) model fit to the log-lynx data. We see that the cycle length variability is slightly more than that of the log-lynx data, and this is reflected by forecasts getting out of sync. However, other realizations from the ARMA(2,3) model displayed forecast behavior very similar to that of the log-lynx data.

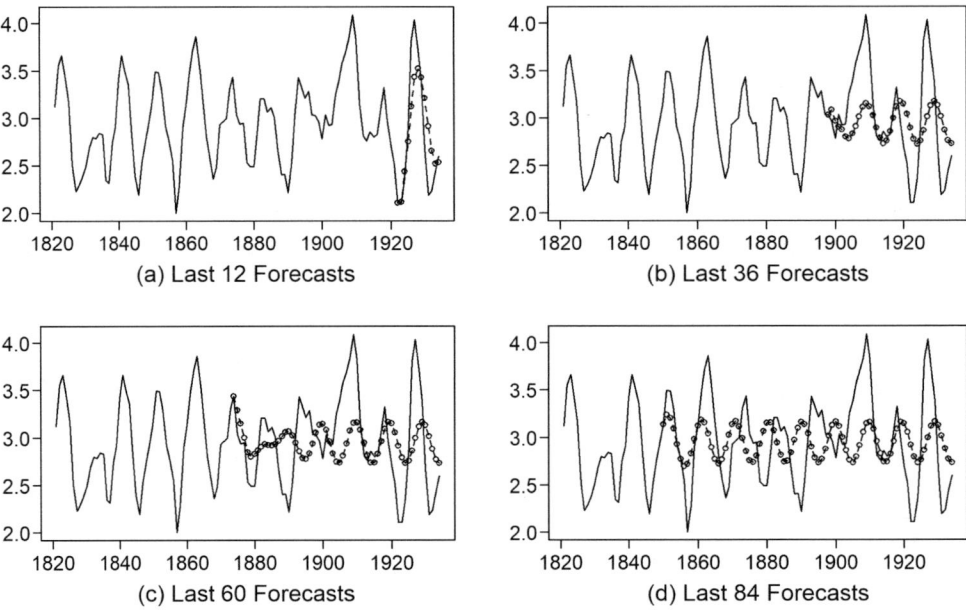

FIGURE 8.20   Realization from the ARMA(2,3) model fit to the log-lynx data with forecasts for last 12, 36, 60, and 84 years.

8.3 CONCLUDING REMARKS

In this chapter we discussed signal+noise models for which the mean changes with time. While many "signals" exist, in this chapter we focused on the line and the sinusoidal (cosine) curve. The crucial (and sometimes difficult!) question that the analyst must answer is whether it is reasonable to expect that the line or sinusoidal pattern is expected to continue. Tests and graphical displays were described which can help answer this question.

Procedures to fit the appropriately chosen model and to derive forecasts were presented in several examples of simulated and real data.

APPENDIX 8A

TSWGE FUNCTIONS

(a) **co.wge(x,maxp)** is a function that performs a Cochrane-Orcutt (CO) test for trend in the time series realization, **x**. The test involves fitting an AR model to the residuals from the fitted line. The order, p, is selected by AIC, and **maxp** specifies the maximum allowed order for p. The autoregressive parameters are estimated using Burg estimates to avoid problems with roots inside the unit circle.

x is a vector containing the time series realization $x_t, t = 1,\ldots,n$.
maxp is the maximum order p for the AR model fit to the residuals from the least squares line.

Output contains the following:

$z are the residuals from the fitted line.
$b0hat is the estimated y-intercept of the fitted line using the CO method.
$b1hat is the estimated slope of the fitted line using the CO method.
$z.order is the order, p, of the model fit to the residuals.
$z.phi are the coefficients of the AR model fit to the residuals.
$pvalue is the p-value of the CO test for the significance of the slope.
$tco is the Cochrane-Orcutt test statistic.

(b) **fore.sigplusnoise.wge(x,linear,method,freq,max.p,n.ahead, lastn,plot, alpha,limits)** is a function that fits a signal+noise model and forecasts **n. ahead** steps ahead for models of the form

(i) $X_t = b_0 + b_1 t + Z_t$, where Z_t satisfies a zero-mean AR process.
(ii) $X_t = C_0 + A\cos(2\pi ft) + B\sin(2\pi ft) + Z_t$, where Z_t satisfies a zero-mean AR process.

Note that signals are restricted to linear or (single) sinusoidal types. You can calculate and plot forecasts **n.ahead** steps beyond the end of the series or you can forecast the last **n.ahead** data values. In the linear model above, all parameters are estimated including the order (p) and associated parameters of the AR(p) model fit to the residuals from the fitted line. In the cosine model the user must specify f. All other model parameters are estimated.

   x = realization
   linear = logical variable. If **linear = TRUE** (default) then the forecasts will be based on the line+noise model above. If **linear = FALSE** then the cosine signal+noise model is fit to the data.
   freq is the variable f that must be specified by the user if **linear=FALSE** (default is **freq = 0**).
   max.p is the maximum order autoregressive model to be fit (based on AIC) to the residuals.
   n.ahead = the number of steps ahead you want to forecast (default = 10).
   lastn=logical variable. If **lastn = FALSE** (default) then the forecasts are for $x(n+1), x(n+2)$, $\ldots, x(n+\text{n.ahead})$ where **n** is the length of the realization. If **lastn = TRUE**, then the program forecasts the last **n.ahead** values in the realization. (default = **FALSE**).

plot = logical variable. If **plot** = **TRUE** (default) then the forecasts will be plotted along with the realization. If **plot** = **FALSE** then no plots are output.

alpha specifies the level of the prediction limits (default=.05).

limits = logical variable. If **limits** = **TRUE** then the 95% forecast limits will be plotted along with the forecasts. (default = **TRUE**)

Example:

Suppose a realization **x** is modeled using a cosine signal+noise model with frequency .0833. The following command will fit the cosine signal+noise model, then fit an AR(p) model to the residuals, and forecast 20 steps beyond the end of the data and plot the results.

```
xfore=fore.sigplusnoise.wge(x,linear=FALSE,freq=0.0833,n.ahead=20,
lastn=FALSE,limits=FALSE)
```

In a real data situation in which you believe the realization may be a cosine signal+noise model, you will typically be able to obtain a good estimate of **freq** using a spectral density estimate or fitting a high order AR model and checking the factor table for the dominant frequency.

(c) **gen.sigplusnoise.wge(n,b0,b1,coef,freq,psi,phi,vara,plot,sn)**   is   a function that generates a realization from the signal-plus-noise model

$$X_t = b_0 + b_1 t + coef[1]\cos(2\pi * freq[1]t + psi[1]) + coef[2]\cos(2\pi * freq[2]t + psi[2]) + Z_t$$

n = length of realization to be generated (t in the above formula specifies the integers from 1 to n)

b0 is the y-intercept

b1 is the slope of a linear regression component (default = 0)

coef is a two-component vector specifying the coefficients (if only one cosine term is desired, define **coef[2]** = **0**) (default = **c(0,0)**)

freq is a two-component vector specifying the frequency components (0 to .5) (default = **c(0,0)**)

psi is a two-component vector specifying the phase shift (0 to 2π) (default = **c(0,0)**)

phi is a vector containing the AR model for the noise process.

vara is the variance of the noise (default = 1).

plot is a logical variable. **plot** = **TRUE** (default) produces a plot of the generated realization.

sn determines the seed number used in the simulation. (default = 0) (see Note below)

Note: **sn=0** (default) produces results based on a randomly selected seed. If you want to reproduce the same realization on subsequent runs of **gen.sigplusnoise.wge** then set **sn** to the same positive integer value on each run.

Example:

The command

```
x=gen.sigplusnoise.wge(n=100,coef=c(1.5,3.5),freq=c(.05,.2),psi=
c(1.1,2.8),
phi=.7,vara=1)
```

calculates

$$X_t = 1.5\cos(2\pi(.05)t + 1.1) + 3.5\cos(2\pi(.2)t + 2.8) + Z_t,$$

where $(1-.7B)Z_t = a_t$, and $\sigma_a^2 = 1$.

(d) **slr.wge(x,tstart)** is a function that uses base R function **lm** to fit a simple linear regression line to the data in vector **x**.

x is a vector containing the time series realization $x_t, t = 1,\ldots,n$

tstart is the index of the first data value in **x**. The independent variable is **time= tstart:length(x)**

Output:

$time contains the values of the independent variable.

That is, the value **tstart, tstart+1, …, tstart+length(x)-1**

$b0hat is the least squares estimate of the y-intercept of the fitted line.
$b1hat is the least squares estimate of the slope of the fitted line.
$pvalue is the p-value for the test of the significance of the slope.
$tstatistic is the t-test statistic for the test of significance of the slope.

(e) **wbg.boot.wge(x,nb,alpha,pvalue=TRUE,sn=0)**

This function performs a Woodward-Bottone-Gray (WBG) test for trend in the time series realization, **x**. The test involves bootstrap replications of the AR model fit to the original data. The maximum value allowed for p is 5. Burg estimates are used to avoid problems with roots inside the unit circle and because of the computer intensive nature of the test.

x is a vector containing the time series realization $x_t, t = 1,\ldots,n$.
nb is the number of bootstrap replications (default is 399).
alpha is the significance level of the test.
pvalue=TRUE prints out the p-value of the test.
sn determines the seed number used in the simulation. (default = 0).

Output contains the following:
$p is the AR order used for the bootstrap simulations.
$phi are the coefficients of the AR model fit to the data.
$pv is the p-value of the test.

EXERCISES

8.1 Generate and plot four realizations from the model

$$\left(1 - .79B - .70B^2 + .255B^3 + .24B^4\right)X_t = a_t,$$

where $\sigma_a^2 = 1$.

   (a) Do any of these realizations have a linear-trending behavior? Explain why or why not, based on the model.
   (b) Use the Cochrane-Orcutt test to test for trend in these datasets.

8.2 Consider the following datasets:
   (a) **dfw.yr**
   (b) **wtcrude2020**
   (c) **tesla**
   (d) **bitcoin**
      (1) Fit an ARMA or ARIMA (not linear trend) model to the data. Explain why you have decided to (or not to) include a unit root in your model.
      (2) Test the data for trend using Cochrane-Orcutt (**co.wge**) and **wbg.boot.wge**.
      (3) For each model decide which model you prefer and why.

8.3 Consider the following datasets:
   (a) **patemp,**
   (b) **freeze**

For each dataset:
   (1) fit a cosine+noise model
   (2) Forecast three "cycles" beyond the end of the dataset.
   (3) "Forecast" the last three, six, and nine cycles. Compare the forecasts with the true values using plots and RMSEs.

Model Assessment

9

In the previous chapters we discussed techniques for fitting a univariate time series model to an observed realization. We discussed two basic types of univariate models: (1) strictly correlation-based models and (2) signal-plus-noise models. Strictly correlation-based models include AR and ARMA models for stationary data along with ARIMA and seasonal models for nonstationary data. Signal-plus-noise models are nonstationary models that include a deterministic signal. We fit these models making decisions about unit roots, seasonal or trend behavior, and obtained final models based on AIC-type methods. If you have carefully followed the model fitting steps, it is tempting to accept the fitted model as the "best you can do" and use it as the final model. However, experienced time series analysts know that *fitting a model is just the beginning point* in obtaining a "final model". After fitting a model to a realization, you should take steps to assure that the model is *suitable*. The first step in this process is to examine the residuals. If the model is appropriate, the residuals should be uncorrelated. This topic will be discussed in Section 9.1. Suppose you have obtained a model by using a model identification technique (such as AIC), estimated the parameters, and examined the residuals to satisfy yourself that they are sufficiently white (that is, uncorrelated).[1] The next step is to determine whether the model *makes sense*. As we examine models, we will evaluate them with regard to the following Key Questions which guide us in assessing whether a given model is appropriate.

Key Questions for Assessing a Model Fit:

1. Does the model "whiten" the residuals?
2. Do realizations and their characteristics behave like the data?
3. Do forecasts reflect what is known about the physical setting?

Section 9.1 discusses methods for testing residuals for white noise and normality. In Section 9.2 we use the analysis strategy outlined above to analyze and compare models fit to the global temperature data. Section 9.3 examines models fit to the sunspot data. Sections 9.2 and 9.3 are case studies using the methods discussed in this book to model the global temperature and sunspot datasets.

9.1 RESIDUAL ANALYSIS

We recommended the use of AIC and its variations for ARMA, ARIMA, seasonal, and signal-plus-noise model identification in Chapters 6–8. AIC and its variations select a model for the stationary component of the model *based on reducing the estimated white noise variance*, $\hat{\sigma}_a^2$, while controlling the number of parameters required to do so. The estimate, $\hat{\sigma}_a^2$, is output from **est.ar.wge** and **est.arma.wge** in the variable **$avar**. In this section we consider residuals from three types of models:

---

1 Refer to Section 7.3 regarding uncorrelated noise for which you might suspect conditional heteroskedasticity.

DOI: 10.1201/9781003089070-9

- ARMA models
- ARIMA or seasonal models
- Signal-plus-noise models

(a) Residuals from a Fitted ARMA(p,q) model

For the ARMA(p,q) model

$$X_t - \hat{\phi}_1 X_{t-1} - \cdots - \hat{\phi}_p X_{t-p} = \hat{a}_t - \hat{\theta}_1 \hat{a}_{t-1} - \cdots - \hat{\theta}_q \hat{a}_{t-q} + \bar{x}\left(1 - \hat{\phi}_1 - \cdots - \hat{\phi}_p\right),$$

the residuals were previously given in (6.5) as

$$\hat{a}_t = X_t - \hat{\phi}_1 X_{t-1} - \cdots - \hat{\phi}_p X_{t-p} + \hat{\theta}_1 \hat{a}_{t-1} + \cdots + \hat{\theta}_q \hat{a}_{t-q} - \bar{x}\left(1 - \hat{\phi}_1 - \cdots - \hat{\phi}_p\right). \tag{9.1}$$

In Section 6.1.1.3 we discussed the calculation of the residuals, the concern about dependence on starting values, and the technique of calculating residuals using *backcasting*. The following were discussed in Section 6.1.1.3.

Key Points:

1   We recommend computing the residuals using *backcasting*.
- The backcast residuals provide a "full set" of n residuals that do not have undue dependence on starting values.
2.   All *tswge* routines that calculate residuals use the backcast procedure.

(b) Residuals from an ARIMA or Seasonal Model

The analysis of ARIMA and seasonal models involves two steps:

(1) Identify the nonstationary components of the data and transform the data to stationarity
(2) Model the transformed data using an ARMA model

As an example, for a fitted model, $(1-B)\hat{\phi}(B)(X_t - \bar{X}) = \hat{\theta}(B)\hat{a}_t$, we

(1) Transform the data to obtain $Y_t = (1-B)X_t$
(2) Fit an ARMA model to the Y_t data
- The backcast residuals in question are those obtained from the modeling of Y_t

(c) Residuals from a Signal-plus-Noise Model

The procedure is similar to the nonstationary case (b) above

(1) Fit a signal (for example, a line) to the data and remove the estimated signal by transforming the data
(2) Fit an AR model to the transformed data

Consider the line+noise model $X_t = a + bt + Z_t$, where Z_t is a zero-mean, stationary process. The analysis steps are as follows:

(1) Find the least squares estimates of a and b, call them $\hat{a}$ and $\hat{b}$, and transform the data to obtain $\hat{Z}_t = X_t - \hat{a} - \hat{b}t$

(2) Fit an AR model to the transformed data, $\hat{Z}_t$
   - The backcast residuals are those obtained from modeling $\hat{Z}_t$

Key Points:

1. Case (a) above involves fitting an ARMA model to a stationary time series
2. In cases (b) and (c), the final step is fitting an ARMA model to the transformed "stationary" data. (In our analyses we restricted our attention to AR models.)

9.1.1 Checking Residuals for White Noise

Recall that the models discussed in Chapters 6–8 all assume that the noise, a_t, is *white* noise. Consequently, if the fitted model is appropriate, then the residuals should be well modeled as white noise.[2] If the residuals of the fitted model contain significant autocorrelation, then the model has not suitably accounted for the correlation structure in the data, regardless of the size of the estimated white noise variance.[3] Another assessment that may be of interest is whether the residuals appear to be *normally distributed*, an assumption of the MLE which is based on the normal likelihood (see Woodward et al. (2017)). Note that the YW and Burg estimation procedures have no underlying distributional assumptions.

Key Points:

1. An appropriate model should "whiten" the residuals.
2. The backcast residuals should have the appearance of white noise.
3. If there is correlation structure remaining in the residuals, then this is an indication that the model is inadequate.

9.1.1.1 Check Residual Sample Autocorrelations against 95% Limit Lines

One method for testing residuals for white noise has already been discussed in Section 6.1.2.1. Specifically, when fitting an ARMA model to a set of data, we recommend first checking the data for white noise. For white noise, the autocorrelations satisfy $\rho_k = 0$, $k \neq 0$, and the method discussed in Section 6.1.2.1 tests $H_0 : \rho_k = 0$ by plotting the sample autocorrelations, $\hat{\rho}_k, k = 1,2,\ldots,K$, against the 95% limits, $\pm 2/\sqrt{n}$, for some K. For a specific lag k, if the data are white noise, there is a 5% chance that $\hat{\rho}_k$ will fall outside the

---

2 We focus the discussion here on the residuals from a *stationary* ARMA(p, q) fit. This discussion is applicable to the fitting of nonstationary models in cases (b) and (c) above, because the final step in those estimation procedures is to fit a stationary AR model to some transformed version of the original data.

3 This is analogous to the examination of residuals in multiple (or simple) linear regression analysis to assess whether the residuals are uncorrelated.

limits $\pm 2 / \sqrt{n}$. Consequently, if the data are white noise then it would not be unusual for about 5% of the sample autocorrelations to fall outside the 95% limit lines. If substantially more than 5% of the sample autocorrelations of the data fall outside these lines, this is evidence against white noise. See Section 6.1.2.1 for a more complete discussion of the use of these limit lines for white noise assessment including the advice to always plot the data.

While we previously used the above method for checking an "original" dataset for white noise, it can also be used to check the (backcast) residuals from a fitted model for white noise. That is, we can compare the *residual* sample autocorrelations against $\pm 2 / \sqrt{n}$ to help decide whether the residuals appear to be "white".

9.1.1.2 Ljung-Box Test

One problem with checking the sample autocorrelations against the 95% limit lines is that the "5% chance of exceeding these lines" applies to the sample autocorrelations separately for each lag k. It would be desirable to have a single *portmanteau*[4] procedure that tests the first K sample autocorrelations "as a group" rather than individually. That is, we test for white noise by testing the hypotheses

$$H_0 : \rho_1 = \rho_2 = \cdots = \rho_K = 0$$

$$H_1 : \text{at least one of the } \rho_k \neq 0 \text{ for } 1 \leq k \leq K. \tag{9.2}$$

Box and Pierce (1970) and Ljung and Box (1978) developed tests of the null hypothesis in (9.2). The test developed by Ljung and Box (1978) is widely used for this purpose. The Ljung-Box test statistic is given by

$$L = n(n+2) \sum_{k=1}^{K} \frac{\hat{\rho}_k^2}{n-k}. \tag{9.3}$$

Ljung and Box (1978) show that L in (9.3) is approximately distributed as χ^2 with K degrees of freedom when the data are white noise. When the data to be tested are residuals from a fitted ARMA(p,q) model, then the test statistic is approximately χ^2 with $K - p - q$ degrees of freedom. Examination of (9.3) shows that the test statistic measures the size of the first K sample autocorrelations as a group, so that large values of L suggest that the data are not white noise. Thus, the null hypothesis of white noise residuals is rejected when L is sufficiently large, and in particular if $L > \chi^2_{1-\alpha}(K - p - q)$ where $\chi^2_{1-\alpha}(m)$ denotes the $(1 - \alpha) \times 100\%$ percentile of the χ^2 distribution with m degrees of freedom. The Ljung-Box test can be implemented using the *tswge* command `ljung.wge`. Although other values of K could be used, in the examples that follow we will use $K = 24$ and $K = 48$, which is consistent with the values of K used by Box et al. (2008). For small sample sizes (n ≤ 100), Box et al. (2008) and Ljung (1986) recommend the use of smaller values of K.

QR 9.1
Ljung-Box Test

---

4 Box and Jenkins used the uncommon term "portmanteau" which can be defined to mean "combining two or more aspects or qualities" to describe a test that considers the sample autocorrelations "as a group".

Example 9.1 Checking Residuals for White Noise

In this example we check the whiteness of residuals for several models fit to datasets in Chapters 6–8.

(a) AR(4) Realization Figure 6.22(a)

Realization 1 in Figure 5.16 and Figure 5.18(g) show a realization from the AR(4) model previously referred to as Model (B). The realization and sample autocorrelations are shown in Figures 9.1(a) and (b), respectively. The code below generates the data, plots the data, sample autocorrelations, and Parzen spectral density estimate. It uses AIC(which selects an AR(4)), and obtains the ML estimates of the AR(4) model fit.

```
modelB=gen.arma.wge(n=150,phi=c(2.6,-3.34,2.46,-.9024),sn=3233)
plotts.sample.wge(modelB)
aic.B=aic.wge(modelB,p=0:10,q=0:4)# AIC selects p=4,q=0
fit40=est.arma.wge(modelB,p=4,q=0)
```

The backcast residuals from the fitted model are contained in **fit40$res**. The residuals are plotted in Figure 9.1(c) and the residual sample autocorrelations are shown in Figure 9.1(d) along with the 95% limit lines. The command

```
plotts.sample.wge(fit40$res,lag.max=48,arlimits=TRUE)
```

will produce Figures 9.1(c) and 9.1(d) along with the Parzen spectral density estimate (not shown).

The residuals appear to be random, and the sample autocorrelations stay within the 95% limit lines, with the exception of $\hat{\rho}_{24} = -.180$ which falls just outside the horizontal lower limit line ($-.163$). The Ljung-Box test results for $K = 24$ and $K = 48$ are obtained using the commands

```
ljung.wge(fit40$res,p=4,q=0)   # K=24 is the default
ljung.wge(fit40$res,p=4,q=0,K=48)
```

Note that p and q are inputs to the **ljung.wge** command because they are involved in calculating the degrees of freedom needed for the chi-square test.[5] The output from these two commands follows.

```
$K 24
$chi.square 25.79192
$df 20
$pval 0.1727946

$K 48
$chi.square 46.69656
$df 44
$pval 0.3622047
```

---

5 If the Ljung-Box test is applied to "raw data" (that are not residuals from a fitted model), then use $p = q = 0$.

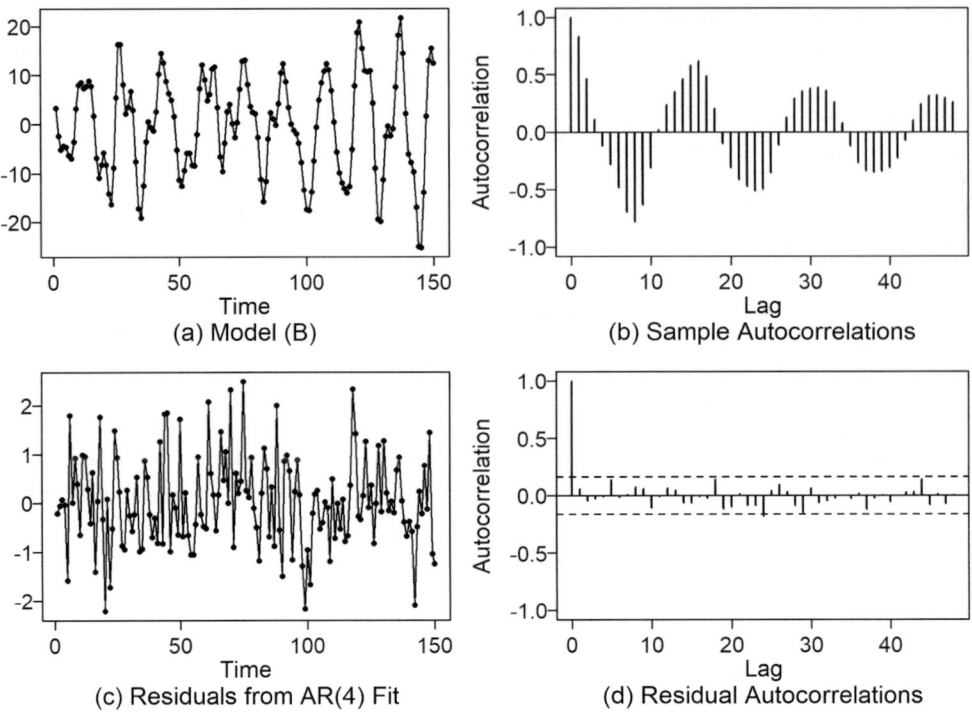

FIGURE 9.1 (a) Realization from Model (B) previously shown in Figure 5.16, (b) sample autocorrelations of the data in (a), (c) residuals from the AR(4) fit to data in Figure 5.16(a), (d) Sample autocorrelations of the residuals in (c) along with 95% limit lines.

The degrees of freedom for the $K = 24$ case, for example, are $K - p - q = 24 - 4 = 20$. For both $K = 24$ and $K = 48$, the p-values are considerably greater than $\alpha = .05$, so we do not have evidence to reject the null hypothesis of white noise. Consequently, based on the plots in Figure 9.1 and the Ljung-Box test, it appears that the fitted model does a good job of "whitening" the residuals.

(b) Log Air Passengers Data
In Example 7.2 we fit a seasonal "airline model" to the log Air Passengers data shown in Figure 7.19(a) (among other places throughout the book). The fitted model using Burg estimates is

$$(1 - B)(1 - B^{12})\phi_{13}(B)X_t = a_t, \tag{9.4}$$

where $\hat{\sigma}_a^2 = 0.0013$ and $\phi_{13}(B)$ is given in (7.15). Recall that the analysis procedure was to difference the data, seasonally difference the "differenced data", and then model the remaining stationary data using an AR(13) model. The residuals to be analyzed and checked for white noise are those from the AR(13) fit. The following commands produce the desired residuals.

```
data(airlog)
d1=artrans.wge(airlog,phi.tr=1) # d1 is the differenced data
d1.s12=artrans.wge(d1,phi.tr=c(rep(0,11),1)) #seasonally difference d1
air.est=est.ar.wge(d1.s12,p=13,type= 'burg')
plotts.sample.wge(air.est$res,lag.max=48,arlimits=TRUE)
ljung.wge(air.est$res,p=13)
ljung.wge(air.est$res,p=13,K=48)
```

The residuals are in `air.est$res` and are plotted in Figure 9.2(a) along with their sample autocorrelations and limit lines in Figure 9.2(b).

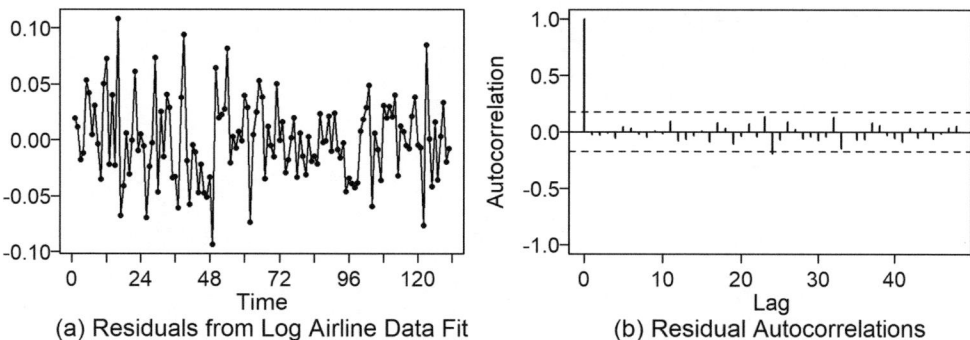

(a) Residuals from Log Airline Data Fit

(b) Residual Autocorrelations

FIGURE 9.2   (a) Residuals from the "airline model" fit in (9.4) to the log airline data and (b) Sample autocorrelations of the residuals in (a) along with 95% limit lines.

The residuals look fairly white with some wandering behavior beginning at about $t = 45$. The sample autocorrelations stay within the limit lines, and the Ljung-Box statistics at $K = 24$ and 48 have p-values of 0.0724 and 0.3504, respectively.[6] The sample autocorrelations are fairly large for k at lags $k = 23$ and $k = 24$. These sample autocorrelations are responsible for the rather small p-value of 0.0724 for $K = 24$.[7] All sample autocorrelations with lags greater than $k = 24$ are inside the limits and most are quite small, which is consistent with the p-value of 0.3504 for $K = 48$. Overall, the model appears to sufficiently whiten the residuals.

(c) DFW Temperature Data

The Dallas-Ft. Worth monthly temperature data from January 2000 through December 2020 are shown in Figure 7.22(a). In Example 7.3 these data were modeled using the nonstationary model

$$\left(1 - 1.732B + B^2\right)\hat{\phi}_{11}(B)\left(X_t - 67.36\right) = \left(1 - .95B\right)a_t, \tag{9.5}$$

where $\hat{\sigma}_a^2 = 9.41$. The model above along with the coefficients of $\hat{\phi}_{11}(B)$ are given in Example 7.3. The residuals from the fitted model along with the plots in Figure 9.3 can be obtained using the following code.

```
data(dfw.mon)
dfw.2000=window(dfw.mon,start=c(2000,1))
tr2.temp=artrans.wge(dfw.2000,phi.tr=c(1.732,-1))
tr2.est=est.arma.wge(tr2.temp,p=11,q=1)
plotts.sample.wge(tr2.est$res,lag.max=48,arlimits=TRUE)
ljung.wge(tr2.est$res,p=11,q=1)
ljung.wge(tr2.est$res,p=11,q=1,K=48)
```

The residuals look fairly white and the sample autocorrelations stay within the limit lines although the sample autocorrelation at $k = 13$ is quite close to the lower limit. The Ljung-Box test at $K = 24$ and 48 has p-values of 0.1044 and 0.1374, respectively. The fact that the sample autocorrelations seem to have a pseudo-cyclic pattern is of some concern. However, the model appears to sufficiently whiten the residuals.[8]

---

6 The degrees of freedom for the Ljung-Box test in this case are $K - 13$ based on the 13th-order stationary AR factor, $\hat{\phi}_{13}(B)$, in (9.4). The factors $1 - B$ and $1 - B^{12}$ were chosen by the investigator to obtain stationarity and are not counted as estimated parameters.

7 If you use the command `ljung.wge(air.est$res,p=13,K=22)` the p-value is .372.

8 The commands `ljung.wge(tr2.est$res,p=11,q=1)` and `ljung.wge(tr2.est$res,p=11,q=1, K=48)` were used because the factor $1 - 1.732 + B^2$ is a nonstationary factor that we "selected" based on examination of the factor tables and our knowledge of the 12-month cycle in the data. It was not "estimated". So, $p = 11$ is based on the 11th-order stationary factor, $\hat{\phi}_{11}(B)$ in (9.5).

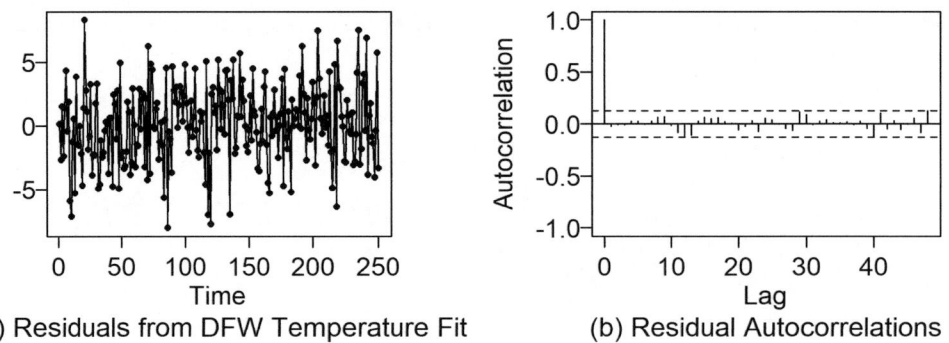

(a) Residuals from DFW Temperature Fit   (b) Residual Autocorrelations

FIGURE 9.3 (a) Residuals from the Example 7.3 model fit to `DFW.2000` temperature data and (b) Sample autocorrelations of the residuals in (a) along with 95% limit lines.

(d) Sunspot Data

In Example 6.15 two models were considered for the `sunspot2.0` data:

 – an AR(2) model
 – an AR(9) model

> (i) Jenkins, and Reinsel (2008) suggested an AR(2) model for a shorter sunspot series based on the partial autocorrelations (see Problem 6.13(a)). By working Problem 6.13(a) you will see that the partial autocorrelations show support for an AR(2) fit to the `sunspot2.0` data. The AR(2) model fit to the `sunspot2.0` data is

$$\left(1-1.38B+.69B^2\right)\left(X_t-78.97\right)=a_t,$$

where $\hat{\sigma}_a^2 = 659.39$.

The following code calculates the residuals and places the backcast residuals in `ss2$res`.

```
data(sunspot.new)
ss2=est.ar.wge(sunspot2.0,p=2)
ljung.wge(ss2$res,p=2)
ljung.wge(ss2$res,p=2,K=48)
```

The backcast residuals in `ss2$res` are plotted in Figure 9.4(a) and they appear to be "somewhat white". Figure 9.4(b) shows the sample autocorrelations where it is seen that 6 of the 48 sample autocorrelations (that is, 12.5%) fall outside the 95% limit lines, suggesting that the AR(2) residuals are not white noise. Further evidence against white noise is the fact that the Ljung-Box p-values at $K = 24$ and $K = 48$ are less than 0.001. Consequently, it does not appear that the AR(2) is a satisfactory fit based on an analysis of the residuals.

> (ii) In Example 6.8 we found that AIC selected the AR(9) model $\phi_9(B)(X_t - 78.97) = a_t$, where

$$\phi_9(B) = 1-1.16B+.41B^2+.13B^3-.10B^4+.07B^5-.01B^6-.02B^7+.05B^8-.22B^9,$$

and $\hat{\sigma}_a^2 = 546.82$. The following code calculates the residuals and places the backcast residuals in `ss9$res`.

```
data(sunspot2.0)
ss9=est.ar.wge(sunspot2.0,p=9)
ljung.wge(ss9$res,p=9)
ljung.wge(ss9$res,p=9,K=48)
```

The backcast residuals in **ss9$res** are plotted in Figure 9.4(c), and these are very similar in appearance to the AR(2) residuals. The sample autocorrelations are shown in Figure 9.4(d) along with the 95% limit lines. The sample autocorrelations stay within the 95% limit lines; however, a few are close to the limits, and the p-values associated with the Ljung-Box test at $K = 24$ and $K = 48$ are 0.0678 and 0.0547, respectively. The plots in Figure 9.4 and the Ljung-Box tests give us some concern about the "whiteness" of the residuals for the AR(9) fit, but we do not "reject white noise" at the $\alpha = .05$ level.

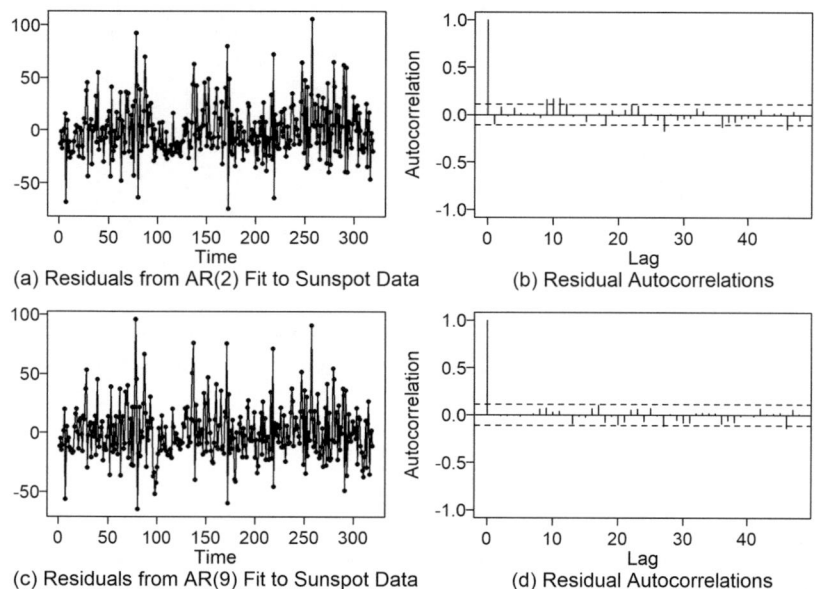

FIGURE 9.4   (a) Residuals from the AR(2) model fit to the sunspot data, (b) sample autocorrelations of the residuals in (a) along with 95% limit lines, (c) residuals from the AR(9) model fit to the sunspot data, and (d) sample autocorrelations of the residuals in (c) along with 95% limit lines.

9.1.2 Checking the Residuals for Normality

The ML estimation procedure assumes that the residuals are normal. Also note that the functions **gen.arma.wge, gen.arima.wge,** and **gen.sigplusnoise.wge** generate realizations based on normal (Gaussian) residuals. For real datasets, you may need to check for normality of the residuals (for example, if you are using ML estimation). Methods for checking normality include the use of histograms, Q-Q plots, or formal tests for normality such as the Shapiro-Wilk test (see Shapiro and Wilk, 1965). In the following we will not examine the AR(2) residuals for normality because the AR(2) model was determined to be a poor fit. The histograms in Figure 9.5 were obtained using the Base R command **hist.** Close examination of the histograms shows that the sunspot model residuals are somewhat skewed to the right.

Base R provides the Shapiro-Wilk test of the null hypothesis of normality via the command **shapiro. test.** Small p-values are an indication of non-normal data. For the residuals from the AR(4) fit that we named **fit40$res,** the command

```
shapiro.test(fit40$res)
```

yields the *p*-value 0.1809, so that normality is not rejected. This is not surprising because the data in Figure 5.16(a) were generated with normal residuals. The *p*-values obtained by applying the Shapiro-Wilk test to the other three datasets are given below:

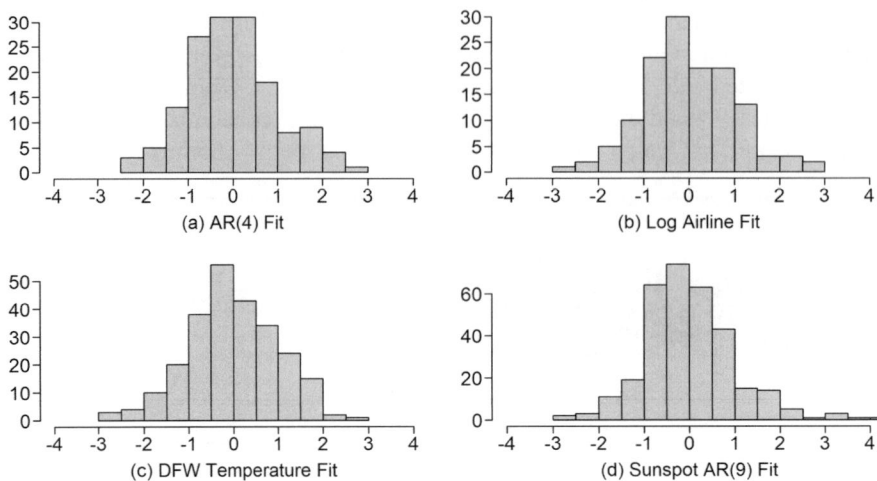

FIGURE 9.5   Histograms of the residuals from the models obtained in Example 9.1: (a) AR(4) fit to data in Figure 5.16, (b) model (9.4) fit to the log airline data, (c) model (9.5) fit to the DFW monthly temperature data, and (d) AR(9) model fit to the sunspot data.

Log airline residuals: `0.7404`
`dfw.2000` temperature residuals: `0.9598`
Sunspot residuals: `7.065e-07`

Consequently, we have reason to doubt the normality of the residuals from the AR(9) fit to the sunspot data.[9] In such cases, because of the normality assumption of the ML estimates, we could use the Burg estimates which make no such assumption. We have previously mentioned that realizations generated (with normal white noise) from the AR(9) sunspot model do not have the asymmetric appearance of the original data (more variability in the heights of the peaks than the troughs).[10] Consequently, nonlinear models (see Tong, 1990) have been proposed for the sunspot data. Another option is to model the log(`sunspot2.0`+10) data as we have done in Example 8.6(b). The log sunspot data will also be analyzed in Section 9.3.

9.2 CASE STUDY 1: MODELING THE GLOBAL TEMPERATURE DATA

We consider modeling the global temperature data as a case study in which we will fit a variety of models to the data and evaluate them with regard to the *Key Questions for Assessing a Model Fit* given in the introduction to this chapter. The temperature data in Figure 9.6(a) are the global average temperature anomalies for the years 1880–2020 from the base period 1901−2000 (ncdc.noaa.gov).

---

9   Other tests for normality, including the Anderson-Darling, Cramer-von Moses, and Lilliefors (Kolmogorov-Smirnov), can be obtained using functions in the CRAN package **nortest**. See Stephens (1974).

10   Even if realizations are generated using a right-skewed noise distribution, the realizations still do not have the dramatic asymmetric appearance observed in the sunspot data.

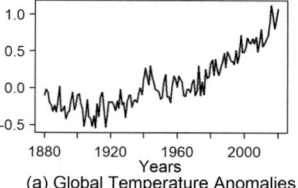
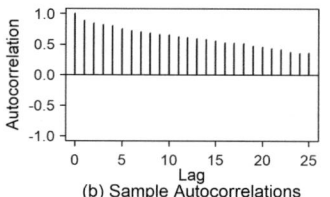
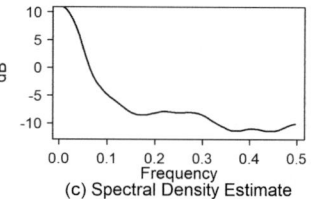

(a) Global Temperature Anomalies　　　(b) Sample Autocorrelations　　　(c) Spectral Density Estimate

FIGURE 9.6   (a) Global average temperature anomalies for the years 1880–2020 from the base period 1901–2000 (ncdc.noaa.gov), (b) sample autocorrelations of the data in (a), and (c) the Parzen spectral density estimate.

The temperature data in Figure 9.6(a) show that there has been a rise in temperatures over the past 60 years. It is well known using proxy data such as tree rings, ocean sediment data, etc., that temperatures have not remained static over time leading up to the rise over the past 60 years, but have tended to vary. There is some historical (non-thermometer-based) evidence that at least in much of the northern hemisphere, the temperatures during the medieval warm period (950–1100) may have been as high as they are today. The medieval warm period was followed by a period known as the Little Ice Age. While not a true ice age, this was a period during which available temperature data indicate there was cooling that lasted into the nineteenth century.

However, the recent increases in global temperature have led many climatologists to conclude that the warming is at least in part due to man-made influences, such as increased atmospheric CO_2, and is a cause for concern. Others have argued that much of the increasing temperatures can be attributed to the fact that we are coming out of the Little Ice Age, and the man-made contribution is not as substantial as others might claim.

CAVEAT:

1.   The topic of global warming is highly charged and political. The analysis and interpretation of the temperature data is a "tricky" and emotionally charged topic.
2.   Our interpretation of the controversy is that climatologists and other scientists believe that the current rise in temperatures is due to two factors:
   −   a natural increase as we come out of the "Little Ice Age" period in the 1800s
   −   warming due to man-made influences such as increased atmospheric CO_2.

The controversy arises concerning the relative contribution of the two sources.[11]

Please Note: In this example, we avoid any controversial issues and simply let the data speak for themselves as we demonstrate the challenges involved in finding an appropriate model for the temperature data and the implications of the selected models for forecasting.

9.2.1 A Stationary Model

In this section we will discuss the topic of modeling the global temperature data using a stationary ARMA model. The exponentially damping sample autocorrelations and wandering (upward) behavior of the data cause us to consider an ARMA model, and we can conjecture that the factored form of the model will have a factor $1 - \alpha B$, where α is close to but less than one. The data are most certainly not white noise, so based

---

11 Our interpretation may be oversimplistic.

on the strategies discussed in Chapter 6, we use AIC-type measures to assist in model identification, and estimate the parameters using ML estimates. We use the following *tswge* commands.

```
data(global2020)
aic.wge(global2020,p=0:10,q=0:4) #AIC selects p=4,q=1
aic.wge(global2020,p=0:10,q=0:4,type='bic') #BIC selects p=3,q=0
global41.est=est.arma.wge(global2020,p=4,q=1)
```

Using the ARMA(4,1) model based on ML estimates we obtain

$$\left(1-.900B+.102B^2-.077B^3-.116B^4\right)\left(X_t-.072\right)=\left(1-.406B\right)a_t, \tag{9.6}$$

where $\hat{\sigma}_a^2 = 0.018$. The associated factor table is given in Table 9.1.

TABLE 9.1   Factor Table for $\left(1-.900B+.102B^2-.077B^3-.116B^4\right)\left(X_t-.072\right)=\left(1-.406B\right)a_t,$

AR-FACTOR	ROOTS	$|r|^{-1}$	f_0
$1-.994B$	1.010	.994	.000
$1-0.289B+.305B^2$	$.473\pm1.748i$	.552	.208
$1+.382B$	-2.618	.382	.500
MA-FACTOR	ROOTS	$|r|^{-1}$	f_0
$1-.406B$	2.460	.406	.000

The AR factor $1-.994B$ is close to the nonstationary factor $1-B$, but the MA factor $1-.406B$ cancels some of the near non stationarity out of the model.

9.2.1.1 Checking the Residuals

The residuals are in vector **global41.est$res**, and these are plotted along with sample autocorrelations in Figure 9.7. There we see no strong evidence against white noise. Additionally, the Ljung-Box p-values for $K=24$ and $K=48$ are .55 and .37, respectively. These are strongly favorable to a decision to not reject white noise. The Shapiro-Wilk p-value is 0.69 suggesting that a normality assumption is reasonable. The *tswge* commands to plot the residuals, their sample autocorrelations with limit lines, and to calculate the Ljung-Box and Shapiro-Wilk tests are

```
plotts.sample.wge(global41.est$res,lag.max=48,arlimits=TRUE)
ljung.wge(global41.est$res,p=4,q=1)
ljung.wge(global41.est$res,p=4,q=1,K=48)
shapiro.test(global41.est$res)
```

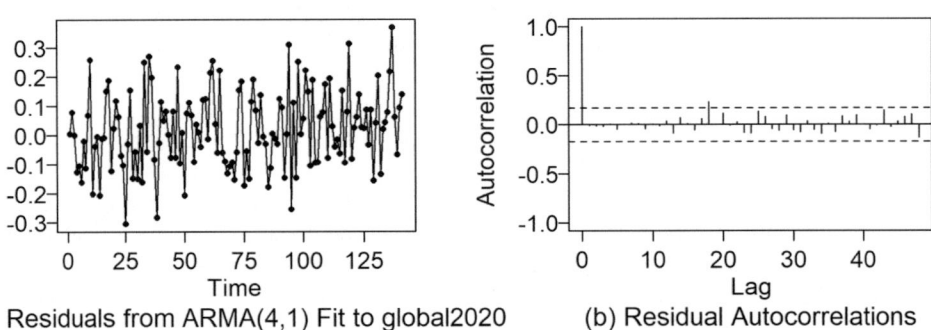

(a) Residuals from ARMA(4,1) Fit to global2020   (b) Residual Autocorrelations

FIGURE 9.7   (a) Residuals from an ARMA(4,1) fit to the global temperature data and (b) sample autocorrelations of the residuals in (a).

9.2.1.2 Realizations and their Characteristics

Figure 9.8(a) shows the global temperature data and Figures 9.8(b)−(f) show five randomly selected realizations of length $n = 141$ from the ARMA(4,1) model in (9.6). The temperature data have a (mostly upward) "wandering or trending behavior along with some high frequency wiggle". The realizations in Figures 9.8(b)−(f) show a variety of behaviors that all could also be described as "wandering or trending behavior along with some high frequency wiggle". Notice that Realization (f) has a strong upward trend while Realization (d) trends down for about 80 time periods and then trends up. However, most of the realizations do not have the primarily monotonic trending behavior of the temperature data in Figure 9.8(a).

Note:

When fitting a correlation-based model (ARMA or ARIMA) to the data, we are tacitly assuming that any observed trends are "random trends" that will eventually abate. The behavior in Figure 9.8(e) is typical in that there is an initial short-lived downward trend, followed by an upward trend that lasts from about $t = 20$ to $t = 100$, followed by a short downward trend. These simulations are designed to help assess whether trending behavior in realizations from the ARMA(4,1) model is likely to be as dramatic as that in the actual temperature data. For this reason, we are interested in downward trends as well as upward trends.

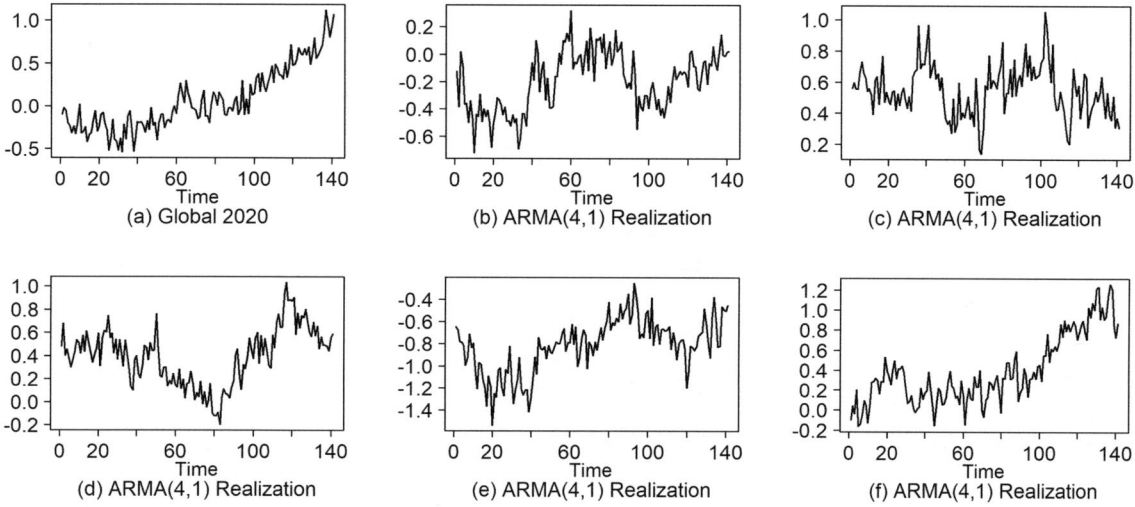

FIGURE 9.8 (a) Global temperature data and (b–f) five realizations of length $n = 141$ from the ARMA(4,1) model in (9.6) which was fit to the global temperature data.

Figure 9.9 compares sample autocorrelations and Parzen spectral density estimates for the global temperature data and the five realizations in Figure 9.8 from the ARMA(4,1) model. Figure 9.9(a) shows that all sample autocorrelations tend to have an exponential-type damping. The sample autocorrelations for the ARMA(4,1) realizations (indicated by open circles connected by solid lines) typically damp more quickly than those for the actual global temperature data. However, the sample autocorrelations for two realizations are quite similar to those for the actual temperature data. The Parzen spectral density estimate for the global temperature data is shown with the bold curve in Figure 9.9(b), and the other five curves are the Parzen spectral density estimates for the five ARMA(4,1) realizations. The Parzen spectral density estimates all have a peak at zero and are similar to the spectral density estimate for the global temperature data. Based on the sample autocorrelations and Parzen spectral estimates in Figure 9.9, the ARMA(4,1) model seems to provide a reasonable fit.

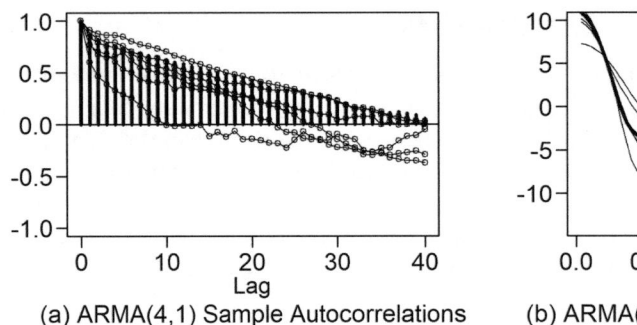 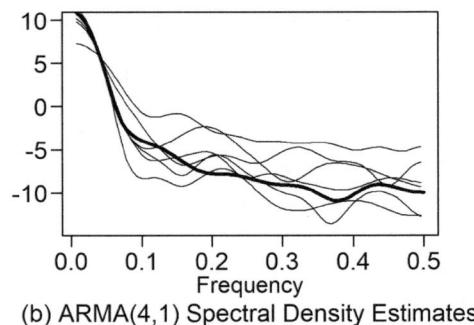

(a) ARMA(4,1) Sample Autocorrelations   (b) ARMA(4,1) Spectral Density Estimates

FIGURE 9.9   (a) Sample autocorrelations for the global temperature data (bold vertical bars) along with sample autocorrelations for the five ARMA(4, 1) realizations in Figure 9.8. (b) Parzen spectral estimate for global temperature (in bold) along with Parzen spectral density estimates for the five ARMA(4, 1) realizations in Figure 9.8.

9.2.1.3 Forecasting Based on the ARMA(4,1) Model

We next use the ARMA(4,1) model to forecast temperatures. Figure 9.10 shows forecasts of the next 50 values for the data in Figure 9.6(a) based on the ARMA(4,1) model.

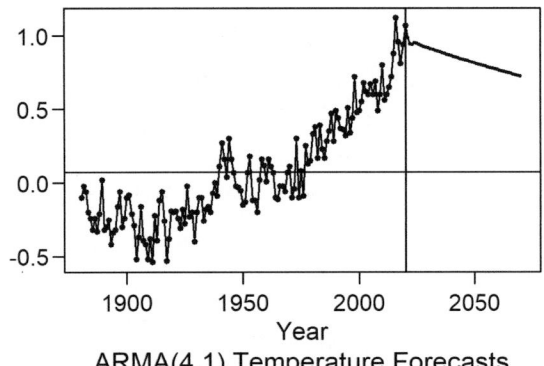

ARMA(4,1) Temperature Forecasts

FIGURE 9.10   Forecasts for global temperature for the years 2021–2070 using the stationary ARMA(4, 1) model in (9.6).

The forecasts were obtained using the command

```
fore.41=fore.arma.wge(global2020,phi=c(.9,-.102,.077,.116),theta=.406,
n.ahead=50,limits=FALSE)
```

(1) Generic Discussion of Forecast Performance

We first discuss the forecasts as if the data in Figure 9.6 were simply some generic dataset. Although the recent data values have been increasing, the ARMA(4,1) model in Equation 9.6 is a stationary model which assumes an equilibrium. Under this model, the interpretation is that the values have wandered far above the mean, and we expect the data values to begin decreasing due to the attraction to a mean level. The eventual forecasts tend to a mean level estimated by the sample mean, $\bar{x} = .072$.

(2) Forecasting Results in the Context of the Problem

While the discussion in the previous paragraph was based on a "generic" time series dataset, the data involved are the global temperature anomalies, and these forecasts result in controversial interpretations in relationship to "global warming". First, we note that based on the appearance of the recent temperature data, the forecasts in Figure 9.10 are not very believable. To our knowledge, no one is predicting that temperatures have reached a maximum and will begin a near immediate decline as predicted by the ARMA(4,1) model. Under this assumption, forecasts that immediately trend downward toward some mean level would be inconsistent with the view of most scientists.

Conclusions Concerning the Stationary ARMA(4,1) Fit

1. The ARMA(4,1) model selected by AIC did a good job of whitening the residuals.
2. Some realizations from the ARMA(4,1) model behave similarly to the temperature data, but most realizations lack the sustained trending behavior. Sample autocorrelations tend to damp more rapidly than in the actual temperature data.
3. Forecasts are quite poor. Because of the equilibrium assumption associated with stationary ARMA models, forecasts tend to trend downward toward an overall mean level.

9.2.2 A Correlation-Based Model with a Unit Root

As an alternative to the ARMA(4,1) model fit, we next consider a model with a unit root. The data in Figure 9.6(a) and the slowly damping sample autocorrelations in Figure 9.6(b) are suggestive of a unit root. Additionally, application of the ADF and KPSS tests discussed in Section 7.1.2.1 conclude that there is a unit root. Therefore, despite our cautions about making decisions solely on the basis of a unit root test, we consider a model for the global temperature data with a unit root. That is, we fit an ARIMA(p,d,q) model by first differencing the data to calculate $Y_t = X_t - X_{t-1} = (1 - B)X_t$, and then modeling Y_t as an ARMA(p,q) process. The differenced data are shown in Figure 9.11(a) along with the sample autocorrelations of the differenced data in Figure 9.11(b). The differenced data actually appear to be white, but we note that the sample autocorrelation at $k = 1$ is outside the limits. AIC and AICC select an MA(2) model for the differenced data, while BIC selects an MA(1). The MA(1) model is more consistent with Figure 9.11(b) because only the first sample autocorrelation appears to be "strongly nonzero".

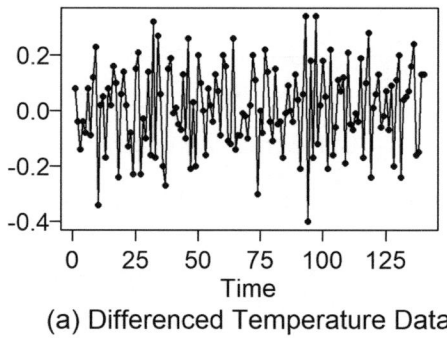
(a) Differenced Temperature Data

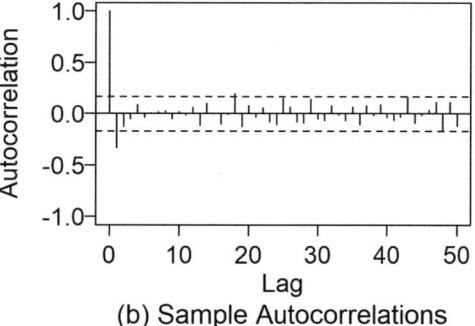
(b) Sample Autocorrelations

FIGURE 9.11   (a) Differenced global temperature data in Figure 9.6(a) and (b) sample autocorrelations of the data in (a).

Fitting the MA(1) model, we obtain

$$(1-B)X_t = (1-.629B)a_t,\tag{9.7}$$

where $\hat{\sigma}_a^2 = 0.018$.

The following are commands that take the first difference of the **global2020** data and plot the first differenced data and sample autocorrelations shown in Figure 9.11. The code also uses BIC to select an MA(1) and obtains the fitted model in (9.7).

```
data(global2020)
y=artrans.wge(global2020,phi.tr=1)
aic.wge(y,p=0:10,q=0:4,type='bic')
# AIC and AICC select an MA(2)while BIC picks MA(1)
# As discussed, we select an MA(1)
y.est=est.arma.wge(y,p=0,q=1)
```

9.2.2.1 Checking the Residuals

Figure 9.12(a) is a plot of the residuals from the MA(1) fit in (9.7) to the differenced data, Y_t. The associated sample autocorrelations are shown in Figure 9.12(b). The residuals appear to be white, and the sample autocorrelations stay sufficiently within the limit lines. The Ljung-Box test (based on $p = 0$ and $q = 1$) reports p-values of 0.598 and 0.367 at $K = 24$ and 48, respectively. The Shapiro-Wilk p-value is 0.252, giving no reason for concern about non-normality of the residuals. Based on analysis of the residuals, the ARIMA(0,1,1) model in (9.7) appears to be a reasonable fit.

The residual plots, their sample autocorrelations with limit lines, the Ljung-Box test and Shapiro-Wilk tests can be obtained using the commands

```
plotts.sample.wge(y.est$res,arlimits=TRUE)
ljung.wge(y.est$res,p=0,q=1)
ljung.wge(y.est$res,p=0,q=1,K=48)
shapiro.test(y.est$res)
```

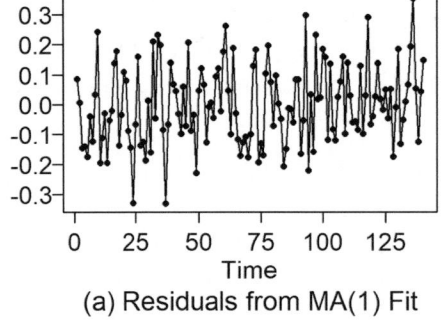
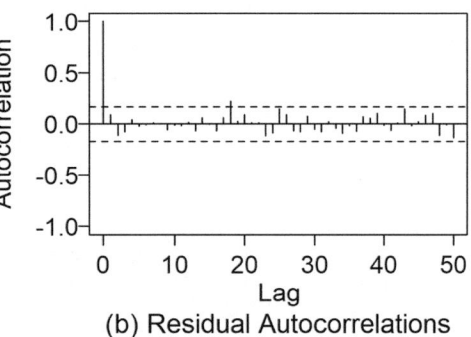

(a) Residuals from MA(1) Fit   (b) Residual Autocorrelations

FIGURE 9.12 (a) Residuals from an MA(1) fit to the differenced global temperature data and (b) sample autocorrelations of the residuals in (a).

Key Points:

1. Models (9.6) and (9.7) both whiten the global temperature data.
2. As discussed in Chapter 7, the decision concerning which model to adopt must be made by the investigator.
3. The decision in (2) will have a major impact on forecasts.

9.2.2.2 Realizations and their Characteristics

Similar to the plots in Figure 9.8 of realizations from the stationary ARMA(4,1) model in (9.6), Figure 9.13(a) shows the `global2020` data and Figures 9.13(b)−(f) are plots of five randomly selected realizations of length $n = 141$ from the ARIMA(0,1,1) model in (9.7). The realizations in Figure 9.13 are similar to those in Figure 9.8 in that they show a variety of behaviors that could all be described as "wandering or trending along with some high frequency behavior." While showing somewhat longer "random" trending behavior than the realizations from the ARMA(4,1) model, as a group they do not have trends that hold up as long as the global temperature data.

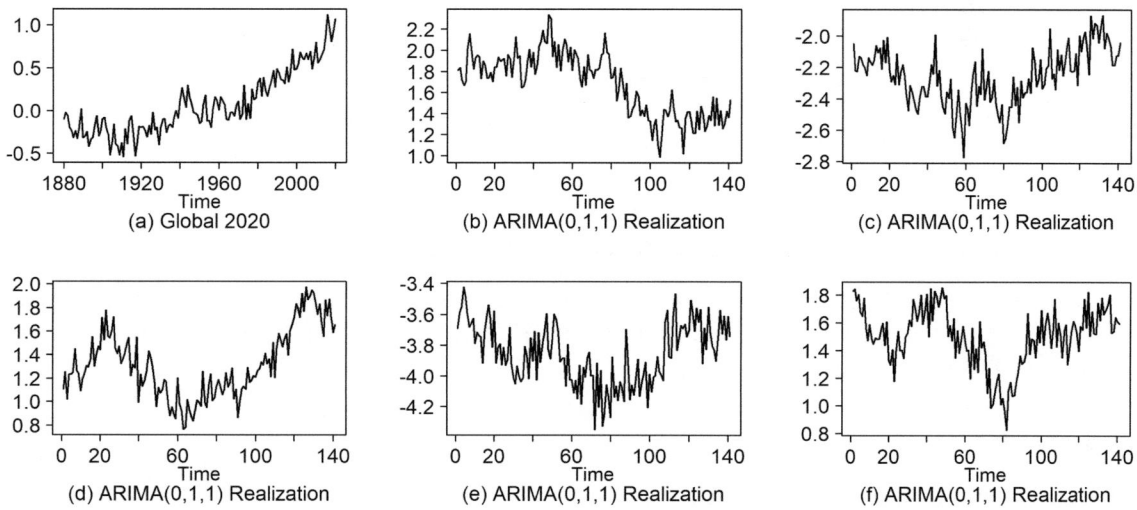

FIGURE 9.13   (a) Global temperature data and (b–f) five realizations of length $n = 141$ from the ARIMA(0, 1, 1) model in (9.7) which was fit to the global temperature data.

Figure 9.14(a) shows that all sample autocorrelations tend to have an exponential-type damping. The sample autocorrelations for the global temperature data (shown using vertical bars) tend to damp more slowly than for most ARIMA(0,1,1) realizations. The Parzen spectral estimates for the global temperature data are shown in bold in Figure 9.14(b). The other five lines represent the Parzen spectral estimates for the five ARIMA(0,1,1) realizations. All simulated realizations have spectral density estimates with a peak at zero and are similar to the spectral density estimate for the temperature data. Based on Figure 9.13, along with the sample autocorrelations and Parzen spectral estimates in Figure 9.14, there is again some cause for concern regarding the ARIMA(0,1,1) model.

Key Points:

1. The ARMA(4,1) and ARIMA(0,1,1) models do not have a "trend component" in the model. However, *some* realizations, e.g. Realization (f) in Figure 9.8 and Realization (b) in Figure 9.13 do show lengthy trends somewhat similar to those in the global temperature data.
2. In general, these "random trends" do not cover as long a time span as the trend observed in the global temperature data.

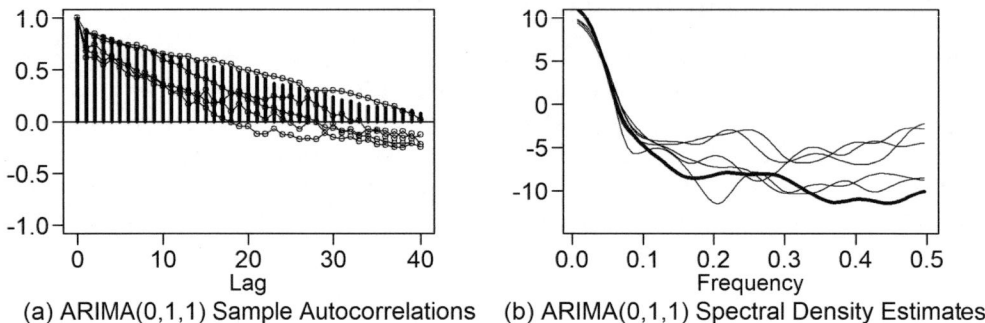

(a) ARIMA(0,1,1) Sample Autocorrelations   (b) ARIMA(0,1,1) Spectral Density Estimates

FIGURE 9.14 (a) Sample autocorrelations for the global temperature data (vertical bars) along with sample autocorrelations for the five ARIMA(0,1,1) realizations in Figure 9.13 (b) Parzen spectral estimate for global temperature (in bold) along with Parzen spectral estimates for the five ARIMA(0,1,1) realizations in Figure 9.13.

9.2.2.3 Forecasting Based on ARIMA(0,1,1) Model

We next use the ARIMA(0,1,1) model to forecast the global temperatures. Figure 9.15 shows forecasts of the next 50 values for the data in Figure 9.6(a) based on the ARIMA(0,1,1) model. The forecasts were obtained using the command

```
fore.011=fore.arima.wge(global2020,d=1,phi=0,theta=.661,
n.ahead=50,limits=FALSE)
```

(1) Generic Discussion of Forecasting Results
While the ARMA(4,1) forecasts in Figure 9.10 trended downward toward a mean level, the ARIMA(0,1,1) model in (9.7) is not stationary and, therefore, under this model, the forecasts do not have an attraction to a mean level. The forecasts follow a horizontal line very similar to the last value observed.[12] The ARIMA(0,1,1) does not predict the current trending behavior to continue (nor does it predict that temperatures will decline).

---

12 The temperature anomaly reading for 2020 is .98 degrees Celsius and the forecasts are a horizontal line at .931 degrees. Recall that forecasts from an ARIMA(0,1,0) model would be equal to the last observed value. However, the moving average term shifts the forecasts slightly.

(2) Forecasting Results in the Context of the Problem

In the context of the problem the forecasts from the ARIMA(0,1,1) model are based on the fact that there is no attraction to a mean level, and the temperatures from 2021 forward are as likely to increase as they are to decrease. This is somewhat in line with those who question global warming although most climatologists believe that there is some "man-made" and some "warming from a colder period" that are impacting global temperatures. This being the case, some continued warming would be expected to persist by most scientists.

Key Points:

1.   The forecasts from an ARIMA(0,1,1) fit are the same as those obtained using exponential smoothing (Shumway and Stoffer, 2017).
2.   ARMA and ARIMA modeling chooses a model before forecasting, while exponential smoothing always uses the same "model" – an ARIMA(0,1,1).

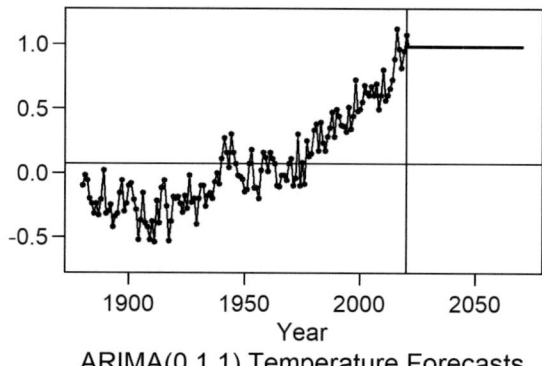

ARIMA(0,1,1) Temperature Forecasts

FIGURE 9.15   Forecasts for global temperature for the years 2021–2070 using the stationary ARIMA(0,1,1) model in (9.7).

(3) Is There a Reasonable Correlation-Based Model that Would Predict the Warming to Continue?

Using modeling techniques discussed in Chapters 6–8, we obtained two reasonable correlation-based models which whiten residuals and have characteristics similar to the temperature data. However, the forecasts from these two models are not consistent with "physical expectations". That is, models (9.6) and (9.7) do not forecast the trend to continue, although (9.7) is "silent" on this issue. In Section 7.1.3 it was noted that a model with one unit root does not predict apparent trends to continue, while ARIMA(p,d,q) models with $d = 2$ predict trending behavior toward the end of the observed realization to continue.[13] Figure 9.11(a) shows the first-differenced global temperature data. There is *absolutely no indication* of a second unit root, and in fact the differenced data differ very little from white noise.

---

13 In Example 7.4 we observed that the correlation-based model, $\phi_{13}(B)(1-B)(1-B^{12})X_t = a_t$, fit to the log airline data predicted the observed trend to continue. We noted that the predicted trending is based on the fact that

$1 - B$ is a factor of $1 - B^{12}$, and, thus, the "airline model" has two factors of $1 - B$.

Key Points:

1. Strictly correlation-based models (ARMA and ARIMA) do not predict the warming to continue because there is absolutely no indication of two unit roots.
2. *Do these models provide conclusive evidence that the increasing trend should not be predicted to continue?* The answer to this question is a resounding "No".
3. If there is a "warming signal" then a signal+noise model would be more appropriate than a strictly correlation-based model. This is the topic of the next section.

9.2.3 Line+Noise Models for the Global Temperature Data

Based on the preceding discussion, we next consider a line+noise model for the global temperature data. That is, we consider a model of the form $X_t = s_t + Z_t$ where s_t is the deterministic "warming signal", and Z_t is a zero-mean stationary noise component that we will model as an AR(p). We consider the linear case in which the signal is a line, $s_t = b_0 + b_1 t$. The line+noise (regression) model is given by

$$X_t = b_0 + b_1 t + Z_t. \tag{9.8}$$

Key Point: If there is a "warming signal" it is almost definitely *not linear* because natural trending patterns are not constrained to behave like a straight line or any other particular mathematical curve. We use a line+noise here because it allows us to use the techniques discussed in Section 8.1.2.

The "beauty" of (9.8) is that it allows for the modeling procedures to "decide" whether the appropriate model is purely correlation based (that is, $b = 0$) or whether including the signal (that is using a model with $b \neq 0$) produces a more suitable model.

Recall that **global2020** contains 141 annual temperature anomalies from 1880–2020. Using the methods of Section 8.1.2, we begin to fit the model (9.8) to **global2020** using the following *tswge* commands:

```
reg=slr.wge(global2020)
t=1:length(global2020)
# residuals from the linear regression line
zhat=global2020-reg$b0hat-reg$b1hat*t
```

$$X_t = -0.4838 + .00782t + \hat{Z}_t t \quad {}^{14} \tag{9.9}$$

Figure 9.16(a) is a plot of the **global2020** dataset, Figure 9.16(b) shows the temperature data along with the least squares regression line in (9.9), and (c) is a plot of the residuals, $\hat{Z}_t$, from the regression line.

---

14 t goes from 1 to **length(global2020)**. To convert the line in (9.9) to years, the formula becomes $-.4838 + .00782$ (year $- 1879$) $= -15.1776 + .00782 *$ year.

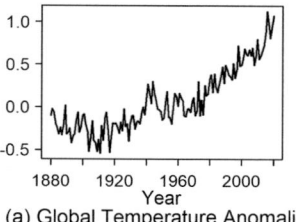 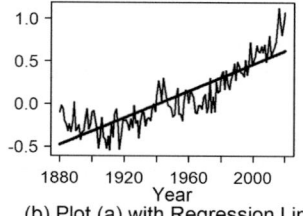 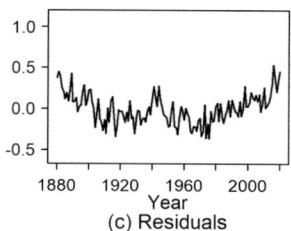

FIGURE 9.16   (a) **global2020** dataset, (b) plot (a) with least squares regression line, and (c) residuals.

The residuals in Figure 9.16(c) do not appear to be white noise. Using the command

```
aic.wge(zhat,p=0:10,q=0:0)
```

AIC selects an AR(4) model. Among the output from **aic.wge** are the coefficients of the AR(4) model along with the estimated white noise variance. The fitted model for the residuals is $\hat{\phi}_4(B)\hat{Z}_t = \hat{a}_t$ where

$$\hat{\phi}_4(B) = \left(1 - .476B - .091B^2 - .109B^3 - .196B^4\right), \tag{9.10}$$

and $\hat{\sigma}_a^2 = .017$.

Thus, the final signal-plus-noise model for the global temperature data, X_t, is

$$X_t = -.4838 + .00782t + \hat{Z}_t. \tag{9.11}$$

where $\hat{\phi}_4(B)\hat{Z}_t = \hat{a}_t$ and $\hat{\sigma}_a^2 = .017$.

(1) Using the Cochrane-Orcutt and WBG Methods to Assess Significance

Because we see an autocorrelation structure in the residuals shown in Figure 9.16(c) are not uncorrelated (which we modeled as an AR(4)) it is not appropriate to test $H_0 : b_1 = 0$ using standard linear regression methods. We will use the Cochrane-Orcutt (CO) and the WBG methods discussed in Section 8.2.3 for this purpose.

```
global.co=co.wge(global2020,maxp=10)
global.wbg=wbg.boot.wge(global2020)
```

The *p*-value for the CO test of $H_0 : b_1 = 0$ is highly significant (less than .0001). Consequently, the CO test indicates that the line in (9.11) should be included in the final model. The WBG test has a *p*-value about .05[15] also suggesting a significant slope but not as strongly as the CO test.

9.2.3.1 *Checking the Residuals, $\hat{a}_t$, for White Noise*

Using the CO and WBG procedures, we have concluded that the line in (9.9) has a slope that is significantly different from zero. The final line-plus-noise model is shown in (9.11). The residuals, $\hat{a}_t$, in our model fit are found using the *tswge* command

```
zhat=est.ar.wge(zhat,p=4)
```

---

15 The WBG test is based on randomly selected bootstrap replications. As such, the result depends on the seed. Using **sn=0** (random seed) and repeating the test five times, the *p*-values were .06, .04, .04, .03, .05.

The residuals, **zhat$res**, and sample autocorrelations with limit lines are shown in Figure 9.17. It can be seen that the sample autocorrelation at lag 19 is outside the 95% limits. The Ljung-Box p-values for $K = 24$ and $K = 48$ are .566 and .435, respectively, so there is insufficient evidence to reject white noise. The Shapiro-Wilk p-value is 0.73, which does not reject normality of the residuals. Based on these criteria, the line+noise fit seems plausible. The *tswge* commands to plot the residuals, their sample autocorrelations with limit lines, and to calculate the Ljung-Box and Shapiro-Wilk tests are as follows:

```
plotts.sample.wge(zhat$res,lag.max=48,arlimits=TRUE)
ljung.wge(zhat$res,p=4)
ljung.wge(zhat$res,p=4,K=48)
shapiro.test(zhat$res)
```

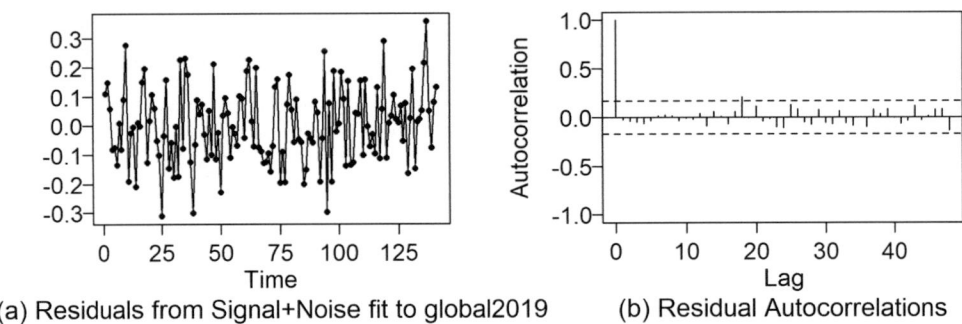

(a) Residuals from Signal+Noise fit to global2019   (b) Residual Autocorrelations

FIGURE 9.17   (a) Residuals from the signal-plus-noise model (9.11) fit to the global temperature data and (b) sample autocorrelations of the residuals in (a).

Key Points: In Equation 9.11 there are two datasets referred to as "residuals".
1. The "residuals" $\hat{Z}_t$ from the regression line are plotted in Figure 9.16(c).
    -   These residuals show an autocorrelation structure, and we model them using ML estimates of the AR(4) model selected by AIC.
2. The "residuals", $\hat{a}_t$, plotted in Figure 9.17(a) are the residuals from the AR(4) model fit to the $\hat{Z}_t$ "residuals".
    -   These residuals should have the appearance of white noise.

9.2.3.2 Realizations and their Characteristics

Figure 9.18(a) shows the **global2020** data and Figure 9.18(b)−(f) are plots of five randomly selected realizations of length $n = 141$ from the line+noise model in (9.11). As a group, the realizations in Figure 9.18 are more similar to the actual temperature data (shown in Figure 9.18(a)) than are the realizations from the two correlation-based models. Because of the linear trend with positive slope, all realizations tend to increase.

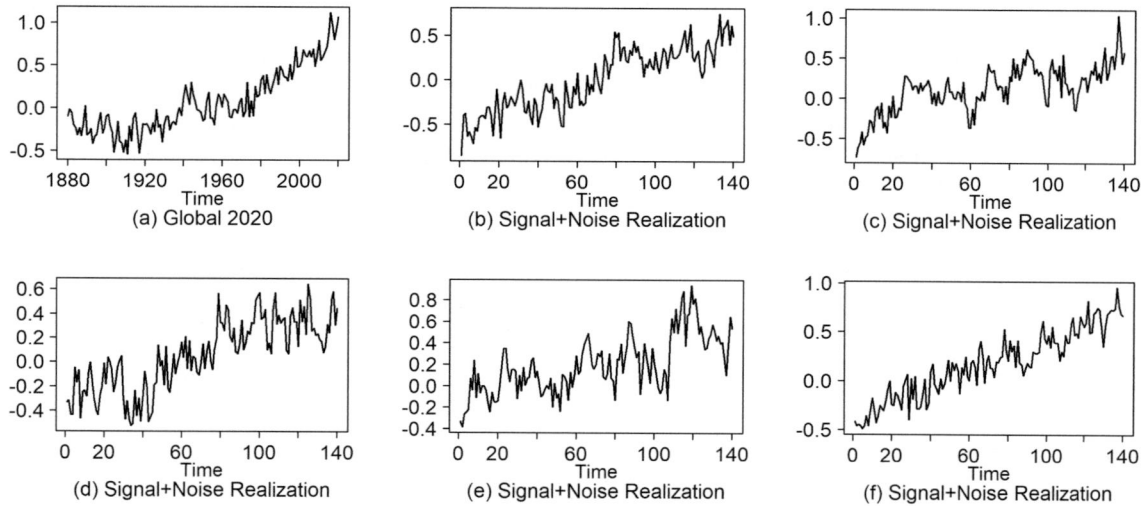

FIGURE 9.18 (a) Global temperature data and (b–f) five realizations of length $n = 141$ from the line+noise model in (9.11) which was fit to the global temperature data.

Figure 9.19(a) plots the sample autocorrelations for the global temperature data (shown using vertical bars) along with the sample autocorrelations shown for the five realizations in Figure 9.18. The sample autocorrelations tend to have an exponential-type damping similar to the sample autocorrelations in Figure 9.6(b) for the actual global temperature data. In fact, three of the five realizations have sample autocorrelations very similar to global temperature sample autocorrelations while the other two damp more quickly. In Figure 9.19(b), the Parzen spectral density estimate for the global temperature data is shown in bold, and the other five curves are the Parzen spectral density estimates for the five signal-plus-noise realizations in Figure 9.18. All spectral estimates have a peak at zero and damp to a similar level by $f = .5$.

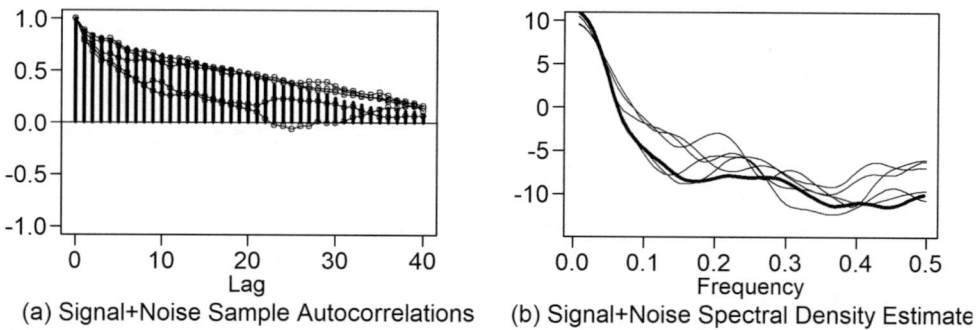

FIGURE 9.19 (a) Sample autocorrelations for the global temperature data (vertical bars) along with sample autocorrelations for the five signal-plus-noise realizations in Figure 9.18. (b) Parzen spectral estimate for global temperature (in bold) along with Parzen spectral estimates for the five signal-plus-noise realizations in Figure 9.18.

9.2.3.3 Forecasting Based on the Signal-plus-Noise Model

Figure 9.20 shows forecasts of the temperature data for the next 50 years using the signal-plus-noise model in (9.11).

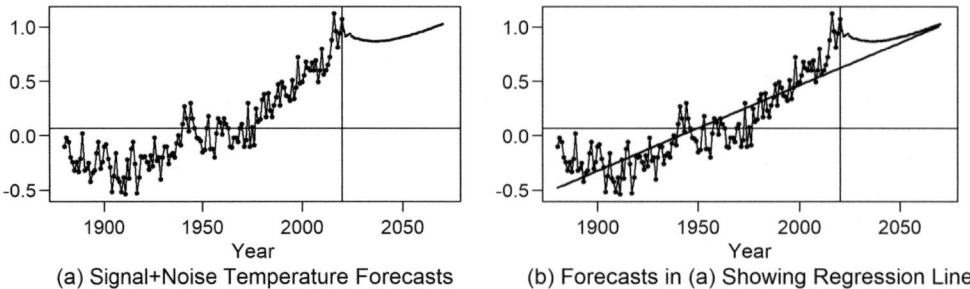

FIGURE 9.20   (a) Signal-plus-noise forecasts for global temperature for the years 2021–2070 and (b) Forecasts in (a) showing fitted regression line.

(1) Generic Discussion of Forecast Performance

The forecasts predict a "near-term" decline with the eventual forecasts following the fitted regression line. The AR(4) model fit to the residuals from the regression line affects the near-term forecasts. However, the eventual forecasts follow the fitted regression line and are not affected by the AR(4) model.

(2) Forecasting Results in the Context of the Problem

As can be seen in Figures 9.20(a) and (b), the forecasts indicate a short decline in temperatures followed by an increase that is at a slower rate than that experienced in the last 45 years. To understand this, note that from 1880–1910 temperatures had a mild decline, followed by a general increase in temperatures from 1910–1940, and this was followed by another mild decline from about 1940–1975. Temperatures have climbed at a relatively high rate after 1975. Consequently, the slope of the regression line fit to the entire dataset is not as steep as the temperature increase for about the last 45 years. The initial dip in the forecasts is reflective of the fact that the forecasts move toward the regression line, which is not as steep as the recent temperature trending behavior.

The following Key Points should be kept in mind.

Key Points Concerning the Line+Noise Fit in (9.11):
1. Realizations from this model show a strong resemblance to the actual temperature data in Figure 9.6(a), and as a whole are more similar to the actual data than are realizations from the strictly correlation-based models.
2. Sample autocorrelations and spectral densities calculated from the realizations are similar to those for the temperature data and are mildly better than those based on realizations from the correlation-based models considered.
3. Forecasts call for an increase in temperatures following an initial short drop.

This leads to the following.

Conclusion: The line+noise model appears to provide a better fit to the global temperature, and thus, based on our analysis of the data, there is reason to believe that temperatures should be predicted to continue to increase.

Before leaving this discussion, we plot the forecasts from the three models along with their 95% limits in Figure 9.21. There it can be seen that the ARMA(4,1) limits would "accommodate" slight increases in temperature as plausible, but these forecasts indicate a decline in temperatures and it appears to be the poorest of the three models. The forecasts for the ARIMA(0,1,1) model in Figure 9.21(b) "allow for the fact" that the temperatures may go up or down, and the limits for the eventual forecasts at, say, 50 steps ahead are wider than those for the signal-plus-noise forecast limits in Figure 9.21(c). One thing to notice about the signal-plus-noise forecasts is that the limits tend to be "tight" (even at longer steps ahead). That is, the uncertainty in the forecasts is smaller *because we fit a deterministic signal as a part of the model*. Forecasting (especially long term) from the linear signal-plus-noise model is an example of *extrapolation* which every good data analyst knows to view with caution. The "small uncertainty" may not be justified because the validity of forecasts depends on whether the process driving temperatures continues into the future. Although the ARIMA(0,1,1) forecasts seem to be poor, these forecasts do "admit" to uncertainty and allow for increasing or decreasing temperatures as being plausible.

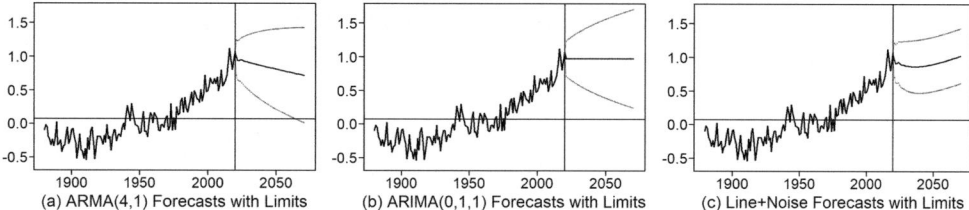

FIGURE 9.21   Forecasts for the years 2021 − 2070 based on (a) the stationary ARMA(4,1) model, (b) the ARIMA(0,1,1) model, and (c) the signal-plus-noise model.

Comments Concerning Temperature Models:

1. Again, if there is a "warming signal", it is almost certain that it is not linear.
2. Beware of extrapolation issues when forecasting with a signal-plus-noise model.
3. Although the ARMA(4,1) and ARIMA(0,1,1) models did not contain a linear trend component, Realization (f) from the ARMA(4,1) model and two of the five realizations from the ARIMA(0,1,1) model had "significant" linear trends according to the CO and WBG tests.
   − Recall that the Cochrane-Orcutt test has inflated significance levels (Section 7.3).
   − The ARIMA(0,1,1) model is "capable" of generating realizations similar to the temperature data and should not be dismissed as a possible model.

9.2.3.4 Other Forecasts

Other possible strategies for forecasting the temperature data include the following:

1. Holt-Winters
2. Only consider data from 1920 on because that frame captures the beginning of carbon emissions. The line+noise forecasts tend toward the fitted regression line in Figure 9.20(b). Because there was no tendency for warming in the early part of the 20th century, the regression line is "flatter" than the current visible temperature increase.

We consider these briefly here.

1. Figure 9.22 shows the Holt-Winters forecasts using a trend but not a seasonal component (solid line) and the line+noise forecasts (dotted line) previously shown in Figures 9.20 and 9.21(c). The commands for the Holt-Winters forecasts follow.

```
data(global2020)
x.hw=HoltWinters(global2020,gamma=FALSE)
x.pred=predict(x.hw,n.ahead=50)
plot(x.hw,x.pred,lty=1:2)
```

Basically, the trend over the last 10–20 years is extended forward. Quite frankly, the Holt-Winters forecasts are frightening! They stand in stark contrast to the more conservative forecasts using the line+noise model.

2. Considering only global temperature data beginning in 1920: Figure 9.22(b) shows the line+noise forecasts (dotted line) and the Holt-Winters forecasts with a trend (solid line) based only on the global temperature data from 1920 through 2020. These forecasts were simply made using the same commands as for Figure 9.22(a) but are based on the dataset `global.1=window(global2020,start=1920)`. In this case, the Holt-Winters and line+noise forecasts are quite similar.

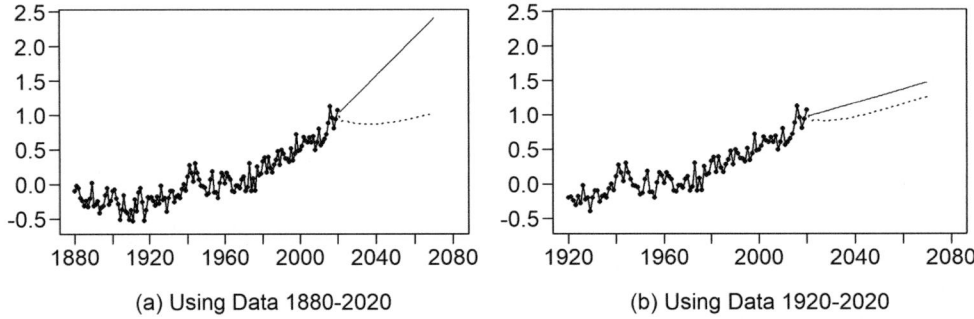

(a) Using Data 1880-2020   (b) Using Data 1920-2020

FIGURE 9.22   Holt-Winters forecasts (solid line) and line+noise forecasts (dotted line) (a) based on the data from 1880–2020 and (b) based on the data from 1920–2020.

Let's be honest:

A *problem with forecasts is that they depend on so many factors that can be controlled by the investigator.*

- If, for example, the investigator wanted to "scare folks" about global warming, then obviously the Holt-Winters forecasts based on the 1880–2020 dataset are the forecasts to use.
- The other forecasts could be used to promote various levels of concern about the global warming issue.

Forecasting Advice: In addition to the comments made regarding forecasts in the Key Points above, we offer the following:

(1) Realize that the model you choose will result in a certain type of forecast, so *carefully choose the type of model* to be used.

Note:
The decision about the "type" of model is more important than the decision to use the AIC or BIC selection of model orders.

(2) It is useful to examine various forecast models before deciding which forecasts to use.
(3) Validate forecasts using techniques described in this book
   −   forecast limits
   −   RMSE and rolling window RMSE are very useful tools

(4) Don't ever provide a forecast value as "the correct answer".
   −   Be sure decision makers are aware of the uncertainties involved and their implications.

(5) As a "consumer" of forecasts, it's important to know "who (or what organization)" produced the forecasts and what assumptions were made.

QR 9.2 Recap

9.3 CASE STUDY 2: COMPARING MODELS FOR THE SUNSPOT DATA

In this section we return to the problem of modeling the sunspot data contained in the *tswge* dataset **sunspot2.0**. As we did with the models for the global temperature data, in this case study we will compare models proposed for the sunspot data based on the three questions posed in the introduction to this chapter:

(1) *Does the model "whiten" the residuals?*
(2) *Do realizations and their characteristics behave like the data?*
(3) *Do forecasts reflect what is known about the physical setting?*

In Chapter 6 we noted two models that have been proposed for the sunspot data: an AR(2) and an AR(9) model. In this case study we will investigate steps involved in a thorough analysis of the data and models selected. Item (2) above is problematic for the sunspot data because we know that realizations from an AR or ARMA model are not going to have the same asymmetric appearance. In Example 8.6(b) we fit a cosine signal+noise model to the data.[16]

```
logss=log(sunspot2.0+10)
```

For convenience we refer to the data calculated above as the *log sunspot data*. Figure 9.23 was previously shown as Figure 8.15, but we repeat it here for completeness. The similarity between the sample

---

16 Recall that the +10 is added to the sunspot numbers because there were instances of zero sunspots, and adding 10 before taking logs produced symmetric behavior similar to that seen in ARMA realizations. The number 10 was arbitrarily chosen for purpose of illustration. However, for example, only adding 1 to each sunspot number did not fully remove the asymmetry

autocorrelations and spectral density estimates of the raw and log sunspot data is striking! In fact, you must look closely to see *any* differences.

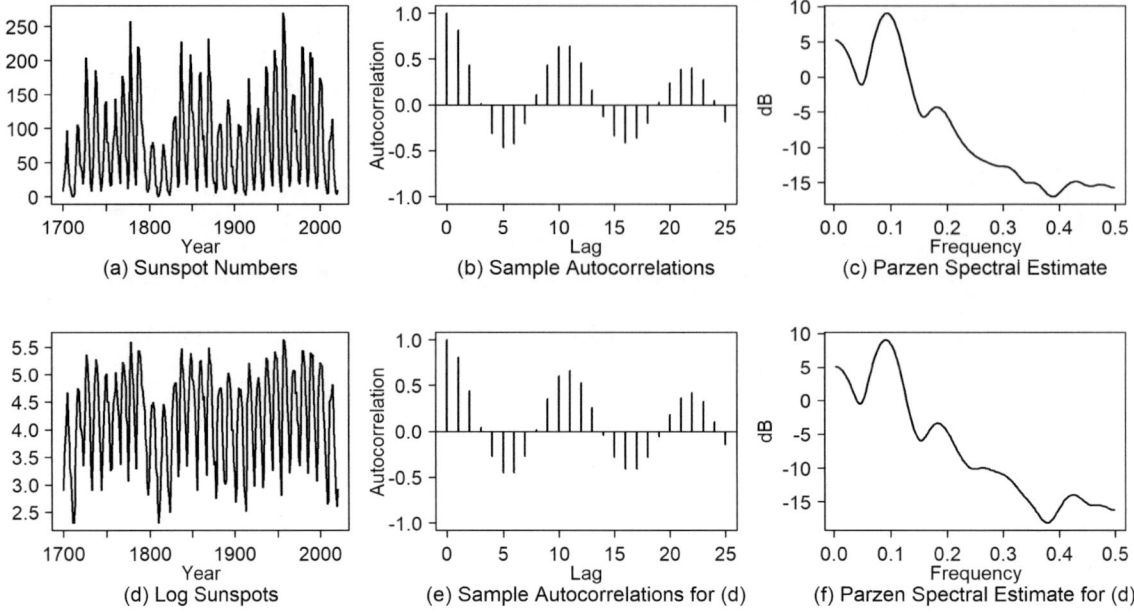

FIGURE 9.23   (a) Sunspot2.0 data, (b) sample autocorrelations and (c) Parzen spectral density estimate of sunspot data in (a), (d) log(sunspot2.0+10), (e) sample autocorrelations and (f) Parzen spectral density estimate of the log sunspot data in (d).

9.3.1 Selecting the Models for Comparison

AR(2) and AR(9) models have been proposed for the "raw" sunspot data, but we will be comparing models for the log sunspot data. Using the `aic5.wge` command below

```
aic.wge(logss,p=0:12,q=0:4)
```

AIC selects an AR(9) model. Using AIC-type tools there is very little support for an AR(2). If you restrict the search to AR models and let p range from 0 to 9, BIC selects an AR(2) as the third choice. Box, Jenkins, and Reinsel (2008) choose an AR(2) (for a different version of the sunspot data) on the basis of the partial autocorrelation function. The *tswge* command

```
pacf.wge(logss)
```

produces the plot in Figure 9.24 which shows two large sample partial autocorrelations suggesting an AR(2).[17]

---

17 Not to be missed is the fact that the `pacf` at lags 6 through 9 are also outside the limits, suggesting an AR(9).

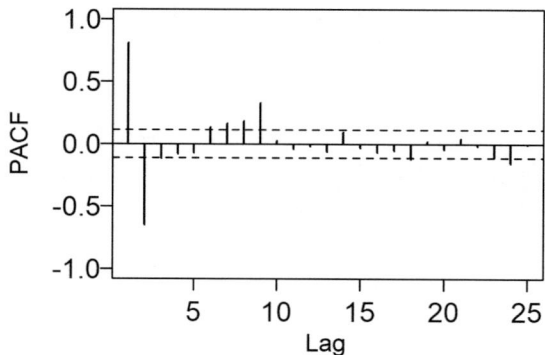

FIGURE 9.24   PACF output for log sunspot data in Figure 9.23(d).

We will use the AR(2) and AR(9) fits for the log sunspot data comparison.

9.3.2 Do the Models Whiten the Residuals?

In Example 9.1 we analyzed the AR(2) and AR(9) models for the raw sunspot data and saw that the AR(9) whitened the residuals while the AR(2) did not. Because we will be comparing models for the log sunspot data, we will repeat the analysis. The following commands estimate the parameters and plot the residuals and their sample autocorrelations for the AR(2) and AR(9) models.

```
data(sunspot2.0)
logss=log(sunspot2.0+10)
ss2=est.ar.wge(logss,p=2)
ss9=est.ar.wge(logss,p=9)
```

The estimated AR(2) model is

$$\left(1-1.37B+.67B^2\right)\left(X_t-4.20\right)=a_t, \tag{9.12}$$

with $\hat{\sigma}_a^2 = .117$. The AR(9) model is

$$\left(1-1.18B+.51B^2+.01B^3-.13B^4+.16B^5-.01B^6-.16B^7+.26B^8-.36B^9\right)\left(X_t-4.20\right)=a_t \tag{9.13}$$

where $\hat{\sigma}_a^2 = .090$. It is informative to find factor tables for AR(2) and AR(9) models for the log data and compare them with the corresponding models for the raw sunspot data.

The residuals and their sample autocorrelations are plotted in Figure 9.25. The AR(2) residuals in Figure 9.25(a) have a "waviness" that doesn't look "white" and several sample autocorrelations are outside the limit lines. The Ljung-Box tests were obtained using the commands

```
ljung.wge(ss2$res,p=2)
ljung.wge(ss2$res,p=2,K=48)
```

and in each case $p < .001$ indicating rejection of the null hypothesis of white noise.

The AR(9) residuals are shown in Figure 9.25(c). These residuals look like white noise and the sample autocorrelations in Figure 9.25(a) stay within the limit lines. The Ljung-Box tests were obtained using the commands

```
ljung.wge(ss9$res,p=9)
ljung.wge(ss9$res,p=9,K=48)
```

and the p-values were .10 and .53 for K=24 and 48, respectively. Consequently, the residuals appear to be white.

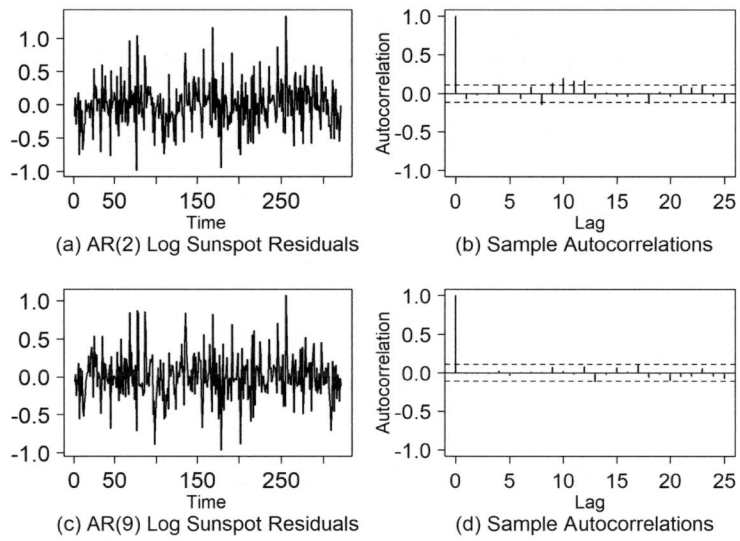

FIGURE 9.25   (a) Residuals from AR(2) fit to log sunspot data, (b) sample autocorrelations of data in (a), (c) residuals from AR(9) fit to log sunspot data, (d) sample autocorrelations of data in (c).

9.3.3 Do Realizations and Their Characteristics Behave Like the Data?

Figures 9.26(a) and 9.27(a) are plots of the log sunspot data. Figures 9.26(b–f) show realizations from the AR(2) model in (9.12) while Figures 9.27(b–f) show five realizations from the AR(9) model in (9.13). The AR(2) model is much more erratic than the log sunspot data and the 10−11 year cycle is barely visible. The AR(9) realizations do a much better job of resembling the log sunspot data. While the behavior is more erratic than the log sunspot data, the 10−11 year cycles are clear.

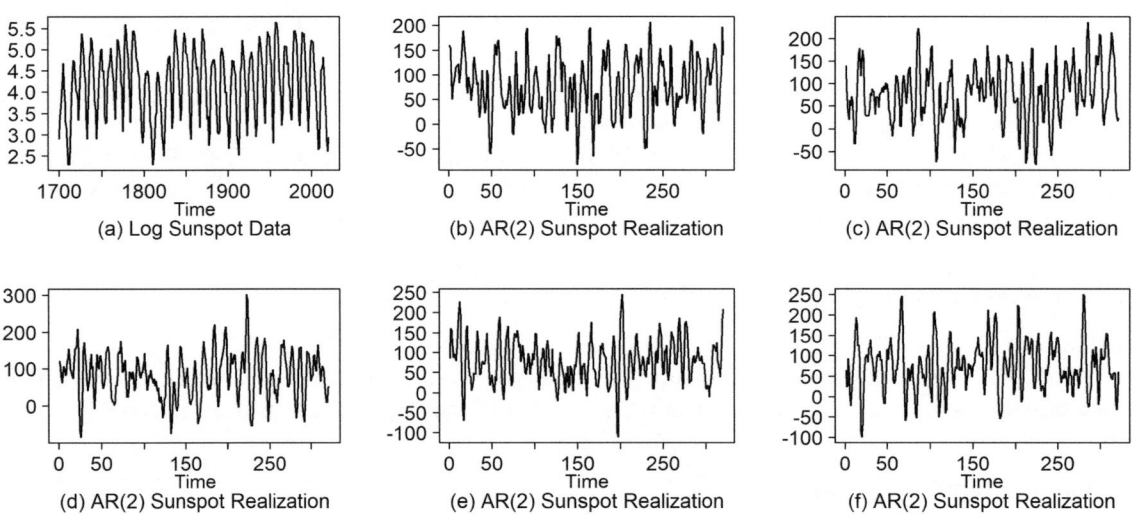

FIGURE 9.26   (a) Log sunspot data and (b–f) realizations from the AR(2) model in (9.12) fit to the log sunspot data.

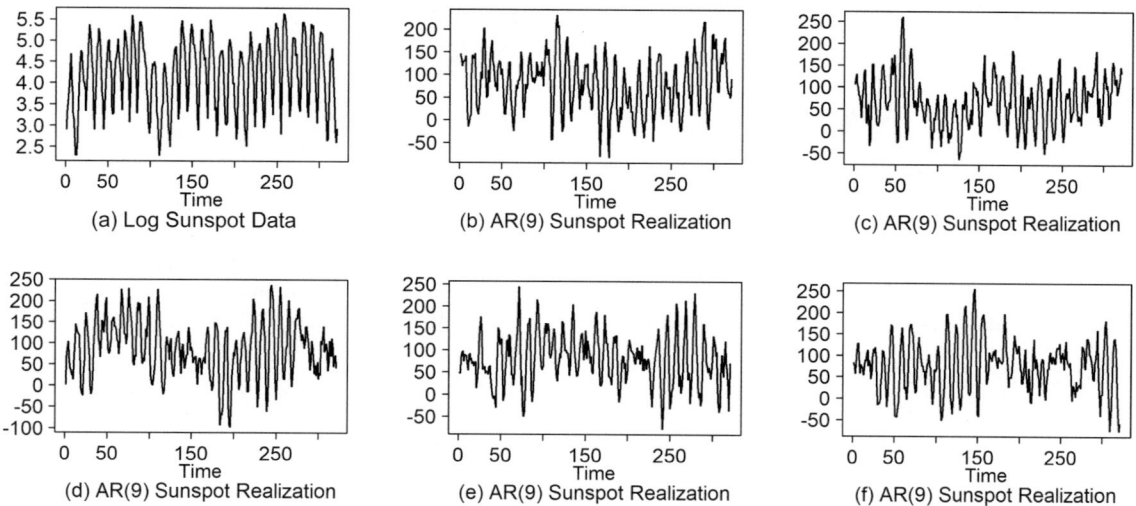

FIGURE 9.27   (a) Log sunspot data and (b–f) realizations from the AR(9) model in (9.13) fit to the log sunspot data.

Figure 9.28(a) shows the sample autocorrelations for the log sunspot data (shown using vertical bars) along with the sample autocorrelations for the five realizations in Figure 9.26 from the AR(2) model. The sample autocorrelations for the realizations have a damped sinusoidal behavior that tend to damp much more quickly than the log sunspot sample autocorrelations. Also, after the first cycle, the sample autocorrelations from the AR(2) model tend to "get off cycle". This is consistent with the lack of distinct cyclic behavior in the AR(2) realizations. The sample autocorrelations in Figure 9.28(b) for the AR(9) realizations in Figures 9.27(b–f) are very similar in appearance to those of the actual log sunspot data.

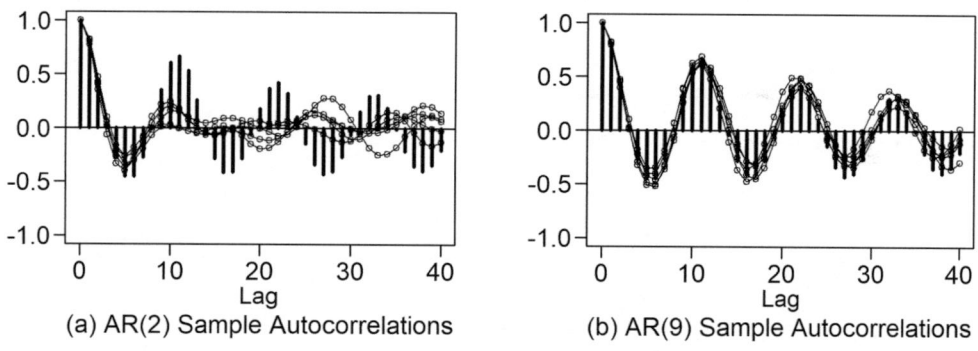

FIGURE 9.28   (a) Sample autocorrelations for log sunspot data (bold vertical bars) and for the five realizations in Figure 9.26 and (b) Sample autocorrelations for actual log sunspot data (bold vertical bars) and for the five realizations in Figure 9.27.

Figure 9.29(a) shows Parzen spectral density estimates for the sunspot data (bold) and for the AR(2) realizations in Figure 9.26. The spectral densities for the AR(2) realizations have peaks at about $f = .1$ that are not as sharp as the $f = .1$ peak for the actual log sunspot data. Noticeably, there is no tendency for the AR(2) spectral density estimates to have a peak at $f = 0$. The spectral density estimates in Figure 9.29(b) for the AR(9) realizations are quite consistent with the spectral density for the sunspot data since they show distinct peaks at $f = 0$ and $f = 0.1$.

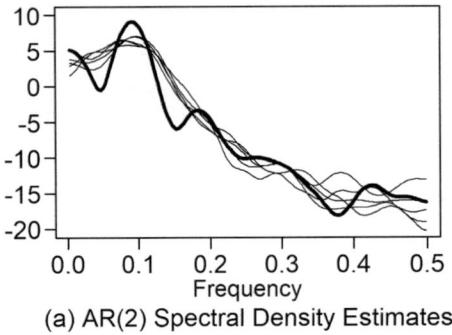 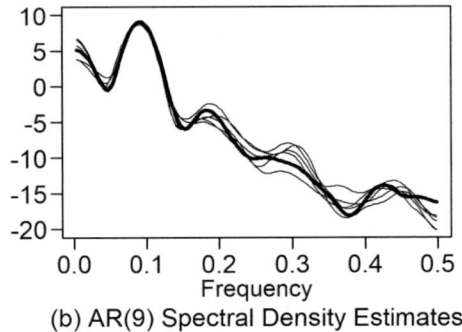

FIGURE 9.29   (a) Parzen spectral density estimate for sunspot data (bold) and for the five AR(2) realizations in Figure 9.26 and (b) Parzen spectral estimate for sunspot data (bold) and for the five AR(9) realizations in Figure 9.27.

9.3.4 Do Forecasts Reflect What Is Known about the Physical Setting?

Example 6.15 went into detail concerning the forecasts from the AR(9) and AR(2) models fit to the raw sunspot data. Specifically, the AR(2) forecasts tended to damp very quickly and simply forecast the mean value for a moderate number of steps ahead. On the other hand, the AR(9) forecasts predicted the cyclic behavior to continue, but because of the fact that the sunspot data does not have fixed cycle lengths, forecasts tend to get off cycle. The behavior of forecasts for the log sunspot data will be similar. The cycles are the same in both datasets. Taking logarithms removed a lot of the asymmetric behavior, but did not alter the cycle lengths.

Conclusion: The AR(2) is a poor model

Although the partial autocorrelations for the sunspot data suggest an AR(2) model:

1.   The residuals are not white noise.
2.   Realizations from the AR(2) model do not have consistent 10–11 year cyclic behavior, and have sample autocorrelations and spectral estimates that are quite different from the corresponding quantities for the actual log sunspot data.
3.   Forecasts do not predict a stable cyclic behavior.

9.3.4.1 Final Comments about the Models Fit to the Sunspot Data

(a)   As mentioned previously, no AR (or ARMA) model will be able to sufficiently account for the asymmetric behavior in the sunspot data. The AR(9) model provides a reasonably good fit with this one exception. Consequently, for this example we used the log sunspot data.
(b)   The Parzen spectral density estimate strongly indicates the existence of two peaks, an obvious one at about $f = .10$ which accounts for the strong 10–11 year cycle in the data, but also a distinct peak at $f = 0$. The factor table for the AR(9) model fit to the log sunspot data is very similar to the one in Table 5.6 for the raw sunspot data, which shows a system frequency of $f = 0$. The AR(2) model only accounts for the 10-11 year cycle.

Key Points:

1. The fact that the Parzen spectral density estimate had strong peaks at $f = 0$ and at about $f = .10$ ***should have told us immediately that an AR(2) model was insufficient.***
2. In order for a model to account for a peak at $f = 0$ and $f = .10$, it ***must*** be at least of order $p = 3$.
    -   Recall that the spectral density for an AR(2) model can have peaks at $f = 0$ and/or $f = .5$, but if it has a peak associated with $0 < f < .5$, then this is the only spectral peak associated with the model. See Section 5.1.3

QR 9.3 Recap and Additional Information

9.4 COMPREHENSIVE ANALYSIS OF TIME SERIES DATA: A SUMMARY

In this chapter, we have discussed issues involved in an analysis of actual time series data. To summarize, given a realization to be modeled, a comprehensive analysis should involve:

1. Examination of the data before fitting a model
    -   is it white noise?
2. Obtaining a model
    -   correlation-based or signal+noise
    -   identifying p, q, d, and seasonal components
    -   ARCH/GARCH
    -   estimating parameters of the signal and of the model for the residuals
3. Checking for model appropriateness
    -   checking residuals
    -   examining realizations and their characteristics
    -   obtaining forecasts, spectral estimates, etc., as dictated by the situation

9.5 CONCLUDING REMARKS

This short (but hopefully useful!) chapter has neatly summarized several of the key time series analysis techniques presented in earlier chapters. The case studies in this chapter, which provide analyses of the global temperature and sunspot datasets, can be a convenient reference for you when you need to quickly review the procedures we recommend for analyzing time series data.

One of the diagnostics for checking the adequacy of a candidate model is the analysis of the residuals. For example, an appropriate time series model should produce residuals which closely resemble white noise. Another attribute of a satisfactory model is that realizations generated from the model should have the overall characteristics of the original data realization. Finally, forecasts given by the model should "make sense" and should be consistent with what is known about the physical setting of the problem.

A variety of examples are provided, and detailed solutions are presented. Interesting comparisons and contrasts are made between various competing models in modeling the global temperature and sunspot data.

It is often the case that you will want to incorporate more than one variable when computing forecasts. This is analogous to the use of several explanatory variables in multiple regression. Such "multivariate" methods will be the topic of Chapters 10 and 11.

APPENDIX 9A

TSWGE FUNCTION

The only new *tswge* function introduced in this chapter is `ljung.wge` which performs the Ljung-Box test on a set of data to test for white noise.

`ljung.wge(x,K,p,q)` performs the Ljung-Box test on the data in vector **x**.

   x is a vector containing the realization to be analyzed for white noise.

   K is the maximum number of lags to be used in the Ljung–Box test statistic formula in Equation 9.3. Box, Jenkins, and Reinsel (2008) recommend running the test using **K = 24** (default) and **K = 48**. Other authors give a variety of recommendations.

   p and **q**: If the data are residuals from an ARMA(p,q) model fit to a set of data, then **p** and **q** specify the model orders of the fit. If **x** is simply a dataset which you want to test for white noise, then use the defaults **p = 0** and **q = 0**.

Output:

   $K is the value of K specified in the input
   $chi.square is the test statistic calculated using (9.3)
   $df is the degrees of freedom associated with the test statistic
   $pval is the p-value of the test $H_0 : \rho_1 = \rho_2 = \cdots = \rho_K = 0$.

BASE R FUNCTION

`shapiro.test(x)` performs a Shapiro-Wilk test of the null hypothesis that the data in **x** are normal.

PROBLEMS

9.1 Consider the data generated using the command

```
ar4=gen.arma.wge(n=100,phi=c(2.76,-3.76,2.6,-.89),sn=463)
```

Find the AR(4) model selected by AIC using ML estimates. Examine the appropriateness of the fitted model using the outline below:

(a) Check the whiteness of the residuals.

(b) Examine the appropriateness of the model in terms of forecasting performance by determining how well the model forecasts the last k steps for whatever value or values you decide k should be.

(c) Generate realizations from the fitted model to determine whether realizations have similar appearance and characteristics (sample autocorrelations and Parzen spectral densities) as the original realization.

9.2 Seasonal model B was generated using the command

```
xB=gen.arima.wge(n=100,phi=c(1.3,-.65),s=4,sn=290)
```

The model fit to the data is given in (7.13). Use the outline in Problem 9.1 to examine the appropriateness of the fitted model.

9.3 For each model below use the outline in Problem 9.1 to assess model appropriateness.

(a) The AR(12) model fit to the log lynx data

(b) The AR(2) model fit to the log lynx data

(c) The ARMA(2,3) model fit to the log lynx data

9.4 In Section 9.1.2 we noted the Shapiro-Wilk test rejected normality for the residuals from the AR(9) fit to the raw sunspot data. Check the residuals of the AR(9) fit to the log sunspot data for normality using the Shapiro-Wilk test. Also, plot a histogram of the residuals and compare it with Figure 9.5(d).

Multivariate Time Series

10

10.1 INTRODUCTION

In previous chapters, we have considered univariate time series, that is, those which involve a single variable. In that scenario, we utilize only the particular time series variable and its past values to fit a model and forecast the future. But in practice, we are often interested in forecasting quantities such as sales or costs which are influenced by other variables as well as past behavior of the dependent variable. In this chapter we will discuss two multivariate time series models:

(1) multiple linear regression (MLR) with correlated errors, and
(2) vector autoregressive (VAR) models.

(1) Multiple linear regression (MLR) with correlated errors

Multiple linear regression with correlated errors is, as its name suggests, very similar to standard multiple linear regression, except that past values of each predictor variable and possibly of the dependent variable can be used as independent variables when predicting the dependent variable. Furthermore, since the error terms are not uncorrelated as is assumed in standard multiple regression, time series methodology must be utilized for meaningful and valid analyses.

(2) Vector autoregressive (VAR) models

Vector autoregressive (VAR) modeling is another technique that is designed to accommodate more than one predictor variable (for example, forecasting sales when considering previous history of both sales *and* advertising). An unusually flexible characteristic of VAR models is that they do not require the analyst to specify which variables are dependent or independent, since there is no distinction between the two.

Examples using simulated and real data will be given to illustrate the two methods.

10.2 MULTIPLE REGRESSION WITH CORRELATED ERRORS

As mentioned, the method of multiple regression with correlated errors is a direct extension of the multiple regression model. However, in the time series setting, either the independent or dependent variables (or some combination of both) may depend on time and occur as realizations of the same length. Because of this dependence on time, it is common for the associated errors to be autocorrelated.

DOI: 10.1201/9781003089070-10

Due to the presence of multiple variables, notation in this chapter will be important. Recall that in univariate modeling, we consider realizations x_t for $t = 1, \cdots, n$ from a time series, X_t. Previous analysis techniques have been presented which use such an observed realization of length n to model the correlation structure within the time series using AR, ARMA, ARIMA, seasonal, etc. models. These models are then used to forecast future values of the univariate variable, that is, $X_{n+1}, X_{n+2}, \ldots$. As mentioned earlier, this basic idea and strategy will be similar for multiple regression with correlated errors.

10.2.1 Notation for Multiple Regression with Correlated Errors

We will denote a multivariate time series regression model (with m independent variables) by[1]

$$Y_t = \beta_0 + \beta_1 X_{t1} + \beta_2 X_{t2} + \ldots + \beta_m x_{tm} + Z_t \tag{10.1}$$

where Z_t is a zero-mean, stationary process. It is also possible for (10.1) to include a trend term and/or previous values of Y_t (denoted Y_{t-k}) and/or previous values of the X_{tj} variables as predictor variables of Y_t. Since each realization is of length n, this model will yield a set of n equations (where $t = 1, \cdots, n$). Because there is one dependent variable and m independent variables, we denote the corresponding $m \times (n+1)$ observations as

$$Y_{1,} X_{11}, X_{21}, \ldots X_{n1}$$

$$Y_{2,} X_{12}, X_{22}, \ldots X_{n2}:$$

$$Y_m, X_{1m}, X_{2m}, \ldots X_{nm}$$

Throughout this chapter, an important notational convention to remember is that the first subscript refers to the time point, while the second subscript refers to the independent variable number.

> **Key Point**: Remember in this chapter that for the term X_{tj}, the first subscript refers to the time point, while the second subscript refers to the independent variable number.

Our approach will be to proceed as in standard multiple regression, with the expectation that the error terms may be correlated, and can be modeled as such. The selection of useful independent variables is similar to that in multiple regression with uncorrelated errors. Then, the resulting correlated error terms, Z_t, are modeled using an AR, ARMA, etc. model, which is included as part of the final model.[2]

To use the independent variables at previous time points as predictors of the dependent variable Y_t, we simply define a separate variable for each independent variable that corresponds to the various lag(s) of interest and enter these variables into the model. For example, if it is hypothesized that both independent variables X_2 and X_3 at a lag of $k = 1$ are important for predicting Y_t (but that X_1 does not have such a lagged relationship with Y_t, and neither do the previous values of Y_t), then the appropriate model corresponding to (10.1) is:

$$Y_t = \beta_0 + \beta_1 X_{t1} + \beta_2 X_{t-1,2} + \beta_3 X_{t-1,3} + Z_t \tag{10.2}$$

---

1   Note that in previous chapters the time series of interest was typically denoted by X_t. However, in the multiple regression section, Y_t will denote the time series of interest (dependent variable) while $X_{tj}s$ are the independent variables.

2   We have chosen to model the residual series, Z_t, using an AR model.

Again, it can be assumed that the residual series Z_t may be correlated and will be modeled accordingly as an AR model.

10.2.2 Fitting Multiple Regression Models to Time Series Data

The procedure we will use to estimate the parameters in (10.2) is as follows:

(a) Use standard MLR procedures to obtain estimates $\hat{\beta}_0, \hat{\beta}_1, \hat{\beta}_2$ and $\hat{\beta}_3$.

(b) Transform the data using $\hat{Z}_t = Y_t - \hat{\beta}_0 - \hat{\beta}_1 X_{t1} - \hat{\beta}_2 X_{t-1,2} - \hat{\beta}_3 X_{t-1,3}$.

(c) Model $\hat{Z}_t$ as an AR(p) model $\hat{\phi}(B)\hat{Z}_t = a_t$ where a_t is well modeled as white noise.

(d) Find the ML estimates of the parameters $\beta_0, \beta_1, \beta_2, \beta_3, \phi_1, \cdots, \phi_p$ in the multiple regression model with correlated errors where the variance-covariance matrix of the error process is based on the covariance structure of the AR fit. The ML estimates are obtained using the Base R command `arima`.

The procedure is illustrated in Example 10.1.

Example 10.1

Suppose that a corporation is interested in forecasting future sales and has evidence or intuition that sales are influenced by the following independent variables: TV advertising expenditures, online advertising expenditures, and the product discount being offered. These data can be found in the *tswge* dataframe **Bsales**. Data points are provided for 100 weeks, meaning that for each of the four variables, there is a corresponding time series realization with length $n = 100$. The resulting notation is as follows.

> Sales (in thousands of dollars): Y_t, for $t = 1, 2, \cdots 100$; that is, $Y_1, Y_2, \cdots Y_{100}$
> TV advertising (in thousands of dollars): $X_{t1} : X_{11}, X_{21}, \ldots X_{100,1}$
> Online advertising (in thousands of dollars): $X_{t2} : X_{12}, X_{22}, \ldots X_{100,2}$
> Discount (in percent): $X_{t3} : X_{13}, X_{23}, \ldots X_{100,3}$

Time series plots for each of these four variables are shown in Figures 10.1(a)–(d). Figure 10.1(a) reveals the cyclic behavior of sales, with a period of noticeably less extreme fluctuation during weeks 40–55; that is, the sales appear to somewhat "flatten out". Figures 10.1(b) and (c) show similar cyclic behavior of TV advertising and online advertising, respectively, but with a less pronounced decrease in fluctuation during weeks 40−55. Figure 10.1(d) is much different from the previous three plots, in that discount remains at zero% or 10% for many consecutive weeks, and all other fluctuations in the plots are quite patterned. Overall, it is not clear from these plots how the independent variables are related to sales, if at all. The code that generates the plots is given below:

```
par(mfrow=c(2,2))
data(Bsales)
sales=Bsales$sales
ad_tv=Bsales$ad_tv
ad_online=Bsales$ad_online
discount=Bsales$discount
plotts.wge(sales,xlab="Week",ylab="Dollars (in thousands)")
plotts.wge(ad_tv,xlab="Week",ylab=" Dollars (in thousands)")
plotts.wge(ad_online,xlab="Week",ylab=" Dollars (in thousands)")
plotts.wge(discount,xlab="Week",ylab=" Dollars (in thousands)")
```

As a first attempt to model this dataset, lag variables will not be included in the model, which can be written as

$$Sales_t = \beta_0 + \beta_1 ad\_tv_t + \beta_2 ad\_online_t + \beta_3 discount_t + Z_t.$$

The following code is used to perform the analysis.

```
data(Bsales)
sales=Bsales$sales
ad_tv=Bsales$ad_tv
ad_online=Bsales$ad_online
discount=Bsales$discount
mlrfit = lm(sales~ad_tv+ad_online+discount)
#Base R function for multiple linear regression
aic.wge(mlrfit$residuals, p=0:8, q=0)
#Selects the optimal p for an AR(p) fit to the residuals- chooses p=7
#the residuals were stored in mlrfit.
# The following computes the ML estimates
fit = arima(sales,order=c(7,0,0),xreg=cbind(ad_tv,ad_online,discount))
fit
```

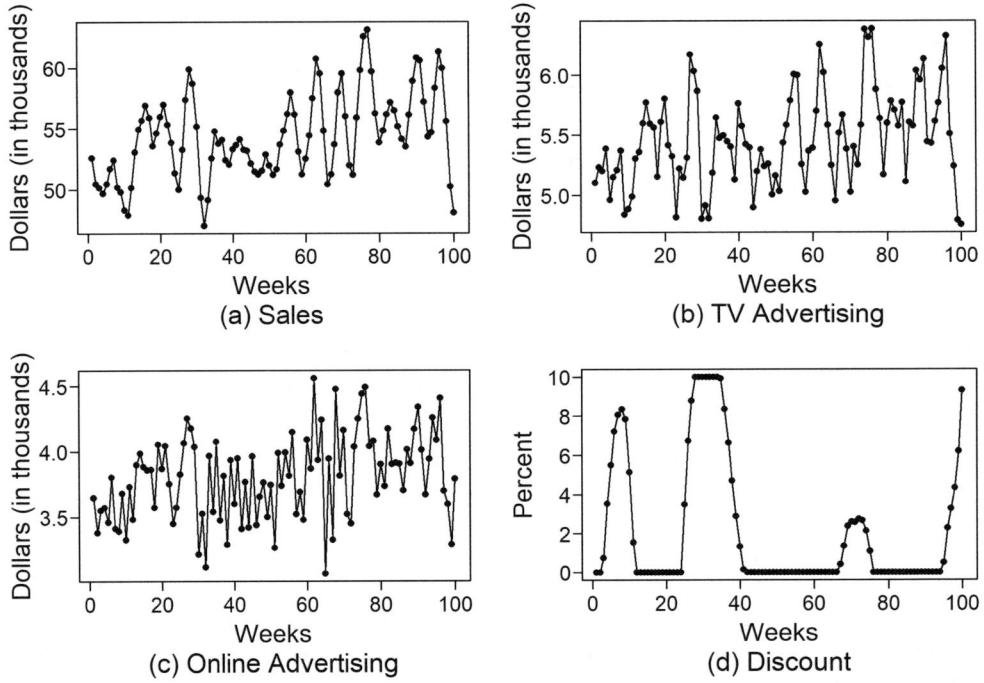

FIGURE 10.1 (a) Sales, (b) TV advertising expense, (c) online advertising expense, and (d) discount.

A summary of the output from the data object **fit** providing parameter coefficients, standard errors, and the ratios of coefficients to parameters is given in Table 10.1.

TABLE 10.1 Summary of Output with No Lagged Independent Variables

	INTERCEPT	AD_TV	AD_ONLINE	DISCOUNT
Coefficients	54.5513	0.0703	−0.0934	−0.1514
S.E	2.204	0.3434	0.2075	0.1315
Ratio	24.75	0.20	−0.45	−1.15

The resulting model is

$$Sales_t = 54.55 + 0.07ad\_tv_t - 0.09ad\_online_t - 0.1514discount_t + Z_t \qquad (10.3)$$

where the error terms Z_t are modeled as an AR(7) process, according to the estimate of the optimal p in `aic.wge`. However, this overall model is unsatisfactory because none of the three independent variables is significant. While p-values are not provided by the function `arima`,[3] we will use a rule of thumb that a variable is significant (at the 5% significance level) if the absolute value of the coefficient exceeds two times the standard error. Consistent with the insignificant predictor variables, further evidence against the adequacy of this model is provided by the diagnostics of the residuals from the AR(7) fit to the residuals of the MLR model.[4] Figure 10.2 shows the sample autocorrelations resulting from the residuals of the AR(7) fit, and reveals that five of the 20 residuals extend beyond the 95% limit lines. Additional confirmation against white noise is suggested by the extremely small p-value from the Ljung-Box test for white noise, where $K = 24$.[5] Corresponding code and output are given below.

```
plotts.sample.wge(fit$resid,arlimits=TRUE)
ljung.wge(fit$resid,p=7)
ljung.wge(fit$resid,p=7,K=48)
```

While the residuals in Figure 10.2(a) from the MLR model fit have the general appearance of white noise, there are several large sample autocorrelations and the p-values from the Ljung-Box test at $K = 24$ and $K = 48$ are both less than .0001. Thus, the MLR has not modeled the data adequately and we need to keep searching for a more appropriate model.

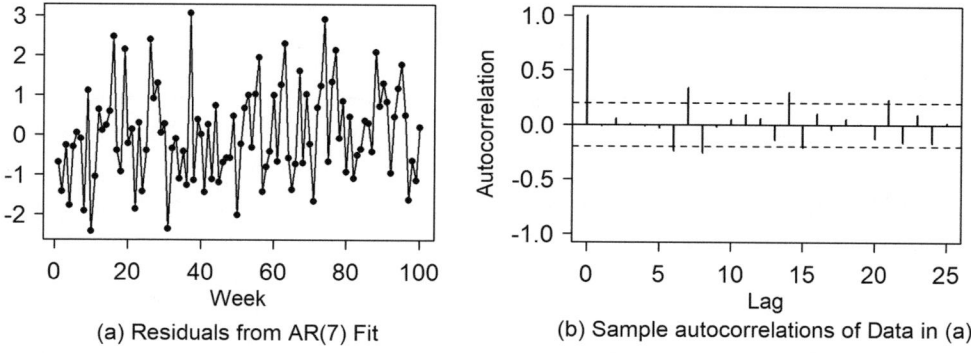

(a) Residuals from AR(7) Fit   (b) Sample autocorrelations of Data in (a)

FIGURE 10.2   (a) Residuals from AR(7) fit in Model (10.1) and (b) sample autocorrelations.

Key Point: It is important to keep in mind that in an MLR with correlated errors analysis, there are two distinct sets of residuals.

1.   The initial set of residuals is a result of the initial MLR fit to the data. These residuals will be modeled as an AR(p) model.

---

3 Recall from Chapter 7 that the Base R function `arima` fits an ARIMA(p,d,q) model, where appropriate estimates of p, d, and q are selected by minimizing AIC.

4 Note that there are two sets of residuals in consideration; one set of residuals is the initial set resulting from the initial MLR fit to the data while the second is the set of are residuals from an AR(p) fit to the residuals remaining after the MLR fit.

5 Recall that the null hypothesis in a Ljung-Box test is that the residuals are white noise, so that small p-values provide evidence against white noise. The default value for number of lags is $K = 24$. We typically use $K = 24$ and $K = 48$.

2. A second set of residuals consists of the residuals (hopefully white noise) resulting from an AR(p) fit to the residuals remaining after the MLR fit. These will be called the "model residuals".

This chapter's diagnostic tests (sample autocorrelation plots and Ljung-Box tests) of the residuals refer to the final set of residuals, that is, from an AR(p) fit to the residuals remaining after the MLR fit.

10.2.2.1 Including a Trend Term in the Multiple Regression Model

In Chapter 8 the idea of including a trend term in a time series model was discussed. A trend term can also be included in the current setting, but the same caveat applies—that is, care must be taken to avoid adding a trend term to model a dataset that does not have a true deterministic trend. In this example, perhaps adding a trend term can improve the previous model. Here, trend will be the week number (from $t = 1, 2, \cdots 100$) and the model of interest is given by

$$Sales_t = \beta_0 + \beta_1 t + \beta_2 ad\_tv_t + \beta_3 ad\_online_t + \beta_4 discount_t + Z_t.$$

We slightly modify the previous code (see code comments above) by adding a trend term, denoted as "t", where $t = 1, 2, \cdots 100$. Note that the addition of the trend term results in AIC choosing an AR(6) model for the residuals. The code is as follows:

```
t = 1:100
mlrfit = lm(sales~ t+ad_tv+ad_online+discount,data=Bsales)
aic.wge(mlrfit$residuals, p=0:8, q=0)
#AIC selects p=6 when fitting the residuals remaining after the MLR fit
fit=arima(Bsales$sales, order=c(6,0,0),xreg=cbind(t,ad_tv,ad_online,discount))
fit
```

The summarized output providing parameter coefficients, standard errors, and the ratios of coefficients to parameters is given in Table 10.2.

TABLE 10.2 Summary of Output with No Lagged Independent Variables, but Including Trend Term

	INTERCEPT	T	AD_TV	AD_ONLINE	DISCOUNT
Coefficients	51.9224	0.0465	0.1123	−0.0508	−0.1701
S.E	2.2242	0.0148	0.3549	0.1939	0.1052
Ratio	23.34	3.14	0.32	−0.26	−1.62

This output suggests the candidate model

$$Sales_t = 51.92 + 0.05t + 0.11ad\_tv_t - 0.05ad\_online_t - 0.17discount_t + Z_t \qquad (10.4)$$

where Z_t is modeled as an AR(6).

In this case, the trend (week number) is significant, but the two advertising variables and discount variable are still insignificant, indicating that this model is probably not adequate. The AR(6) model selected as the best fit to the MLR residuals again produces final residuals that are inconsistent with white noise. Figure 10.3 shows the sample autocorrelations resulting from the residuals of the AR(6) fit to the MLR residuals, and again several autocorrelations fall outside the 95% limit lines. Furthermore, the Ljung-Box test p-value is less than 0.0001 at both $K = 24$ and $K = 48$. Corresponding code and output are given below.

```
plotts.sample.wge(fit$resid,arlimits=TRUE)
ljung.wge(fit$resid,p=6)
ljung.wge(fit$resid,K=48,p=6)
```

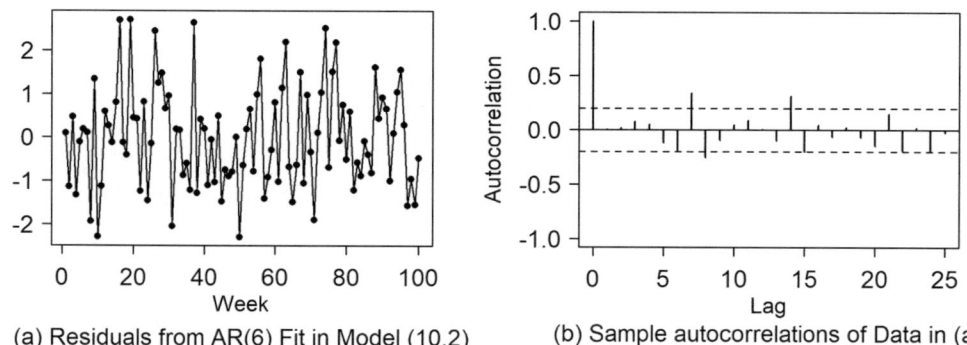

(a) Residuals from AR(6) Fit in Model (10.2)   (b) Sample autocorrelations of Data in (a)

FIGURE 10.3 (a) Residuals from AR(6) fit in model (10.2) and (b) sample autocorrelations.

10.2.2.2 Adding Lagged Variables

The previous two modeling attempts have not taken into consideration the possibility that a relationship may exist between the dependent variable and lagged versions of various independent variables. That is, for example, it is quite possible that the independent variables could have a delayed effect on sales, meaning that advertising in week $t - 1$ may affect sales in week t. So, we create lagged variables (of lag 1) for TV advertising and online advertising, *ad_tv_1* and *ad_online_1*, respectively. The R function used to create lagged variables is **dplyr::lag**. For a detailed tutorial on the **dplyr::lag** function, see the video link referenced by the following QR code. For this model, we will also include the discount variable at time t but will exclude trend, and will fit the model

$$Sales_t = \beta_0 + \beta_1 ad\_tv_{t-1} + \beta_2 ad\_online_{t-1} + \beta_3 discount_t + Z_t.$$

QR 10.1 Business
Sales-Example 10.1

The code is as follows.

```
ad_tv1=dplyr::lag(Bsales$ad_tv,1)#Creating lag 1 for ad_tv
ad_online1=dplyr::lag(Bsales$ad_online,1)#Creating lag 1 for ad_online
discount=Bsales$discount #No lag for discount
Bsales$ad_tv1=ad_tv1 #Add lag (k=1) ad_tv1 to dataset
Bsales$ad_online1=ad_online1 #Add lag (k=1) ad_online1 to dataset
mlrfit=lm(sales~ad_tv1+ad_online1+discount,data=Bsales)#least sqr regression
aic.wge(mlrfit$residuals,p=0:8,q=0) #AIC selects p=7 fit to residuals
fit=arima(Bsales$sales,order=c(7,0,0), xreg=cbind(ad_tv1,ad_online1,discount))
fit
```

Key Point: Creating lagged variables introduces missing data ("NA") values in the new dataset. Some R functions, such as **lm** and **arima**, know by default to omit such a line of data that contains "NA" values. For other functions, such as **VAR** and **VARSelect** (to be introduced in Section 10.3), the analyst must take an extra step to subset the data to exclude the line(s) of data containing "NA" values.

A summary of the output produced by the above code is seen in Table 10.3.

TABLE 10.3 Summary of Output with Lagged Independent Variables, and No Trend Term

	INTERCEPT	AD_TV(T-1)	AD_ONLINE(T-1)	DISCOUNT
Coefficients	4.8382	3.4341	8.1152	−0.0573
S.E	2.827	0.6166	1.2447	0.0281
Ratio	1.71	5.57	6.52	−2.04

From the output, the final model fit is given by

$$Sales_t = 4.84 + 3.43ad\_tv_{t-1} + 8.12ad\_online_{t-1} - 0.06discount_t + Z_t \tag{10.5}$$

where Z_t is fit by an AR(7) model. Interestingly, we find that at the 5% significance level, both advertising variables are now highly significant, and the discount variable is marginally significant. The residuals in Figure 10.4(a) have the appearance of white noise, and all sample autocorrelations in Figure 10.4(b) fall within the 95% limit lines. The Ljung-Box test has a p-value of 0.282 and 0.631, at $K = 24$ and $K = 48$, respectively. This suggests that we have an appropriate model. Corresponding code and output are given below.

```
plotts.sample.wge(fit$resid[2:100],lag.max=50,arlimits=TRUE)
ljung.wge(fit$resid[2:100],p=7)
ljung.wge(fit$resid[2:100],p=7,K=48)
```

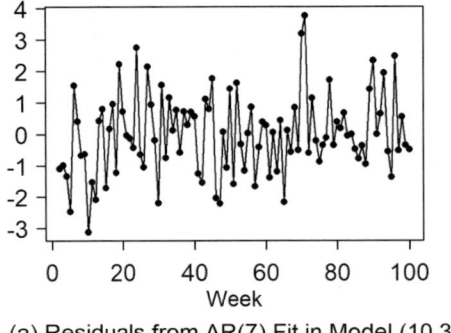

(a) Residuals from AR(7) Fit in Model (10.3)

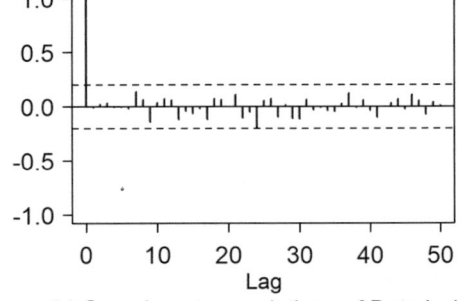

(b) Sample autocorrelations of Data in (a)

FIGURE 10.4 (a) Residuals from AR(7) fit in model (10.3) and (b) sample autocorrelations.

10.2.2.3 Using Lagged Variables and a Trend Variable

The inclusion of lagged variables produces much more satisfactory diagnostic test results overall, but for a final modeling attempt, trend is now included in the previous model containing the significant lagged variables and discount variable to fit the model

$$Sales_t = \beta_0 + \beta_1 t + \beta_2 ad\_tv_{t-1} + \beta_3 ad\_online_{t-1} + \beta_4 discount_t + Z_t.$$

The following code is used.

```
t=1:100 #Adding 100 trend weeks. Remaining code is similar to previous example.
ad_tv1=dplyr::lag(ad_tv,1)
ad_online1=dplyr::lag(ad_online,1)
mlrfit=lm(sales ~ t+ ad_tv1+ad_online1+discount)
aic.wge(mlrfit$residuals,p=0:8,q=0) #AIC selects p=7
fit=arima(sales,order=c(7,0,0),xreg=cbind(t,ad_tv1,ad_online1,discount))
fit
```

The resulting output is seen in Table 10.4.

TABLE 10.4   Summary of Output with Lagged Independent Variables, Including Trend Term

	INTERCEPT	T	AD_TV(T-1)	AD_ONLINE(T-1)	DISCOUNT
Coefficients	6.2215	0.0065	3.318	7.8248	−0.0453
S.E	2.782	0.0038	0.6288	1.302	0.0276
Ratio	2.24	1.71	5.28	6.01	−1.64

From the output we see that

$$Sales_t = 6.22 + 0.0065t + 3.32ad\_tv_{t-1} + 7.82ad\_online_{t-1} - 0.0453discount_t + Z_t, \qquad (10.6)$$

where Z_t is modeled by an AR(7).

Here, we observe again that both advertising variables are highly significant, but in this model, discount is insignificant. The trend variable (week) is technically insignificant (based on significance level .05), but AIC prefers the model including the trend variable over the previous model without trend.[6] The final residuals are modeled by an AR(7), and as was true of the previous model, the diagnostics for these residuals support white noise. In particular, Figure 10.5(a) shows the residuals which have the appearance of white noise, and the sample autocorrelations of the residuals are plotted in Figure 10.5(b) where it can be seen that all sample autocorrelations fall within the 95% limit lines. The Ljung-Box test has p-values 0.264 and 0.617, at $K = 24$ and $K = 48$, respectively. Corresponding code and output are given below.

```
plotts.sample.wge(fit$resid[2:100],lag.max=50,arlimits=TRUE)
ljung.wge(fit$resid[2:100],p=7)
ljung.wge(fit$resid[2:100],p=7,K=48)
```

6 The AIC value is provided in the data object **fit**, and in the previous model was given as 354.86; for this model including trend, the AIC value is 352.87. Remember, however, that it is not recommended to make final model decisions based completely on AIC values.

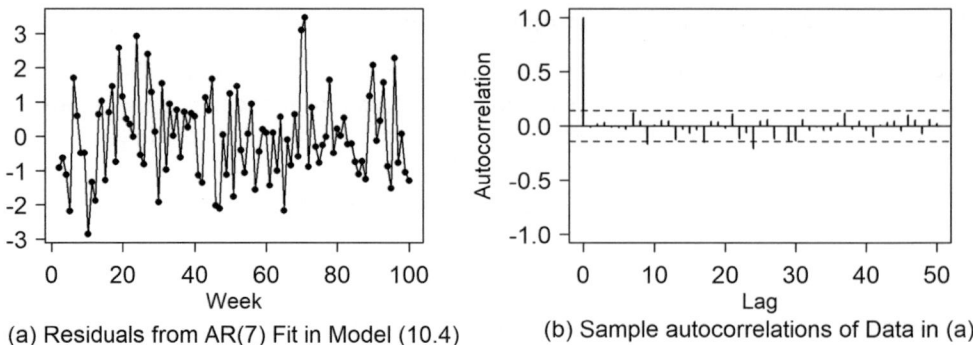

FIGURE 10.5　(a) Residuals from AR(7) fit in model (10.4) and (b) sample autocorrelations.

While including the trend term appears to have improved the model, remember that caution must be taken before including a trend in a model without strong evidence to do so. Because the significance of the trend variable is questionable, this is a case in which the more conservative decision is to not include the trend variable until additional data or domain knowledge gives reason to justify that the trend is real. However, the improvement in the model was noticeable enough that adding the trend term deserves attention and further consideration.

This example emphasizes the added value of using lagged variables for forecasting future values of a time-dependent response variable and also illustrates the strategy used to accommodate correlated residual terms.

10.2.3　Cross Correlation

In Example 10.1 it was shown that when forecasting sales at a given time point t, it was beneficial to use the previous time point $t-1$ of advertising expenses. A question that arises is how one knows, without testing many lags and assessing the significance of each, which lags of which variables should be included in the model. A useful statistical tool for detecting the existence of lagged relationships in time series data is the *cross-correlation function*. The cross-correlation between variables X_{t1} and X_{t2} at lag k is the correlation between $X_{t+k,1}$ and X_{t2}. The general formula for calculating the sample cross-correlations at lag k given a realization of length n for the ordered pairs (X_{t1}, X_{t2}) is given by

$$\hat{\rho}_{ij}(k) = \sum_{t=1}^{n-k} \frac{\left(X_{t+k,i} - \bar{X}_i\right)\left(X_{tj} - \bar{X}_j\right)}{\sqrt{\sum_{t=1}^{n}(X_{ti} - \bar{X}_i)^2}\sqrt{\sum_{t=1}^{n}(X_{tj} - \bar{X}_j)^2}}, \tag{10.7}$$

where i and j represent the variables and k represents the number of lags between the variables. For example, $\hat{\rho}_{12}(k)$ essentially calculates the correlation based on the ordered pairs $(X_{1+k,1}, X_{12}), (X_{2+k,1}, X_{22}), \cdots (X_{n1}, X_{n-k,2})$. Note that if $k = 0$, $\hat{\rho}_{12}(0)$ calculates the correlation from the ordered pairs $(X_{11}, X_{12}), (X_{21}, X_{22}), \cdots, (X_{n1}, X_{n2})$, which is the setting of the familiar Pearson's correlation coefficient. However, when $k \neq 0$, the cross-correlation formula calculates correlation between two variables (either forward or backward). For example, if $k = 2$, then $\hat{\rho}_{12}(2)$ considers the ordered pairs $(X_{31}, X_{12}), (X_{41}, X_{22}), (X_{51}, X_{32}) \cdots (X_{n1}, X_{n-2,2})$, so that the correlation is calculated with variable X_2 being paired with data values of X_1 two time points ahead. However, if $k = -3$, then, $\hat{\rho}_{12}(-3)$ considers the ordered pairs $(X_{11}, X_{42}), (X_{21}, X_{52}), (X_{31}, X_{62}) \cdots (X_{n-3,1}, X_{n2})$, so that the correlation is calculated with variable X_2 being paired with data values of X_1 three time points ago.

Key Point: The cross-correlation between variables X_{t1} and X_{t2} at lag k is the correlation between $X_{t+k,1}$ and X_{t2}. If $k = 0$, the definition is equivalent to the standard Pearson's correlation coefficient introduced in elementary statistics courses. Depending on whether k is positive or negative, the cross-correlation between $X_{t+k,1}$ and X_{t2} is calculated by pairing X_2 values at time t with X_1 values at k lags ahead of, or previous to, time t, respectively.

It is typical to plot the sample cross-correlations for variables X_{ti} and X_{tj} using a plot similar to the sample autocorrelation plot, where the height of a vertical bar indicates the strength of the cross-correlation at a specific lag. In Figures 10.6(a)–(c), we plot the sample cross-correlations from Example 10.1 between the dependent variable sales X_{t1} and each of the independent variables X_{t2} (TV advertising), X_{t3} (online advertising), and X_{t4} (discount). For the cross-correlations between sales and TV advertising (Figure 10.6(a)), and between sales and online advertising (Figure 10.6(b)), we see a positive spike at lag $k = -1$, indicating there is positive correlation between sales at time t and advertising (both TV and online) at time $t - 1$. In other words, there is evidence of substantial positive correlation between sales and advertising lagged by a single week. The cross-correlation between sales and discount (Figure 10.6(c)) reveals no such apparent spike, which suggests that there is no lagged relationship between these two variables. Note that the results suggested by these cross-correlation function ("CCF") plots are consistent with the findings from the modeling procedure in Example 10.1. The Base R function which plots the cross-correlations is **ccf**; the commands to produce the plots in Figures 10.6(a)–(c) are given below.

```
ccf(ad_tv,sales) #Figure 10.6(a)[7]
ccf(ad_online,sales) #Figure 10.6(b)
ccf(discount,sales) #Figure 10.6(c)
```

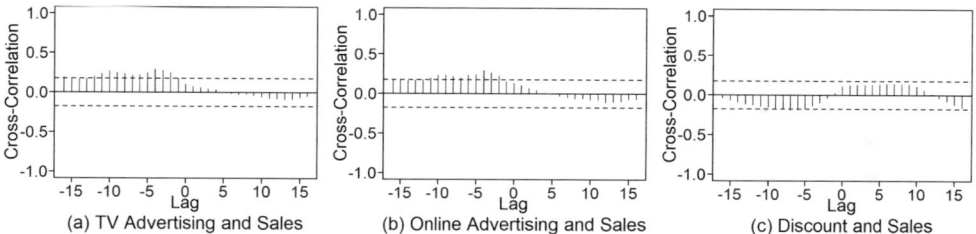

(a) TV Advertising and Sales   (b) Online Advertising and Sales   (c) Discount and Sales

FIGURE 10.6 Cross-correlations between (a) TV advertising and sales, (b) online advertising and sales, and (c) discount and sales.

Caution should be advised due to inconsistencies among authors and software packages in defining the cross-correlation function. To be consistent with Base R, we have defined the cross-correlation function between X_{t1} and X_{t2} at lag k as the correlation between $X_{t+k,1}$ and X_{t2}. That is, the first variable reflects the time shift. The corresponding Base R syntax is **ccf(x1,x2)**. However, some authors (for example, Woodward et al. 2017) have defined the cross-correlation function between X_{t1} and X_{t2} at lag k as the correlation between X_{t1} and $X_{t+k,2}$. This is remedied in R by reversing the syntax as **ccf(x2,x1)**.

QR 10.2
Cross-correlation

---

7 If the order of the variables is reversed, that is, **ccf(sales,ad_tv)**, the spike would appear at lag $k = 1$.

> **Key Point**: Be careful with the definition of cross-correlation among various authors and software packages. While we have defined the cross-correlation function between X_{t1} and X_{t2} at lag k as the correlation between $X_{t+k,1}$ and X_{t2} (as in Base R), it can also be defined as the correlation between X_{t1} and $X_{t+k,2}$. When using R, simply reverse the order of the variables in the syntax to make the definitions match your preference.

10.3 VECTOR AUTOREGRESSIVE (VAR) MODELS

The multiple linear regression with correlated errors method does not take into account the possible correlation structure within and among the independent variables. However, the objective in vector autoregressive (VAR) modeling is to investigate the interrelationships among all variables of interest in order to improve forecasts for one or more of the variables. Since vector autoregressive (VAR) models do not distinguish between dependent and independent variables, the goal is to use all of the variables and to simultaneously forecast all variables.

> **Key Point:** In multiple linear regression (MLR) with correlated errors, the correlation between explanatory variables is somewhat of a liability due to multicollinearity concerns, while VAR leverages such correlation to improve model fitting and corresponding forecasts.

The notation for VAR will remain the same as for MLR with correlated errors, where again the first subscript refers to the time point while the second subscript refers to the variable number.

We first consider the "bivariate VAR(1) process", in which two dependent variables are being modeled but only one time lag is considered for both of the independent predictor variables (i.e. $m = 2$ and $p = 1$). Recall in the univariate case that we often write an AR(1) model as

$$X_t = (1 - \phi_1)\mu + \phi_1 X_{t-1} + a_t.$$

In the univariate AR(1) setting, the model involves the lag 1 value X_{t-1}.

In a bivariate VAR(1) model, there are two variables, X_{t1} and X_{t2}, and these values involve the lag 1 values $X_{t-1,1}$ and $X_{t-1,2}$. The equations for the bivariate VAR(1) model are more complex than for the AR(1) model due to the interrelationships between the two variables and their lag 1 time points. The bivariate VAR(1) model can be expressed as

$$X_{t1} = (1 - \phi_{11})\mu_1 - \phi_{12}\mu_2 + \phi_{11} X_{t-1,1} + \phi_{12} X_{t-1,2} + a_{t1}$$

$$X_{t2} = -\phi_{21}\mu_1 + (1 - \phi_{22})\mu_2 + \phi_{21} X_{t-1,1} + \phi_{22} X_{t-1,2} + a_{t2}, \tag{10.8}$$

where a_{t1} and a_{t2} are the residuals for the models corresponding to Variable 1 and Variable 2, respectively.

Because of the notation complexity, VAR models are often written in a more convenient and abbreviated matrix form. For example, the matrix notation for a bivariate VAR(1) model (with zero mean) is

$$\begin{pmatrix} X_{t1} \\ X_{t2} \end{pmatrix} = \begin{pmatrix} \beta_1 \\ \beta_2 \end{pmatrix} + \begin{pmatrix} \phi_{11} & \phi_{12} \\ \phi_{21} & \phi_{22} \end{pmatrix} \begin{pmatrix} X_{t-1,1} \\ X_{t-1,2} \end{pmatrix} + \begin{pmatrix} a_{t1} \\ a_{t2} \end{pmatrix},$$

QR 10.3 VAR Model

where

$$\begin{pmatrix} \beta_1 \\ \beta_2 \end{pmatrix} = \begin{pmatrix} (1-\phi_{11})\mu_1 - \phi_{12}\mu_2 \\ -\phi_{21}\mu_1 + (1-\phi_{22})\mu_2 \end{pmatrix}$$

If the bivariate VAR(1) model is extended to a bivariate VAR(2) model, there will be two matrices of coefficients ϕ_{ij}. For this reason, we modify the notation so that, for example, ϕ_{11} will be denoted as $\phi_{11(1)}$ for the lag 1 matrix component and denoted as $\phi_{11(2)}$ for the lag 2 matrix component:

$$\begin{pmatrix} X_{t1} \\ X_{t2} \end{pmatrix} = \begin{pmatrix} \beta_1 \\ \beta_2 \end{pmatrix} + \begin{pmatrix} \phi_{11(1)} & \phi_{12(1)} \\ \phi_{21(1)} & \phi_{22(1)} \end{pmatrix} \begin{pmatrix} X_{t-1,1} \\ X_{t-1,2} \end{pmatrix} + \begin{pmatrix} \phi_{11(2)} & \phi_{12(2)} \\ \phi_{21(2)} & \phi_{22(2)} \end{pmatrix} \begin{pmatrix} X_{t-2,1} \\ X_{t-2,2} \end{pmatrix} + \begin{pmatrix} a_{t1} \\ a_{t2} \end{pmatrix}.$$

Here, the vector $\begin{pmatrix} \beta_1 \\ \beta_2 \end{pmatrix}$ is an even more complicated pair of linear combinations of μ_1 and μ_2 than above for the bivariate VAR(1) model. If we multiply the set of matrices, we obtain the expanded version of the bivariate VAR(2) model:

$$X_{t1} = \beta_1 + \phi_{11(1)}X_{t-1,1} + \phi_{12(1)}X_{t-1,2} + \phi_{11(2)}X_{t-2,1} + \phi_{12(2)}X_{t-2,2} + a_{t1}$$

$$X_{t2} = \beta_2 + \phi_{21(1)}X_{t-1,1} + \phi_{22(1)}X_{t-1,2} + \phi_{21(2)}X_{t-2,1} + \phi_{22(2)}X_{t-2,2} + a_{t2} \tag{10.9}$$

Note that both X_{t1} and X_{t2} depend on lagged $t-1$ and $t-2$ values of X_{t1} and X_{t2}.

Obviously, writing the VAR(p) equations in expanded form (and even in matrix form) will quickly become tedious for higher orders and for more than two variables. Next we will establish the forecasting methodology used for VAR(p) models.

10.3.1 Forecasting with VAR(*p*) Models

Forecasting with VAR(p) models is an extension of forecasting with AR(p) models, which was introduced in Chapter 6. Recall that forecasts for an AR(p) model are based on the underlying AR(p) equation and are calculated by

$$\hat{X}_{t_0}(\ell) = \bar{x}\left(1 - \hat{\phi}_1 - \cdots - \hat{\phi}_p\right) + \hat{\phi}_1\hat{X}_{t_0}(\ell-1) + \cdots + \hat{\phi}_p\hat{X}_{t_0}(\ell-p).$$

Specifically, the ℓ-step ahead forecasts, $\hat{X}_{t_0}(\ell)$, for the univariate variable X_t depend on $\ell - 1$ through $\ell - p$ step ahead forecasts, i.e. $\hat{X}_{t_0}(\ell - 1), \ldots, \hat{X}_{t_0}(\ell - p)$, some of which may be actual observed values.

For the sake of simplicity, we will show forecasts for the bivariate VAR(1) model given in (10.8). The forecasts for $\hat{X}_{t_0 1}(\ell)$ and $\hat{X}_{t_0 2}(\ell)$ are given by

$$\hat{X}_{t_0 1}(\ell) = \left(1 - \hat{\phi}_{11}\right)\bar{x}_1 - \hat{\phi}_{12}\bar{x}_2 + \hat{\phi}_{11}\hat{X}_{t_0 1}(\ell - 1) + \hat{\phi}_{12}\hat{X}_{t_0 2}(\ell - 1)$$

$$\hat{X}_{t_0 2}(\ell) = -\hat{\phi}_{21}\bar{x}_1 + \left(1 - \hat{\phi}_{22}\right)\bar{x}_2 + \hat{\phi}_{21}\hat{X}_{t_0 1}(\ell - 1) + \hat{\phi}_{22}\hat{X}_{t_0 2}(\ell - 1).$$

Specifically, the ℓ-step ahead forecasts $\hat{X}_{t_0 1}(\ell)$ for the variable X_{t1} and for X_{t2} depend on $\ell - 1$ step ahead forecasts for both variables X_{t1} and X_{t2}.

QR 10.4 VAR
Model Forecasts

Example 10.2 A Simulated Example

This example illustrates how multivariate techniques detect that one variable is a leading indicator of another variable and how such a relationship is used advantageously in VAR forecasts. The following two time series realizations of length $n = 25$ were generated from AR(2) models and are given below.

```
x1.25=c( -1.03, 0.11, -0.18, 0.20, -0.99, -1.63, 1.07, 2.26, -0.49, -1.54, 0.45,
0.92, -0.05, -1.18, 0.90, 1.17, 0.31, 1.19, 0.27, -0.09, 0.23, -1.91, 0.46,
3.61, -0.03)
```

```
x2.25=c( -0.82, 0.54, 1.13, -0.24, -0.77, 0.22, 0.46, -0.03, -0.59, 0.45, 0.59,
0.15, 0.60, 0.13, -0.04, 0.12, -0.96, 0.23, 1.81, -0.01, -0.95, -0.55, -0.15,
0.71, 0.90)
```

These time series are plotted in Figure 10.7(a), where **x1.25** values are shown as solid dots connected by a solid line and **x2.25** values are represented by open circles connected by dashed lines. From Figure 10.7(a), it is not clear that there is a relationship between the two datasets. However, for $t = 6, 7, \ldots, 20$, the data were created in such a way that $x_{t+5,1}$ is very close to the value $2x_{t2}$. This relationship is shown in Figure 10.7(b). The first five values of **x1.25** and the last five values of **x2.25** are not related in any special way. The cross-correlations between **x1.25** and **x2.25** are shown in Figure 10.7(c). The strong positive cross-correlation at $k = 5$ is the correlation between $X_{t+5,1}$ and X_{t2}, which is high by construction of the datasets.[8] The cross-correlations were obtained using the code

```
ccf(x1.25,x2.25) ## cross-correlation shows the significant lag at 5 ##
```

---

8 Notice that three spikes exceed the limit lines. We conclude that lag five is most influential because it is the most extreme. Just as in a sample autocorrelation plot, it is common for a particular prominent "spike" at a given lag to have several nearby lags which also show relatively strong correlation. It is also important to note that if the **ccf** function in R had reversed the two variables, the most extreme lag would have appeared at $k = -5$. Be sure to pay close attention to this order, as cautioned in the previous Key Point.

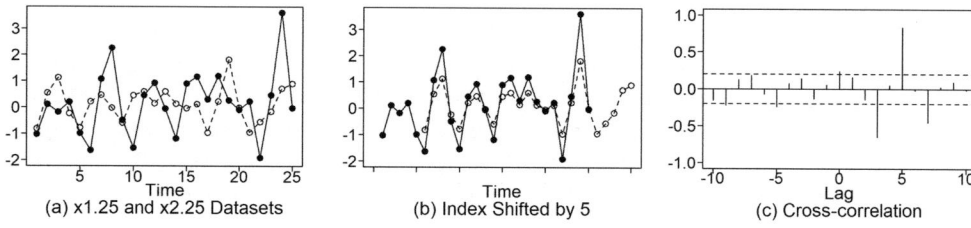

FIGURE 10.7   (a) Datasets x1.25 and x2.25, (b) datasets x1.25 and x2.25 with x2.25 shifted five time units to the right to show relationship, and (c) cross-correlation between x1.25 and x2.25.

10.3.1.1 Univariate Forecasts

We first obtain univariate forecasts for each of the two time series realizations, using techniques discussed in Chapter 6. The following forecasts are obtained using the first 20 data values as a training set on which the forecasts are based, and the last five values are the test set which will be forecast. The training sets are

```
x1=x1.25[1:20]
x2=x2.25[1:20]
```

The univariate forecasts in Figure 10.8 were obtained using the following code.

```
p1=aic.wge(x1,p=0:8,q=0:0) # aic picks p=2
x1.est=est.ar.wge(x1,p=p1$p)
fore.arma.wge(x1.25,phi=x1.est$phi,n.ahead=5,lastn=TRUE,limits=FALSE)
#
p2=aic.wge(x2,p=0:8,q=0:0) # aic picks p=2
x2.est=est.ar.wge(x1,p=p2$p)
fore.arma.wge(x2.25,phi=x2.est$phi,n.ahead=5,lastn=TRUE,limits=FALSE)
#
```

When modeled as univariate models, the forecasts for both **x1.25** and **x2.25** are fairly poor. In Figure 10.8(a), the forecasts for times $t = 21$ to 25 for variable **x1.25** are superimposed on the last five corresponding actual points. The forecasts tend to remain relatively close to the last actual value, and therefore do not predict the strong oscillatory behavior of the actual data.

Similarly, in Figure 10.8(b), the last five forecasts for variable **x2.25** are overlaid on the last five corresponding actual points. These forecasts incorrectly predict the cycle length of the oscillatory behavior; the actual data steadily increase while the forecasts predict a sudden downturn at $t = 24$ which did not occur. Thus, univariate AR(p) forecasts are disappointing for both **x1.25** and **x2.25**.

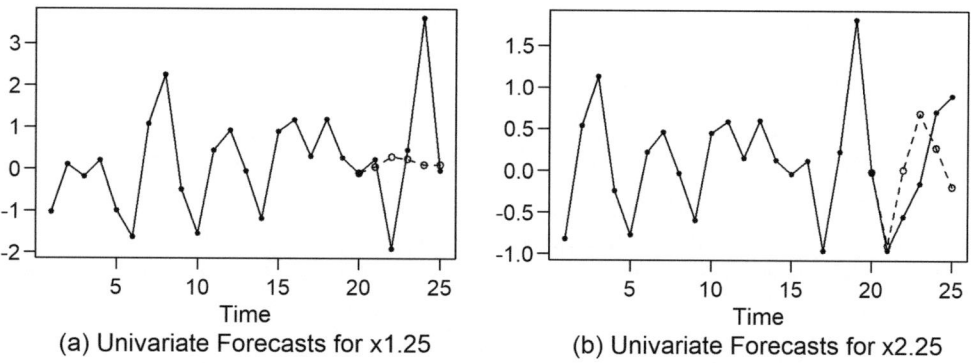

FIGURE 10.8   Univariate Forecasts for (a) x1.25 and (b) x2.25.

10.3.1.2 VAR Analysis

The data in this example are designed to show (in a big way) the value of VAR over univariate forecasting. In this book, we will use the CRAN package *vars* for VAR analysis. The steps involved in VAR forecasting are the same as for the related univariate AR model. Recall that the data on which we will perform the VAR analysis are **x1** and **x2**, which consist of the first 20 values in each realization. We will combine these two datasets (which are stored as column vectors) into a matrix, **X**, using the command

```
X=cbind(x1,x2)
```

1. Model Identification:

The first step is to identify the order p of the VAR(p) model. The function **VARselect** in the *vars* package is analogous in functionality to the **aic.wge** function in *tswge*. **VARselect** will fit models up to a particular lag limit (**lag.max**) and will return AIC and other model identification criteria including BIC (referred to in the output as SC). [9]

```
# You will need to install and load the vars package from CRAN
library(vars)
VARselect(X,lag.max=6,type="const",season=NULL,exogen=NULL)
# AIC and BIC(SC) select p=5
```

The selection of a 5th-order VAR model is expected because there is a "lag 5" relationship built into the data. (Consequently, it was critical that we allowed $p = 5$ as a possible choice.)

2. Parameter Estimation:

Parameter estimation is performed using the *vars* command **VAR** as follows.

```
lsfit=VAR(X,p=5,type="const")
summary(lsfit) ##Note significance of 1 variable, at lag 5 ##
```

The output includes the following

```
Estimation results for equation x1:
===================================
x1=x1.l1+x2.l1+x1.l2+x2.l2+x1.l3+x2.l3+x1.l4+x2.l4+x1.l5+x2.l5+const
```

	Estimate	Std. Error	t value	Pr(>|t|)
x1.l1	-0.0042949	0.0077157	-0.557	0.607
x2.l1	0.0025313	0.0092102	0.275	0.797
x1.l2	0.0040630	0.0081828	0.497	0.646
x2.l2	-0.0077684	0.0103929	-0.747	0.496
x1.l3	-0.0047742	0.0077547	-0.616	0.571
x2.l3	0.0097459	0.0136252	0.715	0.514
x1.l4	0.0044834	0.0053700	0.835	0.451
x2.l4	-0.0079471	0.0140921	-0.564	0.603
x1.l5	-0.0021468	0.0053796	-0.399	0.710
x2.l5	2.0045969	0.0170018	117.905	3.1e-08 ***
const	-0.0001567	0.0055712	-0.028	0.979

The equation in the output is the VAR representation of the first equation in (10.9) for a 5th-order bivariate model. The test results show that $X_{t-5,2}$ has a strong influence on X_{t1} and that there are no

---

9 The BIC (Bayes information criterion) was derived by Gideon E. Schwarz; hence, SC (Schwarz criterion) is an alternate acronym for BIC.

other significant relationships. A second set of tests representing the second equation in (10.9) shows no significant lag relationships.

3. Forecasting:

The model fit information contained in **lsfit** is used to forecast, using the ***vars*** command **predict**. As in Figure 10.8, the forecasts for **x1.25** and **x2.25** for the last five values are shown superimposed on the actual values. Notice that in Figure 10.9(a) the forecasts for **x1.25** (shown as open circles) fall essentially "on top of" the actual data values (shown as smaller black circles). Obviously, **x2.25** is especially helpful in predicting **x1.25** by construction. However, the relationship between the two datasets wasn't obvious in Figure 10.7(a) but was "detected" by the VAR analysis and was used to enhance the forecasts. On the other hand, Figure 10.9(b) reveals that the VAR forecasts for the last five values of **x2.25** are no better than the corresponding univariate forecasts. This again is intuitive because, while **x1.25** was directly calculated from **x2.25**, the converse is not true.

This is an example of how the VAR model can provide excellent results without previously knowing the relationship between the variables. In a multiple regression, it would have been useful to use $X_{t-5,1}$ as an independent variable for predicting X_{t1}. However, this particular independent variable would probably only have been used if the analyst had prior "domain knowledge". In the case in which no domain knowledge was assumed to exist, the VAR model was flexible in the sense that it did not require the analyst to distinguish between independent and dependent variables. Instead, VAR methodology analyzed the relationships among all possible variables, and provided resulting forecasts for the variables. As with univariate model identification, in VAR model identification with **VARselect**, it is important to search for model orders over a sufficiently wide range of possible values. In this case, the range needed to include $p = 5$.

The forecasts and plots in Figure 10.9 can be obtained using the following code.

```
# VAR forecasting
preds=predict(lsfit,n.ahead=5)
```

The output **preds** contains the VAR forecasts for the last five data values. These are in the vectors **f1.12** and **f2.12** defined below.

```
f1.12=preds$fcst$x1[,1] # VAR forecasts for x1
f2.12=preds$fcst$x2[,1] # VAR forecasts for x2

# Plotting Forecasts of x1.25
t=1:25
plot(t,x1.25,type="o",pch=20,cex=1,ylim=c(-2,3.75))
points(t[20:25],c(x1[20],f1.12),type='o',cex=2,pch=1)

# Plotting Forecasts of x2.25
plot(t,x2.25,type="o",pch=20,cex=1,ylim=c(-2,3.75))
points(t[20:25],c(x2[20],f2.12),type='o',cex=2,pch=1)
```

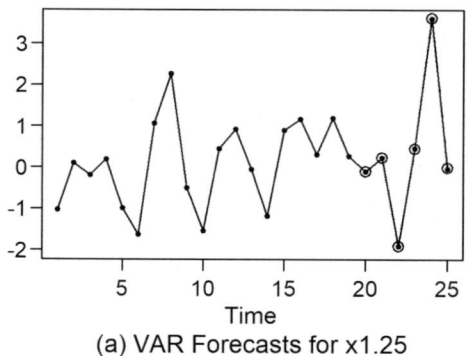
(a) VAR Forecasts for x1.25

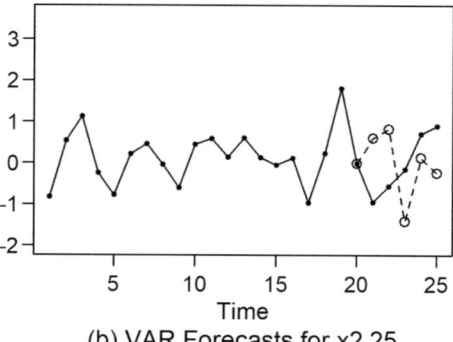
(b) VAR Forecasts for x2.25

FIGURE 10.9   (a): Forecasts for x1.25 and (b) x2.25 using VAR Modeling.

Key Point: In multiple linear regression (MLR) with correlated errors, p refers to the order of the AR fit to the model residuals. In VAR modeling, p refers to the order of the model.

Example 10.3 VAR modeling using sunspot data and melanoma cases

Past research has suggested a likely relationship between sunspot numbers and melanoma cases. The relationship is time-dependent, and the speculation is that the number of melanoma cases[10] is related to the number of recorded sunspots in *previous* years (Houghton, Munster, and Viola, 1978). In particular, it was hypothesized that the lag is two years. In this example, we analyze data from 1936 to 1972 using VAR modeling. Figure 10.10(a) shows the **sunspot2.0** numbers for the years 1936–1972, Figure 10.10(b) is a plot of the melanoma cases during 1936–1972, and Figure 10.10(c) shows the cross-correlations between sunspots and the melanoma incidences. The sunspots show the 10–11 year cycle while the melanoma cases show a rise during this time frame. The cross-correlation at lag $k = -2$ does not give much credibility to the "lag 2" hypothesis. However, we proceed as follows to see if sunspots tend to predict melanoma occurrence.

The data and the code for the cross-correlations are given below.

```
melanoma=c(1.0,0.9,0.8,1.4,1.2,1.0,1.5,1.9,1.5,1.5,1.5,1.6,1.8,2.8,2.5,2.5,
2.4,2.1,1.9,2.4,2.4,2.6,2.6,4.4,4.2,3.8,3.4,3.6,4.1,3.7,4.2,4.1,4.1,4.0,5.2,
5.3, 5.3)
sunspot=c(133,191,183,148,113,79,51,27,16,55,154,215,193,191,119,98,45,20,7,
54,201,269,262,225,159,76,53,40,15,22,67,133,150,149,148,94,98)
t=1:37
year=1935+t
plot(year,sunspot,type='o',pch=20)
plot(year,melanoma,type='o',pch=20)
ccf(sunspot,melanoma,ylim=c(-1,1))
```

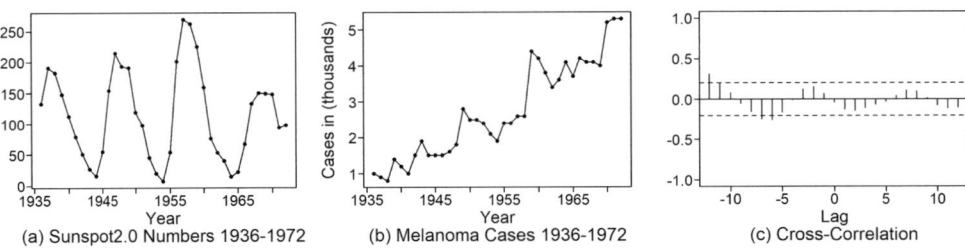

(a) Sunspot2.0 Numbers 1936-1972   (b) Melanoma Cases 1936-1972   (c) Cross-Correlation

FIGURE 10.10   (a) Sunspot numbers and (b) melanoma cases between 1936 and 1972; (c) cross-correlation between melanoma cases and sunspots.

We further explore the possible relationship using VAR techniques to model the data and then use the model for forecasting. As in the simulated data in Example 10.2, we first consider forecasts for both melanoma incidents and sunspot numbers when each is modeled separately as a univariate time series.

For purposes of forecast cross-validation, we used the melanoma and sunspot data for the first 29 years (1936–1964) for model building. These (training) data are contained in the datasets **mel.64** and **sun.64** defined below.

```
mel.64=melanoma[1:29]
sun.64=sunspot[1:29]
```

---

10 Number of incidences of melanoma skin cancer per 100,000 people in Connecticut.

We then use these models to forecast the last eight years (1965–1972). The forecasts for the melanoma and sunspot data are shown in Figures 10.11(a) and 10.11(b), respectively. Forecasts for the last eight years are shown with open circles connected by dotted lines. In each case, the univariate time series is modeled using a stationary model, and therefore the forecasts tend toward the mean. This is particularly evident in the melanoma forecasts where the resulting forecasts of melanoma incidences are quite poor. The univariate forecasts were obtained and plotted using the following code.

```
## Univariate analysis/forecasts for melanoma ##
p.mel=aic.wge(mel.64,p=0:10,q=0:0)
p.mel$p
mel.est=est.ar.wge(mel.64,p=p.mel$p)
pred_m=fore.arma.wge(mel.64,phi=mel.est$phi,n.ahead=8,lastn=FALSE,limits=
FALSE)
plot(year,melanoma,type='o',pch=20)
points(year[29:37],c(melanoma[29],pred_m$f),type='o',lty=2,pch=1)

## Univariate analysis/forecasts for sunspot ##
p.sun=aic.wge(sun.64,p=0:10,q=0:0)
p.sun$p
sun.est=est.ar.wge(sun.64,p=p.sun$p)
pred_s=fore.arma.wge(sun.64,phi=sun.est$phi,n.ahead=8,lastn=FALSE,limits=
FALSE)
plot(year,sunspot,type='o',pch=20)
points(year[29:37],c(sunspot[29],pred_s$f),type='o',lty=2,pch=1)
```

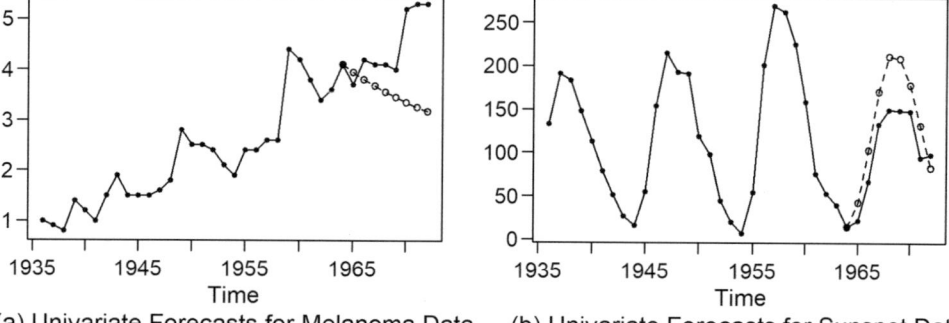

(a) Univariate Forecasts for Melanoma Data   (b) Univariate Forecasts for Sunspot Data

FIGURE 10.11   Univariate forecasts for (a) melanoma data and (b) sunspot data for years 1965–1972.

VAR modeling follows the steps as before:

1. *Use* VARselect *to pick the order* p

The following code combines the melanoma and sunspot training data into a matrix, **x**, and then uses *vars* command **VARselect** to select the order, allowing p to range up to **lag.max=5**. AIC selected five, but all the other criteria, including BIC (SC), selected order four. We will proceed using $p = 4$.

```
X=cbind(mel.64,sun.64)
VARselect(X, lag.max = 5, type = "const",season = NULL, exogen = NULL)
#AIC = 5, BIC picks 4. We go with 4.
```

2. *Use* VAR *to fit the VAR model to the training set*

The following command instructs VAR to fit a 4th-order model on the training set matrix, **x**. You can examine the VAR fit by issuing the command **summary(VARfit)**.

```
VARfit=VAR(X,p=4,type='const') ## This fits 9 parameters ##
```

3. *Forecast using* `predict`

VAR models are complex, and our main goal is to determine whether sunspot information is a "leading indicator" of melanoma incidences. Figure 10.12 shows the VAR-based forecasts. The melanoma forecasts in Figure 10.12(a) no longer tend toward a mean, but in fact are remarkably accurate. The VAR forecasts effectively track the rapid increase in melanoma cases. The forecasts for sunspots are not improved and tend to exaggerate the height of the peak in 1968. However, using melanoma incidences to predict future sunspots does not make sense. It is physically intuitive that the one-directional relationship holds with sunspot numbers being an early indicator of later melanoma cases, but not vice versa. The VAR forecasts for melanoma incidences and sunspot numbers for the years 1965–1972 are found using the following code.

```
preds=predict(VARfit,n.ahead=8)
mel.f=preds$fcst$mel.64[,1]  # VAR forecasts for mel.64
sun.f=preds$fcst$sun.64[,1]  # VAR forecasts for sun.64
```

The overlay plots of the data and forecasts are obtained using the code:

```
t=1:37
year=t+1935
# melanoma forecasts
plot(year,melanoma,type="o",pch=20,cex=1,ylim=c(.5,6))
points(year[29:37],c(melanoma[29],mel.f),type='o',cex=1,pch=1)
# sunspot forecasts
plot(year,sunspot,type="o",pch=20,cex=1)
points(year[29:37],c(sunspot[29],sun.f),type='o',cex=1,pch=1)
```

QR 10.5 VAR Modeling Sunspot and Melanoma- Example 10.3

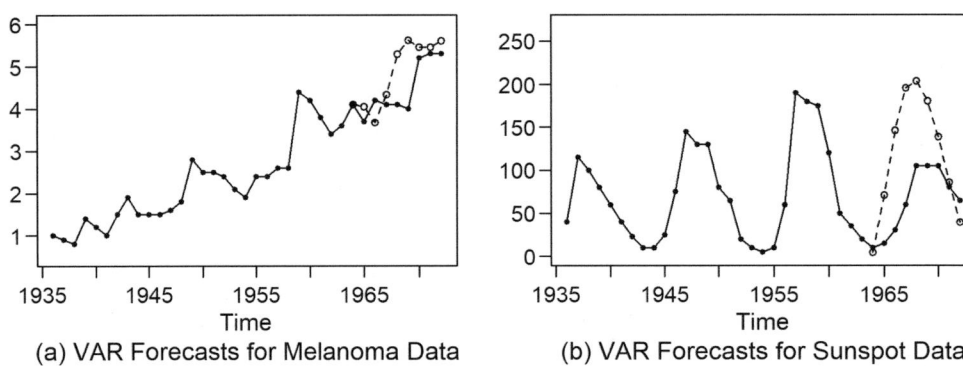

(a) VAR Forecasts for Melanoma Data   (b) VAR Forecasts for Sunspot Data

FIGURE 10.12   VAR forecasts for (a) melanoma data and (b) sunspot data for years 1965–1972.

10.3.1.3 Comparing RMSEs

Obviously, the VAR forecasts for melanoma incidences were far better than the univariate forecasts. To quantify the comparison, we use the RMSEs obtained using the commands

```
RMSE_AR=sqrt(mean((melanoma[30:37]-pred_m$f[1:8])^2))
RMSE_VAR=sqrt(mean((melanoma[30:37]-preds$fcst$mel.64[1:8])^2))
```

The RMSEs for univariate AR and multivariate VAR forecasts of melanoma incidences from 1965–1972 were 1.275 and .7871, respectively.

10.3.1.4 Final Comments

We found that sunspot activity seemed to behave as a leading indicator of melanoma incidences. A couple of points need to be made.

(1) We haven't shown that high sunspot activity causes melanoma. We found an interesting relationship that would require much further investigation.
(2) The results we found did not point directly to a two-year lag relationship. The cross-correlations showed other lagged relationships (4–6 years) that were stronger (and negative).

10.4   RELATIONSHIP BETWEEN MLR AND VAR MODELS

In this chapter, the methodologies have been described and illustrated for multiple linear regression with correlated errors and for VAR, respectively. Partly due to the complexity of the VAR notation, it certainly seems that the two procedures are unrelated, other than the fact they both model multivariate time series. However, a very interesting (and surprising!) relationship between the two exists which may be helpful in better understanding both multivariate methods. For further detail, the reader is encouraged to see Appendix 10B.

10.5 A COMPREHENSIVE AND FINAL EXAMPLE: LOS ANGELES CARDIAC MORTALITY

In this example, we present a classic example (Shumway and Stoffer, 2017) in which the objective is to examine the extent to which cardiac mortality incidences can be predicted from average weekly temperature and air pollution measures. The dataset consists of weekly cardiac mortality, temperatures, and pollution measures for the years 1970–1978 and the first 40 weeks of 1979 in Los Angeles, California. Because it is suspected that cardiac mortality is related to temperature and pollution from *previous* time periods, the use of lagged variables seems appropriate. This example will illustrate the use of VAR modeling to predict cardiac mortality incidences; the modeling steps are shown in detail below. Originally obtained from the package *astsa*, the three variables we will use in our study can be found in the *ts* object **cardiac** in *tswge* and are shown below. To familiarize yourself with the format of the dataset, consider the following code and output.

```
data(cardiac)
head(cardiac)
Time Series:
Start = c(1970, 1)
End = c(1970, 6)
Frequency = 52
          cmort  tempr  part
1970.000   97.85  72.38  72.72
1970.019  104.64  67.19  49.60
1970.038   94.36  62.94  55.68
1970.058   98.05  72.49  55.16
1970.077   95.85  74.25  66.02
1970.096   95.98  67.88  44.01
```

The above output indicates that this multivariate time series dataset is composed of average weekly (**frequency=52**) cardiac mortalities (**cmort**), average weekly temperature (**tempr**), and average weekly number of particulates in the air (**part**). For each of these three variables, plots of the realization, sample autocorrelations, and Parzen spectral density estimate are shown in Figure 10.13. A decreasing trend is evident in the realization for cardiac mortalities over the ten-year period, and the realizations reveal strong evidence of an annual seasonal pattern for all three variables. The sample autocorrelation plots all show a 52-week sinusoidal cycle, and the Parzen spectral density plots all have a peak at approximately $.019 = 1/52$ in each of the spectral densities. The data in Figure 10.14 are the output from a 52nd-order moving average smoother applied to the cardiac mortality data, which shows the downward trend in cardiac mortalities. Noting these behaviors will be vital to the modeling and forecasting procedures that follow.

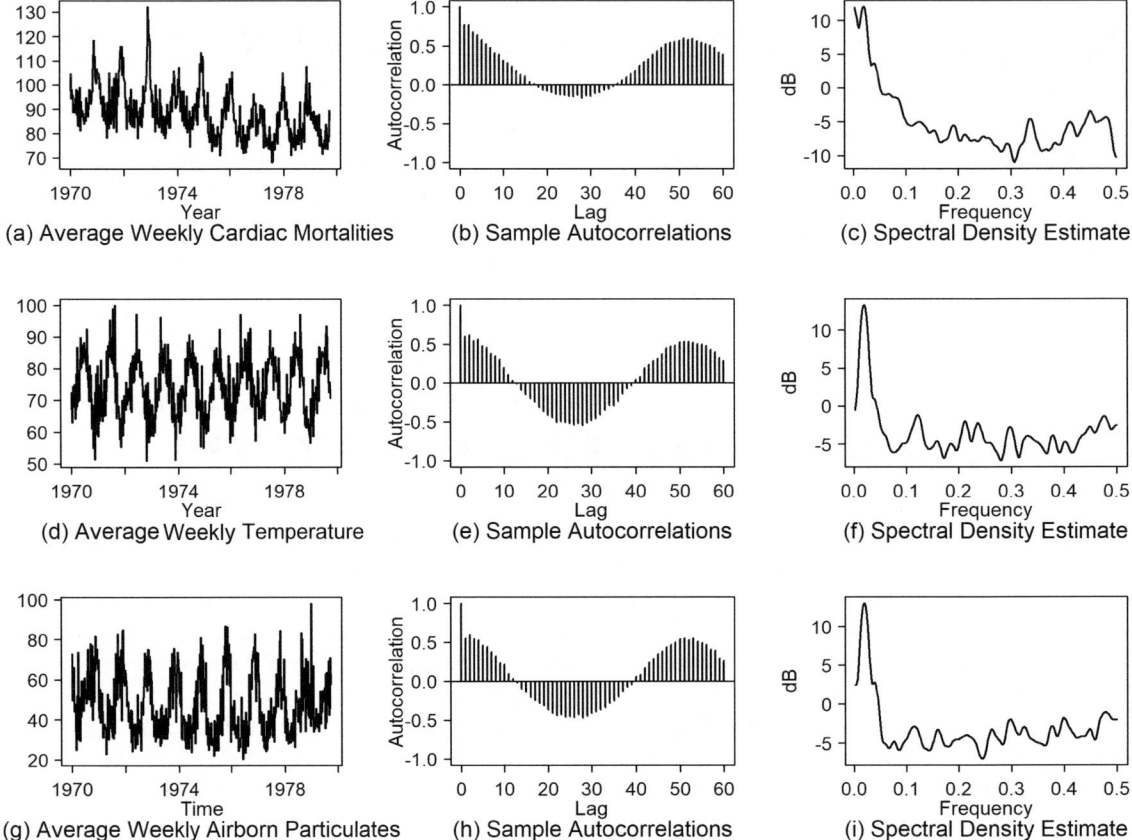

FIGURE 10.13 Realizations, sample autocorrelations, and spectral density estimates for (a)–(c) cardiac mortality data, (d)–(f) weekly temperature data, and (g)–(i) weekly airborn particulates. Note that `trunc=100` for the spectral density plots.

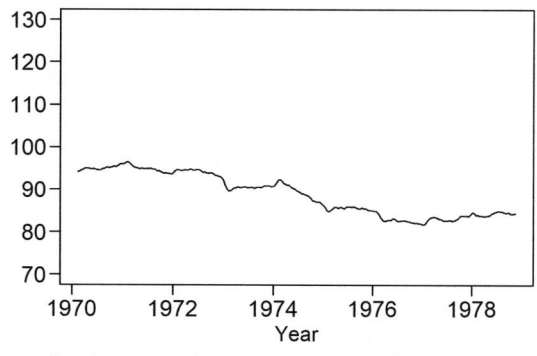

Cardiac Mortality after 52nd Order MA Smoother

FIGURE 10.14 Cardiac mortality data after a 52nd-order centered moving average smoother.

10.5.1 Applying the VAR(*p*) to the Cardiac Mortality Data

(1) Use **VARselect** *to identify candidate model orders*
The first step in generating a "baseline" model for forecasting the cardiac mortalities given the temperature and particulates is to use the AIC and/or BIC(SC) criterion to identify the order of the candidate VAR(*p*) model. Note that due to the previously observed evidence of a decreasing trend in the cardiac mortality series, we have included both an intercept and trend term in the model (**type = "both"**).

```
VARselect(cardiac, lag.max = 10, type = "both")
```

```
$selection
AIC(n)  HQ(n)   SC(n)   FPE(n)
     9      5       2        9
```

```
$criteria
            1        2        3        4        5        6        7        8        9       10
AIC(n)    11.7378 11.3019  1.2679 11.2303 11.1763 11.1527 11.1525 11.1288 11.1192 11.1202
HQ(n)     11.7876 11.3815 11.3774 11.3697 11.3456 11.3518 11.3814 11.3876 11.4078 11.4387
SC(n)     11.8646 11.5048 11.5469 11.5854 11.6076 11.6600 11.7359 11.7883 11.8547 11.9319
FPE(n)   125216.9 80972.3 78268.2 75383.7 71426.1 69758.3 69749.9 68122.4 67477.0 67556.5
```

In the output above, **$criteria** includes the AIC and BIC(SC) for each order, and **$selection** identifies the order that produced the lowest value of each criterion. The BIC(SC) has favored the VAR(2), while the AIC selects a VAR(9). As usual, the analyst is encouraged to consider domain knowledge and any other known and relevant information to make the final decision as to the value of *p* in the model. In addition, a good practice is to select a few values of *p* as candidates and evaluate the competing models by all available measures to choose the most appropriate *p*. This strategy will be implemented next by assessing a visualization and the RMSEs of the forecasts of the last 52 weeks of the series for the VAR(2) and the VAR(9). In addition, the VAR(7) will be included as a candidate model because there is evidence from the cross-correlation function that cardiac mortalities are correlated with particulates after a 7-week lag.

(2) Use `VAR` *to fit the VAR models to the training set*

As in the previous example in this chapter, we will divide the data into a training set, which in this case consists of the first eight years plus the first 40 weeks of 1978. This leaves 52 weeks for the test set (the last 52 weeks of the dataset). [11] The following commands create the training and test sets.

```
cardiacTrain = window(cardiac, start = c(1970,1), end = c(1978,40))
cardiacTest = window(cardiac, start = c(1978,41), end = c(1979,40))
```

The next step is to fit the various candidate models using the following code.

```
CMortVAR2 = VAR(cardiacTrain, type = "both", p = 2)
CMortVAR9 = VAR(cardiacTrain, type = "both", p = 9)
CMortVAR7 = VAR(cardiacTrain, type = "both", p = 7)
```

An initial check of the appropriateness of the models is then conducted using the Ljung-Box test to check the residuals. The Ljung-Box commands for the default $K = 24$ are given below.

```
ljung.wge(CMortVAR2$varresult$cmort$residuals,p=2)
ljung.wge(CMortVAR9$varresult$cmort$residuals,p=9)
ljung.wge(CMortVAR7$varresult$cmort$residuals,p=7)
```

The *p*-values are .554, .215, and .069, respectively. Applying the Ljung-Box test using `K=48` yields *p*-values .779, .431, and .151, respectively. White noise seems to be a reasonable assumption for these lag ranges. (Remember that the analyst should also always plot and visually assess the residuals. (We recommend that you examine the plots.)

(3) Use `predict` *to forecast data values in the test set*

Finally, the forecasts are calculated, using a forecast horizon of 52. The code is as follows:

```
preds2=predict(CMortVAR2,n.ahead=52)
preds9=predict(CMortVAR9,n.ahead=52)
preds7=predict(CMortVAR7,n.ahead=52)
```

A visualization of forecast performance and the RMSE of the forecasts for each of the three models are obtained using the code below. The corresponding results are shown in Figure 10.15 and Table 10.5, respectively.

```
t=1:508
plot(t, cardiac[,"cmort"], type = "l", xlim = c(450,510), ylab = "Cardiac
Mortality", main = "52 Week Cardiac Mortality Forecast", xlab = "Time")
points(t[457:508], preds2$fcst$cmort[,1], type="l", lwd=2, lty=1)
points(t[457:508], preds9$fcst$cmort[,1], type="l", lwd=2, lty=2)
points(seq(457,508,1), preds7$fcst$cmort[,1], type="l", lwd=2,lty=3)

RMSE2 = sqrt(mean((cardiacTest[,"cmort"] - preds2$fcst$cmort[,1])^2))
RMSE9 = sqrt(mean((cardiacTest[,"cmort"] - preds9$fcst$cmort[,1])^2))
RMSE7 = sqrt(mean((cardiacTest[,"cmort"] - preds7$fcst$cmort[,1])^2))
```

---

11 Recall that the cardiac mortality data includes the years 1970–1978 and the first 40 weeks of 1979.

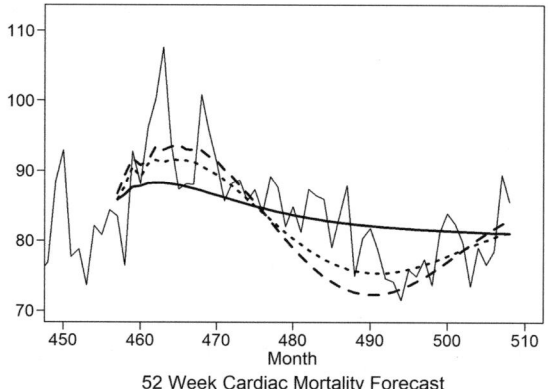

52 Week Cardiac Mortality Forecast

FIGURE 10.15 *Visualization of the 52-week forecast of the VAR(2) (solid line), VAR(9) (dashed line), and VAR(7) (dotted line) models with trend and showing actual values for comparison.*

TABLE 10.5   RMSEs for a Horizon of 52 Weeks for the VAR(2), VAR(9) and VAR(7) with Trend.

MODEL (WITH TREND TERM)	RMSE
Var(2)	5.92
Var(9)	6.08
Var(7)	**5.44**

   The plot in Figure 10.15 shows that the VAR(2) seems to mostly model a decreasing trend in the cardiac mortality rates without much regard to any other behavior. On the other hand, closer inspection of the plots reveals that the VAR(7) and VAR(9) try to model the sharper increases and decreases for the first few weeks in the test set (with some success), and later weeks are predicted with a smooth curve that closely resembles the periodic and trending behavior evident in the series. Ultimately, the VAR(7) appears to most closely model the behaviors in the series which is also reflected by its superior RMSE of 5.44. It is safe to say, George Box would agree, that none of these are the "right" model, although the visualizations and RMSEs suggest that the VAR(7) is the most useful of the three.

10.5.2 The Seasonal VAR(p) Model

(1) Use **VARselect** *to identify candidate seasonal models*
Recall that there was strong evidence of seasonality in the sample autocorrelation functions and spectral densities of each of the cardiac mortality, temperature, and particulate series, with indications of a period of 52 weeks. Analogous to how seasonality can be modeled in AR/ARMA/ARIMA models, the Base R **VAR** function also provides the option to model the seasonal component of a time series by use of seasonal indicator variables (also called "seasonal dummies") with the **season** option. Setting **season = 52** in this case will fit 51 indicator variables for each series to model a mean for each week using the data from the full dataset for variable selection. Code is shown below that will add the seasonal indicator variables (**season=52**) to the **VARselect** call to re-identify the model order, now that seasonal behavior is being considered.

```
VARselect(cardiac, lag.max = 10, season = 52, type = "both")
```

```
$selection
AIC(n)   HQ(n)   SC(n)   FPE(n)
     5       2       2       5
```

```
$criteria
            1        2        3        4        5        6        7        8        9       10
AIC(n)  11.1777  11.0008  10.9939  10.9560  10.9491  10.9673  10.9849  10.9964  11.0005  11.0143
HQ(n)   11.7351  11.5881  11.6111  11.6030  11.6261  11.6741  11.7216  11.7629  11.7969  11.8405
SC(n)   12.5981  12.4973  12.5666  12.6047  12.6740  12.7682  12.8619  12.9495  13.0297  13.1196
FPE(n)  71719.9  60120.7  59744.3  57552.0  57198.8  58289.0  59371.2  60108.7  60410.7  61309.4
```

When seasonality is accounted for, the AIC favors a model with $p = 5$, and the BIC (SC) favors the lag $p = 2$. Consistent with the previous analyses, a seasonal VAR model with p = 7 is also considered.

(2) Use VAR to fit the VAR models to the training set

Each of three models above, seasonal models with lags two, seven, and nine, were fit by using the following code.

```
CMortVAR2 = VAR(cardiacTrain, type = "both", season = 52, p = 2)
CMortVAR9 = VAR(cardiacTrain, type = "both", season = 52, p = 9)
CMortVAR7 = VAR(cardiacTrain, type = "both", season = 52, p = 7)
```

Again, the initial check of the appropriateness of the models is conducted using the Ljung-Box test to assess the residuals. The p-values from this test for each of these models are obtained using the code below.

```
ljung.wge(CMortVAR2$varresult$cmort$residuals,p=2)
ljung.wge(CMortVAR9$varresult$cmort$residuals,p=9)
ljung.wge(CMortVAR7$varresult$cmort$residuals,p=7)
```

The p-values using K=24 (default) were .087, .054, and .011, respectively. Using K=48, the corresponding p-values were .265, .218, and .038. There are concerns about the whiteness of the noise, but we continue to assess the forecast performance of the three models.

(3) Use predict to forecast data values in the test set

The forecasts for the 52 weeks in the test ("hold out") set are calculated as usual:

```
preds2=predict(CMortVAR2,n.ahead=52)
preds9=predict(CMortVAR9,n.ahead=52)
preds7=predict(CMortVAR7,n.ahead=52)
```

Proceeding as before, the models are evaluated first by visualizing the forecasts versus the actual values in the training set and then computing the RMSEs for the last 52 weeks. The appropriate code is given below, and the corresponding output is shown in Figure 10.16 and Table 10.6, respectively.

```
plot(seq(1,508,1), cardiac[,"cmort"], type = "l", xlim = c(450,510), ylab =
"Cardiac Mortality", main = "52 Week Cardiac Mortality Forecast", xlab =
"Time")
```

```
t=1:508
points(t[457:508], preds2$fcst$cmort[,1], type="l", lwd=2, col = "red")
points(t[457:508], preds9$fcst$cmort[,1], type="l", lwd=2, col = "blue")
points(t[457:508], preds7$fcst$cmort[,1], type="l", lwd=2, col = "green")
```

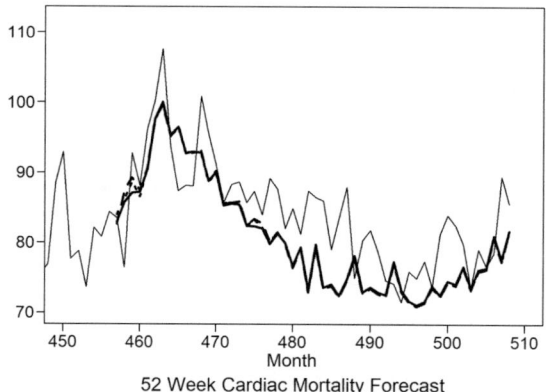

52 Week Cardiac Mortality Forecast

FIGURE 10.16   Visualization of the 52-week forecasts (bold) of the VAR(2), VAR(9), and VAR(7) models with trend and seasonal indicator variables, including actual values for comparison.

```
RMSE2 = sqrt(mean((cardiacTest[,"cmort"] - preds2$fcst$cmort[,1])^2))
RMSE9 = sqrt(mean((cardiacTest[,"cmort"] - preds9$fcst$cmort[,1])^2))
RMSE7 = sqrt(mean((cardiacTest[,"cmort"] - preds7$fcst$cmort[,1])^2))
```

TABLE 10.6   RMSEs for a Horizon of 52 Weeks for the VAR(2), VAR(9) and VAR(7) Models with Trend and Seasonal Indicator Variables.

MODEL	RMSE
(WITH TREND TERM AND SEASONAL INDICATORS)	
Var(2)	6.38
Var(9)	6.44
Var(7)	6.41

In this case, the VAR(2) produces a slightly smaller RMSE than the other two models, although based on the plots in Figure 10.15, the forecasts are very similar and tend to underestimate the actual cardiac mortalities (this fact will be revisited in the next chapter!). Overall, the VAR(7) with trend (and without seasonality) performs better with respect to BIC and RMSE, as summarized in Table 10.7 below.

TABLE 10.7   AIC, BIC, and RMSEs for a Horizon of 52 Weeks for the VAR(7) with Trend and the VAR(2) with Trend and Seasonal Indicators.

MODEL	AIC	BIC	RMSE
VAR(7) with trend	11.15	11.73	5.44
VAR(2) with trend and seasonal indicators	11.00	12.49	6.38

The VAR(2) does have a slightly smaller AIC, which underlines the important fact that these criteria do not tell the analyst which model is "right" or even "better"; rather, these measures are tools the analysts should use along with other information or intuition in order to *make the final decision themselves.*

10.5.3 Forecasting the Future

Suppose that, based on the smaller RMSE of the VAR(7) model with trend (and without seasonality), an analyst decides to forecast the number of cardiac mortalities in the next 52 weeks. Using the entire dataset (**cardiac**), the forecasts and corresponding confidence intervals can be obtained using the code given below. Note that inside the **preds** object is an attribute called **$fcst**, which has a separate set of forecasts for each variable in the vector: **$cmort**, **$tempr** and **$part**. A summary of the data is provided by the function **head**, and the output is shown below.

```
CMortVAR7 = VAR(cardiac, type = "both", p = 7)
preds7=predict(CMortVAR7,n.ahead=52)
#cmort forecasts
head(preds7$fcst$cmort) #cardiac mortality forecasts
```

	fcst	lower	upper	CI
[1,]	86.80259	76.59671	97.00847	10.20588
[2,]	88.77409	77.97191	99.57627	10.80218
[3,]	87.48338	75.78664	99.18012	11.69674
[4,]	88.97024	76.86621	101.07427	12.10403
[5,]	89.56009	77.16757	101.95262	12.39253
[6,]	89.23073	76.49891	101.96254	12.73181

```
head(preds7$fcst$temp) # temperature forecasts
```

	fcst	lower	upper	CI
[1,]	71.36823	59.53554	83.20092	11.83269
[2,]	71.85304	59.82744	83.87865	12.02560
[3,]	68.87960	56.42767	81.33153	12.45193
[4,]	69.47003	56.89573	82.04433	12.57430
[5,]	67.97891	55.08042	80.87740	12.89849
[6,]	67.31557	54.02224	80.60890	13.29333

```
head(preds7$fcst$part) # particulates forecasts
```

	fcst	lower	upper	CI
[1,]	62.35471	42.21521	82.49422	20.13950
[2,]	64.26562	43.88740	84.64384	20.37822
[3,]	61.56504	40.60599	82.52409	20.95905
[4,]	62.94784	41.22299	84.67269	21.72485
[5,]	59.67438	36.89419	82.45457	22.78019
[6,]	58.65509	35.44117	81.86902	23.21392

Because cardiac mortality is the response of interest, we provide the following code to plot the forecasts of cardiac mortalities (**$cmort**) for the next 52 weeks, along with 95% prediction intervals. The output is shown in Figure 10.17.

```
plot(seq(1,508,1), cardiac[,"cmort"], type = "l",xlim = c(450,560), ylim =
c(50,112),xlab = "Time", ylab = "Cardiac Mortality", main = "52 Week Cardiac
Mortality Forecast From a VAR with p = 7")

t=1:560
points(t[509:560],preds7$fcst$cmort[,2], type = "l", lwd = 1, lty = 3)
points(t[509:560],preds7$fcst$cmort[,1] , type = "l", lwd = 1.5, lty = 1)
points(t[509:560],preds7$fcst$cmort[,3] , type = "l", lwd = 1, lty = 3)
```

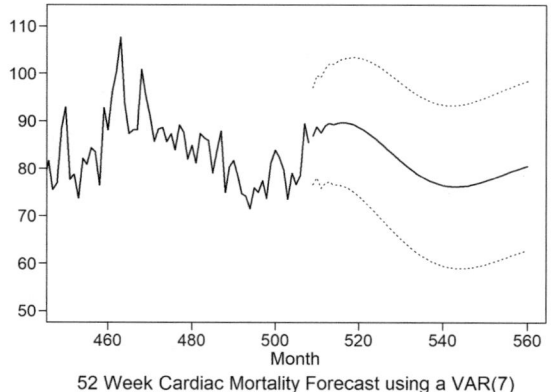

52 Week Cardiac Mortality Forecast using a VAR(7)

FIGURE 10.17   *52-week Cardiac Mortality Forecast (solid) and 95% Prediction Limits (dotted) from VAR(7) Model.*

10.5.3.1 Short vs. Long Term Forecasts

The previous analyses have shown in this particular example that a VAR(7) model without seasonality (including trend) resulted in a superior RMSE based on forecasts of the last year of known data (1979). It is important to note that, similar to the behavior exhibited by an AR model with trend, this VAR(7) model with trend will also gravitate toward the trend line over time. This behavior indeed occurs in the forecasts for the next four years (1980–1984), which are shown in Figure 10.18(a). In contrast, the VAR(2) model with seasonality and trend will continue to perpetuate the periodic behavior in the forecasts into the future, as seen in Figure 10.18(b).

```
# Forecasting next 4 years with VAR(7) Model with Trend

CMortVAR7 = VAR(cardiac, type = "both", p = 7)
preds7=predict(CMortVAR7,n.ahead=208)

plot(seq(1,508,1),  cardiac[,"cmort"], type = "l",xlim = c(450,716), ylim =
c(50,112),xlab = "Time", ylab = "Cardiac Mortality", main = "52 Week Cardiac
Mortality Forecast From a VAR with p = 7")
t=1:716
points(t[509:716],preds7$fcst$cmort[,2], type = "l", lwd = 1, lty = 4)
points(t[509:716],preds7$fcst$cmort[,1] , type = "l", lwd = 1.5, lty = 1)
points(t[509:716],preds7$fcst$cmort[,3] , type = "l", lwd = 1, lty = 3)

# Forecasting next 4 years with VAR(2) Seasonal and Trend Model
CMortVAR2 = VAR(cardiac,season = 52, type = "both", p = 2)
preds2=predict(CMortVAR2,n.ahead=208)

plot(t[1:508],  cardiac[,"cmort"],  type  =  "l",xlim  =  c(450,716),  ylim  =
c(50,112),xlab = "Time", ylab = "Cardiac Mortality", main = "52 Week Cardiac
Mortality Forecast From a VAR(2) with Seasonality and Trend")
points(t[509:716], preds2$fcst$cmort[,2],type="l",lwd=1,lty=4)
points(t[509:716,preds2$fcst$cmort[,1],type="l",lwd =2,lty=1)
points(t[509:716],preds2$fcst$cmort[,3],type = "l",lwd = 1,lty = 3)
```

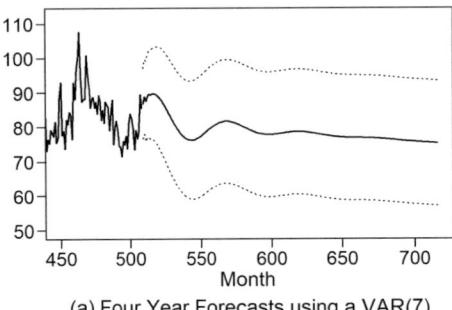

(a) Four Year Forecasts using a VAR(7)

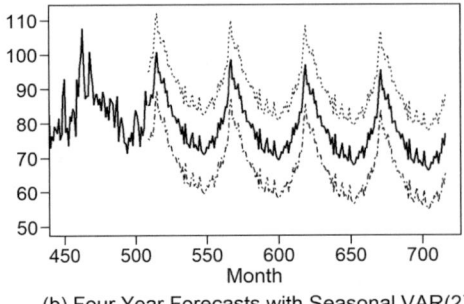

(b) Four Year Forecasts with Seasonal VAR(2)

FIGURE 10.18   (a) Four-year forecasts (solid) for cardiac mortality from VAR(7) with trend model; (b) Four-year forecasts (solid) for cardiac mortality from VAR(2) model with seasonality and trend.

While there seems to be (mild) evidence that the VAR(7) model with no seasonality (trend included) provides superior forecasts for a time horizon of 52, it is not clear, especially given the strong seasonality in the cardiac mortality series, that forecasts from this model will be superior for longer horizons. (Check it out.) Recall also that white noise is questionable for the residuals in the seasonal model. Again, no models are perfect, and the seasonal forecasts are quite good.

10.6 CONCLUSION

This chapter has explored two types of multivariate models and has demonstrated how selecting useful predictor variables and assessing relevant lagged relationships results in better models and improved forecasts. Analogous to how (in the standard statistical setting with independent observations) a multiple regression model can add valuable information to a simple linear regression model, the same holds true for a multivariate time series model. The key for the analyst is to identify related variables that have a lagged relationship with the dependent variable, and then identify the lag. In certain applications such as finance and economics, many types of data have historically been collected and analyzed so that commonly known "leading" indicators are available. For example, when predicting the Gross Domestic Product (GDP), economists often use leading economic indicators which have previously been observed such as unemployment rate, income and wages, corporate profits, etc. This previous information can simplify the process, but in other more obscure or less common applications, finding such related variables may take time, effort, and creativity (and a little luck!).

APPENDIX 10A

BASE R FUNCTIONS

(a) `arima(x, order = c(0L, 0L, 0L),xreg = NULL)` is a multi-purpose function that has many more parameters than listed here. Given the parameters listed here:

(1) If **xref=NULL** then **arima** fits an ARIMA(p,d,q) model to the data in **x** using ML estimates. Note that the coefficients of the moving average parameters will have opposite signs to those used in this book.

(2) If **xreg** specifies variables, then the function computes the ML estimates of a multiple regression with correlated errors in which the parameter order specifies the order of the model fit to the residuals from the MLR fit.

(b) `lm(x~x1+x2+x3)` is a function that performs a standard multiple linear regression assuming uncorrelated errors. In the above, **y** is the dependent variable and **x1**, **x2**, and **x3** are independent

Example 10A.1:

```
x=gen.arma.wge(n=200,phi=c(1.6,-.9),theta=.8,sn=10)
arima(x,order=c(2,0,1))
```

The output is

```
Coefficients:
              ar1       ar2       ma1  intercept
           1.5657   -0.8025   -0.8932    -0.0662
s.e.       0.0436    0.0419    0.0362     0.0325

sigma^2 estimated as 0.9509: log likelihood = -279.86, aic = 569.73
```

Note that the output from the *tswge* command

```
est.arma.wge(x,p=2,q=1)
```

includes:

```
$phi
[1] 1.5657450 -0.8025325
$theta
[1] 0.8932299
```

Note: The *tswge* command `est.arma.wge` calls the function **arima** and changes the sign of the moving average parameters in the output.

Example 10A.2

Example 10A.1 illustrates the use of **arima** for the purpose of multiple regression with correlated errors. The following commands were used in the first multiple regression performed in Example 10.1:

```
mlrfit=lm(sales~ad_tv+ad_online+discount)
arima(sales,order=c(7,0,0),xreg=cbind(ad_tv,ad_online,discount))
```

The command **lm(sales~ad_tv+ad_online+discount)** performs a standard multiple regression with **sales** as the dependent variable and **ad_tv, ad_online,** and **discount** as the independent variables. In Example 10.1 AIC selects an AR(7) model for the residuals.

The command **arima(sales,order=c(7,0,0),xreg=cbind(ad_tv,ad_online, discount))** produces the ML estimates of the parameters of a multiple regression with **sales** as the dependent variable, and **ad_tv, ad_online**, and **discount** as the independent variables, and where the variance-covariance matrix of the residuals is based on the AR(7) model fit to the residuals from the **lm** function call.

CRAN PACKAGE *DPYLR*

dplyr is an R package designed for data manipulation. It was used in the taxicab data example in Chapter 1. In this chapter, we only use the function **lag**.

Example:
```
x=c(1,3,2,4,3,1,5,3)
x1=dplyr::lag(x)
x1
[1] NA 1 3 2 4 3 1 5
```

That is, **x1[1]=NA, x1[2]=x[1], ..., x1[n]=x[n-1]**

CRAN PACKAGE *VARS*

vars is an R package that has functions that perform VAR analysis.

(a) **VARselect(y,lag.max,type,season,exogen)** provides AIC-type measures for identifying the order of the VAR(p) fit.

  y is the multivariate data frame containing the multivariate time series data
  lag.max is the maximum order of p allowed (default=10)
  season is either **NULL** (default) if the data are not seasonal, or the seasonal frequency should be entered. That is, for monthly seasonal data **season=12**.
  type specifies whether the model contains a constant, trend, both, or none using **type='const'**, **'trend', 'both'**, or **'neither'**, respectively.
  exogen is a data frame containing any exogenous variables (default=**NULL**)

(b) **VAR(y,p,type,season,exogen)** fits an AR(p) model to the data in **y**. Other options are available. See documentation. The parameters **type, season**, and **exogen** are the same as in **VARselect**.

(c) **predict(object, n.ahead,ci,dumvar)** uses the VAR object produced by **VAR** and uses it to produce forecasts up to a **n.ahead**, steps ahead. Prediction limits are specified by **ci** (default=.95) and **dumvar** provides for dummy variables (default=**NULL**). See Sections 10.3 and 10.5.

TSWGE DATASETS INTRODUCED IN THIS CHAPTER

1. **Bsales** – a data frame containing 100 weeks of data on the variables:
   sales (in thousands of dollars)
   ad_tv – cost of TV advertising (in thousands of dollars)
   ad_online – cost of online advertising (in thousands of dollars)
   discount – discount (in percent)

2. **cardiac** – a multivariate *ts* object that consists of weekly cardiac mortality, temperatures, and pollution measures for the years 1970–1978 and the first 40 weeks of 1979 in Los Angeles, California. The data were obtained from the *astsa* package. The variables are as follows:

   cmort – average weekly cardiac mortalities
   tempr – average weekly temperatures
   part – average weekly number of particulates in the air

APPENDIX 10B

RELATIONSHIP BETWEEN MLR WITH CORRELATED ERRORS AND VAR

In Section 10.4, it was mentioned that, although not immediately apparent, there is a surprising relationship between MLR with correlated errors and VAR models. To see this, reconsider the simulated example in Example 10.2 (data are given below). Run the following code, and observe that the coefficients for the corresponding variables between the two methods match perfectly!

Three Important Points Should Be Considered:

(1) A VAR model is an MLR model, but not all MLR models are VAR models.
(2) For coefficients to exactly match among the variables for the two methods, the lag order must be the same for each of the variables for both the MLR and for the VAR.
(3) If a trend and intercept term is included for MLR, the parameter **type='both'** must be specified in VAR. If only the intercept term is included for MLR, the function **type='const'** must be specified in VAR.

```
x1 = c(-1.03,0.11,-0.18,0.20,-0.99,-1.63,1.07,2.26,-0.49,-1.54,0.45,0.92,
-0.05,-1.18,0.90,1.17,0.31,1.19,0.27,-0.09,0.23,-1.91,0.46,3.61,-0.03)

x2 = c(-0.82,0.54,1.13,-0.24,-0.77,0.22,0.46,-0.03,-0.59,0.45,0.59,0.15,
0.60,0.13,-0.04,0.12,-0.96,0.23,1.81,-0.01,-0.95,-0.55,-0.15,0.71,0.90)
```

```
XDF = data.frame(x1, x2)

x1Train = x1[1:20]
x2Train = x2[1:20]
XTrainDF = data.frame(x1Train, x2Train)

# MLR Analysis Code
# MLR Lag 2
#Manually Lag x1 and x2 at lags 1 and 2
XTrainDF$X1_l1 = dplyr::lag(XTrainDF$x1,1)
XTrainDF$X1_l2 = dplyr::lag(XTrainDF$x1,2)

XTrainDF$X2_l1 = dplyr::lag(XTrainDF$x2,1)
XTrainDF$X2_l2 = dplyr::lag(XTrainDF$x2,2)

#Show NAs
XTrainDF
```

	x1Train	x2Train	X1_l1	X1_l2	X2_l1	X2_l2
1	-1.03	-0.82	NA	NA	NA	NA
2	0.11	0.54	-1.03	NA	-0.82	NA
3	-0.18	1.13	0.11	-1.03	0.54	-0.82
4	0.20	-0.24	-0.18	0.11	1.13	0.54
5	-0.99	-0.77	0.20	-0.18	-0.24	1.13
6	-1.63	0.22	-0.99	0.20	-0.77	-0.24
7	1.07	0.46	-1.63	-0.99	0.22	-0.77
8	2.26	-0.03	1.07	-1.63	0.46	0.22
9	-0.49	-0.59	2.26	1.07	-0.03	0.46
10	-1.54	0.45	-0.49	2.26	-0.59	-0.03
11	0.45	0.59	-1.54	-0.49	0.45	-0.59
12	0.92	0.15	0.45	-1.54	0.59	0.45
13	-0.05	0.60	0.92	0.45	0.15	0.59
14	-1.18	0.13	-0.05	0.92	0.60	0.15
15	0.90	-0.04	-1.18	-0.05	0.13	0.60
16	1.17	0.12	0.90	-1.18	-0.04	0.13
17	0.31	-0.96	1.17	0.90	0.12	-0.04
18	1.19	0.23	0.31	1.17	-0.96	0.12
19	0.27	1.81	1.19	0.31	0.23	-0.96
20	-0.09	-0.01	0.27	1.19	1.81	0.23

```
# Omit NAs for lm()
XTrainDF2 = XTrainDF[3:20,]

XTrainDF2
```

	x1Train	x2Train	X1_l1	X1_l2	X2_l1	X2_l2
3	-0.18	1.13	0.11	-1.03	0.54	-0.82
4	0.20	-0.24	-0.18	0.11	1.13	0.54
5	-0.99	-0.77	0.20	-0.18	-0.24	1.13
6	-1.63	0.22	-0.99	0.20	-0.77	-0.24
7	1.07	0.46	-1.63	-0.99	0.22	-0.77
8	2.26	-0.03	1.07	-1.63	0.46	0.22
9	-0.49	-0.59	2.26	1.07	-0.03	0.46
10	-1.54	0.45	-0.49	2.26	-0.59	-0.03
11	0.45	0.59	-1.54	-0.49	0.45	-0.59

```
12     0.92     0.15     0.45    -1.54     0.59     0.45
13    -0.05     0.60     0.92     0.45     0.15     0.59
14    -1.18     0.13    -0.05     0.92     0.60     0.15
15     0.90    -0.04    -1.18    -0.05     0.13     0.60
16     1.17     0.12     0.90    -1.18    -0.04     0.13
17     0.31    -0.96     1.17     0.90     0.12    -0.04
18     1.19     0.23     0.31     1.17    -0.96     0.12
19     0.27     1.81     1.19     0.31     0.23    -0.96
20    -0.09    -0.01     0.27     1.19     1.81     0.23
```

#fit the MLR
fit = lm(x1Train ~ X1_l1 + X1_l2 + X2_l1 + X2_l2, data = XTrainDF2)
summary(fit)

```
Call:
lm(formula = x1Train ~ X1_l1 + X1_l2 + X2_l1 + X2_l2, data = XTrainDF2)

Residuals:
     Min       1Q   Median       3Q      Max
-1.42514 -0.35651  0.03649  0.40597  1.76421
```

Coefficients:

```
              Estimate   Std. Error  t value    Pr(>|t|)
(Intercept)    0.14904    0.23394      0.637     0.5351
X1_l1          0.18676    0.22461      0.831     0.4207
X1_l2         -0.58499    0.21574     -2.712     0.0178 *
X2_l1          0.09124    0.35172      0.259     0.7994
X2_l2         -0.07601    0.41576     -0.183     0.8578
---
Signif. codes:  0 '***' 0.001 '**' 0.01 '*' 0.05 '.' 0.1 ' ' 1

Residual standard error: 0.9243 on 13 degrees of freedom
Multiple R-squared: 0.397, Adjusted R-squared: 0.2115
F-statistic: 2.14 on 4 and 13 DF,  p-value: 0.1336
```

 # Compare with VAR Analysis
 # Create ts objects
X_ts = ts(XDF)
XTrain_ts = ts(XDF[1:20,])
XTrain_ts #compare with XTrainDF

```
Time Series:
Start = 1
End = 20
Frequency = 1
        x1     x2
1    -1.03  -0.82
2     0.11   0.54
3    -0.18   1.13
4     0.20  -0.24
5    -0.99  -0.77
6    -1.63   0.22
7     1.07   0.46
8     2.26  -0.03
```

```
9    -0.49  -0.59
10   -1.54   0.45
11    0.45   0.59
12    0.92   0.15
13   -0.05   0.60
14   -1.18   0.13
15    0.90  -0.04
16    1.17   0.12
17    0.31  -0.96
18    1.19   0.23
19    0.27   1.81
20   -0.09  -0.01
```

```
# Fit the model with only the constant
VARfit = VAR(XTrain_ts,p=2, type='const')
VARfit$varresult$x1 #compare with MLR fit at lag 2 above

Call:
lm(formula = y ~ -1 + ., data = datamat)

Coefficients:
x1.l1       x2.l1       x1.l2       x2.l2       const
0.18676  0.09124  -0.58499  -0.07601   0.14904
```

Compare these coefficients with the results of the MLR model fit!

PROBLEMS

10.1 In this chapter multiple linear regression with correlated errors was used to model the **Bsales** data frame. Why does it not make sense to use VAR to model the sales in this analysis?

10.2 Consider the model:

$$Sales_t = \beta_0 + \beta_1 t + \beta_2 ad\_tv_{t-1} + \beta_3 ad\_online_{t-1} + \beta_4 discount_t + Z_t$$

(a) Use multiple linear regression (MLR) assuming independent errors (assume Z_t is iid $N(0, \hat{\sigma}_Z^2)$) to find the RMSE for forecasts of the last 10 sales in the dataset.

(b) Now use multiple linear regression with correlated errors to find the RMSE for the forecasts of the last 10 sales.

(c) Is there a difference between the RMSEs in (a) and (b)? What is your conclusion?

(d) Can you use the model in (b) to forecast sales for week 101? For week 102? Explain.

10.3 Use MLR with correlated errors to model the melanoma count given the sunspot count (use **melanoma2.0**). Find the RMSE for the last 8 years as was done in the text for VAR. Remember to use forecast values for sunspots where appropriate. Does one model seem more useful?

10.4 Variable Selection: Does the addition of temperature and/or particulates variables make a difference?

(a) Fit a VAR(p) model with cardiac mortality (cmort), and particulates (part), but without temperature (tempr). Find and record the AIC of the model with the lowest AIC and find and record the RMSE for forecasts of the last 52 weeks.

(b) Fit a VAR(p) model with cardiac mortality, particulates and temperature. Find and record the AIC of the model with the lowest AIC and find and record the RMSE for the last 52 weeks.

(c) Does your evidence suggest that temperature is an important variable / feature?

10.5 Repeat the analysis in question 4 for the particulate variable.

10.6 Load the *astsa* package in CRAN and load the data `lap`. This is the dataset contained in the dataframe `cardiac` but with many more possible explanatory variables. Create a model or models using additional variables other than temperature and particulates. Can you improve on the model(s) using only these two variables? Explain. Are there any relationships between cardiac mortality and other variables in the `lap` dataset? Provide a thoughtful and well explained (with plots) response.

10.7 *Challenge Problem*: Show that the equation from the VAR(2) model for cardiac mortality with temperature and particulates in the vector with constant and trend components is equivalent to a multiple linear regression with the same lagged variables. Appendix 10B will be helpful in investigating this issue.

Deep Neural Network-Based Time Series Models

11

11.1 INTRODUCTION

Deep Learning and Artificial Intelligence (AI) have recently become very popular and have found applications including self-driving cars, virtual assistants such as Siri and Alexa, fraud detection, and image recognition. As we will see, this exciting technology holds tremendous potential for time series analysis as well.

First conceived in 1958 (Rosenblatt, 1958), the neural network is the most fundamental construct used in building "Deep Learning" applications. In this chapter, we will introduce a basic neural network known as the *perceptron* and see how it can be extended to facilitate the analysis of both univariate and multivariate time series data. In the process, we will make use of functions in the **nnfor** package to construct and forecast from potentially "deep" perceptron-based neural networks.

11.2 THE PERCEPTRON

The first neural networks focused on using initial information to choose between two possible outcomes, that is, binary classification. To illustrate, we will first present a simple non-time series example in which the goal is to classify a particular animal as a dog or a cat, given only the length of each animal from the base of its tail to the tip of its nose. Intuitively, we will first take a sample of *n* animals and record this metric for each animal and whether it is a dog (coded as a 1) or a cat (coded as a 0). The resulting data, in which we know the length of each animal and its classification, are known as the *training set*.

A neural network approach to this problem is as follows:

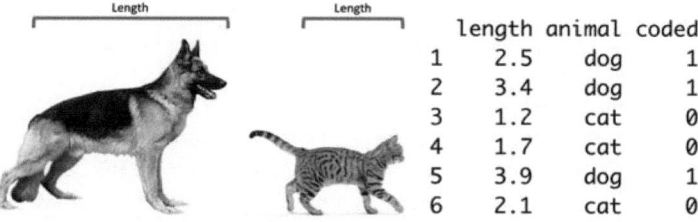

	length	animal	coded
1	2.5	dog	1
2	3.4	dog	1
3	1.2	cat	0
4	1.7	cat	0
5	3.9	dog	1
6	2.1	cat	0

For each observation in the training set:

1. The lengths from each animal are normalized by a form of the "minimax" transformation:

$$length_{t_{minimax}} = \frac{length_t - length_{min}}{length_{max} - length_{min}}$$

where $length_t$ is the length of the tth animal in the sample, and $length_{min}$ and $length_{max}$ are the shortest and longest lengths of all the animals in the sample (dog or cat). The reasons for this transformation will be described later in the chapter.

2. The normalized length is multiplied by a number w (called a model weight) and then added to a constant (aka: "bias term"),θ, which results in the sum

$$f = \theta + w * Length$$

3. f is then used as the input to the sigmoid (or "logistic") function (Figure 11.1):

$$g = \frac{1}{1 + e^{-f}}$$

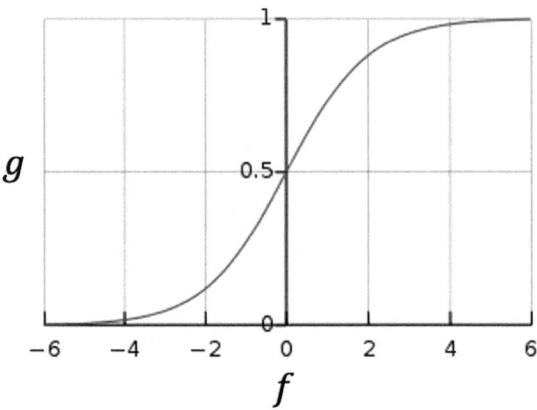

FIGURE 11.1 The logistic (sigmoid) function. This maps a real-valued function f to a value g between 0 and 1

This is known as the "activation function", for which we will have many choices. For binary classification, the sigmoid function is a logical choice as it forces an output between zero and one which corresponds to the probability of a "success" (the outcome coded as a one), in this case, a dog.

4. The final step is then to classify the observation as a dog if g is greater than some threshold, commonly .5, and a cat if it is less than or equal to that threshold.

5. An iterative numerical procedure, known as *gradient descent*, is then used to approximate the "best" w and θ that maximize the number of times the model is correct, its measure of accuracy. This process can be computationally intensive depending on the number of observations and complexity of the model. For this reason, a proxy measure of accuracy that has been shown to

be very efficient is called the "cross-entropy". The cross-entropy measures the disorder that is produced by the model, which will be low if the accuracy is high. Therefore, we call the "cross-entropy" a "loss" function and aim to find the combination of w and θ that will minimize the function

$$Cross\text{-}Entropy = \sum_{i=1}^{n} -p(x_i)\log\big(q(x_i)\big)$$

For each observation i, $p(x_i)$ will either be one if x_i is a dog or zero if x_i is a cat, while $q(x_i)$ is the probability that is output from the activation (sigmoid) function. One reason for our normalization step above is because gradient descent has been shown to be more stable and efficient when the input data is normalized.

The algorithm that includes these five steps combined with the gradient descent procedure to minimize the loss function is known as the *perceptron*. The perceptron for the classification problem is illustrated in Figure 11.2 below, and is broken into an input layer with a single input node (Length) and an output layer with a single output node (Dog/Cat).[1] Going forward, we will call these diagrams the *architecture* of the neural network.

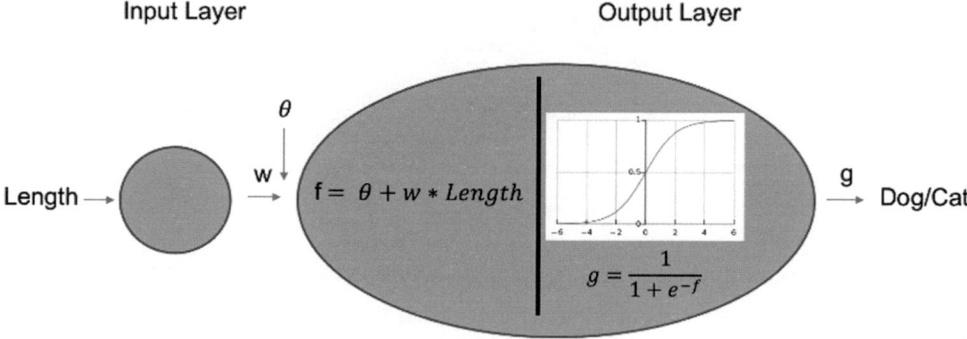

FIGURE 11.2   Architecture of the dog/cat classification perceptron

11.3 THE EXTENDED PERCEPTRON FOR UNIVARIATE TIME SERIES DATA

In this section we will explore how the fundamental perceptron framework described above can be extended to model in the time series setting. The perceptron analog to the AR(p) class of models will be described first at which point "deeper" neural networks will be investigated. Finally, this section will conclude with an example in which these concepts are applied to the **AirPassengers** data.

---

1 If the loss function for the gradient descent is cross-entropy, then this neural network is equivalent to logistic regression.

11.3.1 A Neural Network Similar to the AR(1)

11.3.1.1 The Architecture

As previously mentioned, the perceptron can be adapted to the time series setting by the choice of the *activation function*. For example, suppose we want to model the West Texas Intermediate crude oil prices and we determine that all useful information about this month's price (X_t) is contained in last month's price (X_{t-1}). Note that in contrast to the neural network described previously, X_t, the price, in this particular case is now a continuous variable rather than binary. For this reason, as seen in Figure 11.3, we can modify the perceptron to model time series data by changing the activation function from the sigmoid function to one that will allow for an unbounded response. A commonly used activation function in this setting is the identity function which has output exactly equal to its input: $y = x$.

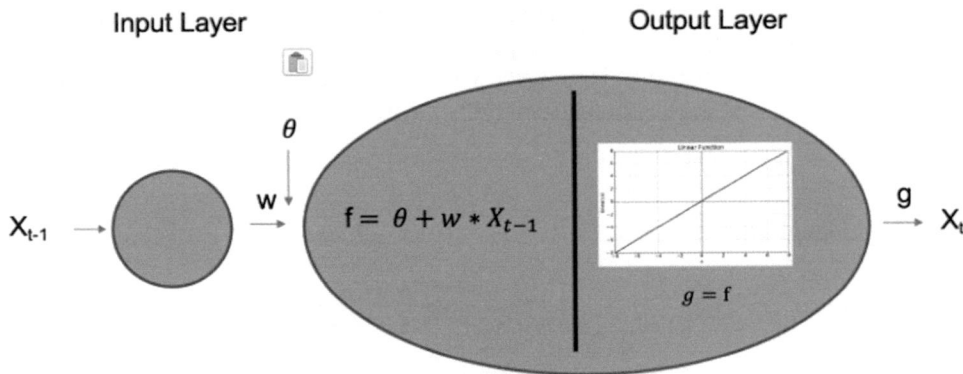

FIGURE 11.3   Architecture of a network similar to the AR(1).

Note that, using the architecture in Figure 11.3, the neural network can be written as

$$X_t = \theta + wX_{t-1} + a_t,$$

which is similar to the AR(1) model,

$$X_t = \mu + \phi_1 X_{t-1} + a_t$$

An additional difference imposed by changing to a continuous response is that of the loss function. We will replace the cross-entropy discussed earlier for classification-based neural networks with the familiar sum of squared error (SSE).[2] The gradient descent algorithm will thus aim to return a set of weights and biases to minimize the SSE.

11.3.1.2 Fitting the MLP

To fit this model in R, we will use the `mlp` function in the ***nnfor*** package. The acronym "MLP" stands for "multilayer perceptron", which pertains to potential multiple *hidden layers* that we will discuss later in this chapter. For the model above, we will have only the input and output layers (no hidden layers: `hd = 0`

---

2 Sum of squared residuals

in the **mlp** function). In this example, we use only one lag (**lags = 1**) and note that the model does not require any pre-differencing (**difforder = 0**) and thus the model can be fit in R using

```
set.seed(1)
AR1Fit = mlp(wtcrude2020, m = 1, hd = 0, lags = 1, difforder = 0, reps = 5)
```

Figure 11.4 displays a simplified diagram (make that "very simplified"!) of the single input (left) and single output node (right) network illustrated in Figure 11.3. This diagram will be more useful (and involved) as the architecture becomes more complex. Obtaining this plot is accomplished using the R code

```
plot(AR1Fit)
```

MLP

FIGURE 11.4   Visualization of the architecture of the AR(1) type neural network from package **nnfor**

Additionally, summary information in (11.1) below about the fitted model can be obtained with the code

AR1Fit[3]

```
MLP fit with 0 hidden node and 5 repetitions.
Univariate lags: (1)
Forecast combined using the median operator.
MSE: 22.2019.
```
(11.1)

The first and second lines of the output in (11.1) summarize the lack of hidden layers (0 hidden) and the inclusion of the X_{t-1} term (Univariate lags: (1)), and also indicate that five repetitions (5 repetitions) were selected to most effectively fit the model. These details are discussed next.

The gradient descent procedure used in estimating the optimal values of w and θ depend on initial weights,[4] or starting values, of w and θ. Depending on the complexity of the neural network, estimates of w may vary considerably based on these starting values. For this reason, we have chosen to fit several neural networks, in this case **reps = 5**, which will be initiated with five different sets of starting values and thus will yield potentially five different estimates of w. (Note that only five repetitions were generated here for illustrative simplicity. In theory, more repetitions will yield better estimates although the increase can also require significantly more time to fit the model; the default is **reps = 20**).[5]

For the West Texas Intermediate Crude example, the five different estimates of θ (the top number in the output) and w (the bottom number in the output) are shown below. They can be accessed using the R command:

AR1Fitnetweights

```
             [,1]
[1,]  -0.00395087
[2,]   0.98539259
```

---

3 Note that all output in this chapter was generated with a seed of 1. To reproduce the output and plots in this chapter, **set.seed(1)** will need to be run immediately before any call to the **mlp** function. If the code is run with a different (random) seed, similar results should be expected. Doing so is often a good exercise.

4 https://stats.stackexchange.com/questions/47590/what-are-good-initial-weights-in-a-neural-network

5 In the simple example case presented here, the optimization problem is "convex", meaning that it only has one (global) minimum. For this reason, the gradient descent algorithm will converge to this single value regardless of the starting value. More complex neural networks that include hidden layers will be very non-convex and thus very dependent on the starting values of w and θ.

```
           [,1]
[1,] -0.003922568
[2,]  0.985443097


           [,1]
[1,] -0.003919923
[2,]  0.985462926


           [,1]
[1,] -0.003844282
[2,]  0.985712695


           [,1]
[1,] -0.003926506
[2,]  0.985443707
```

QR 11.1
MLP for
the AR(1)

Note also that these values of w are very close to the estimate of $\hat{\phi}_1$ in the AR(1) fit using the Burg estimates ($\hat{\phi}_1= .987$) obtained using the code below.

```
estAR1 = est.ar.wge(wtcrude2020,p = 1, method = "burg")
estAR1$phi
```

```
[1] 0.9865561
```

Interestingly, this is the neural network representation of an AR(1)!

11.3.1.3 Forecasting

To calculate the forecasts, the **mlp** function uses the five different weights and θs to generate five separate forecasts of West Texas Intermediate Crude price for the next month (month 373: X_{373}). The forecast for this month (horizon **h = 1**) from each of the five model fits can be calculated and accessed using the code

```
preds = forecast(AR1Fit,h = 1)
preds$all.mean
```

	NN.1	NN.2	NN.3	NN.4	NN.5
Jan 2021	47.16398	47.16486	47.16456	47.16398	47.16455

The columns represent the five neural networks that were fit, and the single row contains the respective forecasts for X_{373} (**wtcrude2020** consists of 372 months of data) from each of the five neural networks. Returning to the third line of the output in (11.1), we note that the final forecast is the median of these five individual forecasts (even though the attribute is always called the "mean"): Forecast combined using the median operator. This final forecast can be accessed using the following code and output. Note that, internal to the *nnfor* package, the attribute is called the "**$mean**" even though for this MLP, the median has been used to combine the forecasts. The value below is indeed the median of the five forecasts.

```
preds$mean # even though this is the median
```

```
        Jan
2021 47.16455
```

Extending this idea to longer horizons, below we see forecasts for a horizon of ten. Again, there is a column for each neural network (for each "repetition") and a row for each of the ten forecasts **(h = 10)**

```
preds = forecast(AR1Fit,h = 10)
preds$all.mean
```

	NN.1	NN.2	NN.3	NN.4	NN.5
Jan 2021	47.16398	47.16486	47.16456	47.16398	47.16455
Feb 2021	47.23303	47.23478	47.23418	47.23305	47.23415
Mar 2021	47.30107	47.30368	47.30279	47.30113	47.30274
Apr 2021	47.36812	47.37158	47.37040	47.36824	47.37034
May 2021	47.43418	47.43849	47.43703	47.43439	47.43695
Jun 2021	47.49928	47.50442	47.50269	47.49960	47.50259
Jul 2021	47.56343	47.56940	47.56740	47.56387	47.56727
Aug 2021	47.62665	47.63343	47.63116	47.62722	47.63102
Sep 2021	47.68894	47.69653	47.69400	47.68967	47.69383
Oct 2021	47.75032	47.75871	47.75592	47.75123	47.75573

As before, the median of each row becomes the ultimate forecast for each month (check it!):[6]

```
preds$mean #even though this is still the median
```

	Jan	Feb	Mar	Apr	May	Jun	Jul	Aug	Sep	Oct
2021	47.16455	47.23415	47.30274	47.37034	47.43695	47.50259	47.56727	47.63102	47.69383	47.75573

Note that while the AR(1) and neural network are similar, a difference between the two models is that the forecasts from the neural network are not attracted to the overall sample mean, $\overline{X}_{wtcrude2020} = 47.47493$, as is the case for the stationary AR(1) model.

Recall that in the AR(1) case, the forecasts are calculated with the eventual forecast function from (6.28):

$$\hat{X}_{t_0}\left(\ell\right) = \hat{\phi}_1 \hat{X}_{t_0}\left(\ell-1\right) + \overline{x}(1-\hat{\phi}_1)$$

On the other hand, forecasts from the neural network model are calculated as

$$\hat{X}_{t_0}\left(\ell\right) = \hat{w}_1 \hat{X}_{t_0}\left(\ell-1\right) + \hat{\theta}_1$$

where in most cases, $\overline{x}(1-\hat{\phi}_1) \neq \hat{\theta}_1$.

This difference is highlighted in Figures 11.5(a)–(c), where the forecasts from both the AR(1) and MLP models are provided (with $h = 300$ to aid in comparison) as well as their respective eventual forecast functions. Figure 11.5(a) displays the forecasts from the end of the **wtcrude2020** series (January 2021), where the divergence is clear. Given $\hat{\phi}_1 = .987$ and $\overline{x}(1-\hat{\phi}_1) = 47.47493(1-.987) = .617$ for the AR(1), and $\hat{w}_1 \cong .986$ and $\hat{\theta}_1 \cong -.004$ for the MLP, the corresponding eventual forecast functions are

$$AR(1): \hat{X}_{t_{372}}\left(\ell\right) = .987\hat{X}_{t_0}\left(\ell-1\right) + .617$$
$$MLP: \hat{X}_{t_{372}}\left(\ell\right) = .986\hat{X}_{t_0}\left(\ell-1\right) - .004$$

The difference in the forecasts in Figure 11.5(a) is the result of the discrepancy between .617 and −.004 in the above eventual forecast function.

---

6 It turns out that the last column (NN.5) is always the median in this simple example; this does not have to be the case.

To further highlight this phenomenon, a new forecast origin was chosen to be April 2014 (observation 292). Using this forecast origin, $\hat{\phi}_1 = .992$ and $\bar{x}(1-\hat{\phi}_1) = 45.52196(1-.992) = .364$ for the AR(1) and for the MLP, $\hat{w}_1 \cong .994$ and $\hat{\theta}_1 \cong .002$. These estimates yield the eventual forecast functions below with corresponding plots in Figure 11.5(b).

$$AR(1): \quad \hat{X}_{t_{292}}(\ell) = .992\hat{X}_{t_0}(\ell-1)+.364$$
$$MLP: \quad \hat{X}_{t_{292}}(\ell) = .994\hat{X}_{t_0}(\ell-1)+.002$$

Figure 11.5(c) displays a comparison of the forecasts from a forecast origin of April 2020 (observation 364) in which again, the eventual forecast functions are different, although this time, there is very little difference that can be seen visually.

$$AR(1): \quad \hat{X}_{t_{364}}(\ell) = .987\hat{X}_{t_0}(\ell-1)+.619$$
$$MLP: \quad \hat{X}_{t_{364}}(\ell) = .988\hat{X}_{t_0}(\ell-1)-.004$$

There is an interesting pattern to these forecasts depending on whether the MLP model's forecast origin is below the sample mean, near the sample mean, or greater than the sample mean.

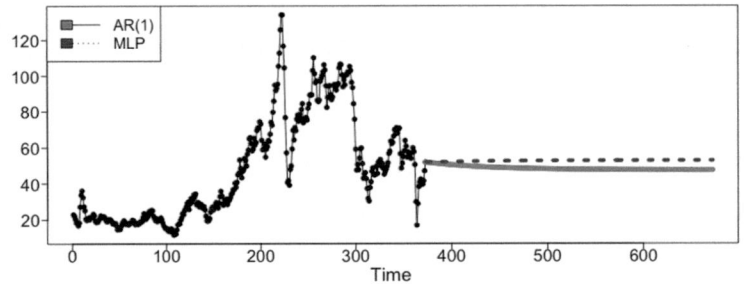

FIGURE 11.5(a)   Plot of AR(1) forecasts from the end of the series (January 2021) with a horizon of 300 (solid) versus forecasts with a horizon of 300 from the MLP model with one lag described above (dashed)

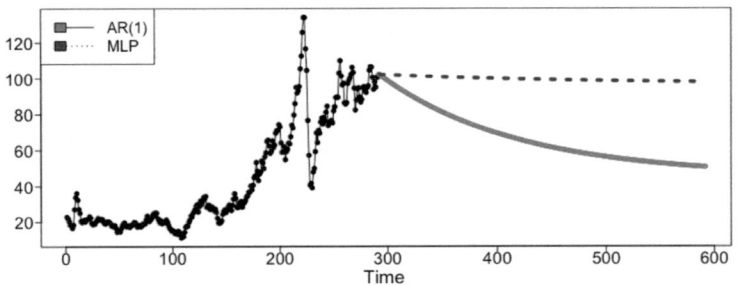

FIGURE 11.5(b)   Plot of AR(1) forecasts from the end of the series (April 2014) with a horizon of 300 (solid) versus forecasts with a horizon of 300 from the MLP model with one lag described above (dashed)

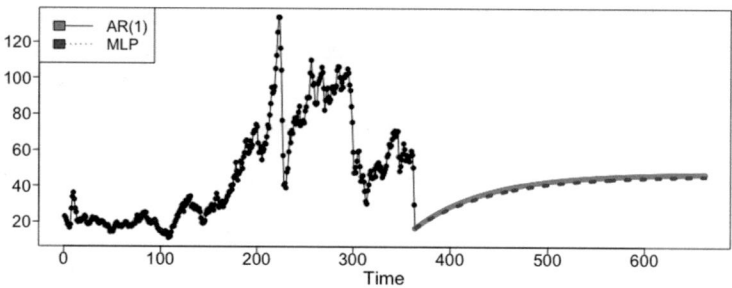

FIGURE 11.5(c) Visualization of a case in which the AR(1) (solid) and MLP (dashed) forecasts are very similar, although closer inspection reveals they are still slightly different

11.3.1.4 Cross Validation Using the Rolling Window RMSE

As in previous chapters, we can produce a rolling window RMSE from the MLP model to assess its fit for a given horizon. In order to do this, we will take the ws and θs from the model fit with the data and pass these estimates along with the desired horizon to the *tswge* function `roll.win.rmse.nn.wge()`. The code and output for the rolling window RMSE of the model contained in `AR1Fit` with a horizon of one are shown below.

```
rwAR1Fit = roll.win.rmse.nn.wge(wtcrude2020,horizon = 1,fit_model = AR1Fit)
"Please Hold For a Moment, TSWGE is processing the Rolling Window RMSE with 363
windows."
"The Summary Statistics for the Rolling Window RMSE Are:"
   Min.   1st Qu.   Median     Mean   3rd Qu.       Max.
0.01466   0.87807   1.88629   3.13809   3.91681   27.08154
"The Rolling Window RMSE is: 3.138"
```

Similar to the rolling window RMSE functions in previous chapters, this function can use the AR1Fit model to make a forecast for nearly every value in the dataset and then calculate the RMSE since the actual values are known.[7] In this case, since the horizon is one, the function calculates 363 (from the first line of the output above) one step-ahead forecasts and then takes the average to produce the "Rolling Window RMSE", which in this case, is 3.138 dollars. The 5-number summary and a histogram (Figure 11.6) are also available in order to describe the distribution of the RMSEs from these 363 windows.

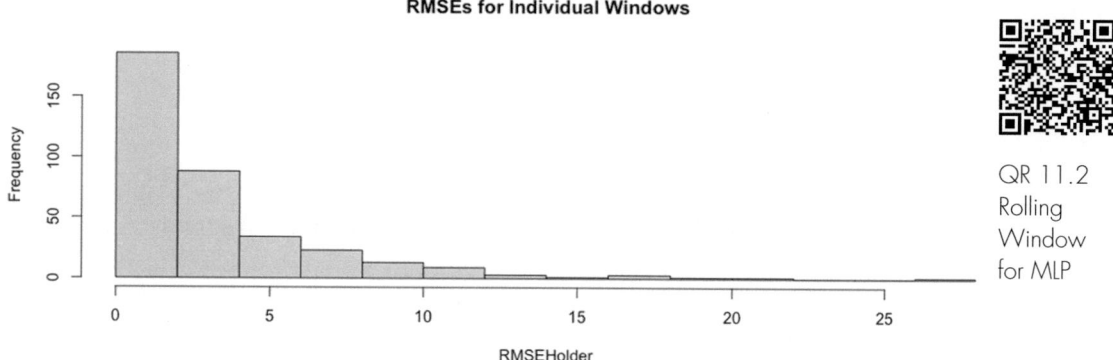

QR 11.2
Rolling
Window
for MLP

FIGURE 11.6 Histogram showing the distribution of the 363 RMSEs calculated from the 363 windows constructed from the **wtcrude2020** series

---

7 The `mlp` function uses the first eight data values and uses a simple exponential smoother to detect trend and thus decides if a first difference is necessary. For this reason, forecasts for the first eight values are not possible.

Intuitively, it is reasonable to expect the forecasting error to increase as we predict further into the future (as the horizon increases). This intuition is supported by considering the rolling window RMSE for the horizon of ten discussed earlier

```
rwAR1Fit = roll.win.rmse.nn.wge(wtcrude2020,horizon = 10,fit_model = AR1Fit)
```

```
"Please Hold For a Moment, TSWGE is processing the Rolling Window RMSE with 354
windows."
"The Summary Statistics for the Rolling Window RMSE Are:"
 Min.     1st Qu.    Median      Mean     3rd Qu.       Max.
1.139       3.974     6.524     9.514      11.222     68.515
"The Rolling Window RMSE is: 9.514"
```

Note that given the larger horizon, we can only compute RMSEs for 354 windows and that the corresponding rolling window RMSE has increased from 3.138 to 9.514 dollars.

11.3.2 A Neural Network Similar to AR(*p*): Adding More Lags

More generally, the perceptron-based neural network can be extended to accommodate up to p lags by adding more input nodes as illustrated in Figure 11.7.

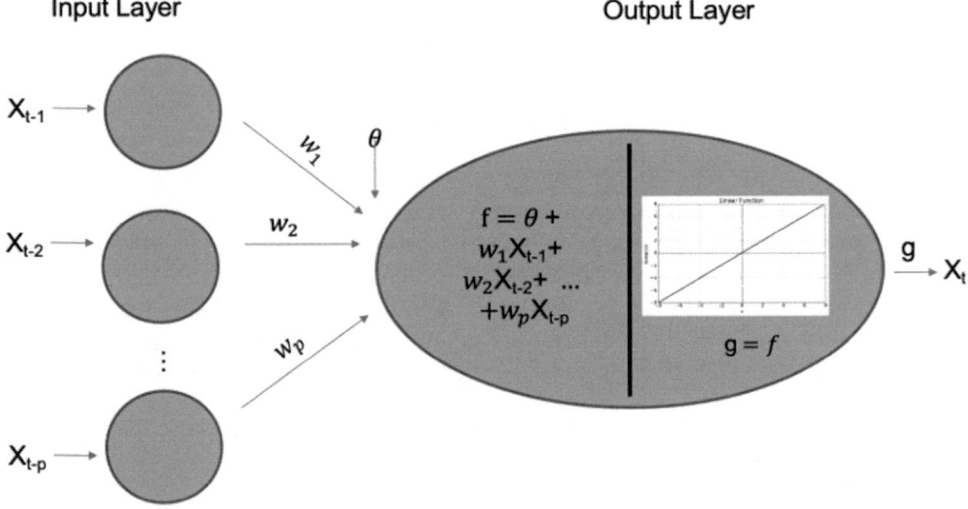

FIGURE 11.7 A Neural Network with input nodes for p lags

To continue the univariate analysis of the West Texas Intermediate Crude oil data, Figure 11.8 shows the architecture for a neural network with inputs for three lags. It will be shown that this is similar, but not equivalent, to the AR(3) fit to these data in Chapter 3.

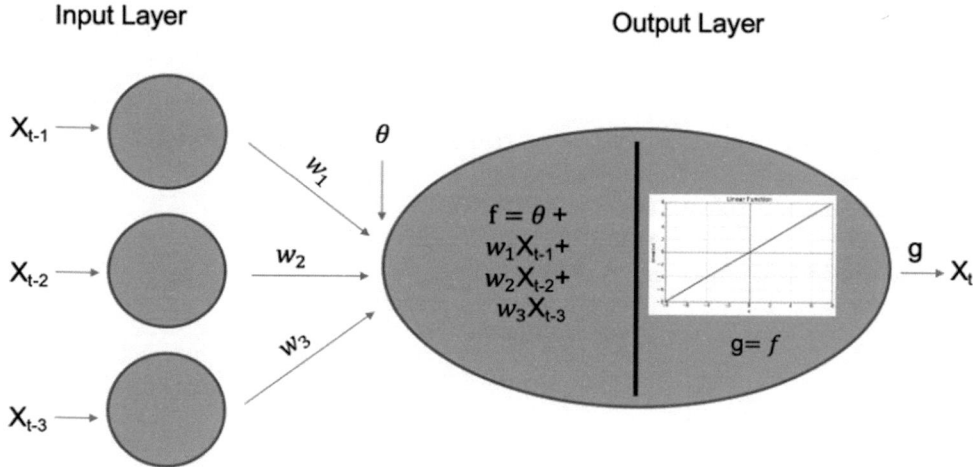

FIGURE 11.8   A Neural Network with input nodes for three lags. This is analogous to an AR(3) model

The **mlp** function will allow for the inclusion of specific lags in the model. In order to include X_{t-1}, X_{t-2}, and X_{t-3} in the model, we must specify **lags = c(1,2,3)**. However, by default, the **mlp** function will automatically select from these lags only those it deems useful (using cross-validation). To override this, the analyst must set the **sel.lag** argument to FALSE (**sel.lag = FALSE**). Below, see the resulting fit and Figure 11.9 shows architecture diagram of this AR(3)-like neural network.

```
set.seed(1)

AR3Fit = mlp(ts(wtcrude2020),hd = 0, lags = c(1,2,3), difforder = 0, reps = 5,
sel.lag = FALSE)

AR3Fit

MLP fit with 0 hidden node and 5 repetitions.
Univariate lags: (1,2,3)
Forecast combined using the median operator.
MSE: 18.8756.

plot(AR3Fit)
```

MLP

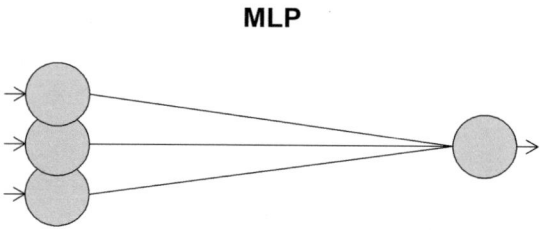

FIGURE 11.9   Architecture of the MLP equivalent of the AR3

As an indication of better fit, note that the MSE for one step-ahead forecasts on the training data has reduced from 22.2 in the model including only X_{t-1}, to 18.9 when adding X_{t-2} and X_{t-3}. While the one step-ahead forecasts may be overly optimistic and should be used with caution, the apples-to-apples comparison here is useful.

As a quick aside, observe what occurs if **mlp** is used to automatically select the lags from X_{t-1}, X_{t-2} and X_{t-3}.

```
set.seed(1)
```

```
AutoFit = mlp(ts(wtcrude2020),hd = 0, lags = c(1,2,3), difforder = 0, reps = 5,
sel.lag = TRUE)
```

```
AutoFit
```

```
MLP fit with 0 hidden node and 5 repetitions.
Univariate lags: (1,2)
Forecast combined using the median operator.
MSE: 18.8279.
```

```
plot(AutoFit)
```

MLP

FIGURE 11.10   Architecture of the MLP equivalent of the AR2

Internally, the **mlp** function fit a model with all combinations of the three lags (c(1), c(1,2), c(1,3), c(2,3) and c(1,2,3)) and returned the model that included only X_{t-1} and X_{t-2}, which results in a one step-ahead MSE of 18.8279. Figure 11.10 above reflects this architecture.

Forecasting from these models is performed as before and forecasts are provided below for reference:

```
preds = forecast(AR3Fit,h = 10)
preds$mean
```

```
Time Series:
Start = 373
End = 382
Frequency = 1
```

t+1	t+2	t+3	t+4	t+5	t+6	t+7	t+8	t+9	t+10
49.28015	50.24284	50.63960	50.79640	50.84923	50.85729	50.84636	50.82766	50.80604	50.78361

```
preds = forecast(AutoFit,h = 10)
preds$mean
```

```
Time Series:
Start = 373
End = 382
Frequency = 1
```

t+1	t+2	t+3	t+4	t+5	t+6	t+7	t+8	t+9	t+10
49.30093	50.16587	50.49673	50.61529	50.64973	50.65107	50.63677	50.61640	50.59460	50.57501

Cross-validation for the horizon of ten is analogous (below) as well, and demonstrates that the RMSEs are close but are slightly smaller for the Lag 2 (**Autofit**) neural network model (Lag 3 Fit: 10.172 versus Lag 2 Fit: 10.099):

```
rwAR3Fit = roll.win.rmse.nn.wge(wtcrude2020,horizon = 10, fit_model = AR3Fit)
```

"Please Hold For a Moment, TSWGE is processing the Rolling Window RMSE with 354
windows."
"The Summary Statistics for the Rolling Window RMSE Are:"

```
 Min.      1st Qu.    Median     Mean    3rd Qu.    Max.
1.220       5.021      7.380    10.172   12.080    62.094
```
"The Rolling Window RMSE is: 10.172"

```
rwAutoFit = roll.win.rmse.nn.wge(wtcrude2020,horizon = 10,fit_model= AutoFit)
```

"Please Hold For a Moment, TSWGE is processing the Rolling Window RMSE with 354
windows."
"The Summary Statistics for the Rolling Window RMSE Are:"

```
 Min.      1st Qu.    Median     Mean    3rd Qu.    Max.
1.305       4.984      7.243    10.099   12.021    62.298
```
"The Rolling Window RMSE is: 10.099"

11.3.3 A *Deeper* Neural Network: Adding a Hidden Layer

We are now in a position to highlight one of the principal differences between neural network models and the AR(p) models we have studied thus far. Remember that "MLP" stands for "multi-layered perceptron" and "multi-layered" refers to the inclusion of a "hidden layer" (that has previously been excluded). This "hidden layer" is inserted between the input and output layer and may be composed of one or more "hidden nodes". While we have technically investigated a few MLP architectures to this point, many in the data science field feel that the term "neural network" implies one or more hidden layers.[8,9]

For simplicity, consider the previous neural network model with X_{t-1}, X_{t-2}, and X_{t-3} except this time a single hidden layer is included, which is composed of two hidden nodes. This architecture is illustrated in Figure 11.11.

---

8 Research has shown that analyzing time series with neural networks usually does not benefit beyond the addition of a single hidden layer. See https://kourentzes.com/forecasting/2019/01/16/tutorial-for-the-nnfor-r-package/

9 The above MLPs were included to introduce the vocabulary, architecture, and functionality of neural networks using a familiar model: the AR(p). As we have seen above, these MLP architectures produce different forecasts and thus are not AR(p) models. If an AR(p) was thought to be the most useful model, which is often the case, AR(p) modeling techniques should be implemented using methods such as those found in *tswge;* forecasting will potentially be much faster and will yield the appropriate forecasts.

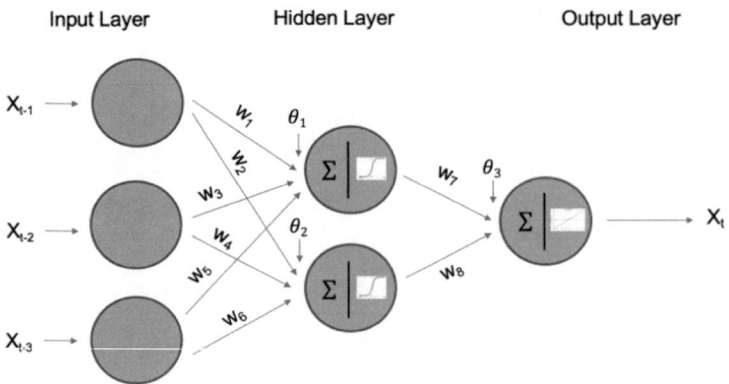

FIGURE 11.11   Architecture with three input nodes in the input layer, two hidden nodes in the hidden layer with logistic activation functions, and a single output node in the output layer with the identity activation function

Although the architecture of this MLP is relatively simple, there is a lot going on here! Note first that each input node is "connected" via a weight to every hidden node and the summation Σ in each hidden node indicates that the inputs, weights, and constants are combined in the same way as described before. For example, the combination of these inputs and parameters facilitated by the first (top) hidden node is displayed in Figure 11.12a. The result is obtained similarly for the second hidden node (Figure 11.12(b)).

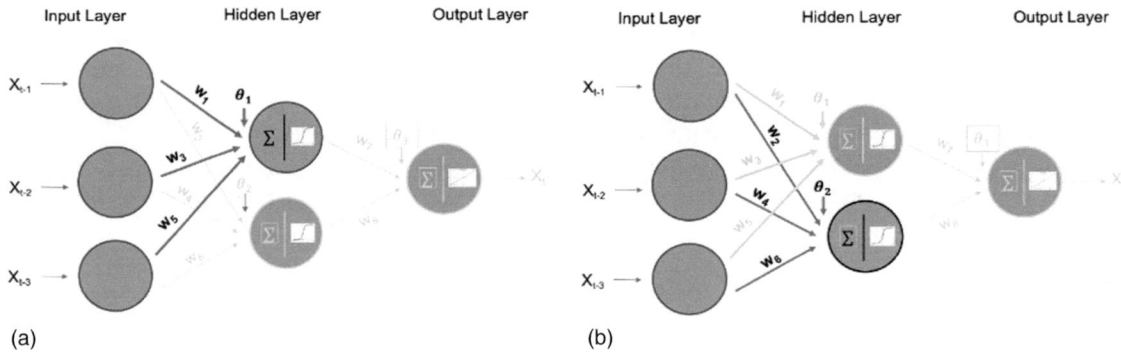

(a)   (b)

FIGURE 11.12   Illustration of the calculations involved when adding a hidden layer. The functional form for (a) f_1 and (b) f_2 are included below each plot

The resulting sum for each hidden node is then acted upon by its corresponding activation function. Figure 11.13 is an enhanced view of the first hidden node in Figure 11.12(a) and reveals that the **mlp** function uses the logistic function for the hidden nodes, in contrast to the identity function used for the output node, which allows for the modeling of nonlinear relationships between X_{t-1}, X_{t-2}, and X_{t-3}.

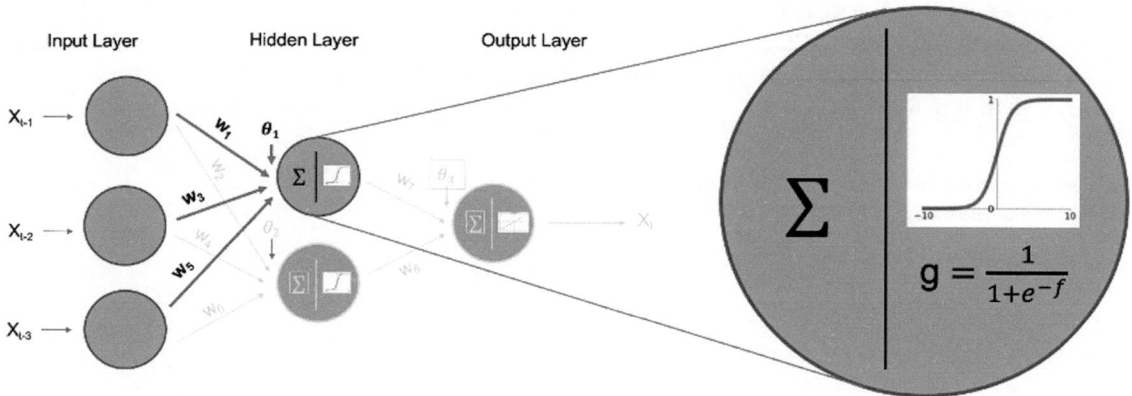

FIGURE 11.13 Close-up of the first hidden node in the single hidden layer highlights the logistic (sigmoid) activation function that maps a real number to a number between 0 and 1

The corresponding function including both hidden nodes is:

$$w_7 g\left(f_1\right) + w_8 g\left(f_2\right) + \theta_3 = X_t,$$

where the w_is and θ_js are again estimated using gradient descent.

The added complexity from the additional weights and the non-linear activation functions degrade the interpretability of these types of models and has even led to these types of models being labeled as "black box"; however, in some settings, they provide more accurate forecasts.

For example, returning to the analysis of the West Texas Crude Oil data, the following R output and Figure 11.14 reveal that the inclusion of a single hidden layer with two nodes (**hd = 2**) has reduced the MSE from 18.8279 to 17.6459.

```
set.seed(1)

DeepFit = mlp(ts(wtcrude2020),hd = 2, lags = c(1,2,3), difforder = 0, reps = 5,
sel.lag = FALSE)

DeepFit

MLP fit with 2 hidden nodes and 5 repetitions.
Univariate lags: (1,2,3)
Forecast combined using the median operator.
MSE: 17.6459.

plot(DeepFit)
```

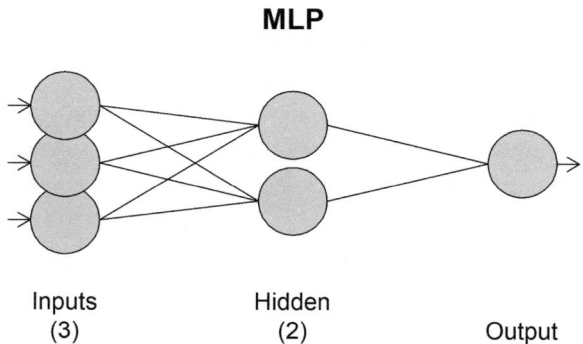

MLP

Inputs
(3)

Hidden
(2)

Output

FIGURE 11.14 Architecture of the "DeepFit" MLP. There are three input nodes corresponding to the three lags, two hidden nodes in the hidden layer, and an output node

However, it is very important to remember that because the loss function of this MLP is the SSE, which is included in the numerator of the MSE itself, *the MSE will always decrease with the inclusion of hidden layers.* A more useful measure of model improvement will be the rolling window RMSE with a horizon larger than 1. Calculating forecasts and the rolling window RMSE is the same as it was with the simpler models discussed above.

```
preds = forecast(DeepFit,h = 10)
round(preds$mean, 2)

Time Series:
Start = 373
End = 382
Frequency = 1

    t+1    t+2    t+3    t+4    t+5    t+6    t+7    t+8    t+9   t+10
  49.19  50.12  50.58  50.87  51.12  51.35  51.59  51.83  52.07  52.30
```

```
rwDeepFit = roll.win.rmse.nn.wge(wtcrude2020,horizon = 10,fit_model= DeepFit)
```

```
"Please Hold For a Moment, TSWGE is processing the Rolling Window RMSE with 354
windows."
"The Summary Statistics for the Rolling Window RMSE Are:"

   Min.   1st Qu.    Median      Mean    3rd Qu.       Max.
 0.6294    3.8252    5.9459    9.0319    12.5727    31.2432

"The Rolling Window RMSE is: 9.032"
```

We note that the MLP with a single hidden layer with two hidden nodes (DeepFit) achieves a lower rolling window RMSE for a horizon of 10, as seen in Table 11.1, than the MLPs with the AR-type architectures fit above.

TABLE 11.1   Comparison of the 4 MLP architectures fit to the wtcrude2020 data.

MODEL	rwRMSE
AR1Fit	9.514
AR3Fit	10.172
AutoFit	10.099
DeepFit	9.032

This indicates that there is evidence to suggest that there may be non-linear relationships in the data and that the hidden layers may be useful in modeling those relationships.

11.3.3.1 Differences and Seasonal "Dummies"

In neural network terminology, the number of lags, hidden layers, hidden nodes per hidden layer, choice of activation function, etc. are called *hyperparameters*. Incidentally, two additional hyperparameters previously studied, differencing and adding dummy (indicator) variables to model seasonal effects, are also useful in MLP-based time series models. We will make use of the **AirPassengers** data to illustrate how to set these additional hyperparameters.

Key Points:

hyperparameter – "A hyperparameter is a parameter that is set before the learning process begins. These parameters are tunable and can directly affect how well a model trains."[10]

Examples of hyperparameters include number of hidden layers, number of hidden nodes, number of differences, presence of seasonal indicator variables, choice of activation function, number of lags, and many more (both a blessing and a curse!)

Recall the monthly `AirPassengers` dataset displayed for convenience in Figure 11.15 (a)-(c).

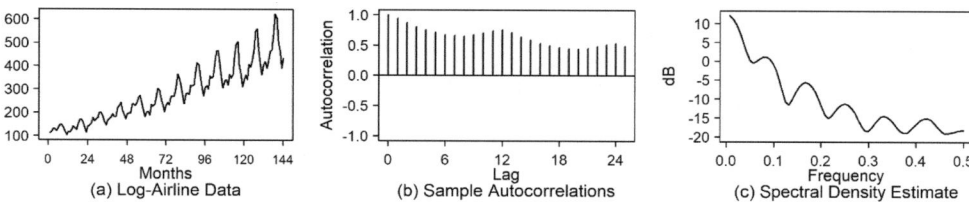

(a) Log-Airline Data   (b) Sample Autocorrelations   (c) Spectral Density Estimate

FIGURE 11.15   (a) The number of airline passengers from 1949 to 1960, (b) the sample autocorrelations of the airline passenger data, (c) the spectral density of the airline passenger data

Domain knowledge of air traffic patterns and visual inspection of the plots in Figure 11.15 provide overwhelming evidence of these data being generated from a non-stationary process. The data clearly exhibit strong evidence of positive trend, which is reflected in the slowly damping sample autocorrelations and the peak at zero in the spectral density. Additionally, the realization, sample autocorrelations, and sample spectral density along with our experience provide equally strong evidence of seasonal behavior with a period of 12. Specifically, sample autocorrelations reveal a sinusoidal pattern with a peak at lag 12, and the spectral density shows a peak at .083 (1/12) with four other peaks that are consistent with a $(1 - B^{12})$ factor. These characteristics support the conclusion that the data exhibit non-stationary seasonal behavior. Previously, in Chapter 7, we modeled these data with a model containing $(1 - B)$ and $(1 - B^{12})$ factors. We called this an "airline model".

We now analyze these data with an MLP by specifying `difforder = c(1,12)`. The following code fits this model by allowing for five hidden nodes (the default) and the automatic selection of the number of lags (also the default). Note that the MLP model will also fit seasonal dummy/indicator variables by default. We will discuss this option later; for now, so we can focus on the differencing, we will turn off the seasonal indicator variable option (`allow.det.season = FALSE`).

```
set.seed(1)

fit.1.12.H5 = mlp(AirPassengers,difforder = c(1,12),allow.det.season = FALSE)
fit.1.12.H5

MLP fit with 5 hidden nodes and 20 repetitions.
Series modeled in differences: D1D12.
Univariate lags: (1,3,4,7,9)
```

---

10 https://deepai.org/machine-learning-glossary-and-terms/hyperparameter

```
Forecast combined using the median operator.
MSE: 53.5823.
```

Figure 11.16 displays the architecture of this model

plot(fit.1.12.H5)

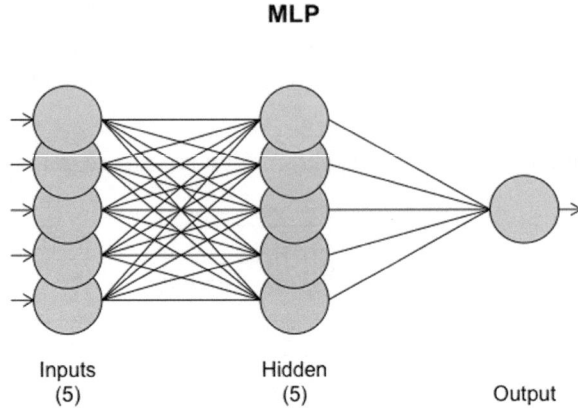

MLP

Inputs
(5)

Hidden
(5)

Output

FIGURE 11.16   Architecture of the MLP to fit the AirPassenger data with a first difference and a twelfth difference. This architecture has five input nodes, one hidden layer with five hidden nodes, and a single output layer

As before, we can see from the output and the plot that lags 1,3,4,7 and 9 make up the inputs and that the architecture includes a single hidden layer with five hidden nodes. However, we also now see in the second line of the output that the first and twelfth difference were also included (Series modeled in differences: D1D12.). While we can see that the MSE for one step-ahead forecasts is 53.5823, assuming we wanted to evaluate this model's performance with a horizon of 36 as in Chapters 2 and 6, we would issue the following commands to calculate the rolling window RMSE

rwfit1.12.H5 = roll.win.rmse.nn.wge(AirPassengers,horizon = 36, fit.1.12.H5)

```
"Please Hold For a Moment, TSWGE is processing the Rolling Window RMSE with 86
windows."
"The Summary Statistics for the Rolling Window RMSE Are:"

 Min.    1st Qu.   Median     Mean   3rd Qu.     Max.
12.32      27.39    34.74    36.47     43.97    68.11
"The Rolling Window RMSE is: 36.474"
```

We will use this model's rolling window RMSE = 36.474 as a baseline and fit a few competing models (architectures). We will fit the first of these competing models by using the **mlp** function to autoselect the difference(s). This can be done by simply *not* specificying the **difforder**

set.seed(1)

fit.1.H5 = mlp(AirPassengers,allow.det.season = FALSE)
fit.1.H5

```
MLP fit with 5 hidden nodes and 20 repetitions.
Series modelled in differences: D1.
Univariate lags: (1,2,3,4,5,6,7,8,9,10,11,12)
Forecast combined using the median operator.
MSE: 20.6619.
```

We see that the function has automatically selected a model with a single difference and lags for each of the previous 12 observations, which reduced the one step-ahead MSE to 20.6619. Again, being cautious when using the in-sample MSE, the code below shows that the rolling window RMSE for this model with a horizon of 36 has been reduced to 12.015. This is strong evidence that the first difference is a useful filter in modeling these data.

```
rwfit.1.H5 = roll.win.rmse.nn.wge(AirPassengers,horizon = 36, fit.1.H5)
```

```
"Please Hold For a Moment, TSWGE is processing the Rolling Window RMSE with 94
windows."
"The Summary Statistics for the Rolling Window RMSE Are:"
```

```
  Min.    1st Qu.   Median     Mean   3rd Qu.     Max.
 8.355     10.682   11.797   12.015   13.458   15.611
"The Rolling Window RMSE is: 12.015"
```

As mentioned earlier, the **mlp** function will also allow for seasonal indicator variables to be added to the model. If we set **allow.det.season = TRUE**, the **mlp** function will automatically add seasonal indicator variables that correspond to the frequency of the series stored in the frequency attribute of the *ts* object. Recall that the **AirPassengers** *ts* object has a frequency of 12 (displayed below) and thus 11 deterministic seasonal indicator (dummy) variables will be added to the model. Note that **difforder** and **lags** are also not specified and therefore these will be automatically selected. The code to fit this MLP is below and the architecture is illustrated in Figure 11.17.

```
frequency(AirPassengers)
```

```
[1] 12
```

```
set.seed(1)
fitSeasonal = mlp(AirPassengers,allow.det.season = TRUE)
# same as fitSeasonal = mlp(AirPassengers)
```

```
fitSeasonal
```

```
MLP fit with 5 hidden nodes and 20 repetitions.
Series modelled in differences: D1.
Univariate lags: (1,2,3,4,5,6,7,8,10,12)
Deterministic seasonal dummies included.
Forecast combined using the median operator.
MSE: 8.7509.
```

```
plot(fitSeasonal)
```

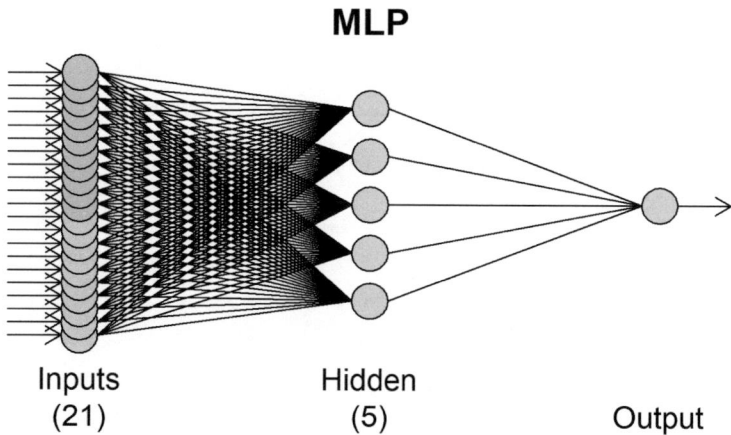

MLP

Inputs
(21)

Hidden
(5)

Output

FIGURE 11.17 Architecture of the MLP to fit the AirPassengers data, with a first difference and seasonal indicator (dummy) variables

Again, we can see that five hidden nodes, a first difference, ten lags (9 and 11 were not included), and seasonal indicators (dummies) were included in the final model. Moreover, with a one step-ahead MSE of 8.7509, we can see that there is evidence that this model is an improvement over the previous two. However, since we are interested in forecasting the next three years, we will evaluate how this model performs with a horizon of 36.

```
rwfitSeasonal = roll.win.rmse.nn.wge(AirPassengers,h = 36, fit_model = fitSeasonal)
"Please Hold For a Moment, TSWGE is processing the Rolling Window RMSE with 1
windows."
"Seasonal MLP models will be evaluated only on the last window."
"The Summary Statistics for the Rolling Window RMSE Are:"
 Min.   1st Qu.   Median    Mean   3rd Qu.    Max.
11.58     11.58    11.58    11.58    11.58    11.58
"The Rolling Window RMSE is: 11.581"
```

Due to the methods the `mlp` function uses to estimate the seasonal indicator variables, note the second line of output from the `roll.win.rmse.nn.wge` call above: "Seasonal MLP models will be evaluated only on the last window". In this case, the last window is the last three years of data, which provides some evidence of improved fit with a *single* window RMSE of 11.581. Although the MLP with the seasonal indicators has the lowest RMSE, it is also based on a single window and this is much more dependent on the window on which it is calculated. Closer inspection of the rolling window code for the D1 model with 12 lags reveals that the median of the 94 rolling window RMSEs generated for this model (11.797) is greater than the single window RMSE generated for the the seasonal model (11.581). Since more than half of the rolling window RMSEs from the D1 with 12 lags model are greater than the RMSE of the last window of the seasonal MLP, we will obtain our final forecasts of the next 36 months with the seasonal indicator MLP model. Table 11.2 summarizes the RMSE results and appear to follow the expected pattern based on the past.

TABLE 11.2   Rolling Window RMSEs for Three Models Fit to the Airline Data

MODEL	rwRMSE	NUMBER OF WINDOWS
"Airline Model" (D1 and D12 Included with 5 Lags)	36.474	86
D1 with 12 Lags	12.051	94
Seasonal Indicators, D1 and 12 Lags	11.581	1

The final step is to forecast the next 36 months of passengers using seasonal indicators with the D1 with 12 Lags model. We will now use *all* the data to make these predictions.

```
set.seed(1)

t = 1:180
fitSeasonal = mlp(AirPassengers,allow.det.season = TRUE)
finalFit = forecast(fitSeasonal, h = 36)
plot(t[1:144], AirPassengers, type = "l", xli = c(0,185), ylim = c(100, 800),
xlab = "Time")
lines(t[145:180], finalFit$mean,lty = 3, lwd = 3)
```

These forecasts are visualized in Figure 11.18 below.

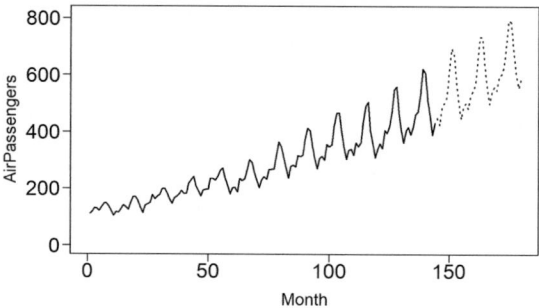

FIGURE 11.18   Final forecasts of the next three years of the AirPassengers series from the MLP with seasonal indicator variables

11.4 THE EXTENDED PERCEPTRON FOR MULTIVARIATE TIME SERIES DATA

11.4.1 Forecasting Melanoma Using Sunspots

11.4.1.1 Architecture

Multivariate time series data are easily accommated by the multilayered perceptron neural network model. Consider the 37 years of melanoma and sunspot data (**melanoma2.0**) studied in Chapter 10. Figure 11.19 is an MLP architecture designed to model melanoma at time t based on melanoma incidences at time t-1 and sunpot activity at time t-1.

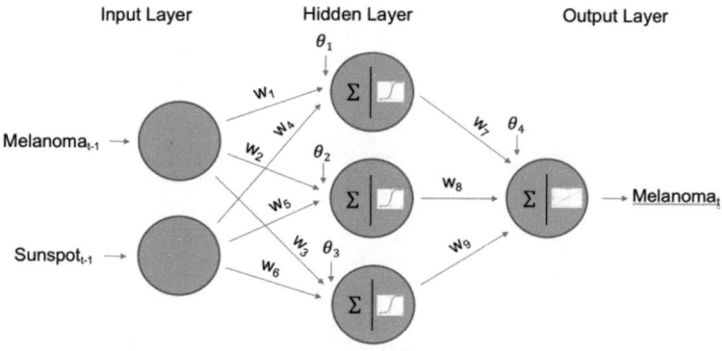

FIGURE 11.19   MLP architecture that models the melanoma count at time t (Melanoma$_t$) given the melanoma and sunspot counts at the previous observation ((Melanoma and Sunspot activity at time t-1), respectively)

11.4.1.2 Fitting the Baseline Model

We will use this model as a baseline and assess it for the purpose of forecasting the next eight years of melanoma counts. Below, a training set is created with the first 29 years of data while the last eight years of data are reserved as a test set for cross-validation. The training data are then used to fit the

MLP architecture in Figure 11.19. Note that the sunspot data are entered into the model through the **mlp** function's **xreg** argument, which requires a data frame.

```
SMtrain = melanoma2.0[1:29,]
SMtest = melanoma2.0[30:37,]

SMtrainDF = data.frame(Sunspot = ts(SMtrain$sunspot))

set.seed(1)

fit.mlp1 = mlp(ts(SMtrain$melanoma), lags = 1,reps = 20,comb = "median",
xreg = SMtrainDF, xreg.lags = 1, allow.det.season = FALSE, hd = 3, difforder = 0,
sel.lag = FALSE)
```

The details of the model can be viewed as usual below.

```
fit.mlp1
```

```
MLP fit with 3 hidden nodes and 20 repetitions.
Univariate lags: (1)
1 regressor included.
- Regressor 1 lags: (1)
Forecast combined using the median operator.
MSE: 0.1492.
```

The output clearly notes that the MLP had two input nodes (melanoma at lag 1 and sunspot at lag 1), three hidden nodes, and resulted in a one step-ahead forecast MSE of .1492. A plot of the architecture (Figure 11.20) is provided as usual and is consistent with the architecture in Figure 11.19.

```
plot(fit.mlp1)
```

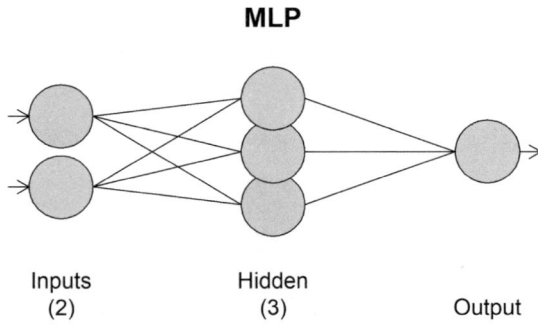

FIGURE 11.20   fit.mlp1 model architecture diagram

11.4.1.3 *Forecasting Future Sunspot Data for Predicting Future Melanoma*

Next, the rolling window RMSE for the MLP is not implemented for the multivariate case, so we will evaluate this model based on the RMSE of the last eight observations (effectively, the last window of a rolling window RMSE). To do so, as is often the case with multivariate time series, it is not realistic that we would know the sunspot number for future years; thus, the sunspot data for the last eight years of the series must be forecast. We will do this with a simple univariate MLP. The code and sunspot forecasts are below, as well as a plot of the forecasts in Figure 11.21.

```
set.seed(1)

fit.mlp.SP = mlp(ts(SMtrainDF$Sunspot),reps = 20, comb = "median")
fore.mlp.SP = forecast(fit.mlp.SP, h = 8)
fore.mlp.SP
```

```
Point             Forecast
30                53.63687
31               161.85283
32               228.67077
33               220.15203
34               204.40068
35               143.16898
36                95.58033
37                62.55229
```

plot(fore.mlp.SP)

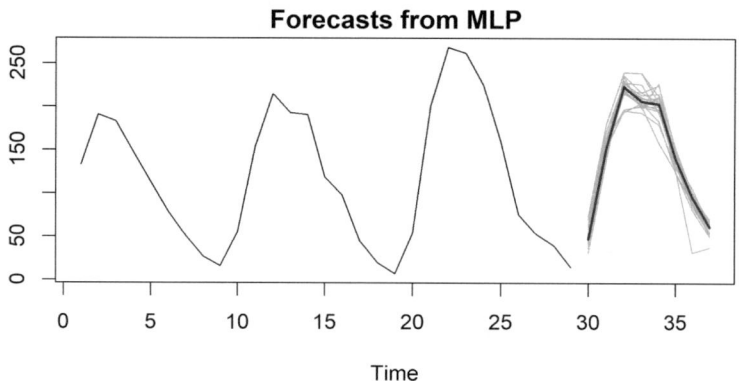

FIGURE 11.21 Plot displaying the 20 forecasts of sunspots for each time period and their median (bold)

It is important to note here when forecasting observations 30–37, it is realistic to assume that we know the actual sunspots for observations 1–29. For this reason, the sunspot forecasts for observations 30–37 must be appended to the assumed known sunspots in the training set and then built into a data frame. Based on the output, it is clear the forecasts start at observation 30, since the known values will be whole numbers and will have decimal values of 0 while the forecasted values from 30–37 (the last column) contain decimals.

SMDF_fore = data.frame(Sunspot = ts(fore.mlp.SP$mean))
SMDF = data.frame(Sunspot = ts(c(SMtrainDF$Sunspot,SMDF_fore$Sunspot)))

```
   Sunspot         10  55.00000      20   54.00000       30   53.63687
1 133.00000        11 154.00000      21 201.00000        31 161.85283
2 191.00000        12 215.00000      22 269.00000        32 228.67077
3 183.00000        13 193.00000      23 262.00000        33 220.15203
4 148.00000        14 191.00000      24 225.00000        34 204.40068
5 113.00000        15 119.00000      25 159.00000        35 143.16898
6  79.00000        16  98.00000      26  76.00000        36  95.58033
7  51.00000        17  45.00000      27  53.00000        37  62.55229
8  27.00000        18  20.00000      28  40.00000
9  16.00000        19   7.00000      29  15.00000
```

11.4.1.4 Forecasting the Last Eight Years of Melanoma

The last eight years of melanoma data can now be forecast. The code and forecasts are below and a plot of the forecasts is in Figure 11.22.

```
fore.mlp = forecast(fit.mlp1, h = 8, xreg = SMDF)
fore.mlp
```

```
Point          Forecast
30             4.058019
31             3.985312
32             3.765170
33             3.857733
34             3.873401
35             3.826858
36             3.645963
37             3.583501
```

```
plot(fore.mlp)
```

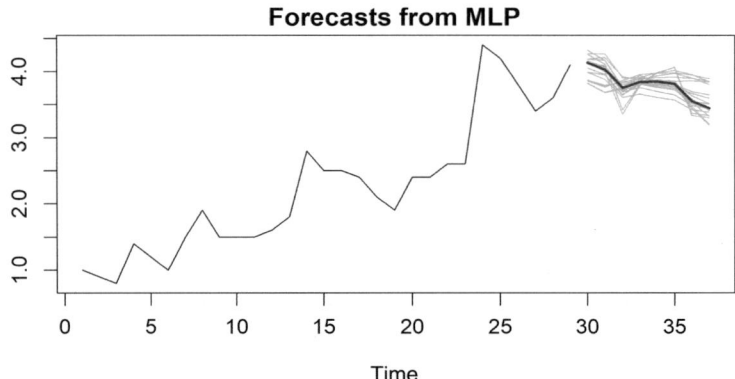

FIGURE 11.22   Melanoma forecasts from fit.mlp1 with sunspot forecasts from fit.mlp.SP

(1) Assessing the Baseline Model on the Last Eight Years of Melanoma
The forecasts seen in Figure 11.21 are not satisfactory because they do not appear to reflect the strong evidence of an upward trend in melanoma counts. Nevertheless, an RMSE can be calculated for the last eight melanoma values of the series, which results in the baseline RMSE of .9955:

```
RMSE = sqrt(mean((SMtest$melanoma - fore.mlp$mean)^2))
RMSE
```

```
0.9955416
```

11.4.1.5 Fitting a Competing Model

Remember that a useful VAR model based on these data in Chapter 10 made use of the melanoma and sunspot data at lag two, so there is hope that this model may have room for improvement. The code below allows the **mlp** function to autoselect the lags and differencing order so as to try and improve the baseline model above.

```
set.seed(1)
```

```
fit.mlp2 = mlp(ts(SMtrain$melanoma), reps = 20,comb = "median", xreg = SMtrainDF,
allow.det.season = FALSE)
```

The resulting model displayed in Figure 11.23 includes a first difference of the melanoma series, which contains melanoma and sunspot data up to lag three and a hidden layer with five nodes. The MSE is not a comparable measure of fit in this case since the data have been differenced, meaning that the resulting MSE (.001) is calculated on the differenced data, rather than the raw data.

fit.mlp2

```
MLP fit with 5 hidden nodes and 20 repetitions.
Series modelled in differences: D1.
Univariate lags: (1,2,3)
1 regressor included.
- Regressor 1 lags: (3,4)
Forecast combined using the median operator.
MSE: 0.001.
```

plot(fit.mlp2)

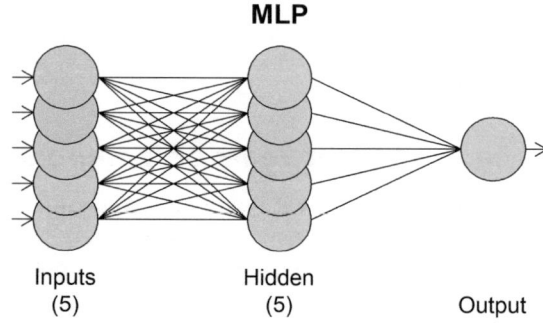

FIGURE 11.23   fit.mlp2 model description and architecture diagram

11.4.1.6 Assessing the Competing Model on the Last Eight Years of Melanoma Data

Forecasts are provided below and visual evidence that this is a more useful model than our baseline is provided in Figure 11.24. The forecasts now appear to be reflecting the trend, and possibly a seasonal component in the melanoma series.

```
fore.mlp = forecast(fit.mlp2, h = 8, xreg = SMDF)
fore.mlp
```

```
30              4.145854
31              4.208989
32              4.191491
33              4.403321
34              5.568593
35              5.317146
36              5.170883
37              4.848404
```

plot(fore.mlp)

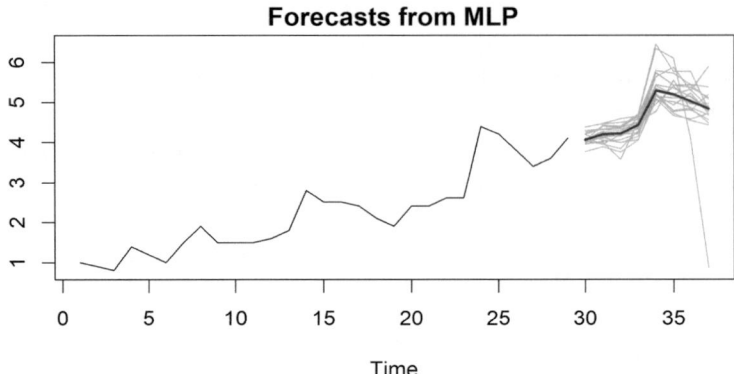

FIGURE 11.24   Melanoma forecasts and a plot summarizing their calculation from model fit.mlp2

Finally, the RMSE of the last eight values is calculated below. The RMSE of .6118 provides evidence that this model is a considerable improvement over the baseline model (RMSE .9955). In fact, this RMSE is also evidence of an improvement over the VAR model fit to these data in Chapter 10 (RMSE = 0.7871)

```
RMSE = sqrt(mean((SMtest$melanoma - fore.mlp$mean)^2))
RMSE
```

```
0.6117655
```

11.4.1.7 Forecasting the Next Eight Years of Melanoma

Now that the final model has been identified, the MLP will need to be refit using *all the data* (**fit.mlp3**) and then melanoma forecasts for the next eight years can be calculated and visualized. The code to perform these steps is shown below, and the output that summarizes the model is in Figure 11.25, while the forecasts are displayed in Figure 11.26.

```
#refit model using all the data
set.seed(1)
fit.mlp3 = mlp(ts(melanoma2.0$melanoma), reps = 20,comb = "median", xreg = data.
frame(Sunspot = melanoma2.0$sunspot), allow.det.season = FALSE)
```

```
fit.mlp3
```

```
MLP fit with 5 hidden nodes and 20 repetitions.
Series modelled in differences: D1.
Univariate lags: (1,2,3)
1 regressor included.
- Regressor 1 lags: (3,4)
Forecast combined using the median operator.
MSE: 0.0046.
```

MLP

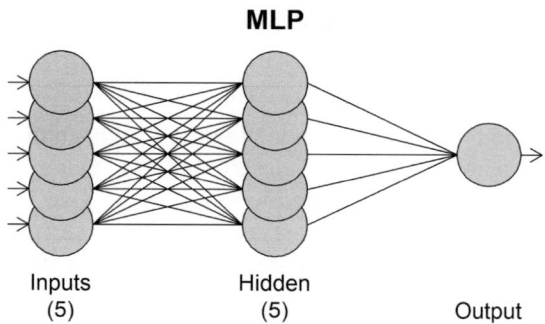

Inputs
(5)

Hidden
(5)

Output

FIGURE 11.25   Architecture of the final MLP fit using the full dataset

```
#forecast the next 8 years of sunspot to use in the forecasts
set.seed(1)
fit.mlp.SP = mlp(ts(melanoma2.0$sunspot),reps = 50, comb = "median")
fore.mlp.SP = forecast(fit.mlp.SP, h = 8)

#put the sunpsots in a data frame so they can be input in mlp()
SMDF_fore = data.frame(Sunspot = ts(fore.mlp.SP$mean))
SMDF = data.frame(Sunspot = ts(c(melanoma2.0$sunspot,SMDF_fore$Sunspot)))

#calculate and visualize the forecasts
fore.mlp = forecast(fit.mlp3, h = 8, xreg = SMDF)
fore.mlp
```

Point	Forecast
38	5.202662
39	5.104457
40	5.527732
41	5.263305
42	5.560360
43	5.701788
44	5.748933
45	5.931066

QR 11.3 Combining
Forecasts with Mean
vs. Median

```
plot(fore.mlp)
```

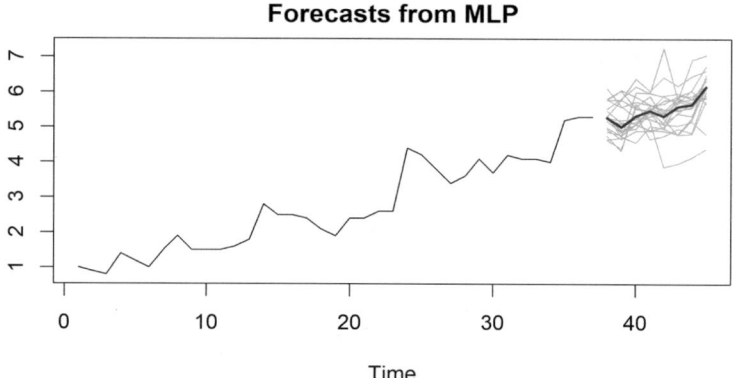

FIGURE 11.26   Forecasts of the next eight years (1980–1987) from the final MLP model fit using the full dataset

11.4.2 Forecasting Cardiac Mortality Using Temperature and Particulates

In this section, we will return to the problem of modeling the weekly cardiac mortality in LA from 1970 to 1979 with temperature and particulates as explanatory variables. Our eventual goal will be to forecast a year in advance; we will compare two candidate models based on their fit (RMSE) to the last year (52 weeks) of the series. Since it is not realistic to know the temperature and particulate counts in the future, these will need to first be forecast for later use in our candidate MLP models aimed at forecasting cardiac mortality. After we have forecast the next year of our explanatory variables (temperature and particulates) we will first fit a model that addresses the periodic behavior in the data through correlatons in the data. Additionally, a competing model will be fit that explicitly models this seasonal behavior using determin-istic seasonal indicator (dummy) variables. These models will be compared using the RMSE of the fit of the cardiac mortality of the *last* 52 weeks of the series; then, the final model will be used to forecast the cardiac mortality for the *next* 52 weeks.

11.4.2.1 General Architecture

Both models considered in this section will contain temperature and particulates as explanatory variables, while only the second model will use seasonal indicator variables to model the mean of each week explicitly. Figure 11.27 displays the general architecture employed in both models. Note that there are potentially many input nodes associated with the lags of the two explanatory variables, tem-perature and particulates. Note, for simplicity, there is only one arrow drawn between each explana-tory variable's input nodes and the hidden nodes. These arrows represent the fact that there is really a unique arrow connecting each node to each hidden layer to form a "fully connected neural network". Also note that the asterisk next to the "Seasonal Indicators" group of nodes indicates that they are pre-sent only in the second model which addresses the periodic behavior with 51 indicator variables (51 additional input nodes.).

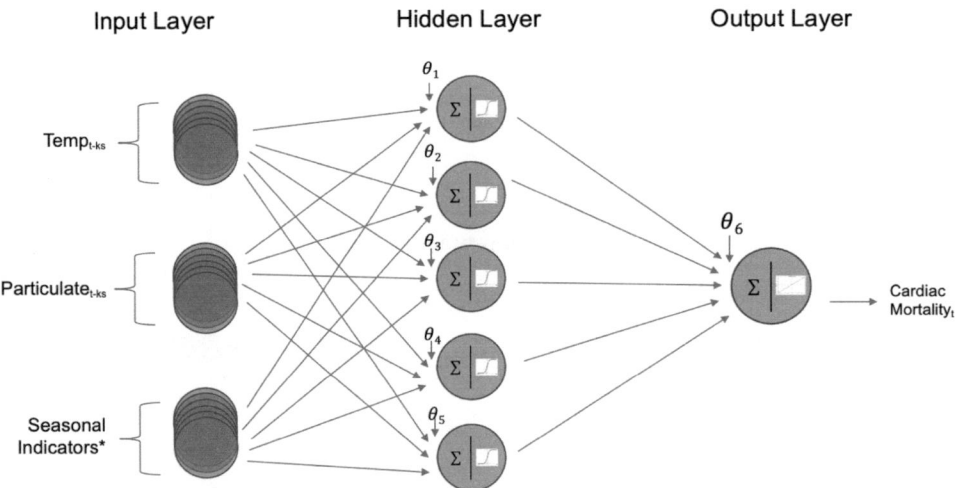

FIGURE 11.27   General archictecure of the MLPs used to model cardiac mortality.
* The seasonal indicator variables will only be present in the second model.

11.4.2.2 Train / Test Split

As seen in previous chapters, a useful first step in forecasting is to divide the dataset into a training and test set. Since the RMSE will be calculated on the last 52 weeks of the dataset, as was done in Chapter 10, the training set will consist of the data between 1970 and the first 40 weeks of 1978 and the test set will consist of the last 52 weeks of the dataset (the last 12 weeks of 1978 and the first 40 weeks of 1979).

```
cardiacTrain = window(cardiac, start = c(1970,1), end = c(1978,40))
cardiacTest = window(cardiac, start = c(1978,41), end = c(1979,40))
```

11.4.2.3 Forecasting Covariates: Temperature and Particulates

Unlike the VAR models we fit in Chapter 10 which forecast the explanatory variables simultaneously with the response, MLP models require the explanatory variables to be forecast in advance if the future values are not known ahead of time. Temperature and particulates are not known in advance, so we will use univariate MLPs to forecast these values for the weeks in the year 1979; these forecasts will be used in the calculation of the RMSE. Later, we will forecast the temperature and particulate values for the weeks of 1980 to calculate our final forecasts.

(1) Temperature
It is intuitive that temperature would have seasonal behavior; the seasonal indicator variables which we have previously fit are available in the MLP model. In this case, 51 (the frequency of the *ts* object minus one) indicator variables corresponding to the week of the year are automatically added if **allow.det. season** is set to **TRUE**, **det.type** is set to **"bin"**, and if their inclusion reduces the overall one step-ahead MSE.[11] The code to fit the MLP is below and the output (11.2), architecture (Figure 11.28(a)), and a plot of the forecasts (Figure 11.28(b)) follow. The output suggests that sufficient fit of the temperature series was achieved using only 16 of the lags from 1 to 52; note that the 51 seasonal indicator variables were left out.

```
# Temperature
set.seed(1)
fit.mlp.temp = mlp(ts(cardiacTrain[,"tempr"],frequency = 52),reps = 50, difforder
= 0, comb = "median", allow.det.season = TRUE, det.type = "bin")
fit.mlp.temp
plot(fit.mlp.temp)
fore.mlp.temp = forecast(fit.mlp.temp, h = 52)
```

```
MLP fit with 5 hidden nodes and 50 repetitions.
Univariate lags: (1,2,4,8,9,12,13,22,28,35,41,42,47,49,50,52)          (11.2)
Forecast combined using the median operator.
MSE: 10.342.
```

```
plot(fore.mlp.temp)
```

---

11 In addition to the binary indicator variables we used in Chapter 10 (**det.type** = **"bin"**), the mlp function allows for a sin/ cos based method as well (**det.type** = **"trg"**). See page 12 and 13 of https://kourentzes.com/forecasting/wp-content/uploads/ 2016/07/Barrow-Kourentzes-2016-impact-special-days.pdf for more information.

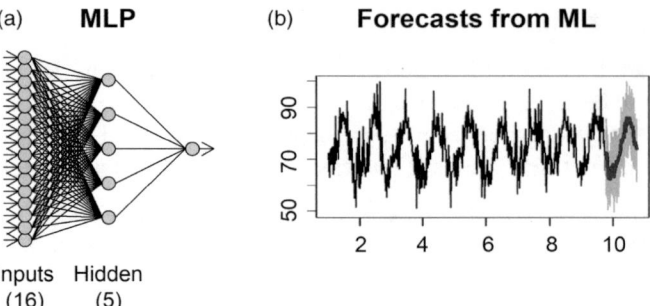

FIGURE 11.28 (a) Architecture of the MLP fit for the temperature series (b)Plot of the forecasts of the last 52 weeks of the temperature data

(2) Particulates
A univariate MLP was used to model the particulates as well. The code to fit this model is below and the output (11.3), architecture (Figure 11.29(a)), and a plot of the forecasts (Figure 11.29(b)) follow. Similar to forecasting the temperatures, a model with seasonal indicators was not deemed to be as useful as a model with only 14 lags (11.3).

```
# Particulates
set.seed(1)
fit.mlp.part = mlp(ts(cardiacTrain[,"part"],frequency = 52),reps = 50,
difforder = 0, comb = "median", allow.det.season = TRUE, det.type = "bin")
```

```
MLP fit with 5 hidden nodes and 50 repetitions.
Univariate lags: (2,3,4,6,12,14,17,28,29,39,43,50,51,52)                    (11.3)
Forecast combined using the median operator.
MSE: 29.8761.
```

```
plot(fit.mlp.part)
```

```
fore.mlp.part = forecast(fit.mlp.part, h = 52)
plot(fore.mlp.part)
```

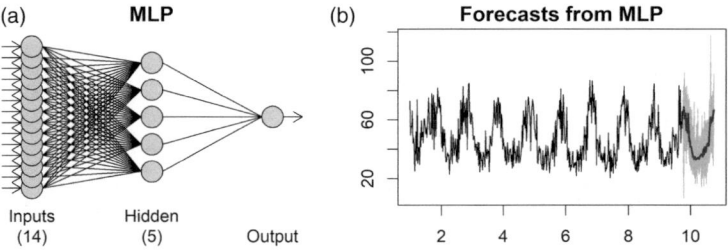

FIGURE 11.29 (a) Architecture of the univariate MLP fit for the particulate series (b) Plot of the forecasts of the last 52 weeks of the particulates series (last 12 weeks of 1978 and first 40 weeks of 1979)

11.4.2.4 Model Without Seasonal Indicator Variables

Now that the temperatures and particulates have been forecast for the last year of the series, models can now be fit to forecast cardiac mortality. The first candidate model considered does not have the option of adding seasonal indicators to the MLP. The code and output below show that a model with a first difference, 60 lagged variables between temperature, particulates and cardicac mortality, and 5 hidden layers was selected. Figure 11.30(a) and (b) show the selected architecture of the model and a plot of the final forecasts, respectively.

```
cardiacDF_xreg = data.frame(temp = ts(cardiacTrain[,"tempr"]),
part = ts(c(cardiacTrain[,"part"])))

set.seed(1)
fit.mlp.cmort = mlp(ts(cardiacTrain[,"cmort"], frequency = 52),reps = 50,
comb = "median",xreg = cardiacDF_xreg, allow.det.season = FALSE)
fit.mlp.cmort
```

```
MLP fit with 5 hidden nodes and 50 repetitions.
Series modelled in differences: D1.
Univariate lags: (1,2,3,4,5,6,7,8,9,10,11,12,13,14,15,16,17,18,19,20,21,22,23,
24,25,26,27,28,29,30,31,32)
2 regressors included.
- Regressor 1 lags: (1,4,14,16,17,18,20,23,27,30,38,44,50,52)
- Regressor 2 lags: (1,7,9,11,16,20,21,26,33,34,44,48,49,50)
Forecast combined using the median operator.
MSE: 0.0621.
```

```
plot(fit.mlp.cmort)
```

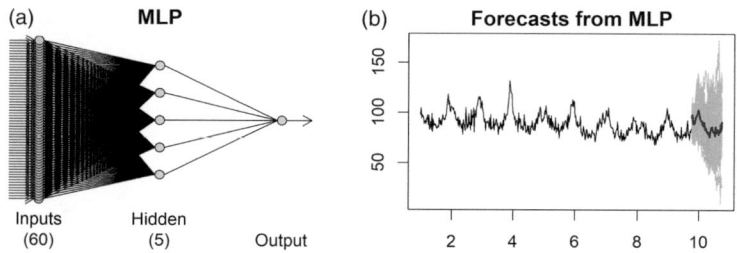

FIGURE 11.30 (a) The architecture of the MLP for cardiac mortality fit.mlp.cmort. (b) the forecasts for the next 52 weeks with the median in bold

Recall that the model above was fit using the data in **cardiacTrain**, which contains only the data from 1970 to the first 40 weeks of 1978. The next step is to evaluate this model by assessing its performance on the hold-out set (the last 52 weeks of the dataset). A realistic scenario would be that we know the actual temperatures and particulates through week 40 of 1978 but would need to use forecast values to predict cardiac mortalities in the remaining 52 weeks. The code below creates a data frame that reflects this structure; the forecast temperatures forecasted earlier (**fore.mlp.temp$mean**) are appended to the known temperatures (**cardiacTrain[,"tempr"]**) in a column called "**temp**", and the same is done for the particulates in a column called "**part**".

```
CMDF_fore = data.frame(temp = ts(c(cardiacTrain[,"tempr"],fore.mlp.temp$mean)),
part = ts(c(cardiacTrain[,"part"],fore.mlp.part$mean)))
```

The final step is to forecast cardiac mortalities for the last 52 weeks of the series (displayed in Figure 11.30(b)) and assess the model by calculating the RMSE of the forecasts using the code below.

```
fore.mlp.cmort = forecast(fit.mlp.cmort, h = 52, xreg = CMDF_fore)
plot(fore.mlp.cmort)
RMSE = sqrt(mean((cardiacTest[,"cmort"][1:52] - fore.mlp.cmort$mean)^2))
RMSE
```

```
7.00397
```

The RMSE for the last 52 weeks of the series was found to be 7.004, which will be compared to the RMSE of a model that allows for seasonal indicator variables; this model will be fit and assessed next.

11.4.2.5 Model With Seasonal Indicator Variables

The code below fits a model that allows for seasonal indicator variables (**allow.det.season = TRUE, det.type = "bin"**). The code and output below (11.4) and the diagram in Figure 11.31(a) indicate that a model with a first difference, 31 total lags for cardiac mortality, 15 total lags for temperature, 15 total lags for particulates, and 51 seasonal indicator variables was selected based on the data in the training set (**cardiacTrain**).

```
cardiacDF_xreg = data.frame(temp = ts(cardiacTrain[,"tempr"]), part =
ts(c(cardiacTrain[,"part"])))

set.seed(1)
fit.mlp.cmortS = mlp(ts(cardiacTrain[,"cmort"],frequency = 52),reps = 50,comb
= "median",xreg = cardiacDF_xreg, allow.det.season = TRUE, det.type = "bin")
fit.mlp.cmortS
```

```
MLP fit with 5 hidden nodes and 50 repetitions.
Series modelled in differences: D1.
Univariate lags: (1,2,3,4,5,6,7,8,9,10,11,12,13,14,15,16,17,18,19,20,21,23,24,
26,31,33,46,47,48,49,51)                                                          (11.4)
2 regressors included.
- Regressor 1 lags: (1,4,5,14,16,17,20,23,27,30,31,38,44,46,50)
- Regressor 2 lags: (2,7,16,18,20,21,26,29,31,32,33,34,44,49,50)
Deterministic seasonal dummies included.
Forecast combined using the median operator.
MSE: 1e-04.
```

```
plot(fit.mlp.cmortS)
```

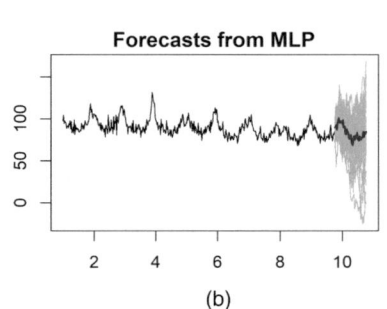

MLP   **Forecasts from MLP**

Inputs   Hidden
(112)   (5)   Output

(a)   (b)

FIGURE 11.31   (a) The architecture of the MLP for cardiac mortality fit.mlp.cmortS (b) the forecasts for the next 52 weeks with the median in bold

As before, a realistic scenario when forecasting the last 52 weeks of cardiac mortality is that the temperatures and particulates for 1970 through the first 40 weeks of 1978 are known, but that these values would need to be forecast for the last 52 weeks (the test set). For this reason, the *forecast* temperatures and particulates for the last 52 weeks of the dataset are appended to the *known* temperatures and particulates for the training set, and then stored in a data frame.

```
CMDF_fore = data.frame(temp =
ts(c(cardiacTrain[,"tempr"],fore.mlp.temp$mean)), part =
ts(c(cardiacTrain[,"part"],fore.mlp.part$mean)))
```

These explanatory variables (**CMDF_fore**) are then passed with the model (**fit.mlp.cmortS**) to the **forecast** function which forecasts the cardiac mortalities for the last 52 weeks. Finally, the model is assessed by calculating the RMSE of the 52 forecasts, which was found to to be 6.4405. The code for the visualization of the forecasts and calculation of the RMSE for the forecasts of the last 52 weeks is below. Figure 11.31(b) is a plot of these forecasts.

```
fore.mlp.cmortS = forecast(fit.mlp.cmortS, h = 52, xreg = CMDF_fore)
plot(fore.mlp.cmortS)
RMSE = sqrt(mean((cardiacTest[,"cmort"][1:52] - fore.mlp.cmortS$mean)^2))
RMSE
```

```
6.440517
```

11.5 AN "ENSEMBLE" MODEL

An "ensemble" model is one that is a combination of two or more individual models. For instance, taking the median forecast for each time point from the 50 repetitions of the MLP (technically, 50 different MLPs) is an example of an ensemble model. Forecasts from both univariate and multivariate models can be "ensembled". In this section, an ensemble of a VAR and a multivariate MLP model is considered.

Figure 11.32(a) displays the 52 forecasts for the VAR with $p = 2$ (including trend and seasonality) that was studied in Chapter 10 and had an RMSE of 6.38. It will be important to note that there is evidence that this model consistently underestimates the cardiac mortality for a large portion of the test set, especially from weeks 466 and later. On the other hand, Figure 11.32(b) displays the forecasts for the previously fit MLP with seasonality (solid)—note that, with the exception of weeks 484–486, these forecasts tend to be greater than the VAR forecasts (dashed). While the MLP model had an RMSE of 6.44, it stands to reason that averaging the forecasts from these two models may yield forecasts that outperform both the individual RMSEs.

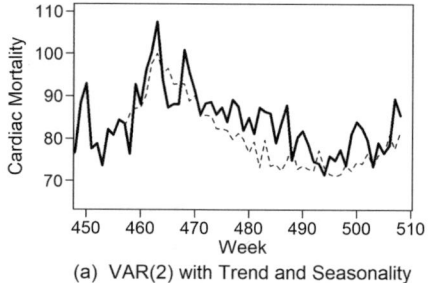

(a) VAR(2) with Trend and Seasonality

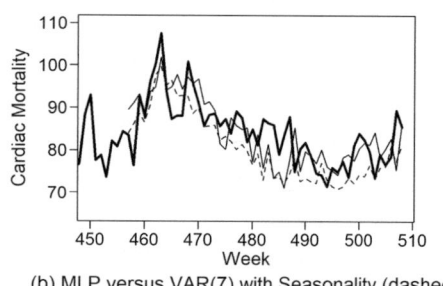

(b) MLP versus VAR(7) with Seasonality (dashed)

FIGURE 11.32   (a) Forecasts from VAR(2) with trend and seasonality and (b) forecasts from MLP (solid) versus VAR(7) with trend and seasonality (dashed)

The VAR model with $p = 2$ (including trend and seasonality) is fit again below for completeness, and the forecasts are stored in the object **preds2S**.

```
CMortVAR2S = VAR(cardiacTrain, season = 52, type = "both", p = 2)
preds2S=predict(CMortVAR2S,n.ahead=52)
```

Next, recall that the forecasts from the MLP with seasonality were stored in the object **fore.mlp.cmortS$mean**. The code below shows the calculation of a very simple ensemble in which the forecasts from these two models are averaged.

```
ensemble = (preds2S$fcst$cmort[,1] + fore.mlp.cmortS$mean)/2
```

Figure 11.33 displays the forecasts produced by the code below and yields an RMSE of 6.059, which is *lower* than the RMSE from either of the individual VAR and MLP forecasts.

```
#Plot
plot(seq(1,508,1), cardiac[,"cmort"], type = "l",xlim = c(450,508), xlab =
"Time", ylab = "Cardiac Mortality", main = "52 Week Cardiac Mortality Forecast
From A VAR/MLP Ensemble")
lines(seq(457,508,1), ensemble, type = "l", lwd = 4, col = "green")

RMSEENSEMBLE = sqrt(mean((cardiacTest[,"cmort"] - as.numeric(ensemble))^2))
RMSEENSEMBLE
```

6.058764

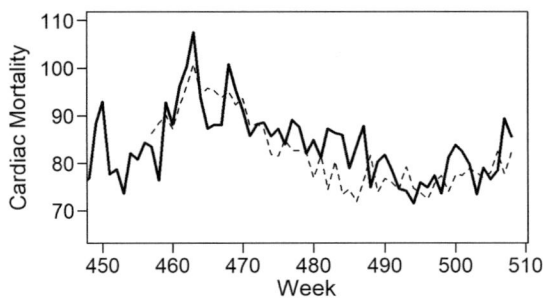

FIGURE 11.33   Forecasts (dashed) from the ensemble model (mean of VAR and MLP models) vs. actual data (solid)

11.5.1 Final Forecasts for the Next Fifty-Two Weeks

Including our analysis in Chapter 10, we have fit several VAR and MLP models to the cardiac mortality data with the goal of making forecasts through the next year. Table 11.2 below summarizes the results with respect to the RMSE of the forecasts on the hold-out set (**cardiacTest**).

TABLE 11.3   Table of VAR, MLP, and Ensemble Performance with a 52-week Horizon.

MODEL	RMSE
VAR(7) with Trend	5.44
VAR(2) with Trend and Seasonality	6.38
MLP with seasonality	6.44
Ensemble: VAR(2) with Trend/Seasonality and MLP with Seasonality	6.06

The VAR(7) with trend from Chapter 10 achieved the lowest RMSE (5.44) on the 52-week hold-out set, outperforming even the ensemble model (RMSE = 6.06) by a notable margin. For this reason, the VAR(7) with trend was selected as the model to provide the forecasts of the next 52 weeks. For convenience and reference, the plot with these forecasts and 95% prediction intervals can be seen in Figure 11.34.

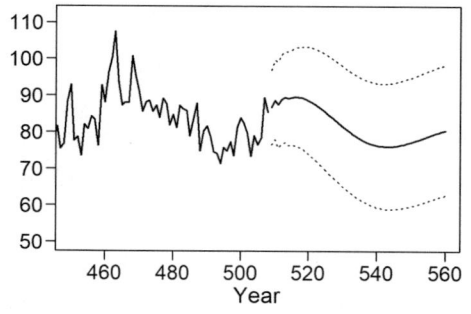

52 Week Cardiac Mortality Forecast using a VAR(7)

FIGURE 11.34   Forecasts (in bold) and 95% prediction intervals (dotted) for the weekly cardiac mortality rates of LA county for the last 12 weeks of 1979 and first 40 weeks of 1980 from the VAR model with p= 7, including trend

11.5.2 Final Forecasts for the Next Three Years (Longer Term Forecasts)

Remember, since there are no seasonality terms, the forecasts from the VAR(7) with trend model will damp to the trend line, and will thus not forecast the strong seasonal behavior evident in the cardiac mortality in the long term. This is illustrated in Figure 11.35, in which the VAR(7) with trend has been used to forecast the next three years (156 weeks) of cardiac mortalities. The RMSE for these 156 forecasts was found to be 6.66.

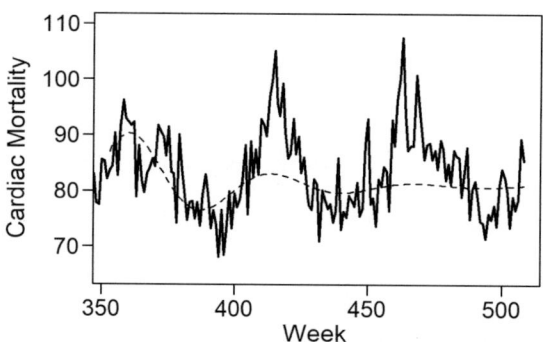

FIGURE 11.35   Three years of forecasts from the VAR(7) with trend. Note the slow, but eventual convergence to the mean / trend line

The VAR(2) with trend and seasonality (Figure 11.36(a)) and the MLP with seasonality (Figure 11.36(b)) were also assessed with respect to the 156-week forecasts, with RMSEs 6.39 and 5.88, respectively. Note that both of these models are not only more visually "satisfying" than the VAR(7) with trend, but have lower RMSEs as well.

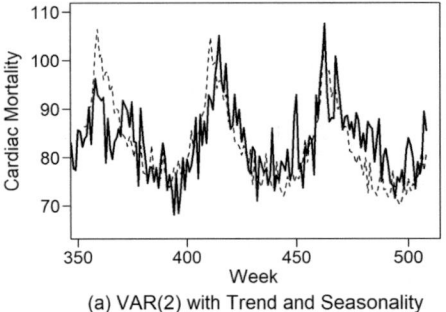

(a) VAR(2) with Trend and Seasonality

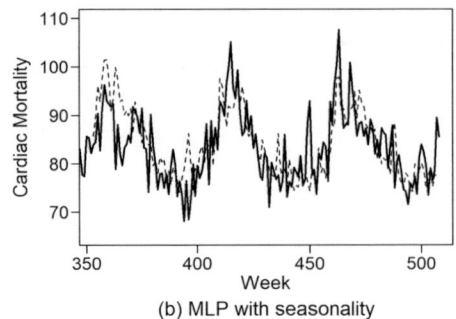

(b) MLP with seasonality

FIGURE 11.36   (a) Three-year forecast using a VAR(2) with trend and seasonality, (b) Three-year forecasts using a MLP model with seasonality

It is visually clear that the MLP forecasts provide closer overall fit to the actual values. We next use the VAR forecasts to construct an ensemble model (Figure 11.37). Averaging the forecasts from these seasonal models results in the ensemble model achieving the lowest RMSE of all the models tested: 5.65. Table 11.4 summarizes these findings. The code for creating Figures 11.36 and 11.37 is included in Appendix 11B.

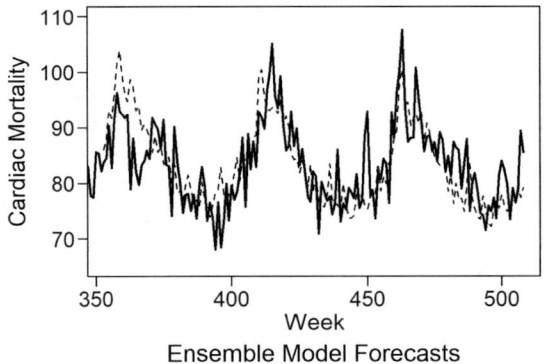

Ensemble Model Forecasts

FIGURE 11.37   Three years of cardiac mortality forecasts from the ensemble model (mean of the VAR and MLP forecasts)

TABLE 11.4   Table of VAR, MLP and Ensemble Performance on a 3-year (156-week) Horizon.

MODEL	RMSE
VAR(7) with Trend	6.66
VAR(2) with Trend and Seasonality	6.39
MLP with Seasonality	5.88
Ensemble: VAR(2) with Trend/Seasonality and MLP with Seasonality	5.65

11.6 CONCLUDING REMARKS

In this chapter, multilayered perceptron (MLP) models were studied with respect to their forecasting of time series data. It has been shown in this chapter, as well as in other areas of research, that MLPs have the *potential* to obtain "better" forecasts with respect to other modeling technologies. This potential stems from the MLP's ability to include hidden layers and various activation functions that allow for the modeling of non-linear relationships that may be missed by other methods. This is certainly a noteworthy potential advantage of MLP models. This advantage, on the other hand, often comes at the cost of significantly longer training times, lengthy hyperparameter tuning, lack of dependable confidence intervals, and black box models that lack interpretability. Many time series applications can tolerate these costs and, in the coming years, increases in computational power as well as new research developments and innovations will almost certainly mitigate some of these concerns.

It is important to underline that currently, the advantage of the MLP model is the *potential* to obtain "better" forecasts. We have witnessed an example of an MLP model outperforming the VAR in forecasting melanoma occurrence. Alternatively, a VAR model was shown to outperform the MLP in the 52-week forecast of cardiac mortality data. Interestingly, an ensemble of the MLP and the VAR was shown to outperform the individual MLP *and* VAR models for 3-year forecasts of cardiac mortality! The take-away is that the time series analysts should make room in their toolbox for MLP, VAR, MLR with correlated errors, ARIMA, ARMA, Holt-Winters, and any or all of the many additional models that are available and will become available in the future.

APPENDIX 11A

TSWGE FUNCTION

```
roll.win.rmse.nn.wge(series,horizon = 1,fit_model)
```

creates as many "windows" as is possible with the data and calculates an RMSE for each window. The resulting "rolling window RMSE" is the average of the individual RMSEs from each window.

> **series** =original realization
> **horizon** = the number of observations to be forecast beyond the forecast origin.
> **fit_model** = the MLP object (model) to be evaluated. This model will have been fit before the call to this function.

CRAN PACKAGE *NNFOR*

The above call statement indicates the parameters available in function **mlp**. See the **mlp** help for more information by typing **help(mlp)** in R.

```
mlp(y, m = frequency(y), hd = NULL, reps = 20, comb = c("median",
  "mean", "mode"), lags = NULL, keep = NULL, difforder = NULL,
```

```
outplot = c(FALSE, TRUE), sel.lag = c(TRUE, FALSE),
allow.det.season = c(TRUE, FALSE), det.type = c("auto", "bin",
"trg"), xreg = NULL, xreg.lags = NULL, xreg.keep = NULL,
hd.auto.type = c("set", "valid", "cv", "elm"), hd.max = NULL,
model = NULL, retrain = c(FALSE, TRUE))
```

APPENDIX 11B

The following is code used to create Figures 11.36 and 11.37.

```
# VAR / MLP / Ensemble Long Term Forecast: 3 years = 156 Weeks
#
# Code for Figure 11.36 1
#
library(astsa)
data(lap) #original cardiac mortality data from schumway
cardiac = window(lap[,c("cmort","tempr","part")],start = c(1970,1))
cardiacTrain = cardiac[1:352,]
cardiacTest = cardiac[353:508,]
seed = 1
#forecast temp and particles
set.seed(seed)
#temp
fit.mlp.temp=mlp(ts(cardiacTrain[,"tempr"],frequency=52),reps=50,difforder=0,
comb = "median", det.type = "bin")
plot(fit.mlp.temp)
fore.mlp.temp = forecast(fit.mlp.temp, h = 156)
plot(fore.mlp.temp)
plot(ts(cardiacTrain[,"tempr"],frequency = 52))
plot(fore.mlp.temp$mean)
set.seed(seed)
#particles
fit.mlp.part=mlp(ts(cardiacTrain[,"part"],frequency=52),reps = 50,difforder=0,
comb = "median", det.type= "bin")
plot(fit.mlp.part)
fore.mlp.part = forecast(fit.mlp.part, h = 156)
plot(fore.mlp.part)
#package them up in data frame. Use the ones you know AND the ones you have to
forecast
CMDF_fore = data.frame(Week = ts(seq(1,508,1)),temp = ts(c(cardiacTrain[,"t
empr"],fore.mlp.temp$mean)),   part   =   ts(c(cardiacTrain[,"part"],fore.mlp.
part$mean)))
CMDF_foreNP = data.frame(Week = ts(seq(1,508,1)), temp = ts(c(cardiacTrain[,"t
empr"],fore.mlp.temp$mean)))
CMDF_foreNT    =    data.frame(Week    =    ts(seq(1,508,1)),    part    =
ts(c(cardiacTrain[,"part"],fore.mlp.part$mean)))
CMDF_foreNW   =   data.frame(temp   =   ts(c(cardiacTrain[,"tempr"],fore.mlp.
temp$mean),frequency   =   52),   part   =   ts(c(cardiacTrain[,"part"],fore.mlp.
part$mean),frequency = 52))
CMDF_fore = data.frame(temp = ts(c(cardiacTrain[,"tempr"],fore.mlp.temp$mean)),
part = ts(c(cardiacTrain[,"part"],fore.mlp.part$mean)),Week = ts(seq(1,508,1)))
#CMDF_fore
```

```
set.seed(seed)
#forecast cmort using mlp with forecasted xreg (don't need to forecast week.)
cardiacDF_xreg     =     data.frame(Week     =     ts(seq(1,352,1)),temp     =
ts(cardiacTrain[,"tempr"]), part = ts(c(cardiacTrain[,"part"])))
cardiacDF_xregNP = data.frame(Week = ts(seq(1,352,1)), temp = ts(cardiacTrain
[,"tempr"]))
cardiacDF_xregNT = data.frame(Week = ts(seq(1,352,1)), part = ts(c(cardiacTra
in[,"part"])))
cardiacDF_xregNW = data.frame(temp = ts(cardiacTrain[,"tempr"], frequency =
52), part = ts(c(cardiacTrain[,"part"]),frequency = 52))
cardiacDF_xreg = data.frame(temp = ts(cardiacTrain[,"tempr"]), part = ts(c(car
diacTrain[,"part"])), Week = ts(seq(1,352,1)))
fit.mlp.cmort  =  mlp(ts(cardiacTrain[,"cmort"], frequency = 52),reps = 50,
difforder = 0, comb = "median",xreg = cardiacDF_xregNW, allow.det.season =
TRUE, det.type = "bin")
fit.mlp.cmort
plot(fit.mlp.cmort)
fore.mlp.cmort = forecast(fit.mlp.cmort, h = 156, xreg = CMDF_foreNW)
plot(fore.mlp.cmort)
RMSE = sqrt(mean((cardiacTest[,"cmort"] - fore.mlp.cmort$mean)^2))
RMSE
#Plot
t=1:508
plot(t[348:508]seq(1,508,1), cardiac[,"cmort"], type = "l",xlim = c(0,510),
ylim = c(0,130), xlab = "Time", ylab = "Cardiac Mortality", main = "52 Week
Cardiac Mortality Forecast")
lines(seq(353,508,1), fore.mlp.cmort$mean, type = "l", col = "red")
#
# Code for Figure 11.37
#
#Ensemble
#VAR p = 7 non seasonal
CMortVAR7 = VAR(cardiacTrain, type = "both", p = 7) #p = 2 from SBC
preds7=predict(CMortVAR7,n.ahead=156)
RMSEVAR7 = sqrt(mean((cardiacTest[,"cmort"] - preds7$fcst$cmort[,1])^2))
RMSEVAR7
#VAR p = 2 seasonal
CMortVAR2S = VAR(cardiacTrain, season = 52, type = "both", p = 2) #p = 2 from
SBC
preds2S=predict(CMortVAR2S,n.ahead=156)
ensemble = (preds2S$fcst$cmort[,1] + fore.mlp.cmort$mean)/2
#Plot
plot(seq(1,508,1), cardiac[,"cmort"], type = "l",xlim = c(350,508), ylim =
c(70,110), xlab = "Time", ylab = "Cardiac Mortality", main = "52 Week Cardiac
Mortality Forecast From A VAR/MLP Ensemble")
lines(seq(353,508,1), ensemble, type = "l", lwd = 4, col = "green")
lines(seq(353,508,1),preds2S$fcst$cmort[,1] , type = "l", lwd = 2, lty = 2, col
= "red")
lines(seq(353,508,1),fore.mlp.cmort$mean , type = "l", lwd = 2, lty = 4, col =
"blue")
lines(seq(353,508,1),preds7$fcst$cmort[,1] , type = "l", lwd = 2, lty = 2, col
= "purple")
RMSEVAR7 = sqrt(mean((cardiacTest[,"cmort"] - preds7$fcst$cmort[,1])^2))
RMSEVAR7
RMSEVAR2S = sqrt(mean((cardiacTest[,"cmort"] - preds2S$fcst$cmort[,1])^2))
```

```
RMSEVAR2S
RMSEMLP = sqrt(mean((cardiacTest[,"cmort"] - fore.mlp.cmort$mean)^2))
RMSEMLP
RMSEENSEMBLE = sqrt(mean((cardiacTest[,"cmort"] - ensemble)^2))
RMSEENSEMBLE
```

CHAPTER 11 PROBLEMS

11.1   Compare the "best" ARMA or ARIMA model with the best MLP model for forecasting the next 12 observations of the global temperature data (**global.temp**).

11.2   Compare the "best" ARMA or ARIMA model with the best MLP model for forecasting the next 12 observations from the Ozona data (**ozona**).

11.3   Pick a stock and download the last three years of closing prices for this stock. Use this dataset to compare the "best" ARMA or ARIMA model with the best MLP model for forecasting the next day's stock price. "Best" here will simply mean whether your model correctly predicted the price to go up or down with respect to the previous day's closing price. Create a clear report or presentation that explains how you approached this problem and obtained your results.

11.4   Fit an MLP with 1 input node, no hidden layers and 1 output node to the **wtcrude2020** data and forecast a horizon of 300 months. Match this result to that of this chapter. Next select two forecast origins below the mean and two that are above the mean and calculate and plot the forecasts for a horizon of 300 months. What do you notice?

11.5   Do your best to model the cardiac mortality (**cmort**) with a univariate MLP model. Compare this model with the multivariate model we fit in this chapter.

11.6   Note that the variable "time" was never input into the melanoma / sunspot models. This is because the trend behavior is addressed with the first difference of the response (**melanoma**). Fit an MLP model that does not take the first difference and explicitly addresses the possible trend in the melanoma count by adding a "year" variable to the MLP as an explanatory variable (in addition to the **sunspot2.0** explanatory variable). Note that you will need to add this variable to the data frame (**melanoma2.0**). How do the results compare with the first difference approach performed in the text?

11.7   Note that the variable "time" was never input into the cardiac mortality models. This is because the trend behavior is addressed using the first difference of the response. Fit an MLP model that does not take the first difference and explicitly addresses the possible trend in the cardiac mortality by adding a "week" variable to the MLP as an explanatory variable. How do the results compare with the first difference approach performed in the text? Note, you will need to add this variable to the **cardiac** data frame.

11.8   With respect to the cardiac mortality study, repeat the short term (52-week) ensemble calculation and analysis using a different seed. Compare your results with the results from the text (**set.seed(1)**). Repeat this process several times using different seeds.

11.9   With respect to the cardiac mortality study, repeat the longer term (156-week) ensemble calculation and analysis using a different seed. Compare your results with the results from the text (**set.seed(1)**). Also, repeat this process several times using different seeds.

11.10 Assess the fit of the MLP on the **AirPassengers** data with and without the seasonal dummy variables. Make the initial assessment based on visualizations of the forecasts for the last three years versus the actual values. Also, compare the results using the RMSEs for forecasts of the last 36 months of the series.

Mini Research Project

There is some research that suggests that the *tanh* (hyperbolic tangent) activation function is better in some sense that the *sigmoid* activation function.

https://towardsdatascience.com/why-data-should-be-normalized-before-training-a-neural-network-c626b7f66c7d

Find out how to change the hidden layer activation functions from the default (sigmoid) to the *tanh* function (this may take some significant research into the `mlp` function and the function it calls). Then think of a way to evaluate whether it is truly "better". Create a report or presentation that explains your findings!

References

Bartlett, M.S. 1946. On the theoretical specification and sampling properties of auto- correlated time series, *Journal of the Royal Statistical Society* B8, 27–41.

Beach, C.M. and MacKinnon, J.G. 1978. A maximum likelihood procedure for regression with autocorrelated errors, *Econometrica* 46, 51–58.

Bhansali, R.J. 1979. A mixed spectrum analysis of the lynx data, *Journal of the Royal Statistical Society* A142, 199–209.

Bollerslev, T. 1986. Generalized autoregressive conditional heteroskedasticity, *Journal of Econometrics* 31, 307–327.

Box, G.E.P. and Jenkins, G.M. 1970. *Time Series Analysis: Forecasting and Control*, Wiley: Hoboken, NJ.

Box, G.E.P., Jenkins, G.M., and Reinsel, G.C. 2008. *Time Series Analysis: Forecasting and Control*. Wiley: Hoboken, NJ.

Box, G.E.P. and Pierce, D.A. 1970. Distribution of the autocorrelations in autoregressive moving average time series models, *Journal of American Statistical Association* 65, 1509–1526.

Brockwell, P.J. and Davis, R.A. 1991. *Time Series: Theory and Methods*, 2nd edn. Springer-Verlag: New York.

Burg, J.P. 1975. Maximum entropy spectral analysis, PhD dissertation, Department of Geophysics, Stanford University: Stanford, CA.

Butterworth, S. 1930. On the theory of filter amplifiers, *Experimental Wireless and the Wireless Engineer* 7, 536–541.

Campbell, M.J. and Walker, A.M. 1977. A survey of statistical work in the Makenzic river series. If annual Canadian lynx trapping for years 1821–1934 and a new analysis. *Journal of the Royal Statistical Society* A140, 411–431.

Clette, F., Cliver, E.W., Lefèvre, L, Svalgaard, L., Vaquero, J.M., Leibacher, J.W. 2016. Preface to Topical Issue: Recalibration of the Sunspot Number, Solar Phys., 291, DOI: 10.1007/s11207-016-1017-8.

Cleveland, W.S. and Devlin, S.J. 1988. Locally Weighted Regression: An Approach to Regression Analysis by Local Fitting. Journal of the American Statistical Association 83: 596–610

Cochrane, D. and Orcutt, G.H. 1949. Application of least squares to relationships containing autocorrelated error terms, *Journal of the American Statistical Association* 44, 32–61.

Cowpertwaite, P.S.P and Metcalfe, A.W. (2009) *Introductory Tie Series with R*. Dordrecht: Springer.

Dickey, D.A. 1976. Estimation and hypothesis testing in nonstationary time series, PhD Dissertation, Iowa State University: Ames, IA.

Dickey, D.A. and Fuller, W.A. 1979. Distribution of the estimators for autoregressive time series with a unit root, *Journal of the American Statistical Association* 74, 427–431.

Durbin, J. 1960. Estimation of parameters in time series regression models, *Journal of the Royal Statistical Society* B 22, 139–153.

Eckner, A. 2014. A Framework for the Analysis of Unevenly Spaced Time Series Data,

Engle, R. 1982. Autoregressive conditional heteroskedasticity with estimates of the variance of United Kingdom inflation, *Econometrica* 50, 987–1007.

Fuller, W.A. (1996). *Introduction to Statistics Time Series,* 2nd edition. Wiley.

Gray, H.L. and Woodward, W.A. 1981. *Application of S-arrays to seasonal data,* Applied Time Series Analysis II. (Ed. D. Findley) Academic Press: New York, 379–413.

Holt, Charles C. (1957). *"Forecasting Trends and Seasonal by Exponentially Weighted Averages".* *Office of Naval Research Memorandum.* 52. reprinted in Holt, Charles C. (January–March 2004).

"Forecasting Trends and Seasonal by Exponentially Weighted Averages". International Journal of Forecasting. 20 (1): 5–10. doi:10.1016/j.ijforecast.2003.09.015.

Houghton, A., Munster, E.W., and Viola, M.V. (1978). Increased incidence of malignant melanoma after peaks of sunspot activity, *Lancet* 1(8067),759–760.

Jenkins, G.M. and Watts, D.G. 1968. *Spectral Analysis and Its Applications*. Holden-Day: San Francisco, CA.

Jones, R.H. 2016. Time series regression with unequally spaced data, *Journal of Applied Probability* 23, Issue A: Essays in Time Series and Allied Processes, 89–98.

Kwiatowski, D., Phillips, P.C.B., Schmidt, P. and Shin,Y. 1992. Testing the null hypothesis of stationarity against the alternative of a unit root, *Journal of Econometrics* 54, (1–3),159–178.

Levinson, N. 1947. The Wiener (root mean square) error criterion in filter design and prediction, *Journal of Mathematical Physics* 25, 262–278.

Ljung, G.M. 1986. Diagnostic testing of univariate time series models, *Biometrika* 73, 725–730.

Ljung, G.M. and Box, G.E.B. 1978. On a measure of a lack of fit in time series models, *Biometrika* 65, 297–303.

Moore, D.S., McCabe, G.P., and Craig, B.A. 2021. Introduction to the Practice of Statistics, 10th edition. W.H. Freeman: New York.

Park, R.E. and Mitchell, B.M. 1980. Estimating the autocorrelated error model with trended data, *Journal of Econometrics* 13, 185–201.

Press, W.H., Teukolsky, S.A., Vetterling, W.T., and Flannery, B.P. 2007. Numerical Recipes: *The Art of Scientific Computing*, 3rd edition. Cambridge University Press: Cambridge, U.K.

Priestley, M.B. 1962. Analysis of stationary processes with mixed spectra-I. *Journal of the Royal Statistical Society B* 24, 215–233.

Priestley, M.B. 1981. *Spectral Analysis and Time Series*. Academic Press: New York.

Quenouille, M.H. 1949. Approximate tests of correlation in time series, *Journal of the Royal Statistical Society, Series B* 11, 68–84.

Riedmiller, M. and Braun, H. 1993. *A direct adaptive method for faster backpropagation learning: The RPROP algorithm.* Proceedings of the IEEE International Conference on Neural Networks (ICNN), 586–591. San Francisco.

Rosenblatt, F. 1958. The perceptron: A probabilistic model for information storage and organization in the brain, Psychological Review 65, 386–408.

Shapiro, S.S. and Wilk, M.B. 1965. An analysis of variance test for normality (com- plete samples), *Biometrika* 52, 591–611.

Shumway, R.H. and Stoffer, D.S. 2017. *Time Series Analysis and Its Applications—With R Examples*, 3rd edn. Springer-Verlag: New York.

Stephens, M.A. 1974. EDF statistics for goodness of fit and some comparisons. Journal of the American Statistical Association 69, 730–737.

Sun, H. and Pantula, S.G. 1999. Testing for trends in correlated data, *Statistics and Probability Letters* 41, 87–95.

Tiao, G.C. and Tsay, R.S. 1983. Consistency properties of least squares estimates of autoregressive parameters in ARMA models, *Annals of Statistics* 11, 856–871.

Tong, H. 1977. Some comments on the Canadian lynx data, *Journal of the Royal Statistical Society A* 140, 432–436.

Tong, H. 1990. *Non-Linear Time Series*. Clarendon Press: Oxford, U.K.

Vogelsang, T.J. 1998. Trend function hypothesis testing in the presence of serial cor- relation, *Econometrica* 66, 123–148.

Wackerly, D., Mendenhall, W. and Scheaffer, R.L. 2007. *Mathematical Statistics with Applications,*7th Ed., Thomson Learning, Inc., USA.

Waldmeier, M. 1961. *The Sunspot-Activity in the Years 1610–1960*, Schulthess and Co.:Zurich, Switzerland.

Wang, Z., Woodward, W.A., and Gray, H.L. 2009. The application of the Kalman filter to nonstationary time series through time deformation, *Journal of Time Series Analysis* 30, 559–574.

Wickham, H. and Grolemund, G. 2017. *R for Data Science: Import, Tidy, Transform, Visualize, and Model Data*. O'Reilly Media, Inc.: Sebastopol, CA.

Winters, P.R. 1960. Forecasting sales by exponentially weighted moving averages. Management Science, 6(3), 324–342.

Woodward, W.A. 2013. Trend detecting. In *Encyclopedia of Environmetrics*, 2nd edition, 6 Volume Set, Abdel H. El-Shaarawi and Walter W. Piegorsch (Editor-in-Chief), John Wiley: Chichester, 2800–2804.

Woodward, W.A., Bottone, S., and Gray, H.L. 1997. Improved tests for trend in time series data, *Journal of Agricultural, Biological and Environmental Statistics* 2, 403–416.

Woodward, W.A. and Gray, H.L. 1983. A comparison of autoregressive and harmonic component models for the lynx data, *Journal of the Royal Statistical Society A* 146, 71–73.

Woodward, W.A. and Gray, H.L. 1993. Global warming and the problem of testing for trend in time series data, *Journal of Climate* 6, 953–962.

Woodward, W.A., Gray, H.L., Haney, J.W., and Elliott, A.C. 2009. Examining factors to better understand autoregressive models, *American Statistician* 63, 335–342.

Woodward, W.A., Gray, H.L., and Elliott, A.C. 2017. *Applied Time Series Analysis with R*, 2nd edition, CRC Press: Boca Raton, FL

Yule, G.U. 1927. On a method of investigating periodicities in disturbed series, with special reference to Wölfer's sunspot numbers, *Philosophical Transactions of the Royal Society* A226, 267–298.

Index

A

Activation function, 456
ADF (Augmented Dickey-Fuller) tests, 289
Aggregating, 45
AI (Artificial intelligence), 455
AIC (Akaike's Information Criterion), 230–241
    BIC, 232
    AICC, 232
    definition, 232
    multiple regression, relationship to, 231–233
    realization length and AIC, 237–239
Airline model, 471
Air passenger data (AirPassengers), *see* Datasets
Akaike's Information Criterion, *see* AIC
Alexa, 455
Aliasing, 128
Aperiodic behavior, 7, 124, 139, 145, 158, 162, 173, 180, 187
ARCH (Autoregressive conditional heteroskedastic) model,
    327–337
    ARCH(1) model, 329–332
    ARCH(p) processes, 332–334
    Dow Jones rates, *see* Datasets
    fitting ARCH/GARCH models to data, 335–339
    return data, daily rates of, 329, 337–338
    simulated realizations, 330–331, 334
ARIMA (Autoregressive integrated moving average), 276–304
    ADF and KPSS tests, 288–291
        Dickey-Fuller test, augmented 288
        KPSS test, 289–290
        simulation study of test performance, 290–291
    autocorrelations, limiting 277–279
    characteristic equations, 277
    decision whether to include 1-B factors, 287–291
    definition, 276
    differencing, 281–282, 291–298
        Dow Jones data, *see* Datasets
    fitting ARIMA models to data, 291–298
    forecasting, 298–304
    parameter estimation, 287–298
    mean, lack of attraction, 279–280
    properties, 276
    unit root tests, 288–291
    US population data, *see* Datasets
    global temperature data, *see* Datasets
    random trends, 280–281
ARMA (Autoregressive-moving average) model, 151,
    191–203, 217–269
    AIC-type measures, *see* AIC (Akaike's Information
        Criterion)
      log-lynx data, *see* Datasets
    AR models, *see* AR (*Autoregressive models*)

AR *vs.* ARMA models, 203–207
    assessing estimation performance, 219–220
    backcast residuals, 222–223
    backward model, 222
    canceling operators, 206–207
    conditional residuals, 221–222
    Dow Jones stock market data, *see* Datasets
    estimating
        mean, 220
        parameters, maximum likelihood, 218–220
        residuals, 220–223
        white noise variance, 220–223
    eventual forecast function, 255
    factor table, 176–178
    fitting ARMA models, 217–223
        estimating parameters, 218–223
        identifying p and q, 218
    forecasting, ARMA (*see also* Forecasting)
        eventual forecast function, 255
        examples, 252–254
        forecasting formula, 251
        probability limits, 255–260
    invertibility conditions, 202
    MA models, *see* MA (*Moving Average model*), 192–201
    maximum likelihood estimation, 218–219, 223
    model identification
        AIC-type measures, 231–232, 239
        types, 230–231
        white noise, checking, 229–230
    probability limits, 255
        forecast errors, facts, 256–260
    π-weights, 199
    ψ-weights, 202
    residual estimates, 221–223
    RMSE, to measure forecast performance, 261–269
AR (Autoregressive) models, 97, 130, 191, 312;
        see also Autoregressive-moving average
    AR(1) model, 153–163
      backshift operator notation, 152
      characteristic equation
        negative roots of, 160–161
        positive roots of, 155–159
      GLP form, 188–190
      spectral density, 155
      stationarity condition, 152
      with root close to +1, 159–160
    AR(2) model, 163
      lynx data (lynx), *see* Datasets
      properties, 164, 173
      autocorrelations, 196
      spectral density, 196

operator notation/characteristic equation, 164–167, 195
  stationary condition, 165
  complex conjugate roots, 168–172
  real roots, 167–168
AR(p) model, 174–190, 223–229, 246–251
  AIC-type identification, 239–240
  Burg estimation, 223–225, 228–229
  forecasting, 246–251
  general linear process, 187–188
  GLP form, 190–191
  linear filters, 187–188
  sunspot data, *see* Datasets
  maximum likelihood estimation, 228
  PACF model identification strategy, 244
  pattern recognition methods, 242
  sunspot data (sunspot2.0, sunspot2.0.month), *see* Datasets
  infinite order autoregressive, 203
Artificial intelligence, *see* AI
Augmented Dickey-Fuller tests, *see* ADF tests
Autocorrelation function, 131, 132
  sample, 112, 113
  theoretical, 111
Autoregressive conditional heteroskedastic model, *see* ARCH model
Autoregressive integrated moving average, *see* ARIMA
Autoregressive models, *see* AR model
Autoregressive-moving average model, *see* ARMA model

B

Backcast residuals, 223, 382–389
Backshift operator, definition, 152
Band-pass filters, 141
Band-stop filters, 141
Bat Echolocation data, *see* Datasets (bat, noctula)
BIC, 231–242, 298
Bitcoin (bitcoin), *see* Datasets
Bivariate data, 80, 82
Bivariate random sample
  assessing association, 82–85
  two random variables, 81–82
Black Monday, 10, 327
Bootstrap test for trend, *see* WBG test for trend
Box, G.E.P., 245
Bureau of Labor Statistics, 58
Burg estimates, 223–225, 228–229, 351, 353
  AR (p) model, 234
  WBG testing, 353
  overfitting, 312
Butterworth filter
  end-effects, 144
  high-pass, 143
  low-pass, 143

C

Canadian lynx data (lynx), *see* Datasets
CCF (Cross-correlation function) plots, 427
Centered moving average smoother, 41, 42
Central limit theorem, *see* CLT
Characteristic polynomial, 153
CLT (Central limit theorem), 78
Cochrane-Orcutt procedure, *see* CO procedure

Coin tossing experiment, 88
Comma-separated values, 20
Conditional heteroskedasticity, 328
Constant/finite variance, 101
CO (Cochrane-Orcutt) procedure, 350–351, 377, 401
  bootstrap-based WBG test, 352–354, 402
Correlation-based models, 381, 393, 395, 399–400, 402, 404
Cosine signal+noise process, 360–376, 407
Covariance stationary, definition, 101
COVID pandemic, 2, 58
  effects on market, 9
  outbreak, 24
  virus, 2
Cross-correlation function plots, *see* CCF plots
Cross-entropy, 457
Cryptocurrency, 2
.csv download option, 32
.csv files, 29
Cyclic/pseudo-periodic data, 122, 124

D

Dallas Ft. Worth, *see* DFW
Datasets,
  AirPassengers, 18, 20, 56, 68
  bat, 317
  bitcoin, 2, 8, 28–32
  Bsales, 419–420, 449
  cardiac, 437–446, 449, 482–491
  chirp, 103
  dfw.mon, 12, 26, 33, 38, 319, 371, 387
  dfw.yr, 26, 38, 43, 66
  dfw.2011, 13–16, 38
  dow.annual, 10, 295
  dow1985, 8, 125, 139, 160, 294–296
  global2020, 390–407
  llynx, 124, 128, 172, 236, 375
  lynx, 15, 18, 124, 128, 172, 236, 375
  mass.mountain, 102
  MedDays, 24
  NAICS, 342
  noctula, 103
  NSA, 342
  ozona, 28, 38
  rate, 328, 337
  sunspot2.0, 24, 38, 44, 124, 187, 241, 254, 262, 372, 388, 407
  sunspot2.0.month, 25, 38, 44
  tesla, 26, 38, 42–44
  tx.unemp.adj, 58
  tx.unemp.unadj, 48, 54, 58, 74
  uspop, 296
  us.retail, 74
  wtcrude2020, 23, 38, 62, 141, 459, 463
  yellowcab.precleaned, 24, 34, 38
Decibels (dB), 130, 237
Decomposition,
  additive, 54–55, 58
  multiplicative, 56–57, 59
DeepFit, architecture of, 470
Deep Learning applications, 455
Deep neural network
  AR(1) model, architecture, 458

AR*(p)*, adding more lags, 464–467
cross validation, using rolling window RMSE, 463–464
"ensemble" model, 487–488
extended perceptron, for univariate time series data, 457
forecasting, 460–463, 489–490
   cardiac mortality, using temperature, 482
   covariates, 483–484
   general architecture, 482
   train/test split, 483
hidden layer, 467–470
mlp function, 458–460
multivariate time series data
   architecture, 475
perceptron, 455–457
seasonal dummies, 470–475
Deviation, standard, 96
DFW (Dallas Ft. Worth), 26, 27
annual temperature data (dfw.yr), 66
monthly temperature data (dfw.mon), 3, 371
national weather service, 33
Dice-rolling process, 86, 114
Dickey-Fuller, 288–289
Differencing, 140–141, 143, 281, 284–285, 291, 395, 471, 478
Dow Jones data (dow.annual, dow1985), *see* Datasets
Durbin Levinson algorithm, 174

E

Earthquake seismic signal, 101–102
mass.mountain, *see* Datasets
Echolocation bat signal (bat, noctula), *see* Datasets
Estimation
   of the autocovariance/autocorrelation, 110–113
   of the mean, 106, 220
   of the spectral density, 132–140
   of white noise variance, 220–223
Estimation of model parameters, *see* Parameter Estimation
Euler's formula, 126–127, 155
Eventual forecast function, 255
Excel format (.xlsx), 21
Expected value, definition, 76
Exponential smoothing, 64
Extrapolation, 83

F

Factor table, 178–179
   AR, 178–179, 186
   ARMA, 203, 206–207
   MA, 193–194, 199
   overfitting, 312–317
   seasonal model, 312–317
Federal Reserve Economic Database, *see* FRED
Filter, *see* Linear filter
Fingent, 10, 11
Forecasting, 60–70
   AirPassengers data, *see* Datasets
   ARIMA, 298–303
   ARMA, 251–303
   assessing accuracy, 68–70, 260–269, 325–326
   cosine signal+noise model, 364–366
   cross-validation, 434
   ensemble model, 487–488

exponential smoothing, 64–65
eventual forecast function, 255
global temperature, *see* Datasets
Holt-Winters
   additive, 66–67
   multiplicative, 67–68
last *k* values, 370
line+noise model, 358–360
multi-layered perceptron, 477, 480
non-model based approaches, 60
prediction intervals, 489
prediction limits, 263
predictive moving average smoother, 62–64
probability limits 255–260
quantifying performance, 263–264
rolling window RMSE, 267–269
univariate forecasts, 435
Fourier representation, 121
Fourier transform pairs, 128
FRED (Federal Reserve Economic Database), 22, 41
Frequency, 123
Frequency domain, 121

G

GARCH model, 332–334
   definition, 333
   fitting ARCH/GARCH models to data, 334–339
   return data, daily rates of, 329, 337–338
Gaussian distribution, 94
Generalized least square, *see* GLS approach
General linear process, *see* GLP
Generating realizations
   ARCH, 330, 334
   ARIMA, 340
   ARMA, 208
   GARCH, 334
   signal+noise, 146
GitHub, 115
Global temperature data (global2020), *see* Datasets
GLP (General linear process), 188, 198
GLS (Generalized least squares), 355
Gradient descent, 456
Great Depression, 10
Great Recession, 48, 58

H

Harmonic model, 360
Hidden nodes, 459, 467–472, 476, 482
High-frequency behavior, 141
High-pass Butterworth filter, 143
High-pass filter, 141
Holt-Winters equations, 66, 67
Holt-Winters forecasting technique, 61, 67, 222, 405, 406
Hyperparameters, 470, 471

I

Infinite-order polynomial, 190
Interpolation, 29–32
   linear, 29–32
   LOCF (Last Observation Carried Forward), 29
Invertibility, 198–203

J

Jarque-Bera test, 335–337

K

KPSS (Kwiatkowski-Phillips-Schmidt-Shin) test, 289–291

L

Last Observation Carried Forward, *see* LOCF
Limiting autocorrelations, 277–278
Linear chirp signal (chirp), *see* Datasets
Linear filter, definition, 187
Linear interpolation, *see* Interpolation
Linear trend, 8, 344–355
Little Ice Age, 391
Ljung-Box test, 335–337, 384–389
Loading tswge, 12
LOESS (Locally estimated scatterplot smoothing), 44
LOCF (Last Observation Carried Forward), 29
Logarithmic transformation, 18
    air passengers data, 50, 310–311, 315–319, 323–326
    lynx data, 18, 116, 138, 172, 236–237
    sunspot data, 407, 409–411
Logistic/sigmoid function, 456
Low-pass Butterworth filter, 143
Low-pass filters, 140, 141
Lynx dataset, *see* Datasets

M

MAD (Mean absolute deviation), 261
MA (Moving average) model, 192
    autocorrelations, 196
    characteristic polynomial, 198
    imposing invertibility, 200
    infinite order MA process, 198, 199
    invertibility, 198–201
    mean, variance, and autocorrelations, 198
    operator notation, 198
    spectral density, 193, 196, 198
    stationary/invertible, 201
    system frequencies, 194, 204
Maximum likelihood, *see* ML
Mean absolute deviation, *see* MAD
Mean square error, *see* MSE
Minitab, 10
Missing data, 29–32
    Linear interpolation, 29–32
    LOCF (Last Observation Carried Forward), 29
ML (Maximum likelihood)
    ARMA model, coefficients of, 218
    estimation procedure, 389
MLP (Multi-layered perceptron)
    analog, 457
    AR3, architecture of, 465, 466
    architecture of, 481
    AR(1) forecasts, 462
    DeepFit, architecture of, 469, 470
    ensemble model, 487–488
    forecasting, 480
    hidden layer, 467, 468

model cardiac mortality, 482
rolling window RMSE, 476
seasonal dummy/ indicator variables, 471, 474
visualization of, 463
MLR (Multiple linear regression), 231, 417, 421
    AR(p) fit, 422
    bivariate VAR(1) process, 428–430, 434
    correlated errors, 428
Model assessment
    assessing forecast performance, 260–261
    examining realizations from the model, 393–394, 402–403
    plotting data , 229–230
    residual analysis, *see* Residual analysis
Model identification
    ARMA(p,q) models
        AIC-type measures, 231–232, 239
        PACF model identification strategy, 244
        pattern recognition methods, 242
        white noise, checking, 229–230
    ARIMA(p,d,q) models
        d, decision concerning differencing order, 287–291
        unit roots, testing for, 288–291
    seasonal models
        overfitting, 312–321
Moving average constant, 152, 279
Moving average model, *see* MA model
MSE (Mean square error), 69
Multi-layered perceptron, *see* MLP
Multiple linear regression, *see* MLR
Multivariate time series, 417–444
    cross correlation function, 426–428
    multiple regression, with correlated errors, 417–428
        adding lagged variables, 423–424
        lagged variables/trend variable, 425–426
        notation, 418–419
        trend term, 422–423
    vector autoregressive (VAR) models, 428–437
        seasonal, 441–444

N

National Weather Service, 22, 26
Neural networks, 455, 460
    AR(3) equivalence, 465
    architecture of, 457
    hidden layer, 467
    logistic/sigmoid activation function, 469
    p lags, input nodes, 464
New York City Taxi & Limousine, 22
Noise process, *see* White noise
Nonstationary models
    ARCH/GARCH, 327–338
    ARIMA. 275–304
    seasonal, 304–327
Normality, testing for
    Jarque-Bera test, 335
    Q-Q plot, 389
    Shapiro-Wilk test, 335, 389
NYCabRaw, *see* Datasets
Nyctalus noctula bat, *see* Datasets
Nyquist frequency, 128–129

O

Observed values, definition, 77
ozona data, *see* Datasets
Order identification, *see* Model Identification
Overfitting, 312
    seasonal factors, checking for, 312–321

P

PACF, 244
Parameter estimation,
    Burg estimation, 223–225, 228–229
    maximum likelihood estimation, 218–219, 223
    Yule-Walker, 223, 226–229
Parsimonious models, 191
Partial autocorrelations, 242–243
Parzen spectral density, 134–135
    sample autocorrelations, vs. 135–137
Parzen window
    log scale, plot in, 137
    model-based spectral density, 135
    sample autocorrelations, 136
Pattern recognition methods, 231, 242
Pearson's correlation coefficient, 426
Perceptron, 455, 457–459, 464, 467, 475
Periodic function, 122
Periodogram, 116, 133–134
Phase shift, 122
π-weights, 199
ψ-weights, 190, 202
Polynomial equation, 176
Probability distribution, 75
Probability limits for forecasting, 255–260
Python, 10

Q

Q-Q plot, 389
QR codes, 3, 11

R

Random sample, definition, 77
Random trending behavior, 7
Random process, 75–77, 85–88, 90, 92, 114
Random variable, 75
Realization
    definition, 90
    length, effect of realization, 97–98, 106–110
    multiple, 90–100
Regression, time series, 343–376
    cosine signal+noise models, 360–361
        cyclic behavior, 369–376
        fitting to data, 361–364
        forecasting, 364–366
        lynx dataset, *see* Datasets
        sunspot data, *see* Datasets
    line+noise models, 343–344
        fitting to data, 355–358
        forecasting, 358–360
        testing for linear trend, 344–355
            bootstrap-based WBG test, 352–354, 402
            CO (Cochrane-Orcutt) test, 350–351, 377, 401

Residual analysis, 381–390
    checking sample autocorrelations, 383–384
    histograms, 390
    Ljung-Box test, 384
    normality, checking, 389–390
    white noise, 383, 385
Residuals, 220–223
    backcasting, 222–223
    normality, checking, 389–390
    variance, estimating, 220–223
    white noise, checking, 229–230, 384–389
RMSE (Root mean square error), 261–269
    cross-validation, 466
    ensemble model, 487–488
    rolling window RMSE, 267–269, 407, 463–464
Root mean square error, *see* RMSE
R programming language, 10–28
    Base R, 12
    loading time series data, 19–22
    plotting time series data, 12–13
    ts object, 13–16
RStudio, 10, 11

S

Sample autocorrelations, 111–113
Sample autocovariances, 110–113
Sample correlation coefficient, 82
Sample spectral density, 132–133
    smoothing, 134–135
Sample standard deviation, 77
SAS software, 10
Seasonal adjustment,
    additive, 58–59, 67
    multiplicative, 59–60
    X13-ARIMA-SEATS, 59
Seasonal data
    additive, 49
    Census-Bureau seasonal adjustment, 59
    multiplicative, 49
Seasonal dummies, 441, 470–475
Seasonal differencing, 309, 315
Seasonal indicators, 482
Seasonal models, 304–327
    factor tables of, 307
    fitting to data, 311–321
        overfitting, 312–321
        seasonally differenced data, 313
    forecasting, 321–327
    properties of, 305
    purely seasonal models, 306
    VAR model, 441–443
Seismic Lg wave, 102
Shapiro-Wilk test, 335, 389
Signal+noise models, *see* Regression, time series
Simple linear regression model, *see* SLR model
Siri, 455
SLR (Simple linear regression) model, 349
    with independent variable time, 343
    line+noise model, 343–358
    standard SLR *t*-test, 349
Smoothing methods, 41

Butterworth filter, low pass, 143
centered moving average smoother, 41, 42
exponential smoothing, 61, 64
Snippet data, 4, 30
Spectral density, 125–128
SPSS software, 10
SSE (Sum of squared error), 458
Standard error, definition, 78
STATA software, 10
Stationarity, covariance, 101
Stationarity conditions
AR(p), 175
ARMA(p,q), 202
GLP (General Linear Process), 188
Stationary time series
autocorrelations, 102
autocovariance, 102
definition, 101
estimation
autocorrelations, 111–113
autocovariance, 110–113
mean, 101, 106–110
spectral density, 125–128, 130–139
variance, 110
spectral density, 125–128
variance, 101–102
Stochastic process, 89
Sum of squared error, see SSE
Sunspot data (sunspot2.0, sunspot2.0.month), see Datasets
System frequency, 169

T

Temperature data (global2020), see datasets
Tesla stock prices (tesla), see Datasets
Testing
for trend, 344–355
for unit roots, 288–291
for white noise, 384–389
Texas unemployment data, see Datasets
Tiao-Tsay result, 312
Time series, applications and relevance, 1–3
Time series regression, see Regression, time series
Tong's model, 374
Trend, testing for, 344–355
Training data, 69, 455
Transformations
autoregressive, 281
difference, 281–282
log, 18, 50, 372
seasonal difference, 309–311

U

Unbiasedness, 77
Unit root tests, 288–291
Univariate time series, 1, 417

V

VAR (Vector autoregressive) models, 417, 428–429
AIC/BIC(SC) criterion, 439
forecasting, 429–437
data values, in test set, 440–441
future, 444–445
MLR, relationship, 437
model identification, 432
parameter estimation, 432–433
RMSEs, comparing, 436
seasonal indicator variables, 443
seasonal VAR(p) model
candidate seasonal models, 441–442
forecasting, 442–443
short vs. long term forecasts, 445–446
training set, 435, 440
trend/seasonality, 487
univariate forecasts, 431
Vector autoregressive models, see VAR models
Videos, see QR codes

W

WBG (Woodward-Bottone-Gray) test, 352, 353, 379
bootstrap-based test, 354
time series realization, 379
Web scraping, 37
West Texas Intermediate Crude Oil (wtcrude2020),
see Datasets
White noise, 86–87
spectral density of, 131
testing for, 298, 335–337, 384–389
White noise variance estimation, see Estimation
Window function, 17
Woodward-Bottone-Gray test, see WBG test
Wrangling data, 37
WTI (West Texas Intermediate Crude Oil), see Datasets

Y

Yahoo!Finance, 25
Yellow Cabs, 27, 35
Yule-Walker estimates, 226, 228
for AR model, 223
equations, 164, 174, 242, 243